Environmental Disaster in China

Xianhua Wu

Environmental Disaster in China

Dilemmas, Economic Loss, and Countermeasures

Springer

Xianhua Wu
School of Economics and Management
Shanghai Maritime University
Shanghai, China

ISBN 978-981-92-0909-5 ISBN 978-981-92-0910-1 (eBook)
https://doi.org/10.1007/978-981-92-0910-1

This work was supported by National Natural Science Foundation of China (72474127) and National Social Science Fund Major Bidding project of China (18ZDA052).

This Springer imprint is published by the registered company Springer Nature Singapore Pte Ltd.
The registered company address is: 152 Beach Road, #21-01/04 Gateway East, Singapore 189721, Singapore

Competing Interests The author has no competing interests to declare that are relevant to the content of this manuscript.

Contents

Part I
Basic Characteristics of Environmental Disasters in China

Chapter 1
Study of Wind Speed and Direction at Yangshan Port

1.1 Introduction

Wind energy, as an infinitely renewable clean energy source, has seen a steady annual increase in generation capacity. By the end of 2023, the cumulative installed capacity of wind power connected to the national grid reached approximately 440 million kilowatts, a year-on-year growth of 20.7% (National Energy Administration, 2023). Wind speed and direction not only affect the generation efficiency and economic benefits of wind farms, but port operations, vessel docking, and transit activities. Therefore, it is of great significance to study the wind speed and direction at ports.

Existing research on wind speed and direction can be broadly categorized into two types. The first category focuses on statistical modeling analysis methods, while the second emphasizes machine learning techniques. In statistical modeling analysis, some scholars treat wind speed data as a stochastic process and utilize time series models for fitting. For example, J.L. Torres and A. Shamshad employed Autoregressive Moving Average (ARMA) models and Markov models, respectively, to model observational data sequences of wind speeds, though their studies failed to consider the wind direction(Torres et al., 2005; Shamshad et al., 2005). Similarly, Youcef developed a first-order Markov chain model for both wind speed and direction; however, this model treats wind speed and direction as independent, neglecting any potential interdependence between the two factors(Youcef Ettoumi et al., 2003). While time series models are effective and reasonable in forecasting, they face challenges in accurately capturing the specific probability distributions of wind speed and direction. Therefore, many researchers have turned to parametric models or non-parametric kernel density estimation models to model and fit the wind vector.

In the existing literature, there are a substantial number of models developed for studying wind speed, among which single distribution function models are commonly employed. These include models such as the Rayleigh distribution, Weibull distribution, Gamma distribution, Inverse Gaussian distribution, Log-Normal distribution, and Extreme Value distribution(Kiss & Jánosi, 2008; Wais, 2017a; Aries et al., 2018;

X. Wu, *Environmental Disaster in China*, https://doi.org/10.1007/978-981-92-0910-1_1

Usta & Kantar, 2012; Celik, 2003). Notably, the Weibull model is frequently applied due to its good fitting capability and straightforward usage(Carrillo et al., 2014). The Weibull model was proposed by Tuller in 1984, who outlined the derivation process of transitioning from a bivariate normal distribution to the Weibull distribution, thereby providing a theoretical foundation for its application in wind speed analysis(Tuller & Brett, 1984). Carta and Ramirez conducted a comparative study of the Log-Normal and Weibull models, highlighting the advantages of the Weibull model over other single distribution models: particularly well-suited for characterizing the occurrence of high wind speeds, demonstrating excellent fitting performance while requiring the estimation of only two parameters(Carta et al., 2009). Precisely owing to the advantages of the Weibull model, researchers have employed it to fit wind speed data from multiple Chinese cities and have compared the performance of various parameter estimation methods(Huo et al., 2024). However, the two-parameter Weibull model demonstrates suboptimal fitting performance when the proportion of low wind speeds (0 ~ 2 m/s) is high, while the three-parameter Weibull model shows improved effectiveness in such cases(Wais, 2017a; Wais, 2017b). Moreover, although single distribution Weibull models perform well when fitting unimodal wind speed data, they struggle with multimodal datasets. To address this issue, some scholars have proposed mixed models composed of multiple single distribution models. For example, Carta utilized a dual Weibull mixed model to analyze average wind speed in Gran Canaria, Spain(Carta & Ramírez, 2007). Ouarda et al. applied ten dual-component mixed models, including Gamma, Weibull, and Gumbel distributions, to study wind speed in Quebec, Canada(Ouarda & Charron, 2018). These mixed models have demonstrated superior fitting performance for complex wind conditions.

Despite the numerous models for studying wind speed, there are relatively few models dedicated to analyzing wind direction. Models used for fitting wind direction include the von Mises distribution, uniform distribution, and wrapped-Cauchy distribution, among others. The von Mises distribution effectively captures the statistical characteristics of wind direction angles and has been widely used for this advantage(Carta et al., 2008a; Masseran et al., 2013). Additionally, mixed von Mises models offer a more flexible fitting approach for complex wind conditions. The model exhibits satisfactory fitting performance in most cases, including scenarios with multiple directional peaks. However, it should be noted that when dealing with wind conditions containing multiple directional peaks, inadequate order selection in the von Mises mixture model may result in compromised fitting effectiveness. Beyond the aforementioned models, other researchers have proposed bivariate joint distribution models that account for both wind speed and direction(Carta et al., 2008b; Li et al., 2019). Meanwhile, in the field of probabilistic wind speed modeling incorporating directional effects, scholars have proposed several innovative approaches. For instance, Xiaowen Ji et al. developed a generalized bivariate mixture (GBM) model for directional wind speed and conducted systematic comparisons with the offset elliptical normal mixture (OENM) model, demonstrating the GBM framework's superior performance. In parallel work, Zihao Yang et al. introduced a semi-parametric trivariate model that simultaneously incorporates wind speed, wind direction and air density within a unified framework for wind energy assessment(Ji et al., 2024; Yang & Dong, 2024).

Shanghai Yangshan Deep-Water Port, also known as Yangshan Port, is one of the busiest sea ports in the world. In 2023, the container throughput at Yangshan Port exceeded 25 million TEUs(Shanghai Maritime Safety Administration, n.d.). In terms of natural conditions, Yangshan Port benefits from abundant wind energy resources, experiencing winds of Force 6 and above on several occasions each month, with an average of 31 to 49 days of strong winds annually. However, most previous research focuses on logistics and freight transportation, with relatively few studies addressing wind speed and direction.

With Yangshan Port as the primary research objective, this study employs a two-parameter Weibull model and a mixed von Mises model to analyze and fit the wind speed and direction data, respectively. Additionally, appropriate Copula functions are selected to derive the bivariate joint probability density function of wind speed and direction. Finally, empirical analysis is conducted using wind data from Yangshan Port.[1]

1.2 Research Methods and Models

1.2.1 Wind Speed Model

Yangshan Port is located in the East China Sea, characterized by significant differences in thermal properties between land and sea, as well as a distinct monsoon climate. The disparities in thermal properties lead to seasonal variations in wind direction, making it unlikely for multiple peaks in wind speed to occur. Furthermore, since Yangshan Port is surrounded by the sea and there is no unique topography that may impede wind flow, the wind speed data at the port exhibits a unimodal distribution. Therefore, a two-parameter Weibull model is employed to study the wind energy characteristics of Shanghai Yangshan Port.

The probability density function (PDF) of the two-parameter Weibull distribution is given by:

$$f(v) = \frac{k}{c}\left(\frac{v}{c}\right)^{k-1} e^{-\left(\frac{v}{c}\right)^k} \tag{1.1}$$

[1] The primary objective of this study is to investigate the long-term aggregate wind speed probability density distribution at Yangshan Port and to compute the corresponding wind energy density. As such, time-dependent wind energy density analysis has not been incorporated within the scope of this research.

The cumulative distribution function is expressed as:

$$F(v) = 1 - e^{-\left(\frac{v}{c}\right)^k} \tag{1.2}$$

Where, v is the wind speed, k is the shape parameter, and c is the scale parameter. Various methods can be used for estimating the parameters of the Weibull model, including method of moments, empirical method, graphical method, maximum likelihood estimation, modified maximum likelihood estimation, and energy pattern factor method(Chang, 2011; Seguro & Lambert, 2000). Studies have shown that the maximum likelihood method yields the best results. In this research, both the least squares method and the maximum likelihood method are used to estimate the model parameters.

1.2.1.1 Least Squares Method

The basic idea of estimating model parameters using the least squares method is to construct a function representing the difference between observed values and predicted values, minimizing the sum of the squared residuals to estimate the parameters of the Weibull function. Taking the logarithm of the two-parameter Weibull distribution function twice results in the following:

$$\ln\{-\ln[1 - M(v)]\} = klnv - klnc \tag{1.3}$$

Wind speed data is divided into certain intervals to perform probabilistic statistics, yielding frequencies $f_1, f_2, \ldots, f_n$. Let $M_1 = f_1$, $M_2 = M_1 + f_1, \ldots$, and $M_n = M_{n-1} + f_n$.

Let

$$x_i = \ln v_i \tag{1.4}$$

$$y_i = \ln[-\ln(1 - M_i)] \tag{1.5}$$

Thus, we obtain:

$$c = \exp\frac{-\left(\sum_{i=1}^n x^2 \sum_{i=1}^n y_i - \sum_{i=1}^n x_i \sum_{i=1}^n x_i y_i\right)}{-\sum_{i=1}^n x_i \sum_{i=1}^n y_i + n\sum_{i=1}^n x_i y_i} \tag{1.6}$$

$$k = \frac{-\sum_{i=1}^n x_i \sum_{i=1}^n y_i + n\sum_{i=1}^n x_i y_i}{n\sum_{i=1}^n x_i^2 - \left(\sum_{i=1}^n x_i\right)^2} \tag{1.7}$$

where v_i is the i-th wind speed sample and n is the number of non-zero wind speed data records.

1.2.1.2 Maximum Likelihood Method

The maximum likelihood function for the two-parameter Weibull distribution is constructed as follows:

$$L(k, c) = \sum_{i=1}^{n} \left[lnk + (k - 1)\ln v_i - klnc - \left(\frac{v_i}{c}\right)^k \right] \quad (1.8)$$

Let

$$F_1 = \frac{\partial L(k, c)}{\partial k} = 0 \quad (1.9)$$

$$F_2 = \frac{\partial L(k, c)}{\partial c} = 0 \quad (1.10)$$

Obtain:

$$F_1 = \sum_{i=1}^{n} \left[\frac{1}{k} - lnc + \ln v_i - \left(\frac{v_i}{c}\right)^k \ln\frac{v_i}{c} \right] = 0 \quad (1.11)$$

$$F_2 = \sum_{i=1}^{n} \left[-\frac{k}{c} + \frac{k}{c}\left(\frac{v_i}{c}\right)^k \right] = 0 \quad (1.12)$$

The Newton-Raphson algorithm is then utilized to iteratively solve the above equations. Starting with initial values, the parameters are subjected to iteration until convergence of the two parameters is achieved, i.e., when $\max(|\Delta k, \Delta c|) < \varepsilon$. Here, ε is a predefined small positive value. Given the specific requirements of the Newton-Raphson algorithm for initial values, the results from the least squares estimation are utilized as the initial values for this process.

1.2.2 Wind Direction Model

The wind direction is fitted using the von Mises distribution. For example, Carta employed a finite mixture of von Mises distributions for the statistical simulation of directional wind speed, utilizing the Levenberg–Marquardt algorithm to estimate the parameters of a multi-order mixture von Mises model and providing methods for initial value settings(Carta et al., 2008a). The probability density function (PDF) of a single von Mises model is expressed as follows:

$$f(\theta; k, u) = \frac{\exp[k\cos(\theta - u)]}{2\pi I_0(k)} \quad (1.13)$$

where θ is the wind direction angle ($0 \leq \theta < 2\pi$). In meteorology, wind direction is typically measured as the clockwise angle from true north, with north defined as 0° (or 360°). Here, u denotes the mean direction ($0 \leq u < 2\pi$), and k is the concentration parameter, indicating the degree of data concentration. The single von Mises function is a unimodal distribution that is symmetric around $\theta = u$. In this equation, $I_0(k)$ represents the modified Bessel function of the first kind of order zero(Carta et al., 2008a).

$$I_0(k) = \frac{1}{\sqrt{2\pi}} \int_0^{2\pi} \exp(k cos\theta) d\theta \tag{1.14}$$

The probability density function and distribution function for the mixture von Mises model can be expressed as follows:

$$f_{vMs}(\theta; k_1, u_1, \omega_1, \ldots, k_N, u_N, \omega_N) = \sum_{i=1}^{N} \frac{\omega_i}{2\pi I_0(k_i)} \exp[k_i \cos(\theta - u_i)] \tag{1.15}$$

$$F_{vMs}(\theta; k_1, u_1, \omega_1, \ldots, k_N, u_N, \omega_N) = \sum_{i=1}^{N} \frac{\omega_i}{2\pi I_0(k_i)} \int_0^{\theta} \exp[k_i \cos(\theta - u_i)] d\theta \tag{1.16}$$

Here, ω_i is the weight of the i-th von Mises model and it adheres to the following constraint:

$$\begin{cases} 0 \leq \omega_i \leq 1 \\ \sum_{i=1}^{N} \omega_i = 1 \end{cases} \tag{1.17}$$

In this study, the LM algorithm is employed to estimate the 3 N parameters of the N-order mixture von Mises model.

1.2.3 Copula Functions

Copula functions are used to describe the degree of correlation between variables. They are considered a class of functions that connect joint distribution functions to their respective marginal distribution functions. Sklar established the relationship between multivariate distributions and Copula functions(Sklar, 1959). The advantage of Copula functions lies in their ability to construct joint distributions with arbitrary marginal distributions without imposing assumptions on the marginal distributions, thereby characterizing nonlinear dependencies between variables. However, given the multitude of Copula functions available, improper selection may result in reduced model accuracy.

Assuming that X_1, X_2, ..., X_n are random variables with marginal distributions F_1, F_2, ..., F_n, then the joint distribution is $F(x_1, x_2, ..., x_n)$, with $u_1 = F_1(x_1)$, ..., $u_n = F_n(x_n)$. If there is a correlation among X_1, X_2, ..., X_n, then a Copula function C exists. For $x \in R$, the relationship between the multivariate distribution $(x_1, x_2, ..., x_n)$ and the Copula function $C(u_1, ..., u_n)$ can be expressed as:

$$\begin{aligned} F(x_1, x_2, \ldots, x_n) &= P(X_1 \le x_1, \ldots, X_n \le x_n) \\ &= C[F_1(x_1), \ldots, F_n(x_n)] \\ &= C(u_1, \ldots, u_n)\# \end{aligned} \tag{1.18}$$

where $F_i(x_i) = u_i$ ($i = 1, 2, ..., n$), with F_i being a continuous function and $u_i \sim U(0, 1)$.

While estimating the joint distribution of wind speed and direction, only a bivariate Copula function is needed since only two variables are involved.

The distribution and probability density functions of the bivariate Copula function are expressed as follows:

$$C(u, v) = F\left(F_1^{-1}(u), F_2^{-1}(v)\right) \tag{1.19}$$

$$c(u, v) = \frac{\partial^2 C(u, v)}{\partial u \partial v} = \frac{f(x_1, x_2)}{f_1(x_1) f_2(x_2)} \tag{1.20}$$

where $u = F_1(x_1)$ and $v = F_2(x_2)$.

Copula functions can be categorized into Archimedean, elliptical, and quadratic types. Archimedean Copulas are particularly useful as they involve only one parameter, making them relatively straightforward to solve. They can generate different adaptable bivariate joint distributions and are widely used in analyses involving wind speed and direction. Among various Archimedean Copulas, Gumbel Copula, Clayton Copula, and Frank Copula are especially typical and frequently employed. Therefore, this study selects these three Copula functions for the joint analysis of wind speed and direction. Their functional expressions are presented in Table 1.1.

The parameters of the Copula functions are estimated using the maximum likelihood method. Let the joint distribution function of the random vector (X, Y) be $F(x, y)$, the joint density function be $f(x, y)$, the marginal distribution functions be $F_1(x)$ and $F_2(y)$, and the marginal density functions be $f_1(x)$ and $f_2(y)$, and the density function of the corresponding Copula function $C(u, v)$ be $c(u, v)$, 1.then:

Table 1.1 Three Copula distribution function expressions were chosen

Copula	Parameter range	Function form
Gumbel	[1,∞)	$e^{\left\{-\left[(-lnu)^{\theta}+(-lnv)^{\theta}\right]^{1/\theta}\right\}}$
Clayton	(0, ∞)	$(u^{-\theta} + v^{-\theta} - 1)^{-1/\theta}$
Frank	R	$-\frac{1}{\theta}\ln\left[1 + \frac{(e^{-\theta u}-1)(e^{-\theta v}-1)}{e^{-\theta}-1}\right]$

$$f(x,y) = c(F_1(x), F_2(y))f_1(x)f_2(y) \tag{1.21}$$

where $c(u,v) = \frac{\partial^2 C(u,v)}{\partial u \partial v}$. The density function $c(u, v)$ reflects the dependence structure of the random variables X and Y.

Given (X_i, Y_i), where $i = 1, 2, \ldots, n$, as a sample of the random vector (X, Y), then the log-likelihood function can be represented as:

$$L = \sum_{i=1}^{n} lnc(F_1(X_i), F_2(Y_i)) + \sum_{i=1}^{n}\left[\ln f_1(X_i) + \ln f_2(Y_i)\right] \tag{1.22}$$

If the marginal distribution functions of the random vector (X, Y) are known, they can be substituted into the equation above. Let $L = 0$, the relevant parameters can be derived.

1.2.4 Angular-Linear Model

Johnson and Wehrly(Johnson & Wehrly, 1978) derived a model representing Angular-Linear distributions based on the principle of maximum entropy, applying it to the study of the joint probability distribution of wind speed and direction. The joint PDF is expressed as:

$$f(v,\theta) = 2\pi g(\xi)f(v)h(\theta) \tag{1.23}$$

where $f(v)$ is the probability density function (PDF) of wind speed, $h(\theta)$ is the PDF of wind direction, and $g(\xi)$ is the PDF of the angular variable ξ. The angular variable ξ is determined by the following equation:

$$\xi_j = \begin{cases} 2\pi\left[F_V(v_j) - F_\Theta(\theta_j)\right], F_V(v_j) \geq F_\Theta(\theta_j) \\ 2\pi\left[F_V(v_j) - F_\Theta(\theta_j)\right] + 2\pi, F_V(v_j) < F_\Theta(\theta_j) \end{cases} \tag{1.24}$$

Where, $F_V(v_j)$ is the cumulative probability distribution of wind speed, and $F_\Theta(\theta_j)$ is the cumulative probability distribution of wind direction. Since the angular variable ξ also lies within the range of $0 \sim 2\pi$, a mixture von Mises distribution can be employed for fitting.

1.2.5 Fitting Evaluation Metrics

To evaluate the goodness of fit of the model, three metrics can be employed: Root Mean Square Error (RMSE), Coefficient of Determination (R^2), and the Akaike Information Criterion (AIC). Among them, the RMSE quantifies the discrepancy between predicted values and observed values; a smaller RMSE indicates a better model fit. The coefficient of determination reflects the degree of correlation between predicted and observed values, with higher values suggesting a stronger fit. Their calculation formulas are as follows:

$$RMSE = \sqrt{\frac{\sum_{i=1}^{n} \left(f_i - \hat{f}_i\right)^2}{n}} \quad (1.25)$$

where f_i is the actual probability for wind speed or direction for the i-th interval, and $\hat{f}_i$ is the estimated value of PDF for the i-th interval.

$$R^2 = 1 - \frac{\sum_{i=1}^{n} \left(f_i - \hat{f}_i\right)^2}{\sum_{i=1}^{n} (f_i - f_m)^2} \quad (1.26)$$

where f_m denotes the average of the actual probabilities across all wind speed and direction data within the i-th interval.

The AIC, also known as the Akaike Information Criterion, is calculated by:

$$AIC = 2k - 2lnL \quad (1.27)$$

where k is the number of parameters, and L is the maximum likelihood function of the model.

This study will utilize RMSE and R^2 to assess the fit of wind speed and direction, while the AIC criterion to evaluate the fit of the bivariate joint distribution of wind speed and direction.

1.3 Empirical Analysis

1.3.1 Data Sources

In this study, 10 m U-wind and V-wind data were obtained from the "ERA5 hourly data on single levels from 1940 to present" dataset available on the European Centre for Medium-Range Weather Forecasts (ECMWF) website. This dataset represents the fifth generation of reanalysis products for global climate and weather over the past 80 years, with a spatial resolution of 0.25° × 0.25° and a temporal resolution of hourly observations. Given that this dataset is the latest generation of global reanalysis products and incorporates ensemble information to characterize the "flow-dependent" covariance matrix of background errors, it is extensively applied in reanalysis. Data from the years 1990 to 2020 were selected, focusing on the region bounded by longitudes 120°E to 125°E and latitudes 25°N to 35°N. Due to the inadequacy of the spatial resolution of 0.25° × 0.25°, linear interpolation was used to enhance the dataset to a resolution of 0.1° × 0.1°. Data from the point located at 122.1°E and 30.6°N was selected as the primary objective of this study, as the coordinates of Yangshan Port are 122.09°E and 30.62°N. The selected point is in close proximity to the port, so the data of this point can well reflect the wind conditions at the port.

1.3.2 Model Application

1.3.2.1 Copula Functions

Figure 1.1 shows the probability distributions across various wind speed intervals, along with two Weibull curves generated using parameters estimated through least squares and maximum likelihood methods. The red solid line represents the Weibull curve estimated using the least squares method, whereas the yellow dashed line represents the Weibull curve obtained via the maximum likelihood method. The blue bars illustrate the distribution of the wind speed data. In existing studies, it is common to use an interval of 1 m/s for wind speed ranges. The maximum recorded wind speed at Yangshan Port does not exceed 22 m/s. Therefore, an interval of 1 m/s was adopted for the wind speed data. It can be observed from the figure that the wind speed distribution at Yangshan Port is unimodal, which aligns with the characteristics of a subtropical monsoon climate. The unimodal distribution suggests that a good fitting can be achieved even without multiple mixed Weibull distributions.

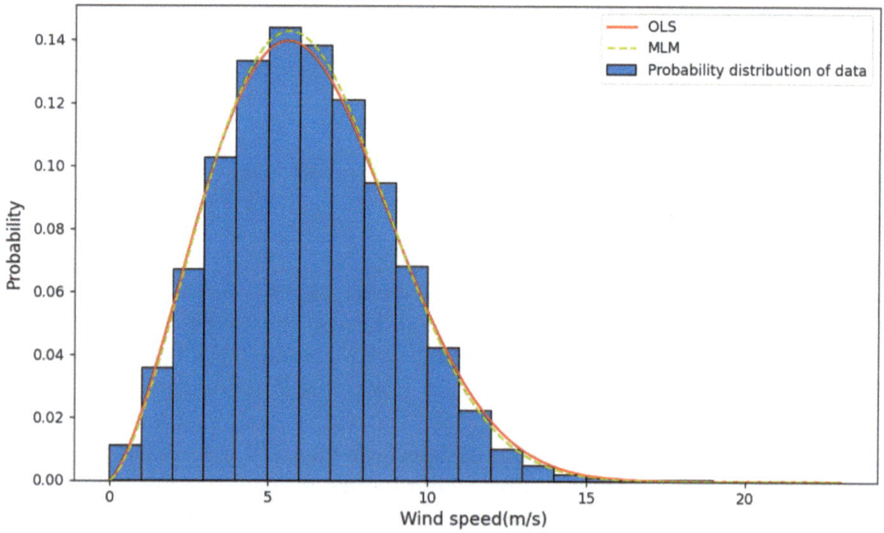

Fig. 1.1 The image of two Weibull functions with different methods of estimating the parameters

For the unimodal two-parameter Weibull model, the maximum likelihood method performs well in estimating parameters. Due to the requirements of the Newton-Raphson algorithm regarding initial values, the two parameters of the model are initially estimated using the least squares method. These parameter values are subsequently employed as initial values of the maximum likelihood method for the Newton-Raphson iteration. This results in two fitted Weibull function curves. The specific results of the parameter estimations are presented in Table 1.2. The results indicate that the parameters estimated using the maximum likelihood method yield a smaller Root Mean Square Error (RMSE) and a larger Coefficient of Determination (R^2) compared to those obtained using the least squares method. This demonstrates that the parameter estimates derived from the maximum likelihood method are superior to those from the least squares method.

Following the modeling of wind speed, the wind direction is modeled. The parameter estimates of the various-order mixture von Mises models are displayed in Table 1.3.

Table 1.4 presents the RMSE and R^2 values for the fitting results of the various-order mixture von Mises models. It is evident that as the order increases, the RMSE values decrease, and the R^2 values increase, indicating improved accuracy and fitting of the model. At the second order, the model exhibits poor fit, indicated by a low R^2 value. The third and fourth-order models show a comparable and relatively good fit, demonstrating a significant improvement over the second order. The fifth and sixth-order models exhibit similar fitting performance and achieve the highest levels of fit, with their R^2 values approaching 1. Compared with the fifth-order mixture von Mises model, the RMSE of the sixth-order model declined by less than 0.001, and the R^2 value only increases by only 0.001, suggesting that the sixth-order von Mises model can well reflect the wind direction distribution. Existing research has shown that when the order exceeds six, the improvement in model fit becomes negligible(Carta et al., 2008a). For this reason, further calculations for the seventh order and above are omitted, and the sixth-order mixture von Mises distribution is selected as the expression function of wind direction for subsequent analyses.

Figure 1.2 presents the function plots of the various-order mixture von Mises distributions, with the blue bars representing the probability distribution histogram of wind direction data. It can be observed that the yellow second-order von Mises distribution, while it roughly reflects the shape of the sample distribution, displays significant deviations at both the peaks and troughs, indicating a poor fitting effect. The third and fourth-order distributions largely overlap, as do the fifth and sixth-order distributions. Notably, the fitting performance of the fifth and sixth-order distributions is significantly superior to that of the third and fourth-order distributions within the

Table 1.2 Different methods for estimating the results of parameters of the Weibull model

Methods	c	k	RMSE	R^2
OLS	7.0156	2.4092	0.00279	0.996833
MLM	6.9824	2.4662	0.00196	0.998559

Table 1.3 Parameter estimation results for mixed von Mises models of all orders

N	ω_i	k_i	u_i
2	0.83608	0.68481	1.93062
	0.16391	9.94129	6.28319
3	0.23472	7.66510	2.85362
	0.55922	0.85888	1.16343
	0.20606	3.52672	6.26602
4	0.55592	0.90742	0.90742
	0.23673	7.57050	2.86158
	0.01375	13.33443	5.26924
	0.19360	4.34406	6.28318
5	0.01243	37.38669	0.11449
	0.32164	1.58472	2.27904
	0.04299	35.19408	3.08192
	0.11168	11.05895	2.83988
	0.51126	1.67253	0.23047
6	0.26824	1.53083	1.42071
	0.08160	22.90602	3.01800
	0.22841	3.74001	0.13878
	0.01146	27.04708	5.56977
	0.19014	4.95621	2.70401
	0.21952	0.57178	6.28319

Table 1.4 Degree of fit of mixed von Mises models of all orders

Order	RMSE	R^2
2	0.05478	0.39704
3	0.01336	0.97607
4	0.01252	0.97888
5	0.00507	0.99657
6	0.00425	0.99763

wind direction ranges of 0° ~ 20°, 90° ~ 110°, and 160° ~ 180°. Therefore, it is reasonable to utilize the sixth-order mixture von Mises distribution to represent the wind direction data. The black line in the figure represents the sixth-order mixture von Mises function, which demonstrates a good fit to the probability distribution of the wind direction data.

After obtaining the expressions for wind speed and wind direction, the Copula function is employed to derive the bivariate joint distribution of wind speed and direction. The probability density expression for wind speed is given by:

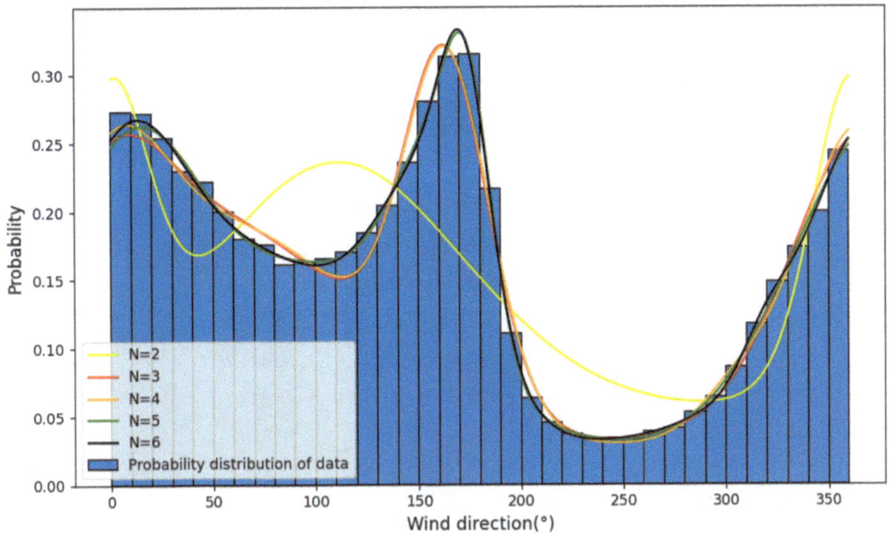

Fig. 1.2 The image of mixed von Mises probability density function of all orders

$$f(v) = \frac{2.4662}{6.9824}\left(\frac{v}{6.9824}\right)^{1.4662} e^{-\left(\frac{v}{6.9824}\right)^{2.4662}} \tag{1.28}$$

The corresponding distribution function is expressed as:

$$F(v) = 1 - e^{-\left(\frac{v}{6.9824}\right)^{2.4662}} \tag{1.29}$$

The PDF for wind direction is expressed as:

$$\begin{aligned} f_{vMs}(\theta) = & \frac{0.268}{2\pi I_0(1.531)} \exp(1.531\cos(\theta - 1.421))\# \\ & + \frac{0.082}{2\pi I_0(22.906)} \exp(22.906\cos(\theta - 3.018)) \\ & + \frac{0.228}{2\pi I_0(3.740)} \exp(3.740\cos(\theta - 0.139)) \\ & + \frac{0.011}{2\pi I_0(27.047)} \exp(27.047\cos(\theta - 5.570)) \\ & + \frac{0.190}{2\pi I_0(4.956)} \exp(4.956\cos(\theta - 2.704)) \\ & + \frac{0.220}{2\pi I_0(0.572)} \exp(0.572\cos(\theta - 6.283)) \end{aligned} \quad \#\#\# \tag{1.30}$$

The distribution function of wind direction is expressed as:

Table 1.5 Parameters and AIC values for each Copula function

	Gumbel	Clayton	Frank
Parameter value	1.0539	0.9696	7.9322
AIC value	−25.822	−25.937	−26.141

$$\begin{aligned} F_{vMs}(\theta) = &\frac{0.268}{2\pi I_0(1.531)}\int_0^\theta \exp(1.531\cos(\theta-1.421))d\theta\# \\ &+\frac{0.082}{2\pi I_0(22.906)}\int_0^\theta \exp(22.906\cos(\theta-3.018))d\theta \\ &+\frac{0.228}{2\pi I_0(3.740)}\int_0^\theta \exp(3.740\cos(\theta-0.139))d\theta \\ &+\frac{0.011}{2\pi I_0(27.047)}\int_0^\theta \exp(27.047\cos(\theta-5.570))d\theta \\ &+\frac{0.190}{2\pi I_0(4.956)}\int_0^\theta \exp(4.956\cos(\theta-2.704))d\theta \\ &+\frac{0.220}{2\pi I_0(0.572)}\int_0^\theta \exp(0.572\cos(\theta-6.283))d\theta \end{aligned} \tag{1.31}$$

The distribution functions for wind speed and direction serve as the two marginal distribution functions. Herein, Gumbel Copula, Clayton Copula, and Frank Copula are applied to construct the joint distribution of wind speed and direction, and the parameter θ of each Copula function is estimated through the maximum likelihood method.

Table 1.5 presents the parameters of the Copula functions along with their AIC values.

From the results of the AIC values, it is evident that there is little variation among the three Copula functions. The Frank Copula has the smallest AIC value of −26.141, followed by the Clayton Copula at −25.937, and the Gumbel Copula with the largest AIC value of −25.822. Each of these three Copula functions is then used as a linking function to generate the bivariate joint PDF of wind speed and direction, utilizing the two marginal PDFs.

1.3.2.2 Angular-Linear Model

The ξ value corresponding to each data record is calculated through Eq. (24), which are then fitted by a mixture von Mises model. Similar to the approach for modeling wind direction, parameter estimation is conducted using the LM algorithm, with the goodness of fit is evaluated based on RMSE and R^2 1.values.

Table 1.6 presents the estimated model parameters for the angular variable, along with their RMSE and R^2 values. Once the mixture von Mises model reaches the third order, it demonstrates a favorable goodness of fit. Therefore, the third-order mixture von Mises distribution is selected as the optimal model for the angular variable. Figure 1.3 illustrates the estimated PDF, revealing that the third-order model well reflects the changes of the angular variable. Thus, based on Eq. (23), the bivariate joint PDF of wind speed and direction under the Angular-Linear (AL) model is obtained.

Table 1.6 Parameter estimation results of the AL model

Order	w	k	u	RMSE	R^2
2	0.521	0.519	1.276	0.0067	0.9286
	0.479	0.649	5.204		
3	0.386	0.478	0.348	0.0022	0.9929
	0.040	7.433	1.764		
	0.574	0.279	5.160		

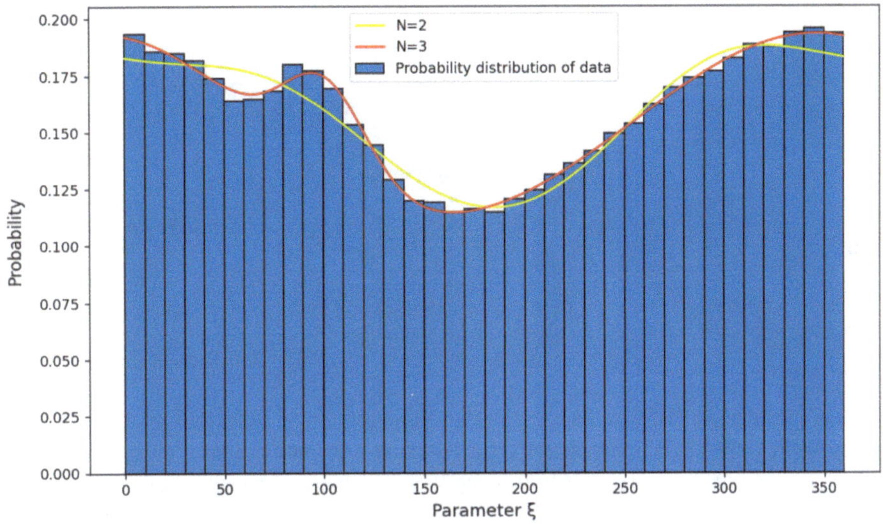

Fig. 1.3 Probability distribution curves of angular variable ξ of different orders

1.3.2.3 Model Comparison

Figures 1.4~1.6 show the joint PDFs of wind speed and direction under the Copula dependence. Figure 1.7 presents the PDF obtained from the Angular-Linear model, and Fig. 1.8 illustrates the case of independence between wind speed and direction.

The bivariate joint PDFs derived from the Copula and the AL model reveal that although both models yield comparable probability densities around wind directions of 180°, the bivariate joint probability density derived from the AL model exhibits two peaks in wind speed at both low and high wind direction angles. This does not align with the actual data. The joint PDF for wind speed and direction produced by the AL model shows unreasonable simulation results in the regions of low and high wind direction angles.

From the PDFs generated by the Copula, it is observed that the PDFs based on the Gumbel Copula dependence are similar to those based on the Frank Copula dependence. However, for the probability density distribution dependent on the Clayton

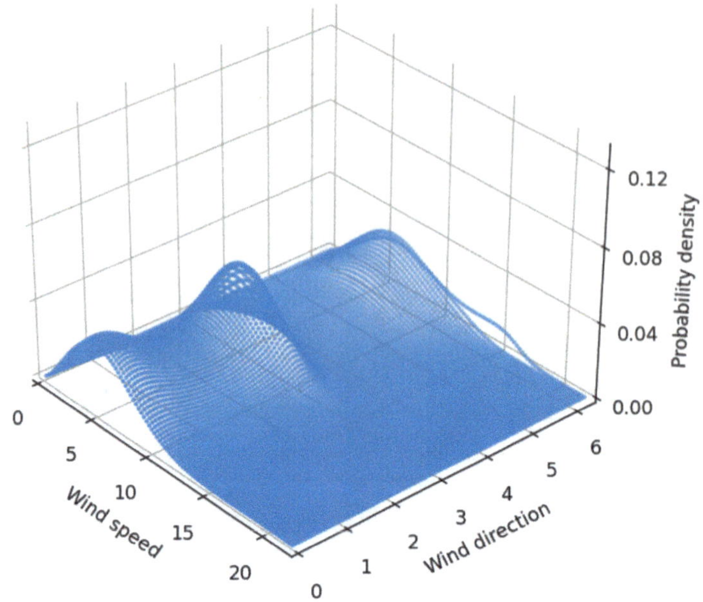

Fig. 1.4 Wind speed and direction are Gumbel-Copula dependent

Copula, significant differences are noted in the wind speed distribution at small wind direction angles when compared to the other two Copulas. Specifically, as seen in Fig. 1.5, the wind speed distribution approaches the x-axis and z-axis when the wind direction angle is small. This indicates that for the bivariate joint PDF of wind speed and direction constructed using the Clayton Copula, the shape parameter k of the two-parameter Weibull distribution is less than 1 in the case of small wind direction angles. This suggests that a significant concentration of wind speed in the low-speed range when the wind direction angle is small. Moreover, it is observed that, at small wind direction angles, wind speeds predominantly concentrate in the range of 0 ~ 5 m/s. However, in the actual data, there are 12,953 records where the wind direction angle is less than 10°, with 2785 of those records having a wind speed between 0 ~ 5 m/s, accounting for approximately 21.5%. For wind direction angles lower than 5°, there are 6303 records, with 1362 falling within the 0 ~ 5 m/s wind speed range, yielding a proportion of 21.6%. Similarly, for angles less than 1°, there are 1324 records, of which 298 correspond to wind speeds of 0 ~ 5 m/s, representing approximately 22.5% of the total. Overall, the proportion of low wind speeds does not exhibit significant variation as the wind direction angle decreases. Furthermore, low wind speed does not account for an overwhelming proportion even when the wind direction angle is small. Consequently, for the bivariate joint PDF of wind speed and direction constructed using the Clayton Copula, significant differences exist at small wind direction angles.

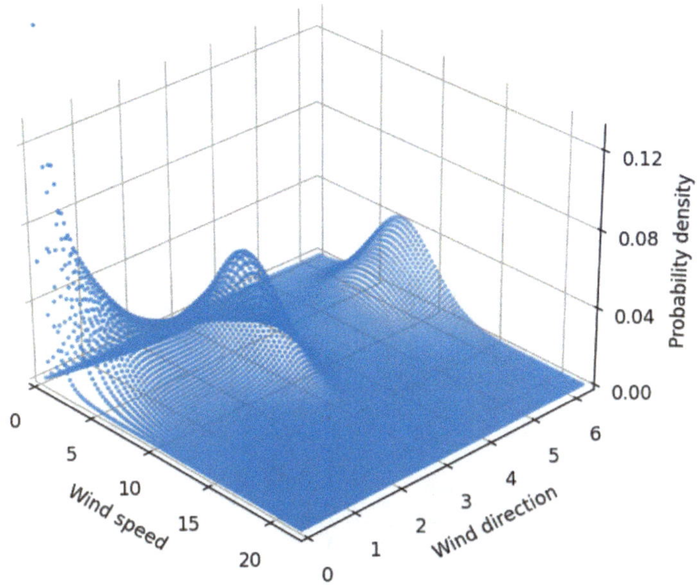

Fig. 1.5 Wind speed and direction are Clayton-Copula dependent

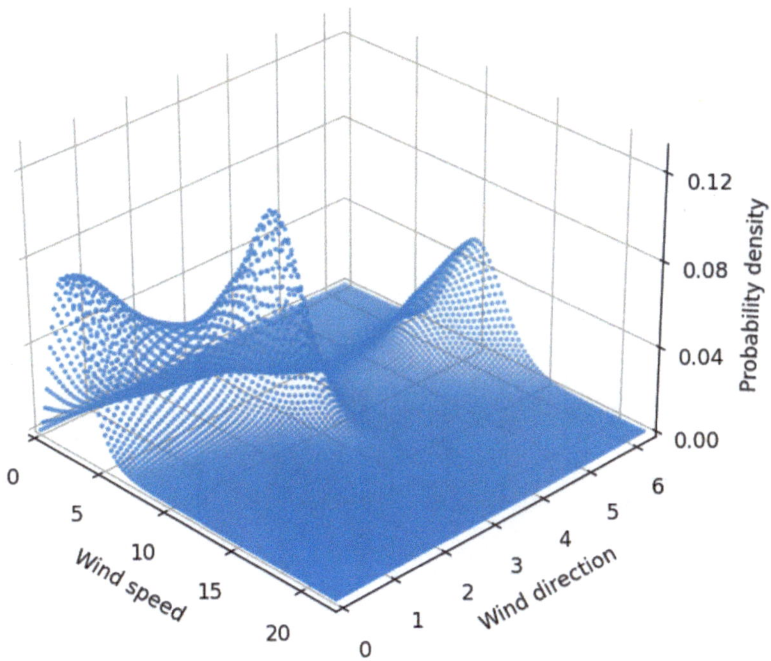

Fig. 1.6 Wind speed and direction are Frank-Copula dependent

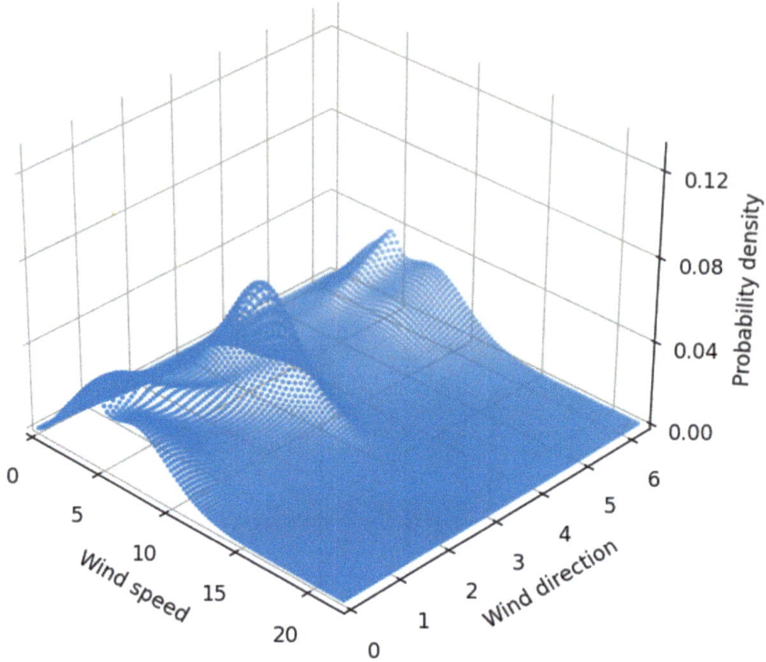

Fig. 1.7 Probability distribution of wind speed and direction under AL model

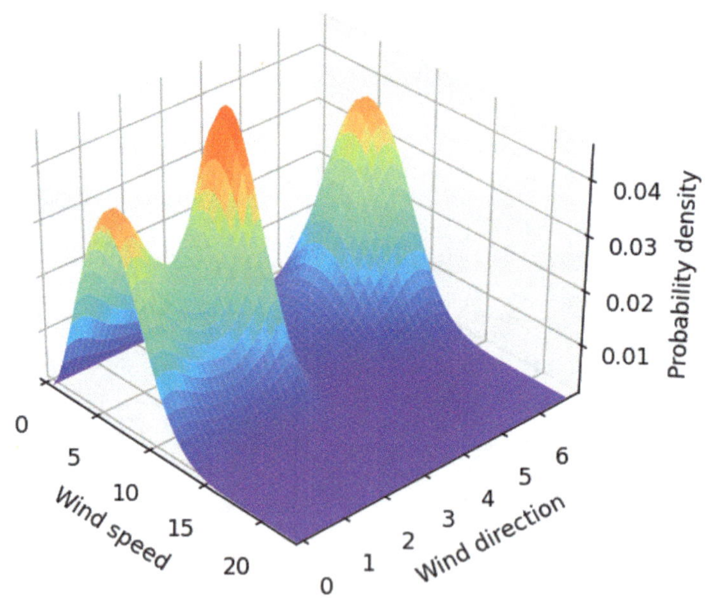

Fig. 1.8 Wind speed is independent of wind direction

Regarding the PDFs dependent on the Gumbel Copula and Frank Copula, it is observed that the bivariate joint PDF based on the Frank Copula exhibits a significant concentration of wind speed data in the range of 0~5 m/s when the wind direction angle is small. By contrast, when the wind direction angle approaches 360°, wind speed data records rarely fall within the 0 ~ 5 m/s range. In the actual data, there are a total of 20,999 data records with wind direction angles greater than 340°, with 4421 of those (approximately 21.1%) corresponding to wind speeds between 0 ~ 5 m/s. In the case of wind direction angles exceeding 350°, there are 11,555 records, of which 2291 (19.8%) exhibit wind speeds in the 0 ~ 5 m/s range. Therefore, compared with the actual data, the calculation results from Frank Copula-based bivariate joint distribution exhibit very few wind speeds below 5 m/s at large wind direction angles, which is unreasonable. In contrast, the PDF graph based on the Gumbel Copula (Fig. 1.4) demonstrates good fitting characteristics regardless of whether the wind direction angle is small or large. Given that the AIC values among the three models do not differ significantly, the Gumbel Copula is ultimately selected as the linking function to derive the bivariate joint probability density of wind speed and direction.

Among the actual data with wind speeds of 8 m/s or above, northwest winds account for 30.7%; in data with wind speeds of 10 m/s or above, northwest winds account for 37.9%; in data with wind speeds of 13 m/s or above, northwest winds account for 40.4%. As wind speed increases, the proportion of northwest winds also escalates, demonstrating a strong dependency relationship between wind speed and direction. In high-wind-speed regions, winds with larger wind direction angles occupy a higher proportion, implying that within the high-wind-speed range, the bivariate joint probability density at large wind direction angles is greater than that at small ones. This observation aligns with the joint probability density depicted in Fig. 1.4. It further underscores the necessity of considering the dependence between wind speed and direction when deriving the bivariate joint probability density function.

As shown in Figs. 1.4~1.8, the wind direction distribution at Yangshan Port exhibits two peaks, near 0° and 180°, respectively. This indicates that the highest wind frequencies in the Yangshan Port area occur from true north and slightly east of true south, followed by winds from the northeast and southeast. The scarcity of wind speed distribution within the wind direction angle range of 200° to 270° indicates that southwest winds rarely occur in the Yangshan Port area. In Fig. 1.4, the peak of the wind speed distribution at larger wind direction angles is slightly shifted toward higher wind speeds compared with that at smaller wind direction angles. This indicates that wind speeds associated with larger wind direction angles tend to be higher than those associated with smaller angles. Given that Yangshan Port is located in a subtropical monsoon climate zone with distinct seasonal variations, the wind directions are noticeably different between summers and winters. Additionally, the landmass to the west and the ocean to the east of Yangshan Port contribute to the regional climate, with southeast winds prevailing in the summer and northwest winds in the winter. Therefore, it can be inferred that the probability of experiencing strong winds in winter is higher than in summer at Yangshan Port. In line with the actual data, the average wind speed from May to July over the past 31 years is 5.94 m/s, while from November to January of the following year, the average wind speed is 6.37 m/s. Additionally, the average wind speed for wind direction angles less than 20° is 7.06 m/s, whereas for angles greater than 340°, the average wind speed is 7.31 m/s. These results are consistent with the observations in Fig. 1.4. The prevalence of northeast winds and the scarcity of southwest winds at Yangshan Port indicate that it is significantly influenced by the northeast trade winds. Since the bivariate joint PDF based on the Gumbel Copula effectively models the wind speed and direction distribution at Yangshan Port, the Copula method employed herein to analyze the bivariate joint probability density of wind speed and direction is deemed reliable.

1.3.3 Wind Energy Resource Assessment

Using the Gumbel Copula function as the linking function, the bivariate joint distribution function of wind speed and direction is constructed:

$$F(v,\theta) = \exp\left\{-\left[(-\ln U_1)^{1.0539} + (-\ln U_2)^{1.0539}\right]^{\frac{1}{1.0539}}\right\} \tag{1.32}$$

where $U_1 = F_{wbl}(v)$ and $U_2 = F_{vMs}\left(\theta;\hat{\Omega}\right)$. By taking the partial derivatives of this distribution function with respect to wind speed v and wind direction θ, the bivariate joint PDF of wind speed and direction is obtained.

Then, the wind energy density at Yangshan Port can be estimated. When ignoring the wind direction, the wind energy density can be calculated using the wind speed PDF as follows:

$$W(v) = \frac{1}{2}\rho\int_0^{\infty} v^3 f_v(v)dv \tag{1.33}$$

where ρ represents the air density, which is taken as the standard value of 1.293 kg/m^3 for this study.

When considering the wind direction, the wind energy density should be computed through the integration of the joint PDF. The formula for this calculation is:

$$W(v,\theta) = \frac{1}{2}\rho\int_0^{2\pi}\int_0^{\infty} v^3 f_{V,\Theta}(v,\theta)dvd\theta \tag{1.34}$$

By substituting the wind speed PDF and the bivariate joint PDF into relevant calculations, the results are obtained, as presented in Table 1.7.

Table 1.7 Wind energy density under different joint distributions unit: W/m^2

	Independent	Gumbel	Clayton	Frank
Wind energy	244.686	244.049	244.541	244.370

Utilizing the bivariate joint PDF based on the Gumbel Copula, the estimated wind energy density at Yangshan Port, considering both wind speed and direction, is approximately 244.049 W/m^2. It can be inferred that if wind direction information is ignored, relying solely on wind speed to estimate the wind energy density at Yangshan Port potentially leads to an overestimation of the wind energy density; however, the difference is negligible. Furthermore, the wind energy density obtained using the other two Copula-based wind speed and direction datasets shows similarly minor differences.

The wind energy density for Yangshan Port can also be referenced from the national standard GB/T 18710–2002, "Methods for Assessment of Wind Energy Resources for Wind Farms," which categorizes the wind energy density level at Yangshan Port as Grade 4. This categorization indicates that the area is classified as having "good" conditions for grid-connected wind power generation. Between 1990 and 2020, the average wind speed at Yangshan Port was approximately 6.2 m/s, exceeding the annual average wind speed reference value of 6.0 m/s provided in GB/T 18710–2002. This suggests that the region possesses abundant wind energy resources, making it suitable for large-scale wind power development.[2]

1.4 Conclusions

Based on Yangshan Port, this study fits the wind speed distribution using a two-parameter Weibull model and investigates the wind direction distribution using a von Mises distribution. After the two marginal distribution functions were obtained, the bivariate joint distribution function of wind speed and direction was derived based on the Copula function, facilitating the assessment of wind energy density at Yangshan Port.

Regarding the models, the two-parameter Weibull model provides a good fit for the wind speed distribution at Yangshan Port, with the parameter estimates obtained through the maximum likelihood method yielding better fitting results compared with those derived from the least squares method. When modeling wind direction, analysis of the root-mean-square error (RMSE) and the coefficient of determination (R^2) indicates that the sixth-order mixed von Mises model outperforms both the fifth-order and lower-order models, achieving a determination coefficient of 0.99763. A joint distribution function of wind speed and direction was constructed using three common Archimedean Copula functions and an Angular-Linear (AL) model. This resulted in three-dimensional probability density plots for the four distributions, as well as a joint probability density plot assuming independence of wind speed

[2] When applying the aforementioned models for wind condition modeling and wind energy density estimation, the computational demands remain relatively low. Specifically, utilizing an NVIDIA GeForce RTX 3060 GPU, parameter estimation for both the Weibull model and Copula functions achieves rapid convergence, while the sixth-order von Mises mixture model parameters can be computed within 20 minutes.

and direction. The AL model exhibited significant deviations in the bivariate joint probability density at both low and high wind direction angles, demonstrating that the Copula function models outperform the AL model. The AIC values for the three Copula models are relatively close. Combining the image results with actual data for comparisons, the Gumbel Copula-based bivariate probability distribution model was ultimately selected. The findings indicate that it is essential to take into account the impact of wind direction in the study of wind speed distribution. As the wind direction angle increases, the overall wind speed at Yangshan Port tends to shift toward higher speeds.

Regarding empirical validation, the wind energy resources at Yangshan Port are relatively abundant. Using the Gumbel Copula function for estimation yields a wind energy density of approximately 244.049 W/m^2; using the Clayton Copula function results in an estimated wind energy density of about 244.541 W/m^2; the Frank Copula function used for estimation leads to an estimated wind energy density of around 244.370 W/m^2.

Acknowledgements Fei Xu and Ruihua Zhang also made great contributions to this manuscript. We express our heartfelt thanks to them. This research also was supported by: National Natural Science Foundation of China (72474127) and National Social Science Fund Major Bidding Project of China (18ZDA052). Part of this chapter was published:Xu F, Wu X, Zhang R. Study of wind speed and direction at Yangshan Port. Renewable Energy 2025;247:122982. https://doi.org/10.1016/j.renene.2025.122982. The publication has been conducted with copyright permission.

References

Aries, N., Boudia, S. M., & Ounis, H. (2018). Deep assessment of wind speed distribution models: A case study of four sites in Algeria. *Energy Conversion and Management, 155*, 78–90. https://doi.org/10.1016/j.enconman.2017.10.082

Carrillo, C., Cidrás, J., Dorado, E. D., & Obando-Montaño, A. F. (2014). An approach to determine the Weibull parameters for wind energy analysis: The case of Galicia (Spain). *Energies, 7*(4), 2676–2700. https://doi.org/10.3390/en7042676

Carta, J. A., Bueno, C., & Ramírez, P. (2008a). Statistical modelling of directional wind speeds using mixtures of von Mises distributions: Case study. *Energy Conversion and Management, 49*(5), 897–907. https://doi.org/10.1016/j.enconman.2007.10.017

Carta, J. A., & Ramírez, P. (2007). Analysis of two-component mixture Weibull statistics for estimation of wind speed distributions. *Renewable Energy, 32*, 518–531. https://doi.org/10.1016/j.renene.2006.05.005

Carta, J. A., Ramírez, P., & Bueno, C. (2008b). A joint probability density function of wind speed and direction for wind energy analysis. *Energy Conversion and Management, 49*, 1309–1320. https://doi.org/10.1016/j.enconman.2008.01.010

Carta, J. A., Ramírez, P., & Velázquez, S. (2009). A review of wind speed probability distributions used in wind energy analysis: Case studies in the Canary Islands. *Renewable and Sustainable Energy Reviews, 13*(5), 933–955. https://doi.org/10.1016/j.rser.2008.05.005

Celik, A. N. (2003). Assessing the suitability of wind speed probability distribution functions based on wind power density. *Renewable Energy, 28*, 1563–1574. https://doi.org/10.1016/s0960-1481(03)00018-1

Chang, T. P. (2011). Performance comparison of six numerical methods in estimating Weibull parameters for wind energy application. *Applied Energy, 88*, 272–282. https://doi.org/10.1016/j.apenergy.2010.06.018

Huo, X., Yang, L., & Li, D. H. W. (2024). Determining Weibull distribution patterns for wind conditions in building energy-efficient design across the different thermal design zones in China[J]. *Energy, 304*, 132013.

Ji, X., Zou, J., Cheng, Z., et al. (2024). Generalized bivariate mixture model of directional wind speed in mixed wind climates[J]. *Alexandria Engineering Journal, 89*, 98–109.

Johnson, R. A., & Wehrly, T. E. (1978). Some angular-linear distributions and related regression models. *Journal of the American Statistical Association, 73*(363), 602–606. https://doi.org/10.1080/01621459.1978.10480062

Kiss, P., & Jánosi, I. M. (2008). Comprehensive empirical analysis of ERA-40 surface wind speed distribution over Europe. *Energy Conversion and Management, 49*, 2142–2151. https://doi.org/10.1016/j.enconman.2008.02.003

H.N. Li, X.W. Zheng, C. Li. Copula-based joint distribution analysis of wind speed and direction. Journal of Engineering Mechanics, 2019, 145 (05): 04019024-., doi:https://doi.org/10.1061/(ASCE)EM.1943-7889.0001600.

Masseran, N., Razali, A. M., Ibrahim, K., & Latif, M. T. (2013). Fitting a mixture of von Mises distributions in order to model data on wind direction in peninsular Malaysia. *Energy Conversion and Management, 72*, 94–102. https://doi.org/10.1016/j.enconman.2012.11.025

National Energy Administration. National Electricity Industry Statistics 2023., https://www.nea.gov.cn/2024-01/26/c_1310762246.htm.

Ouarda, T. B., & Charron, C. (2018). On the mixture of wind speed distribution in a Nordic region. *Energy Conversion and Management, 174*, 33–44. https://doi.org/10.1016/j.enconman.2018.08.007

Seguro, J. V., & Lambert, T. W. (2000). Modern estimation of the parameters of the Weibull wind speed distribution for wind energy analysis. *Journal of Wind Engineering and Industrial Aerodynamics, 85*, 75–84. https://doi.org/10.1016/s0167-6105(99)00122-1

Shamshad, A., Bawadi, M. A., Wan Hussin, W. M. A., Majid, T. A., & Sanusi, S. A. M. (2005). First and second order Markov chain models for synthetic generation of wind speed time series. *Energy, 30*(5), 693–708. https://doi.org/10.1016/j.energy.2004.05.026

Shanghai Maritime Safety Administration. China Communications News: Institutional Innovation to Stimulate the Vitality of First-Class Maritime Services—5th Anniversary of Shanghai Lingang New Area Construction, https://www.sh.msa.gov.cn/mtbd/102098.jhtml.

Sklar, A. (1959). Fonctions de répartition à n dimensions et leurs marges. *Publications de l'Institut de Statistique de l'Université Paris.*

Torres, J. L., García, A., De Blas, M., & De Francisco, A. (2005). Forecast of hourly average wind speed with ARMA models in Navarre (Spain). *Solar Energy, 79*(01), 65–77. https://doi.org/10.1016/j.solener.2004.09.013

Tuller, S. E., & Brett, A. C. (1984). The characteristics of wind velocity that favor the fitting of a Weibull distribution in wind speed analysis. *Journal of Climate and Applied Meteorology, 23*(1), 124–134. https://doi.org/10.1175/1520-0450(1984)023<0124:TCOWVT>2.0.CO;2

Usta, I., & Kantar, Y. M. (2012). Analysis of some flexible families of distributions for estimation of wind speed distributions. *Applied Energy, 89*, 355–367. https://doi.org/10.1016/j.apenergy.2011.07.045

Wais, P. (2017a). A review of Weibull functions in wind sector. *Renewable and Sustainable Energy Reviews, 70*, 1099–1107. https://doi.org/10.1016/j.rser.2016.12.014

Wais, P. (2017b). Two and three-parameter Weibull distribution in available wind power analysis. *Renewable Energy, 103*, 15–29. https://doi.org/10.1016/j.renene.2016.10.041

Yang, Z., & Dong, S. (2024). A semi-parametric trivariate model of wind speed, wind direction, and air density for directional wind energy potential assessment[J]. *Energy Conversion and Management, 314*, 118735.

Youcef Ettoumi, F., Sauvageot, H., & Adane, A.-E.-H. (2003). Statistical bivariate modelling of wind using first-order Markov chain and Weibull distribution. *Renewable Energy, 28*(11), 1787–1802. https://doi.org/10.1016/s0960-1481(03)00019-3

Chapter 2
Shipping Emission Inventory Preparation with AIS Data: A Stratified Random Sampling Method

2.1 Introduction

According to the Fourth Greenhouse Gas Study 2020 of International Maritime Organization, the contribution of shipping emissions to global air pollution has increased from 2.76% in 2012 to 2.89% in 2018.[1] Carbon dioxides, nitrogen oxides and sulfides account for 2.2%, 15% and 13%, respectively (Zhao et al., 2020). The negative externalities of shipping emissions lead to ecological environment problems (Zhang et al., 2023), air pollution (Fu et al., 2023; He et al., 2023) and health problems (Viana et al., 2020; Xiao et al., 2023). In order to reduce shipping emissions, the International Maritime Organization, different countries and regions have actively started corresponding emission restrictions and emission reduction measures. The formulation and implementation of emission reduction policies should be based on high-precision shipping emission inventories. Under such circumstance, the preparation of high-precision shipping emission inventory has become a prerequisite for providing early warning against regional air pollution and formulating pollution prevention and control policies.

Two estimation methods, including the top-down fuel oil method and the bottom-up shipping activity method, are widely used in the preparation of shipping emission inventory. The "top-down" method uses the sales statistics of ship fuel to calculate the total shipping emissions. However, it is difficult for the method to reflect the actual emissions of small and medium scales. The "bottom-up" method uses the navigation status, waterway characteristics, ship load and other AIS (Automatic Identification System) data of every single ship to calculate emissions. This method requires a large volume of data in different types, making it more difficult to obtain complete information. The calculation of total shipping emissions in the study area requires summation of emissions from every single ship. As a result, the accuracy

[1] International Maritime Organization https://www.imo.org/en/MediaCentre/HotTopics/Pages/Fourth-IMO-Greenhouse-Gas-Study-2020.aspx.

X. Wu, *Environmental Disaster in China*, https://doi.org/10.1007/978-981-92-0910-1_2

of shipping emission invention depends to a large extent on the calculation of every single ship. Both the integrity and accuracy of basic data relating to every single ship determine the reliability of emission results (Shu et al., 2023). However, the temporal granularity of AIS data is typically expressed in seconds, and the aggregated data volume is relatively large. The preparation of long-time shipping emission inventories for a region requires longer computing time and more resources. Therefore, this paper establishes a regional stratified sampling emission model based on the emission calculation of single ship and performs stratified calculations according to the waterway characteristics, ship characteristics and engine power in the region so as to obtain a high-efficiency and high-precision regional shipping emission inventory. While improving the preparation efficiency of shipping emission inventory, the uncertainty of the estimation results caused by the lack of static data is reduced accordingly.

2.2 Literature Review

Two methods, including the "top-down" method and the "bottom-up" method, are applied to the preparation of shipping emission inventory according to different calculation paths. The "top-down" method is more commonly used to estimate shipping emissions on a national and international scale. The total emissions are calculated according to the regional fuel sales or consumption data, the pollutant emission factor of different types of fuel oil and fuel consumption data. In the early days, the global atmospheric model was used to calculate the nitrogen and sulfide emissions from international shipping (Capaldo et al., 1999; Corbett et al., 1999). The International Maritime Organization (IMO) uses this method more frequently, and its Fourth Greenhouse Gas Study 2020 calculates the global shipping emissions in the period of 2012 to 2017 based on the natural gas and diesel sales data from the International Energy Agency. In accordance with the *Technical Guide for the Preparation of Air Pollutant Emission Inventory from Non-Road Mobile Sources* compiled by the Ministry of Ecology and Environment of the People's Republic of China of China,[2] the passenger/freight turnover of ships is used as the basis for calculating fuel consumption and preparing the emission inventory. For the "top-down" method, detailed ship data and traffic information are not necessary, thus ensuring a higher calculation efficiency. However, it ignores the characteristics of every single ship, so its precision is lower than that of the "bottom-up" method (Nunes et al., 2017). Based on ship activity data, the "bottom-up" method calculates the total shipping emissions using the ship automatic identification system, engine power, shipping time, unit energy consumption and other data. Jalkanen et al. (2009) established the

---

[2] Ministry of Ecology and Environment of the People's Republic of China: https://www.mee.gov.cn/gkml/hbb/bgg/201501/W020150107594587960717.pdf.

STEAM model (Ship Traffic Emission Assessment Model) based on the information provided by AIS and calculated the emission of various gas pollutants through the shipping data of every single ship. Being highly accurate and real-time, the model has been widely used by scholars as an estimation method (He et al., 2022; He et al., 2023; Yuan et al., 2023). Shu et al. (2023) calculated the emissions under different shipping modes and switching points using the STEAM model and evaluated the accuracy and error of this model. Fu et al. (2023) calculated exhaust emissions using ship activity data and evaluated the impact of emission results on the air quality nationwide based on weather study and forecast mode and multi-scale community air quality mode. Chen et al. (2024) systematically coordinated the relevant results published since 2015, analyzed the cause of shipping emission estimation uncertainty based on AIS data and concluded that data reliability, emission factor and unified engine load calculation methods could improve the accuracy of estimation using emission inventories. The "top-down" fuel method is less accurate in calculating actual shipping emissions on the regional scale and has significant errors in emission estimations on small and medium scales. The "bottom-up" activity method takes into account the type of ships and the state of navigations from the individual characteristics and ensures high accuracy for the exhaust emission calculation results. The division is performed on different scales based on specific study object and scope. Estimation on the global scale: IMO has released four global shipping emission estimation reports so far. The estimation of global shipping emissions accelerates the emission reduction process of shipping industry.[3] Large-scale calculations show higher requirements for data, calculations and human resources. Estimation on the national and river basin scales: The navigation capacity of Arctic route has been further improved, which contributes to the continued growth of shipping in this region. As a result, it suffers more shipping emissions. Due to the large scope and significant impact, scholars start to estimate and study the exhaust emissions from this route (Chen et al., 2021a, 2021b; Chen et al., 2024; Qi et al., 2024). The estimation on the national and river basin scale uses the information of every single ship to improve the shipping emission estimation efficiency and reduce the error of estimation results. Kelmalis et al. (2023) chose the ships on the international coastal short-distance routes as the object and focused on different navigation conditions and pollutant emissions. The results show that nitrogen oxide emissions are the highest, and the contribution rate of energy consumption and emission during berthing are only lower than that in cruising state. Zhou et al. (2022) chose the Yangtze River basin as the object and built a middle-level emission model according to the navigation characteristics of inland ships to estimate the carbon dioxide emissions in the study area. The estimation on the port and regional scale integrates geographical information, ship navigation status, ship type and other data for calculation, and the resulting emission inventory has higher accuracy and resolution. Kwon et al. (2023) considered the differences in navigation

---

[3] International Maritime Organization https://www.imo.org/en/MediaCentre/HotTopics/Pages/Fourth-IMO-Greenhouse-Gas-Study-2020.aspx.

mode, engine output power, ship fuel and other factors, and estimated the emissions of 7 types of gas pollutants at Busan Port in South Korea in 2020. Li et al. (2023) calculated the emissions of main engine, auxiliary engine and boiler using the ship dynamic method, designed the calculation method of high-resolution carbon emission inventory for the ships in Tianjin port area of China and conducted multi-feature prediction and spatial analysis by combining with geographical information. Using the Yangtze River as the study area, Peng et al. (2024) quantified the emission intensity of ships along the Yangtze water by analyzing vessel emissions across entire voyage sections. The calculation integrated emission volumes, cargo throughput, and sailing distance.

In summary, the extensive coverage of ship automatic identification system (AIS) has promoted the popularization of activity-based estimation method, and improved the accuracy of shipping emission inventory preparation. However, the processing of massive ship AIS data requires considerable calculation time and cost. The ship activity method is more applied to the port scale and less applied to the calculation on the global, national and regional scales. Moreover, the collection of detailed parameters poses some challenges to large-scale estimations. Considering the high-precision emission calculation in an enlarged scope study, the cases characterized by large data volume and increasing input parameters, easily cause severe errors to the calculation results. The surge in data volume leads to a significant reduction in calculation efficiency and a significant increase in calculation costs.

2.3 Methods and Models

This paper plans to prepare the shipping emission inventory using the stratified random sampling method to reduce the error caused by missing data during the estimation of shipping emissions, and avoid the high cost and low efficiency of emission calculations based on AIS data. Figure 2.1 shows the framework of the rapid accounting method. This algorithm consists of 4 steps, including data collection and processing, stratified sampling model (sample ship selection), STEAM model (exhaust emission calculation) and emission inventory preparation.

Step 1: Data collection and processing. Coordinate the AIS data of the study area, test and handle abnormal values and add the missing data through interpolation. Use web crawlers to collect static information from the official websites of classification societies of different countries. Check the integrity of the collected ship information and supplement the missing values. At last, classify and match the AIS data and static ship information.

Step 2: Select sample ships and build a stratified sampling model. Perform three-layer sampling according to waterways, ship types and engine power characteristics in the study area. Finish sampling according to the density of ships in each route section of the study area, types of ships in each density layer and power clustering

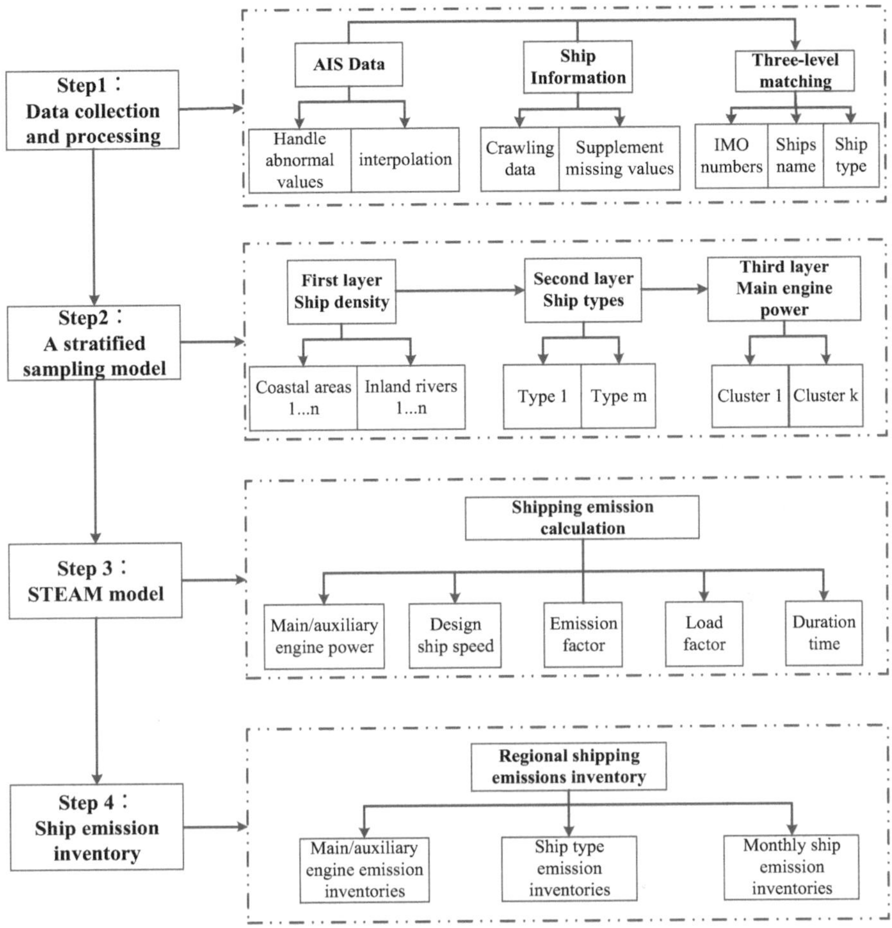

The framework of the rapid accounting method

Fig. 2.1 The framework of the rapid accounting method

results from each type of ship. At last, determine the sample size of ships for the regional exhaust emission calculation.

Step 3: Build the STEAM model and calculate the exhaust emissions of every single ship. Determine the parameters required for the calculation according to the STEAM model. Determine the design ship speed, power of main engine and auxiliary engine, emission factor and load factor according to the area of study object, characteristic data of ships and the existing literature. Calculate the exhaust emissions of every single ship based on the above efforts.

Step 4: Prepare the shipping emission inventory. Use the sample estimation method to calculate the total shipping emissions in the area according to the sample ships in Step 2, and the emissions of every single ship in Step 3. Finally, prepare the emission inventories according to specific main engines, auxiliary engines, different ship types and months, and analyze the emission characteristics.

2.3.1 Data Collection and Processing

2.3.1.1 Data Collection and Processing

In this paper, the registration information of ships were collected from the official websites of China Classification Society (CCS), Japan Classification Society (NK), Korean Register (KR),[4] etc. We read the static data of 16,256 ships using Python programmed crawler. There is no direct correlation between AIS data and static information. Therefore, we searched the corresponding ship IMO number and ship name through MMSI numbers and then matched AIS data with the static information. For the above reason, we obtained the static data of 17,982 ships from a third-party website (chinaports.com).[5] Then, we matched the ships with the corresponding data one by one and supplemented the MMSI information of each ship.

The AIS data of ships is identified as the basis of the crew, shipping companies and maritime supervision authorities to learn about ship status, and track and supervise ships in real time. However, the quality of ship AIS data determines the scientificity and reliability of relevant studies (Ferreira et al., 2023). According to the findings of the related scholars, about 5% of ship AIS data contain errors (Ignasi et al., 2012). In addition to the dynamic information relating to ship navigation, the AIS data set contains a certain proportion of abnormal data and floating object data, which will interfere with the estimation of shipping emissions (Kelly, 2022). The time of data transmission relating to shipping activity information may vary greatly due to equipment errors and failures, data collection cycle settings and other factors. Moreover, environmental interference, equipment factors, improper use, signal transmission and other factors result in the inevitable presence of noise in the AIS data. Therefore, data preprocessing operations, including denoising, error information filtering and data supplementation through interpolation, were performed to improve the data accuracy and provide high-quality basic data for the subsequent studies (Meyers et al., 2022; Yan et al., 2022).

In order to improve the quality of AIS data, it is necessary to pre-process the data set. First, the outliers are eliminated by MMSI code checking. Error types of AIS data include inconsistent MMSI lengths, allocated country codes and ship types.

[4] https://www.ccs.org.cn/, https://www.classnk.or.jp/, https://www.krs.co.kr/kor/.

[5] www.chinaports.com.

The repeated use of MMSI number and incomplete information data are detected by matching the dynamic data (AIS data) with the static data (ship information). At last, the related abnormal data, such as abnormal locations, speeds and steering angles, is checked. The average speed and speed threshold of the adjacent position point are used to judge whether the speed is abnormal and the steering problem is measured by checking whether the adjacent steering angle is smooth in the trajectory. The trajectory data is repaired after the abnormal trajectory points are removed. The same keywords are unavailable between the existing AIS data and static information, failing to establish a direct correlation. Three-level matching is required to correlate the two types of data. First, the first-level matching is performed according to the maritime mobile service identity (MMSI) code in the AIS database. The second-level matching is performed with IMO numbers, MMSI codes and Chinese and English names of ships to find out the missing ships in the first-level matching. Then, the corresponding power, design speed and other information of the ships with missing static information during the second-level matching are obtained by parameter fitting and matching.

2.3.1.2 Parameter Determination

Complete static information and activity data of ships, including engine power, maximum speed, actual speed, pollutant emission factor and shipping time, are necessary for the calculation of shipping emissions. Among the above parameters, shipping time and speed can be directly obtained from AIS data while engine power and maximum speed are available after the successful match of static data. Due to the limited static data sources and sizes of ships, some information is missing, so supplementation should be carried out in an appropriate way. Moreover, it needs to adjust some exogenous parameters, such as pollutant emission factor, load factor and fuel correction factor, because the study objects and scopes are different.

1. Determination of emission factor

Emission factor is an important parameter of shipping emission model and affects the calculation results of shipping emissions. An appropriate emission factor should be selected according to the characteristics of the study area. However, the emission factor is affected by the power, fuel and working conditions of ship engines, and there are significant differences between different types of ships (inland rivers, coastal regions and oceans) and shipping segments. With different types of sample ships and portable emission measurement tools, the related scholars compared different shipping status and found that SO_2, PM and VOC are directly correlated with fuel quality and consumption and free from the impact of ship operating conditions and loads. NO_x, CO_2 and CO are correlated with fuel quality, fuel consumption, ship engine operating conditions and durations (McCaffery et al., 2021; Topic et al., 2023; Yang et al., 2021). In accordance with *Limits and measurement methods for exhaust pollutants from marine engines (CHINA I, II)*, inland ships, river-sea ships and inland fishing ships shall use the diesel that meet the national standard of GB252. In other words, these ships could use the diesel fuel with sulfur content not higher than 3.5% before June 30, 2017, the diesel fuel with sulfur content of 0.5% from July 1, 2017 to January 1, 2018 and the diesel fuel with sulfur content not more than 0.1% after January 1, 2018. For the above reason, we collected the emission factor of inland ships in accordance with the relevant national standards and studies. See Table 2.1 for details.

We also prepared and coordinated the emission factor of coastal ships and ocean ships in accordance with *Regulations for the prevention of air pollution from ships* as Annex VI to MAPROL73/78 Convention of IMO, the Fourth Greenhouse Gas Study released by IMO and the studies on the emission measurement of sea ships and ocean-going ships. See Table 2.2 for details. When a ship is sailing, the power of its main engine will change according to different shipping status. The combustion efficiency of fuel is determined by the real-time load of engine, which directly changes the emission factor of pollutants. See Table 2.3 for the fuel correction factor. When the ship load rate is lower than 20%, a certain impact occurs to the emission factor (Fan et al., 2016; Jinxian et al., 2019; Weng et al., 2020). See Table 2.4 for the correction factor of ship load rate coefficient.

Table 2.1 Fuel correction factor

Fuel type	Sulphur content /%	CO_2	CO	HC	NO_x	SO_2
RO	1.5	1	1	1	1	0.56
MDO	1.5	1	1	1	0.9	0.56
MDO	0.5	1	1	1	0.94	0.18
MDO	0.2	1	1	1	0.94	0.07
MDO	0.1	1	1	1	0.94	0.04

Note RO (residual oil), MDO (marine heavy diesel)

Table 2.2 Emission factors of inland ships

Engines	Fuel types	Sulphur content %	SO_2	NO_x	PM_{10}	$PM_{2.5}$	CO	CO_2	HC
Main	HFO	3.5	10.29	18.1	1.42	1.31	1.4	680	0.5
	MDO	1 ~ 2	3.97	13.2	0.47	0.43	1.1	620	0.5
	MGO	0.5	2.08	13.2	0.31	0.29	1.1	620	0.5
Auxiliary	HFO	3.5	11.98	14.7	1.44	1.32	1.1	620	0.5
	MDO	1 ~ 2	4.24	13.9	0.49	0.45	1.1	620	0.5
	MGO	0.5	2.12	10	0.31	0.29	1.5	683	0.6

Note HFO (Heavy Fuel Oil), MDO (Marine Heavy Diesel Oil), MGO (Marine Gasoline)

Table 2.3 Emission factor of coastal ships and ocean ships

Engines	Fuel types	SO_2	NO_x	PM_{10}	$PM_{2.5}$	HC	CO	Sulphur content %
Main	RO	10.3	18.1	1.4	1.3	0.6	1.4	2.70
Main	MDO	3.6	17	0.5	0.4	0.6	1.4	1.00
Main	MGO	1.8	17	0.3	0.3	0.6	1.4	0.50
Main	RO	11.2	14	1.4	1.3	0.5	1.1	2.70
Main	MDO	4	13.2	0.5	0.4	0.5	1.1	1.00
Main	MGO	2	13.2	0.3	0.3	0.5	1.1	0.50
Auxiliary	RO	12	14.7	1.4	1.3	0.4	1.1	2.70
Auxiliary	MDO	4.2	13.9	0.5	0.5	0.4	1.1	1.00
Auxiliary	MGO	2.1	13.9	0.3	0.3	0.4	1.1	0.50
Main	MGO	2.1	13.2	0.7	0.7	0.5	1.1	0.50
Main	MGO	2.1	13.2	0.3	0.3	0.5	1.1	0.50
Main	MGO	2.1	10	0.3	0.3	0.3	1.5	0.50
Auxiliary	MGO	2.1	10	0.4	0.4	0.3	1.7	0.50
Auxiliary	MGO	2.1	10	0.3	0.3	0.3	1.5	0.50

Note RO (residual oil), MDO (Marine Heavy Diesel Oil), MGO (Marine Gasoline)

2. Determination of design speed

When the ship load rate is higher than 20%, the load factor is the third power of the ratio of actual speed to design speed. The actual speed is provided by AIS data. A large number of design speed data is missing. Therefore, multiple methods are required for estimation. Yau et al. (2012) set the cruising speed after being divided by 0.94 as the maximum speed. Wang et al. (2023) set the average cruising speed after being divided by 0.94 and then rounded up as the maximum cruising speed. Cheng et al. (2016) supplemented the missing data using the maximum speed of ships of the same type and similar tonnage. Starcrest Consulting Group filled the HIS Markit database with the maximum service speed as the missing maximum speed in the database.[6] IMO uses information such as ship length, main engine power, and ship deadweight tons by type to supplement the missing design speed values through multi-linear fitting.

If more static ship data is missing, supplementation is performed by using the data of ships of the same type and similar tonnage. However, only a limited amount of data can be supplemented. If the multi-linear regression method is used and the relevant regression parameters such as main engine power and tonnage data are missing, estimations should be repeatedly performed, which will increase the calculation error of design speed. Each piece of AIS data contains ship speed and the data is

[6] Starcrest Consulting Group, LLC, 2022. Port of Los Angeles Inventory of Air Emissions 2021. Technical Report, September 2022.

Table 2.4 Correction factor of ship load rate coefficient

Load rate	CO_2	CO	HC	NO_x	SO_2	PM_{10}	$PM_{2.5}$
0.01	5.82	19.32	59.28	11.47	1	19.17	19.17
0.02	3.28	9.68	21.18	4.63	1	7.29	7.29
0.03	2.44	6.46	11.68	2.92	1	4.33	4.33
0.04	2.01	4.86	7.71	2.21	1	3.09	3.09
0.05	1.76	3.89	5.61	1.83	1	2.44	2.44
0.06	1.59	3.25	4.35	1.6	1	2.04	2.04
0.07	1.47	2.79	3.52	1.45	1	1.79	1.79
0.08	1.38	2.45	2.95	1.35	1	1.61	1.61
0.09	1.31	2.18	2.52	1.27	1	1.48	1.48
0.1	1.25	1.96	2.18	1.22	1	1.38	1.38
0.11	1.21	1.79	1.96	1.17	1	1.3	1.3
0.12	1.17	1.64	1.76	1.14	1	1.24	1.24
0.13	1.14	1.52	1.6	1.11	1	1.19	1.19
0.14	1.11	1.41	1.47	1.08	1	1.15	1.15
0.15	1.08	1.32	1.36	1.06	1	1.11	1.11
0.16	1.06	1.24	1.26	1.05	1	1.08	1.08
0.17	1.04	1.17	1.18	1.03	1	1.06	1.06
0.18	1.03	1.11	1.11	1.02	1	1.04	1.04
0.19	1.01	1.05	1.05	1.01	1	1.02	1.02
0.2	1	1	1	1	1	1	1

characterized by large size and high completeness and reliability. Therefore, we assumed the main engine power of ships to 85% to 90% of the rated power based on the actual maximum speed of each ship.[7] For the missing design ship speed in the paper, the maximum cruising speed in the AIS dynamic data after being divided by 0.85 is taken as the design speed per hour.

3. Determination of ship power

Main engine power is one of the important parameters for calculating shipping emissions. However, missing data will increase calculation errors. In order to make the emission calculation more accurate, it is necessary to estimate the missing engine data. The missing engine data is supplemented according to the estimation method in the existing studies, reports and references (Chen et al., 2021a, 2021b; Sajjad et al., 2025; Ye et al., 2022; Zhang et al., 2023). To be specific, the model of predicting missing engine power information depends on the key missing information itself; three situations with missing main engine power: ① For the ships without power data

[7] Sheng Zhenbang, Liu Yingzhong, *Principles of Principle of Naval Architecture (Volume II)*, Shanghai Jiaotong University Press.

but with complete width, draft, load and design speed information, linear or nonlinear regression methods are used to estimate the engine data. Classification is performed through the ship type information contained in the static data to determine the fitting model. ② For the ships without engine information and the information necessary for the above estimation process, the missing information is supplemented with the engine data from the ships of similar types and tonnages. ③ For the ships only with MMSI number and ship type information, the average power from the ships with the same types is used as the engine power. First, the static data is classified according to ship types. Then the missing data is supplemented using the data from the ships with the same types and deadweight tons. Linear regression is finally performed according to the ship speed and total ship length and the static ship data with missing main engine power is supplemented. See Fig. 2.2 for the fitting.

2.3.2 *Stratified Sampling Model (Sample Ship Selection)*

This study uses stratified random sampling method to obtain the ship sample data. This method is suitable for the study objects with large differences, such as ships. The sampling framework consists of three layers, and the ships are sampled according to three parameters, including ship density, ship type and main engine power in different shipping segments. The three parameters are identified as the key parameters for the calculation of exhaust emission process based on shipping activities. In the process of parameter selection, the ships with complete static data and dynamic data are selected, and then the emission of every single ship is calculated. Finally the total emissions of all ships in the area are obtained.

The first layer is classified according to ship density. According to the distribution characteristics of waterways in Shanghai, the study area can be divided into inland rivers and coastal areas. Then the study area is further divided into several subareas and each subarea has a similar ship density. The sample data collection method is as follows:

$$n_i = \frac{RT_i}{RT_1 + RT_2 + \cdots + RT_i} * VT \tag{2.1}$$

where, i represents the number of segments in the entire study area, RT_i represents the number of ships in the channel of the i th section, and VT is the total number of ships sampled from the study area. The number of ships sampled from each segmented area is adjusted according to the ship density of this segmented area, with more sampling amount in the higher density segment (Table 2.5).

The second layer is classified according to ship types. As a key parameter for the emission inventory, shipping emission factor is directly correlated with the gross tonnage of ships. According to the study Weng et al. (2020), ship types are correlated with gross tonnage. Ship type data can be obtained more easily than the gross tonnage

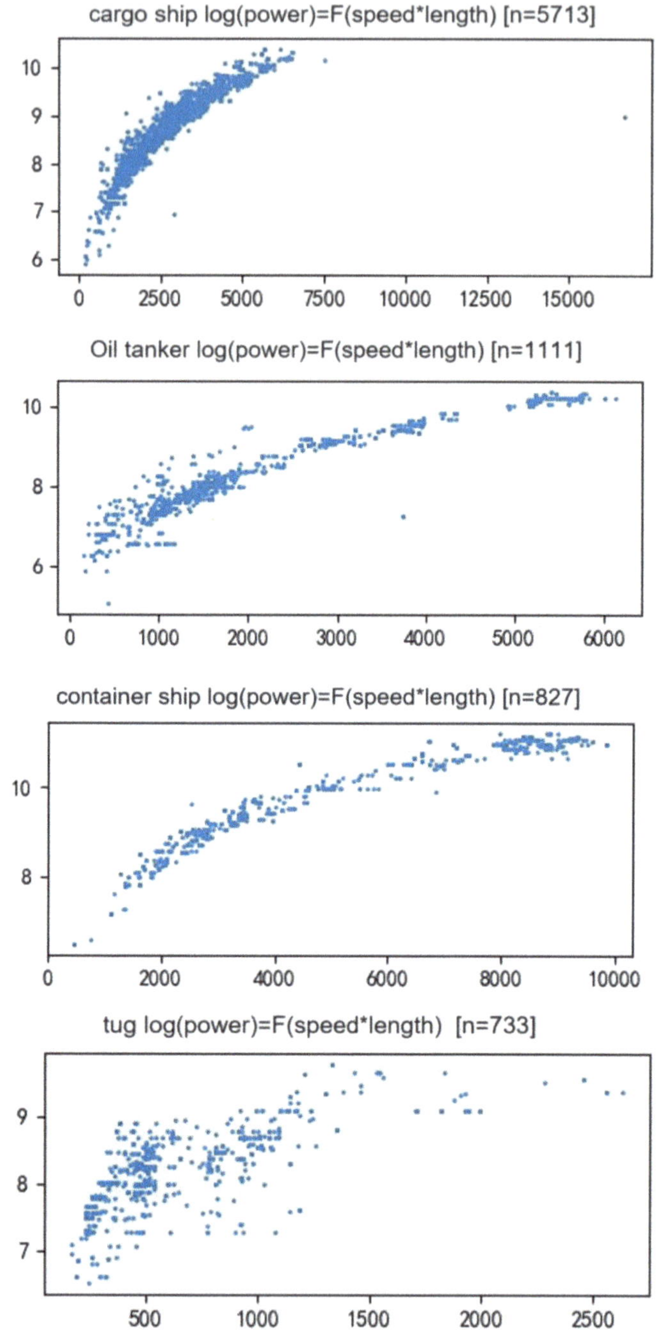

Fig. 2.2 Ship power regression fitting plot

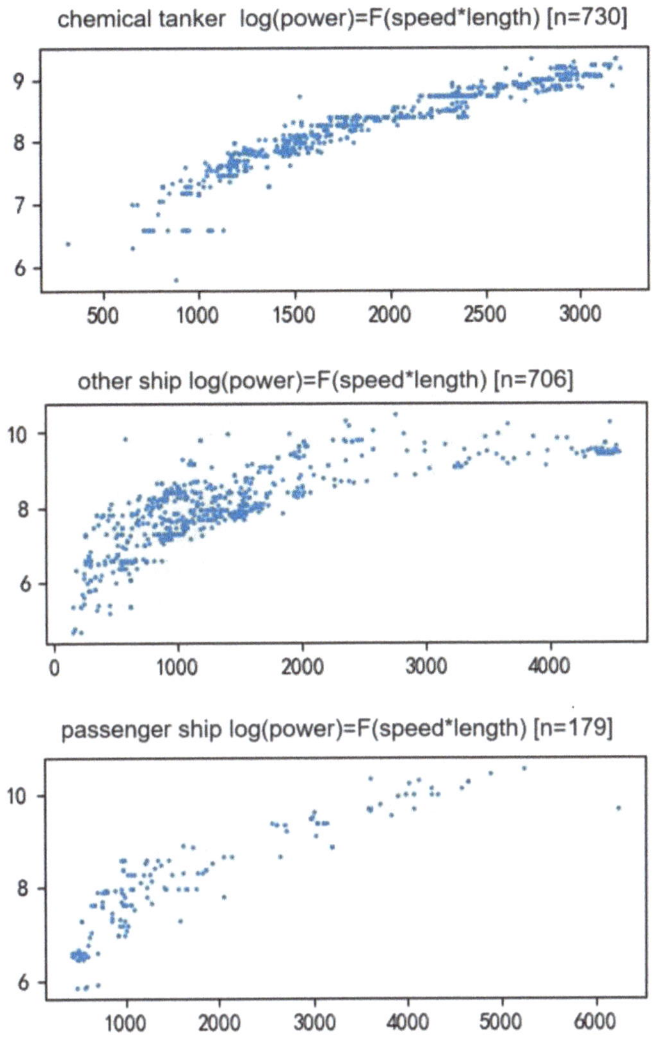

Fig. 2.2 (continued)

data, so ship type is used as the sampling criterion for the second layer. The method of calculating different ship types in each segment is as follows:

$$n_{i-j} = \frac{ST_{i-j}}{ST_{i-1} + ST_{i-2} + \cdots + ST_{i-j}} * n_i \quad (2.2)$$

Where, n_{i-j} is the sample number of the *j*th type of ships in the *i*th segment and ST_{i-j} is total number of the *j*th type of ships in the *i*th segment.

Table 2.5 Ship power regression fitting results

Ship Types	Number	Regression equation	R^2 value
Cargo ship	5713	Log P = −2.8032 + 1.4743 log(S*L)	0.918
Oil tanker	1111	Log P = −2.1824 + 1.4124 log(S*L)	0.898
Container ship	827	Log P = −4.7474 + 1.7377 log(S*L)	0.974
Tugboat	733	Log P = 3.2889 + 0.77812 log(S*L)	0.574
Chemical tanker	730	Log P = −3.7369 + 1.6009 log(S*L)	0.927
Other ship	706	Log P = −0.92332 + 1.2678 log(S*L)	0.707
Passenger ship	179	Log P = −2.6278 + 1.5017 log(S*L)	0.877

Note P(main power), S(speed), L(length)

The third layer is classified according to main engine power. During the navigation of a ship, exhaust emissions vary with the main ship engine power, ship size, the rated ship speed and navigation resistance (meteorological conditions and navigation channels) and energy efficiency. According to the study of (Zhang et al., 2019a, 2019b), the factors affecting shipping emissions have a strong correlation with the main ship engine. Therefore, the power characteristics of main engine are used as the classification basis.

$$\mathrm{dis}(X_i, C_j) = \sqrt{\sum_{t=1}^{m} (X_{it} - C_{jt})^2} \tag{2.3}$$

Where, X is the data sample and C is the cluster center point. According to the principle of shortest distance, the best cluster center point is obtained as per the distance formula of each data point to each cluster center. With the new center point, we continue iteration until the maximum set number of iterations is reached or we stop the calculation when no new center point is generated.

2.3.3 STEAM Model (Shipping Emission Calculation)

In this study, the STEAM model (Lei et al., 2021; Weng et al., 2020) is used to estimate the shipping emissions (Jalkanen et al., 2009). Based on the AIS activity data of ships, this method combines the static ship of every single ship to calculate the emissions of various pollutants. The formula of calculating the emissions of every single ship is as follows:

$$E = E_{main} + E_{auxiliary} \tag{2.4}$$

$$E_{main} = P_{main} \times LF_{main} \times T_{main} \times LEF \tag{(2.5)}$$

$$E_{auxiliary} = P_{auxiliary} \times LF_{auxiliary} \times T_{auxiliary} \times LEF \tag{2.6}$$

E is the exhaust emissions of a single ship, and main, auxiliary and boiler represent the emissions of main engine, auxiliary engine and boiler. P is the power of the ship power equipment; LF is the load factor; T is the operation (duration) time; LEF is the localized emission factor.

$$LF_{main} = \left(v_{actual}/v_{max}\right)^3 \tag{2.7}$$

v_{actual} represents the actual ship speed; some factors, such as meteorological conditions and geographical environments, are added according to the real-time ship speed. v_{max} represents the maximum ship speed, and this parameter is substituted into Formula (2.5) to obtain:

$$E_{main} = P_{main} \times \left(v_{actual}/v_{max}\right)^3 \times T_{main} \times EF_{main} \tag{2.8}$$

$$LEF = EF \times ECF \times FCF \times CF \tag{2.9}$$

EF is the benchmark shipping pollutant emission factor, ECF is the engine load correction factor, FCF is the ship fuel correction factor based on the sulfur content in the local fuel, and CF is the control factor of the emission reduction measures or policies. The emission factors of various pollutants depend on ship types, and types, status and fuel types of ship engines. If the basic emission factor is used, other parameters are set to 1.

According to the relationship specified by IMO between main engine power and auxiliary engine power,[8] we estimated the auxiliary engine power using the formula below.

When the main engine power is less than or equal to 10,000 kW:

$$P_{auxiliary} = 0.05P_{main}/0.85 \tag{2.10}$$

When the main engine power is more than 10,000 kW:

$$P_{auxiliary} = 0.025P_{main}/0.85 \tag{2.11}$$

[8] IMO. Designation of an Emission Control Area for Nitrogen Oxides, Sulphur Oxides and Particulate Matter. Marine Environment Protection Committee, 2010.

2.3.4 Total Shipping Exhaust Emissions

For the preparation of shipping emission inventory, it is necessary to calculate the total exhaust emissions from all the samples using the stratified sampling model.

$$\mathrm{Emission_N} = \mathrm{emission_N} \pm \Delta\bar{e} \tag{2.12}$$

$\mathrm{Emission_N}$ is the total shipping emissions, $\mathrm{emission_N}$ is the emissions without taking into account errors and $\Delta\bar{e}$ is the error term.

$$\mathrm{emission_N} = \sum \left(\mathrm{Emission_k}/\mathrm{n_k} \right) \tag{2.13}$$

$\mathrm{Emission_K}$ is the shipping emissions in the sample cluster k and $\mathrm{N_K}$ is the size of samples in the *k*th cluster.

$$\Delta\bar{e} = \mathrm{t} \times \mu\bar{e} \tag{2.14}$$

t is the mean confidence interval value and $\mu\bar{e}$ is the mean sample error value.

$$\mu\bar{e} = \sqrt{\frac{s^2}{n-1} \times \left(\frac{N-n}{N-1} \right)} \tag{2.15}$$

n is the number of ships in the sample; N is the total number of ships.

$$s^2 = \frac{\sum(\Delta s_k \times n_k)}{n} \tag{2.16}$$

n is the number of samples, Δs_k is the sample variance of cluster k and n_k is the number of samples in cluster k.

2.4 Empirical Analysis

2.4.1 Shipping Emission Inventory

According to the stratified sampling model and exhaust emission model, the preparation of shipping emission inventory involves Formula (2.1) to (2.16). We completed the selection of sample ships, calculated the emissions of every single ship and all ships and obtained the shipping emissions of Shanghai from January to December 2019. First, we read the AIS data for the whole year and coordinated the number of ships sailing in the area of Shanghai in 2019. We used ship MMSI number to query the ships one by one at http://ship.chinaports.com, and obtained the IMO number and name of these ships. Then a ship database covering the period of the study area is thus established through IMO number and ship name as well as the static information of ships in the classification societies of various countries. According to the stratified sampling steps, the sampling process contains Formula (2.1) to (2.3) and the waterway is divided into 12 sampling segments, as shown in Fig. 2.3.

In this study, the sampling continues for 10 days every month according to the calculation efficiency and sampling result stability (4.2 Stability analysis). Sample ships are selected according to the sampling framework. The first layer consists of 12 areas based on the ship density per unit area in the study area. The number of ships in these areas is 8051, 8116, 8849, 8363, 6792, 3848, 2222, 1523, 1133, 941, 778 and 278 respectively. 4,922,293 pieces of AIS data are sampled from the target areas. In the second layer, ships are divided into five types, including cargo ships, oil tankers, tugboats, passenger ships and other ships. Each type of ships has complete static data and AIS data. The number of ships of each type varies depending on sampling segments. According to the proportion of different types of ships to the total number of ships, we calculated the number of each type of ships sampled from each segment.

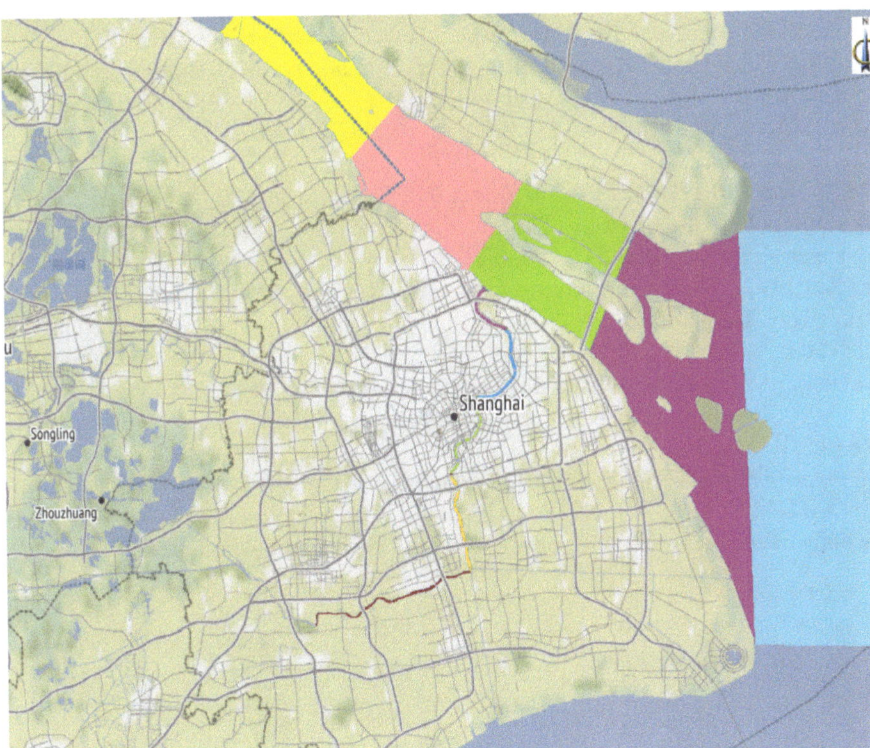

Fig. 2.3 Ship density

Table 2.6 Emission inventories by ship type (main engine emissions/t)

Ship type	CO_2	SO_2	NO_x	$PM_{2.5}$	PM_{10}	CO	VOC
Cargo ships	98265.84	1510.68	1870.36	172.65	215.81	201.42	71.94
Other ships	24045.42	369.66	457.67	42.25	52.81	49.29	17.60
Oil tankers	18468.14	283.92	351.52	32.45	40.56	37.86	13.52
Tugboats	739.99	11.38	14.09	1.30	1.63	1.52	0.54
Passenger ships	1899.82	29.21	36.16	3.34	4.17	3.89	1.39
Total	143419.20	2204.83	2729.79	251.99	314.98	293.98	104.99

In the thrid layer, the ships of the same type are divided into different groups with similar main engine power through the K-means clustering method for the engine power of the same types of ships. The engine power of the same type of ships is divided into three categories using the power estimation model of static data. Then the number of sampled ships under each category is calculated.

The emissions of sampled ships are calculated according to the shipping emission model. Based on the sampling frequency of 10 days/month (the average results of 10 times of sampling), the emission of various pollutants from the main engine and auxiliary engine of ships are respectively obtained, as shown in Tables 2.6, 2.7, 2.8 and 2.9. In 2019, the total amount of major types of shipping emissions in Shanghai is 154,977.147 tons, of which the amounts of CO_2, SO_2, NO_x, $PM_{2.5}$, PM_{10}, CO and VOC are 148,853.033 tons, 2288.37 tons, 2833.22 tons, 261.523 tons, 326.91 tons, 305.116 tons and 108.97 tons respectively, accounting for 96.5%, 1.48%, 1.83%, 0.17%, 0.21%, 0.2% and 0.07%. According to the analysis of ship types, the pollutant emissions of cargo ships, other ships, oil tankers, tugboats and passenger ships are 106,339.346 tons, 26,010.575 tons, 19,749.922 tons, 785.947 tons and 2091.357 tons, accounting for 68.62%, 16.78%, 12.74%, 0.51% and 1.35% respectively. Cargo ships are the main source of shipping emissions in Shanghai, followed by oil tankers. Tugboat emissions are less compared to the total emissions. The amount of pollutants from January to December is 9962.958 tons, 7220.073 tons, 9002.458 tons, 7507.819 tons, 7428.384 tons, 8086.984 tons, 8986.070 tons, 19,830.210 tons, 7423.345 tons, 20,081.397 tons, 19,931.828 tons and 19,515.622 tons respectively, accounting for 6.43%, 4.66%, 5.81%, 4.84%, 4.79%, 5.22%, 5.8%, 12.8%, 11.24%, 12.96%, 12.86% and 12.59%. The total emissions from August to December account for 62.45% of the total emissions in the whole year.

2.4.2 *Stability Analysis*

This section analyzes the impact caused by the number of sampling days on the calculation results. According to the data of cargo ships and the proportion of emissions from cargo ships, cargo ships play an important role in the shipping sector

Table 2.7 Emission inventories by month (main engine emissions/t)

Month	CO_2	SO_2	NO_x	$PM_{2.5}$	PM_{10}	CO	VOC
January	9273.34	142.56	176.51	16.29	20.37	19.01	6.79
February	6720.10	103.31	127.91	11.81	14.76	13.78	4.92
March	8364.68	128.59	159.21	14.70	18.37	17.15	6.12
April	6995.93	107.55	133.16	12.29	15.36	14.34	5.12
May	6906.19	106.17	131.45	12.13	15.17	14.16	5.06
June	7526.34	115.71	143.25	13.22	16.53	15.43	5.51
July	8350.13	128.37	158.93	14.67	18.34	17.12	6.11
August	18291.69	281.21	348.16	32.14	40.17	37.49	13.39
September	16037.56	246.55	305.25	28.18	35.22	32.87	11.74
October	18554.66	285.25	353.16	32.60	40.75	38.03	13.58
November	18407.12	282.98	350.36	32.34	40.43	37.73	13.48
December	17991.47	276.59	342.44	31.61	39.51	36.88	13.17
Total	143419.20	2204.83	2729.79	251.98	314.98	293.98	104.99

Table 2.8 Emission inventories by ship type (auxiliary engine emissions/t)

Ship type	CO_2	SO_2	NO_x	$PM_{2.5}$	PM_{10}	CO	VOC
Cargo ships	3871.38	59.52	73.69	6.80	8.50	7.94	2.83
Other ships	937.31	14.41	17.84	1.65	2.06	1.92	0.69
Oil tankers	501.34	7.71	9.54	0.88	1.10	1.03	0.37
Tugboats	14.90	0.23	0.28	0.03	0.03	0.03	0.01
Passenger ships	108.89	1.67	2.07	0.19	0.24	0.22	0.08
Total	5433.83	83.54	103.43	9.55	11.93	11.14	3.98

of Shanghai. The stability of sampling results with cargo ships as typical ships is analyzed according to the sampled data, calculation time, calculation resources and result interpretability. We have also plotted the curve of estimated CO_2 emissions from cargo ships by comparing and analyzing the sampling results on the third day to the 12th day. See Fig. 2.4 for details. With the increase of sampling days, the curve of total CO_2 emissions from cargo ships tends to be stable. As shown by the curve, the total CO_2 emissions of cargo ships tend to be consistent when the sampling rate is 23% to 40% on the seventh day to the 12th day. Considering the calculation efficiency, cost and error, we have set 10 days as the sampling benchmark and compared the relative error of different sampling rates, as shown in Fig. 2.5. The relative error rate of the estimation results is less than 20% in the sampling interval from the seventh day to the 12th day, and the relative error rates are 1.15% and 2.96% on the 11th day and the 12th day respectively. Therefore, the number of shipping emission sampling days is set to 10 days in this paper.

Table 2.9 Emission inventories by month (auxiliary engine emissions/t)

Month	CO_2	SO_2	NO_x	$PM_{2.5}$	PM_{10}	CO	VOC
January	295.93	4.55	5.63	0.52	0.65	0.61	0.22
February	214.66	3.30	4.09	0.38	0.47	0.44	0.16
March	282.03	4.34	5.37	0.50	0.62	0.58	0.21
April	215.21	3.31	4.10	0.38	0.47	0.44	0.16
May	228.66	3.52	4.35	0.40	0.50	0.47	0.17
June	241.07	3.71	4.59	0.42	0.53	0.49	0.18
July	280.85	4.32	5.35	0.49	0.62	0.58	0.21
August	754.90	11.61	14.37	1.33	1.66	1.55	0.55
September	697.28	10.72	13.27	1.23	1.53	1.43	0.51
October	733.20	11.27	13.96	1.29	1.61	1.50	0.54
November	737.08	11.33	14.03	1.30	1.62	1.51	0.54
December	752.96	11.58	14.33	1.32	1.65	1.54	0.55
Total	5433.83	83.54	103.43	9.55	11.93	11.14	3.98

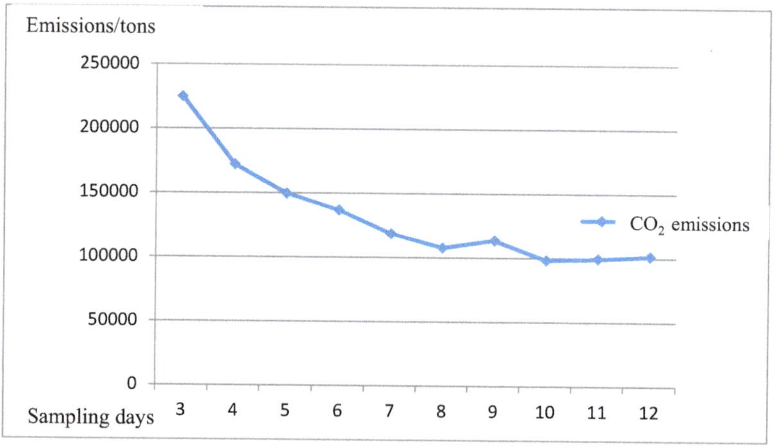

Fig. 2.4 CO_2 emission curves of cargo ships with different sampling rates

2.5 Conclusion and Prospect

This study calculates shipping emissions using ship AIS data and stratified sampling emission model according to specific waterways, ship types, main engine power and other characteristics. Based on complete ship information, this calculation method reduces the uncertainty caused by the absence of single ship parameters during estimation. When the ship sampling ratio is not lower than 1/3, the relative error of calculation results is less than 3%. When any ship information is missing and the scope of this study is enlarged, the rapid accounting method significantly improves

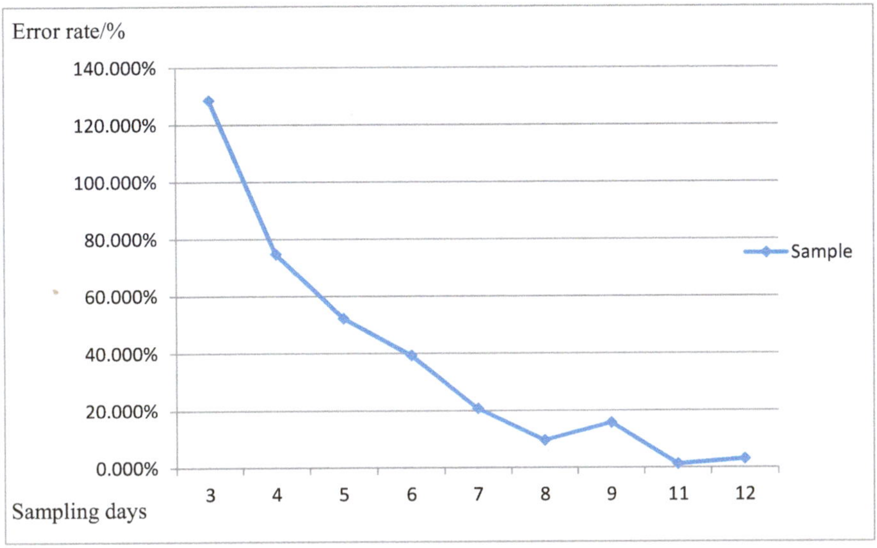

Fig. 2.5 Relative error rates for different sampling durationsSampling days

the calculation efficiency and helps prepare a high-precision shipping emission inventory.

1. According to the stratified sampling emission model, ships in the area of Shanghai emitted a total of 148,853.033 tons of CO_2, 2288.37 tons of SO_2, 2833.22 tons of NO_x, 261.528 tons of $PM_{2.5}$, 326.910 tons of PM_{10}, 305.116 tons of CO and 108.970 tons of VOC in 2019. The shipping emissions are dominated by CO_2, SO_2 and NO_x. The Marine Emission Control Area Plan for Pearl River Delta, Yangtze River Delta and Bohai Rim Area (Beijing, Tianjin and Hebei) only restricts the sulfur content in fuel oil. According to the results of estimation, the consumption of fuel oil with sulfur content of 0.5% and 0.1% would reduce the SO_2 emissions by 81% and 96% in 2020 compared with that in 2015 (Liu et al., 2018), showing a significant effect in emission reduction. Therefore, it is recommended to add restrictive policies on the fuel containing carbon and nitrogen elements to reduce the emissions of CO_2, CO and NO_x.
2. According to the statistics of shipping emissions by type, the total emissions from cargo ships account for 68.62% of the emissions from all ships. International freight plays an important role in Shanghai and the cargo ships mainly include container ships, conventional ships and refrigerated ships. If some emission reduction plans are made for cargo ships (for example, installing more energy-saving and emission reduction equipment), the emission reduction effect will far exceed that of other types of ships. Moreover, it is recommended to increase the coverage of shore power and use electric tugboats to reduce the emissions of cargo ships when they are going into a port.

3. According to the monthly emission inventories, the pollutant emissions of ships from August to December account for more than 68%. However, the surge in shipping demand has caused congestion when ships are going into or out of a port, thereby resulting in ship detention, causing a longer waiting time and increasing pollutant emissions to a certain extent. The introduction of cargo incentive plans for the off-season to increase the shipping demand can reduce the emissions caused by ship congestion.
4. In order to minimize the uncertainty caused by any missing data, this study tries to make the calculation results of emission inventory close to the true values. The data is subject to preprocessing and numerical supplementation. Most ships are included in the statistics. However, some ships, such as ultra-low-power fishing ships, are not arranged with AIS system according to the regulations of IMO. Therefore, they are not included in the statistics. Small inland rivers are not installed with AIS system due to installation cost and other factor. The above actual situations will cause uncertain deviations to the calculation results.

Acknowledgements Shaoli He, Jun Wang, and Yumeng Lu also made great contributions to this manuscript. We express our heartfelt thanks to them. This research also was supported by: National Natural Science Foundation of China (72474127) and National Social Science Fund Major Bidding Project of China (18ZDA052). Part of this chapter was published: He S, Wu X, Wang J, Lu Y. Shipping emission inventory preparation with AIS data: A stratified random sampling method. Transportation Research Part D: Transport and Environment 2025; 138:104520. https://doi.org/10.1016/j.trd.2024.104520. The publication has been conducted with copyright permission.

References

Capaldo, K., Corbett, J. J., Kasibhatla, P., Fischbeck, P., & Pandis, S. N. (1999). Effects of ship emissions on sulphur cycling and radiative climate forcing over the ocean. *Nature, 400*, 743–746. https://doi.org/10.1038/23438

Chen, J., Kang, S., Wu, A., & Chen, L. (2024). Projected emissions and climate impacts of Arctic shipping along the Northern Sea route. *Environmental Pollution, 341*, 1–8. https://doi.org/10.1016/j.envpol.2023.122848

Chen, Q., Lau, Y.-Y., Ge, Y.-E., Dulebenets, M. A., Kawasaki, T., & Ng, A. K. Y. (2021). Interactions between Arctic passenger ship activities and emissions. *Transportation Research Part D, 97*, 1–19. https://doi.org/10.1016/j.trd.2021.102925

Chen, S., Meng, Q., Jia, P., & Kuang, H. (2021). An operational-mode-based method for estimating ship emissions in port waters. *Transportation Research Part D, 101*, 1–15. https://doi.org/10.1016/j.trd.2021.103080

Cheng, L., Yuan, Z., Ou, J., Fan, X., Ye, S., & Xiao, T. (2016). An AIS-based high-resolution ship emission inventory and its uncertainty in Pearl River Delta region, China. *Science of the Total Environment, 573*, 1–10. https://doi.org/10.1016/j.scitotenv.2016.07.219

Corbett, J. J., Fischbeck, P. S., & Pandis, S. N. (1999). Global nitrogen and sulfur inventories for oceangoing ships. *Journal of Geophysical Research, 104*(D3), 3457–3470. https://doi.org/10.1029/1998jd100040

Fan, Q., Zhang, Y., Ma, W., Ma, H., Feng, J., & Yu, Q. (2016). Spatial and seasonal dynamics of ship emissions over the Yangtze River Delta and East China Sea and their potential environmental. *Environmental Science & Technology, 50*(3), 1322–1329. https://doi.org/10.1021/acs.est.5b03965

Ferreira, M. D., Campbell, J., Purney, E., Soares, A., & Matwin, S. (2023). Assessing compression algorithms to improve the efficiency of clustering analysis on AIS vessel trajectories. *International Journal of Geographical Information Science, 37*(3), 660–683. https://doi.org/10.1080/13658816.2022.2163494

Fu, X., Chen, D., Wang, X., Li, Y., & Lang, J. (2023). The impacts of ship emissions on ozone in eastern China. *Science of the Total Environment, 903*, 1–12. https://doi.org/10.1016/j.scitotenv.2023.166252

He, S., Wu, X., & Wang, J. (2023). A calculation algorithm for ship pollutant gas emissions and diffusions based on real-time meteorological conditions and its application. *Ocean Engineering, 287*, 1–12. https://doi.org/10.1016/j.oceaneng.2023.115825

He, S., Wu, X., Wang, J., & Guo, J. (2022). A ship emission diffusion model based on translation calculation and its application on Huangpu River in Shanghai. *Computers & Industrial Engineering, 172*, 1–15. https://doi.org/10.1016/j.cie.2022.108569

Ignasi, P.-O., Ismael, C.-U., Pere, B.-R., Xenofontas, D., & Josep, S.-P. (2012). Practical anomaly detection based on classifying frequent traffic patterns. *Proceedings—IEEE INFOCOM, 49–54*. https://doi.org/10.1109/INFCOMW.2012.6193518

Jalkanen, J.-P., Brink, A., Kalli, J., Pettersson, H., Kukkonen, J., & Stipa, T. (2009). A modelling system for the exhaust emissions of marine traffic and its application in the Baltic Sea area. *Atmospheric Chemistry and Physics, 9*(23), 9209–9223.

Jinxian, W., Kun, S., Xiafan, G., Guorong, L., & Zhi, H. (2019). Ship emission estimation with high spatial-temporal resolution in the Yangtze River estuary using AIS data. *Journal of Cleaner Production, 248*, 1–14. https://doi.org/10.1016/j.jclepro.2019.119297

Kelly, P. (2022). A novel technique to identify AIS transmissions from vessels which attempt to obscure their position by switching their AIS transponder from normal transmit power mode to low transmit power mode. *Expert Systems with Applications, 202*, 1–12. https://doi.org/10.1016/j.eswa.2022.117205

Kelmalis, A., Lekkas, D. F., Moustakas, K., & Vakalis, S. (2023). Assessing the emissions of short sea international shipping: A case study of the Mytilini-Ayvalik route. *Environmental Science and Pollution Research, 30*, 115496–115505. https://doi.org/10.1007/s11356-023-30595-5

Kwon, J.-W., Yeo, S., & Lee, W.-J. (2023). Assessment of shipping emissions on Busan port of South Korea. *Marine Science and Engineering, 11*(4), 1–18. https://doi.org/10.3390/jmse11040716

Lei, Y., Qijun, Z., Yanjie, Z., Zongyan, L., Yanan, W., Lin, W., & Hongjun, M. (2021). An AIS-based emission inventory and the impact on air quality in Tianjin port based on localized emission factors. *Science of the Total Environment, 783*. https://doi.org/10.1016/j.scitotenv.2021.146869

Li, Y., Xie, W., Yang, Y., Mei, Q., & Wang, Z. (2023). Research on the carbon emissions traceability inventory and multi-horizon prediction of ship carbon emissions: A case study of Tianjin port. *Frontiers in Marine Science, 10*, 1–19. https://doi.org/10.3389/fmars.2023.1174411

Liu, H., Meng, Z., Shang, Y., Lv, Z., Jin, X., Fu, M., & He, K. (2018). Shipping emission forecasts and cost-benefit analysis of China ports and key regions' control. *Environmental Pollution, 236*, 49–59. https://doi.org/10.1016/j.envpol.2018.01.018

McCaffery, C., Zhu, H., Karavalakis, G., Durbin, T. D., Miller, J. W., & Johnson, K. C. (2021). Sources of air pollutants from a tier 2 ocean-going container vessel: Main engine, auxiliary engine, and auxiliary boiler. *Atmospheric Environment, 245*, 1–9. https://doi.org/10.1016/j.atmosenv.2020.118023

Meyers, S. D., Yilmaz, Y., & Luther, M. E. (2022). Some methods for addressing errors in static AIS data records. *Ocean Engineering, 264*, 1–11. https://doi.org/10.1016/j.oceaneng.2022.112367

Nunes, R. A. O., Alvim-Ferraz, M. C. M., Martins, F. G., & Sousa, S. I. V. (2017). The activity-based methodology to assess ship emissions—A review. *Environmental Pollution, 231*, 87–103. https://doi.org/10.1016/j.envpol.2017.07.099

Peng, X., Ding, Y., Yi, W., Laroussi, I., & He, T. (2024). The inland waterway ship emission inventory modeling: The Yangtze River case. *Transportation Research Part D: Transport and Environment, 129*, 1–13. https://doi.org/10.1016/j.trd.2024.104138
Qi, X., Li, Z., Zhao, C., Zhang, Q., & Zhou, Y. (2024). Environmental impacts of Arctic shipping activities: A review. *Ocean and Coastal Management, 247*, 1–13. https://doi.org/10.1016/j.ocecoaman.2023.106936
Sajjad, M., Chu-Van, T., & Shahariar, G. M. H. (2025). Ship emissions and fuel economy under transient conditions: Revisiting the propeller law. *Marine Pollution Bulletin, 221*, 1–15. https://doi.org/10.1016/j.marpolbul.2025.118535
Shu, Y., Hu, A., Zheng, Y., Gan, L., Xiao, G., Zhou, C., & Song, L. (2023). Evaluation of ship emission intensity and the inaccuracy of exhaust emission estimation model. *Ocean Engineering, 287*, 1–11. https://doi.org/10.1016/j.oceaneng.2023.115723
Topic, T., Murphy, A. J., Pazouki, K., & Norman, R. (2023). NOx emissions control area (NECA) scenario for ports in the North Adriatic Sea. *Journal of Environmental Management, 344*, 1–15. https://doi.org/10.1016/j.jenvman.2023.118712
Viana, M., Rizza, V., Tobías, A., Carr, E., Corbett, J., & Sofiev, M. (2020). Estimated health impacts from maritime transport in the Mediterranean region and benefits from the use of cleaner fuels. *Environment International, 138*, 1–11. https://doi.org/10.1016/j.envint.2020.105670
Wang, P., Hu, Q., Xie, W., Wu, L., Wang, F., & Mei, Q. (2023). Big data–driven carbon emission traceability list and characteristics of ships in maritime transportation—A case study of Tianjin port. *Environmental Science and Pollution Research, 30*, 71103–71119.
Weng, J., Shi, K., Gan, X., Li, G., & Huang, Z. (2020). Ship emission estimation with high spatial-temporal resolution in the Yangtze River estuary using AIS data. *Journal of Cleaner Production, 248*, 1–14. https://doi.org/10.1016/j.jclepro.2019.119297
Xiao, G., Wang, T., Luo, Y., & Yang, D. (2023). Analysis of port pollutant emission characteristics in United States based on multiscale geographically weighted regression. *Frontiers in Marine Science, 10*, 1–13. https://doi.org/10.3389/fmars.2023.1131948
Yan, R., Mo, H., Yang, D., & Wang, S. (2022). Development of denoising and compression algorithms for AIS-based vessel trajectories. *Ocean Engineering, 252*, 1–29. https://doi.org/10.1016/j.oceaneng.2022.111207
Yang, L., Zhang, Q., Zhang, Y., Lv, Z., Wang, Y., Wu, L., & Feng, X. (2021). An AIS-based emission inventory and the impact on air quality in Tianjin port based on localized emission factors. *Science of the Total Environment, 783*, 1–11. https://doi.org/10.1016/j.scitotenv.2021.146869
Yau, P. S., Lee, S. C., Corbett, J. J., Chengfeng, W., Cheng, Y., & Ho, K. F. (2012). Estimation of exhaust emission from ocean-going vessels in Hong Kong. *Science of the Total Environment, 431*, 299–306.
Ye, J., Chen, J., Wen, H., Wan, Z., & Tang, T. (2022). Emissions assessment of bulk carriers in China's east coast-Yangtze River maritime network based on different shipping modes. *Ocean Engineering, 249*, 1–12. https://doi.org/10.1016/j.oceaneng.2022.110903
Yuan, Y., Zhang, Y., Mao, J., Yu, G., & Xu, K. (2023). Diverse changes in shipping emissions around the Western Pacific ports under the coeffect of the epidemic and fuel oil policy. *Science of the Total Environment, 879*, 1–9. https://doi.org/10.1016/j.scitotenv.2023.162892
Zhang, L., Meng, Q., & Fwa, T. (2019a). Big AIS data based spatial-temporal analyses of ship traffic in Singapore port waters. *Transportation Research Part E, 129*, 287–304. https://doi.org/10.1016/j.tre.2017.07.011
Zhang, X., v, d., R., Ding, J., Zhang, X., & Yin, Y. (2023). Significant contribution of inland ships to the total NOx emissions along the Yangtze River. *Atmospheric Chemistry and Physics, 23*, 5587–5604. https://doi.org/10.5194/acp-23-5587-2023
Zhang, Y., Fung, J. C. H., Chan, J. W. M., & Lau, A. K. H. (2019b). The significance of incorporating unidentified vessels into AIS-based ship emission inventory. *Atmospheric Environment, 203*, 102–113. https://doi.org/10.1016/j.atmosenv.2018.12.055

Zhao, J., Zhang, Y., Patton, A. P., Ma, W., & Kan, H. (2020). Projection of ship emissions and their impact on air quality in 2030 in Yangtze River delta, China. *Environmental Pollution, 263*, 1–11. https://doi.org/10.1016/j.envpol.2020.114643

Zhou, C., Ding, Y., Huang, H., Huang, L., & Lu, Z. (2022). Meso-level carbon dioxide emission model based on voyage for inland ships in the Yangtze River. *Science of the Total Environment, 838*, 1–10. https://doi.org/10.1016/j.scitotenv.2022.156271

Chapter 3
A Calculation Algorithm for Ship Pollutant Gas Emissions and Diffusions Based on Real-Time Meteorological Conditions and Its Application

3.1 Introduction

Ships undertake about 90% of the transport volume arising from global trade (Meng et al., 2022). With the continuous growth of domestic and foreign trade, for China's waterway transportation industry, compared to 2011, the cargo volume and turnover in 2021 have risen by 93.44% and 53.24%, respectively. At the same time, ships emit massive atmospheric pollutants. According to the Fourth IMO (International Maritime Organization) GHG Study Report (2020), CO_2 emissions from ships were 0.962 billion tonnes in 2012 and 1.056 billion tonnes in 2018 (increased by 9.3%), and it is estimated that the emissions in 2050 will reach 90% to 130% of the volume in 2008. According to the prediction of China's *Green Transportation Development Planning for the 14th Five-Year Plan Period*, the demand of the waterway transportation industry will still be very huge, with a moderate and growth rate. The rocketing demand for ship transportation leads to an increase in the emission of atmospheric pollutants, which will cause continuous air pollution in the areas along waterways near the emission source and bring health risks to the residents nearby (Ramacher et al., 2019; Sorte et al., 2018). After long-term accumulation, the atmospheric pollutants emitted by ships will result in early mortality and increase labor losses (Yin, 2022). Now the government, scholars and the public have attached great importance to the reduction of atmospheric pollutants emitted by ships.

Identified as a scientific basis to reduce ship emissions, the calculation of atmospheric pollutant emissions helps formulate targeted pollution prevention and control programs. At present, the calculation of ship pollutant emission and diffusion mostly depends on the Gaussian puff model. This model is based on the hypothesis of a stable meteorological environment to study the emission concentration distribution of single ship in a short-time and small-scale range (Bai et al., 2020; Juan et al., 2020). The stable meteorological environment helps reduce the workload of real-time meteorological data collection, and avoid the model calculation problem caused by the addition of parameters. Under the circumstance of single emission

X. Wu, *Environmental Disaster in China*, https://doi.org/10.1007/978-981-92-0910-1_3

source and short time, the sequential cyclic superimposition calculation is typically applied, which needs a longer calculation time and ignores calculation cost and efficiency. In fact, multi-period, multi-ship and long-term observations in a waterway are more common. However, the calculation of ship pollutant emission and diffusion under such circumstance still follows the traditional model, which will reduce the calculation efficiency, increase the calculation time and cost and disable large-scale calculations. Therefore, it is necessary to propose a new fast algorithm, and use the algorithm to analyze the rule of ship atmospheric pollutant emission and diffusion under unstable meteorological conditions, such as wind direction and wind speed, according to real-time meteorological data.

3.2 Literature Review

3.2.1 Ship Pollutant Gas Emissions

Most current studies on ship atmospheric pollutant emissions pay attention to the compilation of emission inventory, the analysis of emission characteristics and the research of factors affecting emissions. Fuel statistics and activity estimation are two typical methods for the compilation of emission inventory. The fuel statistics method estimates the total emissions from different types of fuels based on sales data, and the emissions of various components of pollutants after unit fuel combustion (Ernest et al., 2021). For example, this method is adopted by the Fourth IMO GHG Study Report (2020) and Air Pollutant Emission Inventory of Marine in China. The fuel statistics method is suitable for large-scale ship pollutant emission calculation, which can reflect the emission characteristics from the macro level. For example, Kwon et al. (2019) performed comparative analysis using the fuel statistics method and activity estimation method, demonstrating that the calculation results could provide support for macro-policy formulation, when the fuel statistics method is applied to large-scale areas. However the activity method is more accurate in estimating emissions in small-scale areas such as ports. As for the activity estimation method, it calculates the emission of atmospheric pollutants based on the relevant data, such as the activity data of each ship, ship engine power, navigation time and emission from energy consumption per unit kilowatt-hour. For example, Weng et al. (2020) estimated ship emissions in the Yangtze River Estuary (Shanghai Municipality and Jiangsu Province) region using the STEAM model. Yang et al. (2022) calculated the emission inventory 2020 of Qingdao using AIS (Automatic Identification System) data and evaluated the implementation effect of the policy for restricting the atmospheric pollutant emissions from ships in the region. Fan et al. (2022) established a real-time measurement system with portable detection equipment, detection information platform and cloud server to obtain real-time ship pollutant emission data.

The emission inventory can be used to analyze the emissions and distribution characteristics of various pollutants. For example, Zhang et al. (2019) calculated the atmospheric component pollutant emissions according to ship types, fuel types and ship navigation status. Liu et al. (2018) prepared emission inventories at the port and regional levels to assess the costs and benefits of domestic sulfur emission control areas. The results show that regional control has a better cost–benefit ratio than port control. Cao et al. (2018) compiled an emission inventory based on the data of inland ships of the Beijing-Hangzhou Grand Canal in Jiangsu Province. The emission characteristics of atmospheric pollutants are analyzed from the aspects of pollutant classification, ship type, ship navigation status and channel segment. Using the Yangtze River as the study area, Peng et al. (2024) quantified the emission intensity of ships along the Yangtze water by analyzing vessel emissions across entire voyage sections. The calculation integrated emission volumes, cargo throughput, and sailing distance. Dai et al. (2025) developed high spatiotemporal-resolution ship emission inventory for Shanghai in 2023 using AIS data, analyzing the temporal-spatial characteristics of CO2 emissions, spatial heterogeneity, and consistency. Furthermore, they projected future emission scenarios based on current policy trends.

Ship atmospheric pollutant emissions are affected by many factors, including fuel type, engine power and revolution, ship navigation status, ship load, combustion rate, etc. For example, Yang et al. (2022) measured the emission factors of organic compounds (VOCs), SO_2, and PM from large ocean-going ships using a portable measurement system. The results show that the carbides reach the peak when ships are moored and anchored while the emissions of nitrogen oxides are larger in the cruise mode. Zhang et al. (2022) measured the gas emissions of China's inland ships. The results show that the concentration values of CO, THC, PN and PM under the status of startup and idling are significantly higher than other ship statuses, and the concentration values of NOx and SO_2 emissions during cruising are the highest. Zhang et al. (2021) tested the emissions under different ship load modes using HFO (Heavy Fuel Oil) fueled low-speed marine engines as the object. The results show that the CO emissions are the highest under the load mode of 50%. Taking the air quality of Tianjin Port as an example, Yang et al. (2021) established a ship pollutant emission inventory using local emission factors. The results show that the fuel quality has a significant impact on emissions.

With the in-depth study on ship atmospheric pollutant emissions, emission inventories will be compiled in a more precise and localized way. Emission inventories provide data support for improving air quality, optimizing living environment, reducing social and economic losses, and formulating regional pollution prevention & control and emission reduction programs. High-precision emission calculation results can effectively improve the research quality and help obtain more scientific and reliable conclusions.

3.2.2 Ship Pollutant Gas Diffusions

The diffusion of atmospheric pollutants includes the transportation, diffusion and conversion of pollutants. The typical atmospheric diffusion models evolve from the first generation of air quality models (Gaussian model, Lagrangian model and Box model). The Gaussian model has the following advantages: fewer parameters, lower difficulty in data acquisition, concise and clear model, simple calculation process, highly accurate calculation results, adjustable model and parameters according to research requirements, and good flexibility. It is suitable for continuous or intermittent pollution source emissions, and compatible with the simulation and prediction range up to several hundred kilometers. Moreover, the model structure is more mature and has a wider range of applications. Several commonly used diffusion models, such as CALPUFF model, AERMOD model and ADMS model, derive from the Gaussian model. *Technical guidelines for environmental impact assessment - Atmospheric environment* (HJ 2.2–2018) issued by the Ministry of Ecology and Environment of China recommends these types of Gaussian model as applicable ambient air quality models.

With respect to the calculation of ship pollutant diffusions, the Gaussian puff model is mostly used for the research on how pollutants affect air quality. For example, Peng et al. (2021) used two container ships as research objects in the emission control area at Yantian Port of Shenzhen City, and compared the SO_2 diffusion calculated through the Gaussian puff model with the SO_2 value measured through UAV (Unmanned Aerial Vehicle). The analysis result shows that the simulation method has met the requirements of fuel monitoring in the emission control area, and reduced the monitoring cost. Juan et al. (2020) conducted single-ship real-time diffusion simulation at Xiamen Port using the Gaussian puff model, Jave Web technology, GIS (Geographic Information System) technology. They also performed a comparative analysis of NO_2 diffusions under inbound, cruising and outbound statuses. Xin et al. (2020) created the Gaussian puff diffusion model for mobile ship pollution sources, and simulated the distribution of single-ship emission and diffusion concentrations using the control variable method. With reference to the WRF/CALLPUFF model, He et al. (2021) conducted single-ship emission diffusion simulations for four types of typical ships, and studied the layout of onshore monitoring points at Yantian Port. Toscano et al. (2021) compared and analyzed the diffusion of atmospheric pollutants from ships using single or combination of CALLPUFF model, CFD model and wind tunnel model. The results show that the Gaussian puff diffusion process combined with CFD has reduced the uncertainty of numerical prediction, and improved the accuracy of the model. Tseng et al. (2024) employed the CALPUFF atmospheric dispersion model to simulate the spatiotemporal distribution characteristics of $PM_2.5$ concentrations in the Kaohsiung Port area and adjacent urban zones. The results indicate that afternoon sea breezes significantly influence $PM_2.5$ dispersion, with ship emissions contributing substantially to local pollution levels. Specifically, main engines accounted for 2.9%–5.3% of $PM_2.5$ concentrations, while auxiliary engines contributed 3.8%–8.3%.

According to the findings of the above research on ship pollutant emission and diffusion, most of the current research results are associated with small-scale single-source diffusion, and seldom involved in multi-source emissions. In fact, it is more common for multiple ships to sail in the same area and time period. In reality, multi-source emission and diffusion are the main type of pollutant emission and diffusion. At the same time, the diffusion process of ship atmospheric pollutants is closely related to the ship navigation status, environmental factors (wind, wave and marine current), geographical location, meteorological and climatic conditions and other factors (Huang et al., 2018). Ship navigation status data can be obtained from AIS. AIS data is characterized by short data interval and huge data size. Data sets of multiple ships will significantly reduce the calculation efficiency. The real-time acquisition of environmental and meteorological data is a difficult and costly process. In this sense, the calculation involving multiple parameters and data sets will not only increase the model complexity and analysis cost, but also reduce the calculation efficiency.

3.2.3 Reduction of Ship Pollutant Gas Emissions

The research on ship atmospheric pollutant emission and diffusion includes the compilation of emission inventory, simulation of diffusion process, calculation of pollutant concentration value and visual analysis. Based on the above research, we analyze the impact of ship pollutant emissions on air quality and the health hazards caused by such emission to the residents along waterways and in port areas.

Many countries have established emission control areas to regulate marine fuel and emission standards, thus reducing ship atmospheric pollutant emissions. The establishment of emission control areas is more convenient, feasible and effective than other measures in emission reductions. Yang et al. (2022) prepared an inventory of ship pollutant emissions for Qingdao in 2016 and 2020. They performed a comparative analysis to the emission situations before and after the implementation of the emission control policy in the same region. The results show that after the implementation of the policy, the emission reduction effect is significant, and especially the emissions of SO_2 and $PM_{2.5}$ are significantly reduced. Taking the emission control areas in the Yangtze River Delta region as the research object, Zhao et al. (2020) compared three different emission reduction scenarios, and evaluated the emission reduction effect of emission control areas. The results show that the emission control area policy effectively reduces ship pollutant emissions. However, high-standard marine fuel will increase ship operating costs. By analyzing the ship trajectory in the emission control area, Weng et al. (2022) found that merchant ships outside the control area would make adjustments according to such factors as ship size, voyage distance and fuel price difference to achieve emission reductions.

In addition to the establishment of emission control areas, the relevant scholars have proposed emission reduction plans, such as ship speed reduction, clean energy substitution, ship route optimization and ship technology upgrade. Chou et al. (2021) used the energy efficiency operating index (EEOI), and compared the efficiency and emissions of the ferries fully powered by diesel and the diesel-electric hybrid ferries, with CO_2 as the object. Diesel-electric hybrid ferries can improve energy efficiency by 12% and reduce emissions by 10%. In urban ports, the use of small medium-speed passenger ships instead of small high-speed passenger ships and large passenger ships can reduce emissions by 11% to 26%. Shukai et al. (2021) estimated ship pollutant emissions near ports based on different operation modes of ships, and found that tugboats eased the main engine load of large ships by 10% and reduced sulfur oxide emissions by 2% to 6% during operation. Based on port ship scheduling and decelerated ship navigation, Xia et al. (2021) proposed the scheme of optimizing ship scheduling and reducing ship speed for emission reductions. As proven by the empirical data of different ports, the scheme can reduce the ship pollutant emissions by 8% to 11.9% at all stages of the port, and improve the traffic efficiency by 3.8% to 6.2%.

3.2.4 Literature Comments

In light of the above research, the high-precision ship pollutant emission diffusion results help formulate ship pollutant prevention and control measures. Most of the current research calculates emissions and diffusions based on a stable meteorological environment, ignoring the influence of meteorological factor variations on ship pollutant emissions and diffusions. In fact, changing meteorological conditions will change the speed of ship navigation, pollutant emissions, pollutant concentration and diffusion process. In addition, the calculation of atmospheric pollutant emission and diffusion based on ship AIS data sets has higher accuracy. With the increasing application of ship automatic system, the ship activity estimation method will replace the fuel statistics method, and become an important method to study the calculation of ship pollutant emissions. However, the ship AIS data size is huge. If the number of ships and the duration of study increase, the amount of data to be processed in the calculation process will increase exponentially. For the above reason, the traditional algorithms will be inefficient and costly and also unable to realize real-time calculations when the data size is large. Based on the current research findings, this paper proposes a new calculation method, and simulates and calculates the real-time emissions and diffusions of ships by adding emission sources, increasing observation time and enlarging observation distance.

3.3 Ship Pollutant Gas Emission and Diffusion Model

The ship traffic emission assessment model (STEAM) (Jalkanen et al., 2009) based on AIS data is one of the most widely used methods for estimating ship atmospheric pollutant emissions (Weng et al., 2020; Xin et al., 2020; Yang et al., 2021). The calculation formula is as follows:

$$\mathrm{Q} = \sum_i Q_i = \sum_i \mathrm{P_i} \times \mathrm{LF_i} \times T_i \times \mathrm{EF_i},\ \mathrm{i} \in \text{main, auxiliary, boiler} \tag{3.1}$$

Q is the total emission of atmospheric pollutants from a single ship (g). i represents the three types of emission equipment, including ship's main engine, auxiliary engine and boiler respectively. P is the power (kW) of various types of equipment of the ship. LF is the load factor. The determination of load factor requires considering the actual ship speed and the calculation formula is shown in (2). T is the time (h) of (continuous) ship operation. EF is the emission factor, and its value shall be determined according to the ship engine type, the ship maneuvering status, and the type of fuel consumed by the ship.

$$LF_i = (v/v_{max})^3 \tag{3.2}$$

v is the actual ship speed (kn/h) and v_{max} is the maximum design ship speed (kn/h), which are substituted into Eq. (3.1) to obtain:

$$Q = \sum_i P_i \times (v/v_{max})^3 \times T_i \times EF_i \tag{3.3}$$

First, the geographical coordinates of the ship's location were converted into the coordinates of sailing wind direction, thus establishing the coordinates system of ship pollutant emissions and diffusions. Then, the ship pollutant diffusion in the navigation status was simulated using the Gaussian puff diffusion model. The calculation formula is as follows:

$$C(x, y, z, H_e) = \frac{Q}{2\psi_y \sigma_z u} e^{\frac{-y^2}{2\sigma_y^2}} \left(e^{\frac{-(z-H_e)^2}{2\sigma_z^2}} + e^{\frac{-(z+H_e)^2}{2\sigma_z^2}}\right) \tag{3.4}$$

where, $C(x, y, z, H_e)$ is the concentration of pollutants at any position; Q is the pollutant emission of the emission source within a unit time of period; u refers to wind speed; x, y and z respectively represent the downwind distance from the origin of emission source coordinates, the distance along the wind direction and the distance perpendicular to the wind direction. H_e refers to the height of ship pollutant emission point. σ_y and σ_z respectively represent the diffusion coefficient in the y-axis and z-axis directions.

As shown by the emission diffusion Eq. (3.3) and (3.4), the emission amount, diffusion speed and direction of air pollutants generated by a ship during activities are closely related to the wind direction during the ship navigation. Therefore, any wind direction will influence the actual ship speed. The factors influencing ship speed roughly fall into two categories: (1) external factors (environmental factors), including weather (wind), sea/river conditions (visibility), water area (ship traffic) and so on. (2) Internal factors (ship factors), including ship functions (power), ship type, load and so on.

According to the calculation formula (3.4) of the traditional Gaussian puff model. To calculate the pollution concentration in the target area for a mobile point source, t + 1 recursive calls should be performed from the starting time t and ending the observation time (T) is exceeded. If there are multiple continuous emission sources in a large observation area, the traditional model needs to superimpose the calculation of each puff group to obtain the multi-source emission and diffusion results in the same space and period of time. The calculated time of the single emission source for the superimposed diffusion calculation is the square of the observation time (T). If the number of emission sources is N, the calculation time of multi-source emission will be $N*T^2$. If only the observation space is expanded, the calculation difficulty will be significantly increased and the calculation efficiency will be greatly reduced. If the traditional Gaussian puff model is directly applied to the diffusion research of multi-variable and multi-source emissions, the research will be more costly and difficult. Therefore, this paper intends to optimize the original model. First, the variable wind direction is used as an unstable meteorological parameter to discuss the emission and diffusion changes of ship atmospheric pollutants in this situation. Then, the fast calculation of ship pollutant emission and diffusion with variable wind direction is realized by the translational diffusion algorithm.

3.4 Fast Algorithm of Ship Pollutant Emission and Diffusion with Variable Wind Direction

3.4.1 Ship Pollutant Gas Emissions with Variable Wind Direction

The coordinates of ship navigation under the influence of wind are shown in Fig. 3.1, and XOY is the common coordinates system with X as the due north and Y as the due east. $xo\prime y$ is the geographical coordinates of the ship location, $O\prime$ is the center of gravity of the ship, x is the bow direction, and y is the starboard of the ship. V_{actual}, V_T and V_{AIS} respectively represent the corrected ship speed, the ship drift speed caused by wind, and the ship speed collected by AIS. φ and ω respectively represent the ship heading angle in AIS and the wind angle in weather forecast. u_{AIS} and v_{AIS} are the component of V_{AIS}. u_T and v_T are the component of V_T. β is the drift angle. θ_T is the angle between wind direction and bow direction, where, $v_T = v$.

$$\overrightarrow{V_{AIS}} = \overrightarrow{V_{actual}} + \overrightarrow{V_T} \tag{3.5}$$

$$\begin{cases} u_{AIS} = V_{AIS} - V_T * \cos\theta_T \\ u_{AIS} = -V_T * \sin\theta_T \\ V_{AIS} = \sqrt{{u_{AIS}}^2 + {v_{AIS}}^2} \end{cases} \tag{3.6}$$

$$V_{actual} = \sqrt{{V_{AIS}}^2 - {u_{AIS}}^2} + V_T * \cos\theta_T = \sqrt{{V_{AIS}}^2 - (-V_T * \sin\theta_T)^2} + V_T * \cos\theta_T \tag{3.7}$$

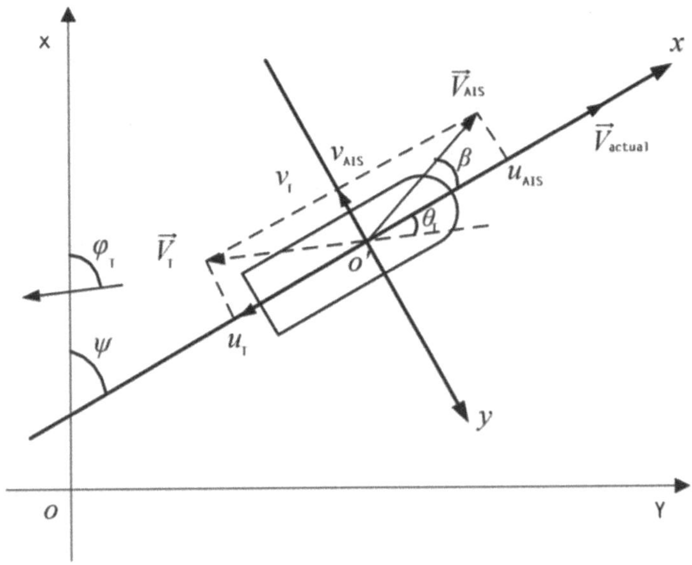

Fig. 3.1 A schematic diagram of ship movement under the influence of wind

The influence of wind is substituted into the process of ship navigation and pollutant diffusion. The ship speed calculated during actual ship navigation using Eq. (3.7) is substituted into (3) to obtain the emission amount of ship atmospheric pollutants under the influence of wind. The pollutant emission is obtained accordingly when the wind direction changes during the ship navigation. See the equation below for details:

$$Q_i = P_i \times (v_{actual}/v_{max})^3 \times T_i \times EF_i \tag{3.8}$$

3.4.2 *Ship Pollutant Gas Diffusions with Variable Wind Direction*

The influence of wind direction change on the diffusion of atmospheric pollutants from ships involves diffusion amount (Q_i) and spatial location. The change of emissions is calculated according to the corrected emission Eq. (3.8), and the results are substituted into the Gaussian puff model. The change of spatial location caused by the change of wind direction is transformed through coordinate transformation. When the wind speed is used as a control variable, the equation for calculating the concentration of any point in the space is the same when the wind direction changes, only the area affected by pollutants will change. Wind direction is an important basis for determining cross-wind direction and downwind direction. After coordinate transformation, the y-axis is taken as the cross-wind direction and the x-axis as the downwind direction. Then, the concentration of each point in the space after puff diffusion is calculated.

The coordinates transformation is performed in the plane with the emission point as the source point O (0, 0) and the due north direction as the y axis (Fig. 3.2). Position transformation is performed to the diffusion process after emission to obtain the diffusion process when the wind direction changes. x and y respectively represent the coordinates after the rotation angle of the position of spatial point before the wind direction changes, and x_1 and y_1 respectively represent the coordinates of the rotation β relative to the spatial point after the wind direction changes. After the angle transformation around point O (0, 0), the new location of point A (x, y) will turn into point C (x1, y1).

$$\mathrm{dist}|AO| = \frac{y}{sin\alpha} = \frac{x}{cos\alpha} \tag{3.9}$$

$$\mathrm{dist}|CO| = \frac{y1}{\sin(\alpha + \beta)} = \frac{x1}{\cos(\alpha + \beta)} \tag{3.10}$$

where, x and y respectively represent the horizontal axis and vertical axis coordinates of point A. × 1 and y1 respectively represent the horizontal axis and vertical axis coordinates of point C. dist|AO| and dist|CO| respectively represent the distance between A and O and the distance between C and O. α represents the original wind angle. β represents the angle changed after the rotation relative to point A. At the same time, the radius r, dist|AO| and dist|CO| for the circle of rotation have an equal distance with details given as follows:

$$r = \frac{y}{sin\alpha} = \frac{x}{cos\alpha} = \frac{y1}{\sin(\alpha + \beta)} = \frac{x1}{\cos(\alpha + \beta)} \tag{3.11}$$

The following equation is obtained according to the trigonometric function and difference formula:

$$x1 = r * \cos(\alpha + \beta) = r * cos\alpha * cos\beta - r * sin\alpha * sin\beta = x * cos\beta - y * sin\beta \tag{3.12}$$

$$y1 = r * \sin(\alpha + \beta) = r * sin\alpha * cos\beta + r * cos\alpha * sin\beta = y * cos\beta + x * sin\beta \tag{3.13}$$

The clockwise rotation variation formula is as follows:

$$\begin{cases} x_1 = x * cos\beta + y * sin\beta \\ y_1 = y * cos\beta - x * sin\beta \end{cases} \tag{3.14}$$

Counterclockwise rotation variation formula:

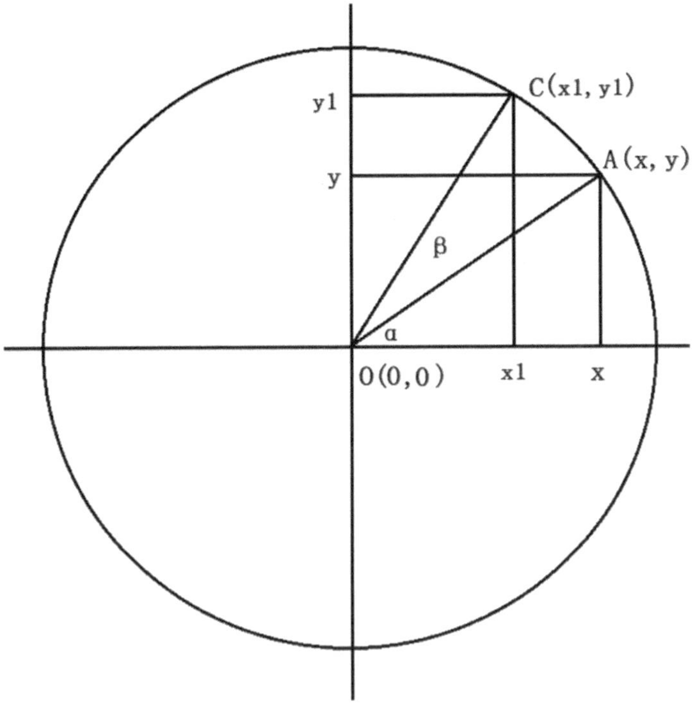

Fig. 3.2 Diagram of wind direction rotation coordinates

$$\begin{cases} x_1 = x * \cos\beta - y * \sin\beta \\ y_1 = y * \cos\beta + x * \sin\beta \end{cases} \tag{3.15}$$

In the Gaussian puff diffusion model, set the downwind direction as the horizontal axis x, and the cross-wind direction as the vertical axis y, and then calculate the concentration value of each point at different moments. The wind direction is only related to the coordinate axes and has nothing to do with the concentration calculation. In other words, the wind direction only affects the puff diffusion direction rather than the puff diffusion pattern. Therefore, the wind direction can be taken as the diffusion direction. However, the continuous effect of different wind directions will cause the puff to shift as a whole, as shown in Fig. 3.3. Under the influence of wind speed U, the pollutants at the emission source point $P_0(x_0, y_0)$ actually diffuse to $P_n(x_n, y_n)$ from t_0 to t_n and the wind direction changes from w_0 to w_n. In order to save the calculation cost, the diffusion position $P_{n'}(x_{n'}, y_{n'})$ of pollutants is obtained by directly substituting the wind direction w_n into the moment $P_{n'}(x_{n'}, y_{n'})$. The offset position at different moments is calculated by the virtual diffusion position and the actual diffusion position at the moment t_n, and the final offset from t_0 to t_n with the calculation formula as follows:

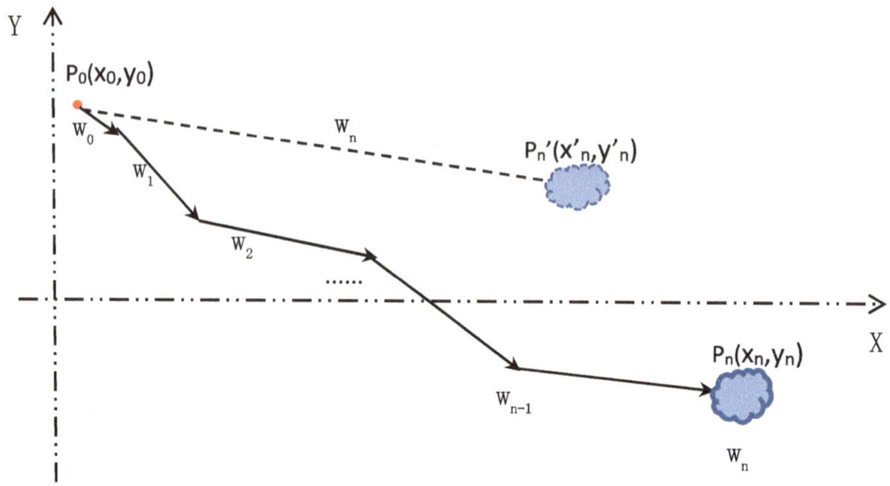

Fig. 3.3 Offset diagram

$$\begin{cases} x_n = x_0 - U * t_n sin(w_n) \\ y_n = y_0 - U * t_n cos(w_n) \end{cases} \quad (3.16)$$

$$\begin{cases} x_n' = x_0 - \sum_i U * (t_{i+1} - t_i)\sin(w_i) \\ y_n' = y_0 - \sum_i U * (t_{i+1} - t_i)\cos(w_i) \end{cases} \quad (3.17)$$

$$\begin{cases} \Delta x = x_n' - x_n \\ \Delta y = y_n' - y_n \end{cases} \quad (3.18)$$

As the wind direction changes, all the puff diffusion points will shift with the wind direction. Therefore, the horizontal and vertical coordinates of puff diffusion are processed according to the offset formula, and the Gaussian puff diffusion model under the influence of variable wind direction can be obtained.

$$\begin{cases} C = C_{t_i} \\ x\prime = x + \Delta x \\ y\prime = y + \Delta y \end{cases} \quad (3.19)$$

3.4.3 Fast Algorithm of Ship Pollutant Gas Diffusions with Variable Wind Direction

When directly used for simulating atmospheric pollutant diffusions, the Gaussian puff model has lower calculation efficiency, and requires a high cost. The original concentration calculation method based on superimposition is changed since the puffs are independent of each other during diffusion. The matrix translation method not only enables to flexibly set the calculation accuracy but also greatly improves the calculation efficiency (He et al., 2022). Taking two puffs of instantaneous emission as an example, under the same atmospheric environment, and according to the principle of constant diffusion function, the change of wind directly affects the final spatial position and diffusion distance of puffs. The pattern of instantaneous puff at the origin (0, 0) in the period from t_i to t_k, when the wind speed is $\overline{u}$ is shown in Fig. 3.4 (a). All puffs from a continuously moving emission source at different moments diffuse according to the standard pattern in Fig. 3.4 (a) to constitute the puff group as shown in Fig. 3.4 (b). In other words, for the instantaneous emission at any point, the puff diffusion at a particular moment can be considered as the coordinate translation of the standard puff at the same moment.

First, the emission diffusion coordinate system is established, and the ship position coordinate is converted into wind direction coordinate. Through the Gaussian puff model, the continuous emission source is calculated for emission. Assuming that the location of a certain emission source at the moment t_0 is (x_0, y_0, z_0), the formula for calculating the concentration at any point (x, y, z) at the moment t_i after puff diffusion is as follows:

$$C_{t_i} = \rho(x, y, z, t_i) = \frac{Q_{t_0}}{(2\pi)^{1.5}\sigma_{x_0}\sigma_{y_0}\sigma_{z_0}} e^{\frac{-[x_0 - u*(t_i - t_0)]^2}{2\sigma_{x_0}^2} + \frac{-y_0^2}{2\sigma_{y_0}^2}} \left(e^{\frac{-(z_0 - H_e)^2}{2\sigma_{z_0}^2}} + e^{\frac{-(z_0 + H_e)^2}{2\sigma_{z_0}^2}}\right) \tag{3.20}$$

Ship atmospheric pollutant emission is a continuous and dynamic process. However, the diffusion and emission of each puff are independent of each other. According to the different characteristics and correlation of emission and diffusion

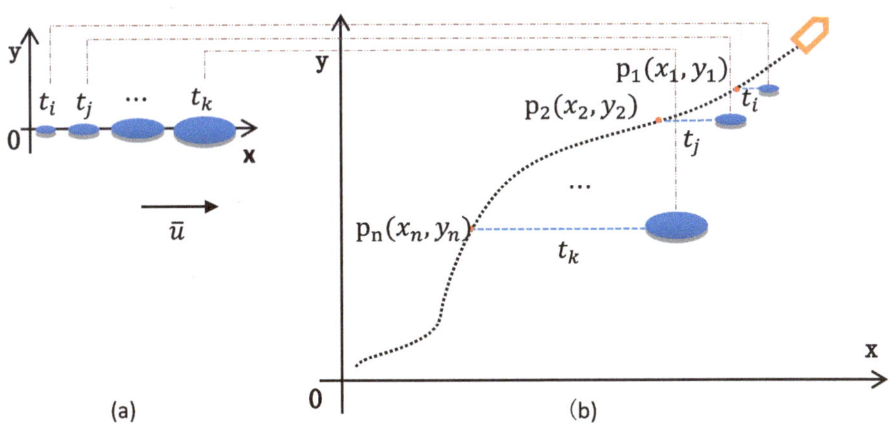

Fig. 3.4 Translational calculation of standard puff diffusion

process, continuous emissions are assumed to be an infinite instantaneous puff superimposition process ignoring the time intervals (Pan et al., 2021; Rui et al., 2021; Ying et al., 2019). In other words, the pollutant concentration at any point (x, y, z) in the air is C_t at the observation moment t:

$$C_t = \sum_i \rho(x, y, z, t - t_i) = \sum_i \frac{Q_{t_i}}{(2\pi)^{1.5}\sigma_{x_i}\sigma_{y_i}\sigma_{z_i}} e^{\frac{-[x_i - u*(t-t_i)]^2}{2\sigma_{x_i}^2} + \frac{-y_i^2}{2\sigma_{y_i}^2}} \left(e^{\frac{-(z_i - He)^2}{2\sigma_{z_i}^2}} + e^{\frac{-(z_i + He)^2}{2\sigma_{z_i}^2}}\right) \tag{3.21}$$

Similarly, the observation range is extended to assume multiple emission sources in the area. It is only necessary to sum the concentration values of each independent puff to obtain the total multi-source continuous emission gas concentration data in the whole area.

$$C_t = \sum_k \sum_i \rho(x, y, z, t - t_i) \tag{3.22}$$

Then, the concentration at the observation plane is expressed as a data matrix. In other words, **n** scales on the axis x correspond to n columns of the matrix, and **n** scales on the axis y correspond to n rows of the matrix. The value of $C_i(x, y)$ is expressed as the element of the intersection between the column x and the row y of the matrix, thus expressing the puff diffusion concentration matrix as follows:

$$C_i = \begin{bmatrix} c_{11}^i & \cdots & c_{1n}^i \\ \vdots & \vdots & \vdots \\ c_{n1}^i & \cdots & c_{nn}^i \end{bmatrix} \tag{3.23}$$

Move the unit matrix having the same dimension as the puff diffusion concentration matrix C_i (N x N) to the right for r columns to obtain the position translation matrix M_r:

$$M_r = \left[\begin{array}{c|c} \begin{matrix} 0 \\ 0 \end{matrix} & \mathbf{E} \\ \hline \underbrace{}_{r} & 0 \end{array}\right]_{NN} \}\, r \tag{3.24}$$

Where r represents the number of columns (distance) moved to the right and E is the unit matrix.

According to the diffusion formula (3.20), the emission Q does not change the puff diffusion pattern, and it only affects the concentration. In other words, the concentration of any point in the air is Q times the concentration of the corresponding point for the unit emission. Assuming that any puff is emitted by a continuous emission source at point p (x, y), the diffusion concentration value $F_{x,y}(i)$ of the puff is obtained by translating all the data of C_i by x rows and y columns and can be expressed as follows:

$$F_{x,y}(i) = Q * {M_y}^T * C_i * M_x \quad (3.25)$$

Since emission sources are independent of puff diffusions, the total diffusion of puffs from multiple continuous emission sources are expressed using the sum of diffusions from each emission source. The details are as follows:

$$F = \sum_k \sum_i \left(Q_{ki} * {M_{y_{ki}}}^T * C_{ki} * M_{kx_i}\right) \quad (3.26)$$

3.4.4 *Time Saving Analysis of the Fast Algorithm*

Assuming that there are n emission sources and the observation time is T, there are n*T puffs in the diffusion process of continuous emission. According to the Gaussian puff model, the Eq. (3.20) should be continuously called when using the Eq. (3.21) to measure the diffusion of pollutants in the observation interval. The time and memory resources required by the traditional algorithm in the running process are $n*T^2$, and the algorithm complexity is S ($n*T^2$) accordingly. According to the fast algorithm, the Eq. (3.20) is called once at first when using the Eq. (3.25) for calculation, and then all puffs are translated one by one, so the complexity of the translation algorithm is S (T + n*T), which can be simplified to S(n* T). According to the comparative analysis relating to the complexity of the two algorithms, the fast algorithm S (n*T) is more efficient than the traditional algorithm S ($n*T^2$).

According to Table 3.1, Fig. 3.5 (Time consumption of the algorithm) is plotted for comparisons in three dimensions, including observation time, emission source and observation radius.

Figure 3.5 (a) shows the ratio of time consumption for different emission sources, and observation radii with the observation time of 2 h. The time consumption increases with the number of emission sources, and the change of observation radius has a slight effect on the time consumption ratio of the algorithm. There is only a slight difference of time consumptions for calculating a single emission source with different observation radii. However, the fast algorithm has higher calculation efficiency. For the radius of 2,500 m, the traditional algorithm consumes 1,056 s while the translation algorithm only consumes 1.3 s. For the radius of 12,500 m, the traditional algorithm consumes 20,963 s while the translation algorithm only consumes 28.8 s.

Figure 3.5 (b) shows the comparative results of single emission sources. With the observation time of 2 h, the algorithm consumes almost the same time for three different radii. However, with the increasing observation time, the time consumption ratio increases with radius. For example, for a 4-h observation period, the traditional algorithm and translation algorithm consume 4,270 s and 2.67 s at a radius of 2500 m,

Table 3.1 Time-consuming comparison of emission diffusion calculations

Radius/m	Duration/h	Emission sources	Time consumption comparison/s		
			Traditional algorithm	Translation algorithm	Save time
2500	2	1	1056.148	1.334	1054.814
		5	5030.998	1.328	5029.670
		10	9849.848	1.339	9848.509
	4	1	4270.377	2.675	4267.702
		5	20,323.741	2.688	20,321.053
		10	40,160.727	2.689	40,158.038
	6	1	9410.295	4.028	9406.267
		5	46,046.686	4.043	46,042.643
		10	93,308.793	4.045	93,304.748
5000	2	1	3401.053	4.090	3396.963
		5	17,396.832	4.096	17,392.736
		10	30,845.958	4.098	30,841.860
	4	1	12,635.757	8.336	12,627.421
		5	74,368.099	8.340	74,359.759
		10	138,940.388	8.357	138,932.031
	6	1	31,206.4926	12.642	31,193.851
		5	165,975.658	12.780	165,962.878
		10	479,323.5834	12.808	479,310.775
12,500	2	1	20,963.496	28.873	20,934.623
		5	99,851.577	29.164	99,822.413
		10	197,700.902	29.175	197,671.727
	4	1	311,143.2253	57.975	311,085.250

Note The running environment is: Processor Intel(R) Xeon(R) Gold 5118 CPU @ 2.3 GHz (24 processors); RAM 48 GB; Python 3.8.4

and the calculation time for a radius of 12,500 m is 31,143 s and 57.9 s respectively. When both the observation time and radius increase, the calculation time required by the fast algorithm is shorter.

Figure 3.5 (c) shows the comparison based on 5,000-m observation window. For a single emission source, the advantages of the fast algorithm are not obvious when the observation time and the observation window are increased. However, the fast algorithm can finish the calculation within 8.36 s under the situation of 10 emission sources, 5,000-m observation window and 4-h observation time, saving by 54 h compared with traditional algorithms. The comparative analysis in the three dimensions shows that the fast algorithm has short calculation time and high efficiency under the situation of multiple emission sources, changing observation time and changing distance.

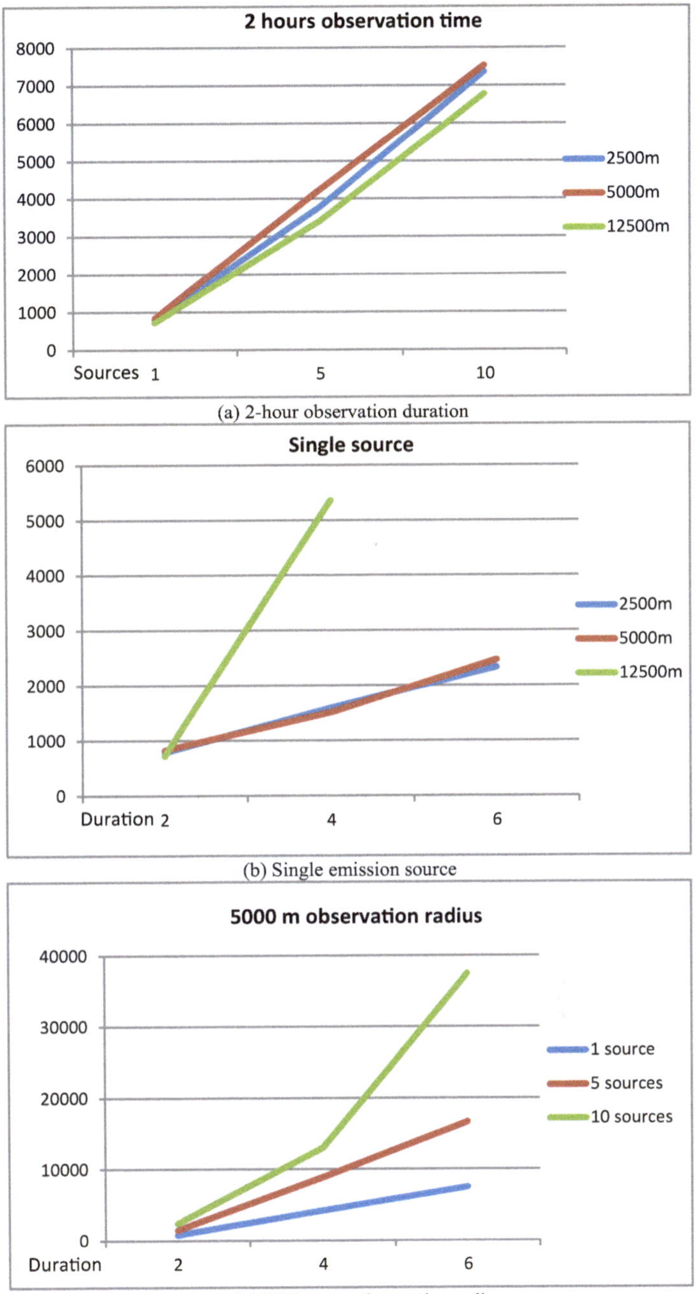

(a) 2-hour observation duration

(b) Single emission source

(c) 5000 meter observation radius

Fig. 3.5 Time consumption of the algorithm

3.5 Calculation and Comparison of Ship Pollutant Gas Emission and Diffusion

3.5.1 *Study Area and Data Source*

In this paper, Huangpu River in Shanghai is taken as an example with the study area from 121°27'E to 123°00'E and 29°05'N to 30°29'N. The collection period of ship AIS data and meteorological data is from 00:00 on October 28, 2022 to 00:00 on October 29, 2022. The AIS data set is collected from ship.chinaport.com, involving a total of 10 ships, and the relevant information of each ship is shown in Table 3.2. The meteorological data is collected from Shanghai Water Authority (http://bm.swj.sh.gov.cn/), and the webpage data is captured once every 10 min. The distribution of meteorological stations is shown in Fig. 3.6. The pollutant emissions are calculated using the ship speed after wind direction changes. Based on the translational diffusion model, the wind speed is set as a control variable to simulate the diffusion, and distribution of atmospheric pollutants in the real-time wind direction.

Table 3.2 Vessels information

MMSI	Start time	End time	Main engine	Max speed
412,047,220	1:26	22:40	1471	13
412,047,690	9:46	22:02	2070	12
412,428,730	10:58	18:48	1471	11.5
412,437,030	1:26	18:30	809	9
413,313,720	0:00	22:22	2427	12.5
413,352,360	0:00	6:00	2206	12
413,358,110	11:36	22:44	15,540	15
413,370,020	0:00	18:04	2426	12.2
413,432,970	0:08	18:02	1630	16.5
413,440,560	0:00	19:22	735	11
413,444,770	0:56	6:46	7200	14
413,546,000	0:00	22:12	2480	12
413,553,160	0:00	22:44	3285	14.5
414,783,000	0:00	21:00	7948	13.9

Note Missing vessels for static data are replaced using vessels of the same type, tonnage and draught

Fig. 3.6 Regional map of meteorological sites

3.5.2 Comparison of Ship Pollutant Emissions with Variable Wind Direction

9 wind angles, including real-time wind direction, north wind (0°), northeast wind (45°), east wind (90°), southeast wind (135°), south wind (180°), southwest wind (225°), west wind (270°) and northwest wind (315°), are set according to the ship pollutant emission Eq. (3.8) after wind direction changes. The wind speed is set to the range of 0 (0 m/s) to 7 (13.9 m/s) on the Beaufort scale, and then the carbon dioxide emissions at the same wind speed and different wind directions are calculated. Table 3.3 shows the pollutant emission simulation results of some ships, including the maximum emission, minimum emission and the percentage of emission difference. It can be seen from the table that the difference in emissions from the ship (MMSI: 413,352,360) is 34.05%, 36.85% and 5.42% in the situations of downwind, upwind and real-time wind respectively. The difference in emissions from the ship (MMSI: 413,546,000) is 9.79%, 9.82% and 4.43% in the situations of downwind, upwind and real-time wind respectively. When the wind speed is 0 m/s and 13.9 m/s, the emission

ratio of the ship decreases by 34.5% under the condition of downwind and increases by 36.85% under the condition of upwind, and the real-time emission difference is up to 19.97%.

3.5.3 *Empirical Analysis of Ship Pollutant Emissions and Diffusions*

Figure 3.7 (a-d) shows the spatial characteristics of ship pollutant emission and diffusion and concentration distribution in four scenarios. Carbon dioxide emission and diffusion are taken as an example, the period from 03:00 to 06:00 is the observation period and there are 8 ships with a sailing speed greater than or equal to 1 nautical mile (1kn). When the sailing speed is less than 1kn, the berthing state is the default state, and considered not to emit. The coordinate range of the observation window is (-20000 m, 20000 m). It can be seen from the figure that at different wind speeds, the wind direction changes in the same period, and the final pollutant distribution position is roughly the same. After controlling the wind direction variable, the pollutant concentration near the ship trajectory is the highest when the wind speed is low under the same wind direction change. The ship pollutants diffuse slowly around the point source over time. When the wind speed increases gradually, the pollutant diffusion will be accelerated and the scope of influence will be expanded. If the pollutant emission remains unchanged, the pollutant concentration value in the affected area will decrease. The change of pollutant position shows that the influence of pollutant emission and diffusion from ships near the estuary decreases with the increasing wind speed. The wind in the opposite direction to residential areas helps diffuse air pollutants to open areas and reduce the direct damage caused by air pollution to residents. Some residential areas are distributed along inland waterways. They have no areas for pollutant diffusion and transfer to marine directions. Therefore, the pollutant emissions from ships will cause a longer period of air pollution to these residential areas.

3.6 Conclusion and Prospect

(1) This paper proposes an algorithm to measure the emission of atmospheric pollutants that takes into account the effects of wind direction changes during ship navigation. The ship exhaust gas emission estimation model under variable wind direction is established according to ship activity data, ship static information and real-time wind direction change. This model has improved the calculation accuracy of pollutant emissions during ship activities, and provided data support for ship atmospheric pollution prevention and control.

Table 3.3 Part of the CO2 emissions(kg)

MMSI	0° (downwind)			180° (upwind)			Real-time		
	Maximum emission	Minimum emission	Difference Percentage	Maximum emission	Minimum emission	Difference Percentage	Maximum emission	Minimum emission	Difference Percentage
412,047,220	4106.76	3487.41	17.76%	3487.41	2973.31	17.29%	3569.11	3435.72	3.88%
413,553,160	3040.20	2446.56	24.26%	2446.56	1949.87	25.47%	2681.55	2235.24	19.97%
413,313,720	4218.70	3581.78	17.78%	3581.78	3018.59	18.66%	3689.75	3496.27	5.53%
413,546,000	11,055.72	10,069.68	9.79%	10,069.68	9169.59	9.82%	10,300.58	9863.51	4.43%
413,440,560	1100.30	901.45	22.06%	901.45	732.42	23.08%	934.26	876.61	6.58%
413,352,360	1660.94	1239.01	34.05%	1239.01	905.39	36.85%	1284.53	1218.53	5.42%
413,370,020	6788.22	6160.84	10.18%	6160.84	5577.41	10.46%	6335.23	5990.11	5.76%

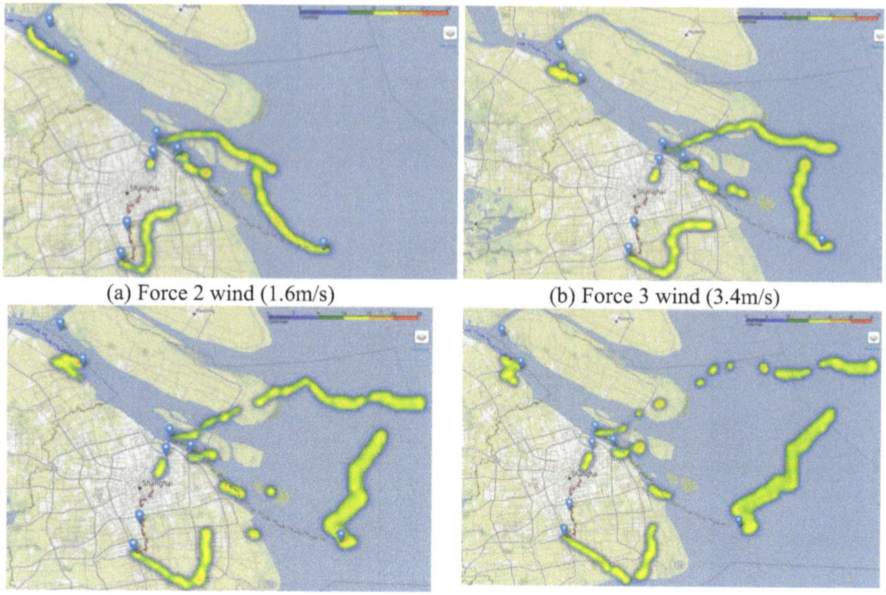

(a) Force 2 wind (1.6m/s) (b) Force 3 wind (3.4m/s)

(c) Force 4 wind (5.5m/s) (d) Force 5 wind (8.0m/s)

Fig. 3.7 Spatial characteristics of ship emissions diffusion

(2) A fast translational diffusion algorithm is established based on the Gaussian puff model. Meanwhile, spatial position transformation is performed using the coordinate transformation method. The fast calculation of ship atmospheric pollutant emission and diffusion with variable wind direction is realized accordingly. A 20 km * 20 km observation window is set to efficiently simulate the real-time diffusion scenario of multiple emission sources. In comparison with the traditional algorithms, the fast algorithm has the same calculation accuracy but its calculation efficiency has been improved significantly.

(3) This paper presents the spatial distribution of multi-ship pollutant emissions and diffusions. Taking the emission source point during ship navigation as the center, this paper simulates the changes of pollutant emission and diffusion under the influence of variable wind direction. Atmospheric pollutants from ships produce a long-term effect on the air quality around waterways. In particular, the downwind residential areas are affected for a longer time. Wind speed determines the speed and scope of pollutant diffusion. Higher wind speed will make pollutants farther from the emission source and stay for a shorter time around the emission source in the same period.

In this paper, the real-time meteorological information of onshore stations is used to simulate the process of carbon dioxide emissions and diffusions during the navigation of multiple ships. The wind direction information is collected from onshore monitoring stations. Therefore, it is different from the meteorological environment actually faced by the ships. If the real-time meteorological data of ships is available, the calculation accuracy will be further improved.

Acknowledgements Shaoli He and Jun Wang also made great contributions to this manuscript. We express our heartfelt thanks to them. This research also was supported by: National Natural Science Foundation of China (72474127) and National Social Science Fund Major Bidding Project of China (18ZDA052). Part of this chapter was published:He S, Wu X, Wang J. A calculation algorithm for ship pollutant gas emissions and diffusions based on real-time meteorological conditions and its application. Ocean Engineering 2023;287:115825. https://doi.org/https://doi.org/10.1016/j.oceaneng.2023.115825. The publication has been conducted with copyright permission.

References

Bai, S., Wen, Y., He, L., Liu, Y., Zhang, Y., Yu, Q., & Ma, W. (2020). Single-vessel plume dispersion simulation: Method and a case Study Using CALPUFF in the Yantian Port Area, Shenzhen (China) [Single-Vessel plume dispersion simulation: Method and a case study using CALPUFF in the Yantian Port Area, Shenzhen (China)]. *International Journal of Environmental Research and Public Health, 17*(21), 7831. https://doi.org/10.3390/ijerph17217831

Cao, Y., Wang, X., Yin, C., Xu, W., Shi, W., Qian, G., & Xun, Z. (2018). Inland vessels emission inventory and the emission characteristics of the Beijing-Hangzhou grand canal in Jiangsu province. *Process Safety and Environmental Protection, 113*, 4898–4506. https://doi.org/10.1016/j.psep.2017.10.020

Chou, C.-C., Hsu, H.-P., Wang, C.-N., & Yang, T.-L. (2021). Analysis of energy efficiencies of in-port ferries and island passenger-ships and improvement policies to reduce CO2 emissions. *Marine Pollution Bulletin, 172*,. https://doi.org/10.1016/j.marpolbul.2021.112826

Dai, Y., Shi, X., & Deng, Q. (2025). Synergistic effects of CO2 and air pollutants from ship emissions in Shanghai, China: Spatial-temporal characteristics, prediction assessment, policy implications. *Journal of Environmental Management, 376*,. https://doi.org/10.1016/j.jenvman.2025.124417

Ernest, C., T, C. G., Aneta, O.-J., Barbara, P., & Theo, N. (2021). An energy consumption approach to estimate air emission reductions in container shipping. *Energies, 4*(2). https://doi.org/10.3390/en14020278

Fan, Z., Jing, L., Hang, Z., Xiaodong, Y., & Yunli, F. (2022). A real-time measurement-modeling system for ship air pollution emission factors. *Journal of Marine Science and Engineering, 10*(6). https://doi.org/10.3390/jmse10060760

He, L., Wang, J., Liu, Y., Zhang, Y., He, C., Yu, Q., & Ma, W. (2021). Selection of onshore sites based on monitoring possibility evaluation of exhausts from individual ships for Yantian Port, China. *Atmospheric Environment, 247*,. https://doi.org/10.1016/j.atmosenv.2021.118187

He, S., Wu, X., Wang, J., & Guo, J. (2022). A ship emission diffusion model based on translation calculation and its application on Huangpu River in Shanghai [A ship emission diffusion model based on translation calculation and its application on Huangpu River in Shanghai]. *Computers & Industrial Engineering, 172*,. https://doi.org/10.1016/j.cie.2022.108569

Huang, L., Wen, Y., Geng, X., Zhou, C., & Xiao, C. (2018). Integrating multi-source maritime information to estimate ship exhaust emissions under wind, wave and current conditions [Integrating multi-source maritime information to estimate ship exhaust emissions under wind, wave and current conditions]. *Transportation Research Part D, 59*, 148–159. https://doi.org/10.1016/j.trd.2017.12.012

Jalkanen, J.-P., Brink, A., Kalli, J., Pettersson, H., Kukkonen, J., & Stipa, T. (2009). A modelling system for the exhaust emissions of marine traffic and its application in the Baltic Sea area [A modelling system for the exhaust emissions of marine traffic and its application in the Baltic Sea

area]. *Atmospheric Chemistry and Physics, 9*(23), 9209–9223. https://doi.org/10.5194/acpd-9-15339-2009
Juan, L., Wenbin, X., & Jutao, C. (2020). Simulation of emission and diffusion of air pollutants from ships sailing in port [Simulation of emission and diffusion of air pollutants from ships sailing in port]. *Journal of Xiamen University(Natural Science), 59*(6), 1016–1024. https://doi.org/10.6043/j.issn.0438-0479.202001026
Kwon, Y., Lim, H., Lim, Y., & Lee, H. (2019). Implication of activity-based vessel emission to improve regional air inventory in a port area. *Atmospheric Environment, 203*, 262–270. https://doi.org/10.1016/j.atmosenv.2019.01.036
Liu, H., Meng, Z.-H., Shang, Y., Lv, Z.-F., Jin, X.-X., Fu, M.-L., & He, K.-B. (2018). Shipping emission forecasts and cost-benefit analysis of China ports and key regions' control. *Environmental Pollution, 236,* 49–59. https://doi.org/10.1016/j.envpol.2018.01.018
Meng, L., Liu, K., He, J., Han, C., & Liu, P. (2022). Carbon emission reduction behavior strategies in the shipping industry under government regulation: A tripartite evolutionary game analysis Journal of Cleaner Production, 378. https://doi.org/10.1016/j.jclepro.2022.134556
Pan, K., Lim, M. Q., Kraft, M., & Mastorakos, E. (2021). Development of a moving point source model for shipping emission dispersion modelling in EPISODE-CityChem v1.3 [Development of a moving point source model for shipping emission dispersion modelling in EPISODE-CityChem v1.3]. https://doi.org/10.5194/gmd-2021-47
Peng, X., Ding, Y., Yi, W., Laroussi, I., & He, T. (2024). The inland waterway ship emission inventory modeling: The Yangtze River case. *Transportation Research Part D: Transport and Environment, 129*, 1–13. https://doi.org/10.1016/j.trd.2024.104138
Peng, X., Huang, L., Wu, L., Zhou, C., Wen, Y., Chen, H., & Xiao, C. (2021). Remote detection sulfur content in fuel oil used by ships in emission control areas: A case study of the Yantian model in Shenzhen. *Ocean Engineering, 237*,. https://doi.org/10.1016/j.oceaneng.2021.109652
Ramacher, M. O. P., Karl, M., Bieser, J., Jalkanen, J.-P., & Johansson, L. (2019). Urban population exposure to NOx emissions from local shipping in three Baltic Sea harbour cities—A generic approach [Urban population exposure to NOx emissions from local shipping in three Baltic Sea harbour cities—A generic approach]. *Atmospheric Chemistry and Physics, 19*(14), 9153–9179. https://doi.org/10.5194/acp-19-9153-2019
Rui, X., Xiaoling, D., Hang, W., Yanpeng, C., & Xipeng, P. (2021). A deep learning method to repair atmospheric environmental quality data based on Gaussian diffusion [A deep learning method to repair atmospheric environmental quality data based on Gaussian diffusion]. *Journal of Cleaner Production, 308*,. https://doi.org/10.1016/j.jclepro.2021.127446
Shukai, C., Qiang, M., Peng, J., & Haibo, K. (2021). An operational-mode-based method for estimating ship emissions in port waters. *Transport Research Part D-Transport and Environment, 101*,. https://doi.org/10.1016/j.trd.2021.103080
Sorte, S., Rodrigues, V., Ascenso, A., Freitas, S., Valente, J., Monteiro, A., & Borrego, C. (2018). Numerical and physical assessment of control measures to mitigate fugitive dust emissions from harbor activities [Numerical and physical assessment of control measures to mitigate fugitive dust emissions from harbor activities]. *Air Quality, Atmosphere & Health, 11*, 493–504. https://doi.org/10.1007/s11869-018-0563-7
Toscano, D., Murena, F., Quaranta, F., & Mocerino, L. (2021). Assessment of the impact of ship emissions on air quality based on a complete annual emission inventory using AIS data for the port of Naples [Assessment of the impact of ship emissions on air quality based on a complete annual emission inventory using AIS data for the port of Naples]. *Ocean Engineering, 232*(5), Article 109166. https://doi.org/10.1016/j.oceaneng.2021.109166
Tseng, Y., Cheng, W., & Yuan, C. (2024). Impacts of ship emissions and sea-land breeze on urban air quality using chemical characterization, source contribution and dispersion model simulation of PM2.5 at Asian seaport. *Environmental Pollution, 347*(15), 123663. https://doi.org/10.1016/j.envpol.2024.123663
Weng, J., Han, T., Shi, K., & Li, G. (2022). Impact analysis of ECA policies on ship trajectories and emissions. *Marine Pollution Bulletin, 179*,. https://doi.org/10.1016/j.marpolbul.2022.113687

Weng, J., Shi, K., Gan, X., Li, G., & Huang, Z. (2020). Ship emission estimation with high spatial-temporal resolution in the Yangtze River estuary using AIS data. *Journal of Cleaner Production, 248*,. https://doi.org/10.1016/j.jclepro.2019.119297

Xia, Z., Guo, Z., Wang, W., & Jiang, Y. (2021). Joint optimization of ship scheduling and speed reduction: A new strategy considering high transport efficiency and low carbon of ships in port. *Ocean Engineering, 233*,. https://doi.org/10.1016/j.oceaneng.2021.109224

Xin, P., Yuanqiao, W., Changshi, X., liang, H., Chunhui, Z., Fan, Z.,…Yimeng, Z. (2020). Improved calculation model for ship exhaust emission dispersion [Improved calculation model for ship exhaust emission dispersion]. *Journal of Safety and Environment, 20*(1), 255–264. CNKI:SUN:AQHJ.0.2020-01-034

Yang, L., Zhang, Q., Zhang, Y., Lv, Z., Wang, Y., Wu, L.,…Mao, H. (2021). An AIS-based emission inventory and the impact on air quality in Tianjin port based on localized emission factors. Science of the Total Environment, 783. https://doi.org/10.1016/j.scitotenv.2021.146869

Yin, S. (2022). Decadal changes in PM2.5-related health impacts in China from 1990 to 2019 and implications for current and future emission controls [Decadal changes in PM2.5-related health impacts in China from 1990 to 2019 and implications for current and future emission controls]. *Science of the Total Environment, 834*. https://doi.org/10.1016/j.scitotenv.2022.155334

Ying, Z., Xiaxia, Y., Xiangming, K., Quanling, T., Yongping, L., Guohe, H., & Yexin, L. (2019). Industry-environment system management based on an uncertain Gaussian diffusion optimization model for coal-dependent cities in ecologically fragile areas [Industry-environment system management based on an uncertain Gaussian diffusion optimization model for coal-dependent cities in ecologically fragile areas]. *Journal of Cleaner Production, 234*, 832–857. https://doi.org/10.1016/j.jclepro.2019.06.142

Zhang, F., Chen, Y., Cui, M., Feng, Y., Yang, X., Chen, J.,…Liu, H. (2019). Emission factors and environmental implication of organic pollutants in PM emitted from various vessels in China. *Atmospheric Environment, 00*, 302–311. https://doi.org/10.1016/j.atmosenv.2018.12.006

Zhang, F., Chen, Y., Su, P., Cui, M., Han, Y., Matthias, V., & Wang, G. (2021). Variations and characteristics of carbonaceous substances emitted from a heavy fuel oil ship engine under different operating loads. *Environmental Pollution, 284*,. https://doi.org/10.1016/j.envpol.2021.117388

Zhang, Y., Zhao, K., Lou, D., & Fang, L. (2022). Study on the real-world emission characteristics of gaseous and particulate pollutants from an inland ship using a portable emission measurement system. *Marine Pollution Bulletin, 184*,. https://doi.org/10.1016/j.marpolbul.2022.114205

Zhao, J., Zhang, Y., Patton, A. P., Ma, W., Kan, H., Wu, L.,…Walker, K. (2020). Projection of ship emissions and their impact on air quality in 2030 in Yangtze River delta, China. *Environmental Pollution, 263*. https://doi.org/10.1016/j.envpol.2020.114643

Chapter 4
A Ship Emission Diffusion Model Based on Translation Calculation and Its Application on Huangpu River in Shanghai

4.1 Introduction

Shipping logistics accounts for two thirds of the world's total trade logistics (Weng et al., 2019). The global maritime trade volume in 2020 was 10.648 billion tons and is expected to increase by 4.3% in 2021 (UNCTAD, 2021). According to the Report of the International Maritime Organization (IMO) (Zincir et al., 2019), the annual global consumption of petroleum by ships is 300 million tons, which generates 2.2% of the world's carbon dioxide (CO_2) (Bilgili & Celebi, 2018). A large amount of nitrogen oxides (NOx), sulfur oxides (SOx) and suspended particulate matter (PM) are also produced during ships' activities (Cao et al., 2022; Nakatsubo et al., 2019; Sharafian et al., 2019). With the growing demand for ship transportation, air pollutants discharged by shipping and port activities have become the third major source of pollution following industrial exhaust gas and road traffic exhaust gas, which has an increasing negative impact on air quality and residents' health in coastal areas along the Yangtze River (Liu et al., 2017). Therefore, the issue of ship exhaust emissions has attracted extensive attention. For example, IMO and some governments have issued policies to establish emission control areas (Chen et al., 2018) to reduce the impact of ship exhaust emissions on the surrounding ecological environment.

The existing research on ship exhaust emissions mainly includes the calculation of pollutant emissions, the impact of pollutant diffusion on air quality, and the socio-economic impact. Among them, the calculation of pollutant emissions from ships is mainly studied from different regional scales (port level and regional level) (Huang et al., 2020; Zhang et al., 2019) and global scale (Viana et al., 2014). In terms of the socio-economic impact of exhaust emissions, the external costs of health and other damage caused by ship emissions are calculated (Progiou et al., 2020; Tichavska & Tovar, 2015). For the quality of regional air caused by exhaust emissions, the existing air quality models are often used, or the concentration of air pollutants is estimated based on a small scale such as a ship (Bai et al., 2020) and a port (Fu & Li, 2018; Liu et al., 2020; Wen et al., 2017; Xie et al., 2025). Among them, the widely used

X. Wu, *Environmental Disaster in China*, https://doi.org/10.1007/978-981-92-0910-1_4

commercial model like AERMOD (Ekmekçioğlu et al., 2020) and CALPUFF (Bai et al., 2020; Tseng et al., 2024) require strict basic data of geographic and meteorological data, which increases the cost of data collection and computational difficulty. Moreover, such models are highly complex and multi-modular, and thus are difficult to be widely used in real-time application analysis. In addition, they have high requirements on the scale of application. Because shipping activities are mainly in coastal and riverine areas, the area within the emission range of air pollutants is densely populated and developed. Considering above characteristics of the atmospheric diffusion model, this paper attempts to reach a balance among the calculation scale, accuracy and cost according to the dynamic multi-source ship emissions as well as the strong flow, high dispersion, and easy diffusion of pollutants. Then, based on the Gaussian CALPUFF model, this paper, through the big data of AIS shipping activities and the environment data of ship activities, simulates the diffusion of ship emissions under different meteorological conditions. Finally, based on existing research, a data translation model for analyzing ship emission diffusion is established so as to improve the calculation accuracy and efficiency and further reduce the calculation cost.

4.2 Literature Review

4.2.1 Ship Emissions

The research on ship exhaust emissions includes the compilation of inventory of ship emissions (Huang et al., 2020), the impact of exhaust gas on social, economic, and ecological environment, residents' health, etc. The compilation of ship emission inventory can be divided into actual measurement and statistical evaluation according to the degree of accuracy. The actual measurement was first studied based on laboratory engine measurement, ship certification level, combustion rate data of different fuels, sampling measurement of specific ships and geographical locations, etc. (Huang et al., 2020; Kasper et al., 2007; McLaren et al., 2012). This method however requires perfect equipment and facilities, and needs to determine the accuracy of data according to the research content. Although it has high estimation accuracy in a small-scale range, the calculation cost is high, restricting its large-scale applications. According to different accounting paths, it can be divided into top-down fuel methods and bottom-up activity methods. The fuel method is mainly based on fuel sales data, fuel types and corresponding fuel emission factors to form emission inventory (Corbett & Koehler, 2003). This method lacks ship-related technical information, which will increase the uncertainty of emission inventory in the case of regional scale division. By contrast, the activity method is based on ship characteristics and navigation characteristics. Ship characteristics include the number of ship engines, load factor, fuel type, fuel consumption for sailing, required fuel consumption rate, etc. Navigation characteristics include navigation state, speed (Jiang et al., 2019), travel time and distance under conditions such as entering and leaving ports, cruising, and berthing (Chu-Van et al., 2017). This method has high requirements on basic data, not only the dynamic information of during navigation, but also the complete static data of the ship's archives.

With the progress of information technology, mobile internet devices and the Automatic Identification System (AIS) data, the emission estimation models based on AIS data have become a new measurement method, which not only improves the estimation accuracy of emissions, but also achieves accurate statistics at different spatial scales (Huang et al., 2020). In order to measure (Yang et al., 2021; Zhang et al., 2019) and predict air pollution emissions, and analyze the impact of emissions on the environment (Zhuang et al., 2025), economy and residents, an emission control zone can be set up to real-time monitor pollutant emissions (Chen et al., 2018; Huang et al., 2020).

4.2.2 *Diffusion Model of Ship Emissions*

The atmospheric diffusion model is mainly used to simulate the diffusion process of exhaust pollutants in the neighboring area centered on the emission source. Air quality or pollutant concentrations at different locations downwind are predicted based on emission source intensity and meteorological conditions. According to the development stage, the models can be divided into three generations. The first generation is the air quality model, including Gaussian model, Box model and Lagrangian model (Holmes & Morawska, 2006). The second generation is the photochemical pollution model, which simulates the chemical transformation process in the diffusion process. The third generation is the comprehensive regional-scale diffusion model (Liu et al., 2017) which considers interactions between pollutants. Different atmospheric diffusion models are developed for different purposes, with large differences in principles, theories (K-mode, Taylor statistical model, Gaussian diffusion model) and application scope (small scale (Wu et al., 2018), regional scale, and global model (Bindle et al., 2021)).

There are few studies on the diffusion of ship exhaust gas at home and abroad. Because the comprehensive regional scale diffusion model is strict about the basic data of meteorological conditions and pollution sources, it has flexible functions but low operability, a large amount of calculation, and high requirements on computing equipment. Some studies based on Gaussian model (Stockie, 2011) to analyze the atmospheric diffusion of single and multiple ships are as follow. Liu et al. (2020) adopted Gaussian CALPUFF model, Jave Web technology, and geographic information system technology to simulate the diffusion of a single cargo ship in Xiamen port. Peng et al. (2020) studied the simulation calculation method of ship emission dispersion based on the Gaussian CALPUFF model, and conducted empirical tests through single ship application and shore-based monitoring stations. Shubin Bai et al. combined the WRF/CALMET/CALPUFF model with ship AIS data to simulate the diffusion of air pollution within 2 kilometers of a single ship. Since the single ship model has a small amount of data in a certain area, there is no need for research on computational efficiency or cost. If the multi-ship calculation is carried out directly according to the single-ship model, the calculation cost will be greatly increased and the calculation efficiency will be reduced. In terms of multi-vessel atmospheric dispersion, Murena et al. (2018) used Gaussian CALPUFF model to assess the impact of cruise emissions on air quality based on fixed locations, with hourly averages as ship activity data. Toscano et al. (2021) simulated the Gaussian CALPUFF diffusion of multi-ship emissions by clustering ships and simplifying ship activity trajectories, and evaluated the impact of exhaust gas on regional air quality. Fu et al. (2025) employed the Weather Research and Forecasting (WRF)-Community Multiscale Air Quality (CMAQ) modeling system to comprehensively assess the impacts of ship emissions on $PM_{2.5}$ and ozone (O_3) pollution in the Yangtze River Delta (YRD) region. Their study revealed the spatiotemporal distribution of these impacts and elucidated the atmospheric evolution mechanisms of ship-emitted pollutants. These findings provide critical scientific support for developing targeted emission control policies in maritime-influenced areas. In the multi-vessel atmospheric dispersion model study, the AIS data is compressed by the increase of time granularity and means the value of vessel activity. Even if the error is small, it will still affect the accuracy of the diffusion concentration to a certain extent. Both single-ship or multi-ship diffusion modes are based on multiple calculations of emission sources, and then aggregated and superimposed, so it cannot meet the requirements of calculation efficiency and cost in a small time granularity and a large regional range.

4.3 Real-Time Estimation of Ship Emissions

The air pollutant emissions from ships are estimated through the AIS data-based Ship Traffic Emissions Assessment Model (STEAM) (Jalkanen et al., 2009), including the air pollutants (SO_2, NOx, CO_2, PM, etc.) emitted by a single ship during its voyage. This calculation method based on ship activities has been widely applied (Tichavska & Tovar, 2015; Wen et al., 2017; Weng et al., 2019; Yang et al., 2021). The real-time estimation formula of air pollutant emissions from ships of this paper is as follows:

$$\mathrm{E} = \mathrm{E}_{\mathrm{main}} + \mathrm{E}_{\mathrm{auxiliary}} + \mathrm{E}_{\mathrm{boiler}} \tag{4.1}$$

$$\mathrm{E}_{\mathrm{main}} = \mathrm{P}_{\mathrm{main}} \times \mathrm{LF}_{\mathrm{main}} \times T_{main} \times \mathrm{EF}_{\mathrm{main}} \tag{4.2}$$

$$\mathrm{E}_{\mathrm{auxiliary}} = \mathrm{P}_{\mathrm{auxiliary}} \times \mathrm{LF}_{\mathrm{auxiliary}} \times T_{\mathrm{auxiliary}} \times \mathrm{EF}_{\mathrm{auxiliary}} \tag{4.3}$$

$$\mathrm{E}_{\mathrm{boiler}} = \mathrm{P}_{\mathrm{boiler}} \times \mathrm{LF}_{\mathrm{boiler}} \times T_{\mathrm{auxiliary}} \times \mathrm{EF}_{\mathrm{boiler}} \tag{4.4}$$

E is individual ship exhaust emissions (g), main, auxiliary, and boiler denote the main engine, auxiliary engine, and boiler respectively. P is ship power equipment power (kW); LF is load factor; T is operating (duration) time (h); EF is emission factor.

$$LF_{main} = \left(v_{actual}/v_{max}\right)^3 \tag{4.5}$$

v_{actual} represents the actual speed, based on the real-time speed (kn/h) acquired by the ship in the real sailing environment, adding the influence factors of meteorological conditions and geography. v_{max} is the maximum speed (kn/h) of the ship is sailing. The substitution of the load factor into Eq. (4.2) gives:

$$E_{main} = P_{main} \times \left(v_{actual}/v_{max}\right)^3 \times T_{main} \times EF_{main} \tag{4.6}$$

Emission factors for air pollutants depend on the ship type, type and state of ship engine, and fuel type. Therefore, in the empirical study, the data should be selected according to the specific situation.

4.4 Diffusion Model of Ship Emissions

4.4.1 Modelling Assumptions

The concentration of air pollutants is affected by the emission pattern, geographical location and environmental factors of the pollution source. Ship emissions are a typical mobile point source emission models that differs significantly from the emissions and dispersion of factories and motor vehicles. Ship navigation is a dynamic process. The emission concentration of exhaust pollutants is closely related to sailing conditions (cruising, anchoring and berthing), sailing speed, ship equipment (main engine, auxiliary engine, boiler and quantity). The spatial location determines the diffusion characteristics of exhaust gas. Ships sail in open water where there are no tall buildings nearby, so they are mainly affected by meteorological conditions during emission diffusion.

Therefore, the following conditional assumptions need to be considered when building the diffusion model of ship air pollutants by Gaussian model.

1. During the time interval of dispersion simulation, the pollutants emitted by the ship are characterized by continuity, stability and uniformity.
2. The pollutants discharged on water surface are totally reflected without absorption effect.
3. The total amount of pollutants remains unchanged during the diffusion process, only the diffusion of exhaust gas emitted from the source is considered, and there is no chemical transformation during the diffusion.

4. The diffusion model is based on the downwind direction of the puffs, and the diffusion in other directions is not considered herein.

4.4.2 *Dispersion Model of Ship Emissions*

4.4.2.1 Wind Direction Coordinate System and Diffusion Model

To solve the exhaust gas diffusion, a wind direction coordinate system is established, where the coordinate origin is set as the exhaust gas emission source, the downwind direction is horizontal axis x and the cross wind direction is vertical axis y. Since ship emissions change along with the navigational state, the origin of emissions is not always located at the origin of coordinates, and the wind direction may also deviate from the horizontal axis. It is therefore necessary to convert the navigation coordinates into wind direction coordinates, and convert the origin of the ground coordinate $A'(x', y')$ to the position of the ship's emission source (X_s, Y_s). The new origin of the wind direction coordinate system is $A'(x', y')$, as shown in Fig. 4.1.

The ship emission diffusion coordinate system is established based on the wind direction coordinates converted during the ship's navigation, and the Gaussian CALPUFF diffusion model is adopted to simulate the distribution of the ship's emission diffusion concentration in sailing state. The calculation formula of pollution concentration under continuous discharge and total reflection of diffusion of point source is:

$$C(x, y, z, H_e) = \frac{Q}{2\psi_y\sigma_z u} e^{\frac{-y^2}{2\sigma_y^2}} \left(e^{\frac{-(z-H_e)^2}{2\sigma_z^2}} + e^{\frac{-(z+H_e)^2}{2\sigma_z^2}} \right) \tag{4.7}$$

The instantaneous emission of point source takes the spatial position of emission source at time i as the origin, and x_i', y_i' and z_i' are the downwind direction, crosswind direction and elevation distance of the coordinate system, respectively.

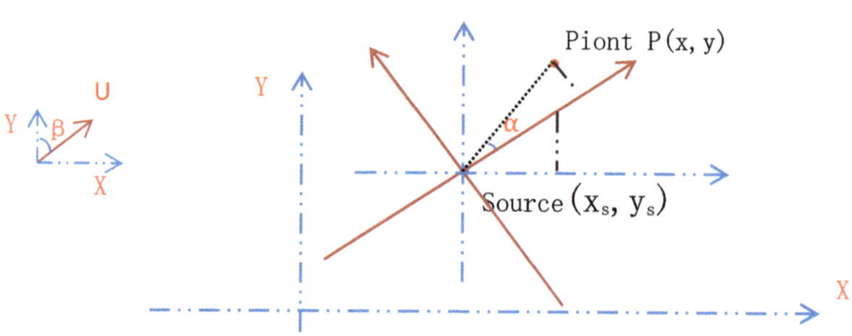

Fig. 4.1 Coordinate system conversion

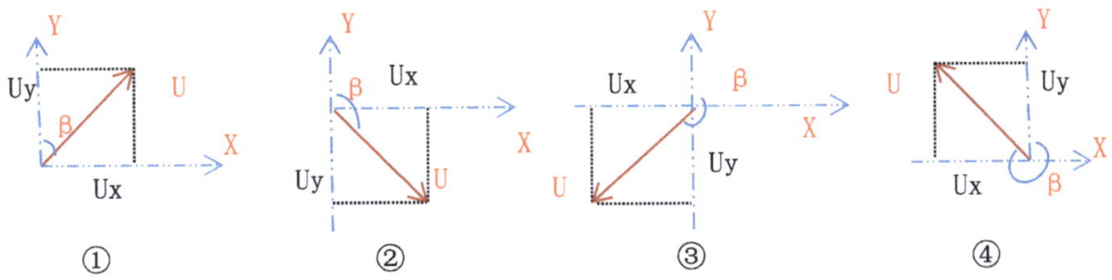

Fig. 4.2 Wind projection map

$$\rho_i(x, y, z, t_n) = \frac{Q_i}{(2\pi)^{1.5}\sigma_{x_i'}\sigma_{y_i'}\sigma_{z_i'}} e^{\frac{-[x_i'-u(t_n-t_i)]^2}{2\sigma_{x_i'}^2}+\frac{-y_i'^2}{2\sigma_{y_i'}^2}} \left(e^{\frac{-(z_i'-H_e)^2}{2\sigma_{z_i'}^2}} + e^{\frac{-(z_i'+H_e)^2}{2\sigma_{z_i'}^2}} \right) \quad (4.8)$$

Where $\rho_i(x, y, z, t_n)(ug/m^3)$ is the mass concentration of the exhaust gas emitted by the emission source at time i diffused to any point (x, y, z) at time n; Q_i (g) is the emission intensity at time i; u (m/s) is the wind speed; $\sigma_{x_i'}$, $\sigma_{y_i'}$ and $\sigma_{z_i'}$ are the diffusion coefficient on the directions of x_i', y_i' and z_i' respectively.

The puff diffusion is always along the downwind direction. If the wind direction is set to θ, then the downwind direction is β= θ − 180°. Therefore, the projection of the downwind direction on the x and y axes is calculated by Fig. 4.2.

1. $0° \le \theta - 180° < 90°$

$$u_x = u * \sin(\theta - 180°) = -u * \sin\theta \; u_y = u * \cos(\theta - 180°) = -u * \cos\theta \quad (4.9)$$

2. $90° \le \theta - 180° < 180°$

$$u_x = u * \sin[180° - (\theta - 180°)] = -u * \sin\theta \; u_y = u * \cos[180° - (\theta - 180°)] = u * \cos\theta \quad (4.10)$$

3. $180° \le \theta - 180° < 270°$

$$u_x = u * \sin[(\theta - 180°) - 180°] = u * \sin\theta \; u_y = u * \cos[(\theta - 180°) - 180°] = u * \cos\theta \quad (4.11)$$

4. $270° \le \theta - 180° < 360°$

$$u_x = u * \sin[360° - (\theta - 180°)] = u * \sin\theta \; u_y = u * \cos[360° - (\theta - 180°)] = -u * \cos\theta \quad (4.12)$$

That is to say, $u_x = -u * \sin\theta$ while $0° \le \theta - 180° < 180°$, $u_x = u * \sin\theta$ while $180° \le \theta - 180° < 360°$, $u_x = -u * \cos\theta$ while $270° \le \theta - 180°$ or $\theta - 180° < 90°$, and $u_x = u * \cos\theta$ while $90° \le \theta - 180° < 270°$.

Based on the formula for the angle between two lines:

$$\cos(\alpha) = \frac{U \bullet P}{\sqrt{U} \bullet \sqrt{P}} = \frac{u_x \bullet (x - x_0) + u_y \bullet (y - y_0)}{\sqrt{u_x{}^2 + u_y{}^2} \bullet \sqrt{(x - x_0)^2 + (y - y_0)^2}} \tag{4.13}$$

Given the angle α between the wind direction and the ray point P (x', y') in the new coordinate system, we have wind projection map:

$$x' = \sqrt{x^2 + y^2} \bullet \cos\left\{\arccos\left[\frac{u_x \bullet (x - x_0) + u_y \bullet (y - y_0)}{\sqrt{u_x{}^2 + u_y{}^2} \bullet \sqrt{(x - x_0)^2 + (y - y_0)^2}}\right]\right\} \tag{4.14}$$

$$y' = \sqrt{x^2 + y^2} \bullet \sin\left\{\arccos\left[\frac{u_x \bullet (x - x_0) + u_y \bullet (y - y_0)}{\sqrt{u_x{}^2 + u_y{}^2} \bullet \sqrt{(x - x_0)^2 + (y - y_0)^2}}\right]\right\} \tag{4.15}$$

Let t_0 moment a certain emission source run to the position (x_0, y_0, z_0) when the emission of smoke mass at t_i moment diffusion to any point (x, y, z) mass concentration is calculated public as follows:

$$C_{t_i} = \rho(x, y, z, t_i) = \frac{Q_{t_0}}{(2\pi)^{1.5}\sigma_{x_0}\sigma_{y_0}\sigma_{z_0}} e^{\frac{-[x_0 - u*(t_i - t_0)]^2}{2\sigma_{x_0}^2} + \frac{-y_0{}^2}{2\sigma_{y_0}^2}} \left(e^{\frac{-(z_0 - He)^2}{2\sigma_{z_0}^2}} + e^{\frac{-(z_0 + He)^2}{2\sigma_{z_0}^2}}\right) \tag{4.16}$$

For the emission source that emits continuously in constant movement, it is assumed that the puffs emitted at different times are independent of each other during the diffusion. So the continuous emission can be regarded as puff clusters formed by the superposition of numerous instantaneous puffs at very short time intervals (Pan et al., 2021; Xu et al., 2021; Zhu et al., 2019). Then at the observation time t, the formula for the mass concentration of emissions at any point (x, y, z) in the air is:

$$C_t = \sum_i \rho(x, y, z, t - t_i) = \sum_i \frac{Q_{t_i}}{(2\pi)^{1.5}\sigma_{x_i}\sigma_{y_i}\sigma_{z_i}} e^{\frac{-[x_i - u*(t - t_i)]^2}{2\sigma_{x_i}^2} + \frac{-y_i{}^2}{2\sigma_{y_i}^2}} \left(e^{\frac{-(z_i - He)^2}{2\sigma_{z_i}^2}} + e^{\frac{-(z_i + He)^2}{2\sigma_{z_i}^2}}\right) \tag{4.17}$$

Similarly, in the case of continuous emission by multiple emission sources in the observation space, the total emission mass concentration there can be obtained by the sum of their respective puff clusters:

$$C_t = \sum\nolimits_k \sum\nolimits_i \rho(x, y, z, t - t_i) \tag{4.18}$$

4.4.2.2 Fast Translation Algorithm for Diffusion Model

According to the puff diffusion model of instantaneous emission of a single emission source (Eq. 4.16), the diffusion of puff after emission is only related to the initial emission conditions (emission place and emission amount), wind (wind scale and wind direction), atmospheric diffusion parameters (diffusion coefficients on x, y and z axes) and time. That is, the puff diffuses independently of the emission source after emission. The wind and atmospheric diffusion parameters of the same type of atmospheric pollutant emissions are relatively stable within a relatively small observation range. Therefore, it is considered that the puff diffusion is related to the initial emission source and time parameters.

Under the influence of the same atmospheric diffusion parameters and wind, the two instantaneous emission puffs in the movement of the emission source follow the same diffusion function, the differences are the puff location and the diffusion duration. Affected by wind $\overline{u}$, the puff emitted instantaneously at the origin (0, 0) presents different shapes at different times, as shown in Fig. 4.3a. Then, under the influence of the same wind $\overline{u}$, the puffs emitted at different points (places) by the continuously moving emission source all follow the standard diffusion model in Fig. 4.3a, forming the puff clusters shown in Fig. 4.3b. In other words, the instantaneous emission of the diffused puff at any point at a specific time is equivalent to the result of coordinate translation of the standard puff at the corresponding diffusion time.

Under specific atmospheric diffusion parameters and wind, for the puff emitted in unit emission amount at the origin (0, 0) of the XY coordinate system at time 0, its diffusion at a different time is expressed by matrix C_i:

$$C_i = \frac{1}{(2\pi)^{1.5}\sigma_{x_0}\sigma_{y_0}\sigma_{z_0}} e^{\frac{-[u*i]^2}{2\sigma_{x_0}^2}} \left(e^{\frac{-(z_0-H_e)^2}{2\sigma_{z_0}^2}} + e^{\frac{-(z_0+H_e)^2}{2\sigma_{z_0}^2}} \right) \tag{4.19}$$

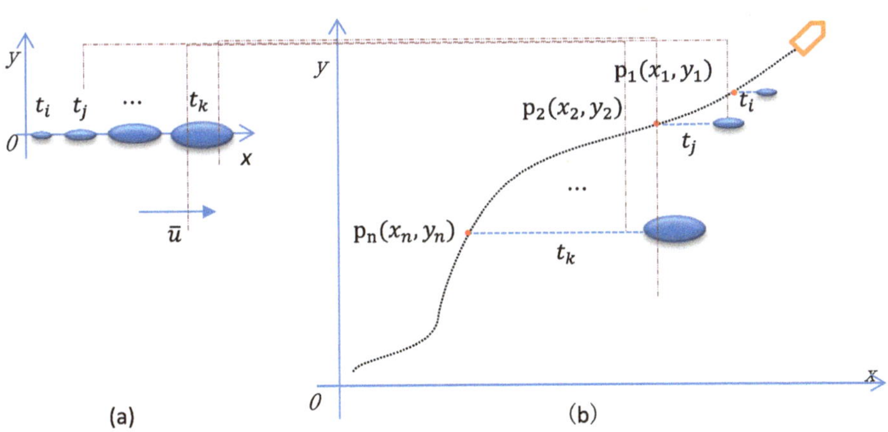

Fig. 4.3 Calculation of continuous emission clusters based on standard cluster diffusion advection

Where Eq. (4.19) shows the special case of Eq. (4.16), i.e., $x_0 = 0$, $y_0 = 0$ and $t_0 = 0$.

1. Matrix translation algorithm

The concentration of C_i on the observation surface is represented as a data matrix, that is, the n scales on the x-axis correspond to n columns of the matrix, and the n scales on the y-axis to n rows of the matrix. The value of $C_i(x, y)$ is factor at the intersection point of the matrix's x column and y row, so the matrix of puff diffusion is expressed as:

$$C_i = \begin{bmatrix} c^i_{11} & \cdots & c^i_{1n} \\ \vdots & \vdots & \vdots \\ c^i_{n1} & \cdots & c^i_{nn} \end{bmatrix} \tag{4.20}$$

Let the matrix obtained by translating r columns to the right of a unit array with the same dimension as the matrix C_i (N x N) be M_r, i.e.:

$$M_r = \left[\begin{array}{c:c} 0 & \\ 0 & \mathbf{E} \\ \hdashline \underbrace{\quad}_{r} & 0 \end{array}\right]_{NN} \begin{matrix} \\ \\ \}\, r \end{matrix} \tag{4.21}$$

Where r is the number of columns (distance) moving to the right, and E is the identity matrix.

It can be seen from Eq. (4.7) that the instantaneous emission amount Q does not affect the diffusion form of puff, but only the concentration, that is, the concentration at any point in the air is Q times that at the corresponding point under unit emission. Assuming that a certain puff is emitted by a continuous emission source at point p (x, y), and it has just diffused for i units of time at the time of observation, then the diffusion $F_{x,y}(i)$ of the puff at the observation time is the translation of all data of C_i by x rows and y columns:

$$F_{x,y}(i) = Q * {M_y}^T * C_i * M_x \tag{4.22}$$

At time i, one continuous emission source has a total of i puffs diffusing in the air, so the total diffusion F_i of the puff clusters can be expressed as:

$$F_i = \sum_i \left(Q_i * {M_{y_i}}^T * C_i * M_{x_i}\right) = Q_i * {M_{y_i}}^T * C_i * M_{x_i} + {M_{y_i - y_{i-1}}}^T * F_{i-1} * M_{x_i - x_{i-1}} \tag{4.23}$$

Under the influence of north wind (5 m/s), a ship starts from the origin (0, 0) with an emission intensity of 40 g/s, and sails on the course shown in Fig. 4.4a. Its sailing trajectory and puff diffusion concentration contours are shown in Fig. 4.4b, and c, respectively. After the puff diffuses to a certain extent, the emission concentration is greatly reduced until it is invisible to naked eyes.

In the x-y coordinate system, it is difficult to know the specific location of the ship's emission diffusion, which needs to be transformed into the corresponding geographic coordinate system. Assuming that the ship is sailing in a sea area near Dalian port, the geographical coordinates corresponding to the initial observation point (origin) are set to [38.9666,121.6727], so each point in the x-y coordinate system can be transformed into the geographical coordinates expressed in latitude and longitude according to the conversion rules between x-y and geographical coordinate systems. Under the restriction of concentration threshold 1e-5, the ship's sailing track and puff cluster diffusion during sailing are shown in Fig. 4.4d.

Since each puff is diffused independently, the total diffusion of the puff cluster from multiple continuous emission sources can be directly expressed as the sum of the diffusion of each emission source, i.e.:

$$F = \sum_{k} \sum_{i} \left(Q_{ki} * {M_{y_{ki}}}^{T} * C_{ki} * M_{kx_{i}} \right) \tag{4.24}$$

It is assumed that in the same observation area, there are multiple ships navigating in the ways shown in Table 4.1 under identical environmental conditions. According to the Eq. (4.24), the puff clusters of the emission sources are added. The puff cluster diffusion and concentration contour lines observed after a period are shown in Fig. 4.5a, and b. Figure 4.5a represents the instantaneous multi-source emission dispersion superimposed concentrations in the observation area. The shade of the color represents the value of the shade. However, this picture can't compare the concentrations of adjacent points. Figure 4.5b represents the concentration contour map in the observation area. The figure shows the pollutant concentration values, the concentration line distribution, and the spatial distribution of concentration after superposition diffusion of multiple smoke clusters.

The concentration and contour diagrams reflect the diffusion patterns of puff clusters emitted by various continuous emission sources at a certain observation time, but they cannot reflect the influence process of puff clusters. Figure 4.5c represents the superimposed diffusion average concentration of multi-source emissions in the observation period. The figure allows us to observe and evaluate the impact of multi-source emissions and the level of effect (concentration), thus providing a basis for formulating and implementing relevant policies.

The emission process of the x-y coordinate system is mapped to the geographic coordinate system to obtain the trajectory of each ship. Under the action of wind, the puff clusters emitted continuously by multiple emission sources are superimposed in the observation space, and the final emission diffusion shown in Fig. 4.6.

2. Point translation algorithm

Puff diffusion accounts for a small part of the whole observation area. In other words, when the matrix is used to represent the diffused puff cluster and its concentration, the

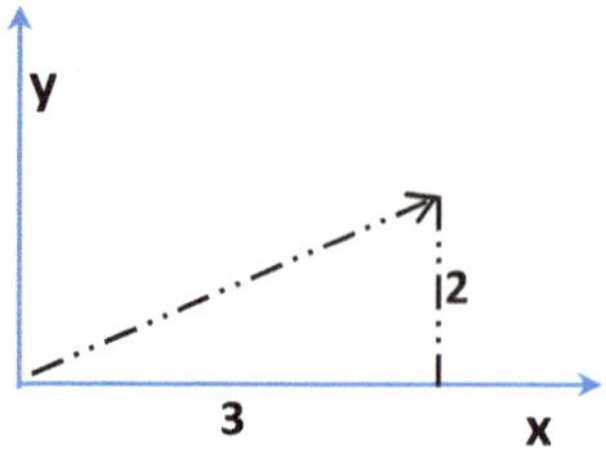

(a) Direction of navigation(m/s)

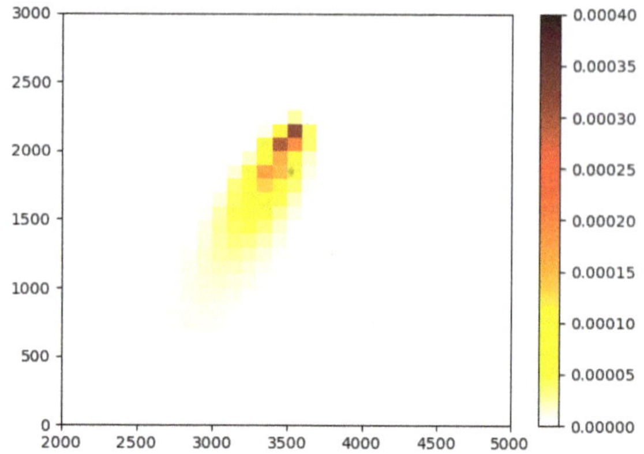

(b) Puff cluster diffusion map during navigation

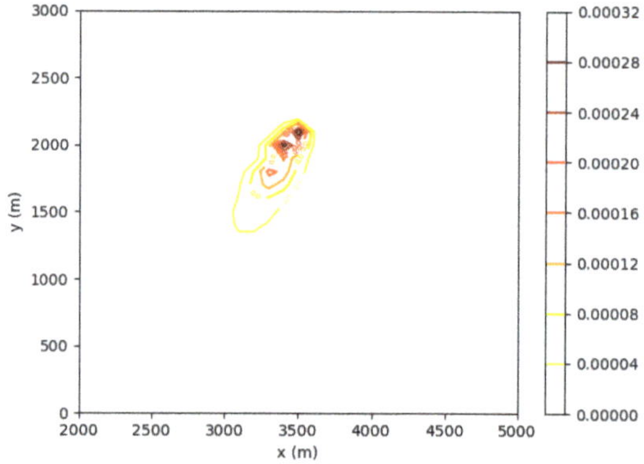

(c) Contour map of puff cluster diffusion during navigation

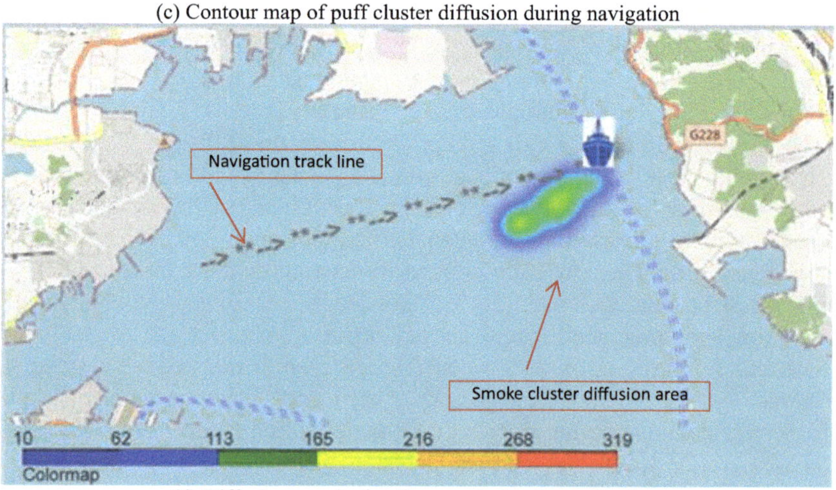

(d) Diffusion map in the diagram during the navigation

Fig. 4.4 Atmospheric pollutant diffusion diagram. **a** Direction of navigation (m/s). **b** Puff cluster diffusion map during navigation. **c** Contour map of puff cluster diffusion during navigation. **d** Diffusion map in the diagram during the navigation

Table 4.1 Ship discharge location and diffusion

Number	Initial Points	X-axis direction speed (m/s)	Y-axis direction speed (m/s)	Emissions (g/s)
1	(0, 0)	3	3	40
2	(−100, 250)	4	2	40
3	(−200, −500)	5	3	40
4	(200, 200)	3	2	50
5	(−400, 600)	4	2	60
6	(300, −200)	4	2	40

matrix is a sparse matrix because there are non-zero values in a small part of positions. In the calculation of puff diffusion, the size and granularity of the observation area will directly affect the size of the sparse matrix. In addition, the time interval and duration of the observations will directly affect the number of sparse matrices that need to be stored. With the expansion of the scale and number of sparse matrices, the storage space required for calculation will also increase rapidly, accompanied by a large number of invalid translation calculations containing zero. Taking the horizontal and vertical axes of the x-y coordinate system with a granularity of 10 meters, an observation window of 5000 meters, and 1000 observation moments as an example, the dimension of each sparse matrix is 500x500, and the whole observation process requires 2.5×10^8 data units. Obviously, it will be a huge burden for both data storage and diffusion computing.

Therefore, in the case of large data size or long observation duration, point translation method can be adopted for calculation. That is, the data whose concentration value is greater than a certain threshold (for example 1e-5) is used for storage and translation calculation, but the storage of corresponding coordinate value should be increased accordingly. Thus, the data satisfying the threshold requirement in matrix (Eq. 4.25) is stored as:

$$C_i = \begin{bmatrix} p_{11}^i & p_{21}^i & v_1^i \\ p_{12}^i & p_{22}^i & v_2^i \\ \vdots & \vdots & \vdots \\ p_{1m}^i & p_{2m}^i & v_m^i \end{bmatrix} \tag{4.25}$$

where (p_{j1}^i, p_{j2}^i) are the coordinates of observation space corresponding to observation time i, and v_j^i is the puff diffusion concentration of the coordinate point. Then the translation formula for data matrix C_i is:

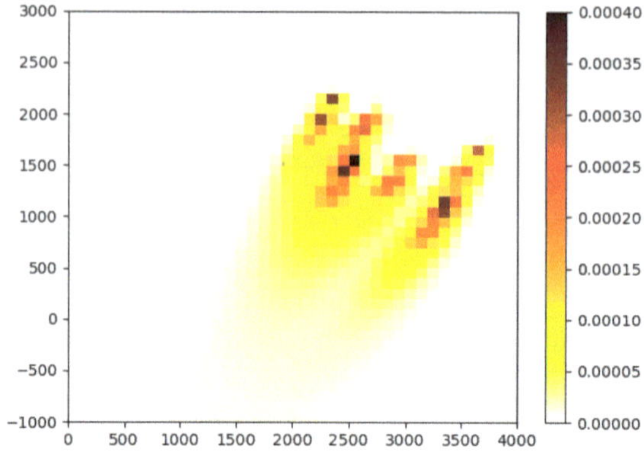

(a) Multi-emission source puff cluster diffusion map

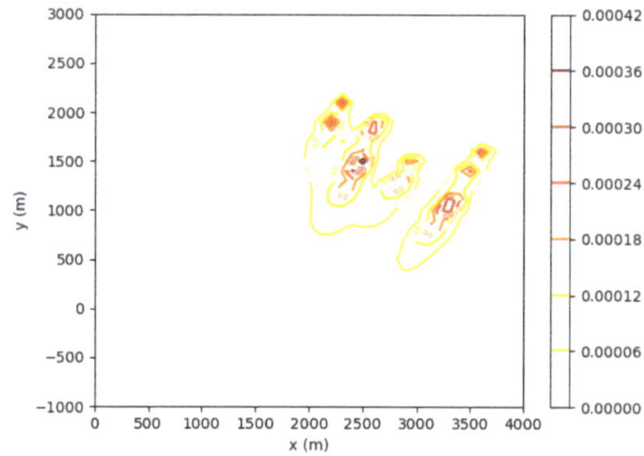

(b) Contour map of puff cluster diffusion from multiple emission sources

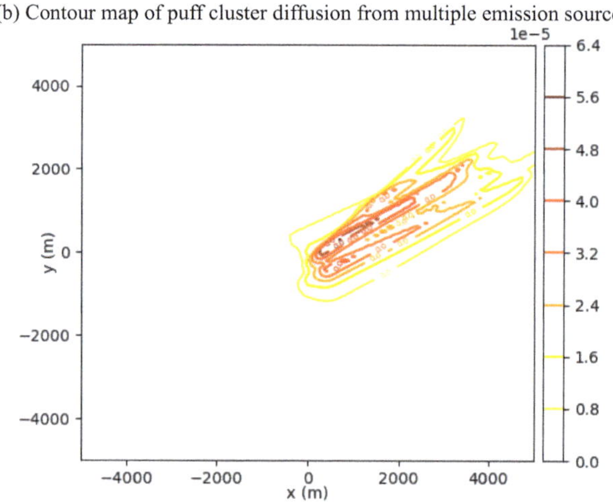

(c) Contour map of the average diffusion concentration of puff clusters from multiple emission sources

Fig. 4.5 Multi-source continuous emission puff cluster diffusion map. **a** Multi-emission source puff cluster diffusion map. **b** Contour map of puff cluster diffusion from multiple emission sources. **c** Contour map of the average diffusion concentration of puff clusters from multiple emission sources

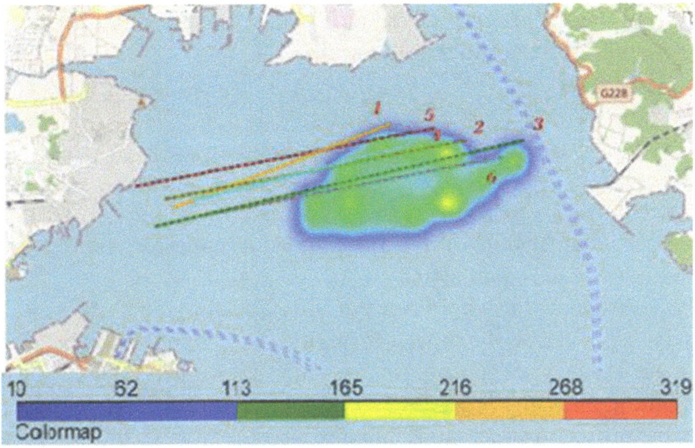

Fig. 4.6 Diffusion map of continuous emissions from multiple sources based on geographical coordinates

$$\begin{cases} C_i^{'} = \begin{bmatrix} p_{11}^i + \Delta p_1 & p_{21}^i + \Delta p_2 & v_1^i \\ p_{12}^i + \Delta p_1 & p_{22}^i + \Delta p_2 & v_2^i \\ \vdots & \vdots & \vdots \\ p_{1m}^i + \Delta p_1 & p_{2m}^i + \Delta p_2 & v_m^i \end{bmatrix} \\ \Delta p_1 = p_{j1}^0 - p_1^0 \\ \Delta p_2 = p_{j2}^0 - p_2^0 \end{cases} \tag{4.26}$$

where (p_1^0, p_2^0) are the coordinates of the instantaneous emission point of the reference puff and (p_{j1}^0, p_{j2}^0) are the coordinates of the instantaneous emission point of the target puff. Thus, $(\Delta p_1, \Delta p_2)$ are the translation amount of the target puff relative to the reference puff, and the puff translation can be directly obtained through the matrix linear calculation.

The accuracies of matrix translation algorithm and point translation algorithm are both determined by the granularity of reference puff. A large granularity means that the puff calculation is rough; yet a small granularity will result in a large data size and further increase computing time and storage space. Therefore, a moderate calculation granularity is often selected in actual calculation. The matrix translation algorithm is based on the number of translations in rows and columns, and moving a row or column means that the horizontal or vertical axis is shifted by 1 granularity. Generally, the distance between the emission point and the emission point of the reference puff is not exactly an integer multiple of the granularity, leading to certain errors. But the data size of the puff will remain unchanged, so it is easy to obtain the puff superposition. By contrast, the point translation algorithm directly adds and subtracts the coordinate difference between the emission point and the reference puff emission point on the original coordinates, and there is no calculation error.

However, due to the uncertainty of the puff data scale, the puff superposition will lead to a continuous increase in the data scale.

4.4.3 Algorithm Complexity Analysis

The whole-process diffusion of the instantaneous emission source puffs corresponding to Eq. (4.16) is calculated by:

It can be seen from the whole-process diffusion flow chart (Fig. 4.7) that the algorithm has one cycle, so the time complexity of the algorithm is O (T). There are k*T puffs in the whole process of continuous emission from k sources. The traditional calculation formula (Eq. 4.17) needs to constantly call the diffusion model (Eq. 4.16) to calculate the diffusion concentration in the target region, so the complexity of the whole algorithm is $O(kT^2)$. Eq. (4.19) is a special form of Eq. (4.17) with the same complexity. When the translation formula (Eq. 4.22) is adopted to calculate the diffusion of the puff clusters from continuous emission sources, the single diffusion model is called first to calculate the concentration matrix (Eq. 4.19) in the target area, and then the translation calculation of puff is performed one by one. Thus, the complexity of the whole algorithm is $O(T + kT)$, which can be simplified to $O(kT)$. It can be concluded that the translation algorithm O (kT) has an advantage over the traditional algorithm $O(kT^2)$ in time complexity, and can realize the rapid calculation of the puff diffusion.

Matrix translation algorithm and point translation algorithm are the same in basic time complexity, both O (kT). The major difference is that the former uses matrix product calculation, while the latter adopts matrix linear calculation, which will result in a certain difference with the change of data scale.

According to Table 4.2, the point translation fast algorithm takes sharply less time compared to the conventional algorithm, both in terms of the number of emission sources and the observation window radius for comparison. In a multi-source emission dispersion algorithm, the time-efficient variation of the number of emission sources is an important factor to measure the efficiency of the algorithm. Therefore, the average time consumption ratio between the traditional algorithm and the new algorithm was used to indicate the degree of time efficiency improvement of the new algorithm. If R is close to 1, indicating that the two algorithms were comparable in time efficiency. If R is greater than 1, the new algorithm is more time-efficient than the traditional algorithm, and vice versa, the new algorithm is less time-efficient than the traditional algorithm. According to the statistics in Table 4.2, $Min(R) > 145$, which means that the new algorithm has at least 140 times more time efficiency improvement relative to the traditional algorithm. To observe the relationship between the efficiency improvement and the number of emission sources, a trend diagram of the variation of R with the number of emission sources (N) is constructed, as shown in Fig. 4.8.

As can be seen, the average time-consuming ratio curve divided into two stages. In the first stage, with 1 to 5 emission sources, the average efficiency decreases slightly

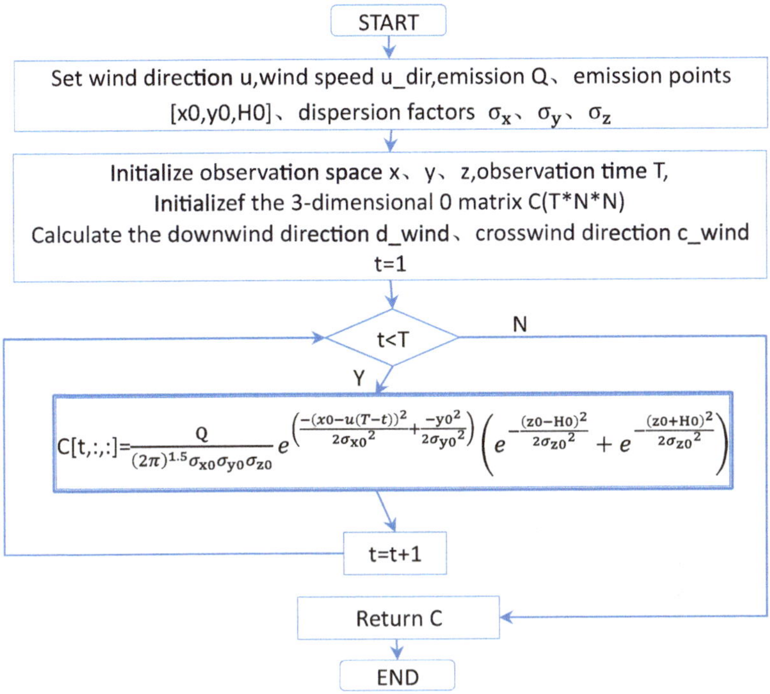

Fig. 4.7 The whole process diffusion flow chart

Table 4.2 Comparison of algorithm time consumption

Comparison group	Emission sources	Radius of observation interval (m)	Time consumption (s)		
			Traditional Algorithm	Matrix translation algorithm	Point translation algorithm
Group 1	1	2500	630.1904	1.2219	5.9744
	1	5000	1983.9246	7.1500	13.2212
	1	10,000	10292.4156	20.9899	30.3435
Group 2	5	2500	3097.1956	2.1394	9.2505
	5	5000	9512.6411	19.8338	26.2761
	5	10,000	45317.9116	40.4356	76.1639
Group 3	10	2500	6200.2693	3.2858	12.0316
	10	5000	18974.2314	35.5446	46.3610
	10	10,000	91764.1653	63.5102	131.4490
Group 4	20	2500	12289.9316	5.8665	18.3387
	20	5000	37748.4156	67.5919	89.4610
	20	10,000	181893.5057	112.0139	252.3658

Note The running environment are mainly the Processor Intel(R) Xeon(R) Gold 5118 CPU @ 2.3GHz (24 processors), RAM 48GB and Python 3.8.4

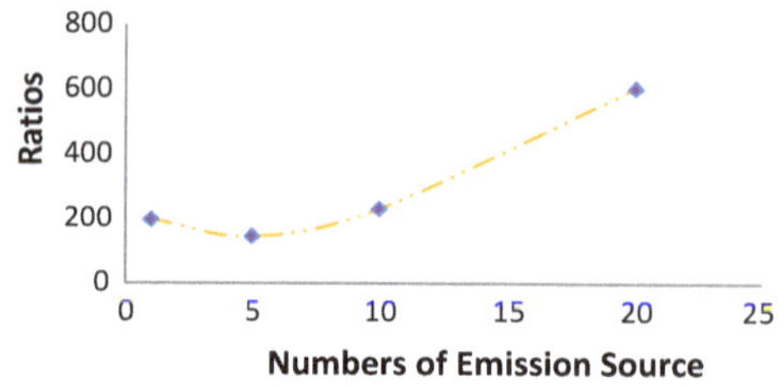

Fig. 4.8 Average time-consuming ratio of traditional algorithm to new algorithm

from 200 to 145. There was no advantage for the new algorithm when the number of emission sources was not sufficient. However, the efficiency of the new algorithm increased linearly as the number of emission sources increased, in the second stage with 5 to 25 sources. Thus, from the curve, the new algorithm has a linear increase in time efficiency relative to the traditional algorithm when the number of emission sources is relatively large.

In conclusion, the translation algorithm has a huge advantage over traditional algorithm in time complexity, which becomes increasingly obvious with the increase of data size. Moreover, in reality, it is usually necessary to analyze the diffusion of puff clusters emitted continuously by multiple ships in adjacent areas at the same time. In this case, the translation algorithm needs several minutes for computation, which is more efficient than the traditional time-consuming algorithm.

4.5 Numerical Simulation and Analysis

4.5.1 Determination of Model Parameters

Sulfides, nitrogen oxides, carbon oxides, and PM emitted by ships are major components of pollutants, and are important factors causing air pollution and affecting residents' health. Carbon emissions are closely related to the carbon content and combustion power of fuel. With the implementation of the "carbon peaking and carbon neutrality" and ship carbon emission control policies, CO_2 emissions have become the main concern following sulfur dioxide and carbon dioxide. Therefore, this paper takes carbon dioxide pollutants as the analysis object to simulate the diffusion model. The emission diffusion process is affected by factors such as wind speed, wind direction, the height of the ship's chimney, and atmospheric stability.

The mainstream of the Huangpu River is 82.5 kilometers. The river runs through Shanghai and flows into the Yangtze River at Wusong Mouth. The river is about 300 meters wide. When the water level is 3.5 meters, the cross-section of the water is 3540~4100 square meters, and the river bottom elevation is −8.0~−20.0 meters. Huangpu River belongs to the subtropical monsoon climate, with high temperature and rain in summer and mild and little rain in winter. The Huangpu River, a tributary of the Yangtze River in Shanghai, is an important port connecting inland and harbor.

Huangpu River has important research significance and value as an important part of China's inland river shipping and Shanghai's international shipping distribution system.

The ship AIS data selected based on the following three considerations. First, the emission dispersion model requires AIS activity data and static data of a ship. The loss of any type of data will reduce the scientific validity of the measurement results. Therefore, the AIS data should have complete MMSI and IMO information and the ability to obtain relevant static data through each classification society database. Second, AIS data of navigating vessels at similar times and in adjacent areas, such data can facilitate the dynamic simulation demonstration of emission dispersion processes. Third, the study area needs to be typical, with a high level of vessel traffic flow and activity data. For this purpose, the voyage data of vessels with a considerable length of voyage trajectory on a certain day intercepted from the public data website (http://ship.chinaport.com/) as the study object, using the Huangpu River inlet section as the observation window. The ship files were from China Classification Society (https://www.ccs.org.cn/) and Japan Classification Society (ClassNK). The AIS data was collected from 50 randomly selected ships sailing in the Yangtze River Basin (Huangpu River region) from 00:00 to 24:00 on January 28, 2022. According to the requirements of AIS data in the empirical process, eight vessels from 8:00 to 20:00 selected for the empirical study, and the vessel information shown in Table 4.3. Based on the translational diffusion model and the real-time calculation model of air pollution emissions, the diffusion distribution of air pollutants from ships with different wind directions and speeds was simulated. Figure 4.9 shows the trajectory of a ship. S is the starting point and E is the end within the timeline. Figure 4.10 shows the start and end times of the voyage of ships. Due to the difference in starting and ending time, speed and heading, the track lines of eight ships are also different (Table 4.4).

Table 4.3 Empirical ship-related information

MMSI	IMO	Main engine model	Main engine power/KW	Machine Speed/rpm	Built time	Design speed/kn
413,228,080	9,842,360	6RT-flex58T-E	12,900	105	2020	18.9
413,303,210	9,194,646	6 L28/32	1320	775	1999	12
413,358,220	9,536,014	6S42MC7	6480	136	1997	14.6
413,376,840	9,638,898	8DKM-28	2500	750	2012	11.6
413,660,000	524,288	6S50MC-C	9480	127	2007	14.2
413,890,000	9,538,347	6S50MC-C	9480	127	2007	14.2
477,222,100	9,632,313	6S50MC-C	9480	127	2013	14.2
538,007,143	9,432,347	6S50MC-C	9480	127	2010	14.2

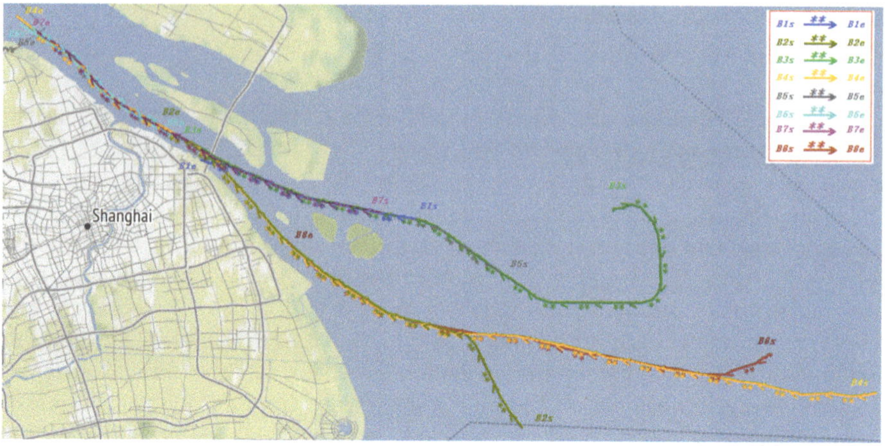

Fig. 4.9 Track line

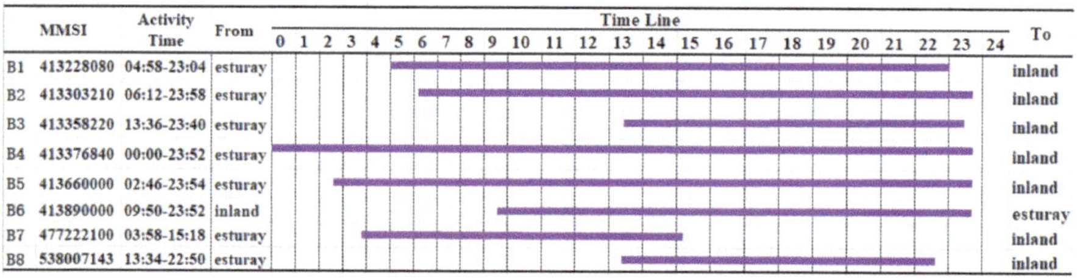

Fig. 4.10 Time line

Table 4.4 Ship emission factors (g/kWh)

Ship type	Built time	Nox	SO2	PM2.5	PM10	CO	CO2	VOC
Low-speed diesel engine	≤1999	18.1	10.5	1.2	1.5	1.4	620	0.6
Medium-speed diesel engine	≤1999	14	11.5	1.2	1.5	1.1	683	0.5
Low-speed diesel engine	2000–2010	17	10.5	1.2	1.5	1.4	620	0.6
Medium-speed diesel engine	2000–2010	13	11.5	1.2	1.5	1.1	683	0.5
Low-speed diesel engine	2011–2021	15.3	10.5	1.2	1.5	1.4	620	0.6
Medium-speed diesel engine	2011–2021	11.2	11.5	1.2	1.5	1.1	683	0.5

Note Low speed is less than 50% of the design speed, medium speed navigation is greater than or equal to 50% of the design speed

4.5.2 Case Simulation Analysis

According to the statistical data of the Meteorological Bureau and related studies, the wind direction in this paper was set as constant. First, the study area is dominated by seasonal winds, which vary from season to season, but remain constant in specific seasons (e.g. southeast wind in summer and northwest wind in winter). Therefore, the constant wind direction is consistent with the climatic characteristics of the study area. Meanwhile, the diffusion time in this paper is less than 24 hours, and the constant wind direction is in line with the actual meteorological conditions. Second, their impact is negligible for winds that change direction in the short term. Data are difficult to collect and study for winds that change direction over a long time. All of these can be the content of future research. Huangpu River in Shanghai, where southeasterly winds dominate in summer and northwesterly winds in winter (Lu et al., 2022). Therefore, emission diffusion simulation was performed for two different wind directions. In order to know the diffusion distribution area in the case of different wind directions, the spatial state of CO_2 pollutant diffusion was simulated at the wind scale of 5 (8.0 m/s) according to the actual sailing state of the ship. When the pollutant concentration is lower than 10^{-9}, it will be invisible. The observation window was set to 10000*10000 meters, as shown in Fig. 4.11a, and b. A total of 5,060,639 puffs were calculated, which took 319.3644 seconds. The influence of seasonal wind direction on the coastal area was significantly different, and the diffusion showing a central pattern according to the distance from the coast. In winter, northwest winds affected these areas: Pudong New Area (Gaodong Town, Caolu Town, Heqing Town), Baoshan District (Luojing Town, Songnan Town), and Chongming District (Changxing Town). In summer, southeast wind affected these areas: Baoshan District (Luojing Town, Yuepu Town, Youyi Road Street), Chongming District (Changxing District), Taicang City (Fuqiao Town, Liuhe Town). When wind speed and total air pollutants are constant, the area of pollution changes following wind direction. Pollutants spread from inland to the coast in winter and move from coast to inland in summer.

The diffusion of air pollutants was not only impacted by wind direction, but also wind speed. Figure 4.12a, b, and c show the diffusion characteristics of southeaster winds of scales 4(5.5 m/s), 5(8.0 m/s), and 6(10.8 m/s), respectively. The number of puffs was 8,823,681, 5,060,639, and 3,173,181, respectively, and the corresponding calculation time was 431.5450, 319.3644, and 282.3567 seconds. An increase in the number of puffs by millions did not dramatically increase the computational efficiency and cost of the diffusion process. As seen in Fig. 4.12a, b, and c the contaminated areas include Baoshan, Pudong, Chongming District in Shanghai, and Taicang city in Jiangsu province. When the wind speed is level 4 (5.5 m/s), pollutants disperse will relatively slowly and affect long on the waterway. When the wind speed is level 5 (8 m/s), the polluted area will be reduced. When the wind speed is level 6 (10.8 m/s), the pollution area will be reduced and moved inland. The faster the wind, the quicker the pollutants spread. When the total amount of pollutants and wind direction are constant, the higher the wind level is, the less pollution will be in the waterway area.

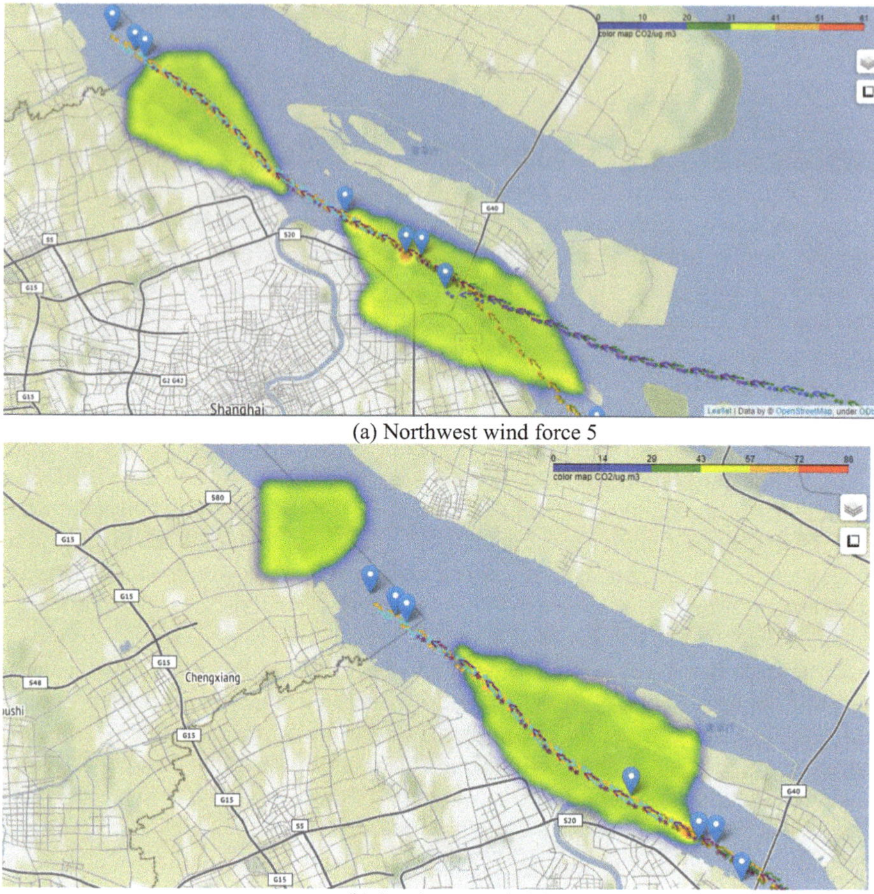

(a) Northwest wind force 5

(b) Southeaster wind force 5

Fig. 4.11 **a** Northwest wind force 5. **b** Southeaster wind force 5

4.6 Conclusion and Prospect

1. Based on the Gaussian CALPUFF diffusion model, a fast translation algorithm for the diffusion model is established by using the big data of ship AIS. With the popularization of the ship AIS, this algorithm has the advantage of time complexity, which can investigate the impact of ship emission diffusion on air quality in real time.
2. The matrix translation algorithm is suitable for local spatial scale data, such as ports and watersheds. The algorithm accuracy is determined by the calculation granularity of the benchmark puff. Therefore, selecting the appropriate granularity according to the actual application requirements and target expectations can effectively save computing time and storage space

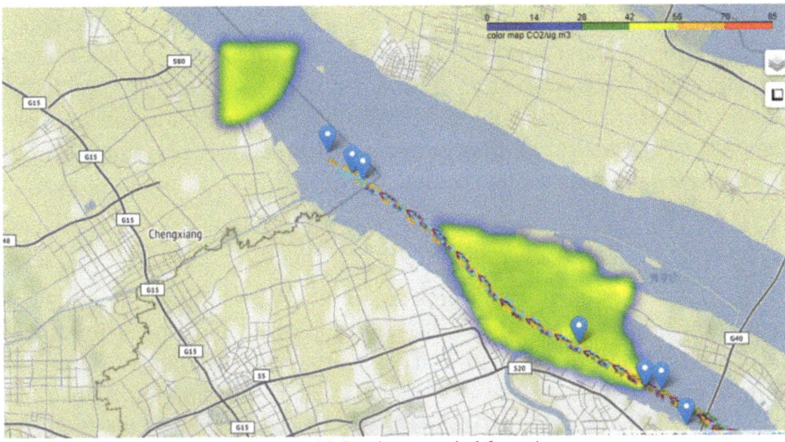

(a) Southeaster wind force 4

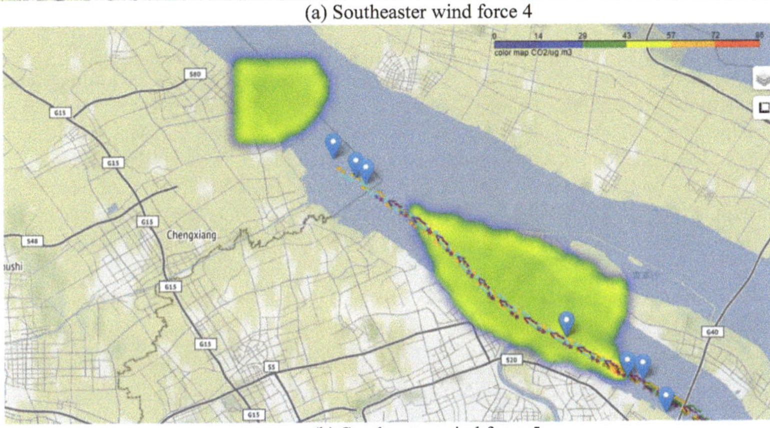

(b) Southeaster wind force 5

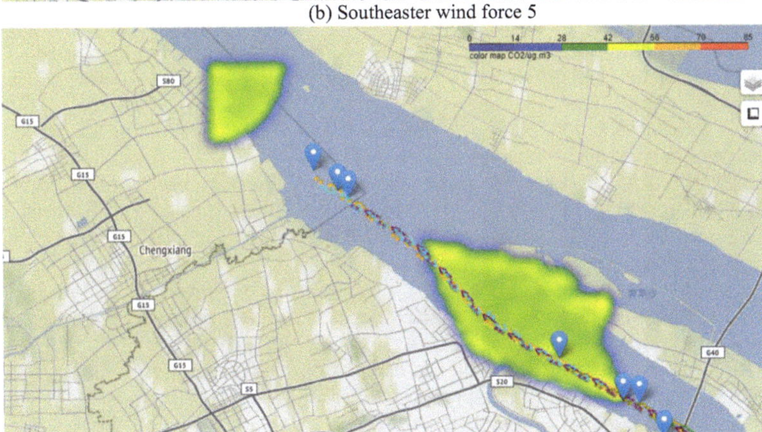

(c) Southeaster wind force 6

Fig. 4.12 **a** Southeaster wind force 4. **b** Southeaster wind force 5. **c** Southeaster wind force 6

3. The point translation algorithm is suitable for regional spatial scale date, such as cities, provinces and countries. The algorithm directly adds and subtracts the coordinate difference between the emission point and the reference puff emission point on the original coordinates, so there is no calculation error. Furthermore, compared with traditional algorithms, it has no loss of accuracy. Due to the uncertainty of the data scale, the amount of data will continue to grow in the superposition process.

In order to improve the calculation efficiency and reduce the calculation cost, this study focuses on the rapid calculation of the emission and diffusion of atmospheric pollutants from ships with AIS big data. Nevertheless, it fails to carry out in-depth analysis on the accuracy of calculation. Thus, in future study, data integration can be applied to improve the data processing speed without affecting the diffusion calculation accuracy in the case of sharp increase in data size caused by the superposition of multiple puffs.

Acknowledgements Shaoli He and Jun Wang also made great contributions to this manuscript. We express our heartfelt thanks to them. This research also was supported by: National Natural Science Foundation of China (72474127) and National Social Science Fund Major Bidding Project of China (18ZDA052). Part of this chapter was published: He S, Wu X, Wang J, Guo J. A ship emission diffusion model based on translation calculation and its application on Huangpu River in Shanghai. Computers & Industrial Engineering 2022;172:108569. https://doi.org/10.1016/j.cie.2022.108569. The publication has been conducted with copyright permission.

References

Bai, S., Wen, Y., He, L., Liu, Y., Zhang, Y., Yu, Q., & Ma, W. (2020). Single-vessel plume dispersion simulation: Method and a case study using CALPUFF in the Yantian port area, Shenzhen (China). *International Journal of Environmental Research and Public Health, 17*(21), 7831. https://doi.org/10.3390/ijerph17217831

Bilgili, L., & Celebi, U. B. (2018). Developing a new green ship approach for flue gas emission estimation of bulk carriers. *Measurement, 120*, 121–127.

Bindle, L., Martin, R. V., Cooper, M. J., Lundgren, E. W., Eastham, S. D., Auer, B. M., et al. (2021). Grid-stretching capability for the GEOS-Chem 13.0.0 atmospheric chemistry model. *Geoscientific Model Development, 14*(10), 5977–5997. https://doi.org/10.5194/gmd-14-5977-2021

Cao, M., Liu, Y., Gai, T., Zhou, M., Fujita, H., & Wu, J. (2022). A comprehensive star rating approach for cruise ships based on interactive group decision making with personalized individual semantics. *Journal of Marine Science and Engineering, 10*, 1–20.

Chen, L., Yip, T. L., & Mou, J. (2018). Provision of emission control area and the impact on shipping route choice and ship emissions. *Transportation Research Part D, 50D*, 280–291.

Chu-Van, T., Ristovski, Z., Pourkhesalian, A. M., Rainey, T., Garaniy, V., Abbassi, R., et al. (2017). On-board measurements of particle and gaseous emissions from a large cargo vessel at different operating conditions. *Environmental Pollution, 237*, 1–10. https://doi.org/10.13140/RG.2.2.33675.80167

Corbett, J. J., & Koehler, H. W. (2003). Updated emissions from ocean shipping. *Journal of Geophysical Research-Atmospheres, 108*(D20), 4650. https://doi.org/10.1029/2003JD003751

Ekmekçioğlu, A., Kuzu, S. L., Ünlügençoğlu, K., & Çelebi, U. B. (2020). Assessment of shipping emission factors through monitoring and modelling studies. *Science of the Total Environment, 743*, 140742. https://doi.org/10.1016/j.scitotenv.2020.140742

Fu, J., & Li, Y. (2018). Study on ship's exhaust-gas diffusion based on Gaussian plume model. *Marine Science Bulletin, 37*(2), 1–6. https://doi.org/10.11840/j.issn.1001-6392.2018.02.016

Fu, X., Wang, X., & Chen, D. (2025). Process analysis of the impacts of ship emissions on PM2.5 and O3 in the Yangtze River Delta. *China. Journal of Environmental Sciences., 161*, 372–383. https://doi.org/10.1016/j.jes.2025.05.001

Holmes, N., & Morawska, L. (2006). A review of dispersion modelling and its application to the dispersion of particles: An overview of different dispersion models available. *Atmospheric Environment, 40*(30), 5902–5928.

Huang, L., Wen, Y., Zhang, Y., Zhou, C., Zhang, F., & Yang, T. (2020). Dynamic calculation of ship exhaust emissions based on real-time AIS data. *Transportation Research Part D, 80*, 102277. https://doi.org/10.1016/j.trd.2020.102277

Jalkanen, J.-P., Brink, A., Kalli, J., Pettersson, H., Kukkonen, J., & Stipa, T. (2009). A modelling system for the exhaust emissions of marine traffic and its application in the Baltic Sea area. *Atmospheric Chemistry and Physics, 9*(23), 9209–9223. https://doi.org/10.5194/acpd-9-15339-2009

Jiang, H., Wu, G., Li, T., & He, P. (2019). Characteristics of particulate matter emissions from a low-speed marine diesel engine at various loads. *Environmental Science and Technology, 53*(19), 11552–11559. https://doi.org/10.1021/acs.est.9b02341

Kasper, A., Aufdenblatten, S., Forss, A., Mohr, M., & Burtscher, H. (2007). Particulate emissions from a low-speed marine diesel engine. *Aerosol Science and Technology, 41*(1), 24–32. https://doi.org/10.1080/02786820601055392

Liu, J., Xie, W., & Chen, J. (2020). Simulation of emission and diffusion of air pollutants from ships sailing in port. *Journal of Xiamen University (Natural Science), 59*(6), 1016–1024. https://doi.org/10.6043/j.issn.0438-0479.202001026

Liu, Z., Lu, X., Feng, J., Fan, Q., Zhang, Y., & Yang, X. (2017). Influence of ship emissions on urban air quality: A comprehensive study using highly time-resolved online measurements and numerical simulation in Shanghai. *Environmental Science and Technology, 51*(1), 202–211.

Lu, S., Wenyu, Z., Chaoshun, L., Wenshou, T., & Chungang, F. (2022). Characteristics of sea-land breezes and their impact on PM2.5 and O3 in Shanghai and adjacent areas. *Acta Scientiae Circumstantiae, 42*(9), 339–350. https://doi.org/10.13671/j.hjkxxb.2022.0111

McLaren, R., Wojtal, P., Halla, J. D., Mihele, C., & Brook, J. R. (2012). A survey of NO2:SO2 emission ratios measured in marine vessel plumes in the strait of Georgia. *Atmospheric Environment, 46*, 655–658.

Murena, F., Mocerino, L., Quaranta, F., & Toscano, D. (2018). Impact on air quality of cruise ship emissions in Naples, Italy. *Atmospheric Environment, 187*, 70–83. https://doi.org/10.1016/j.atmosenv.2018.05.056

Nakatsubo, R., Oshita, Y., Aikawa, M., Takimoto, M., Kubo, T., Matsumura, C., et al. (2019). Influence of marine vessel emissions on the atmospheric PM2.5 in Japan's around the congested sea areas. *Science of the Total Environment, 702*(5), 134744.

Pan, K., Lim, M. Q., Kraft, M., & Mastorakos, E. (2021). Development of a moving point source model for shipping emission dispersion modelling in EPISODE-CityChem v1.3. *Geoscientific Model Development, 14*(7), 4509–4534. https://doi.org/10.5194/gmd-2021-47

Peng, X., Wen, Y., Xiao, C., Huang, L., Zhou, C., Zhang, F., & Zhang, Y. (2020). Improved calculation model for ship exhaust emission dispersion. *Journal of Safety and Environment, 20*(1), 255–264.

Progiou, A. G., Bakeas, E., Evangelidou, E., Kontogiorgi, C., Lagkadinou, E., & Sebos, I. (2020). Air pollutant emissions from Piraeus port: External costs and air quality levels. *Transportation Research Part D Transport and Environment, 91*(2017), 102586. https://doi.org/10.1016/j.trd.2020.102586

Sharafian, A., Blomerus, P., & Mérida, W. (2019). Natural gas as a ship fuel: Assessment of greenhouse gas and air pollutant reduction potential. *Energy Policy, 131*, 332–346. https://doi.org/10.1016/j.enpol.2019.05.015

Stockie, J. M. (2011). The mathematics of atmospheric dispersion modeling. *SIAM Review, 53*, 349–372. https://doi.org/10.1137/10080991X

Tichavska, M., & Tovar, B. (2015). Environmental cost and eco-efficiency from vessel emissions in Las Palmas port. *Transportation Research Part E, 83*, 126–140.

Toscano, D., Murena, F., Quaranta, F., & Mocerino, L. (2021). Assessment of the impact of ship emissions on air quality based on a complete annual emission inventory using AIS data for the port of Naples. *Ocean Engineering, 232*(5), 109166. https://doi.org/10.1016/j.oceaneng.2021.109166

Tseng, Y., Cheng, W., & Yuan, C. (2024). Impacts of ship emissions and sea-land breeze on urban air quality using chemical characterization, source contribution and dispersion model simulation of PM2.5 at Asian seaport. *Environmental Pollution, 347*(15), 123663. https://doi.org/10.1016/j.envpol.2024.123663

UNCTAD. (2021). *Review of Maritime Transport 2021* [Review of Maritime Transport 2021]. https://unctad.org/en/PublicationsLibrary/rmt2021_en.pdf

Viana, M., Hammingh, P., Colette, A., Querol, X., Degraeuwe, B., Vlieger, I. D., & Aardenne, J. V. (2014). Impact of maritime transport emissions on coastal air quality in Europe. *Atmospheric Environment, 90*, 96–105. https://doi.org/10.1016/j.atmosenv.2014.03.046

Wen, Y., Geng, X., Huang, L., Zhou, C., & Liu, Y. (2017). Calculation model for ship exhaust emissions on a count of the impacts of wind, wave and water current included. *Journal of Safety and Environment, 17*(5), 1969–1974. https://doi.org/10.13637/j.issn.1009-6094.2017.05.067

Weng, J., Shi, K., Gan, X., Li, G., & Huang, Z. (2019). Ship emission estimation with high spatial-temporal resolution in the Yangtze River estuary using AIS data. *Journal of Cleaner Production, 248*, 119297. https://doi.org/10.1016/j.jclepro.2019.119297

Wu, H., Zhang, Y., Yu, Q., & Ma, W. (2018). Application of an integrated weather Researchand forecasting (WRF)/CALPUFF modeling tool forsource apportionment of atmospheric pollutantsfor air quality management: A case study in theurban area of Benxi, China. *Journal of the Air & Waste Management Association, 68*(4), 347–368. https://doi.org/10.1080/10962247.2017.1391009

Xie, R., Xing, Q., Gao, J., & Wang, X. (2025). Investigation on the effect of ship emissions on the air quality: A case study in Hainan Island, China. *Journal of Environmental Sciences, 156*, 114–125. https://doi.org/10.1016/j.jes.2024.08.010

Xu, R., Deng, X., Wan, H., Cai, Y., & Pan, X. (2021). A deep learning method to repair atmospheric environmental quality data based on Gaussian diffusion. *Journal of Cleaner Production, 308*, 127446. https://doi.org/10.1016/j.jclepro.2021.127446

Yang, L., Zhang, Q., Zhang, Y., Lv, Z., Wang, Y., Wu, L., et al. (2021). An AIS-based emission inventory and the impact on air quality in Tianjin port based on localized emission factors. *Science of the Total Environment, 783*, 146869. https://doi.org/10.1016/j.scitotenv.2021.146869

Zhang, F., Chen, Y., Cuia, M., Feng, Y., Yang, X., Chen, J., et al. (2019). Emission factors and environmental implication of organic pollutants in PM emitted from various vessels in China. *Atmospheric Environment, 200*(MAR), 302–311. https://doi.org/10.1016/j.atmosenv.2018.12.006

Zhu, Y., Yan, X., Kong, X., Tong, Q., Li, Y., Huang, G., & Li, Y. (2019). Industry-environment system management based on an uncertain Gaussian diffusion optimization model for coal-dependent cities in ecologically fragile areas. *Journal of Cleaner Production, 234*, 832–857. https://doi.org/10.1016/j.jclepro.2019.06.142

Zhuang, B., Zhou, Y., & Hu, Y. (2025). Influence of ship emitted sulfur and carbonaceous aerosols on east Asian climate in summer. *Atmospheric Environment, 344*, 121035. https://doi.org/10.1016/j.atmosenv.2025.121035

Zincir, B., Deniz, C., & Tunér, M. (2019). Investigation of environmental, operational and economic performance of methanol partially premixed combustion at slowspeed operation of a marine engine. *Journal of Cleaner Production, 235*, 1006–1019.

Chapter 5
Study on Spatial Correlation of Air Pollution and Control Effect of Development Plan for the City Cluster in the Yangtze River Delta

5.1 Introduction

In past decades, with the rapid industrialization and urbanization, urban air quality in China has deteriorated. Air pollution, such as haze, has been a particularly acute problem, which has become the focus of attention of the government and the public (Hu et al., 2020). The Yangtze River Delta (YRD), a cluster of 41 cities including Shanghai, is no exception. According to *China's ecology and environment gazette 2018*, the air quality of more than 1/4 days in 2018 in the YRD failed to meet the air quality standards (http://www.h2o-china.com/news/292237.html).

Due to the spatial spillovers, joint efforts among cities are called to effectively control air or water pollution (D'Amato et al., 2018; Portnov et al., 2009). Indeed, the cities in the YRD have experimented and implemented "integrated" planning and governance on issues related to air pollution. For example, in June 1992, a joint committee of directors of the YRD was established. In May 2016, China State Council issued *The Development Plan for the City Cluster in the YRD* (hereinafter referred to as *Plan*). The *Plan* is one of the most important documents to jointly control environmental pollution for cities in the Yangtze River Delta. For example, the *Plan* focuses on the control of water pollution, air pollution and soil pollution, promoted joint prevention and treatment of the environment, worked together to prevent air pollution and improve the quality of the ecological environment in all regions (National Development and Reform Commission, 2016). Since 2016 few, if any, systematic studies have been completed on the effectiveness of the *Plan* on air pollution. In this paper, with the implementation of *Plan* taken as natural experiment, the Spatial Difference-in-difference Model (hereinafter referred to as SDID) model is constructed. Panel data of 41 cities in the Yangtze River Delta from 2011 to 2018 has been collected to study the effect of the *Plan* on air pollution.

The characteristics of this paper include: firstly, the Spatial Autoregression Model (hereinafter referred to as SAR) model is used to analyze the air pollution problems in Jiangsu-Zhejiang-Shanghai-Anhui area. The SAR model can not only reflect the

X. Wu, *Environmental Disaster in China*, https://doi.org/10.1007/978-981-92-0910-1_5

specific causes of pollution, but also consider the spatial correlation of air pollution between regions. Secondly, the implementation effect of the *Plan* is evaluated by the difference-in-difference model (hereinafter referred to as DID). Under the conclusion that the policy has a significant impact, the SDID model is further introduced, and then the placebo test is carried out.

The rest of paper is arranged as follows. The second part is literature review. The third part is the analysis of spatial measurement model of air pollution in Jiangsu-Zhejiang-Shanghai-Anhui area. The fourth part includes the construction of DID model and SDID model, parallel trend test, placebo test and empirical analysis. The fifth part is the conclusion and the suggestion.

5.2 Literature Review

5.2.1 Study on Spatial Spillover Effect of Air Pollution

At present, there are rich studies on spatial spillover effect of air pollution. For example, Tao (2018) studied and analyzed the air quality of city cluster in Yangtze River Delta from two dimensions of space and time. Xu (2019) analyzed the spatial and temporal characteristics of air pollution in the Yangtze River Delta. According to the study of Niu (2019), air pollution in China does have a certain agglomeration effect, and factors such as industrial structure, population size, scientific and technological innovation, greening level and urbanization process have different effects on air pollution in different regions. Gong (2016) found the spatial transmission effect of air pollution in surrounding cities on Beijing. Tang (2019) found that air pollution in Nanjing had a significant superposition effect, which was influenced by the combined effect of local pollution and air pollution in surrounding areas. Xie (2015) found that trans-regional air pollution transmission control is an important approach for reducing air pollution. Wang et al. (2020) discussed the main influencing factors of air pollution from the perspective of space. According to the above research, the spatial transmission effect of air pollution exists between cities. Therefore, in order to study the spillover effect of air pollution, it is necessary to introduce a spatial econometric model.

5.2.2 Research Status of SDID Model

The SDID model is intended to evaluate the effect of project or policy implementation. For example, Sunak and Madlener (2015) studied the potential depreciation effect of wind farms on German real estate. Heckert and Mennis (2012) measured the impact of the "space greening" implemented in Philadelphia on the value of nearby residential property. Heckert (2015) evaluated the impact of space greening scheme on the value of nearby property. Dubé et al. (2014) assessed the impact of public transport system on real estate value. Taking the opening of Singapore's new loop as a natural experiment, Diao et al. (2017) explored the impact of urban rail transit network on the value of private residences. Han (2018) explained the possible spatial spillover effect of IKEA entry. Yan et al. (2025) found that China's clean heating policies significantly improved energy efficiency in pilot cities. Zhou et al. (2025) observed that regional science and technology planning positively influenced the development of local green generic technologies, yet generated negative geographical spillover effects on neighboring cities. Casquet et al. (2025) examined the impact of transport policies on carbon emissions through a case study of bus fleet retrofitting and electrification in the Piedmont region.

Many Chinese scholars evaluated the effectiveness of a major infrastructure project or a new policy. For example, Fan (2018) evaluated the impact of the approval of bonded areas on the trade of dependent provinces, and the results showed that the DID and SAR model could well evaluate the trade development effect of bonded areas. By the SDID model, Wang (2018) found rail transit has a significant impact on housing price. Using the SDID model and spatial quantile regression model, Yu and Zhao (2019) showed the Beijing-Shanghai high-speed railway significantly promoted the economic growth of the cities where it was located and the economic development of neighboring cities. By using the DID space autoregressive dynamic model, Zhang et al. (2019) found that the economic trade zone could significantly promote economic growth and the free trade zone has significant positive spatial spillover effect on the surrounding cities. Zheng et al. (2019) found the opening of high-speed railway has a significant spatial spillover effect on the development of residents' consumption. Jiang et al. (2025) observed that the opening of high-speed rail lines stimulated entrepreneurial activity not only in local cities but also in neighboring urban centres. Li and An (2025) found that the construction of logistics hubs in China substantially improved urban green logistics efficiency through both direct and spillover effects.

From the above studies, if there is an obvious spatial association among the subjects, and the spatial effect is ignored in the traditional regression, it is easy to underestimate the impact of projects or policies. Therefore, the SDID model can be used to analyze the effect of policies on air pollution control.

5.3 Spatial Econometric Model in Jiangsu-Zhejiang-Shanghai-Anhui Area

5.3.1 Variable Selection and Data Source

1. Treatment group and control group. According to the *Plan*, a total of 26 cities in Jiangsu-Zhejiang-Shanghai-Anhui area are listed as city clusters in Yangtze River Delta, then these 26 cities are treatment groups. In order to avoid the

misestimation of the model due to the large difference in other control factors between the treatment group and the control group, 15 other cities in Jiangsu-Zhejiang-Shanghai-Anhui area are selected in this paper, which are in the same province and close to each other, as the control group.

2. Dependent variable. Through data searching and availability of index data, this paper selected the annual average $PM_{2.5}$ of each city as the dependent variable to measure the pollution of each region.
3. Independent variables. Zhang (2017) found that the civil automobile ownership in the Beijing-Tianjin-Hebei region had a significant impact on the urban $PM_{2.5}$ concentration. For every 1% increase in resident car ownership, the $PM_{2.5}$ concentration increases by 13.3ug/m^3, so the number of civil car ownership (vehicles) is taken as an independent variable. Generally, the larger the industrial scale is, the more serious the air pollution will be. Therefore, industrial power consumption (one billion kilowatt-hours) is selected as the independent variable. Wu et al. (2019) found that industrial SO_2 concentration had a significant effect on air pollution. Xiao et al. (2012) found that the air in Taiyuan city in warm season was greatly affected by industrial smoke and dust, which contributed significantly to the $PM_{2.5}$ concentration of the city. Therefore, industrial SO_2 emissions and industrial smoke and dust emissions are taken as explanatory variables. In addition, there is a certain relationship between the $PM_{2.5}$ concentration and dust, so the construction area is selected. Borge et al. (2017) found significant negative correlations between wind speed and air pollutant concentration, so the days with wind speed greater than category 3 are taken as an independent variable. Finally, human economic activity is an important factor affecting environmental conditions, and GDP per capita is taken as an independent variable. The variables involved are shown in Table 5.1.

5.3.2 *Spatial Metrology Model Based on Spatial Weight Matrix*

5.3.2.1 Establishment of Spatial Weight Matrix

Geographical location and distance are important factors that need to be considered. For the sake of comparison, adjacent weight matrix and inverse distance weight matrix are selected, and their definitions are as follows.

Set w_{ij} as the distance between region *i* and region *j*, and.

$$w_{ij} = \begin{cases} 1, Region\ a\ is\ adjacent\ to\ region\ b \\ 0, Region\ a\ is\ not\ adjacent\ to\ region\ b \end{cases} \tag{5.1}$$

Table 5.1 Variables and data sources

Type of variables	Name of variables	Unit of measurement	Sources of data
Dependent variables	Annual average $PM_{2.5}$ concentration (y)	ug/ m^3	Socioeconomic Data and Applications Center(SEDAC) (https://sedac.ciesin.columbia.edu/), China regional statistical yearbook (2011–2018), China urban statistical yearbook (2011–2018), Sina News Center (http://news.sina.com.cn/o/2019-01-23/doc-ihqfskcn9723756.shtml)
	civil automobile ownership (x_1)	10,000 vehicles	China regional statistical yearbook (2011–2018)
	industrial power consumption (x_2)	100 million kwh	China regional statistical yearbook (2011–2018)
Independent variables	industrial SO_2 concentration (x_3)	10,000 tons	China regional statistical yearbook (2011–2018)
	Industrial dust concentration (x_4)	10,000 tons	China regional statistical yearbook (2011–2018)
	construction area (x_5)	10,000 square meters	China regional statistical yearbook (2011–2018)
	wind speed greater than category 3 (x_6)	day	2345 history weather (http://tianqi.2345.cn/wea_history/58141.html)
	GDP per capita (x_7)	10,000 yuan	China regional statistical yearbook (2011–2018)

Therefore, $W = \begin{bmatrix} w_{11} & \cdots & w_{1n} \\ \vdots & \vdots & \vdots \\ w_{n1} & \cdots & w_{nn} \end{bmatrix}$ is the adjacent weight matrix of n cities, and the main diagonal elements are all 0 (the distance between the same city is 0).

Define: $w_{ij} = \frac{1}{d_{ij}}$ d_{ij} is defined as geographical distance.

$W_d = \begin{bmatrix} w_{11} & \cdots & w_{1n} \\ \vdots & \vdots & \vdots \\ w_{n1} & \cdots & w_{nn} \end{bmatrix}$ is the inverse distance weight matrix, in which the main diagonal elements are also 0.

The rows of the spatial weight matrix need to be standardized. Each element in the matrix (denoted as $\tilde{w}_{ij}$) is divided by the sum of the elements in its row. Ensure that the sum of the elements in each row is 1, and the formula is:

$$w'_{ij} \equiv \frac{\tilde{w}_{ij}}{\sum_j \tilde{w}_{ij}} \tag{5.2}$$

5.3.2.2 Selection of Optimal Spatial Weight

The following three basic assumptions are given to select the optimal weight matrix (Fan & Shi, 2018):

1. The data structure matching between the optimal weight matrix and the research object matrix is the highest.
2. The data structure matching degree of the object matrix population and the optimal matrix population is the highest.
3. The elements of the optimal weight matrix have the highest correlation with the elements of the research object matrix.

According to the above three basic assumptions, the research object matrix is constructed as follows:

$$ty = tril\left\{kron\left[\left(Y^{1/2}Y^{1/2}\right)^{1/2}, \left(Y^{1/2}Y^{1/2}\right)^{1/2}\right]\right\} \tag{5.3}$$

In formula (5.3), *tril* represents the trigonometric matrix function extracted from matlab; *kron* stands for matrix multiplication function; *Y* represents the column 1, row *nt*matrix of the independent variable (*n*represents the number of cities, and *t* represents the number of periods).

The correlation coefficient between the object matrix and the elements of the spatial weight matrix is:

$$r = \frac{\sum_{v=1}^{nt_1 \times nt_1}\left(ty_v'' - mean\left(ty''\right)\right)\left(tw_v'' - mean\left(tw''\right)\right)}{\sqrt{\sum_{v=1}^{nt_1 \times nt_1}\left(ty_v'' - mean(ty'')\right)^2}\sqrt{\sum_{v=1}^{nt_1 \times nt_1}\left(tw_v'' - mean(tw'')\right)^2}} \tag{5.4}$$

In formula (5.4), *mean ()* mean value; *ty* is research object matrix.

Fischer T statistics:

$$|r|\sqrt{(nt_1 \times nt_1 - 2)\left(1 - r^2\right)} \sim T(nt_1 \times nt_1 - 2) \tag{5.5}$$

The selection principles of the optimal weight matrix are:

1. The spatial weight matrix is optimal when the row and column are statistical significant in the test of the mean difference and variance ratio of the structure matching of the spatial weight matrix and the research object matrix are the largest.
2. The larger the correlation coefficient is, the better the corresponding space weight matrix will be.
3. When the structure matching principle conflicts with the effective correlation principle in selecting the optimal weight, the structure matching principle should be moderately weakened, and the optimal spatial weight matrix should be selected mainly based on the effective correlation principle.

5.3.2.3 Spatial Autocorrelation Test

The $PM_{2.5}$ transmission effects between neighboring cities reflect the spatial interdependence (Chen, 2013), so it is essential to conduct the spatial autocorrelation test on the $PM_{2.5}$ data of the studied cities. Moran's I was used to measure the autocorrelation of spatial sequences $\{x_i\}_{i=1}^{N}$. The formula is as follows:

$$I = \frac{\sum_{i=1}^{N}\sum_{j=1}^{N} W_{ij}(x_i - \bar{x})(x_j - \bar{x})}{S^2\sum_{i=1}^{N}\sum_{j=1}^{N} W_{ij}} \tag{5.6}$$

$$S = \frac{\sum_{i=1}^{N}(x_i - \bar{x})^2}{N} \tag{5.7}$$

S^2 is the sample variance; W_{ij} is the adjacent weight matrix or the inverse distance weight matrix.

The value range of Moran's I is generally $[-1,1]$. If it is greater than 0, it represents a positive correlation, that is, the high value is adjacent to the high value, and the low value is adjacent to the low value. If it is smaller than 0, it represents a negative correlation, that is, the high value is adjacent to the low value. If the Moran is close to 0, it indicates that the spatial distribution is random and there is no spatial correlation.

5.3.2.4 Construction of SAR

The general SAR is as follows:

$$y_{it} = \begin{cases} \tau y_{i,t-1} + \rho w_i^T y_t + x_{it}^T \beta + d_i^T X_i \delta + u_i + \gamma_t + \varepsilon_{it} \\ \varepsilon_{it} = \lambda m_i^T \varepsilon_t + v_{it} \end{cases} \tag{5.8}$$

To reflect the spillover effect of air pollution in the Yangtze River Delta from 2011 to 2018, the model is as follows:

$$y_{it} = \beta_0 + \rho W y_{it} + \beta ln x_{it} + \beta' x'_{it} + u_i + \varepsilon_{it} \tag{5.9}$$

y_{it} represents the annual mean of $PM_{2.5}$ of the i-th city in period t, and it is used to measure the air quality level; β_0 represents the intercept term; ρ represents the spatial autoregressive coefficient; W represents the spatial weight matrix (W_1 is used to represent a row-normalized adjacent weight matrix, and W_d represents the row-normalized inverse distance weight matrix); lnx_{it} is the logarithm value of the number of civilian car ownership, industrial electricity consumption, construction area and days with wind speed greater than category 3 in the i-th city in the period of t; x'_{it} represents the industrial sulfur dioxide emissions, industrial smoke and dust emissions and per capita GDP in the i-th city in period t; u_i is the individual effect of city i; ε_{it} is the random disturbance term.

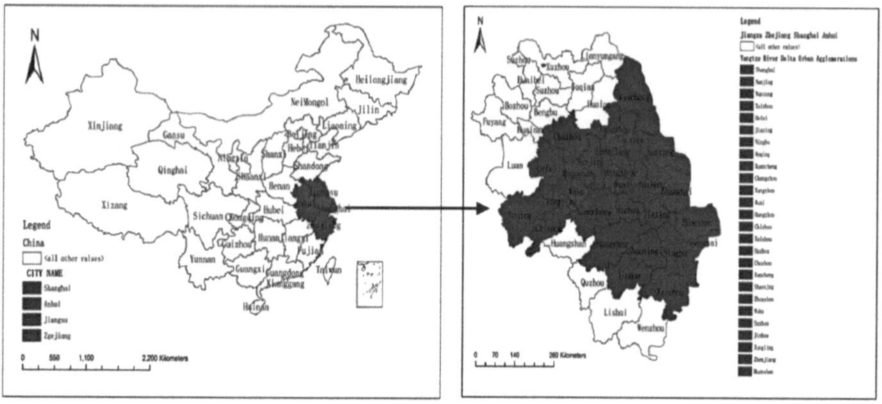

Fig. 5.1 Location of Yangtze river delta in China

5.3.3 *Empirical Analysis of Spatial Econometric Model*

5.3.3.1 Research Objects

Jiangsu-Zhejiang-Shanghai-Anhui area are the combined names of Jiangsu, Zhejiang, Shanghai and Anhui provinces. Its economic development ranks highly in China, but air pollution is also a long-standing problem left over from history as it used to be dominated by heavy industry. According to the *12th five-year Plan for the prevention and control of air pollution in key areas*, the Yangtze River Delta has been listed as a key pollution control area. Therefore, a SAR model was established based on the data of 41 cities in Jiangsu-Zhejiang-Shanghai-Anhui area from 2011 to 2018 for quantitative analysis, so as to analyze the causes of pollution and the correlation of regional pollution and jointly control air pollution. The geospatial map of the Yangtze River Delta is as follows (Fig. 5.1):

5.3.3.2 Spatiotemporal Analysis of $PM_{2.5}$ in Jiangsu-Zhejiang-Shanghai-Anhui Area

The annual average $PM_{2.5}$ concentration in Jiangsu-Zhejiang-Shanghai-Anhui area in 2018, 2017 and 2016 were showed in Figs. 5.2, 5.3 and 5.4, separately. From the perspective of space, air pollution in the northwest of Jiangsu-Zhejiang-Shanghai-Anhui area is heavier than that in the central part and south. In terms of cities, air quality in Zhejiang province is generally better than that in other provinces, followed by Shanghai. Air pollution in Jiangsu and Anhui provinces is relatively serious. From the point of urban distribution, the areas with the most serious air pollution in Jiangsu province are Xuzhou, Suqian, Zhenjiang and Changzhou, followed by Huaian, Yangzhou, Taizhou and Yangzhou. The cities with poor air quality in Anhui province are concentrated in the Huaibei area of Anhui province. The heavily polluted cities include Suzhou, Huaibei, Fuyang, Bengbu, Bozhou, and Huainan; Chuzhou, Hefei, Wuhu and Tongling also have serious air pollution. The air quality of Zhejiang Province is generally better than that of others. The cities with more serious pollution include Hangzhou, Shaoxing and Jiaxing. From the perspective of time, the air pollution situation in 2018 was significantly better than that in 2016, and the heavily polluted areas were significantly reduced. However, in some areas, such as Jiangsu, Zhejiang, Shanghai, Anhui, and several cities in the northwest, the pollution situation has increased significantly.

The trend of the annual average $PM_{2.5}$ concentration in Jiangsu-Zhejiang-Shanghai-Anhui area was showed in Fig. 5.5. It can be seen that the annual average $PM_{2.5}$ concentration in each city was the highest in 2015, which began to decline in 2016, and the concentration was significantly improved from 2017 to 2018. *Urban development Plan of Yangtze River Delta* was complimented in 2016, which is the basis for the following research on the restraining effect of the urban development plan of Yangtze River Delta on air pollution in Jiangsu-Zhejiang-Shanghai-Anhui area. Further analysis from the spatial level shows that the annual average $PM_{2.5}$ concentration is generally "agglomerated." For example, the annual average $PM_{2.5}$ concentration in Changzhou—Lu 'an and other regions is between 45 and 65 ug/m^3, while the annual average $PM_{2.5}$ concentration in Anqing-Huzhou is significantly lower than that in other cities. It can be speculated that the annual average $PM_{2.5}$

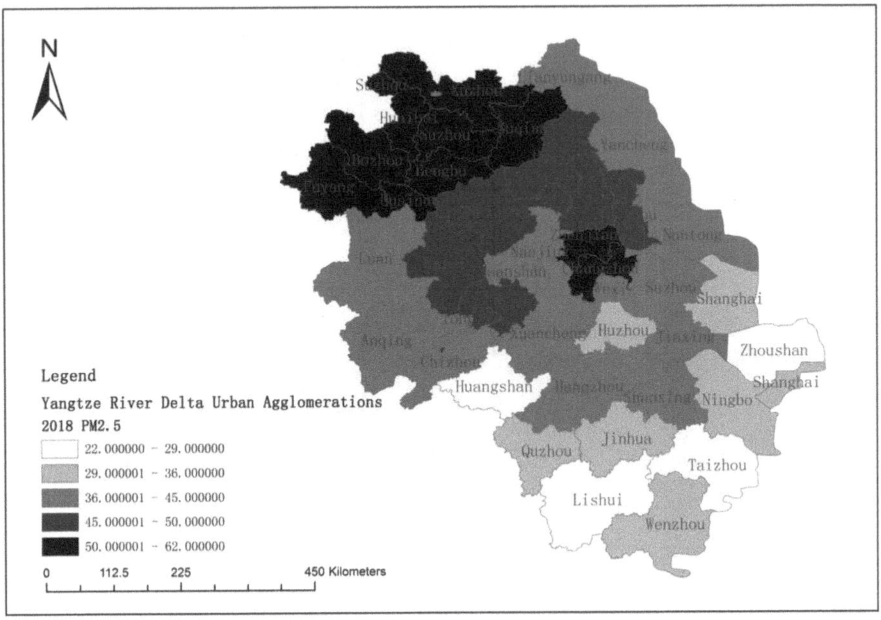

Fig. 5.2 The $PM_{2.5}$ concentration of Jiangsu, Zhejiang, Shanghai and Anhui in 2018

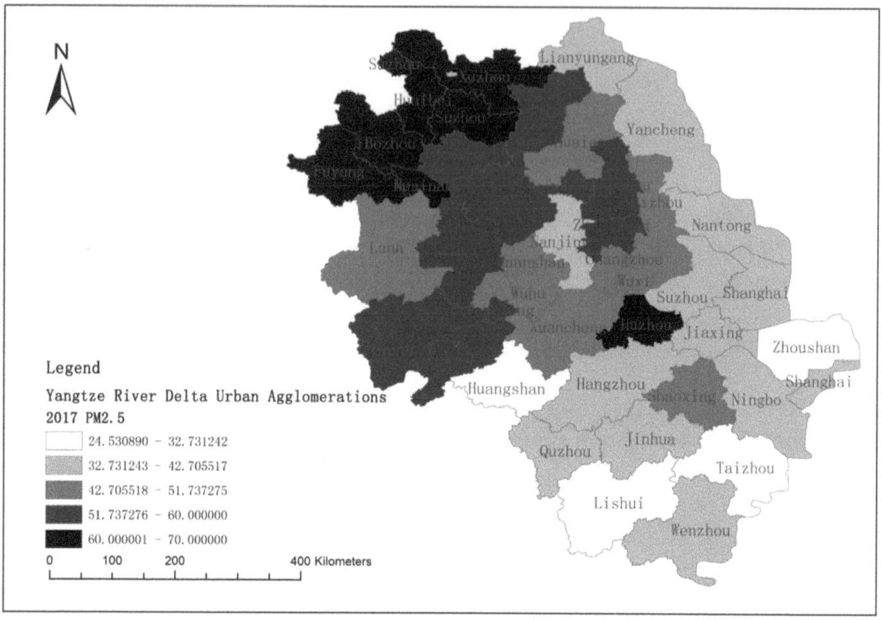

Fig. 5.3 The $PM_{2.5}$ concentration of Jiangsu, Zhejiang, Shanghai and Anhui in 2017

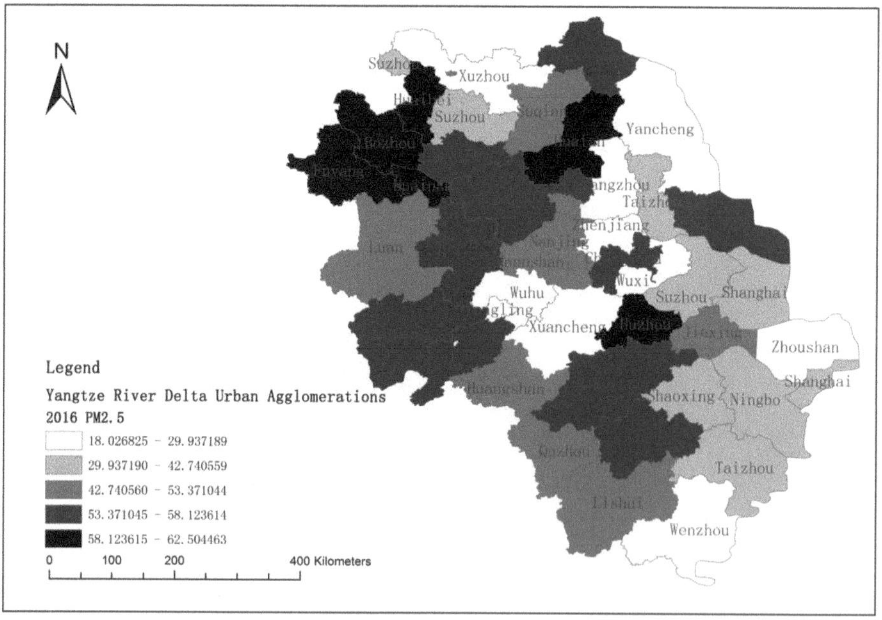

Fig. 5.4 The $PM_{2.5}$ concentration of Jiangsu, Zhejiang, Shanghai and Anhui in 2016

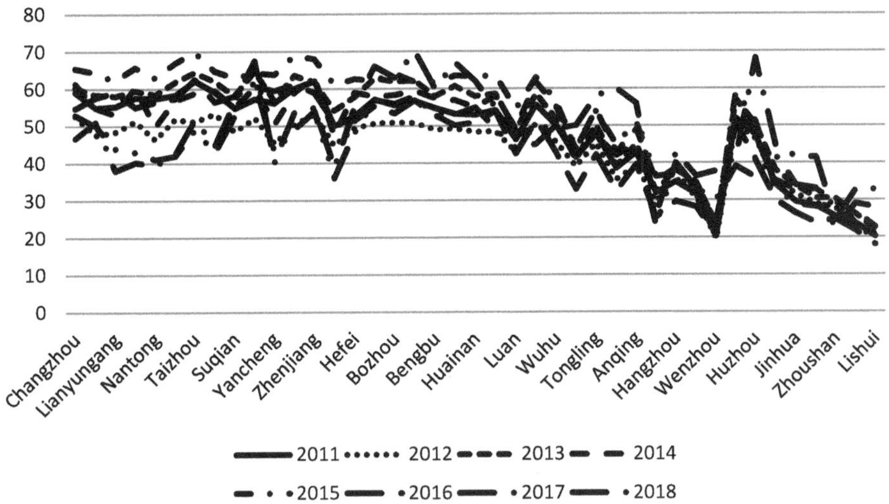

Fig. 5.5 Annual average $PM_{2.5}$ concentration of cities. *Notes* (1) The labels for the vertical axe are annual average $PM_{2.5}$ concentration (ug/ m^3); (2) The labels for the horizontal axes are names of cities in Jiangsu, Zhejiang, Shanghai and Anhui

concentration may be highly correlated with the annual average $PM_{2.5}$ concentration between cities, that is, there may be a significant spatial autocorrelation.

From the perspective of region, the annual average $PM_{2.5}$ concentration of Hangzhou, Wenzhou, Jinhua, Zhoushan, Lishui and other cities in Zhejiang province is significantly lower than that of Shanghai, Anhui and Jiangsu, indicating that air pollution in Shanghai, Anhui and Jiangsu is more serious than that in Zhejiang.

Figure 5.6 depicts the year to year variation of annual average $PM_{2.5}$ concentration in Jiangsu-Zhejiang-Shanghai-Anhui area. We can see that 2016 was a watershed year for the annual average $PM_{2.5}$ concentration of Jiangsu-Zhejiang-Shanghai-Anhui area. The annual average $PM_{2.5}$ concentration of some cities decreased significantly after 2016, while the annual average $PM_{2.5}$ concentration of other cities increased from 2017 to 2018 and then declined sharply. It is found that the cities whose $PM_{2.5}$ concentration has been declining since 2016 are in city clusters in Yangtze River Delta belonging to the urban development plan of Yangtze River Delta, which shows that the government is effective in air pollution control. It can also be seen from Fig. 6 that the annual average $PM_{2.5}$ concentration of Zhejiang province is significantly lower than that of other cities, reflecting that the air quality of Zhejiang province is better than that of Shanghai, Jiangsu and Anhui, which is consistent with the results in Fig. 5.5. And we can know that the air pollution in Jiangsu province is the most serious in Jiangsu-Zhejiang-Shanghai-Anhui area, and its annual average $PM_{2.5}$ concentration is significantly higher than that of other provinces and cities.

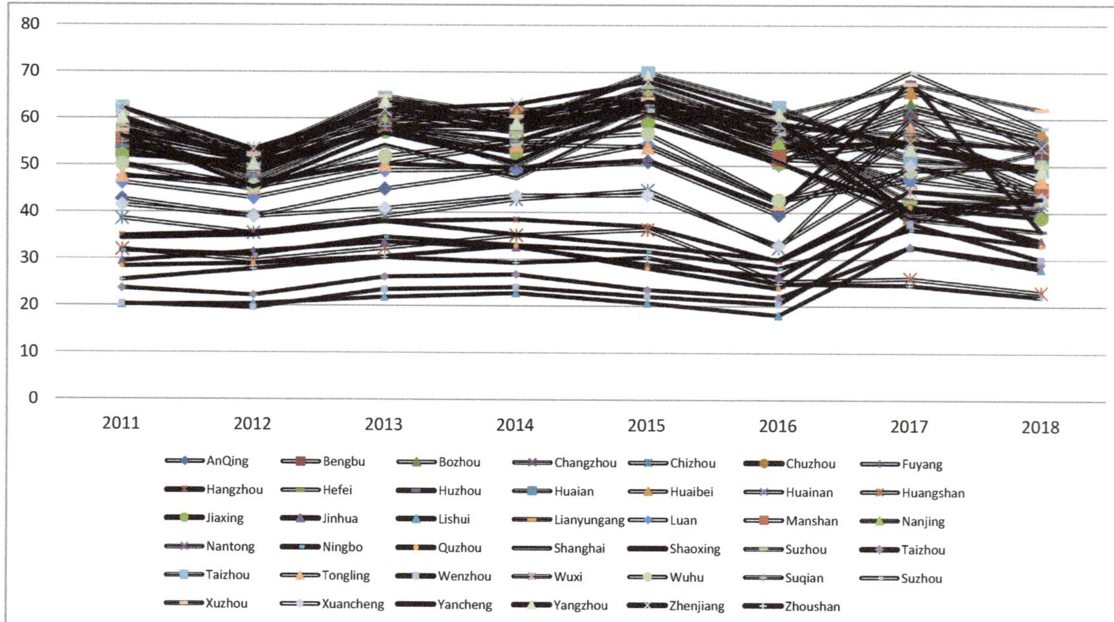

Fig. 5.6 Annual variation of $PM_{2.5}$ concentration of cities in Jiangsu, Zhejiang, Shanghai and Anhui. *Notes* (1) The labels for the vertical axe are annual average $PM_{2.5}$ concentration (ug/ m^3); (2) The labels for the horizontal axes are the names of cities in Jiangsu, Zhejiang, Shanghai and Anhui

5.3.3.3 $PM_{2.5}$ Spatial Autocorrelation Test

The variation of the annual average $PM_{2.5}$ concentration in Jiangsu-Zhejiang-Shanghai-Anhui areas were showed from Figs. 5.2, 5.3, 5.4, 5.5, and 5.6. We guess that there may be a spatial relationship between the annual average $PM_{2.5}$ concentrations. The spatial autocorrelation test is carried out below to further verify it.

The Moran's of annual average $PM_{2.5}$ concentration of 41 cities in Jiangsu-Zhejiang-Shanghai-Anhui areas from 2012 to 2018 was showed in Table 5.2. The Moran's are all positive and have passed the significance test at the level of 1%, indicating that there is a significant positive spatial correlation between 41 cities in Jiangsu-Zhejiang-Shanghai-Anhui areas. That is, for cities with higher annual average $PM_{2.5}$ concentration, one or more cities with higher annual average $PM_{2.5}$ concentration are clustered around them, and conversely, one or more cities with lower annual average $PM_{2.5}$ concentration are clustered around cities with lower annual average $PM_{2.5}$ concentration.

From the perspective of the adjacent weight matrix, the positive relationship fluctuated around 0.8 and reached the lowest value of 0.501 in 2017, indicating that the spatial correlation remained stable and at a high level. From the perspective of the inverse distance weight matrix, the inverse distance weight matrix adds the distance information. Although this positive relation is somewhat weaker than that of adjacent weight matrix, there is still a significant positive relation, indicating that the spatial econometric model is applicable.

Figure 5.7 show the Moran scatter diagram of the annual average $PM_{2.5}$ concentration in Jiangsu-Zhejiang-Shanghai-Anhui areas from 2012 to 2018. The horizontal axis represents the standardized annual average $PM_{2.5}$ concentration, and the vertical axis represents the spatial lag value of the standardized annual average $PM_{2.5}$ concentration. It can be seen from the figures that from 2012 to 2018, almost all cities are in the first and third quadrants, representing "high-high" and "low-low" regions. The scatter diagram shows the stability of the positive correlation of the annual average

Table 5.2 Global Moran's statistical index of annual average $PM_{2.5}$ concentration in 41 cities of Jiangsu-Zhejiang-Shanghai-Anhui from 2012 to 2018

Weights	Year	Morans'I	year	Morans'I
0–1 weight	2012	0.802* * *	2016	0.847* * *
	2013	0.819* * *	2017	0.501* * *
	2014	0.825* * *	2018	0.682* * *
	2015	0.863* * *	/	
Inverse distance weight	2012	0.248* * *	2016	0.270* * *
	2013	0.266* * *	2017	0.194* * *
	2014	0.266* * *	2018	0.26* * *
	2015	0.269* * *	/	

$PM_{2.5}$ concentration from the internal structure. From the perspective of time, cities with low $PM_{2.5}$ concentration, such as Wenzhou (33), Quzhou (38), Zhoushan (39) and Taizhou (40), tend to be stable from 2011 to 2018, and the air pollution has been kept at a low level. At the same time, it can be clearly seen that the air quality had been improved. In the Moran scatter diagram, the cities in the first quadrant accounted for the majority in 2012, but as time passed, pollution was controlled, and the cities in the third quadrant slowly began to increase, showing low-low benefits in cities, that is, the pollution situation in low-polluted cities and surrounding cities has been improved.

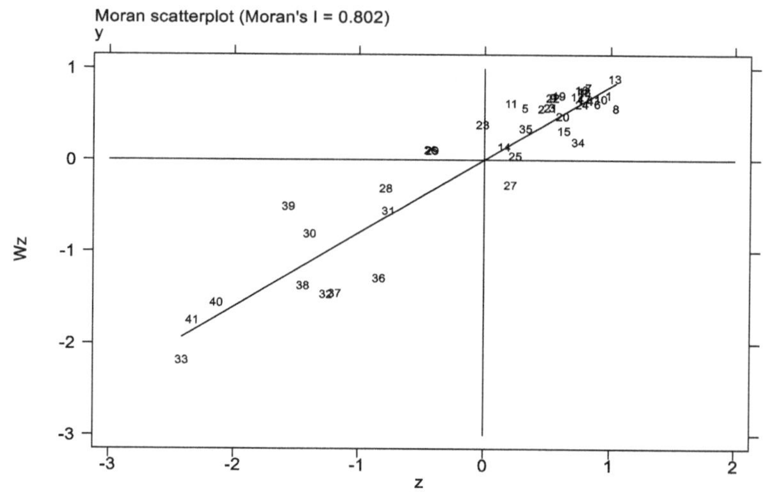

(a)

Fig. 5.7 **a** Moran scatter diagram of annual average $PM_{2.5}$ concentration in cities of Jiangsu, Zhejiang, Shanghai and Anhui in 2012. **b** Moran scatter diagram of annual average $PM_{2.5}$ concentration in cities of Jiangsu, Zhejiang, Shanghai and Anhui in 2013. **c** Moran scatter diagram of annual average $PM_{2.5}$ concentration in cities of Jiangsu, Zhejiang, Shanghai and Anhui in 2014. **d** Moran scatter diagram of annual average $PM_{2.5}$ concentration in cities of Jiangsu, Zhejiang, Shanghai and Anhui in 2015. **e** Moran scatter diagram of annual average $PM_{2.5}$ concentration in cities of Jiangsu, Zhejiang, Shanghai and Anhui in 2016. **f** Moran scatter diagram of annual average $PM_{2.5}$ concentration in cities of Jiangsu, Zhejiang, Shanghai and Anhui in 2017. **g.** Moran scatter diagram of annual average $PM_{2.5}$ concentration in cities of Jiangsu, Zhejiang, Shanghai and Anhui in 2018. *Note* Cities numbered 1 to 41 in Fig. 5.7 are: Changzhou, Huaian, Lianyungang, Nanjing, Nantong, Suzhou, Taizhou (belong to Jiangsu Province), Wuxi, Suqian, Xuzhou, Yancheng, Yangzhou, Zhenjiang, Shanghai, Hefei, Huaibei, Bozhou, Suzhou, Bengbu, Fuyang, Huainan, Chuzhou, Luan, Maanshan, Wuhu, Xuancheng, Tongling, Chizhou, Anqing, Huangshan, Hangzhou, Ningbo, Wenzhou, Jiaxing, Huzhou, Shaoxing, Jinhua, Quzhou, Zhoushan, Taizhou (belong to Zhejiang Province), Lishui

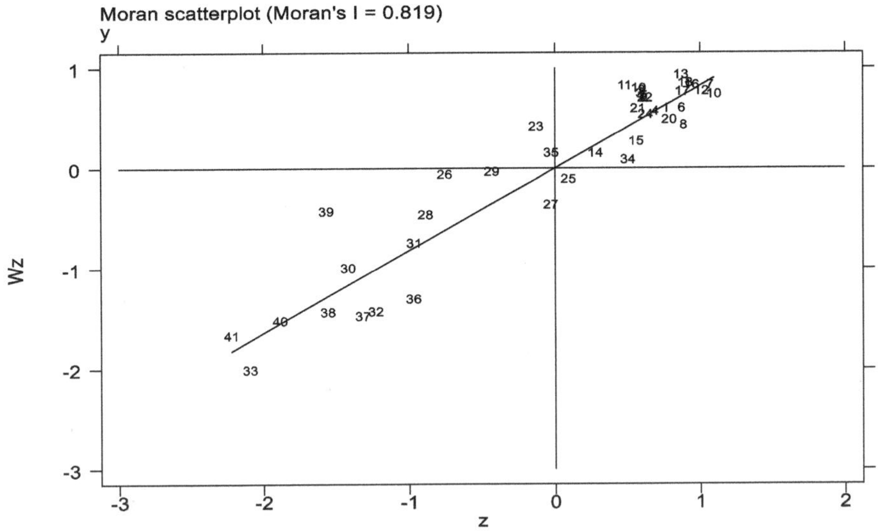

(b)

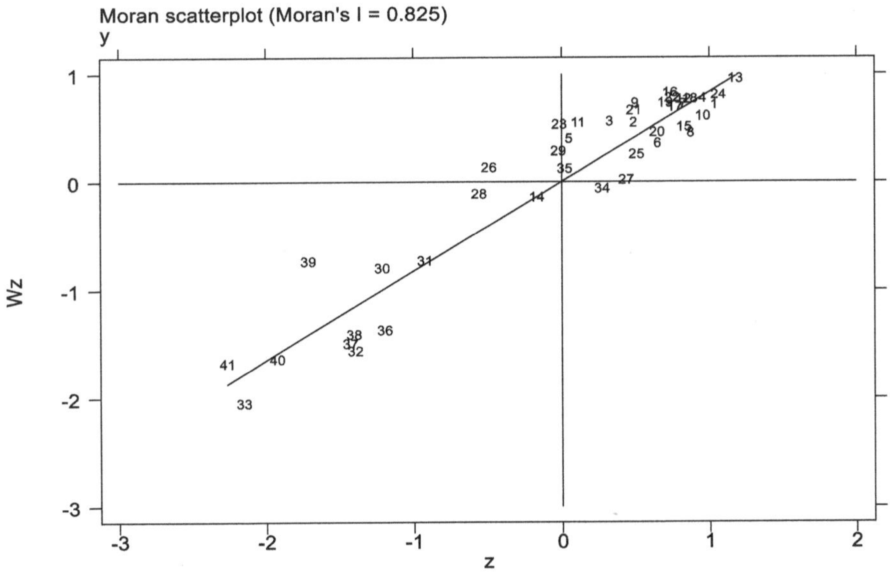

(c)

Fig. 5.7 (continued)

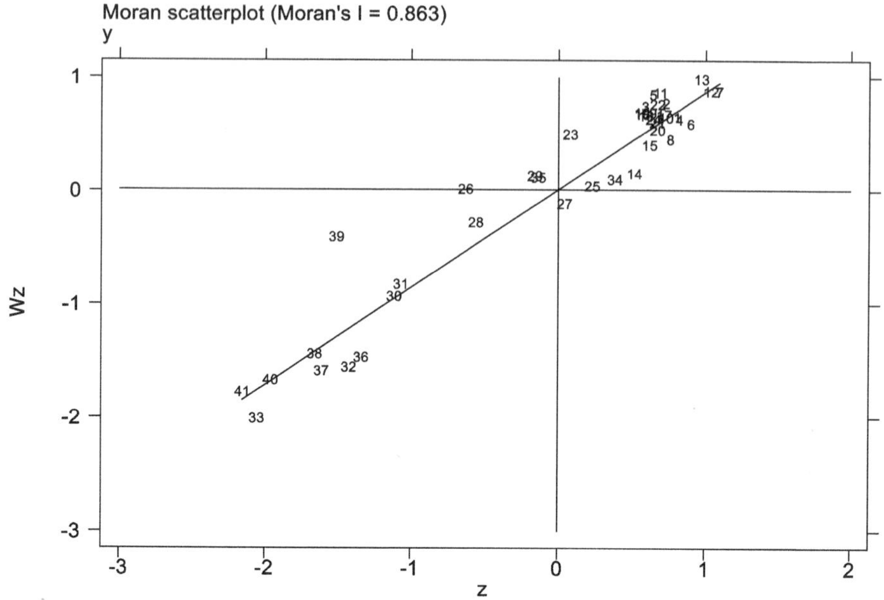

(d)

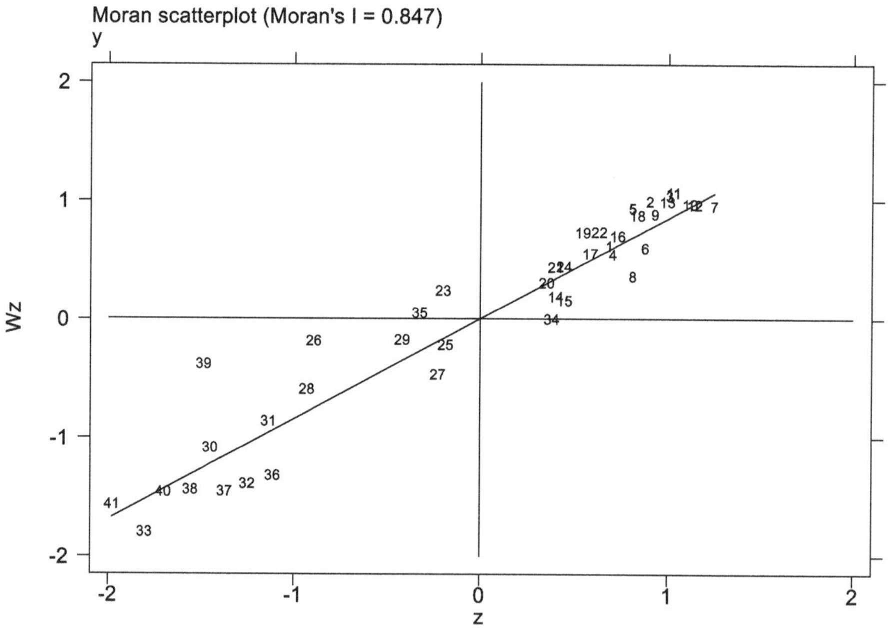

(e)

Fig. 5.7 (continued)

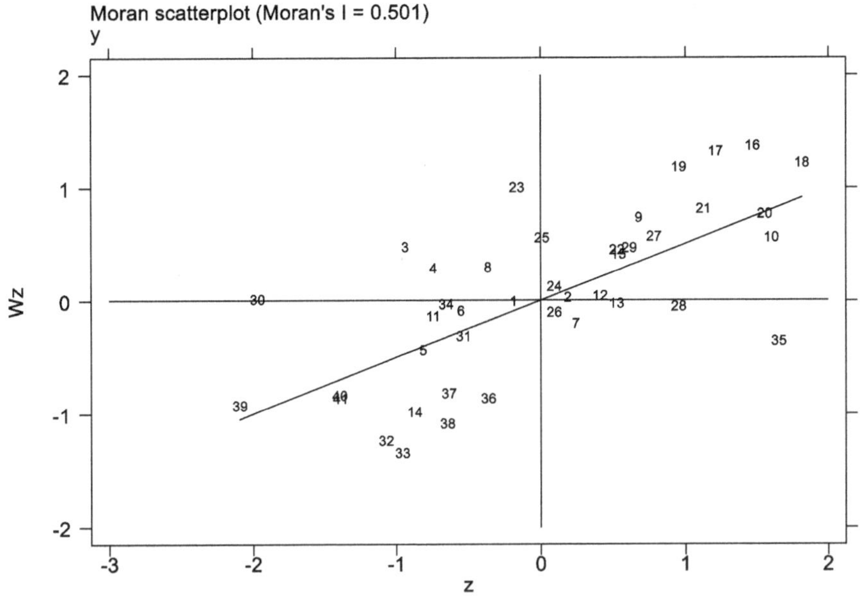

(f)

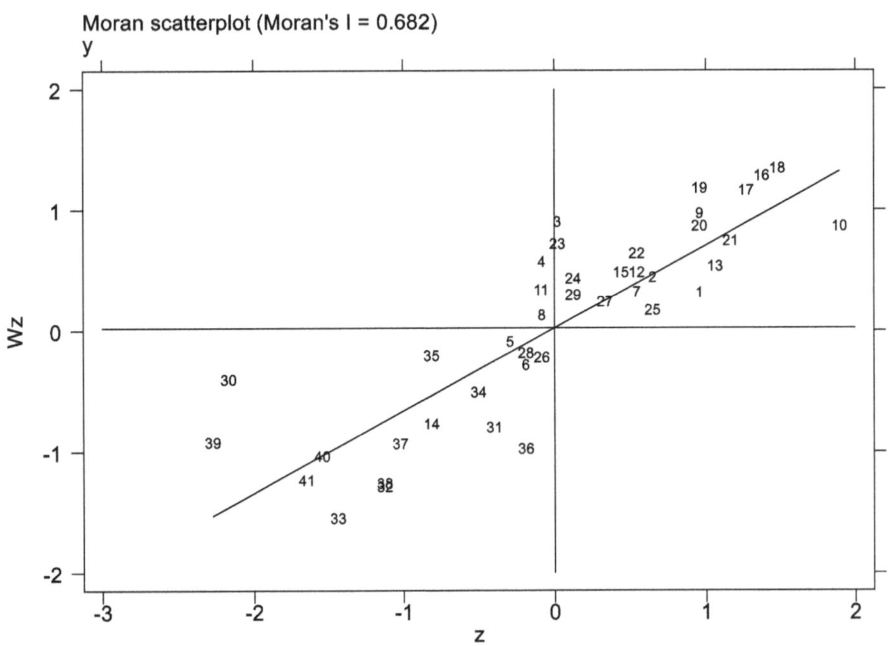

Fig. 5.7 (continued)

5.3.3.4 SAR Estimation and Result Analysis in Jiangsu-Zhejiang-Shanghai-Anhui Areas

This paper establishes a SAR model based on the data of 41 cities in in Jiangsu-Zhejiang-Shanghai-Anhui areas from 2011 to 2018. The spatial weight matrix here is the adjacent spatial weight matrix and the inverse distance weight matrix (The distance between cities is considered in the adjacent spatial weight matrix and the inverse distance weight matrix, so it is more reliable to study the air pollution correlation between cities). In order to eliminate the influence of heteroscedasticity, the logarithms of variables such as civil automobile ownership, industrial electricity consumption, building construction area and days with wind speed greater than category three are selected, and the SAR model estimation results are shown in Table 5.3. It can be seen from Table 5.3 that the spatial autoregressive coefficients are all significant at the 1% level, indicating the existence of spatial autoregressive effect in the annual average $PM_{2.5}$ concentration in Jiangsu-Zhejiang-Shanghai-Anhui area. It is consistent with the above conclusions in Fig. 5.7. Then Hausman test is conducted on the model. For the adjacent weight matrix, the p-value of the Hausman test is 0.000, so the original hypothesis of random effects is rejected, and a fixed-effect SAR model is selected. For the inverse distance weight matrix, Hausman statistics is negative, and the original hypothesis of random effect is accepted. Therefore, according to different weight choices, the model results in column (2) and column (5) in Table 5.3 are used for analysis.

In Table 5.3, the spatial autoregressive coefficients of column (2) and (5) are greater than zero, indicating that there is a significant spatial spillover effect in the annual average $PM_{2.5}$ concentration in Jiangsu-Zhejiang-Shanghai-Anhui area and all of them play a positive spillover effect. For the adjacent weight matrix, for every 1% increase in the annual average $PM_{2.5}$ concentration in the surrounding area, the annual average $PM_{2.5}$ concentration will increase by 0.803% in the region. For the inverse distance weight matrix, every 1% increase in the annual average $PM_{2.5}$ concentration in the surrounding area will cause the annual average $PM_{2.5}$ concentration in the area to increase by 0.879%. The spatial spillover effect is obvious.

The annual average $PM_{2.5}$ concentration in Jiangsu-Zhejiang-Shanghai-Anhui area was significantly affected by the civil car ownership, industrial smoke and dust emissions and per capita GDP. The civil car ownership has a significant positive effect on the annual average $PM_{2.5}$ concentration in Jiangsu-Zhejiang-Shanghai-Anhui area. The two weight matrix results show that for every 1% increase in civil car ownership, the annual average $PM_{2.5}$ concentration in Jiangsu-Zhejiang-Shanghai-Anhui area increased by 1.75% and 2.729%, indicating that the car exhaust contributes greatly to the $PM_{2.5}$ concentration. Under the two spatial weight matrices, the emission of industrial smoke and dust has a positive effect on the $PM_{2.5}$ concentration. In the inverse distance weight matrix, the coefficient passed the significance test of 10%. For every 1% increase in the emission of industrial smoke and dust, the annual average $PM_{2.5}$ concentration in Jiangsu-Zhejiang-Shanghai-Anhui area increased by 0.323%. Furthermore, it was found that per capita GDP had a restraining effect on the $PM_{2.5}$ concentration, indicating that when the economy developed to a high level,

Table 5.3 Estimation results of spatial autoregressive model (SAR)

variables (1)	Fixed effect (W_1) (2)	Random effect (W_1) (3)	Fixed effect (W_d) (4)	Random effect (W_d) (5)
Civil automobile ownership (lnx_1)	1.75***	0.971*	3.244***	2.729***
Industrial power consumption (lnx_2)	−2.41	1.244	−2.174	−0.007
Industrial concentration (x_3)	−0.037	0.0044	−0.017	0.0122
Industrial dust concentration (x_4)	0.235	0.243	0.256	0.323*
Construction area (lnx_5)	0.866	−0.089	0.5882	−0.312
Wind speed greater than category 3 (lnx_6)	−0.107	0.091	−0.304*	−0.0697
GDP per capita (x_7)	−0.45**	−0.333**	−0.733***	−0.592**
ρ	0.803***	0.842***	0.874***	0.879***
Hausman Test χ^2	79.06		−2.35	
	0.0000			

people's awareness of environmental protection would be improved, and people's investment in environmental protection would also increase, so as to achieve better environmental quality than that of less developed cities.

5.4 Effectiveness of Urban Development Plan of Yangtze River Delta on Air Pollution Control

5.4.1 The DID Model

Plan was approved and issued by the state council in May 2016, specifying that the city cluster in Yangtze River Delta includes 26 cities such as Shanghai, Jiangsu and Zhejiang. The city cluster in Yangtze River Delta is an important intersection zone between "The Belt and Road" and the Yangtze river economic belt and plays a decisive strategic role in the overall situation of China's national modernization and the all-round opening pattern. *Plan* is a policy project, and its implementation will inevitably have a different impact on the air pollution in the cities where it is implemented and in the neighboring cities (cities in Jiangsu-Zhejiang-Shanghai-Anhui area). The net effect of *Plan* on the treatment group can be better obtained by using the DID model. Therefore, the DID model is set as follows:

$$\begin{aligned} y_{it} = \beta_0 + \beta_1 DT + \alpha_1 \ln x_{1it} + \alpha_2 \ln x_{2it} + \alpha_3 \ln x_{5it} + \alpha_4 \ln x_{6it} \\ + \alpha_5 x_{3it} + \alpha_6 x_{4it} + \alpha_7 x_{7it} + \delta_i + \varepsilon_{it} \end{aligned} \tag{5.10}$$

In formula (5.10), subscript i is city, and t is time; y_{it} represents the annual mean value of $PM_{2.5}$ in the $i-th$ city in period t, and it is used to measure the air quality level; β_0 represents the regression intercept term; D stands for grouped dummy variable. If it belongs to the city cluster in Yangtze River Delta, D = 1; otherwise, D = 0; T stands for time group dummy variable. Before 2016, T = 0, and T = 1 from 2016. DT stands for the multiplication term of D and T, and β_1 represents the net effect of the *Plan*. lnx_{it} is the logarithm value of the number of civil car ownership, industrial electricity consumption, construction area and days with wind speed greater than category 3 in the period of t in the i-th city; x_{it} represents the industrial sulfur dioxide emissions, industrial smoke and dust emissions and per capita GDP of i cities in period t; δ_i is the individual effect of city i; ε_{it} is the random error term. β_1 reflects the net effect of the *Plan* on local air pollution.

5.4.2 SDID Model Construction of the Governance Effectiveness of Planning

The DID model considers the net effect of the *Plan* on air pollution but fails to consider the spatial correlation of air pollution in various cities. Ignoring the connection can lead to bias. Therefore, a SDID model is constructed by combining the DID model with SAR to evaluate the impact of the implementation of the *Plan* on air pollution.

5.4.2.1 SDID Spatiotemporal Effect Model Construction

Same as the dummy variable introduced in the DID model, the samples are divided into treatment group, control group, pre-change sample and post-change sample. D is the grouping dummy variable. If it belongs to the city cluster in Yangtze River Delta, D = 1, otherwise D = 0. T is the time grouping dummy variable, T = 0 before 2016 and T = 1 after 2016. See Fig. 5.8.

Formula (5.10) is further modified to establish the following SAR—SDID model:

$$\begin{aligned} y_{it} = \beta_0 + \beta_1 DT + \rho W y_{it} + \alpha_1 \ln x_{1it} + \alpha_2 \ln x_{2it} + \alpha_3 \ln x_{5it} + \alpha_4 \ln x_{6it} \\ + \alpha_5 x_{3it} + \alpha_6 x_{4it} + \alpha_7 x_{7it} + \delta_i + \varepsilon_{it} \end{aligned} \tag{5.11}$$

In formula (5.11), subscript i is city, and t is time; y_{it} represents the annual mean value of $PM_{2.5}$ in the $i-th$ city in period t, and it is used to measure the air quality

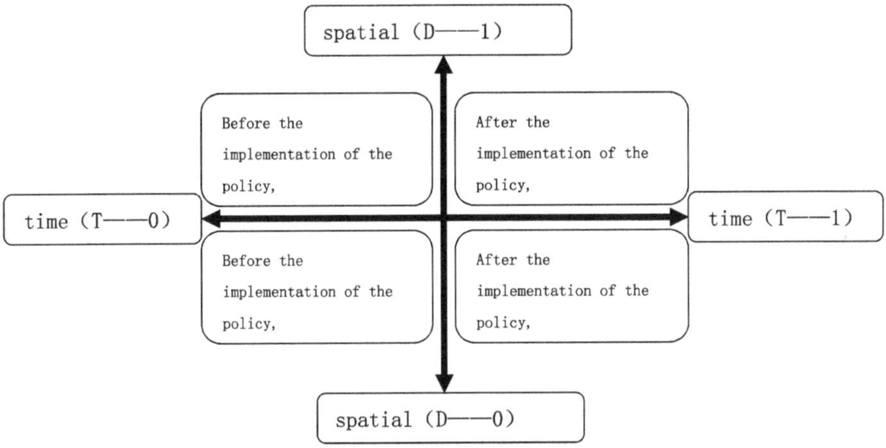

Fig. 5.8 Group setting diagram of virtual variables D and T

level; β_0 represents the regression intercept term; D is the grouping dummy variable, and if it belongs to the city cluster in Yangtze River Delta, $D = 1$, otherwise $D = 0$; T is a time grouping dummy variable. Before 2016, T = 0, and T = 1 after 2016. DT represents the multiplication term of D and T, and β_1 represents the net effect of *Plan*. ρ represents the spatial autoregressive coefficient; W represents the spatial weight matrix W_1 is used to represent the row-normalized adjacent weight matrix, and W_d is used to represent the row-normalized inverse distance weight matrix); lnx_{it} is the logarithm value of the number of civil car ownership, industrial electricity consumption, construction area and days with wind speed greater than category 3 in the i-th city in the period of t; x_{it} represents the industrial sulfur dioxide emissions, industrial smoke and dust emissions and per capita GDP of i cities in period t; δ_i is the individual effect of city i; ε_{it} is the random error term.

This paper focuses on the DT coefficient β_1, which represents the net effect of the implementation of the *Plan* on the annual average $PM_{2.5}$ concentration in Jiangsu-Zhejiang-Shanghai-Anhui area.

5.4.2.2 Assumptions and Basic Conditions

In combination with the DID model and the study in this paper, the SDID model should satisfy the following assumptions: first, the parallel trend hypothesis, that is, the treatment group and the control group had the same development trend before the implementation of *Plan*; Secondly, the implementation of *Plan* only has an impact on the treatment group, but not on the control group. Thirdly, before and after the implementation of the *Plan's* policy, other control variables have the same effect on the treatment group and the control group. Fourthly, before and after the implementation of the *Plan's* policy, some important characteristics of the treatment group and the

Table 5.4 Division of treatment group and control group

City	Grouping	Prefecture level city
Belong to Yangtze River Delta Urban Agglomerations	Treatment	Shanghai, Nanjing,Wuxi, Changzhou, Suzhou, Nantong, Yancheng, Yangzhou, Zhenjang, Taizhou, Hangzhou, Ningbo, Jiaxing, Huzhou, Shaoxing, Jinhua, Zhoushan, Taizhou, Hefei, Wuhu, Maanshan, Tongling, Anqing, Chuzhou, Chizhou, Xuancheng
Not belong to Yangtze River Delta Urban Agglomerations	Controlled	Huaian, Lianyungang, Suqian, Xuzhou, Huaibei, Bozhou, Suzhou, Bengbu, Fuyang, Huainan, Luan, Huangshan, Wenzhou, Quzhou, Lishui

control group are distributed stably during the period. Fifthly, there is a significant spatial correlation of $PM_{2.5}$ between the cities in Jiangsu-Zhejiang-Shanghai-Anhui area.

Basic conditions to be met:

1. Taking the implementation of *Plan* as a key time node, a panel data set corresponding to at least two years (one year before and after the change) is needed, and the dummy variable T is assigned 0 and 1 respectively.
2. The implementation area and unimplemented area of the *Plan* are found and divided into treatment group and control group. Dummy variable D is set to be 1 and 0, respectively.

5.4.2.3 Division of Treatment Group and Control Group

The 41 cities in Jiangsu-Zhejiang-Shanghai-Anhui area are divided into treatment group and control group. The results are shown in Table 5.4.

5.4.3 Empirical Analysis of SDID

5.4.3.1 Parallel Trend Test

The premise of the application of DID model is to satisfy the parallel trend test, so the parallel trend test is conducted first.

1. The basic regression method is employed. $PM_{2.5}$ is selected as the dependent variable, and the period dummy variable T, grouping dummy variable D and intersection term DT of T and D are set as independent variables to see whether DT is significant.
2. The first-order difference of the dependent variable y is set as the dependent variable and grouping dummy variable D as the independent variable. Control other variables for regression to see if the D coefficient is significant.

3. The yearly dummy variable and the interaction term of the dummy variable of treatment group are set as the independent variable, and P as the dependent variable for regression, and the regression coefficient of five years is plotted.

The results of the first two test methods are shown in Table 5.5, and the result of the third test method is shown in Fig. 5.9.

The basic regression results show that the DT coefficient is −4.24, and the corresponding p value is 0.002, indicating that the net effect of the policy implementation is significant. In the difference score method, the coefficient of D is −1.55 and it fails to pass the significance test, indicating that before the implementation of the

Table 5.5 Parallel trend test based on 41 cities

Variables	Regression (y)	Difference-in-difference (y)
DT	−4.24*** (0.002)	
D	0.713 (0.849)	−1.55 (0.208)
T	−0.056 (0.96)	
Other independent variables	Uncontrolled	Controlled
Year	Controlled	Controlled
N	328	164

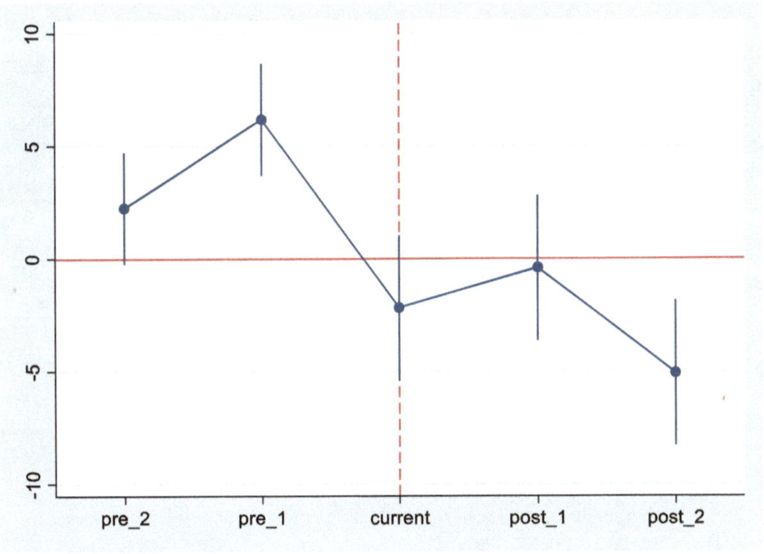

Fig. 5.9 Comparison of treatment groups in two years before and after the implementation of the policy. *Notes* (1) The labels for the vertical axe are annual average $PM_{2.5}$ concentration (ug/m^3); (2) The labels for the horizontal axes are: pre_2 = 2014, pre_1 = 2015, current = 2016, post_1 = 2017, and post_1 = 2018

policy, there was no significant difference in the decreasing trend of air pollution concentration between the treatment group and the control group.

As can be seen from Fig. 5.9, the coefficient was always positive in the first two years before the implementation of *Plan*, and the coefficient was significantly negative after implementing policy, indicating that the division of the treatment group and the control group is meaningful. It can also be seen from Fig. 5.9 that the implementation of the *Plan* has a significant negative effect on air pollution concentration. In other words, the implementation of *Plan* effectively controlled air pollution, playing a significant role at the beginning of the implementation. Although there was a rebound one year after the implementation of the *Plan*, it is effective.

5.4.3.2 Model Estimation Results and Analysis

According to the model established in Sect. 4.2.1, the estimation results of the model are shown in Table 5.6. In the third part, although the SAR model is used to consider the spillover effect between regions, reflecting the spillover effect in the implementation areas of the *Plan* during the whole period (from 2011 to 2018), but it cannot reflect the spillover effect in different periods. Therefore, the SAR model is used to compare the changes of the spatial autocorrelation coefficients ρ for the data from 2011 to 2015 and 2016 to 2018 and investigate the change of spillover effect before and after the implementation of the *Plan*. The reason for using the inverse distance weight matrix is that the inverse distance weight matrix takes into account the distance between cities, which is equivalent to considering the transmission distance in air pollution, and it is more effective.

Table 5.6 Results of model estimations

Variables (1)	DID (2)	SDID (W_d) (3)	SAR (W_d) (4)	SAR (W_d) (5)
Year	2011 ~ 2018	2011 ~ 2018	2011 ~ 2015	2016 ~ 2018
ρ		0.864***	0.918***	0.81***
DT	−3.59**	−2.219**		
D	−0.241	1.9209		
T	−3.87**	0.642		
lnx_1	6.132***	2.878***	−1.023	7.54***
lnx_2	−0.168	0.2759	2.378*	−4.141
x_3	−0.247**	−0.0352	−0.131*	1.085
x_4	0.697***	0.2431	0.195	1.534
lnx_5	−1.763*	−0.662	−0.296	−2.058
lnx_6	0.119	−0.0355	−0.334**	−1.091*
x_7	−0.163	−0.473*	0.131	−0.0969

First of all, column (2) in Table 5.6 shows that the DT regression coefficient in the DID model has passed the significance test and is significantly negative, indicating that the implementation of *Plan* has effectively controlled air pollution in these areas. The results show that the implementation of *Plan* has reduced the average annual $PM_{2.5}$ concentration by 3.59 units on average of city cluster in the Yangtze River Delta. The main independent variables affecting the air pollution concentration are the civil automobile ownership, the industrial sulfur dioxide emission and the industrial smoke and dust emission. The annual average $PM_{2.5}$ concentration is positively affected by the number of civil vehicles ownership and the emission of industrial smoke and dust. Although this conclusion may violate our daily cognition, it is consistent with the conclusion of Li et al. (2019).

Column (3) in Table 5.6 shows the estimated results obtained by using the inverse distance weight matrix. As can be seen from column (3), the spatial autoregressive coefficient is significantly positive with a high value of 0.864, indicating that there is a significant spatial spillover effect in the whole area of Jiangsu-Zhejiang-Shanghai-Anhui area, that is, air pollution between cities is highly correlated. The DT coefficient passes the significance test at the level of 5%, which is consistent with the results obtained by the DID model, indicating that the implementation of *Plan* could indeed curb air pollution. Moreover, the per capita GDP plays the negative influence on $PM_{2.5}$ concentration, and the SDID model also passes the significance test at the level of 10%. The possible causes of this phenomenon that is when a regional economic develops to a high level, high-polluted industries are transferred to the other places, energy structure is adjusted, and technological innovation is strengthened, resulting in a situation in which the higher the per capita GDP, the lighter the air pollution.

Column (4) and (5) in Table 5.6 adopt spatial autoregression model to study the air pollution in Jiangsu-Zhejiang-Shanghai-Anhui area from 2011 to 2015 and 2016 to 2018 respectively, that is, the changes of air pollution before and after the implementation of the policy. It can be seen from the column (4) and (5), the spatial autoregressive coefficient is significantly positive with high value of 0.918 and 0.81, respectively, showing high regional air pollution correlation before and after the implementation of policy and indicating that from 2011 to 2015, industrial power consumption, industrial sulfur dioxide emissions and the number of days with wind speed greater than category 3 have a significant impact on air pollution. The greater the industrial power consumption, the more serious the air pollution is; while the industrial sulfur dioxide and the number of days with wind speed greater than category 3 have a negative impact on air pollution indicators. Generally speaking, the higher the wind speed is, the more conducive it is to the diffusion movement of air pollution, thus exerting a negative impact. The explanation of the negative impact of industrial sulfur dioxide emissions on $PM_{2.5}$ can be explained by referring to that of DID model. From 2016 to 2018, the significant influencing variables is the number of civil car ownership and the number of days with wind speed more than category 3, and the number of civil car ownership has significant positive influence on air pollution. Nowadays, more and more families own private cars, and the environmental pollution problems caused by automobile exhaust are becoming more and more serious. The conclusions in columns (5) and (6) actually reflect from the side that with the development of the times, the causes of air pollution are also constantly changing, which reminds us to timely adjust the prevention and control strategies of air pollution according to specific situation.

5.4.3.3 Placebo Test

One method to test placebo: change the pilot time. The *Plan* was implemented in 2016. In this paper, 2014, 2015 and 2017 are proposed as the policy implementation time respectively, and the $PM_{2.5}$ is set as the dependent variable, multiplication term DT of time dummy variable T and grouping dummy variable D is set as the independent variable for regression. The results are shown in Table 5.7. Table 5.7 shows that if the pilot time is changed to 2014 and 2015, the coefficient of key independent variables is not significant, indicating that the year of 2014 and 2015 had no significant impact on air pollution. In 2017, the coefficient of key independent variables was significant, and the coefficient was significantly negative, indicating that the implementation of the *Plan* in 2016 played a significant role in controlling air pollution.

The second placebo test method: replace treatment group, that is, the *Plan* was not implemented, that is, the 26 cities in Yangtze River Delta were not "planned". The treatment group and the control group were shuffled in order, and the treatment group and the control group were randomly selected for double-difference basic regression. According to the estimation formula (4.1), the estimated value $\hat{\beta}_1$ of DT coefficient can be expressed as:

$$\hat{\beta}_1 = \beta_1 + \gamma \frac{\text{cov}(D_i \times T_t, \varepsilon_{it}|z)}{\text{var}(D_i \times T_t|z)} \tag{5.12}$$

In formula (5.12), z represents all the control variables involved in the paper. If γ =0, the $\hat{\beta}_1$ estimation is unbiased, that is, non-observed factors will not interfere with the estimation results. However, it is not possible to directly test whether γ is 0. Therefore, in this paper, the treatment group and the control group are randomly scrambled to randomly generate treatment group and the control group. Then, this process is performed for basic regressions for 1000 times. As can be seen from Fig. 5.10, all the coefficients of $\hat{\beta}_1$ are concentrated around 0, so it can be considered that γ =0 without influencing the estimated results. The estimated results are robust.

Table 5.7 Placebo test—change of pilot time

	2014	2015	2017
DT	0.1183	−0.974	−3.829
P value	(0.892)	(0.248)	(0.000)

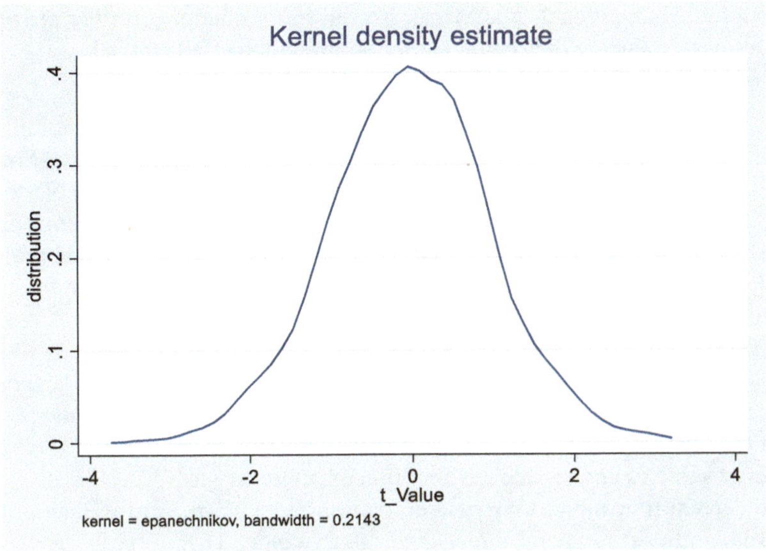

Fig. 5.10 Placebo test chart

5.5 Conclusion and Suggestion

The conclusions of this paper mainly include: first, the implementation of the *Plan* significantly reduce the urban annual average $PM_{2.5}$ concentration in Jiangsu-Zhejiang-Shanghai-Anhui area. Secondly, there is a significant spillover effect of air pollution in the Yangtze River Delta, and the correlation value of annual average $PM_{2.5}$ concentration between regions is relatively high. Therefore, the correlation of air pollution between regions should be considered in the treatment of urban air pollution, so as to achieve cross-regional coordinated treatment. Thirdly, from 2011 to 2015 and 2016 to 2018, the factors affecting air pollution in Jiangsu-Zhejiang-Shanghai-Anhui area have changed. From 2011 to 2015, industrial power consumption, industrial sulfur dioxide emissions and the number of days with wind speed greater than category 3 had a significant impact on air pollution. The greater the industrial power consumption was, the more serious the air pollution was, while industrial sulfur dioxide emissions and the number of days with wind speed greater than category 3 had a negative impact on air pollution indicators. From 2016 to 2018, significantly influencing variables included civil car ownership and the number of days with wind speed greater than category 3, and the civil car ownership had significant positive influence on air pollution, showing the economic development in China and the fact that automobiles have become an increasingly common means of transportation in China from the side and that automobile exhaust will cause serious air pollution problems. Fourthly, the SDID model can separate the net effect of the implementation of *Plan* on air pollution in Jiangsu-Zhejiang-Shanghai-Anhui area. The spatial autoregressive coefficients are all significant positive at the 1% confidence level and highly significant. The Moran is used for verification, further proving the necessity of introducing the spatial lag term based on the DID model.

Based on the above studies, the following suggestions are proposed:

1. Promote measures to jointly prevent and control environmental pollution. There is a significant spatial correlation of air pollution in Jiangsu-Zhejiang-Shanghai area, showing a high-high, low-low aggregation effect. It is possible to build multiple city "clusters" for pollution control with several highly polluted cities as the core and surrounding cities as the auxiliary circle. Environmental protection policies suitable for their own characteristics can be implemented in their cities.
2. Raise pollution control emission standards of vehicle exhaust. Automobile exhaust is an important factor of air pollution. On the one hand, it is necessary to control exhaust emissions. In other words, stricter standards should be set to limit the emission of vehicles that fail to meet environmental standards. On the other hand, starting from the energy utilization, we should develop new energy vehicles and reduce the emission of polluting gas.
3. Build a real-time monitoring system for air pollution. The causes of air pollution have significant temporal effects and are dynamic. Therefore, the government departments should establish a more complete real-time monitoring system for air quality, and deal with pollution more intuitively and scientifically.

Finally, this paper constructs a SDID model to evaluate the impact of the implementation of *Plan*. The differences between cities in Jiangsu-Zhejiang-Shanghai-Anhui area are selected. Cities outside these four regions can be selected as the control group in the future. The Yangtze River Delta region has successively introduced and implemented a number of policies to control environmental pollution and air pollution. How to use multiple DID models to separate the effects of these policies when data are available is also worthy of further study.

Acknowledgements Zhiqing Tian, Yun Kuai, Shunfeng Song, and Stephen M Marson also made great contributions to this manuscript. We express our heartfelt thanks to them. This research also was supported by: National Natural Science Foundation of China (72474127) and National Social Science Fund Major Bidding Project of China (18ZDA052). Part of this chapter was published: Wu X, Tian Z, Kuai Y, Song S, Marson S. M. Study on spatial correlation of air pollution and control effect of development plan for the city cluster in the Yangtze River Delta. Socio-Economic Planning Sciences 2022;83:101213. https://doi.org/10.1016/j.seps.2021.101213. The publication has been conducted with copyright permission.

References

Borge, R., Lumbreras, J., & Vardoulakis, S. (2017). Analysis of long-range transport influences on urban PM10 using two-stage atmospheric trajectory clusters. *Atmospheric Environment, 41*(21), 4434–4450.

Casquet, M. Á., Sciullo, A., Wang, H., & Quiroga, S. (2025). Electrification and modal shift in decarbonizing transportation: Impact evaluation in Piedmont region. *Transportation Research Part D: Transport and Environment., 144*, 104757.

Chen, Q. (2013). *Advanced econometrics and Stata application* (pp. 578–582). Higher Education Press.

D'Amato, A., Mazzanti, M., Nicolli, F., & Zoli, M. (2018). Illegal waste disposal: enforcement actions and decentralized environmental policy. *Socio-Economic Planning Sciences, 64*(DEC.), 56–65.

Diao, M., Leonard, D., & Sing, T. F. (2017). Spatial-difference-in-differences models for impact of new mass rapid transit line on private housing values. *Regional Science and Urban Economics, 67*, 64–77.

Dubé, J., Legros, D., Thériault, M., & Des, R. F. (2014). A spatial Difference-in-Differences estimator to evaluate the effect of change in public mass transit systems on house prices. *Transportation Research Part B: Methodological, 64*, 24–40.

Fan, Q. (2018). Has the establishment of free trade port promoted the trade development of territorial provinces?—Based on the analysis of dual vehicle space measurement model. *Journal of customs and trade., 39*, 112–124.

Fan, Q., & Shi, M. J. (2018). Selection method of endogenous spatiotemporal weight matrix based on structure matching and effective correlation. *The Journal of quantitative economics., 9*(2), 115–134.

Gong, P. P. (2016). *Study on the influencing factors of air quality based on spatial statistical method—Taking Beijing as an example*. Capital University of Economics and Business.

Han, M. (2018). Comparison and one-stop shopping after big-box retail entry: A spatial difference-in-difference analysis. *Journal of Retailing and Consumer Services, 40*, 175–187.

Heckert, M. (2015). A Spatial Difference-in-Differences Approach to Studying the Effect of Greening Vacant Land on Property Values. *City, 17*(1), 51–60.

Heckert, M., & Mennis, J. (2012). The Economic Impact of Greening Urban Vacant Land: A Spatial Difference-In-Differences Analysis. *Environment and Planning A, 44*(12), 3010–3027.

Hu, C. Y., Huang, K., Fang, Y., Yang, X. J., Ding, K., Jiang, W., Hua, X. G., Huang, D. Y., Jiang, Z. X., & Zhang, X. J. (2020). Maternal air pollution exposure and congenital heart defects in offspring: A systematic review and meta-analysis. *Chemosphere, 253*, 126668.

Jiang, W., Jiang, N., & Wang, K.-L. (2025). The local effects and neighborhood effects of high-speed railway on urban entrepreneurial vitality: Evidence from China. *Socio-Economic Planning Sciences, 98*, 102172.

Li, K., Daniel, J. J., Liao, H., & Zhu, J. (2019). A two-pollutant strategy for improving ozone and particulate air quality in China. *Nature Geoscience, 12*(11), 906–911.

Li, Z., & An, J. (2025). The role of China's national logistics hubs in promoting green logistics efficiency: spillover effects and endogenous heterogeneity. *Applied Economics., 58*(17), 3342–3357.

National Development and Reform Commission, Development planning of urban agglomeration in Yangtze River Delta. 2016. https://www.ndrc.gov.cn/xxgk/zcfb/ghwb/201606/t20160603_962187.html.

Niu, Y. F. (2019). *Reserch on the Spatial Agglomeration and Regional Differences of Air Pollution in China*. China University of Geosciences.

Portnov, B. A., Dubnov, J., & Barchana, M. (2009). Studying the association between air pollution and lung cancer incidence in a large metropolitan area using a kernel density function. *Socio-Economic Planning Sciences, 43*(3), 141–150.

Sunak, Y., & Madlener, R. (2015). The impact of wind farm visibility on property values: A spatial difference-in-differences analysis. *Energy Economics, 55*, 79–91.

Tang, Z. K. (2019). Analysis on PM2.5 regional transport and budget of a winter heavy haze pollution event in Nanjing. *Act Science Circumstantiae., 38*(12), 4606–4611.

Tao, T. H. (2018). *Temporal and Spatial Patterns and Influencing Factors of Air Quality in urban Agglomeration of the Yangtze River Delta*. Jiangxi University of Science And Technology.

Wang, F. (2018). *SDID-Based Spatial-temporal Effects of Rail Transit on Surrounding Housing Prices*. Zhejiang University.

Wang, S. J., Hua, G., & Zhou, H. (2020). What Are the Key Factors Affecting Air Pollution? Research on Jiangsu, China from the Perspective of Spatial Differentiation. *Sustainability., 12*(6), 1–17.

Wu, X., Chen, Y., Zhao, P., Guo, J., & Ma, Z. X. (2019). Study of haze emission efficiency based on new coopetition data envelopment analysis. *Expert Systems, 3*, 1–21.

Xiao, Z. M., Bi, X. H., & Feng, Y. C. (2012). Source Apportionment of Ambient PMand PMin Urban Area of Ningbo City. *Study Methodology of Environmental Science, 25*(5), 549–555.

Xie, W. (2015). On the legislation of transregional air pollution transmission control in China. *Social Scientist, 8*, 98–102.

Xu, Y. W. (2019). *Analysis of Pollution Synoptic Patterns and Impacts of boundary layer characteristics on the distribution of PM2.5 over the Yangtze River Delta in winter*. Nanjing University of Information Science and Technology.

Yan, W., Chen, Y., & Wang, Y. (2025). Efficiency improvement effect of clean energy transformation —A quasi-natural experiment based on China's clean heating policy. *Energy, 334*, 137798.

Yu, L., & Zhao, J. M. (2019). Disentangling treatment effect, spatial spillover effect and quantile effect: An empirical study of Beijing-Shanghai High Speed Rail. *World Regional Studies, 28*(1), 47–57.

Zhang, J., Yan, D. S., Feng, Z. X., & Li, C. (2019). *Study on the Economic Growth Effect of Free Trade Zones-A Dynamic Analysis Based on Double Difference Space Autoregressive Model., 36*(4), 71–77.

Zhang, S. H. (2017). *The study on regional air pollution linkage control under consideration of intercity transmission*. Shanghai University.

Zheng, L., Long, F., Chang, Z., & Ye, J. (2019). Ghost town or city of hope? the spatial spillover effects of high-speed railway stations in China. *Transport Policy, 81*(SEP.), 230–241.

Zhou, Y., Zhang, Q., Xu, G., & Wang, Y. (2025). Regional S&T planning and green generic technologies: A spatial differences-in-difference analysis. *Energy Economics., 150*, 108815.

Chapter 6
A Multi-Scale Periodic Study of $PM_{2.5}$ Concentration in the Yangtze River Delta of China Based on Empirical Mode Decomposition-Wavelet Analysis

6.1 Introduction

In recent years, $PM_{2.5}$ pollution has become a critical challenge faced by all countries in the world in the process of industrial development. In China, $PM_{2.5}$ pollution is particularly severe in the economically developed Beijing-Tianjin-Hebei region, the Yangtze River Delta and the Pearl River Delta. $PM_{2.5}$ contains a variety of toxic and harmful substances, which greatly impact the respiratory health of the public and directly harm people's physical health (Bao & Fang, 2007; Li et al., 2007; Wang & Wang, 2014). Moreover, $PM_{2.5}$ has a significant impact on atmospheric visibility (Malm et al., 1994; Shen et al., 2005) and climate change (Hansen et al., 2005). The scope of influence of $PM_{2.5}$ on the environment and the harm to human health is more severe than that of PM_{10} and PM_{100}. The United States, the European Union and other developed countries have listed $PM_{2.5}$ as a typical air pollutant (Yang et al., 2012). The factors leading to and influence the concentration of $PM_{2.5}$ in the atmosphere are very complex. Studies have proved that the variation of $PM_{2.5}$ concentration is closely related to the law of human production activities and local meteorological changes (Liang et al., 2007), and these factors tend to be cyclical. For example, the seasonal characteristics of the concentration of the main chemical components of $PM_{2.5}$ in autumn and winter are significantly higher than those in spring and summer (Zhang et al., 2013), and the time series data of $PM_{2.5}$ should also have an implicit periodicity.

To control PM, one of the feasible ways is to find the periodic law of PM time series, which will help government management departments to implement targeted measures before the peak of PM, such as shutting down key pollution sources and reminding the public to deal with the negative impact of PM. For example, Some of China's northern cities, such as Beijing and Xi 'an, have implemented "coal-to-gas" projects, which wanted to avoid the peak PM emission in winter and spring caused by the burning of coal for heating. In 2015, Shanghai implemented a "temporary air quality control scheme in winter" to avoid the peak of PM emissions in winter.

X. Wu, *Environmental Disaster in China*, https://doi.org/10.1007/978-981-92-0910-1_6

However, few researchers currently use common methods in the information and communication technologies (ICT) field to study the PM cycle. This article combines classical methods in the ICT field, such as EMD and WA, to study the periodic characteristics of PM from different scales, which can master the periodic operation characteristics of $PM_{2.5}$, controlling $PM_{2.5}$ concentration, and introducing targeted governance measures to optimize air quality.

The decomposition method of EMD (Empirical Mode Decomposition, hereinafter referred to as EMD) does not need to be given a basis function in advance. It carries out adaptive decomposition based on the data sequence, and the results can better reflect the fluctuation characteristics of the original sequence in different time scales. This paper uses the empirical mode decomposition and wavelet analysis to construct the EMD-WA (Empirical Mode Decomposition-Wavelet Analysis, hereinafter referred to as the EMD-WA) model, and multi-scale periodicity of $PM_{2.5}$ in the Yangtze River Delta is studied with the method of measurement after decomposition. The innovation is that: (1) Previous studies on the periodicity of $PM_{2.5}$ often used filtering methods directly. In wavelet analysis, it is necessary to manually set parameters and give basis functions, so it is difficult to avoid subjectivity. The EMD decomposition method applied in this paper does not need to set parameters and a given basis function, but it is an adaptive decomposition process based on the data. The results of decomposition can well reflect the local time characteristics of the original data. (2) When selecting the period length of the Morlet wavelet analysis sequence, the Morlet wavelet analysis has the characteristics of the period study in Fourier analysis, accurately reflecting the original sequence's multi-time-scale periodic fluctuation.

The rest is as follows. The second part is a literature review; The third part introduces the principle and characteristics of EMD decomposition and wavelet analysis. The fourth part constructs EMD-WA model and explains the specific steps. The fifth part is the empirical analysis of EMD-WA model and cycle measurement of each city. The last part is the conclusion and the research prospect.

6.2 Literature Review

Scholars have conducted extensive studies in the fields of EMD decomposition, wavelet analysis and period research of $PM_{2.5}$, but there are relatively few studies on combining EMD decomposition and wavelet analysis to analyze the multi-scale period of $PM_{2.5}$. The following part will summarize the study of $PM_{2.5}$ period and the quantitative analysis method of time series period.

6.2.1 Periodic Study of $PM_{2.5}$

At present, there are in-depth studies on the time series of $PM_{2.5}$ concentration. For example, Zhao (2008) studied the seasonal and daily variation characteristics of $PM_{2.5}$ concentration in urban areas and suburbs of Beijing, and Li et al. (2011) pointed out the variation characteristics of $PM_{2.5}$ concentration in urban areas of Beijing, etc. Xia et al. (2024) found evident periodic oscillation patterns in $PM_{2.5}$ concentration in Hefei, with a greater number of decreasing variables compared to rising variables in the sequence, and the change points exhibit a distinct "phased" characteristic. Studies in the Yangtze River Delta region mainly analyze the correlation between $PM_{2.5}$ and meteorological factors. For example, Gu et al. (2015) studied the relationship between the variation characteristics of $PM_{2.5}$ concentration and meteorological conditions in Chongming island of Shanghai in the past three years, and pointed out that the seasonal characteristics of $PM_{2.5}$ concentration are obvious. Liu et al. (2025) collected hourly $PM_{2.5}$ concentration data from 168 monitoring stations across the Yangtze River Delta and North China Plain between 2014 and 2024, identifying periodic behavioural characteristics of $PM_{2.5}$. However, in terms of research methods, conventional statistical analysis is used. For example, Mao et al. (2017) studied the spatial and temporal pattern of $PM_{2.5}$ in cities in the Yangtze River Delta region in 2015 and its influencing factors, and pointed out that the seasonal changes of $PM_{2.5}$ concentration generally showed a high seasonal cycle in spring and winter and low in summer and autumn. By time series models, Cekim (2020) forecasted PM_{10} concentrations in most polluted cities in Turkey. Used Generalized Additive Models, Fiffer et al. (2020) assessed the Long-term impact of $PM_{2.5}$ mass and sulfur reductions on ultrafine particle trends in Boston from 1999 to 2018. Zhou et al. (2018) analyzed the periodic changes of $PM_{2.5}$ in Taiyuan City from January 1, 2014 to May 31, 2016. The results show that there are significant cycles of 2–8 d and 10–16 d in the $PM_{2.5}$ concentration change in Taiyuan. Li et al. (2025) constructed daily $PM_{2.5}$ transport flux data for the central and eastern regions of China during the winter seasons from 2015 to 2019. By extracting the dynamic spatial distribution of $PM_{2.5}$ transport in this area, they identified quasi-annual oscillations in regional $PM_{2.5}$ transport and characterised their spatiotemporal variations across central and eastern China. However, because each research city is different and the research period is also different, the period of each time series is not exactly the same. But on the whole, from the seasonal trend, the PM cycle is more significantly affected by the season, and there are generally high data in winter and spring, and low data in summer and autumn. Looking at the micro data again, the "Sunday" effect is widespread, such as the cycle being about 7–8 days. This shows that people emit more PM in their production and life, and that the reduction in production and life on weekends will also reduce PM. This paper finds that the PM time series in the Yangtze River Delta has more abundant features from other scales in addition to the above "seasonality" and "Sunday" characteristics.

For the systematic study of variation period of $PM_{2.5}$ concentration, the method of wavelet analysis is mostly applied directly. For example, Li et al. (2017) used Morlet continuous wavelet to study the period of $PM_{2.5}$ at different time scales in the urban area of Beijing, and concluded that the primary period was about 24 h, 8d and 14d. Wang et al. (2017) used Morlet continuous wavelet to study and found that the change periods of the PM_{10} and $PM_{2.5}$ time series are mainly short periods of 4–8 days. From the perspective of the authors, the current literatures on the periodic research of PM time series are not comprehensive enough. This article innovatively combines the commonly used methods in the ICT field, such as EMD and WA technology, to study the characteristics of PM in terms of season, short cycle, medium cycle and long cycle, and try to find the law, not only enriching existing research, but also providing countermeasures and suggestions for PM governance.

6.2.2 *Quantitative Analysis Method of Periodicity of Time Series*

In the past, the main idea of analyzing the periodicity of time series was to process the original sequence, extract the sub-sequence containing periodic fluctuations, usually called the period term, and then calculate the period length of the sub-sequence. This process of extracting periodic items is mainly realized by filtering or decomposition. Pedregal (2003) designed the HP (hodrick-prescott) filter model and analyzed the trend part obtained by the Gross National Product (GNP) serial low-pass filter and the periodic part obtained by the band-pass filter. Mohr (2006) used a transconductin-capacitor (TC) filter to analyze the cycles and trends of gross domestic product (GDP) in various European countries, and verified that the TC filter was more convenient than the HP filter through empirical analysis. He and Song (2013) used HP filtering and BP (band-pass) filtering to study the periodic variation of the marine economy. Cendejas et al. (2017) concluded that after economic globalization, all economies have a common business economic cycle based on the 4–7 year cycle determined by HP filter model. Carvalho et al. (2012) decomposed GDP of U.S. by singular spectrum analysis. Sella et al. (2010) verified the synchronicity of economic cycle based on singular spectrum analysis (SSA for short), Multi Taper Method (MTM for short) and Maximum Entropy Method (MEM for short). The study of Yogo (2008) showed that wavelet decomposition could decompose GDP time series into three parts: white noise, period and trend. Most of the above studies extract the period term from the original time series through the methods of sequence noise reduction, trend term filtering or sequence decomposition, but these methods also have their own limitations. For example, the filtering method is often subjective in the selection of parameters, resulting in a large gap in the cycle measurement results of the same time series; a basis function is needed in the decomposition method, and it is easy to appear false fluctuation scale in the decomposition process.

The research on the calculation of the length of the fluctuation cycle has evolved several times, from the initial analysis of variance to the spectrum analysis (Kuo & Lee, 1988), and then into the field of wavelet analysis (Cui, 1995). It is a development process from time domain to frequency domain. Baubeau and Cazelles (2009) used continuous wavelet transform analysis. Addo et al. (2014) proved that the wavelet analysis can analyze hidden period fluctuations of time series more effectively. Berdiev and Chang (2015) adopted wavelet analysis to discuss the synchronicity of economic development cycles in China, Japan, the United States and other Asia–pacific countries. Marczak and Gomez (2015) extracted the original sequence period term through wavelet analysis, and studied the periodicity of behavior of producers and wages of consumers. The above method for measuring the period length indicates that wavelet analysis solves the problem of low time–frequency resolution of Fourier transform spectrum method to a certain extent, and can effectively measure the time scale of the period term.

The Empirical mode decomposition (EMD) method was proposed by Huang et al. (1998, 1999). This method has advantages in dealing with nonlinear and non-stationary time series. Flandrin et al. (2004) proved that the empirical mode decomposition method could be used as a method similar to wavelet decomposition, which is an adaptive high-pass filtering process and suitable for the analysis of nonlinear time series. It can be seen that from the EMD decomposition principle, its filtering process is based on the characteristics of the analyzed time series instead of artificial function. The decomposed IMF (Intrinsic Mode Function) can represent fluctuations on different time scales, so that each IMF component can characterize the original sequence more accurately and comprehensively. Based on the above advantages, the empirical mode decomposition method is popular in the fields of time series prediction (Li et al., 2012; Wang et al., 2010; Yu et al., 2008; Zhang et al., 2008) and multi-scale periodic analysis (Wang et al., 2020; Wu et al., 2018; Zhang, 2014).

In recent years, rapid developments and applications of modern ICT have created more and more opportunities for air pollution control (Wu et al., 2018). For example, online monitoring instruments can more accurately monitor and predict the emission of air pollutants. The modern information analysis technologies used in this article, such as EMD and WA technologies, are commonly used to reduce noise and find rules for data information. This article combines the two and analyzes the periodic characteristics of PM, which is the application of ICT technology in environmental governance and has a good typical demonstration significance.

6.3 Research Methods

6.3.1 *Empirical Mode Decomposition*

The core objective of empirical mode decomposition is the process of decomposing a nonlinear non-stationary sequence into a finite number of IMF components and a trend term. The specific steps of EMD decomposition are as follows:

(1) Search for all local extremum (maximum and minimum) points of the original data sequence.
(2) Choose a curve interpolation method (cubic spline interpolation method is applied in this paper), connect all local maxima to get the upper envelope $x_{\max}(t)$, and connect the local minima to get the lower envelope $x_{\min}(t)$. It can be known

that the upper and lower envelope should cover all local maxima and minima of $x(t)$, namely

$$x_{\min}(t) \le x(t) \le x_{\max}(t) \tag{6.1}$$

where, t is time. In the part of "5. Empirical analysis", t is the day data.

(3) The upper and lower envelope $x_{\max}(t)$, $x_{\min}(t)$ and mean value $m_{11}(t)$ can be calculated, namely

$$m_{11}(t) = \frac{x_{\max}(t) + x_{\min}(t)}{2} \tag{6.2}$$

(4) Then subtract $m_{11}(t)$ from sequence $x(t)$, and the new sequence $h_{11}(t)$ is obtained, namely

$$x(t) - m_{11}(t) = h_{11}(t) \tag{6.3}$$

(5) Finally, check whether the new sequence $h_{11}(t)$ meets the two conditions in the definition of the eigenmode function. If satisfied, $h_{11}(t)$ is selected as the first IMF component $c_1(t)$ of $x(t)$; If not, the new sequence $h_{11}(t)$ should be used as the original data sequence $x(t)$ to repeat steps (1) to step (4) until the envelope mean of the new sequence approaches zero. Repeat the operation for k times to get the first IMF component of $x(t)$, namely

$$c_1(t) = h_{1k}(t) = h_{1(k-1)}(t) - m_{1k}(t) = x(t) - \sum_{i=1}^{k} m_{1k}(t) \tag{6.4}$$

Taking the difference between the original data sequence $x(t)$ and the first IMF component $c_2(t)$ as the original signal, repeat steps (1) to (5) to get the remaining IMF components $c_2(t)$, $c_3(t)$. .. $c_n(t)$. After repeating for n times, the last margin $r_n(t)$ is obtained, which represents the trend information of the original data sequence $x(t)$, in which there is only one extreme point. $r_n(t)$ is called the trend term of $x(t)$.

$$r_1(t) = x(t) - c_1(t) \tag{6.5}$$

$$c_n(t) = r_{n-1}(t) - r_n(t) \tag{6.6}$$

In steps (1) to (5), the process of decomposition of IMF from the original data sequence $x(t)$ is called the "filtering" process of EMD method.

The condition for EMD algorithm to stop filtering is determined by the difference between two consecutive filtering results, Standard deviation (SD), and the threshold value of SD is set as the judging basis for stopping the filtering process. The formula of SD is as follows

$$SD = \sum_{t=0}^{T} \frac{\left|h_{1(k-1)}(t) - h_{1k}(t)\right|^2}{h_{1(k-1)}^2(t)} \tag{6.7}$$

The selection of SD threshold value is subjective to some extent, and the selection is mostly determined by experience. When the magnitude of the original sequence data is 10, the value of SD is usually [0.2, 0.3] (Zheng, 2010). The length of the data sequence in the study is about 2000, so the value of SD is 0.003.

From the above empirical mode decomposition principle, it can be known that the decomposed IMF can fully and clearly reflect the different time scales of the original sequence, so the study of the periodic characteristics of each IMF component can clearly understand the periodic characteristics of the original sequence.

6.3.2 *Wavelet Analysis*

In the analysis of time series, we usually start from the two directions of time domain and frequency domain. The time domain exists objectively in reality and develops linearly according to time, while the frequency domain is a description of frequency, which can be obtained by the time domain through the Fourier transform. Common time domain analysis and frequency domain analysis are usually suitable for stationary time series. However, the fluctuation of $PM_{2.5}$ concentration has a multi-time-scale structure. In the early 1980s, the wavelet analysis with time–frequency multi-resolution function proposed by Jean Morlet provided a solution to the problem of multi-time-scale periodic change hidden in time series.

6.3.2.1 Wavelet Function

The definition of the wavelet function is as follows: suppose $\psi(t)$ is a square integrable function in the real number domain, that is $\psi(t) \in L^2(R)$; if its Fourier transform $\hat{\psi}(t)$ satisfies formula (6.8), $\psi(t)$ is called a basic wavelet function

$$\int_R \frac{|\hat{\psi}(t)|^2}{\omega} d\omega < 0 \tag{6.8}$$

It can be seen that the value of $\hat{\psi}(t)$ is zero when $\omega = 0$, so it means that the integral of the wavelet function in its time domain is also zero, namely:

$$\int_{-\infty}^{+\infty} \psi(t) d_t = 0 \tag{6.9}$$

The volatility shown in formula (6.9), combined with the characteristic of rapid attenuation to zero in the definition of wavelet function, indicates that the wavelet function is convergent.

$\psi(t)$ can form a family of continuous wavelet functions through translation and scaling, also known as subwavelet, denoted as $\psi_{ab}(t)$, and expressed as follows:

$$\psi_{ab}(t) = |a|^{\frac{1}{2}}\psi(at - b) \tag{6.10}$$

In the formula, a is the scale scaling factor, and the change of a can reflect the period length of different time scales; b is the translation factor, and represents the translation in time.

6.3.2.2 Wavelet Transform

The definition of wavelet transform is as follows:

If there are a sequence $x(t)$ and a wavelet $\psi_{ab}(t)$, then

$$W_x(a, b) = \int_{-\infty}^{\infty} x(t)\overline{\psi}_{ab}(t)\mathrm{d}t = |a|^{\frac{1}{2}} \int_{-\infty}^{\infty} x(t)\overline{\psi}(at - b)\mathrm{d}t \tag{6.11}$$

is the continuous wavelet transform of the sequence $x(t)$, $W_x(a, b)$ is the wavelet coefficient, $\overline{\psi}_{ab}(t)$ is the complex conjugate function of $\psi_{ab}(t)$.

From the above definition of wavelet transform, it can be known that wavelet analysis is an integral transformation that maps a one-dimensional linear time series to a two-dimensional time-scale plane, which is conducive to extracting the hidden multi-time-scale information.

The modulus $\|\psi_{ab}(t)\|_0^2$ of $\psi(t)$ calculated according to formula (6.10) can be obtained as follows:

$$\begin{aligned}\|\psi_{ab}(t)\|_0^2 &= \int_R |\psi_{ab}(t)|^2\mathrm{d}t = \int_R \left|a^{1/2}\psi(at - b)\right|^2\mathrm{d}t \\ &= \int_R |\psi(at - b)|^2\mathrm{d}(at - b) \\ &= \int_R |\psi(t)|^2\mathrm{d}t = \|\psi(t)\|_0^2\end{aligned} \tag{6.12}$$

It indicates that after translation and scaling, the modulus of $\psi(t)$ is unchanged after being transformed into Eq. (6.10). Therefore, the multi-time scale of the original sequence can be grasped by scaling and changing the scale parameters a to obtain the information of high and low frequencies in the sequence.

At the same time, under different values of a, continuous wavelet transform has the adaptive ability of time-window and frequency-window, that is, the wavelet function has the adaptive short-term analysis effect on both high and low frequency sequences. The specific analysis is as follows:

$$\begin{aligned} \hat{\psi}_{ab}(\omega) &= \int_R a^{1/2}\psi(at-b)\mathrm{e}^{\mathrm{i}at}\mathrm{d}t \\ &= a^{-1/2}\int_R \psi(at-b)\mathrm{e}^{-i\frac{\omega}{a}(at)-b}\mathrm{e}^{-i\omega\frac{b}{a}}\mathrm{d}(at-b) \\ &= a^{-1/2}\mathrm{e}^{-i\omega\frac{b}{a}}\hat{\psi}_{ab}\left(\frac{\omega}{a}\right) \end{aligned} \tag{6.13}$$

From formula (6.10) and formula (6.12), it can be seen that, when $a < 1$, from the perspective of time domain, the time window is widened and the fluctuation amplitude is compressed after transforming $\psi(t)$ to $\psi_{ab}(t)$. From the perspective of frequency domain, the frequency window is narrower than before, and the time–frequency window is flatter than before after transforming $\psi(t)$ to $\psi_{ab}(t)$, so it can adapt to the localization requirements of the lower frequency sequence. When $a > 1$, from the perspective of time domain, after transforming $\psi(t)$ to $\psi_{ab}(t)$, the time window is compressed and the fluctuation amplitude is enlarged; From the perspective of frequency domain, after transforming $\psi(t)$ to $\psi_{ab}(t)$, the frequency window is wider than before, and the time–frequency window is taller and thinner than before, so it can adapt to the localization requirements of higher frequency sequences.

6.3.2.3 Wavelet Variance

The wavelet coefficient can demonstrate the variation characteristics of the original sequence at a specific time scale. Different wavelet coefficients correspond to different time scales, and the larger the absolute value of the wavelet coefficient is, the more significant the time scale is. Therefore, the square of all wavelet coefficients related to a in the time domain is integrated in the b domain to obtain the wavelet variance:

$$\mathrm{Var}(a) = \int_{-\infty}^{\infty} |W_x(a,b)|^2 \mathrm{d}b \tag{6.14}$$

Since a is a scaling factor, its change can reflect the periodicity of the original sequence at different time scales, and $\mathrm{Var}(a)$ can reflect the strength of the fluctuation period at this time scale. Therefore, the main time scale, namely the primary period, can be determined through the peak of its wavelet variance. In order to avoid the multiple local peaks of wavelet variance when determining the period with the peak of wavelet variance, the primary period is usually analyzed by combining the wavelet variance graph.

6.4 Model Design and Data Description

6.4.1 Model Design

Firstly, the empirical mode decomposition method is used to decompose the variation sequences of $PM_{2.5}$ concentration of 41 cities, and then the decomposed IMF is reconstructed in groups. Finally, the wavelet analysis method is used to calculate the reconstructed IMF period as the period of the original sequence.

In terms of grouping reconstruction, although the empirical mode decomposition method can be widely used to analyze multi-scale time series, it has limitations. Firstly, in the process of applying the empirical mode decomposition method, spurious modes are easy to appear, that is, the waveform superposed by the real mode and error, and periodic calculation of the spurious mode will obtain the false cycle. Secondly, generally speaking, the main fluctuation information of the original sequence is often concentrated in a few decomposed IMF. If each of the decomposed IMF is subjected to period measurement, it cannot accurately reflect the multiscale period features implied in the original sequence. In view of the above limitations, the orthogonality of empirical mode decomposition is used to reconstruct the IMF groups.

To sum up, the EMD-WA model applied in this paper can be summarized into four steps: decomposition, reconstruction, measurement and analysis. In the decomposition step of the model, the empirical mode decomposition method enables the decomposed sequence to cover all the local time domain features of the original sequence, which effectively overcomes the defect of ignoring some local time domain features of the sequence due to the subjective selection of specific wavelet basis functions in the study of periodicity of time series by wavelet analysis.

The steps of decomposition, reconstruction, measurement and analysis of EMD-WA model are as follows:

(1) Decomposition: the $PM_{2.5}$ concentration data sequence $x(t)$ is decomposed by EMD. The specific steps of EMD decomposition are shown in steps (1) to (6) in 2.1. to obtain the IMF sequence $\{c(t)\}$ and trend item sequence $\{r(t)\}$.

(2) Reconstruction: the IMF decomposed by the $PM_{2.5}$ concentration sequence of each city is applied to wavelet analysis for periodic analysis. The morlet wavelet is selected as the basis function in the form of $\psi(t) = e^{kt}e^{\frac{-t^2}{2}}$. After calculating the wavelet coefficient $W_x(a, b)$, the primary period of each IMF component in each city is calculated with the wavelet variance $\text{Var}(a)$ in formula (6.12). Then, clustering analysis is conducted according to the main cycle length of each IMF, and clustering results are used as the basis for grouping. According to the orthogonality of the EMD method, the same group of the IMF is added as the reconstruction result, so as to obtain more accurate and comprehensive multi-scale periodic result of the sequence.

(3) Measurement: wavelet analysis method is used to calculate the cycle length of IMF series in various cities after reconstruction.

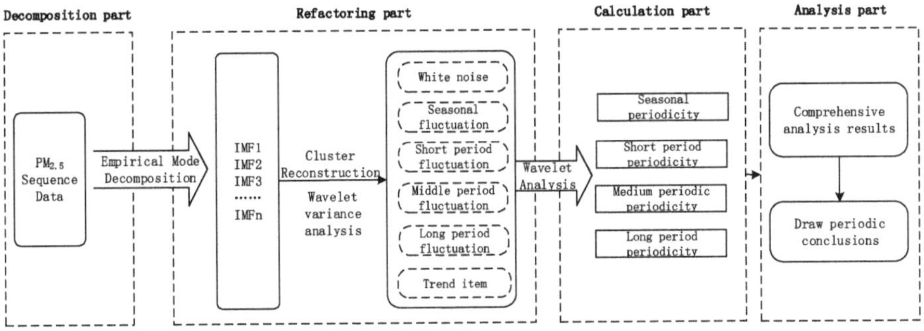

Fig. 6.1 Cycle decomposition and analysis process of $PM_{2.5}$

(4) Analysis: combined with geographical location, high concentration period and other information, the results of various cities are compared to comprehensively analyze the multi-scale period characteristics of $PM_{2.5}$ in the Yangtze River Delta.

The periodic decomposition and prediction analysis process of $PM_{2.5}$ are shown in Fig. 6.1.

6.4.2 Data Description

In October 2019, when the *19th Yangtze River Delta urban economic coordination meeting* was held, 7 cities including Huangshan, Bengbu, Lu'an, Huaibei, Suzhou, Bozhou, and Fuyang in Anhui were recruited as new members of the Economic Coordination Committee of the Yangtze River Delta, and 41 prefecture-level cities in Shanghai, Jiangsu, Zhejiang, and Anhui all became new members in the Yangtze River Delta. Therefore, the Yangtze River Delta includes: Shanghai, Nanjing, Wuxi, Xuzhou, Changzhou, Suzhou, Nantong, Lianyungang, Huaian, Yangzhou, Zhenjiang, Taizhou, Suqian, Hangzhou, Ningbo, Wenzhou, Jiaxing, Huzhou, Shaoxing, Jinhua, Quzhou and Zhoushan, Taizhou, Lishui, Hefei, Wuhu, Huainan, Maanshan, Tongling and Anqing, Chuzhou, Xuancheng, Chizhou, Huangshan, Bengbu, Luan, Huaibei, Suzhou, Bozhou, Fuyang.

This paper collected 1,795 hourly $PM_{2.5}$ concentration data of 41 cities in the Yangtze River Delta from January 1, 2015, to November 30, 2019 from Website of https://www.aqistudy.cn/historydata/. In the study, the average concentration of $PM_{2.5}$ from 0 AM to 23 Afternoon in each city was taken as the $PM_{2.5}$ concentration of that day. $PM_{2.5}$ concentration data is measured as ug/m^3 and can be input the models directly.

6.5 Empirical Analysis

6.5.1 Periodic Analysis of Daily $PM_{2.5}$ Concentration in the Yangtze River Delta Region

The daily $PM_{2.5}$ concentration data of 41 prefecture-level cities in Shanghai, Jiangsu, Zhejiang and Anhui in the Yangtze River Delta from January 1, 2015 to November 30, 2019, and the EMD-WA model constructed above are used to conduct periodic analysis of daily $PM_{2.5}$ concentration data of various cities. In order to show the analysis process in detail while avoiding repeated descriptions, this section takes the $PM_{2.5}$ daily concentration data in Shanghai as an example to show the empirical process in detail, and summarizes the results of other cities (the conclusions please see Part of Sect. 6.1) (Fig. 6.2).

First, decompose step (1) of the daily $PM_{2.5}$ concentration in Shanghai processed by the EMD-WA model, and steps (1) to (6) of the empirical mode decomposition method are specifically used. The fluctuations or trends of different scales in the Shanghai's daily $PM_{2.5}$ concentration data sequence $x(t)$ are decomposed one by one to obtain the new series containing a trend term $r_n(t)$, namely formula (6.5), and the intrinsic modal function $c_n(t)$, which is formula (6.6).

Figure 6.3 shows the obtained IMF after the daily $PM_{2.5}$ concentration sequence $x(t)$ in Shanghai was decomposed by EMD. Among them, a total of 11 IMF sequences and 1 trend item sequence were decomposed. For the convenience of description, the 12 sequences were denoted as $c(i)$, $i = 1, 2, \ldots, 12$. As can be seen from Fig. 6.3, the frequency of each IMF sequence is arranged from high to low. Therefore, compared with the earlier IMF sequence $c(i-n)$, $i > n$, the later IMF sequence $c(i)$ has the characteristics of slower fluctuation and longer cycle. Until it is decomposed to the last sequence, it can be called the residual component or the trend term, which is characterized by the extreme slowdown of fluctuations and the absence of periodicity. It can represent the long-term trend of the daily concentration of $PM_{2.5}$ in Shanghai.

After completing the decomposition of EMD-WA, the IMF sequences $c(i)$ decomposed from the daily concentration sequences $x(t)$ of $PM_{2.5}$ in Shanghai were reconstructed to eliminate the false modes in the eigenmode function, so as to make the

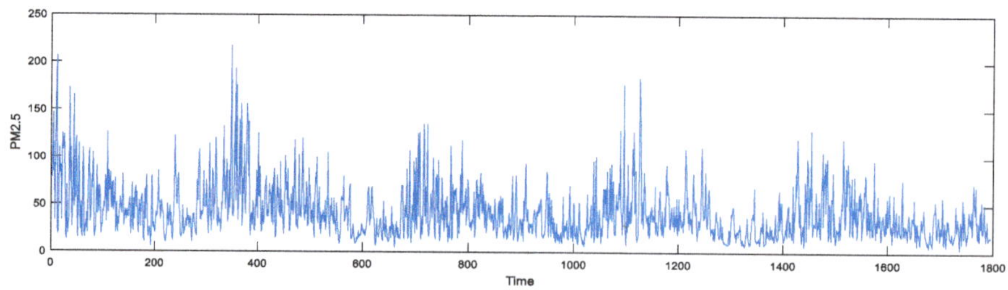

Fig. 6.2 Daily concentration sequence of $PM_{2.5}$ in Shanghai

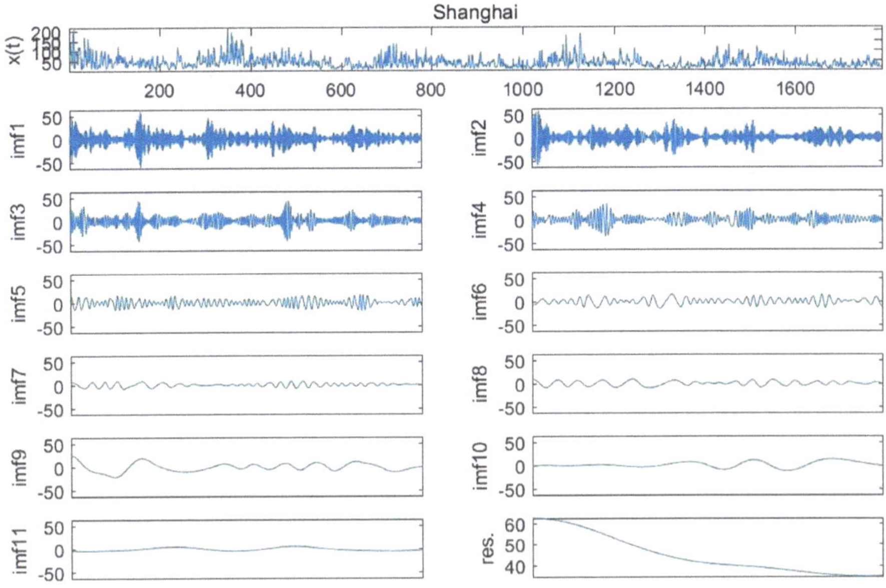

Fig. 6.3 $PM_{2.5}$ EMD decomposition results of concentration series in Shanghai

cycle more accurate. Therefore, the morlet wavelet in form of $\psi(t) = e^{kt}e^{\frac{-t^2}{2}}$ is used as the basic wavelet function, and wavelet analysis method is used to obtain the wavelet coefficient $W_x(a, b)$ according to formula (6.11), and then calculate the wavelet variance $\mathrm{Var}(a)$ according to formula (6.14). When determining the primary period with the wavelet variance, the time when the maximum value appears is selected as the primary period. In practice, it is often analyzed by combining the wavelet variance graph to avoid the problem of partial and unobjective selection of the primary period caused by maxima in multiple positions. After the wavelet analysis of IMF sequences $c(i)$ decomposed by step (1) of the EMD-WA model, the wavelet variance graph obtained is shown in Fig. 6.4.

It can be seen intuitively from Fig. 6.4, the wavelet variance map corresponding to $c(1), \cdots, c(5)$ is not smooth with violent, dense and no obvious trend of fluctuations. The wavelet variance can reflect the strength of the fluctuation cycle under the time scale, for $c(1), \cdots, c(5)$ without obvious maximum value, it can be identified that the first five IMFs $c(1), \cdots, c(5)$ decomposed by $x(t)$ through EMD method have the random fluctuation characteristic of white noise, so there is no obvious primary period.

In this paper, after morlet wavelet analysis, one or more IMF sequences that are relatively "front" (that is, relatively high frequency) will be converted into the white noise part of the original sequence if the corresponding wavelet variance has a sharp fluctuation and no obvious maximum value. The periodic characteristics will not be analyzed. The first five IMFs $c(1), \cdots, c(5)$ decomposed from the original sequence $x(t)$ are the white noise part of the daily $PM_{2.5}$ concentration in Shanghai.

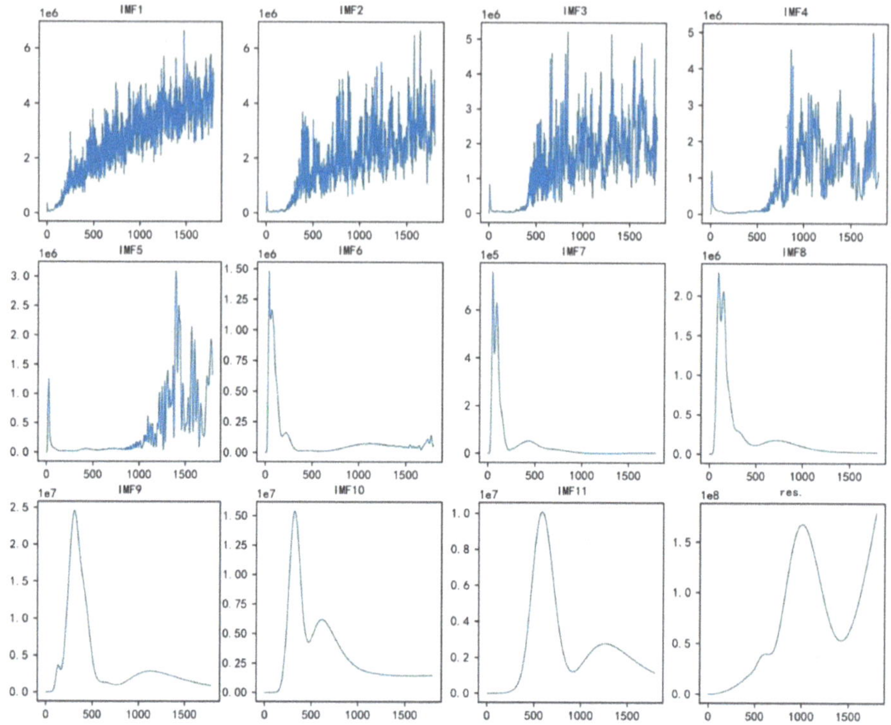

Fig. 6.4 IMF wavelet variance in Shanghai

Starting from $c(6)$, the wavelet variance graphs obtained by morlet wavelet analysis in subsequent sequences $c(i), i \geq 6$ all show obvious maximum values. However, there are still some unsmooth parts in the wavelet variance graphs of $c(6)$, $c(7)$, and $c(8)$, indicating that relatively high-frequency IMF sequences other than white noise may have the limitation of empirical mode decomposition (that is, spurious modes). The main cycle results of each IMF sequence are shown in the Table 6.1.

The reconstruction of the decomposed IMF sequence can eliminate the spurious mode generated by the EMD decomposition of the daily $PM_{2.5}$ concentration

Table 6.1 Main period of IMF series in Shanghai

Name	Cycle/Day	Name	Cycle/Day
$c(1)$	No main period	$c(7)$	48
$c(2)$	No main period	$c(8)$	91
$c(3)$	No main period	$c(9)$	305
$c(4)$	No main period	$c(10)$	318
$c(5)$	No main period	$c(11)$	576
$c(6)$	35	$c(12)$	1794

sequence $x(t)$ in Shanghai and centralize the similar cycle information hidden in each IMF. In combination with Yogo (2008)'s study on the periodicity of time series, the multi-scale periodicity of $PM_{2.5}$ was divided into four categories: seasonal cycle, short cycle, medium cycle and long cycle. Among them, the seasonal cycle refers to the repeatability of a trend line within a short cycle of time, so it is defined as the shortest period in this paper.

It can be seen from Table 6.1 that $c(6)$ and $c(7)$ are relatively close in Shanghai's results, with a period difference of about 13 days. Their shape of the wavelet variance graphs are also very similar; $c(8)$ is unique, and it is not close to the period before and after; the period of $c(9)$ and $c(10)$ are relatively close and the difference is also 13 days; $c(11)$ is a unique category with the longest period; $c(12)$ is trend item and shows a long-term trend, so the period length is almost equal to the sequence length. It can be seen from wavelet variance graph that, it also presents an obvious rising trend. It can not find longer cycle characteristics due to the length of data set in this paper, which is also a limitation of this paper.

The classification above is based on the analysis of the wavelet variance graph and the subjective judgment of the difference in spans between cycles. However, in order to avoid the reconstruction caused by too subjective classification, K-means clustering method is used to cluster the cycle length of $c(6)$ to $c(11)$ of IMF series except white noise and trend items, and the objective is to divide it into four categories for reconstruction. The purpose of clustering is to verify the subjective classification results and provide references for the reconstruction of IMF sequences. The clustering used in this paper is based on the principle of minimization of Euclidean distance. Finally, $c(6)$ and $c(7)$ are put into a category, $c(8)$ is put into a category, $c(9)$ and $c(10)$ are put into a category, and $c(11)$ is put into a category. The result of K-means clustering is consistent with that of subjective classification, verifying the subjective classification result combined with the analysis of the wavelet variance graph. Therefore, we have decomposed 12 IMF sequences from the daily $PM_{2.5}$ concentration in Shanghai, and divided them into 6 categories according to the characteristics of different cycle scales:

(1) The part with no obvious primary period of $c(1), \cdots, c(5)$ is defined as the white noise of the sequence $x(t)$, which is defined as the white noise item after reconstruction.
(2) $c(6)$ and $c(7)$ have the shortest period length measured by wavelet analysis. They are defined as the seasonal cycle of the sequence $x(t)$ after reconstruction.
(3) $c(8)$ is defined as the short cycle term of the sequence $x(t)$.
(4) After $c(9)$ and $c(10)$ are reconstructed, they are defined as the medium cycle of the sequence $x(t)$.
(5) $c(11)$ is defined as the long cycle item of the sequence $x(t)$.
(6) According to the decomposition principle of EMD, $c(12)$ shows the long-term trend of the sequence $x(t)$, so it is defined as the trend term of the sequence $x(t)$.

As shown in Fig. 6.5, the seasonal cycle term, short cycle term, medium cycle term and long cycle term formed by the reconstructed $c(6)$ to $c(11)$ can reflect the

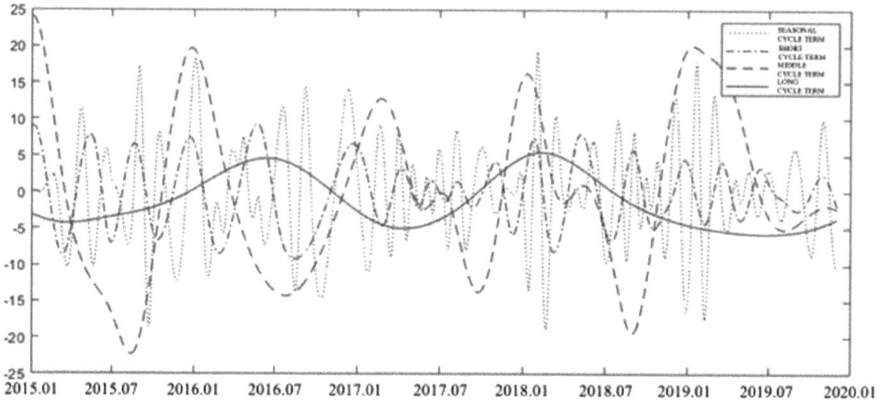

Fig. 6.5 The seasonal, short, medium and long-term terms of IMF reconstruction in Shanghai

multi-scale periodic fluctuation characteristics hidden in the original sequence $x(t)$ as scheduled from high to low in frequency. Therefore, it is reasonable to take the period length of these periodic terms as the multi-scale periodic characteristics of the original sequence $x(t)$.

After the decomposition and reconstruction of steps (1) and (2) of the EMD-WA model, four sequences with cycle characteristics of different time scales, namely, seasonal cycle term, short cycle term, medium cycle term and long cycle term, are obtained. The cycle measurement of step (3) of the EMD-WA model is carried out in the following part. The wavelet analysis is used to analyze the four sequences to obtain the wavelet coefficients $W_x(a, b)$ according to formula (6.11), and the wavelet variance $\mathrm{Var}(a)$ is calculated according to formula (6.14). The time scale corresponding to the maximum point is found by combining the wavelet variance graph, which is the primary period of the sequence. The wavelet variance graph is shown below.

According to Fig. 6.6 and wavelet analysis, the following conclusions can be drawn:

(1) The seasonal cycle of the daily $PM_{2.5}$ concentration sequence $x(t)$ in Shanghai is about 49 days.
(2) The short cycle of the daily $PM_{2.5}$ concentration sequence $x(t)$ in Shanghai is about 91 days.
(3) The medium cycle of the daily $PM_{2.5}$ concentration sequence $x(t)$ in Shanghai is about 305 days.
(4) The long cycle of the daily $PM_{2.5}$ concentration sequence $x(t)$ in Shanghai is about 575 days.

After the cycle measurement of step (3) of the EMD-WA model is completed, a comprehensive analysis is needed to reach a conclusion. Moreover, in order to verify the validity of the conclusion, Therefore, it is of more practical value to study the multi-scale periodicity of $PM_{2.5}$ concentration in the Yangtze River Delta.

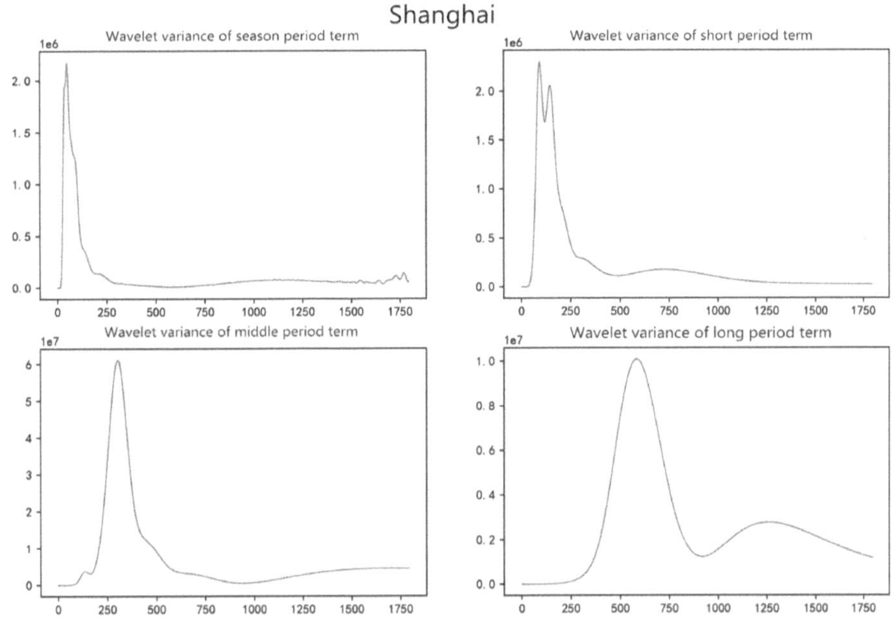

Fig. 6.6 Wavelet variance of seasonal periodic, short-term, mid-period and long-period terms of Shanghai $PM_{2.5}$ sequence

6.5.2 Comprehensive Analysis of Empirical Results Based on the EMD-WA Model

In the previous section, taking Shanghai as an example, the operation process of the EMD-WA model was elaborated to analyze the multi-scale periodicity of the daily $PM_{2.5}$ concentration in Shanghai. This paper aims to analyze the multi-scale periodicity of $PM_{2.5}$ in the Yangtze River Delta. In the previous studies on periodicity of $PM_{2.5}$, its periodicity of a city is only analyzed. This paper analyzes the periodicity of a region, because $PM_{2.5}$ exists objectively, and its concentration distribution is not discrete, but the values obtained by each monitoring station are discrete due to measurement errors, recording errors, missing data, etc. Since $PM_{2.5}$ in neighboring areas is correlated to a certain extent, this paper intends to eliminate the errors and one-sidedness of the results of the periodic study on $PM_{2.5}$ in a single city through the periodic results of neighboring areas. The results of 41 prefecture-level cities in the Yangtze River Delta through the EMD-WA model are shown below.

As shown in Table 6.2, the daily $PM_{2.5}$ concentration sequence of 41 prefecture-level cities in the Yangtze River Delta region was decomposed by the EMD-WA model, and morlet wavelet analysis was used to measure the primary period of the decomposed IMF sequence of each city.

According to Fig. 6.7, the number of IMF in each city ranges from 11 to 14, and the number of IMF in most cities is 12 or 13. To some extent, the number of IMF

Table 6.2 Main period of IMF sequence in 41 cities of three provinces and one city

IMF Cycle /Day City	1	2	3	4	5	6	7	8	9	10	11	12	13	14
Shanghai	*	*	*	*	*	35	48	91	305	318	576	1794	None	None
Nanjing	*	*	*	*	*	38	62	93	240	303	662	1794	None	None
Zhenjiang	*	*	*	*	*	31	50	65	238	303	814	1794	None	None
Yangzhou	*	*	*	*	*	29	50	149	303	325	586	1794	None	None
Taizhou	*	*	*	*	*	28	51	137	300	300	637	1003	None	None
Changzhou	*	*	*	*	*	28	50	70	253	297	671	1021	None	None
Wuxi	*	*	*	*	*	50	60	101	296	459	894	1794	None	None
Suzhou	*	*	*	*	*	53	57	93	238	305	1794	None	None	None
Nantong	*	*	*	*	*	36	78	156	305	319	1064	1794	None	None
Huaian	*	*	*	*	*	28	51	96	212	297	612	799	1794	None
Yancheng	*	*	*	*	*	33	43	53	169	307	1794	None	None	None
Lianyungang	*	*	*	*	*	59	38	50	106	287	320	456	1033	1794
Suqian	*	*	*	*	*	41	48	206	297	311	542	782	1794	None
Xuzhou	*	*	*	*	*	33	62	126	290	463	1404	1794	None	None
Hangzhou	*	*	*	*	*	32	60	95	155	296	492	1020	1794	None
Huzhou	*	*	*	*	*	49	69	101	310	448	566	1794	None	None
Jiaxing	*	*	*	*	*	39	107	262	318	488	1794	None	None	None
Shaoxing	*	*	*	*	*	36	70	104	300	419	960	1794	None	None
Jinhua	*	*	*	*	*	*	46	62	125	292	475	1794	None	None
Quzhou	*	*	*	*	*	36	57	70	111	290	637	1794	None	None
Lishui	*	*	*	*	*	38	91	134	294	945	1794	None	None	None
Wenzhou	*	*	*	*	*	36	51	106	280	309	927	1794	None	None
Taizhou	*	*	*	*	*	55	73	113	295	448	879	1794	None	None
Ningbo	*	*	*	*	*	37	68	95	137	305	320	501	766	1794
Zhoushan	*	*	*	*	*	35	58	101	287	301	431	851	1794	None
Hefei	*	*	*	*	*	41	65	160	292	419	1794	None	None	None
Chuzhou	*	*	*	*	*	35	54	101	290	484	744	1794	None	None
Maanshan	*	*	*	*	*	35	57	90	318	327	735	1794	None	None
Bengbu	*	*	*	*	*	46	47	147	297	532	1794	None	None	None
Suzhou	*	*	*	*	*	27	57	101	291	525	595	1124	1794	None
Huaibei	*	*	*	*	*	33	51	101	300	656	879	1794	None	None
Haozhou	*	*	*	*	*	34	57	96	307	599	677	1213	1794	None
Huainan	*	*	*	*	*	27	108	96	354	521	624	595	1794	None
Fuyang	*	*	*	*	*	*	51	101	404	320	519	960	1794	None
Liuan	*	*	*	*	*	36	58	107	280	788	799	735	1794	None

(continued)

Table 6.2 (continued)

IMF Cycle /Day City	1	2	3	4	5	6	7	8	9	10	11	12	13	14
Anqing	*	*	*	*	*	55	74	96	297	311	547	1794	None	None
Chizhou	*	*	*	*	*	36	55	101	287	525	1794	None	None	None
Tongling	*	*	*	*	*	28	60	96	262	357	287	814	1794	None
Huzhou	*	*	*	*	*	*	57	101	150	300	548	814	1794	None
Xuancheng	*	*	*	*	*	32	60	72	106	290	799	1794	None	None
Huangshan	*	*	*	*	*	41	70	125	291	559	973	None	None	None

Note * Indicates that the white noise part of the IMF is no main period

can reflect the fineness of EMD decomposition in filtering the original sequence. The number of IMF can be controlled by setting the SD value in the EMD method (formula 6.7). Too many decomposed IMF will generate more spurious modes, resulting in increased difficulty of grouping during IMF reconstruction. However, if the number of decomposed IMF is too small, it means that the filtering original sequence fluctuation is not refined enough. It is impossible to study the periodicity of the time series on multiple scales. The number of IMF obtained in this paper is relatively moderate, indicating that the setting of SD = 0.003 is reasonable.

Fig. 6.7 Frequency histogram of IMF number in different cities

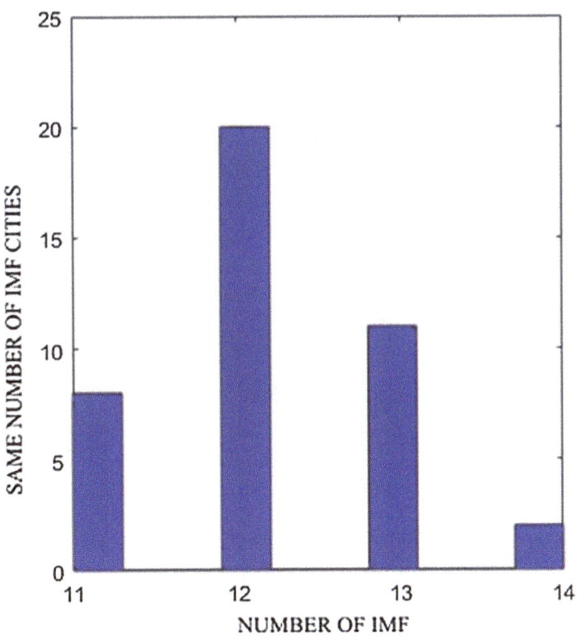

What is shown in Table 6.3 is as follows: (1) In addition to the white noise and the sequence trend item, the IMF sequence of 41 cities is classified by K-means clustering for its primary period length; (2) The reconstructed seasonal cycle item, short cycle item, medium cycle item, long cycle item are calculated according to the classification results.

According to the multi-scale cycle length of various cities shown in Table 6.4, the boxplot of seasonal period item, short period item, medium period item and long period item is made. As shown in Fig. 6.8, there is more overlap between each period item, and the length of each scale period is not concentrated enough, indicating that there may be cases where different time scale period items are grouped into a group during IMF reconstruction, so it is not effective to take clustering results as the basis for reconstruction grouping. Therefore, combined with the wavelet variance graph of each city's IMF sequence and the principle of cycle length concentration, after several reasonable attempts, the reconstruction grouping was corrected, and the results are shown in the following table.

As shown in Fig. 6.9, after the correction and reconstruction grouping based on the wavelet variance graph, the overlap between seasonal cycle items, short cycle items, medium cycle items and long cycle items is significantly reduced, and the cycle length of seasonal cycle items, short cycle items and medium cycle items are quite concentrated in various cities, showing good effectiveness. Although the concentration degree of the long-cycle term is slightly improved, its upward and downward range is still large, so it is not effective enough. Because the insufficient data length used in this paper leads to a longer period fluctuation hidden in the trend term during the EMD operation, resulting in a higher degree of dispersion in the range of long periods and poor effectiveness.

The research objective of this paper is to study the multi-scale period characteristics of $PM_{2.5}$ in the Yangtze River Delta region. In order to take into account the overall macroscopicity and local specificity, the above cycle results are analyzed in two parts: (1) The cycle length of 41 prefecture-level cities in the Yangtze River Delta is analyzed; (2) Taking the province as the unit, statistical analysis is carried out for different provinces to study the local multi-scale periodic characteristics of the Yangtze River Delta region more clearly.

The above table provides a statistical summary of the seasonal cycle items, short cycle items, medium cycle items, and long cycle items in all prefecture-level cities in the Yangtze River Delta from six aspects: including median, upper quartile, and lower quartile, average, minimum and maximum, all indicators are rounded. In combination with Fig. 6.9, in order to objectively measure the period length and reduce the influence of extreme values, the median is adopted as the period length of each period item. The following conclusions are drawn:

(1) The seasonal cycle of daily $PM_{2.5}$ concentration in the Yangtze River Delta is 46 days (about 1.5 months).
(2) The short cycle of daily $PM_{2.5}$ concentration in the Yangtze River Delta is 101 days (about 3.5 months).

Table 6.3 IMF clustering results of 41 cities in three provinces and one city and the main period after IMF reconstruction

City	Seasonal cycle term		Short cycle term		Middle cycle term		Long cycle term	
	Cluster results	Cycle/ Day	Cluster results	Cycle/ Day	Cluster results	Cycle/ Day	Cluster results	Cycle/ Day
Shanghai	[6,7]	49	[8]	91	[9,10]	305	[11]	576
Nanjing	[6,7,8]	62	[9]	240	[10]	303	[11]	662
Zhenjiang	[6,7,8]	50	[9]	238	[10]	303	[11]	814
Yangzhou	[6,7]	39	[8]	149	[9,10]	295	[11]	586
Taizhou	[6,7]	41	[8]	137	[9,10]	299	[11]	637
Changzhou	[6,7,8]	57	[9]	253	[10]	297	[11]	671
Wuxi	[6,7,8]	53	[9]	296	[10]	459	[11]	894
Suzhou	[6,7]	36	[8]	93	[9]	238	[10]	305
Nantong	[6,7]	36	[8]	156	[9,10]	296	[11]	1064
Huaian	[6,7,8]	49	[9,10]	294	[11]	612	[12]	799
Yancheng	[6]	33	[7,8]	50	[9]	169	[10]	307
Lianyungang	[6,7,8,9]	101	[10,11]	300	[12]	456	[13]	1033
Suqian	[6,7]	41	[8,9,10]	294	[11]	542	[12]	782
Xuzhou	[6,7,8]	96	[9]	290	[10]	463	[11]	1404
Hangzhou	[6,7,8,9]	96	[10]	296	[11]	492	[12]	1020
Huzhou	[6,7,8]	58	[9]	310	[10]	448	[11]	566
Jiaxing	[6]	39	[7]	107	[8,9]	310	[10]	448
Shaoxing	[6,7,8]	93	[9]	300	[10]	419	[11]	960
Jinhua	[7,8]	65	[9]	125	[10]	292	[11]	475
Quzhou	[6,7,8]	45	[9]	111	[10]	290	[11]	637
Lishui	[6]	38	[7,8]	104	[9]	294	[10]	945
Wenzhou	[6,7]	41	[8]	106	[9,10]	295	[11]	927
Taizhou	[6,7,8]	95	[9]	295	[10]	448	[11]	879
Ningbo	[6,7,8,9]	101	[10,11]	297	[12]	501	[13]	766
Zhoushan	[6,7,8]	99	[9,10]	305	[11]	431	[12]	851
Hefei	[6,7]	46	[8]	160	[9]	292	[10]	419
Chuzhou	[6,7,8]	95	[9]	290	[10]	484	[11]	744
Maanshan	[6,7]	55	[8]	90	[9,10]	299	[11]	735
Bengbu	[6,7]	44	[8]	147	[9]	297	[10]	532
Suzhou	[6,7,8]	55	[9]	291	[10,11]	431	[12]	1124
Huaibei	[6,7,8]	50	[9]	300	[10]	656	[11]	879
Haozhou	[6,7,8]	95	[9]	307	[10,11]	381	[12]	1213
Huainan	[6,7,8]	62	[9]	354	[10]	521	[11,12]	331
Fuyang	[7,8]	53	[9,10]	294	[11]	519	[12]	960

(continued)

Table 6.3 (continued)

City	Seasonal cycle term		Short cycle term		Middle cycle term		Long cycle term	
	Cluster results	Cycle/Day	Cluster results	Cycle/Day	Cluster results	Cycle/Day	Cluster results	Cycle/Day
Liuan	[6,7]	55	[8]	107	[9]	280	[10,11,12]	311
Anqing	[6,7]	62	[8]	96	[9,10]	292	[11]	547
Chizhou	[6,7]	41	[8]	101	[9]	287	[10]	525
Tongling	[6,7,8]	60	[9,10]	305	[11]	287	[12]	814
Huzhou	[7,8,9]	150	[10]	300	[11]	548	[12]	814
Xuancheng	[6,7,8]	62	[9]	106	[10]	290	[11]	799
Huangshan	[6,7]	43	[8]	125	[9]	291	[10]	559

Note The number in [] indicates the IMF serial number decomposed by EMD of PM2.5 daily concentration series of the city; the period is the main period calculated after IMF reconstruction in []

(3) The medium cycle of the daily $PM_{2.5}$ concentration in the Yangtze River Delta is 294 days (about 10 months).
(4) The long cycle of daily $PM_{2.5}$ concentration in the Yangtze River Delta is 671 days (about 22.5 months).

Among all the periodicity scales, the medium cycle is the most significant. Not only is the cycle length concentration high in all cities, but the results are basically consistent with that of 298 days of cycle results of Chen et al. (2020)'s study on $PM_{2.5}$ in Chengdu.

After obtaining the overall multi-scale periodicity in the Yangtze River Delta region, we still want to know whether there is consistency or specificity in the local multi-scale periodicity in the Yangtze River Delta region. In order to facilitate the research, elaboration and analysis of relevant results, this paper divides the Yangtze River Delta region into four parts according to the provincial administrative regions: Jiangsu, Shanghai, Zhejiang and Anhui, which can roughly correspond to the north, east, south and west of the Yangtze River Delta region. Except for Shanghai, the statistical analysis is carried out in the other three provinces according to the multi-scale cycle results of cities in Table 6.4 based on the statistical caliber in Table 6.5. The statistical results are shown in Tables 6.6, 6.7 and 6.8.

Table 6.6 Multi scale periodic statistical analysis of the Northern Yangtze River Delta (Jiangsu)

Periodic scale	Median/day	Upper quartile/day	Lower quartile/day	Average/day	Min/day	Max/day
Seasonal cycle term	45	39	53	45	33	58
Short cycle term	101	93	137	111	50	206

(continued)

(continued)

Periodic scale	Median/day	Upper quartile/day	Lower quartile/day	Average/day	Min/day	Max/day
Middle cycle term	295	292	297	282	169	302
Long cycle term	666	606	894	742	305	1404

According to the above three tables and the results of multi-scale cycle in Shanghai (east of Yangtze River Delta), it can be found that the period length is roughly consistent on the seasonal, short and medium cycle scales; however, there are large differences on the long cycle scale: (1) The median and average of the long cycle term in the south of the Yangtze River Delta (Zhejiang) are higher than that in other regions of the Yangtze River Delta; (2) In the west of the Yangtze River Delta (Anhui), the dispersion degree of the long period term is relatively large, and the minimum value of the long period term even overlaps with that of the middle period term.

To sum up, the EMD-WA model was applied to carry out an empirical study of multi-scale periodic of $PM_{2.5}$ in the Yangtze River Delta region to comprehensively analyze its overall and local characteristics. The results are summarized in Table 6.9.

6.5.3 Short Periodic Analysis in Yangtze River Delta During Heavy Haze

This section intends to study the periodic characteristics of hourly $PM_{2.5}$ concentration within 10 days in the most severe haze days (240 h) in the Yangtze River Delta region. The data set selected is from January 1, 2015 to November 30, 2019. The moving average maximum value of hourly $PM_{2.5}$ concentration in cities in the Yangtze River Delta during the 240 window periods. The results of 41 prefecture-level cities are shown in Table 6.10.

As shown in Table 6.10, the moving average value of the 240 window periods of the hourly $PM_{2.5}$ concentration in Xuzhou city is the largest in all cities. The most polluted period is from 0:00 on January 13, 2018 to 23:00 on January 22, 2018, with a total length of 240 h. The results of using the EMD-WA model are as follows.

6.5.4 Short Periodic Analysis in Yangtze River Delta During Heavy Haze

The length of the data set applied in this section is only 240, so the stop filter condition of the EMD method is set as SD = 0.03. As shown in Fig. 6.10, IMF $c(1), \cdots, c(5)$

Table 6.4 IMF restructuring groups and the main period after IMF restructuring

City	Seasonal cycle term		Short cycle term		Middle cycle term		Long cycle term	
	Cluster results	Cycle/Day	Cluster results	Cycle/Day	Cluster results	Cycle/Day	Cluster results	Cycle/Day
Shanghai	[6,7]	49	[8]	91	[9,10]	305	[11]	576
Nanjing	[6,7]	58	[8]	93	[9,10]	297	[11]	662
Zhenjiang	[6,7]	45	[8]	65	[9,10]	300	[11]	814
Yangzhou	[6,7]	39	[8]	149	[9,10]	295	[11]	586
Taizhou	[6,7]	41	[8]	137	[9,10]	299	[11]	637
Changzhou	[6,7]	48	[8]	70	[9,10]	293	[11]	671
Wuxi	[6,7]	53	[8]	101	[9,10]	297	[11]	894
Suzhou	[6,7]	36	[8]	93	[9]	238	[10]	305
Nantong	[6,7]	36	[8]	156	[9,10]	296	[11]	1064
Huaian	[6,7]	47	[8]	96	[9,10]	294	[11,12]	666
Yancheng	[6]	33	[7,8]	50	[9]	169	[10]	307
Lianyungang	[6,7,8]	53	[9]	106	[10,11,12]	302	[13]	1033
Suqian	[6,7]	41	[8]	206	[9,10]	292	[11,12]	606
Xuzhou	[6,7]	58	[8]	126	[9]	290	[10,11]	1404
Hangzhou	[6,7]	57	[8,9]	95	[10,11]	297	[12]	1020
Huzhou	[6,7]	49	[8]	101	[9,10]	300	[11]	566
Jiaxing	[6]	39	[7]	107	[8,9]	310	[10]	448
Shaoxing	[6,7]	36	[8]	104	[9,10]	294	[11]	960
Jinhua	[7,8]	65	[9]	125	[10]	292	[11]	475
Quzhou	[6,7,8]	45	[9]	111	[10]	290	[11]	637
Lishui	[6]	38	[7,8]	104	[9]	294	[10]	945
Wenzhou	[6,7]	41	[8]	106	[9,10]	295	[11]	735
Taizhou	[6,7]	70	[8]	113	[9,10]	294	[11,12]	879
Ningbo	[6,7]	38	[8,9]	137	[10,11]	297	[12,13]	735
Zhoushan	[6,7]	46	[8]	101	[9,10,11]	303	[12]	851
Hefei	[6,7]	46	[8]	160	[9]	292	[10]	419
Chuzhou	[6,7]	41	[8]	101	[9,10]	294	[11]	744
Maanshan	[6,7]	55	[8]	90	[9,10]	299	[11]	735
Bengbu	[6,7]	44	[8]	147	[9]	297	[10]	532
Suzhou	[6,7]	55	[8]	101	[9,10,11]	292	[12]	1124
Huaibei	[6,7]	35	[8]	101	[9]	300	[10,11]	542
Haozhou	[6,7]	33	[8]	96	[9,10,11]	297	[12]	1213
Huainan	[6,7]	59	[8]	96	[9]	354	[10,11,12]	297
Fuyang	[7]	51	[8]	101	[9,10,11]	294	[12]	960

(continued)

Table 6.4 (continued)

City	Seasonal cycle term		Short cycle term		Middle cycle term		Long cycle term	
	Cluster results	Cycle/ Day	Cluster results	Cycle/ Day	Cluster results	Cycle/ Day	Cluster results	Cycle/ Day
Liuan	[6,7]	55	[8]	107	[9]	280	[10,11,12]	311
Anqing	[6,7]	62	[8]	96	[9,10]	292	[11]	547
Chizhou	[6,7]	41	[8]	101	[9]	287	[10]	525
Tongling	[6,7]	57	[8]	96	[9,10,11]	287	[12]	814
Huzhou	[7]	57	[8,9]	149	[10,11]	301	[12]	814
Xuancheng	[6,7]	58	[8,9]	96	[10]	290	[11]	799
Huangshan	[6,7]	43	[8]	125	[9]	291	[10]	559

Note The number in [] indicates the IMF serial number decomposed by EMD of PM2.5 daily concentration series of the city; the period is the main period calculated after IMF reconstruction in []

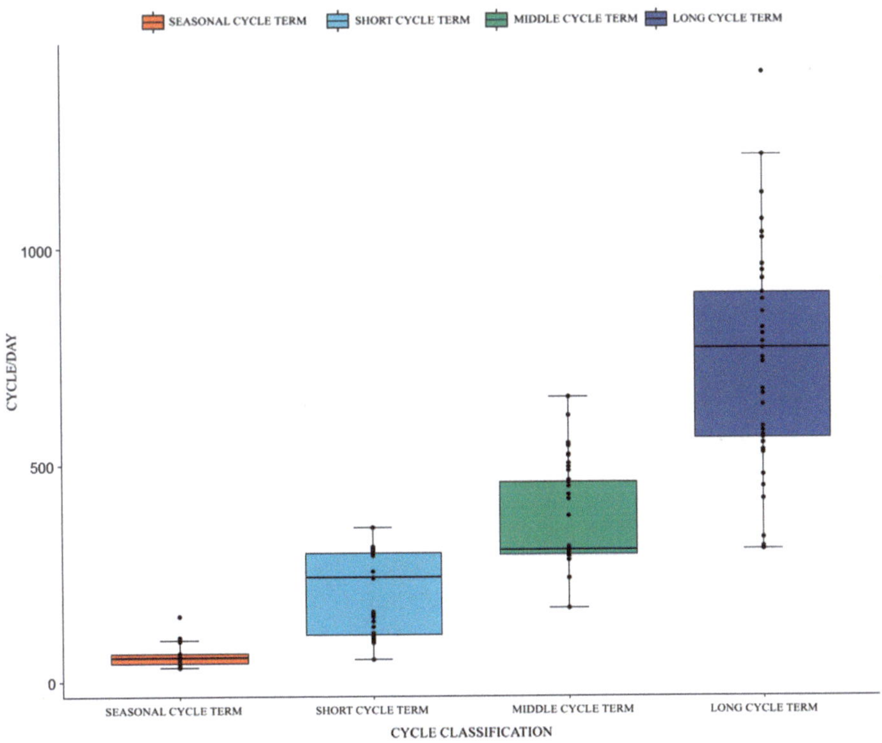

Fig. 6.8 Boxplot of each period determined by clustering results

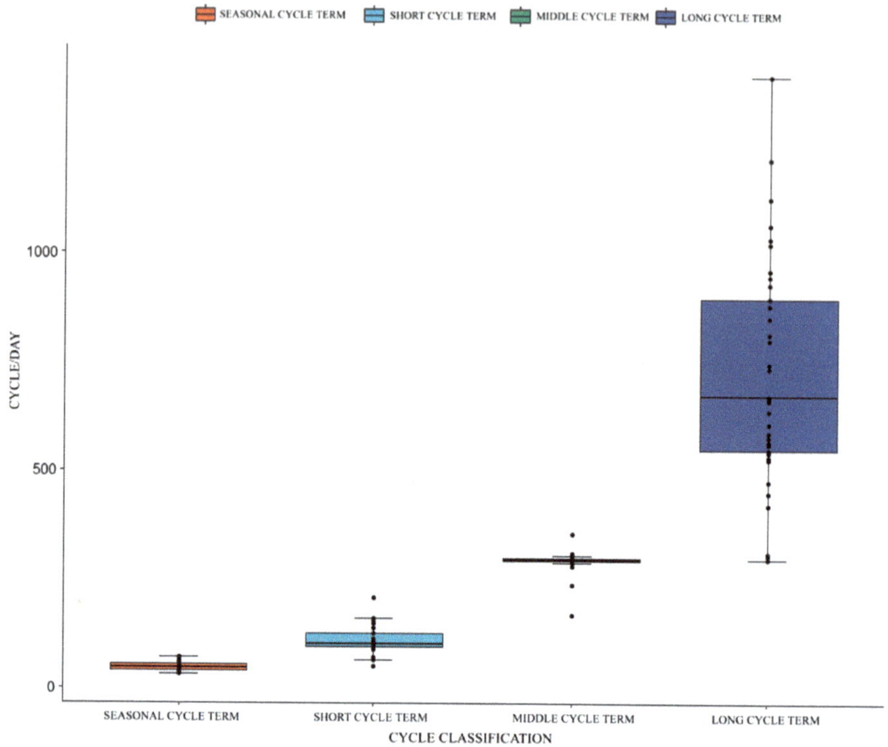

Fig. 6.9 Box line diagram of each period determined after modification and reconstruction grouping

Table 6.5 Multi scale periodic overall statistical analysis of all cities in the Yangtze River Delta

Periodic scale	Median/day	Upper quartile/day	Lower quartile/day	Average/day	Min/day	Max/day
Seasonal cycle term	46	41	55	48	33	70
Short cycle term	101	96	125	110	50	206
Middle cycle term	294	292	299	292	169	354
Long cycle term	671	547	894	722	297	1404

and trend item $c(6)$ are decomposed through the EMD method, and their cycle length is calculated by morlet wavelet analysis for reconstruction.

The results are shown in Table 6.11.

Table 6.7 Multi scale periodic statistical analysis of cities in the south of Yangtze River Delta (Zhejiang)

Periodic scale	Median/day	Upper quartile/day	Lower quartile/day	Average/day	Min/day	Max/day
Seasonal cycle term	45	39	53	48	36	70
Short cycle term	106	103	112	109	95	137
Middle cycle term	295	294	299	297	290	310
Long cycle term	851	602	936	768	448	1020

Table 6.8 Multi scale periodic statistical analysis of Western Yangtze River Delta (Anhui)

Periodic scale	Median/day	Upper quartile/day	Lower quartile/day	Average/day	Min/day	Max/day
Seasonal cycle term	53	43	57	50	33	62
Short cycle term	101	96	112	110	90	160
Middle cycle term	293	291	298	297	280	354
Long cycle term	647	530	814	683	297	1213

Table 6.9 Summary of multi-scale periodic results in the Yangtze River Delta and its parts

Periodic scale	Yangtze River Delta (overall)/day	Northern Yangtze River Delta (Jiangsu)/day	Eastern Yangtze River Delta (Shanghai)/day	Southern Yangtze River Delta (Zhejiang)/day	Western Yangtze River Delta (Anhui)/day
Seasonal cycle term	46	45	49	45	53
Short cycle term	101	101	91	106	101
Middle cycle term	294	295	305	295	293
Long cycle term	671	666	575	851	647

Table 6.10 Average maximum value of PM 2.5 hourly concentration in 240 window periods in different cities

City	PM 2.5	City	PM 2.5	City	PM2.5	City	PM2.5
Shanghai	107	Lianyungang	142	Taizhou	100	Fuyang	171
Nanjing	131	Suqian	167	Ningbo	100	Liuan	128
Zhenjiang	145	Xuzhou	210	Zhoushan	65	Anqing	128
Yangzhou	118	Hangzhou	118	Hefei	130	Chizhou	130
Taizhou	127	Huzhou	135	Chuzhou	139	Tongling	119
Changzhou	129	Jiaxing	106	Maanshan	138	Huzhou	154
Wuxi	123	Shaoxing	117	Bengbu	147	Xuancheng	118
Suzhou	114	Jinhua	121	Suzhou	167	Huangshan	66
Nantong	122	Quzhou	98	Huaibei	161		
Huaian	147	Lishui	89	Haozhou	160		
Yancheng	128	Wenzhou	96	Huainan	152		

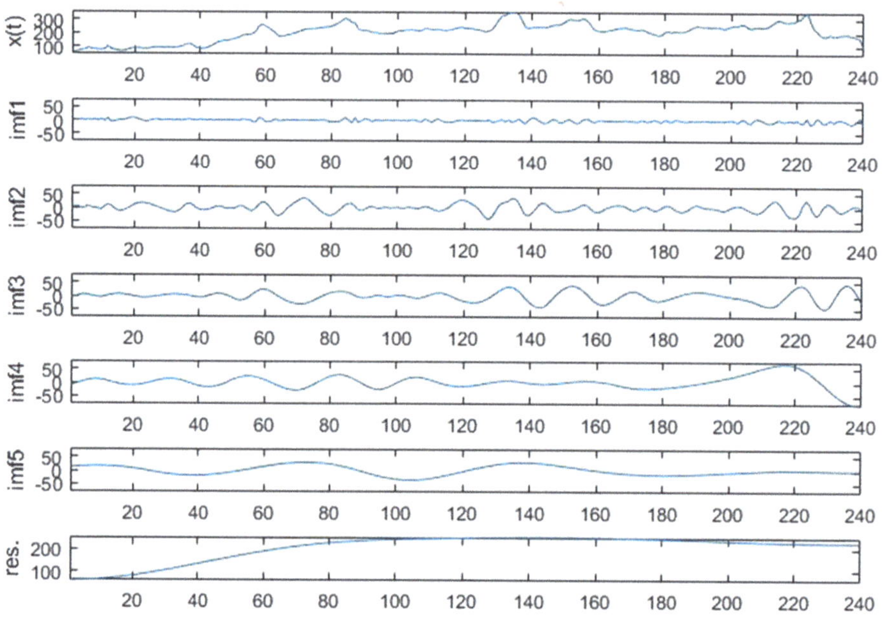

Fig. 6.10 Timing chart of $PM_{2.5}$ concentration in Xuzhou City and IMF

Table 6.11 Main cycle of IMF sequence in Xuzhou

Name	Cycle/hour	Name	Cycle/hour
$c(1)$	No main period	$c(4)$	47
$c(2)$	9	$c(5)$	55
$c(3)$	14	$c(6)$	240

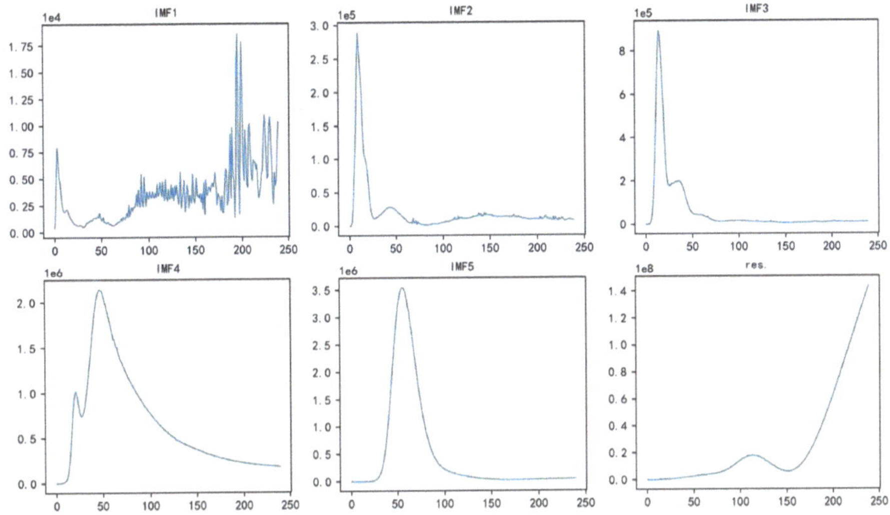

Fig. 6.11 IMF wavelet variance in Xuzhou

According to the cycle length of each IMF in Table 6.11, combined with Fig. 6.11 corresponding to the wavelet variance graph of IMF sequence, it is divided into two groups of short cycle term and long cycle term for reconstruction, and $c(2)$, $c(3)$ are short cycle term, and $c(4)$, $c(5)$ are long cycle term.

After the reconstruction, the period is calculated and the results are as follows: (1) During the 240 h with the most severe haze in Xuzhou, the short cycle is 14 h. (2) During the 240 h with the most severe haze in Xuzhou, its long cycle is 55 h.

Therefore, in the short period during severe haze in the Yangtze River Delta, the variation of $PM_{2.5}$ concentration still has a significant periodicity. However, due to its short sequence, the periodicity may be more inclined to the oscillation period, and the trend period is not obvious.

6.6 Conclusion and Outlook

In this article we attempt to apply ICT technology to PM time series, and verifies the feasibility and practicability of these methods in PM time series analysis.

In this paper, EMD and WA are combined to analyze the following two data sets: *a.* the daily $PM_{2.5}$ concentration in 41 prefecture-level cities in the Yangtze River Delta from January 1, 2015 to November 30, 2019; *b.* The hourly $PM_{2.5}$ concentration sequence during the heavy haze region of the Yangtze River Delta from 0 on January 13, 2018 to 23 on January 22, 2018. The multi-scale periodicity of $PM_{2.5}$ in the Yangtze River Delta is analyzed and the main conclusions are as follows:

(1) In the analysis of data set *a*, through the decomposition and reconstruction of the EMD-WA model, cycle characteristics of four scales from short to long can be obtained, which are seasonal, short, medium and long cycle terms respectively.
(2) The $PM_{2.5}$ concentration in the Yangtze River Delta region has an obvious multi-scale periodicity on the whole, with a seasonal cycle of 46 days (about 1.5 months), a short cycle of 101 days (about 3.5 months), a medium cycle of 294 days (about 10 months), and a long cycle of 671 days (about 22.5 months).[1]
(3) There is good consistency in terms of seasonal, short and medium cycle scales in the north (Jiangsu), east (Shanghai), south (Zhejiang), and west (Anhui) of the Yangtze River Delta, but there are significant differences in terms of long cycle scales.
(4) Based on the analysis of data set b, it is proved that in the 240 h with severe haze in the Yangtze River Delta region, $PM_{2.5}$ concentration still has a relatively obvious periodicity, showing a short cycle of 14h and a long period of 55h.

According to the findings above, policy suggestions can be given as follows:

(1) According to the short-term periodic characteristics of the PM sequence, it is possible to strengthen the detection of key pollution sources, shut down and reduce pollution sources before the arrival of PM peaks such as winter and spring or working days; Early warning of high-polluting weather is issued to remind the public to prevent the impact of haze on health, to remind the transportation industry to deal with the possible impact of severe haze on vehicle driving and aircraft travel, and to remind hospitals of the negative effects of the surge in respiratory diseases caused by PM.
(2) According to the long-term periodic characteristics of the PM sequence, the most important thing is to reduce pollution emissions in industrial production, improve the environmental standards of fuels used, reduce vehicle exhaust emissions, and strengthen urban greening work; It is necessary to advocate the implementation of a low-carbon lifestyle, and to gradually eliminate the use of fireworks or firecrackers to celebrate holidays. Only by starting with systems and regulations and persevering, can we achieve sound haze reduction effects (Wu et al., 2020).

Future researches include:

(1) During the analysis of the long-cycle term, it is found that there are abnormal points in the period length. A possible reason is that the insufficient length of the selected sequence makes it impossible to find fluctuation trends with longer cycle lengths (Song & Wang, 2018).
(2) When judging the period of multi-scale period results of the Yangtze River Delta region at present, we find that the data of only one city may not be able to

[1] These conclusions are basically consistent with the other study. For example, Wang et al. (2020) found hourly concentrations of $PM_{2.5}$ from 2015 to 2018 in the cities of the Yangtze River Delta have two dominant periods: an annual cycle on the time scale of 250–480 days and a semi-annual cycle on the time scale of 130–220 days.

represent the Yangtze River Delta region, but this paper has not found a better method to study this point-to-face problem.

(3) When analyzing the multi-scale period of $PM_{2.5}$, little information about the causes of haze is mentioned, but this period characteristic is closely related to a series of factors such as geographical conditions, weather, climate, human production activities (Wu et al., 2019a). It is necessary to consider the joint analysis of the geographic characteristics, climatic characteristics, economic composition, and policy orientation of the research area under the premise of studying the concentration sequence of $PM_{2.5}$ to study the periodicity of $PM_{2.5}$ more accurately.

Acknowledgements Shaoli He and Weiheng Sun also made great contributions to this manuscript. We express our heartfelt thanks to them. This research also was supported by: National Natural Science Foundation of China (72474127) and National Social Science Fund Major Bidding Project of China (18ZDA052). Part of this chapter was published:Wu X, He S, Guo J, Sun W. A multi-scale periodic study of PM2.5 concentration in the Yangtze River Delta of China based on Empirical Mode Decomposition-Wavelet Analysis. Journal of Cleaner Production 2021;281:124853. https://doi.org/https://doi.org/10.1016/j.jclepro.2020.124853. The publication has been conducted with copyright permission.

References

Addo, P. M., Billio, M., & Guégan, G. (2014). *Nonlinear dynamics and wavelets for business cycle analysis.* Springer International Publishing.

Bao, C., & Fang, C. L. (2007). Water resources constraint force on urbanization in water deficient regions: A case study of the Hexi Corridor, arid area of NW China. *Ecological Economics, 62*(3–4), 508–517.

Baubeau, P., & Cazelles, B. (2009). French economic cycles: A wavelet analysis of French retrospective GNP series. *Cliometrica, 3*(3), 275–300.

Berdiev, A. N., & Chang, C. P. (2015). Business cycle synchronization in Asia-Pacific: New evidence from wavelet analysis. *Journal of Asian Economics, 37*, 20–33.

Carvalho, M. D., Rodrigues, P. C., & Rua, A. (2012). Tracking the us business cycle with a singular spectrum analysis. *Economics Letters, 114*, 32–35.

Cendejas, J. L., Muñoz, F. F., & Fernández-de-Pinedo, N. (2017). A contribution to the analysis of historical economic fluctuations (1870–2010): Filtering, spurious cycles, and unobserved component modeling. *Cliometrica, 11*(1), 93–125

Cekim, H. O. (2020). Forecasting PM10 concentrations using time series models: A case of the most polluted cities in Turkey. *Environmental Science and Pollution Research., 27*, 25612–25624.

Chen, X. B., Yin, L. R., Fan, Y. L., Song, L. H., Ji, T. T., Liu, Y., Tian, J. W., & Zheng, W. F. (2020). Temporal evolution characteristics of PM2. 5 concentration based on continuous wavelet transform. *Science of the Total Environment, 699*, 134244.

Cui, J. T. (1995). *Wavelet analysis.* Xi'an Jiaotong University Press.

Fiffer, M., Kang, C.-M., Requia, M. J., & Koutrakis, P. (2020). Long-term impact of $PM_{2.5}$ mass and sulfur reductions on ultrafine particle trends in Boston, MA from 1999 to 2018. *Journal of the Air & Waste Management Association, 70*(7), 700–707.

Flandrin, P., Rilling, G., & Gonçalves, P. (2004). Empirical mode decomposition as a filter bank. *IEEE Signal Processing Letters, 11*(2), 112–114.

Gu, K. H., Shi, H. X., Zhang, S., Fan, S. X., Xu, J. M., & Tan, J. G. (2015). Variation characteristics of $PM_{2.5}$ levels and the influence of meteorological conditions on Chongming Island in Shanghai. *Resources and Environment in the Yangtze Basin, 24*(12), 2108–2116.

Hansen, J., Sato, M. K. I., Ruedy, R., Nazarenko, L., Lacis, A., Schmidt, G. A., & Bell, N. (2005). Efficacy of climate forcings. *Journal of Geophysical Research: Atmospheres, 110*(D18).

He, J. L., & Song, W. L. (2013). Measurement and analysis of ocean economic cyclical changes based on filtering method. *Marine Science Bulletin, 32*(1), 1–7.

Huang, N., Shen, Z., Long, S. R., & Wu, M. C. (1998). The empirical mode decomposition and the Hilbert spectrum for nonlinear and non-stationary time series analysis. *Proceedings of the Royal Society A-Mathematical Physical & Engineering Sciences,* 454(1971), 903–995.

Huang, N. E., Shen, Z., & Long, S. R. (1999). A new view of nonlinear water waves: The Hilbert spectrum. *Annual Review of Fluid Mechanics, 31*(1), 417–457.

Kuo, S. M., & Lee, B. H. (1988). *Fast Fourier transform and its applications.*

Li, C. G., Tian, Y. X., & He, J. R. (2012). Prediction model of AC algorithm based on EMD decomposition combined with GMDH and its application. *Journal of Systems & Management, 21*(1), 105–110.

Li, Q., Wei, W., Zhang, H., Sun, M., Wu, B., & Liu, J. (2025). Impacts of turbulent intermittency and pressure oscillation on persistent heavy haze pollution. *Atmospheric Research, 326*, Article 108305.

Li, X. F., Chu, J. H, Yu L. D., Zhu, G. H., & Wang, G. F. (2011). Variational characteristics of $PM_{2.5}$ concentration in a monitoring site in Beijing urban area. *Journal of Beijing Normal University(Natural Science), 47*(03), 258–261.

Li, X. Y., Jin, M. J., Chen, K., Xiang, H. Q., & Liu, Q. M. (2007). The case-crossover studies of air particulate matter pollution and cardiovascular disease death. *China Environmental Science, 27*(5), 657–660.

Li ZiMing, L. Z., Sun ZhaoBin, S. Z., Shao Xie, S. X., Liao XiaoNong, L. X., Zhang XiaoLing, Z. X., Xiong YaJun, X. Y., & Ma XiaoHui, M. X. (2017). Using Morlet wavelet analysis to analyze multiple time scale periodically in PMzs in Beijing. *China Environmental Science, 37*(02), 407–415.

Liang, M. Y., Dong, L., & Tao, J. (2007). Pollution level of the Airborne Particulate Matter ($PM_{2.5}$) during the Haze period in winter in Guangzhou. *Environmental Monitoring in China,* (05), 52–54+70.

Liu, Y., Adamu, A. T., & Tan, J. (2025). Spatial characterization of periodic behaviors of ground PM2.5 concentration across the Yangtze River Delta and the North China Plain during 2014–2024: A new insight on driving processes of regional air pollution. *Environmental Research, 277*, 121648.

Mao, W. L., Xu, J. H., Lu, D. B., Yang, D. Y., & Zhao, J. N. (2017). An analysis of the spatial-temporal pattern and influencing factors of $PM_{2.5}$ in the Yangtze River Delta in 2015. *Resources and Environment in the Yangtze Basin, 26*(02), 264–272.

Pedregal, D. J. (2003). Filter-design and model-based analysis of economic cycles. *Documento de trabajo*, 13.

Shen, L., Cheng, S., Gunson, A. J., & Wan, H. (2005). Urbanization, sustainability and the utilization of energy and mineral resources in China. *Cities, 22*(4), 287–302.

Sella, L., Vivaldo, G., Ghil, M., & Groth, A. (2010). Economic cycles and their synchronization: Spectral analysis of macroeconomic series from Italy, The Netherlands, and the UK. In *EGU General Assembly conference abstracts* (Vol. 12, p. 11847).

Song, M., & Wang, S. (2018). Market competition, green technology progress and comparative advantages in China. *Management Decision, 56*(1), 188–203.

Malm, W. C., Sisler, J. F., Huffman, D., Eldred, R. A., & Cahill, T. A. (1994). Spatial and seasonal trends in particle concentration and optical extinction in the United States. *Journal of Geophysical Research: Atmospheres, 99*(D1), 1347–1370.

Marczak, M., & Gomez, V. (2015). Cyclicality of real wages in the USA and Germany: New insights from wavelet analysis. *Economic Modelling, 47*, 40–52.

Mohr, M. F. (2006). The missing cycle in the HP filter and the measurement of cyclically-adjusted budget balances. *SSRN Electronic Journal*, 73–111.

Wang, F., Han, Y. L., & Zhao, Y. (2017). Spatial-temporal variations of PM_{10} and $PM_{2.5}$ on different time-scales in Taiyuan. *Ecology and Environmental Sciences, 26*(09), 1521–1528.

Wang, G. C., & Wang, P. C. (2014). $PM_{2.5}$ Pollution in China and Its Harmfulness to Human Health. *Science & Technology Review, 32*(26), 72–78.

Wang, J. J., Lu X. M., Yan Y. T., Zhou, L. G., & Ma, W. C. (2020). Spatiotemporal characteristics of PM 2.5 concentration in the Yangtze River Delta urban agglomeration, China on the application of big data and wavelet analysis. *Science of The Total Environment,* 724.

Wang, W. B., Fei, P. S., & Yi, X. M. (2010). Prediction of China stock market based on EMD and neural network. *Systems Engineering Theory & Practice, 30*(6), 1027–1033.

Wu, H. H., Kuang, H. B., Meng, B., & Feng, W. W. (2018). Study on the periodic characteristics of BDI index based on EMD-WA model. *Systems Engineering c Characteristics o, 38*(06), 1586–1598.

Wu, X. H., Cao, Y. L., Xiao, Y., & Guo, J. (2018). Finding of urban rainstorm and waterlogging disasters based on microblogging data and the location-routing problem model of urban emergency logistics. *Annals of Operations Research,* 1–32.

Wu, X. H, Chen, Y. F, Zhao, P., Guo, J., & Ma, Z. X. (2019a). Study of haze emission efficiency based on new co-opetition data envelopment analysis. *Expert Systems,* (3), 1–12.

Wu, X. H., Xu, Z., Liu, H., Guo, J., & Zhou, L. (2019b). What are the impacts of tropical cyclones on employment? An analysis based on meta-regression. *Weather, Climate, and Society*, 259–275.

Wu, X. H., Wang, Z. J., Gao, G., Guo, J., & Xue, P. P. (2020). Disaster probability, optimal government expenditure for disaster prevention and mitigation, and expected economic growth. *Science of the Total Environment, 709*, Article 135888.

Xia, M., Dong, L., Jiang, L., & Zeng, M. (2024). Research on time series change point detection and influencing factors under machine learning: Based on PM2.5 concentration data in Hefei city. *Earth Science Informatics, 17*(1), 351–64.

Yang, X. X., Feng, L. H., & Wei, P. (2012). Air particulate matter $PM_{2.5}$ in Beijing and its harm. *Frontier Science, 6*(01), 22–31.

Yogo, M. (2008). Measuring business cycles: A wavelet analysis of economic time series. *Economics Letters, 100*, 208–212.

Yu, L., Wang, S., & Lai, K. K. (2008). Forecasting crude oil price with an EMD-based neural network ensemble learning paradigm. *Energy Economics, 30*(5), 2623–2635.

Zhang, W. A. (2014). Study on the dynamic and mechanism of regional economical disparity in Western China—Multi-scale analysis based on EMD method. *Journal of Applied Statistics and Management, 33*(6), 951–964.

Zhang Z. S., Tao, J., Xie, S. D., Zhou, L. D., Song, D. L., Zhang, P., Cao, J. J., & Luo, L. (2013). Seasonal variations and source apportionment of $PM_{2.5}$ at urban area of Chengdu. *Acta Scientiae Circumstantiae, 33*(11), 2947–2952.

Zhao, X. J. (2008). Seasonal and daily variation characteristics of $PM_{2.5}$ concentration in urban and suburban areas of Beijing. *China Meteorological Society: China Meteorological Society,* 11.

Zhang, X., Lai, K. K., & Wang, S. Y. (2008). A new approach for crude oil price analysis based on empirical mode decomposition. *Energy Economics, 30*(3), 905–918.

Zheng, Z. G. (2010). *Empirical modal analysis and wavelet analysis* (pp. 1–2). Meteorological Publishing House.

Zhou, J., Zhang Y. J., Xiang, D., & Han, Z. Y. (2018). The periodicity and cause analysis of PM2.5 in Taiyuan. *Ecology and Environmental Sciences, 27*(3), 527–532.

Chapter 7
A Spatiotemporal XGBoost Model for $PM_{2.5}$ Concentration Prediction and Its Application in Shanghai

7.1 Introduction

The fluctuation of $PM_{2.5}$ is affected by many factors. There are more and more researches on $PM_{2.5}$ prediction using machine learning methods, especially the extreme gradient boosting model (XGBoost). Random forest model (Stafoggia et al., 2019), enhanced long short-term memory neural network model (Li et al., 2017) and LightGBM model (Zhong et al., 2021) performed effective in prediction of short-term $PM_{2.5}$ concentration changes, which indicates an advantage of using machine learning algorithms to predict short-term changes in $PM_{2.5}$, especially those data sets with high spatial and temporal resolution.

The methods used in current research are mainly divided into two kinds: conventional statistical methods, and machine learning methods. Early research on prediction models mainly focused on methods related to regression analysis, like data-driven ordinary differential equation (ODE) model established by Wang et al. (2020) and an improved grey correlation analysis method called grey multivariate convolution model (GMCN(1,*N*)) applied by Zhang et al. (2020a). What's more, the relationship between $PM_{2.5}$ and other pollutants came to the fore (Yahya et al., 2014), while sequence information be extracted to improve the precision by using complementary integrated empirical mode decomposition method– after reconstructing, the predicted values of the decomposed subsequences could perform better in predicting time sequence like $PM_{2.5}$ concentration (Liu & Sun, 2019). Parameter optimization method like principal component analysis (PCA) used by (Sun & Sun, 2017) was gradually added to prediction model like long-short-term memory (LSTM) used by Xu and Yoneda (2021), in order to compose hybrid model, which could extremely improve the predicting capability.

As a relatively advanced gradient boosting algorithm, XGBoost has attracted much attention in recent years. With the development of machine learning technology, scholars began to apply XGBoost to predict $PM_{2.5}$ concentration. Early studies mainly used XGBoost alone for empirical analysis, and compared it with

X. Wu, *Environmental Disaster in China*, https://doi.org/10.1007/978-981-92-0910-1_7

other algorithms to verify its advantages (Luo et al., 2021). In short-term prediction, especially when we get the data set of hourly $PM_{2.5}$ mass concentration (Ma et al., 2020a), XGBoost has a smaller percentage error than the LSTM network (Chang et al., 2020), which may be ascribed to the difference in training mechanism between them. When taking more stations into consideration (Gui et al., 2020), the prediction results of constructed XGBoost network always show strong robustness and high accuracy. Combined other predicting model and XGBoost to establish a hybrid model for forecasting of $PM_{2.5}$ could largely improve the accuracy in some extreme point—in these days the $PM_{2.5}$ concentration is largely higher or lower than the day before or later (Fan et al., 2020). The establishment of a reliable $PM_{2.5}$ prediction system based on distributed interactive platform is conducive to real-time prediction (Shang et al., 2019).

Optimization of the XGBoost prediction data itself can improve the prediction performance of the model. There are two methods for scholar to do the research when preprocessing $PM_{2.5}$ sequence before predicting. How to fill in these missing values by improving XGBoost is the focus of scholars. The handling of missing values is a common method in predicting $PM_{2.5}$ concentration, especially the problem of non-random missing data (Chen et al., 2019; Han et al., 2022). For example, Wang et al. (2022) used wavelet decomposition (WD) to solve the Aerosol Optical Depth (AOD) missing problem in $PM_{2.5}$ prediction.

Another method to improve the prediction performance of the model is analyze the $PM_{2.5}$ sequence before predicting. This study uses empirical modal decomposition (EMD) to do the preprocessing (Zhu et al., 2017). EMD for $PM_{2.5}$ sequences preprocessing can improve the prediction accuracy of machine learning models. Furthermore, with the EMD method for $PM_{2.5}$ prediction, Jin et al. (2020) reconstructed the decomposed sequences by frequency, then predicted them separately, and finally merged the frequency sequences to obtain the predicted values. By doing so, the root mean square error on the test set is significant reduced. What's more, the prediction efficacy of neural network models could also be enhanced by EMD model (Huang et al., 2021). Thus, we apply EMD method to improve the predicting accuracy of our model.

The spatial distribution of $PM_{2.5}$ has long been a matter of interest to scholars. The model prediction of $PM_{2.5}$ distribution can help to discover its spatial characteristics. Liu et al. (2009) found that the spatial distribution of $PM_{2.5}$ was significantly different between urban and rural areas. The division of the study area according to administrative areas is a common practice of regional partitioning (Wang & Fang, 2016) when analyzing the spatial–temporal distribution of $PM_{2.5}$. What should also be noted in partitioning the study area is the altitude differences among each sub-area (e.g. plain area, river delta, the bottom of basin etc.) will also affect the distribution of $PM_{2.5}$ concentration (Ning et al., 2018). Thus, we innovatively utilize the inverse distance weighting (IDW) method to deal with data sets and we use kriging interpolation method to show the difference of $PM_{2.5}$ concentration in 16 districts in Shanghai. The combination of these two methods performs well.

The current research on XGBoost mainly focuses on how to build a combined model to improve XGBoost. One way is to combine multiple methods. For example, ridge regression (Liu et al., 2021) and three-dimensional landscape pattern index (TDLPI) (Zhang et al., 2022) both being added to XGBoost could make up for the low data monitoring accuracy of the air quality detector. A hybrid simulation model was developed by integrating the Multiscale Geographically Weighted Regression model (MGWR) and XGBoost (Li et al., 2025). Multi-source data including spatiotemporal, periodic, meteorological, vegetation, human and topological features could be integrated into the hybrid predicting model (Feng et al., 2020; Just et al., 2018; Zhang et al., 2020b), which could significantly contribute to the prediction accuracy if all variables were independent with each other. Another way is to adjust the XGBoost model parameters by using grid search method or genetic algorithm to optimize the predictions (Pan et al., 2022). In addition to its good performance in the prediction of continuous variables, XGBoost is also widely used in image recognition and classification (Ma et al., 2020b; Sheridan et al., 2016). However, up to date, the literature concerning searching the best weighing between each part of the combined model is still lacking. One innovation of us is that we efficiently find the best parameter of this gradient boosting algorithm and get the best weighing.

In general, machine learning algorithms tend to replace traditional statistical methods in the prediction of $PM_{2.5}$ (Camilleri et al., 2025; Cordero, 2025). Among them, gradient boosting algorithms are increasingly used for $PM_{2.5}$ prediction. In terms of variables affecting $PM_{2.5}$, the relationship between AOD data and $PM_{2.5}$ has become a research hotspot, but there is little discussion on other variables except meteorological factors and other pollutants. Thus, one of the contributions is that we put air pollutant indicators, meteorological environmental factors and macroeconomic indicators into XGBoost model in order to fully acquire the relationship between $PM_{2.5}$ and all these possible variables. Numerous studies have shown that fusion models can improve the accuracy of predictions. Therefore, the focus of this paper is to select a suitable model to fuse with XGBoost to improve the prediction effect. What's more, we find that the fusion of gradient boosting algorithms LightGBM, GBDT and XGBoost can enhance the complementarity of the first-layer prediction model to a certain extent. Furthermore, we weighted the predicted $PM_{2.5}$ values by the three gradient lifting algorithms, and found that the new sequence formed was closer to the actual $PM_{2.5}$ value.

7.2 Research Methods and Modeling

7.2.1 Important Parameter Choice of Main Model

7.2.1.1 Revised Empirical Mode Decomposition

$PM_{2.5}$ concentration data are nonlinear non-stationary time series with daily fluctuations. When there are too many subsequences decomposed by EMD, or modal aliasing occurs after decomposition, the SSA method can be applied to decompose the sequence. After the window width is specified, the decomposed submatrix is reconstructed according to Eq. (7.1):

$$e_k = \begin{cases} \frac{1}{K} \cdot \sum_{P=1}^{K} e^*_{P,K-P+1}, & 1 \leq K < m \\ \frac{1}{m} \cdot \sum_{P=1}^{m} e^*_{P,K-P+1}, & m \leq K < n \\ \frac{1}{N-k+1} \cdot \sum_{P=K-n+1}^{N-n+1} e^*_{P,K-P+1}, & n \leq K < N \end{cases} \tag{7.1}$$

where, e_P is the decomposed subsequence; e_K is the reconstructed high-frequency, low-frequency and residual sequences; m and n are the given widths. The sequences with low correlation (correlation coefficients between all the decomposed subsequences and the original $PM_{2.5}$ sequence less than 0.15) will be removed. According to the index value, the subsequences are classified into three categories: high frequency, low frequency and residual. After that, the three reconstructed sequences are predicted respectively.

7.2.1.2 The Parameter Choice of XGBoost Modeling

Given that XGBoost uses a second-order Taylor expansion, a quadratic function can improve the accuracy of the approximation. At this time, the default regular term selected is L_2, that is, the square of the norm of the second fundamental form. Chen and Guestrin (Chen & Guestrin, 2016) once gave the regular term of the gradient boosting algorithm.

$$\Omega(f(t)) = \gamma T + \frac{1}{2}\bar{\lambda} w^2 \tag{7.2}$$

where $\Omega(f(t))$ is the complexity of the model in the t-th round iteration, w is the weight of the leaf node, γ is the learning rate, T is the number of the leaf nodes, and $\bar{\lambda}$ is the L_2 regularization term for the weight.

At the same time, in the Eq. (7.2), the L_1 regularization method makes it easier for the parameter $\bar{\lambda}$ to be 0, so it is not as effective as L_2 for non-sparse dependent variables. In addition, since the conditions for L_1 to obtain the extreme value at point 0 are looser than that for L_2, the number of optimal parameters will increase, or it will fall into a local optimal solution. Thus, the XGBoost model should choose the square term as the regular term.

The innovation lies that we remove those sequences which have lower relationship with original sequence and complete incomplete reconstruction in EEMD-XGBoost model when prediction. What's more, we analyze non-sparse $PM_{2.5}$ data set and sparse $PM_{2.5}$ data at different air quality monitoring stations by using L2 regularization method and L1 regularization method separately, which will affect the learning

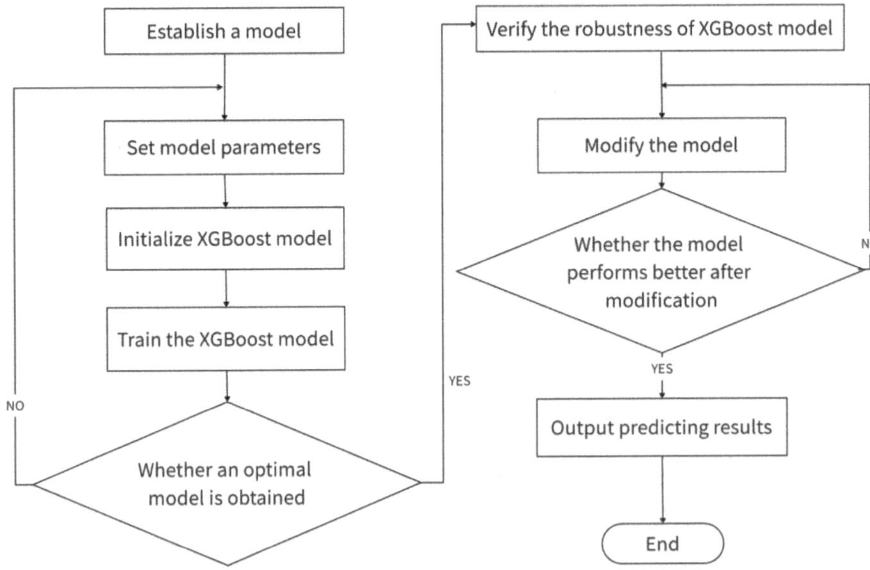

Fig. 7.1 XGBoost modeling flow chart

rate γ in XGBoost model. What we talked above could improve the accuracy of prediction.

The specific modeling process is shown in Fig. 7.1.

7.2.2 *Modeling Steps*

7.2.2.1 Establishment of a Spatiotemporal XGBoost Model Modified by Empirical Mode Decomposition

First, the parameters of the model are adjusted to determine the optimal parameters of XGBoost. Next, the original $PM_{2.5}$ concentration sequence after signal decomposition is substituted into the model as a dependent variable for prediction. The EMD-based modeling process is shown as in Fig. 7.2.

In Fig. 7.2, EEMD is to perform empirical mode decomposition on the original sequence, and the set of sub-sequences obtained is denoted as IMF. Then, the correlation between sub-sequences and the original $PM_{2.5}$ sequence is checked. Each subsequence is predicted and analyzed by XGBoost, and finally the predicted sequence is recombined to obtain the predicted value of $PM_{2.5}$ sequence. Next, a fusion model of the gradient boosting algorithm is constructed to enhance the stability of the XGBoost model, as set in Fig. 7.3

As shown in Fig. 7.3, the three gradient boosting methods are integrated, and then combined with the multiple linear regression for secondary prediction. The

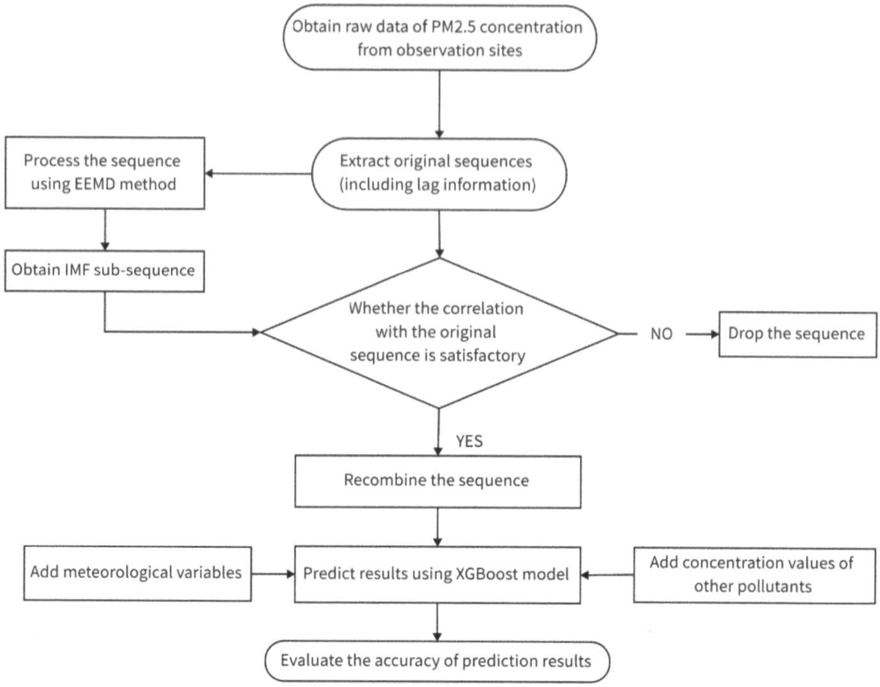

Fig. 7.2 EEMD-XGBoost modeling flow chart

prediction results of the integrated model and XGBOOST are compared. Herein, fivefold cross-validation is adopted. According to the time sequence, 20% of the data is set aside each time, and the remaining data is used for training. The XGBoost, LightGBM, and GBDT are respectively used as the first layer of the integrated model for prediction, and the error on the training set and test set is calculated. Then the first total sequence prediction value is obtained by weighted average. The method of variable weight assignment (Eqs. (7.6)–(7.8)) is adopted to determine the weight to correct the prediction error at the extreme point.

$$f(X_i) = \frac{1}{n}\sum_{i=1}^{3} W(t) f_i(X_i) \tag{7.3}$$

where $W_i(t)$ is the weight of the i-th model at time t,

$$W_i(t) = \frac{\frac{1}{\varepsilon_l(t)}}{\frac{1}{\sum_{i=1}^{3}\overline{\overline{\varepsilon_l(t)}}}} = \frac{\sum_{i=1}^{3}\overline{\overline{\varepsilon_l(t)}}}{\varepsilon_l(t)} \tag{7.4}$$

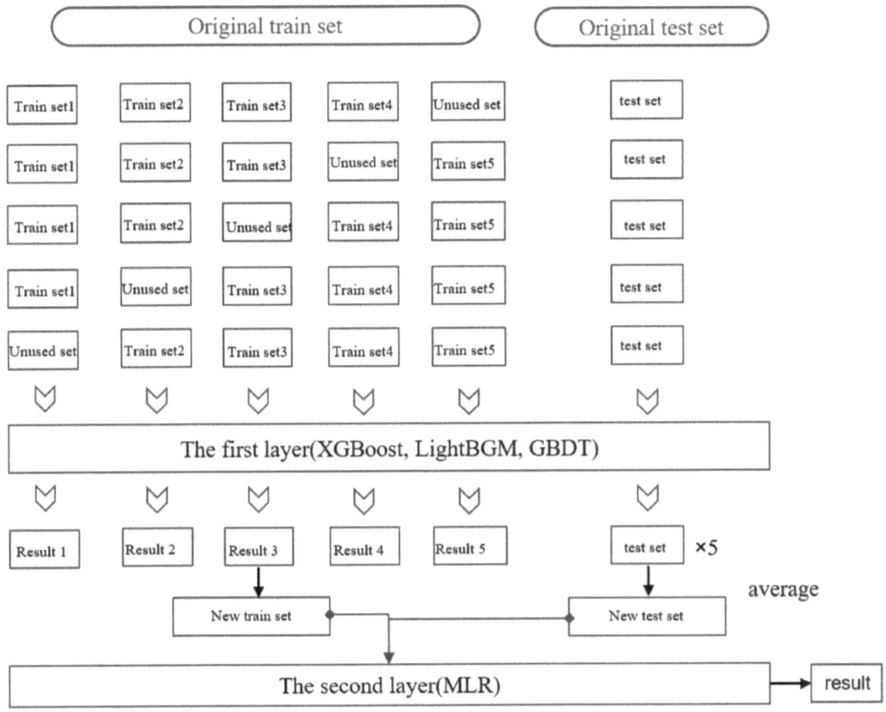

Fig. 7.3 Improved XGBoost model by integration

where $\varepsilon_l(t)$ is the prediction error of the l-th model at time t and it satisfies $W_i(t) > 0$. The prediction results of XGBoost model with variable weight modification can be obtained by substituting Eq. (7.4) into Eq. (7.3),

$$s.t. \sum_{i=1}^{3} W_i(t) f_i(X_i) \tag{7.5}$$

In view of the fact that a weight can be obtained at each observation moment (there may be errors), the weight at each moment is adjusted by means of residual weighting. Then the corresponding optimal weight is calculated by

$$W_j(t) = \frac{1}{m} \sum_{k=1}^{m} W_j(t-k),\ m = 3 \tag{7.6}$$

where the weight before the correction at time t is obtained by weighting the weight coefficients of the three adjacent periods. At the same time, the absolute error between the predicted value and the actual value at this time and the next period of this time can be calculated, i.e.,

$$e_{i,t} = \sum_{i=1}^{n} W_i(t) \cdot f_i(X_i) - f(t) \tag{7.7}$$

$$e_{i+1,t} = \sum_{i=1}^{n} W_{i+1(t)} \cdot f_{i+1}(X_{i+1}) - f(t) \tag{7.8}$$

The values of residual errors $e_{i,t}$ and $e_{j,t}$ are compared. If $e_{i,t} < e_{j,t}$, the weight combination of the previous moment with small error replaces the weight combination of the current period; if $e_{i,t} \geq e_{j,t}$, the weight of the current period is retained. The weight combination of three periods is considered, which enhances the stability of the model. In this way, EEMD-XGBoost completes fusion and adaptive weight adjustment.

7.2.2.2 Importance Ranking and Spatial Effect Analysis

The XGBoost model can solve classification and regression problems, and explore the influence of various factors on $PM_{2.5}$ concentration. Since $PM_{2.5}$ concentration is a continuous variable, the dependent variable needs to be classified first. In the process of outlier processing, the observation points greater than or less than 1.5 standard deviations from the mean are regarded as outliers Therefore, the values calculated from Eqs. (7.9) and (7.10) are the upper and lower thresholds for determining whether a certain point is an outlier,

$$y_L = \bar{y} - 1.5.\delta_y \tag{7.9}$$

$$y_U = \bar{y} + 1.5.\delta_y \tag{7.10}$$

where y_L and y_U are the upper and lower limits of the reasonable estimation interval of y, respectively, and δ_y is the standard deviation of the $PM_{2.5}$ concentration sequence. All values in the concentration sequence are traversed to determine whether the value is an outlier. Outliers greater than y_U are marked as 1, the outlier smaller than y_L as − 1, and the normal values as 0. This categorical variable is regarded as a dependent variable.

In order to analyze the interaction between measured $PM_{2.5}$ concentration values at each observation point, the predicted $PM_{2.5}$ concentration values are modified by inverse distance weighting. The steps to construct the distance weight matrix are as follows: the longitude and latitude coordinates of 19 monitoring stations in Shanghai are obtained by accurate positioning with the help of Google Map. The latitude and longitude can be converted into the arc length of the corresponding spherical surface through the spherical coordinate system. Let the coordinates of a point in space be $P(X,Y,Z)$, where X and Y are the coordinates of the observation point in the space cartesian coordinate system, and Z is the height of the point from the horizontal

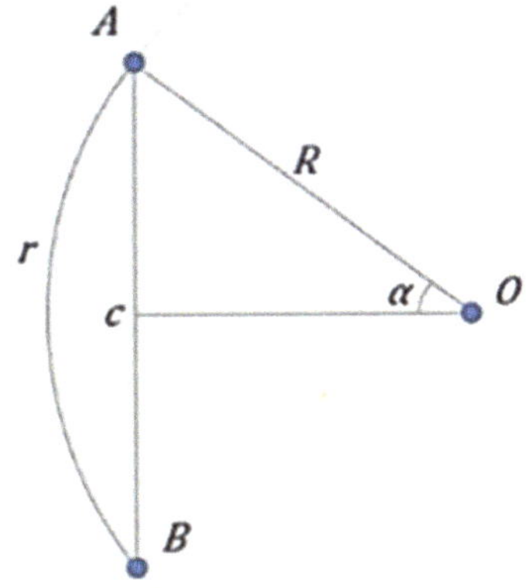

Fig. 7.4 Calculation of the real distance between two points *A* and *B*

ground. Let (*e*,*n*) be the latitude and longitude of a point *A* on the earth's surface and the position *P* can be obtained from Eqs. (7.11)–(7.13):

$$X = R \cdot \cos(n) \cdot \cos(e) \tag{7.11}$$

$$y = R \cdot \cos(n) \cdot \cos(e) \tag{7.12}$$

$$z = R.\sin(n) \tag{7.13}$$

The radius of the earth is 6371 km, denoted as *R*. The length of the minor arc corresponding to the string *AB* (the relationship of each variable is shown in Fig. 7.4).

Namely, the distance between stations *A* and *B*, can be obtained by Eqs. (7.14)–(7.16):

$$c = 2 \cdot R \cdot \sin(\alpha) \tag{7.14}$$

$$\alpha = \arcsin(\frac{\mathrm{c}}{2 \cdot \mathrm{R}}) \tag{7.15}$$

$$\mathrm{r} = 2 \cdot \alpha \cdot \mathrm{R} \tag{7.16}$$

Wu et al. (2003) utilized the inverse distance weighting (IDW) method to determine the weight of the pollution from a wildfire in the studied area contributed to overall $PM_{2.5}$ concentration of rest of the region. Next, the inverse distance weighting method is used to reprocess the predicted values of all stations, and the process is shown in Eqs. (7.17) and (7.18):

$$P_d = \frac{\sum_{i=1}^{n} \frac{P_i}{d_i^2}}{\sum_{i=1}^{n} \frac{1}{d_i^2}} \tag{7.17}$$

$$P_t = t \tag{7.18}$$

where *i* is the *i*-th station around and *t* is the observation time. The estimation accuracy of the model can be improved by introducing spatial factors. Regarding the research on distance weight, some scholars proposed the bandwidth selection method (Shang et al., 2019):

$$W_{i,j} = \begin{cases} \left[1 - \left(\frac{d_{ij}}{b_j}\right)^2\right]^2, & d_{ij} < b_j \\ 0, & d_{ij} > b_j \end{cases} \tag{7.19}$$

$$PM_{2.5}(u_i, v_i) = \sum_{j=1}^{n} W_{i,j} \cdot F_j(X_i) \tag{7.20}$$

where *n* is the total number of stations; (u_i, v_i) are the spatial coordinates; *i* and *j* are stations; b_j is the bandwidth, kilometer; $W_{i,j}$ is the weight. The $PM_{2.5}$ concentration at station *i* is obtained by weighting. When the distance between two stations is smaller than the given bandwidth, the station does not participate in the weighting. Wu et al. (Wu et al., 2003) applied Kriging interpolation to study the differences in $PM_{2.5}$ concentrations, which provides a method for studying the spatial network distribution of $PM_{2.5}$ concentrations. In our research, the revised Kriging interpolation method is used to fill in the predicted value of the whole Shanghai, and the main steps are as follows:

Let Z_0^* be the $PM_{2.5}$ concentration to be estimated, and Z_0 and Z_i be the $PM_{2.5}$ concentrations at the known and unknown stations, respectively. First, the cost function *J* is introduced, and then the calculation formula of $PM_{2.5}$ concentration at unknown station is substituted into *J* to obtain Eq. (7.22):

$$Z_0^*(x_0, y_0) = \sum_{i=1}^{n} \lambda_i \cdot Z(x_i, y_i) \tag{7.21}$$

$$J = Var(Z_0^* - Z_0) \tag{7.22}$$

$$J = \sum_{i=0}^{n} \sum_{j=0}^{n} \lambda_i \cdot \lambda_j Cov(Z_i, Z_j) - 2 \cdot \sum_{i=1}^{n} \lambda_i \cdot Cov(Z_i, Z_0) + Var(Z_0) \tag{7.23}$$

Therefore, the concept of semi-variance is introduced as in Eq. (7.23), where λ_i is the combination coefficient which needs to satisfy the constraints of Eq. (7.24):

$$r_{ij}= \sigma^2 - Cov(x_i, y_i) \tag{7.24}$$

$$\sum_{i=1}^{n} \lambda_i = 1 \tag{7.25}$$

Based on the obtained distance between any two monitoring points, Gaussian function is chosen to determine the semi-variance between any two points. The relationship between distance and semi-variance is shown in Eq. (7.26):

$$Cov(Z_i, Z_j) = f\left(\sqrt{(x_i - x_j)^2 + (y_i - y_j)^2}\right) = f(x) = a \cdot \exp\left(-\frac{(x - b)^2}{2 \cdot c^2}\right) \tag{7.26}$$

where *a*, *b*, and *c* are all constant.

Equation (7.26) is used as the Lagrangian operator to derive the cost function in Eq. (7.27), and the equation system is obtained as below:

$$\begin{cases} r_{i0} - \lambda_i \cdot r_{ij} = 0 \\ \sum_{i=1}^{n} \lambda_i = 1 \end{cases} \tag{7.27}$$

The combination coefficients $\lambda_1, \lambda_2, \ldots, \lambda_n$ under the minimum estimated variance can be obtained by solving equation system (7.27). By substituting these coefficients into Eq. (7.21), the $PM_{2.5}$ concentration at any location in Shanghai can be estimated. Then the $PM_{2.5}$ predictions of all stations in Shanghai can be obtained.

The solution of the Eq. (7.30) must satisfy the two constraints (7.28) and (7.29) (Kriging interpolation conditions):

$$\underbrace{\min}_{\lambda_i} Var\left(Q_d(x_0) - \sum_{i=1}^{n} \lambda_i \cdot Q(x_i)\right) \tag{7.28}$$

$$E\left(\overline{Q_d(x_0)} - \sum_{i=1}^{n} \lambda_i \cdot \overline{Q(x_i)}\right) = 0 \tag{7.29}$$

where $Q_d(x_0)$ is the spatial $PM_{2.5}$ attribute of Shanghai, and λ_i is the weighted combination coefficient of five observation stations.

After filling by Kriging interpolation, the predicted concentration of $PM_{2.5}$ in Shanghai at each time can be obtained.

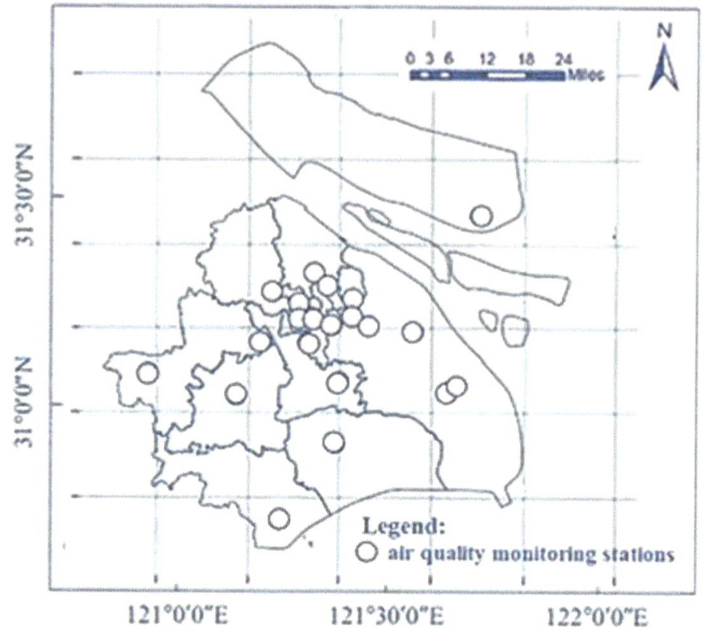

Fig. 7.5 Distribution of air quality monitoring stations (only) in Shanghai

7.2.3 Data Preparation

7.2.3.1 Introduction to the Study Area

In this paper, the data of 19 environmental quality monitoring stations in Shanghai is selected for analysis. Shanghai is located in the Yangtze River Delta region along the east coast of China with small topographic relief. It is a typical megacity with an area of about 6340 km^2 and has a subtropical monsoon climate. The air quality monitoring stations in China and Shanghai are shown in Fig. 7.5.

The whole city is divided according to the administrative boundaries, and the meteorological data obtained are rasterized by inverse distance weighting method. Then the data will be projected and unfolded at each point. After that, the hourly concentrations of other pollutants and meteorological data are converted into daily data.

7.2.3.2 Data Acquisition and Preparation

The dataset in this paper can be obtained from the website by using python. The data can be divided into 3 categories: concentrations of other pollutants, meteorological data, and macro variables. Specifically, the concentrations of other pollutants (AQI,

SO_2, CO, PM_{10}, O_3, and NO_2) can be obtained from Zhenqi.com (https://www.zq12369.com). The meteorological data (2013–2021) includes the sunshine duration, humidity and wind speed can be obtained from China Meteorological Administration (http://www.cma.gov.cn), and the precipitation data (2013–2021) is available from European Centre for Medium-Range Weather Forecasts (https://www.ecmwf.int/), and the temperature data (2013–2021) from the National Oceanic and Atmospheric Administration (https://www.noaa.gov). The macro variables (population density, car ownership, coal use, sunny days, etc.) can be obtained from the statistical yearbook of Shanghai over the years. Other pollutant concentrations and meteorological data include daily and hourly data, macro variables are annual data, and car ownership and electricity consumption are monthly data. The higher the proportion of coal in energy consumption, the more serious the air pollution caused by the secondary industry. What's more, the correlation between coal and energy consumption is lower than 10% showing that it's necessary to take both variables into consideration. While GDP are important indicators of the economy, we find that the high correlation between GDP, population density and private car ownership could bring multicollinearity (month by month data) which will lead to the failure of significant test when analyzing the relationship of $PM_{2.5}$ and these 3 variables.

The variables and symbols involved in the empirical analysis of this paper are listed in Table 7.1.

The calculation of the average daily $PM_{2.5}$ concentration in the city is based on the *Technical Specifications for Ambient Air Quality Assessment (Trial)* issued by the Ministry of Environmental Protection (Ministry of Environmental Protection, 2013). The daily average concentration of $PM_{2.5}$ at an observation station is the arithmetic mean of the hourly $PM_{2.5}$ concentration of the station from 0:00 to 24:00 every day. The proportion of missing values in the observed $PM_{2.5}$ concentration data of all observation stations in the city at a certain time t is calculated. When the proportion is less than 25%, the arithmetic mean of all observation stations is the $PM_{2.5}$ concentration in Shanghai at that time. When there are many missing data, that is, the proportion is $\geq$ 25%, the maximum among all observed values is taken as the $PM_{2.5}$ concentration in Shanghai (Huang et al., 2021). For example, Jung et al. (Jung et al., 2021) adopted this method when dealing with the missing $PM_{2.5}$ concentration values in Japan, and achieved good results.

7.3 Empirical Analysis

7.3.1 Screening of Important Stations in Shanghai

A hierarchical analysis method was applied by Liang et al. (Liang et al., 2019) for an index system in evaluating environmental pollution levels in an urbanization process. As based on entropy weight method and principle of minimum information entropy, such method has an important reference in determining the impact weights

Table 7.1 Variables and symbols involved in this paper

Index	Variable name	Symbol
Air pollutant indicators (Type I)	Inhalable micro particulate matter	$PM_{2.5}$
	Inhalable particulate matter	PM_{10}
	Carbon monoxide	CO
	Nitrogen dioxide	NO_2
	Sulfur dioxide	SO_2
	Ozone	O_3
Meteorological environmental indicators (Type II)	Precipitation	X_1
	Sunshine duration	X_2
	Humidity	X_3
	Wind speed	X_4
	Average temperature	X_5
	Evaporation	X_6
Macroeconomic indicators (Type III)	Private car ownership	Cars
	Population density	density_of_people
	Annual coal use	coal_amount
	Energy loss	loss_of_energy
	Number of sunny days	number_of_good_weather
	Air quality rate	rate_of_excellent_whethe
	Energy consumption	Energy
	Total population	people

of different regions. The 19 observation stations in Shanghai are scored by entropy weight. Let X_{ij} be the j-th evaluation index of the i-th observation station, and Y_{ij} be the index after standardization. P_{ij} is the proportion of the j-th pollution gas to the i-th observation station, E_j is the entropy of each indicator, W_j is the weight of each indicator, and Z_i is the score of each station.

$$P_{ij} = \frac{X_{ij}}{\sum_{i=1}^{m} X_{ij}},\ i = 1, 2, \ldots, m,\ j = 1, 2, \ldots, n \tag{7.30}$$

$$E_j = -\frac{1}{\ln(m)} \cdot \sum_{i=1}^{m} P_{ij} \cdot \ln\left(P_{ij}\right), j = 1, 2, \ldots, n \tag{7.31}$$

$$W_j = \frac{1 - E_j}{\sum_{i=1}^{m} (1 - E_j)},\ j = 1, 2, \ldots, n \tag{7.32}$$

$$Z_i = \sum_{j=1}^{n} W_j \cdot Y_{ij}, j = 1, 2, \ldots, n \tag{7.33}$$

The score Z_i of each station is converted, ranging between 60 and 100. The higher the score, the larger the pollution index. The results of pollution level are shown in Table 7.2.

It can be seen from Table 7.2 that the heavily polluted areas are concentrated in Yangpu, Jingan, Hongkou, Xuhui and other urban areas, while Qingpu,

Table 7.2 Air quality results evaluation at 19 observation stations in Shanghai

Station name[a]	$PM_{2.5}$	Industrial waste gas (sulfur nitride)	Others	Comprehensive index of air quality	Score
Yangpu Sipiao	30.360	30.546	20.199	18.895	100.000
Jing'an Observation Station	31.438	23.597	20.606	19.431	95.072
Putuo	30.657	25.038	19.455	19.197	94.347
Hongkou	28.913	29.067	18.447	17.769	94.195
Shanghai Normal University, Xuhui Campus	29.853	25.824	20.023	17.977	93.676
Nanxiang Town, Jiading	26.897	28.635	17.742	17.036	90.310
Miaohang Town, Baoshan	30.349	22.287	17.467	19.799	89.902
Shiwuchang Station	29.357	21.170	20.646	17.665	88.837
Zhangjiang, Pudong	28.736	22.821	18.706	17.422	87.684
Pudong New Area Observation Station	27.957	22.795	19.298	17.500	87.550
Xujing Town, Qingpu	29.559	19.377	17.641	18.809	85.385
Chuansha Town, Pudong	26.866	17.124	19.228	16.799	80.017
Songjiang Library	27.997	16.143	18.005	17.750	79.895
Xianxia Street, Changning	29.499	13.060	17.606	19.062	79.228
Huinan Town, Pudong	25.402	14.733	18.407	17.102	75.645
Pujiang Town, Minhang	28.134	10.720	17.722	17.949	74.525
Fengxian Nanqiao New City	26.759	9.858	19.013	17.952	73.581
Jinshan New Town	26.662	1.685	19.501	17.500	65.348
Dongtan School Affiliated to Shanghai Experimental School, Chongming	22.297	5.417	19.057	15.186	61.956

[a] The evaluation ranking is sorted by score of each observation station. Higher scores indicate worse air quality

Pudong, Songjiang, Jinshan, and Fengxian—districts located in the suburbs—are less polluted, with relatively good air quality. Among the five new urban districts, Jiading has slightly worse air quality, which is related to the proportion of industry and population density.

According to the $PM_{2.5}$ concentration, the pollution can be divided into three levels: severe pollution, mild pollution, and light pollution, and the corresponding score ranges of the station are 90–100, 80–90, and 60–80, respectively. To ensure selected stations are representative, after the importance of each station is calculated by XGBoost model, the top two sites with the highest importance are selected respectively from the three pollution levels. It should be pointed out that the importance of the low pollution group is generally low, so only Minhang station is retained. Finally, the five most important observation stations, Pudong (Huinan Town), Jiading (Nanxiang Town), Qingpu (Dianshan Lake), Hongkou and Minhang (Pujiang Town) are selected to construct the weight matrix.

7.3.2 Analysis of the Importance of Meteorological Variables

All matched variables are independent variables, $PM_{2.5}$ concentration is the dependent variable. The importance of independent variables is ranked by XGBoost and the results are shown in Fig. 7.6. Generally, when the importance of a variable is lower than 20%, the variable is considered to have a weak correlation with $PM_{2.5}$, so the variable should be eliminated. Next, stepwise regression is performed on the remaining variables that have passed the importance screening, and variables that fail the significance test are removed.

For annual data such as cumulative precipitation, evaporation, and number of sunny days, it is necessary to convert them into monthly data according to frequency, and then match them with the monthly average $PM_{2.5}$ concentration to expand the sample size, and further study their relationship with $PM_{2.5}$. Herein, the start time of 2003 is $t = 1$, and the end time of 2021 is $t = 18$. Each year is an interval, and the cubic spline interpolation in Eq. (7.34) is used for data expansion. The value of α_{ij} is determined by the chosen fitting function and can be given by Eviews fitting

$$f(t) = \alpha_{1i} + \alpha_{2i} \cdot (t - t_i) + \alpha_{3i} \cdot (t - t_i)^2 + \alpha_{4i} \cdot (t - t_i)^3 \tag{7.34}$$

where $t_i \leq t \leq t_{i+1}$ and $i = 1, 2, N - 1$.

There is a lag correlation between some variables, such as air temperature and wind speed, and $PM_{2.5}$. For these data, we first substitute them into Eq. (7.35), and the lag period is determined by multiple regression. Then Eq. (7.34) is applied to calculate the correlation of each variable with $PM_{2.5}$.

$$PM_{2.5} = \beta_0 + \beta_1 . A_{t-i} + \beta_2 . B_t \tag{7.35}$$

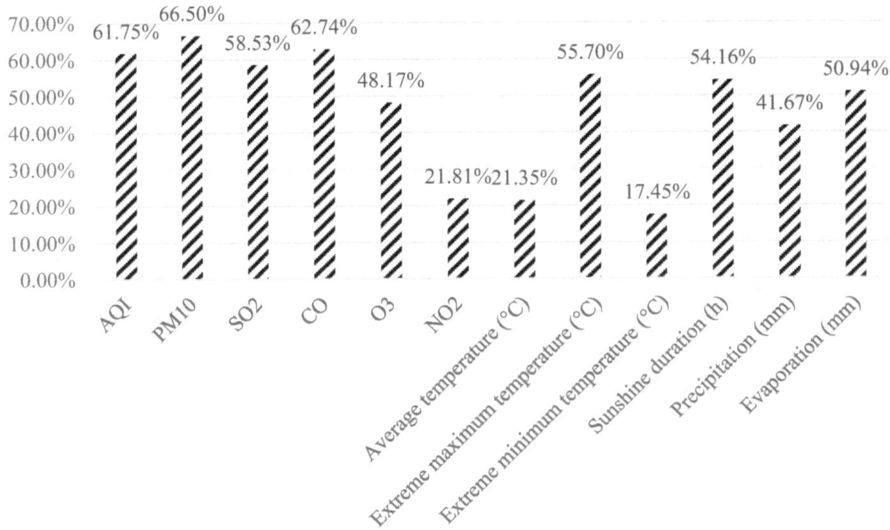

Fig. 7.6 Importance ranking of environmental factors affecting $PM_{2.5}$ concentration

where, $A_{t-i} = (X_1, X_2, X_3, \ldots)^T$ represents the column vector composed of variables with lag compared with PM2.5; $B_t = (B_1, B_2, B_3, \ldots)^T$ represents the column vector composed of variables with synchronization compared with $PM_{2.5}$; and $\beta_1 = (\alpha_1, \alpha_2, \alpha_3, \ldots)$ and $\beta_2 = (\gamma_1, \gamma_2, \gamma_3, \ldots)$ are the coefficient matrices of each independent variable.

$$\mathrm{VIF} = \frac{1}{1 - R^2} \tag{7.36}$$

Forming together vectors A_{t-i} and B_t retrieved from Eq. (7.35), we obtain the matrix $X = \left[A_{t-i}\ B_t\right]$ and calculate the eigenvalues of the matrix X^TX. The smaller the absolute value of the eigenvalues of a variable, the higher possibility of multicollinearity between this variable with other independent variables would be. Using Eq. (7.36), we can then determine whether the variable should be retained.

After frequency conversion, the importance of variables is calculated, as shown in Fig. 7.6

Among the variables in Fig. 7.6, extreme minimum temperature fails the significance test, and other variables are finally determined as independent variables. The importance of independent variables is shown as in Fig. 7.6. Among all indicators, the PM_{10} concentration ranks first in importance for predicting $PM_{2.5}$ concentration, with a correlation of 66.5%, and AQI, CO and SO_2 has a correlation of about 60%. Precipitation and monthly evaporation have a great impact (negative) on $PM_{2.5}$ concentration. The reason is that the increase in air humidity will take away part of the pollutants and cause part of the gas particles to settle, which is conducive to improving air quality.

7.3.3 Analysis of Factors Affecting $PM_{2.5}$

7.3.3.1 Ranking of the Importance of Factors Affecting $PM_{2.5}$

The correlation between each variable and $PM_{2.5}$ is determined according to the XGBoost importance score (AUC value). Its disadvantage is that dependent variables need to be transformed into categorical variables before screening. Note that the data collection frequency of macroeconomic factors cannot be measured in terms of days and thus needs to be converted by Eq. (7.34). Therefore, compared with meteorological factors, the sample size of macroeconomic counterparts is relatively small. We notice the effectiveness of grey correlation analysis in this case, as can solve the probable error caused by insufficient sample size, and is more helpful to explore the relationship between macroeconomic factors and $PM_{2.5}$. Therefore, Eqs. (7.37) and (7.38) can be used to investigate the relationship of each time series. This method considers the sequence of each element and can make up for the error caused by the absence of sequence.

$$\varepsilon_i = \frac{\underbrace{\min}_{i}\underbrace{\min}_{k}|X_0(k) - X_i(k)| + \rho \cdot \underbrace{\max}_{i}\underbrace{\max}_{k}|X_0(k) - X_i(k)|}{|X_0(k) - X_i(k)| + \rho \cdot \underbrace{\max}_{i}\underbrace{\max}_{k}|X_0(k) - X_i(k)|} \tag{7.37}$$

$$r_i = \frac{\sum_{k=1}^{n} \varepsilon_i(k)}{n} \tag{7.38}$$

where ρ is the resolution coefficient, which it is set to 0.5, the commonly used value in literature; ε_i is the grey correlation coefficient between each variable and $PM_{2.5}$ concentration at a certain time; $\underbrace{\min}_{i}\underbrace{\min}_{k} X_0(k) - X_i(k)$ and $\underbrace{\max}_{i}\underbrace{\max}_{k} X_0(k) - X_i(k)$ are the minimum and maximum differences; r_i is the grey correlation index between the independent variable and $PM_{2.5}$ concentration.

The frequency conversion of macro variables uses Eq. (7.34), and the conversion function is the quadratic interpolation curve in the Eviews (this function can be modified until the fitting error is minimal). For data such as population density and industrial structure distribution in Shanghai, the fluctuation of data is not obvious after changing to daily data, so the constant method is adopted for conversion. After conversion, Fig. 7.7 can be obtained by Eqs. (7.37) and (7.38).

It can be seen from Fig. 7.7 that the private car ownership, population density, coal consumption, and total population in Shanghai are negatively correlated with $PM_{2.5}$ concentration, and the correlation coefficients are − 58.7%, − 45.15%, − 27.55%, and −68.52%. Among them, private car ownership and total population is highly correlated with $PM_{2.5}$ concentration, and their correlation coefficients both exceed 55%. In recent years, the $PM_{2.5}$ concentration has not shown a fluctuating upward trend with the popularization of private cars and the massive influx of population in Shanghai. The number of days with severe pollution was 46 between 2013 and

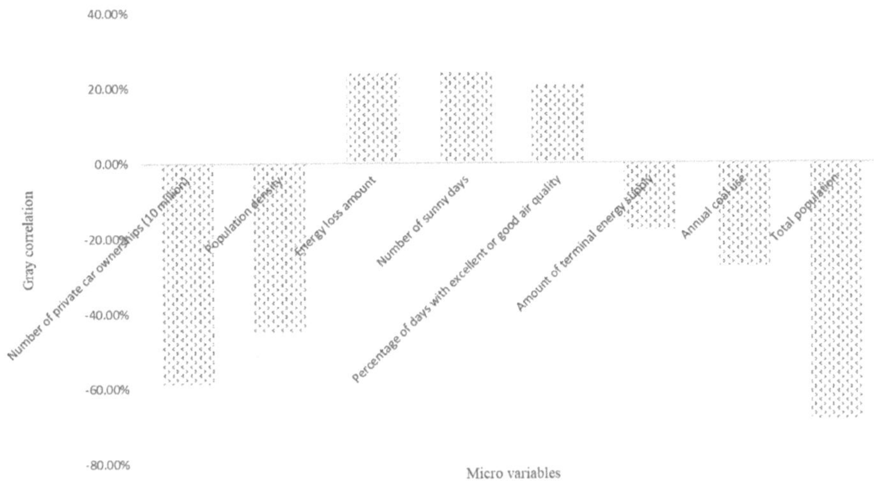

Fig. 7.7 Importance of economic factors

2015, while the total number of days with severe pollution in the six years from 2016 to 2021 was less than 40, indicating that Shanghai's air quality is improving. Due to Shanghai's special license plate policy, the choice of new energy vehicles is a favorable means for migrants to obtain license plates, resulting in the growth of private cars mainly concentrated in the new energy field, which explains the negative correlation. In addition, Pudong, Songjiang and other new areas are developing high-tech industries, and the secondary industry is also moving to the surrounding areas of Shanghai. Such factors are possible influencing factors of $PM_{2.5}$ concentration.

7.3.3.2 Prediction After Introduction of Different Variables

This section will discuss whether to add these three types of variables (air pollutant indicators, meteorological environmental indicators, and macroeconomic indicators), and compare the prediction accuracy and corresponding error of the XGBoost model before and after introduction of different independent variables, as shown in Table 7.3.

After introduction of variables of type II, the average decision error and root mean square error of the model are slightly reduced and the goodness of fit is slightly improved on both the training set and the test set. The only environmental variable that has a greater impact on $PM_{2.5}$ concentration is rainfall. After introduction of environmental variables, the error term is improved by less than 5%, and the goodness of fit is not significantly enhanced. The goodness of fit does not decrease after removing the environmental variables with low correlation such as the highest temperature, lowest temperature and sunshine duration, so they can be eliminated.

Table 7.3 Prediction results comparison after variables are added

Variable added	Group	MAE	Goodness of fit (R^2)	RMSE	RPE
Type I	Trained	0.8295	99.08%	1.179	2.11%
	Test	3.7490	89.45%	4.898	9.05%
Types I and II	Trained	0.7796	99.13%	1.149	2.06%
	Test	3.6735	89.89%	4.801	8.86%
Types I and III	Trained	0.7980	99.14%	1.140	2.04%
	Test	3.7737	89.32%	4.927	9.11%
Types I, II, and III	Trained	0.7726	99.17%	1.122	2.01%
	Test	3.8586	88.60%	5.091	9.41%

MAE: Mean absolute prediction error; RMSE: Root mean square error; RPE: Relative prediction error

Also shown in Table 7.2, the performance of the model on the training set is improved to a certain extent after the introduction of the processed type I variables. Moreover, the goodness of fit is increased by about 1%, and the relative error is reduced. By contrast, on the test set, the effect of these four evaluation indicators decline.

Similarly, when the three types of variables are introduced at the same time, the evaluation indices on the training set are further improved. However, on the test set, the relative prediction error increases by 0.36% and the mean square error drops by 2.28%. When only precipitation is added to the model, the error on the test set and the training set are reduced, and the performance on the training set and the testing set is consistent. It is therefore can be concluded that the prediction accuracy of the XGBoost model cannot be improved by simply increasing the number of variables.

7.3.4 Empirical Results

7.3.4.1 Signal Decomposition Results

The EEMD model is used to decompose the $PM_{2.5}$ sequence, and the results are shown in Fig. 7.8.

Among the 11 subsequences decomposed, the first five IMFs are high-frequency sequences, the 6–11 sequences are low-frequency sequences, and the last item is the residual term. Generally, when the number of integrations is set to around 100 and the original sequence is processed with white noise, the decomposition effect can be improved and the number of iterations required for convergence can be effectively reduced. First, the mean square error of the original sequence is calculated, and then the variance of the white noise is obtained by introducing the coefficient factor for correction. Then, white noise is added to the original sequence to get a new sequence. It is found that when the standard deviation of white noise is 0.1, the effect is better.

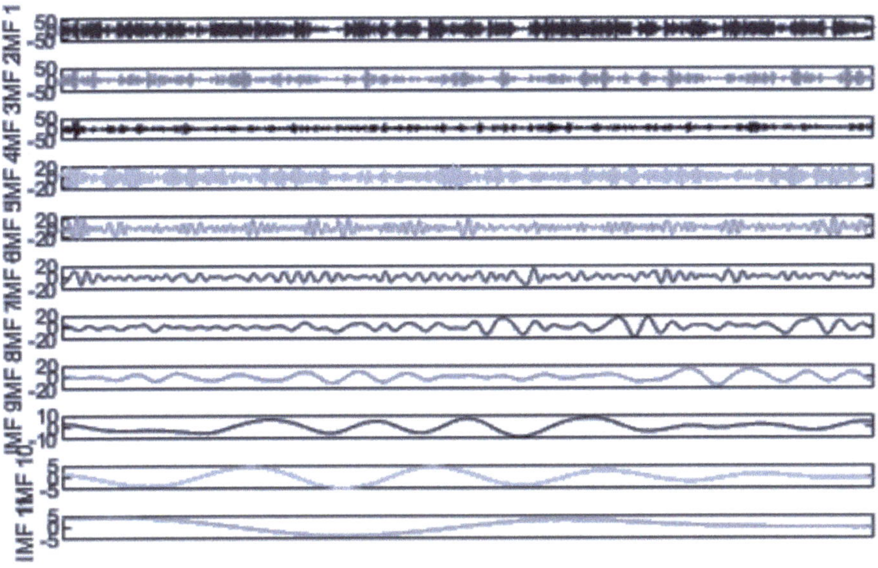

Fig. 7.8 Sequence decomposition effects using EEMD method

Therefore, noise_width is set to 0.1, which is added to the 11 IMF subsequences and the 1 residual sequence obtained after decomposition. The residual sequence shows the original variation trend of the daily average concentration of $PM_{2.5}$ in Shanghai. The characteristics of the decomposed subsequences are shown in Table 7.4.

Table 7.4 Characteristics of decomposed sequences using EEMD method

Sequence	Index						
	Kendall's correlation coefficient	Variance contribution rate	Pearson's correlation coefficient	Spearman's correlation coefficient	Period (day)	Sample entropy	Number of run
IMF1	0.5280	48.34%	0.7326	0.7239	1.4309	1.2538	1496
IMF2	0.3243	8.13%	0.4609	0.4704	2.5209	0.4688	1114
IMF3	0.2960	4.75%	0.4215	0.4350	3.5185	0.8775	760
IMF4	0.2405	4.63%	0.3505	0.3556	6.2147	0.6269	431
IMF5	0.1926	3.52%	0.2789	0.2854	11.4419	0.5679	244
IMF6	0.1527	1.7%	0.2322	0.2275	21.2374	0.4532	127
IMF7	0.1733	2.61%	0.2559	0.2528	42.1714	0.1868	60
IMF8	0.1593	2.35%	0.2378	0.2362	86.8235	0.0994	32
IMF9	0.0967	1.19%	0.1518	0.1444	184.5	0.0413	28
IMF10	0.0675	0.5%	0.0989	0.0994	295.2	0.0263	10
IMF11	0.0369	0.43%	0.0641	0.5513	984	0.0069	4
R	0.0256	0.58%	0.0485	0.0373	2952	1.1324×10^{-4}	2

From the decomposed sequence, it can be seen that from IMF1 to IMF11, the frequency of subsequences decreases and the number of periods increases. The average number of periods per sequence is obtained by dividing the total number of observation days by the number of extreme points. The average period of IMF3 is 6.2147 days, which is about one week; the period of IMF5 is 21.2374 days, more than half a month; the period of IMF6 is 42.1714 days, which is one and a half months; the average period of IMF8 is 184.5 days, which is about half a year. It can be seen that the $PM_{2.5}$ concentration in Shanghai changes roughly on a weekly, monthly and quarterly basis.

IMF1 is the most important subsequence with a variance contribution rate of 48.34%. The variance contribution rates of IMF2–IMF4 are 8.13%, 4.75%, 6.63% and 3.52%, respectively. The rest of the subsequences have low contribution rates. The cumulative variance contribution rate of the first five items reaches 69.37%, nearly 70%. The variance contribution rate of all subsequences is 78.73%. Overall, IMF1–IMF4 have strong correlation with the original sequence.

The mean of each subsequence is calculated respectively. The data points less than the mean in each subsequence are denoted as " − ", and the points larger than the mean are denoted as " + ". Therefore, 12 symbol sequences (11 components and a residual vector) can be obtained. Then, the number of runs of the sequence is calculated. The larger the number of runs, the stronger the randomness and fluctuation of the sequence. It can be found that the number of runs of the sequence plunges from IMF1 to R. The "run_test" proves that all subsequences pass the run test. The time span of the original $PM_{2.5}$ sequence is 2587 days, and the sample entropy of each subsequence (including the original sequence) is calculated. In general, the tolerance for a sequence should be set to 0.2 times its standard deviation.

Before reconstruction, the PE of subsequence should be calculated. After comparison, when the dimension of embedding number is set to 5 and the time delay parameter τ to 1, the PE of each component is 0.913, 0.457, 0.456, 0.358, 0.173, 0.163, 0.137, 0.108, 0.008, 0.006, 0.001, and 0 respectively. Thus, components 1–4 are reconstructed as high-frequency components, 5–7 as low-frequency components, and 8–11 as residual components, and the PE of residual R is 0. Table 7.5 presents the sequence prediction results after reconstruction.

The reconstruction process follows the steps of "reconstruction by frequency, first prediction, and then superimposition". Herein, high_fre, low_fre and R represent high frequency sequence, low frequency sequence and residual sequence respectively. For the original daily data $PM_{2.5}$ concentration, the correlation coefficient between the decomposed subsequence and the original sequence is calculated. The sequence whose absolute value of correlation coefficient is greater than 0.2 is selected for reconstruction according to the frequency. The subsequence is considered to contain less information of the original sequence when its correlation coefficient is less than 0.2, so the sequence is not involved in reconstruction. Three reconstructed sub-sequences are obtained: high-frequency subsequence, low-frequency subsequence, and residual sequence. XGBoost prediction is performed on these three subsequences. On the training set, the goodness of fit of the residual sequence is the best, and that of low-frequency sequences and high-frequency sequences can reach about 90%. On the test

Table 7.5 Sequence prediction results after reconstruction

Sequence	Group	MAE	Goodness of fit (R^2)	RMSE	RPE
Reconstructed $PM_{2.5}$ sequence	—	0.745	85.74%	0.929	8.80%
high_fre	Trained	0.744	95.24%	0.920	1.65%
	Test	1.147	88.70%	1.418	12.1%
low_fre	Trained	0.710	90.32%	0.9160	20.07%
	Test	0.866	85.6%	1.117	29.63%
R	Trained	1.297	95.93%	1.634	50.66%
	Test	5.495	38.38%	6.324	40.34%

MAE: Mean absolute prediction error; RMSE: Root mean square error; RPE: Relative prediction error

set, the goodness of fit of low-frequency sequences and high-frequency sequences is about 85%, but that of the residual sequence is low, only 38.38%. This is reasonable since the residual sequence itself only contains a little short-term fluctuation information of the original sequence. On the test set, the relative prediction errors of high-frequency sequences, low-frequency sequences, and residual sequences are 12%, 38%, and 40%, respectively, but the low-frequency sequence has the largest mean absolute error of 6.7%.

Next, these three subsequences are superimposed and compared with the original $PM_{2.5}$ sequence. The mean absolute error of the reconstructed sequence is 0.745, the root mean square error is only 0.929, and the relative prediction error is 8.8%. These indicate that the reconstructed sequences well reflect the inherent fluctuations and changes of the $PM_{2.5}$ sequence, and achieve the purpose of reducing the prediction error of the XGBoost model through signal decomposition. The predictions of the high-frequency sequence, the low-frequency sequence, and the residual sequence are compared to obtain the reconstructed prediction sequence of $PM_{2.5}$ concentration. The comparison of the prediction curve on the test set with the original curve is shown in Fig. 7.9.

The reconstructed sequence has poorer prediction performance in the three periods of mid-August 2020, late November 2020, and late December 2020, when $PM_{2.5}$ concentration increased significantly (intensified pollution), and the prediction effect of the remaining periods all reaches the expectation. The wild fluctuations in $PM_{2.5}$ since 2020 also reflect the impact of urban development on environment in recent years. This method of selecting the subsequences with high correlation for first reconstruction and then performing the second reconstruction after prediction is an "incomplete reconstruction" at the cost of partial information. If all subsequences are allowed to participate in the reconstruction, it is difficult to avoid the problem of "modal aliasing".

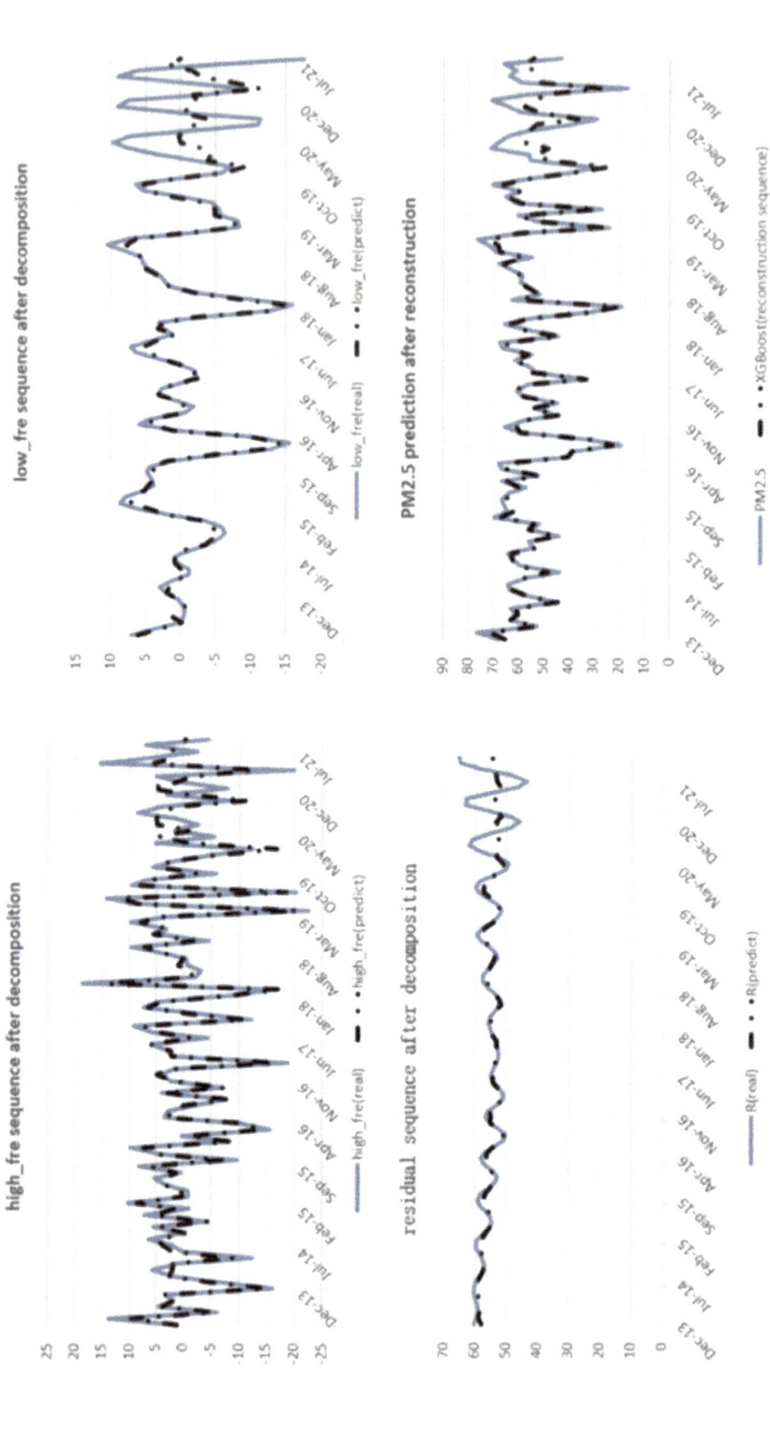

Fig. 7.9 $PM_{2.5}$ concentration predicting results generated by sub-sequences of different frequencies after first prediction and sequence reconstruction

7.3.4.2 Prediction Results of Modified Spatiotemporal XGBoost Model

The three methods of XGBoost, LightGBM, and GBDT after parameter adjustment and signal decomposition are used as the first layer of the integrated model (these three methods are all gradient boosting algorithms, which are conducive to model fusion). The error on the test set of the training set is calculated respectively, and the average of the three prediction sequences is calculated to obtain the first prediction sequence. In order to avoid overfitting, the second layer model should not be complicated. The purpose of adding the second layer model is to make the trend extrapolation of the fused model easier. Since the second layer model conducts is the second prediction based on the first layer of prediction sequence, the prediction accuracy will be reduced if the second layer model is complex. Therefore, a simple linear regression model is selected for prediction, and the final fusion model obtained is called the spatiotemporal XGBoost model. Comparison of prediction results among the three models is shown in Table 7.6.

It can be seen from Table 7.6 that the prediction accuracy of the fusion model is around 90%. The 8 variables with the strongest correlation with $PM_{2.5}$ concentration are used as independent variables, and the high-frequency, low-frequency, and residual sequences reconstructed after signal decomposition as dependent variables. XGBoost is used for prediction, and then the prediction sequence is superimposed to obtain the prediction result of XGBoost in the first layer of the fusion model. Similarly, the above operations are performed with LightBGM and GBDT respectively to obtain their prediction sequences. As shown in Table 7.6, the goodness of fit of each layer model is above 90%. The mean absolute error of the fusion model is 3.3485, and the goodness of fit is 98.1%, which is significantly improved compared with the single model. The performance of XGBOOST processed by the adaptive weighting method on the test set is shown in Fig. 7.10.

The goodness of fit of the predicted value and the actual value regression of the Stacking fusion model, XGBoost, LightBGM, GBDT, and the XGBOOST model modified by adaptive weighted method are 0.977, 0.993, 0.989, 0.988, and 0.993,

Table 7.6 Errors of stacking fusion models

Model	Group	MAE	Goodness of fit (R^2)	RMSE	RPE
First layer: LightBGM	Trained	3.99	96.67%	6.49	10.90%
	Test	5.8041	92.51%	8.53	16%
First layer: XGBoost	Trained	1.9775	99.41%	2.535	4.92%
	Test	3.8256	98.07%	4.94	8.03%
First layer: GBDT	Trained	2.7400	98.12%	3.54	6.36%
	Test	5.7336	95.04%	7.69	14.92%
Second layer: Multiple regression	—	3.3485	98.10%	6.59%	9.06%

MAE: Mean absolute prediction error; RMSE: Root mean square error; RPE: Relative prediction error

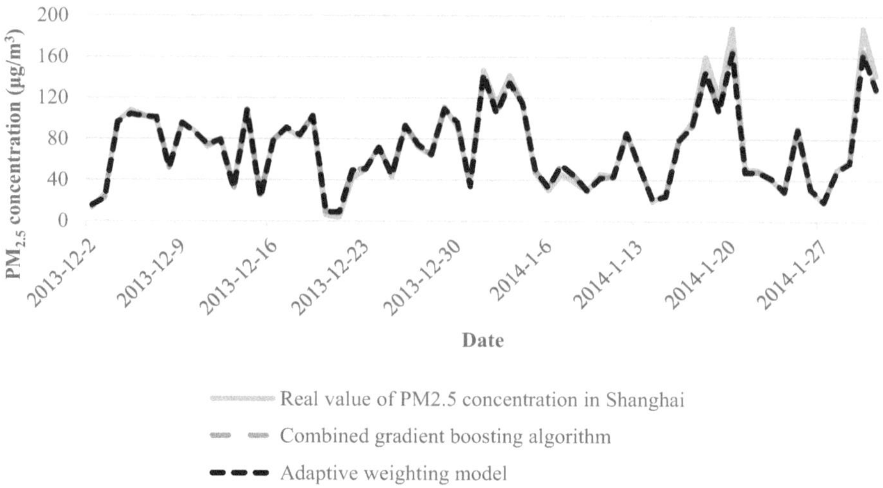

Fig. 7.10 Accuracy comparison of adaptive weighting method against the traditional method

respectively. This indicates that the adaptive weighting method improves the goodness of fit and reduces the prediction error. It can be seen from Fig. 7.11 that the performance of the model at extreme points is also improved. The fitting equation between the actual value and the predicted value of the final fusion model is

$$y = 1.026 \cdot x - 1.548 \tag{7.39}$$

The spatiotemporal XGBoost model is modified by the adaptive weight adjustment method. The fitting equation between the predicted value of the modified model and the actual value is $y = 0.9709 \cdot x + 1.4527$ (Eq. (7.39) is the fitting equation before the model modification). R^2 is 0.993, which achieves the desired effect. Finally, the established model is used to forecast the $PM_{2.5}$ sequence in Shanghai, and Fig. 7.11 is obtained.

In the 10-year observation data, March, April, and May represent spring; June, July, and August represent summer; September, October, and November represent autumn; December, January, and February represent winter. The XGBoost model after parameter adjustment has better predictions in spring and autumn. The possible reason is that the pollution is more serious in winter and there are many extreme points of $PM_{2.5}$ concentration, which impact the prediction accuracy. The error of each evaluation indicator in April (spring) is far smaller than that in other seasons. The active atmosphere in summer leads to the accelerated migration rate of air pollutant particles, so it is not easy for more $PM_{2.5}$ particles to accumulate in local areas. The prediction performance of the XGBoost model differs in different seasons, which also reflects the difference in meteorological conditions among seasons.

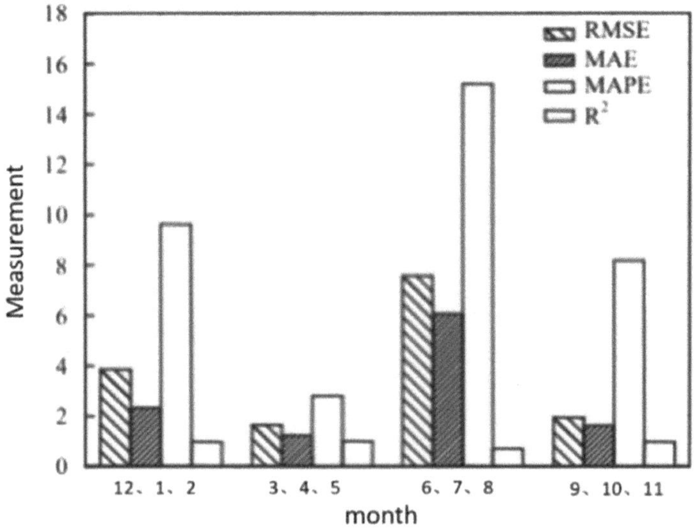

Fig. 7.11 XGBoost prediction results for different seasons

7.3.4.3 Model Performance After Introduction of Observations from Each Station

The surface arc length distance *r* between two places is obtained. One observation station is reserved in each administrative region, so there are a total of 12 observation stations in Shanghai, as shown in Fig. 7.12.

The top 5 observation stations of XGBoost importance are selected: Hongkou, Minhang (Pujiang), Pudong (Huinan), Putuo and Hongkou, which are used as distance matrix to participate in weighting. The mutual influencing factors of stations are introduced for secondary prediction. There are two processing methods for the distance between two stations, namely the inverse distance weighting method and the bandwidth selection method. The empirical analysis results of these two distance processing methods are given. The optimal bandwidth combination of each station is selected, and the mean square error of the observation station under the bandwidth is calculated with the weight. The optimal bandwidths of Fengxian, Pudong New Area, Putuo, Yangpu and Xujiahui are 42 km, 41 km, 22 km, 8 km and 62 km respectively. The corresponding root mean square errors are 16.1453, 16.8092, 16.8145, 17.2938, and 17.4082, respectively.

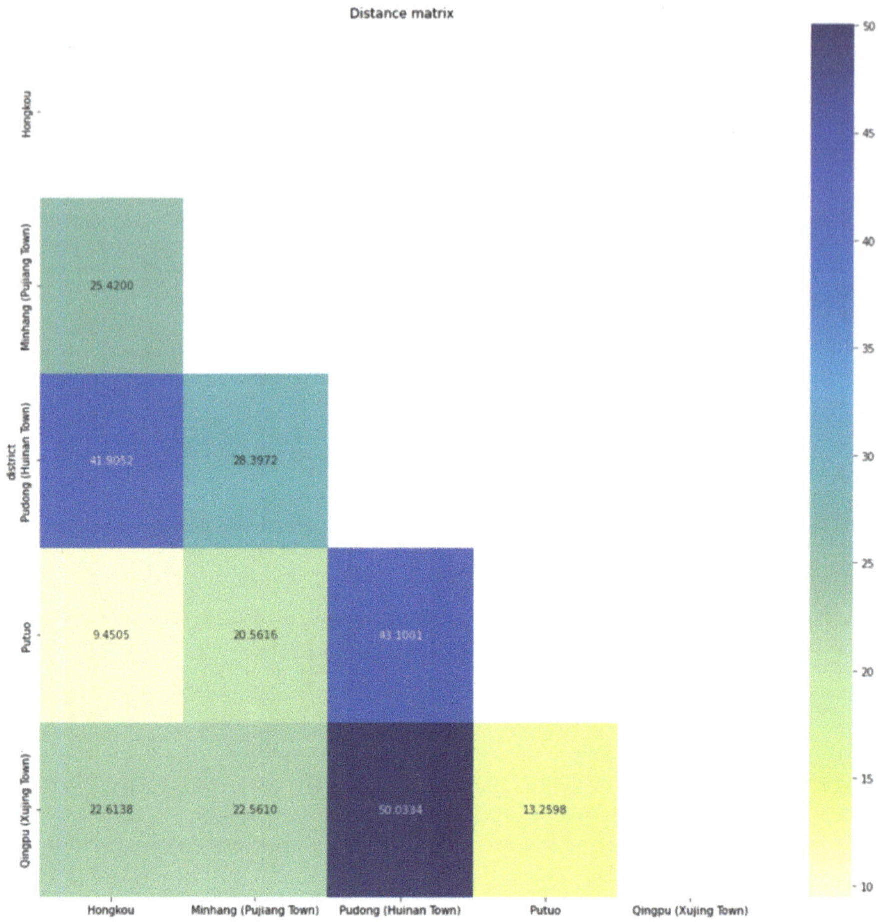

Fig. 7.12 The heatmap of distance matrix between every two of the top five observation stations of highest correlation with each other

7.3.5 Extending XGBoost's Spatial Prediction by Kriging Interpolation

The modified XGBoost model is used to predict the $PM_{2.5}$ concentration of 19 stations in Shanghai. The predicted result series filled with missing values are summarized to calculate the annual $PM_{2.5}$ concentration of each station. The time span of the predicted result is from 1/1 to 12/31 in 2021. After inverse distance weighted correction, the distribution of $PM_{2.5}$ concentration in Shanghai is obtained, as shown in Fig. 7.13.

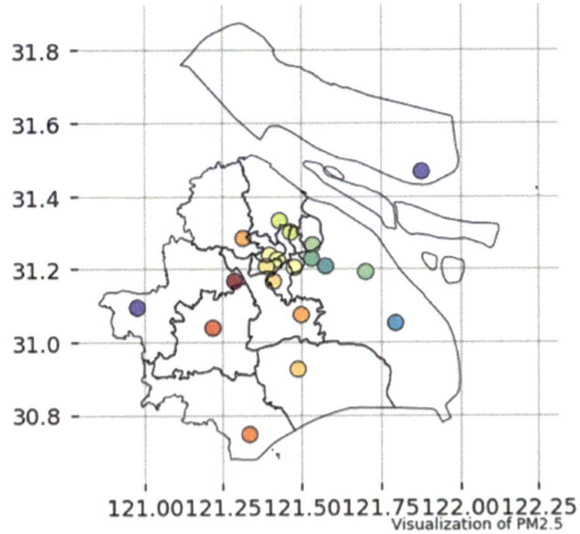

Fig. 7.13 Distribution of $PM_{2.5}$ concentration (μg/m^3) in each district in Shanghai

It can be seen that the $PM_{2.5}$ concentration in Putuo, Xujiahui, Yangpu and other urban areas is at a high level, especially in the Yangpu Industrial Zone. The distribution of $PM_{2.5}$ concentration is quite similar to that of industrial areas and population density. The accuracy of spatial prediction is consistent with the traditional understanding of air pollution distribution. Seen from Fig. 7.15, the $PM_{2.5}$ concentration in winter is at a high level, and air pollution is more serious, which may be resulted from the climatic characteristics of winter. The frequent rainfall in summer is an important reason for the improvement of air pollution, as shown in Fig. 7.14.

Next, Kriging interpolation method is used to establish a spatial distribution network for the predicted values of each observation station. And the kernel density

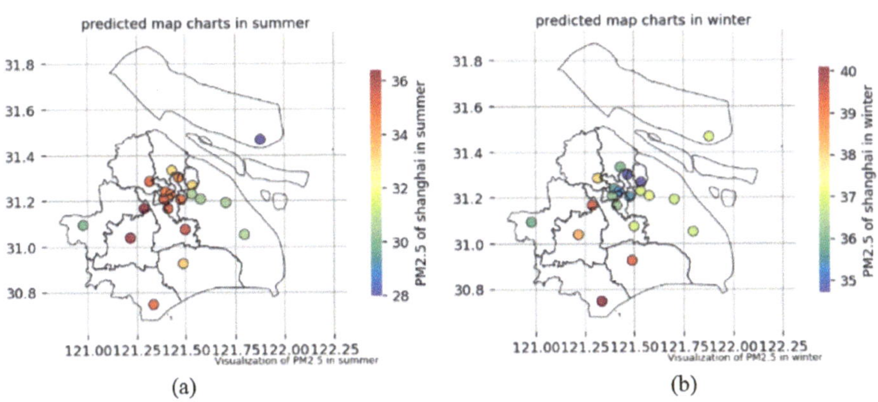

Fig. 7.14 Distribution of $PM_{2.5}$ concentration (μg/m^3) in each district in Shanghai by time: In summer (a) and in winter (b), respectively

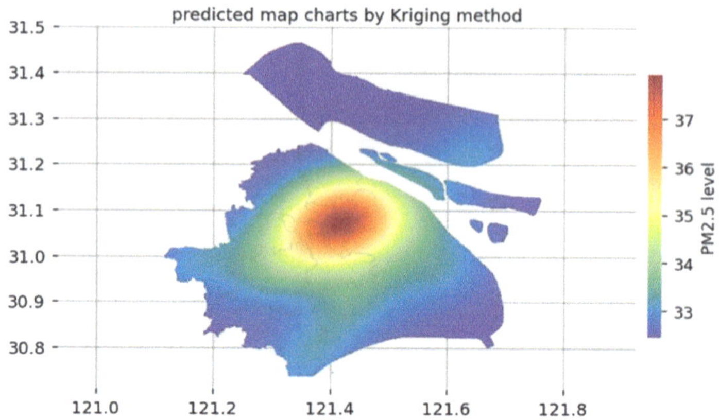

Fig. 7.15 Distribution of $PM_{2.5}$ concentration predicted values (μg/m^3) of across Shanghai, using Kriging interpolation method

function is a Gaussian function. When the Kriging conditions are met, the $PM_{2.5}$ concentration value of the adjacent station is filled in. During the filling process, the actual value of the observation station is used for training and correction to ensure the accuracy of the established spatial $PM_{2.5}$ network. Finally, by reading the boundary data (.json file) of each administrative district in Shanghai, the spatial $PM_{2.5}$ gridded data is cut and divided, and the predicted $PM_{2.5}$ concentration of each station is obtained, as shown in Fig. 7.15.

It is further confirmed in Fig. 7.15 that the areas with the most serious air pollution in Shanghai are the old urban areas and industrial areas. In urban areas, $PM_{2.5}$ concentration is on the decline as it approaches to suburban areas. As for the real-time evaluation system for the observation stations, the spatial dynamic change diagram of the predicted $PM_{2.5}$ concentration can be established, from which the predicted $PM_{2.5}$ concentration at any station in Shanghai can be obtained.

7.3.6 *Result Analysis*

The $PM_{2.5}$ concentration data of several observation stations in Shanghai are added to establish a spatial XGBoost model for prediction, through which the prediction level of $PM_{2.5}$ concentration in each district is obtained. Finally, the kriging interpolation method is used to fill the adjacent areas of each observation station so as to obtain the prediction results of each observation station in Shanghai.

Table 7.7 Prediction effect comparison between XGBoost models before and after modification, taking the predicted results of Xujiahui as an example

Model	Group	MAE	Goodness of fit (R^2)	RMSE	RPE
XGBoost	Trained	0.1801	90%	3.2643	0.1587
	Test	0.1912	82.37%	3.9	0.2065
Modified XGBoost	Trained	0.0785	98.05%	2.5289	0.0821
	Test	0.1757	87.40%	3.2976	0.1746

MAE: Mean absolute prediction error; RMSE: Root mean square error; RPE: Relative prediction error

7.4 Results Comparison and Robustness Analysis

7.4.1 Comparison of Prediction Effects Before and After XGBoost Model Modification

7.4.1.1 Prediction Effect of XGBoost Model Before and After Modification

After parameter adjustment, the optimal parameter settings of the XGBoost model can be determined as follows: min_child_weight is set to 3, n_estimate to 600, colsample_bytree to 0.9, max_dept to 3, gamma to 0.1, subsample to 0.8, and learning_rate to 0.1; the additional objective parameter selects the default term linear. The scores of the optimal values of these parameters are all over 80%. The other models in the gradient boosting algorithm are selected as the comparison models of XGBoost. Taking the observation data set of Xujiahui as an example, the prediction effect of XGBoost model before and after modification is compared to demonstrate the superiority of the improved XGBoost model, as shown in Table 7.7.

It can be seen that the modified XGBoost model has better performance in all evaluation dimensions.

7.4.1.2 Comparison of XGBoost with Other Gradient Boosting Algorithms

Gradient boosting algorithms can be roughly divided into three types: LightGBM, XGBoost, and GBDT. Among them, XGBoost has the smallest average prediction error (0.1045), and LightGBM has the largest prediction error (0.1114). The difference between them is small. The goodness of fit is over 95%, and the prediction effect is good. The prediction results of the three models are compared, as shown in Table 7.8.

In general, XGBoost is superior in all evaluation dimensions, and the root mean square error of GBDT prediction is large. The performance of XGBoost on the

Table 7.8 Error comparison among three models after introducing EMD decomposed sequences

Model	MAE	Goodness of fit (R^2)	RMSE	RPE
LightBGM	0.1114	98.86%	3.62	6.53%
GBDT	0.1669	98.53%	4.1076	7.40%
XGBoost	***0.1045***	99.25%	***2.9247***	***5.27%***

MAE: Mean absolute prediction error; RMSE: Root mean square error; RPE: Relative prediction error

test set is even better than that of the combined modified model, which indicates that the prediction ability of XGBoost has been greatly improved after parameter adjustment. Compared with GBDT, XGBoost improves the accuracy of the loss function. It replaces the first-order with the second-order Taylor expansion, and is superior to GBDT in the improvement of the regularization term. Moreover, the objective function is directly used as the loss function, which improves the flexibility of the model. In addition, XGBoost allows cross-validation in each iteration, and the later round can continue training on the training results of the previous round, which improves the operation efficiency. LightBGM is mainly to lower the value of loss function through the segmentation point algorithm, so as to achieve the purpose of improving training efficiency and reducing memory (Ke et al., 2017). Thus, it is reasonable that the prediction accuracy of the XGBoost model after parameter adjustment is higher than that of LightBGM.

According to the comparison in Sects. 4.1.1 and 4.1.2, the modified XGBoost significantly outperforms the gradient boosting model before optimization and the XGBoost model without parameter adjustment.

7.4.2 *Robustness Analysis*

In order to verify the robustness of the improved model, the data observed in Fengxian District are used again for model comparison with other conditions being unchanged. The results are listed in Table 7.9.

Table 7.9 Prediction effect comparison between XGBoost models before and after modification, taking the predicted results of Fengxian District as an example

Model	Group	MAE	Goodness of fit (R^2)	RMSE	RPE
XGBoost	Trained	0.2584	83.68%	3.1887	0.2108
	Test	0.3381	73.63%	4.6310	0.3021
Modified XGBoost	Trained	0.1164	98.15%	2.3647	0.0851
	Test	0.3001	86.36%	3.3305	0.2173

MAE: Mean absolute prediction error; RMSE: Root mean square error; RPE: Relative prediction error

The results demonstrate that improved XGBoost model is robust and has better performance on two randomly selected datasets. Next, the number of observation stations is added to the model to investigate whether the prediction accuracy at a certain observation station can be improved. Table 7.10 is obtained by extending the distance matrix from 5 to 8 dimensions.

The 5 most representative stations are selected by importance with XGBoost. When the matrix dimension increases to 8, the prediction effect for Xujiahui decreases (Fengxian rises slightly). With only 5 stations, the workload can be greatly reduced. Subsequently, the predictions of the spatial XGBoost model in urban and suburban areas are compared to see if there are differences.

Data in Table 7.11 shows that the prediction effect is not necessarily related to whether it is located in the urban area: the prediction result at Xujiahui is good, but that at Putuo is poor; the prediction result at Fengxian is good, while that at Qingpu is poor. Therefore, it is not necessary to consider the specific location of each observation station (urban or suburban) when using XGBoost to screen important observation stations.

Table 7.10 Comparison of predicted results accuracy among extended distance matrices with different dimensions on test sets

Matrix type	Area	MAE	Goodness of fit (R^2)	RMSE	RPE
5 × 5	Xujiahui	0.1193	93.37%	4.0797	0.1501
	Fengxian	0.3715	82.14	14.2832	0.5036
8 × 8	Xujiahui	1.7822	33.41%	18.86	1.7977
	Fengxian	0.1923	78.74%	7.2674	0.2562

MAE: Mean absolute prediction error; RMSE: Root mean square error; RPE: Relative prediction error

Table 7.11 Comparison of test sets accuracy for between urban and suburb areas

Observed area		MAE	Goodness of fit (R^2)	RMSE	RPE
Urban	Xujiahui	0.1569	89.44%	5.1481	0.1894
	Putuo	0.2916	61.47%	10.3795	0.3697
Suburb	Fengxian	0.1019	93.99%	3.8623	0.1362
	Qingpu	0.4081	51.58%	11.1775	0.4465

MAE: Mean absolute prediction error; RMSE: Root mean square error; RPE: Relative prediction error

7.5 Conclusions and Limitations

This study innovatively adopted the adjusted XGBoost model associated with EEMD and Kriging interpolation to quantify and spatialized $PM_{2.5}$ pollution. Based on the dataset, indicators representing air pollutant indicators, meteorological environmental indicators and macroeconomic indicators were analyzed for 16 case districts in Shanghai over a long-term period (2013–2021). Then, this study analyzes the seasonal effect and the $PM_{2.5}$ distribution in Shanghai by using the model we constructed. The main findings can be summarized as follows:

(1) Compared with the original model, the goodness of fit of the modified XGBoost model on the test set increased by 17%, and the root mean square error decreased by 28%, indicating that the performance of XGBoost model after parameter adjustment is significantly improved. Meanwhile, the improved model outperforms the other gradient boosting algorithms. In addition, the error of the constructed XGBoost fusion model at extreme value points is reduced. The reconstructed high-frequency sequence contains the most information of the original sequence, especially the fluctuation information, which means that the signal decomposition method can effectively obtain most information of original $PM_{2.5}$ sequence when getting rid of useless information from some extreme points.

(2) The variation of $PM_{2.5}$ concentration in Shanghai has a significant seasonal (cyclical) effect, and its fluctuation period is 3 months (a quarter). The empirical results show that the $PM_{2.5}$ level in Shanghai has shown a downward trend with fluctuations in recent years. The $PM_{2.5}$ level in Shanghai is low in summer, but higher in winter. The air pollution shows obvious seasonal effect and presents a certain periodicity.

(3) In terms of spatial distribution, the $PM_{2.5}$ concentration in Putuo, Xujiahui, Yangpu and other urban areas is at a high level, and the pollution in Yangpu Industrial Zone is the most severe, and the $PM_{2.5}$ concentration gradually decreases from center city to the surrounding areas.

This study has several limitations. For example, when the dimension of the distance matrix increases, the prediction accuracy of the model cannot be improved by either the method of inverse distance weighting or bandwidth selection. This indicates that before adding the data of each observation station, it is necessary to judge whether the data of this observation station is significantly different from those of other stations. Thus, further study can be carried out on these aspects.

Acknowledgements Zidong Wang, and You Wu also made great contributions to this manuscript. We express our heartfelt thanks to them. This research also was supported by: National Natural Science Foundation of China (72474127) and National Social Science Fund Major Bidding Project of China (18ZDA052). Part of this chapter was published:Wang Z, Wu X, Wu Y. A spatiotemporal XGBoost model for $PM_{2.5}$ concentration prediction and its application in Shanghai. Heliyon 2023;9:e22569. https://doi.org/https://doi.org/10.1016/j.heliyon.2023.e22569. The publication has been conducted with copyright permission.

Abbreviations

EEMD	Ensemble Empirical Mode Decomposition
EMD	Empirical Mode Decomposition
GIR	Grid Importance Rank
GMCN	Grey Multivariate Convolution Model
KNN	*K*-Nearest Neighbors
LSSVM	Least Squares Support Vector Machine
LSTM	Long-Short-Term Memory
MAE	Mean Absolute Prediction Error
ODE	Ordinary Differential Equation
PCA	Principal Component Analysis
PE	Permutation Entropy
RMSE	Root Mean Square Error
RPE	Relative Prediction Error
SVD	Singular Value Decomposition
IDW	Inverse Distance Weighting
TDLPI	Three-Dimensional Landscape Pattern Index
WD	Wavelet Decomposition
XGBoost	Extreme Gradient Boosting Model
AOD	Aerosol Optical Depth

References

Camilleri, R., Harrison, R. M., & Aquilina, N. J. (2025). Application of machine learning algorithms in predicting indoor residential $PM_{2.5}$ concentrations. *Atmospheric Pollution Research, 16*, 102609. https://doi.org/10.1016/j.apr.2025.102609

Chang, Y. S., Chiao, H. T., Abimannan, S., Huang, Y. P., Tsai, Y. T., & Lin, K. M. (2020). An LSTM-based aggregated model for air pollution forecasting. *Atmospheric Pollution Research, 11*(8), 1451–1463. https://doi.org/10.1016/j.apr.2020.05.015

Chen, T., & Guestrin, C. (2016, Aug 13). XGBoost: A scalable tree boosting system. In *Proceedings of 22nd ACM SIGKDD International Conference on Knowledge Discovery and Data Mining (KDD 2016)*, New York, NY, USA. pp. 785–794. https://doi.org/10.1145/2939672.2939785

Chen, Z. Y., Zhang, T. H., Zhang, R., Zhu, Z. M., Yang, J., Chen, P. Y., Ou, C. Q., & Guo, Y. (2019). Extreme gradient boosting model to estimate $PM_{2.5}$ concentrations with missing-filled satellite data in China. *Atmospheric Environment, 202*, 180–189. https://doi.org/10.1016/j.atmosenv.2019.01.027.

Cordero, J. M. (2025). A two-stage algorithm to estimate ground-level $PM_{2.5}$ concentration levels in Madrid (Spain) from AOD satellite data and surface proxies. *Atmospheric Pollution Research.* https://doi.org/102678.10.1016/j.apr.2025.102678

Fan, Z., Zhan, Q., Yang, C., Liu, H., & Bilal, M. (2020). Estimating $PM_{2.5}$ concentrations using spatially local xgboost based on full-covered SARA AOD at the urban scale. *Remote Sensing, 12*(20), 3368. https://doi.org/10.3390/rs12203368.

Feng, L., Li, Y., Wang, Y., & Du, Q. (2020). Estimating hourly and continuous ground-level $PM_{2.5}$ concentrations using an ensemble learning algorithm: The ST-stacking model. *Atmospheric Environment, 223*, 117242. https://doi.org/10.1016/j.atmosenv.2019.117242

Gui, K., Che, H., Zeng, Z., Wang, Y., Zhai, S., Wang, Z., Luo, M., Zhang, L., Liao, T., Zhao, H. and Li, L. (2020). Construction of a virtual $PM_{2.5}$ observation network in China based on high-density surface meteorological observations using the Extreme Gradient Boosting model. *Environment International, 141*, 105801. https://doi.org/10.1016/j.envint.2020.105801.

Han, M., Jia, S., & Zhang, C. (2022). Estimation of high-resolution $PM_{2.5}$ concentrations based on gap-filling aerosol optical depth using gradient boosting model. *Air Quality, Atmosphere & Health, 15*(4), 619–631. https://doi.org/10.1007/s11869-021-01149-w

Huang, G., Li, X., Zhang, B., & Ren, J. (2021). $PM_{2.5}$ concentration forecasting at surface monitoring sites using GRU neural network based on empirical mode decomposition. *Science of the Total Environment, 768*, 144516. https://doi.org/10.1016/j.scitotenv.2020.144516

Jin, X. B., Yang, N. X., Wang, X. Y., Bai, Y. T., Su, T. L., & Kong, J. L. (2020). Deep hybrid model based on EMD with classification by frequency characteristics for long-term air quality prediction. *Mathematics, 8*(2), 214. https://doi.org/10.3390/math8020214

Jung, C. R., Chen, W. T., & Nakayama, S. F. (2021). A national-scale 1-km resolution $PM_{2.5}$ estimation model over Japan using MAIAC AOD and a two-stage random forest model. *Remote Sensing, 13*(18), 3657. https://doi.org/10.3390/rs13183657

Just, A. C., De Carli, M. M., Shtein, A., Dorman, M., Lyapustin, A., & Kloog, I. (2018). Correcting measurement error in satellite aerosol optical depth with machine learning for modeling $PM_{2.5}$ in the Northeastern USA. *Remote Sensing, 10*(5), 803. https://doi.org/10.3390/rs10050803

Ke, G., Meng, Q., Finley, T., Wang, T., Chen, W., Ma, W., Ye, Q., & Liu, T. Y. (2017, Dec). Light GBM: a highly efficient gradient boosting decision tree. In *Proceedings of 31st Conference on Neural Information Processing Systems (NIPS 2017)*, Long Beach, CA, USA. pp. 3149–3157. https://doi.org/10.5555/3294996.3295074

Li, D., Jin, X., Xu, F., Liang, J., & Wang, X. (2025). Fine simulation and spatiotemporal analysis of urban PM2. 5 using Mobile monitoring data. *Urban Climate, 62*, 102498. https://doi.org/10.1016/j.uclim.2025.102498

Li, X., Peng, L., Yao, X., Cui, S., Hu, Y., You, C., et al. (2017). Long short-term memory neural network for air pollutant concentration predictions: Method development and evaluation. *Environmental Pollution, 231*, 997–1004. https://doi.org/10.1016/j.envpol.2017.08.114

Liang, L., Wang, Z., & Li, J. (2019). The effect of urbanization on environmental pollution in rapidly developing urban agglomerations. *Journal of Cleaner Production, 237*, Article 117649. https://doi.org/10.1016/j.jclepro.2019.117649

Liu, B., Tan, X., Jin, Y., Yu, W., & Li, C. (2021). Application of RR-XGBoost combined model in data calibration of micro air quality detector. *Science and Reports, 11*, 15662. https://doi.org/10.1038/s41598-021-95027-1

Liu, D., & Sun, K. (2019). Short-term $PM_{2.5}$forecasting based on CEEMD-RF in five cities of China. *Environmental Science and Pollution Research, 26*, 32790–32803. https://doi.org/10.1007/s11356-019-06339-9

Liu, Y., Paciorek, C. J., & Koutrakis, P. (2009). Estimating regional spatial and temporal variability of $PM_{2.5}$ concentrations using satellite data, meteorology, and land use information. *Environmental Health Perspectives, 117*(6), 886. https://doi.org/10.1289/ehp.0800123

Luo, Z. G., Wang, Z. Y., Wang, H. W., He, H. D., & Peng, Z. R. (2021). Characterizing spatiotemporal distributions of black carbon and $PM_{2.5}$at a toll station: Observations on manual and electronic toll collection lanes. *Building and Environment, 199*, 107933. https://doi.org/10.1016/j.buildenv.2021.107933

Ma, J., Cheng, J. C. P., Xu, Z., Chen, K., Lin, C., & Jiang, F. (2020b). Identification of the most influential areas for air pollution control using XGBoost and Grid Importance Rank. *Journal of Cleaner Production, 274*, Article 122835. https://doi.org/10.1016/j.jclepro.2020.122835

Ma, J., Yu, Z., Qu, Y., Xu, J., & Cao, Y. (2020). Application of the XGBoost machine learning method in $PM_{2.5}$prediction: A case study of Shanghai. *Aerosol and Air Quality Research, 20*(1), 128–138. https://doi.org/10.4209/aaqr.2019.08.0408

Ministry of Environmental Protection, Specification HJ 663–2013: Technical regulation for ambient air quality assessment (on trial), Beijing, China, Ministry of Ecology and Environment, 2013.

Ning, G., Wang, S., Ma, M., Ni, C., Shang, Z., Wang, J., & Li, J. (2018). Characteristics of air pollution in different zones of Sichuan Basin, China. *Science of the Total Environment, 612*, 975–984. https://doi.org/10.1016/j.scitotenv.2017.08.205

Pan, S., Zheng, Z., Guo, Z., & Luo, H. (2022). An optimized XGBoost method for predicting reservoir porosity using petrophysical logs. *Journal of Petroleum Science and Engineering, 208*, Article 109520. https://doi.org/10.1016/j.petrol.2021.109520

Shang, Z., Deng, T., He, J., & Duan, X. (2019). A novel model for hourly $PM_{2.5}$concentration prediction based on CART and EELM. *Science of the Total Environment, 651*, 3043–3052. https://doi.org/10.1016/j.scitotenv.2018.10.193

Sheridan, R. P., Wang, W. M., Liaw, A., Ma, J., & Gifford, E. M. (2016). Extreme gradient boosting as a method for quantitative structure–activity relationships. *Journal of Chemical Information and Modeling, 56*, 2353–2360. https://doi.org/10.1021/acs.jcim.6b00591

Stafoggia, M., Bellander, T., Bucci, S., Davoli, M., De Hoogh, K., De'Donato, F., Gariazzo, C., Lyapustin, A., Michelozzi, P., Renzi, M., & Schwartz, J. (2019). Estimation of daily PM_{10} and $PM_{2.5}$ concentrations in Italy, 2013–2015, using a spatiotemporal land-use random-forest model. *Environment International*, 124, 170–179. https://doi.org/10.1016/j.envint.2019.01.016

Sun, W., & Sun, J. (2017). Daily $PM_{2.5}$ concentration prediction based on principal component analysis and LSSVM optimized by cuckoo search algorithm. *Journal of Environmental Management, 188*, 144–152. https://doi.org/10.1016/j.jenvman.2016.12.011

Wang, J., He, L., Lu, X., Zhou, L., Tang, H., Yan, Y., & Ma, W. (2022). A full-coverage estimation of $PM_{2.5}$ concentrations using a hybrid XGBoost-WD model and WRF-simulated meteorological fields in the Yangtze River Delta Urban Agglomeration, China. *Environmental research, 203*, 111799. https://doi.org/10.1016/j.envres.2021.111799.

Wang, Y., Wang, H., & Zhang, S. (2020). Prediction of daily $PM_{2.5}$concentration in China using data-driven ordinary differential equations. *Applied Mathematics and Computation, 375*, 125088. https://doi.org/10.1016/j.amc.2020.125088

Wang, Z. B., & Fang, C. L. (2016). Spatial-temporal characteristics and determinants of $PM_{2.5}$ in the Bohai Rim Urban Agglomeration. *Chemosphere, 148*, 148–162. https://doi.org/10.1016/j.chemosphere.2015.12.118

Wu, J., Winer, A. M., & Delfino, R. J. (2006). Exposure assessment of particulate matter air pollution before, during, and after the 2003 Southern California wildfires. *Atmospheric Environment, 40*, 3333–3348. https://doi.org/10.1016/j.atmosenv.2006.01.056

Xu, X., & Yoneda, M. (2021). Multitask air-quality prediction based on LSTM-auto encoder model. *IEEE Trans. Cybern., 51*, 2577–2586. https://doi.org/10.1109/tcyb.2019.2945999

Yahya, K., Zhang, Y., & Vukovich, J. M. (2014). Real-time air quality forecasting over the southeastern United States using WRF/Chem-MADRID: Multiple-year assessment and sensitivity studies. *Atmospheric Environment, 92*, 318–338. https://doi.org/10.1016/j.atmosenv.2014.04.024

Zhang, P., Yang, L., Ma, W., Wang, N., Wen, F., & Liu, Q. (2022). Spatiotemporal estimation of the $PM_{2.5}$ concentration and human health risks combining the three-dimensional landscape pattern index and machine learning methods to optimize land use regression modeling in Shaanxi, China. *Environmental Research, 208*, 112759. https://doi.org/10.1016/j.envres.2022.112759

Zhang, Y., Zhang, R., Ma, Q., Wang, Y., Wang, Q., Huang, Z., et al. (2020b). A feature selection and multi-model fusion-based approach of predicting air quality. *ISA Transactions, 100*, 210–220. https://doi.org/10.1016/j.isatra.2019.11.023

Zhang, Z., Wu, L., & Chen, Y. (2020). Forecasting $PM_{2.5}$and PM_{10} concentrations using GMCN (1, N) model with the similar meteorological condition: Case of Shijiazhuang in China. *Ecological Indicators, 119*, 106871. https://doi.org/10.1016/j.ecolind.2020.106871

Zhong, J., Zhang, X., Gui, K., Wang, Y., Che, H., Shen, X., Zhang, L., Zhang, Y., Sun, J., & Zhang, W. (2021). Robust prediction of hourly $PM_{2.5}$ from meteorological data using LightGBM. *National Science Review, 8*, nwaa307. https://doi.org/10.1093/nsr/nwaa307

Zhu, S., Lian, X., Liu, H., Hu, J., Wang, Y., & Che, J. (2017). Daily air quality index forecasting with hybrid models: A case in China. *Environmental Pollution, 231*, 1232–1244. https://doi.org/10.1016/j.envpol.2017.08.069

Chapter 8
Visual Analysis of Global Air Pollution Impact Research: A Bibliometric Review (1996–2022)

8.1 Introduction

Natural resources play an important role in the development of a country (Song et al., 2022). The atmosphere belongs to natural resources. Natural resources can be divided into biological resources, agricultural resources, climate and meteorology, water resources, etc. The atmosphere is one of the climatic and meteorological resources. The intensification of globalization has led to an increase in the attention paid to resource security, which, as one of the most important material bases for human survival and development, constitutes the key factor affecting overall resource security (Song et al., 2020). The impact of air pollution on social economy has been examined by scholars from various countries since the Industrial Revolution. Western developed countries, such as the United Kingdom and the United States, suffered from severe smog and paid a heavy price in their early development. For example, more than 4000 people died of respiratory diseases and heart disease as a result of the Great Smog of London on December 5–9, 1952; the Donora Smog in the United States on October 26–31, 1948, inflicted eye pain, sore throat, headache, chest tightness, vomiting, and diarrhea on more than 6000 people, and resulted in over 20 deaths. Since the middle of the last century, developing countries such as China and India have begun to face similar air pollution issues. Air pollution in China has become an urgent environmental problem to be solved, given the great prominence of its regional compound air pollution characterized by ozone, $PM_{2.5}$, and acid rain which poses a serious threat to people's health and sustainable socio-economic development (Huang et al., 2014; Xue & Zhao, 2017). In 2006, the economic loss caused by enironmental pollution accounted for about 10% of China's GDP. Although China has actively taken measures to address existing environmental problems, issues such as rising greenhouse gas emissions is still likely to accelerate in some regions of the country. In India, air pollution has been identified as a main health risk, and the mortality rate of its residents caused by air pollution remains high (Oliveira et al., 2019). According to the BP Statistical Review of World

X. Wu, *Environmental Disaster in China*, https://doi.org/10.1007/978-981-92-0910-1_8

Energy 2020 report (Company, 2020): China, the United States and India were the top three countries in primary energy consumption in 2019, with growth rates of 4.4%, −1.0%, and 2.3%, respectively, compared with the previous year (Company, 2020). Quantitative evidence demonstrates the economic magnitude of theseimpacts global crop yield losses from air pollution range from 11–18 billion US Dollarsannually, with ozone pollution alone causing wheat yield losses of 9.6% globally and up to17.4% in South Asia (Avnery et al., 2011; Saroj et al., 2025). Therefore, it is of great theoretical value and practical significance to systematically summarize the impact of air pollution from multiple dimensions using bibliometric methods. This can not only inform the academia about the frontiers in the research field, but also help government departments take targeted prevention and control measures.

The rest of this paper is organized as follows: The second part is an overview of relevant literature; the third part introduces the data source and research method; the fourth part presents the statistical analysis of the research topics; the fifth part further analyzes the research focuses; the sixth part illustrates the development trend of the research themes using strategic diagram; the final part summarizes the work done in this paper.

8.2 Literature Review

The following is a literature review from two aspects: research methods and content. Methodologically, statistic analysis and bibliometric analysis are frequently adopted by scholars. Regarding the research content, the major themes can be divided into the following categories, most of which involve the impact of air pollution on human health.

1. The impact of air pollution on human health (including human cognition, nervous system, psychology, disease, and death.) Lelieveld et al. (2015), using a global atmospheric chemistry model and in accord with the Global Burden of Disease Study 2010, found that outdoor air pollution dominated by $PM_{2.5}$ caused 3.3% premature deaths in Asia every year. The study also discovered that the emissions from residential energy use (e.g., heating and cooking) prevalent in India and China had the greatest impact on global premature mortality. The model predicted under a business-as-usual emission scenario that the contribution of outdoor air pollution to premature death may double by 2050.
2. The impact of air pollution on the health of residents in a region or a country. Crouse and Peters (2012) studied the risk of death associated with long-term exposure to fine particles and identified the link between mortality and long-term exposure to $PM_{2.5}$. Via investigating the relationship between cardiovascular mortality among Canadian adults and their long-term exposure to $PM_{2.5}$. The author concluded that every 10 μg/m3 increase in $PM_{2.5}$ concentration would lead to a risk ratio of 1.15 resulted from non-accidental causes, and of 1.31 from

ischemic heart disease. After analyzing subjects in 11 cities, the author found a similar association between non-accidental mortality and $PM_{2.5}$.

3. The impact of certain air pollutants on the incidence rates of certain diseases of human beings, such as the effect of particulate pollutants on asthma. Pope et al. (2002) examined the relationship between lung cancer, cardiorespiratory mortality, and long-term exposure to fine particulate air pollution, and found the correlation of fine particulate matter and sulfur oxide-related pollution with all-cause mortality, lung cancer, and cardiorespiratory mortality. Cohen et al. (2018) estimated the global population-weighted mean concentrations of particle mass with an aerodynamic method, and also the relative risk of mortality from ischemic heart disease, cerebrovascular disease, chronic obstructive pulmonary disease, lung cancer, and lower respiratory infections using non-linear exposure–response functions. In addition, it was calculated that ambient $PM_{2.5}$ was the fifth-ranking mortality risk factor in 2015, and environmental air pollution accounted for a large and increasing proportion of global disease burden. Regarding the impact of particles on public health, Brook et al. (2010) also drew relevant conclusions: there was a causal relationship between $PM_{2.5}$ exposure and cardiovascular morbidity and mortality, so $PM_{2.5}$ exposure was deemed a modifiable factor leading to cardiovascular morbidity and mortality.
4. The impact of air pollution on special populations such as children and pregnant women. Undertaking a systematic review and meta-analysis, Bowatte et al. (2015) identified the influence of traffic-related air pollution exposure on risk of asthma, allergy, and sensitization in children. The study also found that childhood exposure to traffic pollutants $PM_{2.5}$ and black carbon was associated with an increased risk of subsequent asthma in childhood. Specifically, early childhood exposure to traffic-related air pollutants was related to development of asthma from childhood up to 12 years old. The magnitude of these associations increased with age, and the pattern was the most prominent for $PM_{2.5}$. The results revealed a correlation between autism and prenatal air pollution exposure, which was primarily related to traffic sources.

In terms of the research methods for analyzing air pollution impact, statistical and bibliometric analyses are mostly used, but relevant studies are few. Lim et al. (2012) summarized the risk of burden of disease and injury attributable to 67 risk factors and risk factor clusters in 21 regions from 1990 to 2010. Quantifying the disease burden caused by different risks could inform disease prevention. The authors found that the three major risk factors of global diseases in 2010 were high blood pressure, tobacco smoking (including second-hand smoke), and household air pollution from solid fuels. Cheng et al. (2019) reviewed the impact of volatile organic compounds (VOCs) on human health in the past 16 years using a bibliometric method. Han et al. (2021) used CiteSpace software to study the frontiers in research on the influence of $PM_{2.5}$ on health. The visual analysis discovered that relevant literature has increased steadily since 2000; in recent five years, the number of publications in China has increased rapidly, the United States enjoyed the largest number of publications, and the research on the impact of $PM_{2.5}$ on health has gradually deepened.

Sun et al. (2020) bibliometrically analyzed the impact of global air pollution on children from 1999 to 2018, and drew a trend chart. Smith and MedSc (2009) reviewed the literature published in Archives of Environmental & Occupational Health from 1975 to 2004, analyzed the number of publications, the number of citations, the immediacy index, and the journal impact factor, and believed that citation analysis could provide a useful lens for journal development.

To sum up, most of the existing studies qualitatively summarize the effect of air pollution on human health, with few using bibliometric methods to analyze air pollution impact. Moreover, relevant review articles to date mostly focus on some countries or regions, or take some specific types of air pollution as the theme, rarely adopting a global perspective and covering a long time span. In view of this, the present paper systematically presents the status quo, hotspots, and development trends of air pollution impact research over the past 26 years utilizing bibliometric analysis. Its possible contribution lies in the overall bibliometric analysis of the impact of air pollution research, the discussion of key issues, and the identification of current research gaps, which can provide reference for subsequent inquiries.

8.3 Data Source and Research Method

8.3.1 Data Source

The data were sourced from the Web of Science (WOS) Core Collection database developed by the Institute for Scientific Information (ISI). The data from 1996 to April 2022 were used for analysis. First of all, the determination of a retrieval strategy is crucial during data collection for ensuring the authenticity and integrity of the research results. Through multiple discussions and comparisons of various retrieval strategies, the topic search method was finally adopted for literature retrieval. Noteworthily, the studies on air pollution prevention or treatment were excluded during the topic search due to their lack of direct relevance to the research objective. The specific retrieval strategy and literature selection range are shown in Table 8.1.

8.3.2 Method

Bibliometric methods are used for analyzing the evolution of academic literature over time, aiming to acquire an understanding of scientific facts through the retrieval, identification, sorting, and analysis of literature. Knowledge mapping can facilitate the vivid demonstration of the development profile of a research field or discipline, and help people to comprehensively review the research hotspots or other information. Co-word analysis is a content analysis technique which, via analyzing the

Table 8.1 Keywords selection and results of the paper topic search

The process flow		Results
(1) The process of theme search		
a. The process of literature identification		42,176
Data sources	Web of Science Core Collection	
Indexed format	TS = air pollution AND TS = (impact OR affect OR influence OR effect OR lead OR result OR cause OR bring OR generate)NOT TS = (prevent OR precaution OR guard against OR prevention OR jointly prevention) NOT TS = (manage OR govern OR control OR curb OR treat OR reduce OR mitigation)	
Paper language	English	
Literature type	Article	
Time span	1996–2022	
b. The process of irrelevant subject exclusion		0
(2) The process of removing duplicates		0
(3) The process of title, keywords and abstract selection		42,176

terminology in literature reflecting research topics or directions within a certain discipline, identifies the relationships between the researched topics within the discipline to illustrate its research structure. This technique allows not only for detecting high-frequency keywords reflecting the research topics or directions within a discipline, but also for obtaining co-occurrence matrices. Burst terms refer to single words or phrases extracted from the title, abstract and other sections of literature, whose frequency changes suddenly in a certain period. Cluster analysis is a method for classifying things based on their quantitative characteristics. In the clustering process, the distance between words in the co-word relationship network is first calculated, the subject terms closer to each other are then gathered so that the terms within a cluster share high attribute similarity, and finally, clusters with relatively independent concepts are formed. Derived from the co-word analysis, strategic diagram is capable of visually representing the research structure of a certain field, and different clusters have differed representations in different quadrants.

8.3.3 Research Tools and Ideas

To illustrate the current situation and development trend of air pollution assessment research, this paper uses the visualization software CiteSpace to statistically analyze keyword frequency, identify high-frequency keywords, conduct social network analysis and hierarchical cluster analysis, and obtain the keywork strategic diagram. Data processing tools and research ideas are shown in Fig. 8.1.

8.4 Statistical Analysis of Data

8.4.1 Spatio-Temporal Analysis of Publications

The temporal and spatial variation in literature quantity is an important indicator to evaluate the development of a field, and it is of vital significance to evaluate the development trend and dynamic evolution of a field. This paper first described the number and changing trend of literature published in the past 26 years (Fig. 8.2). In general, literature quantity has continuously increased from 400 in 1997 to 4827 in 2021, indicating that the impact of air pollution is attracting rising scholarly attention and gradually becoming a research hotspot.

Secondly, CiteSpace was used to draw the spatial evolution map of the literature recorders and the burst country/region maps (Figs. 8.3, 8.4 and Tables 8.2, 8.3). (1) Temporally, the research on this topic started in 1996. (2) Spatially, scholars from 171 countries/regions have studied this topic. (3) Judging from the centrality and number of publications, the US ranks first, indicating that it has the largest academic influence in this field and is the world's research center in the field. China ranks second in terms of the number of publications, which considerably outnumbers those of the UK which ranks third. However, the centrality of China's papers is zero, indicating that research on this topic in China is in a booming state, but the depth and influence are still weak. It can be seen that China and the US are the most active in the research field and boast the highest output, which is 11,235 and 8902 respectively, together accounting for around half of the total literature quantity. In terms of centrality, ranked second to tenth are all European countries except Canada. This reveals that apart from American and Canadian scholars, European scholars also play an important role in this field. In addition, the centrality of other countries, with the exception of China, is closely related to their number of publications. (4) Regarding the burst intensity of the publications, China ranks the highest, and the burst intensity of two of its periods entered the top ten. The first, with a value of 371.88, was from 2018 to 2022; the second, with a value of 228.51, was from 1998 to 2007. Germany ranks second (with a burst intensity of 310.59 from 1996 and 2007); the UK (with a burst intensity of 269.72 from 1996 to 2007) and Italy (with a burst intensity of 227.93 from 1996 to 2007) rank third and fourth respectively, indicating that the research in these four countries played an academically leading role in a specific period of time.

In a nutshell, the spatio-temporal analysis of publications reveals that the US started early in the field of air pollution assessment, and its academic leadership has lasted a long time. Seven European countries (the UK, Germany, France, Italy,

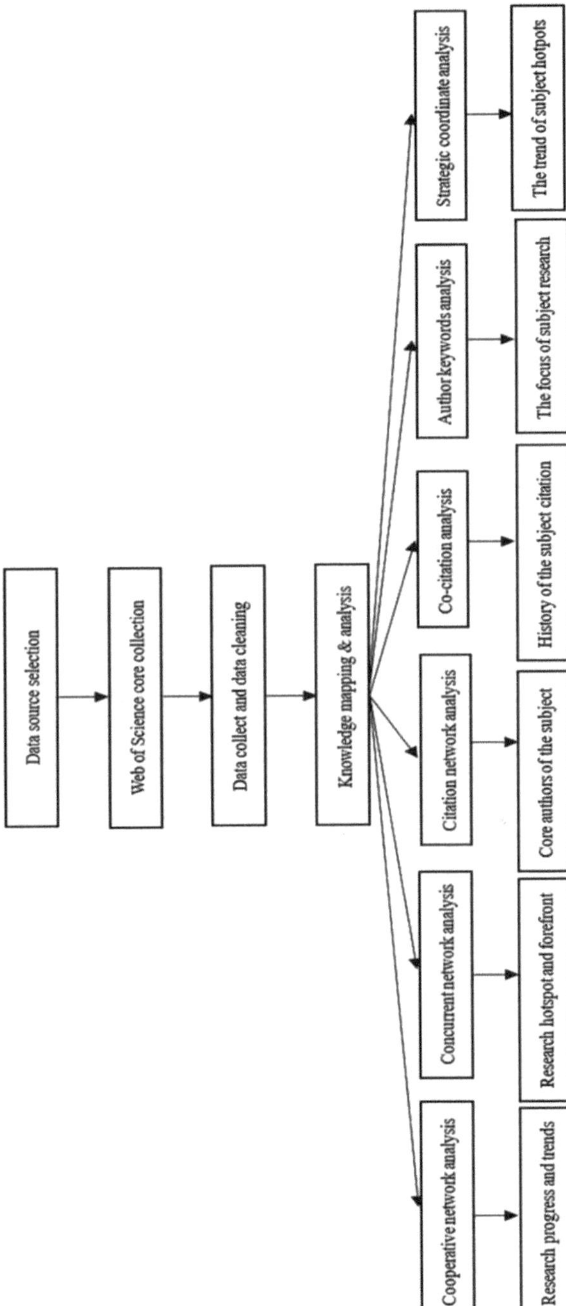

Fig. 8.1 The research idea and data processing flow

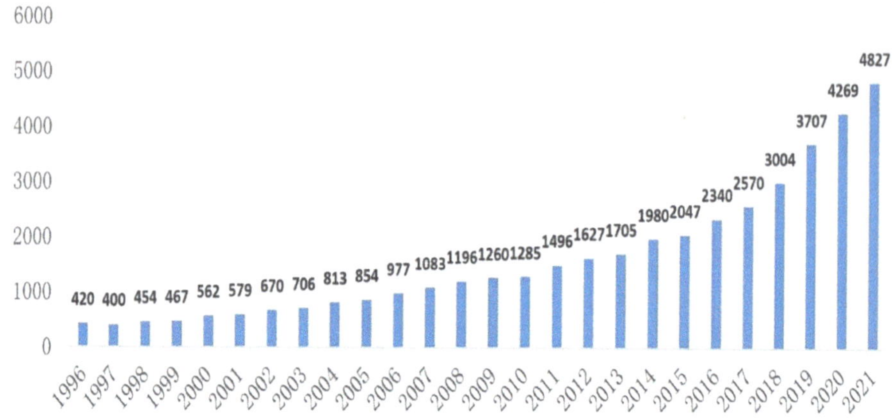

Fig. 8.2 Time distribution of publications

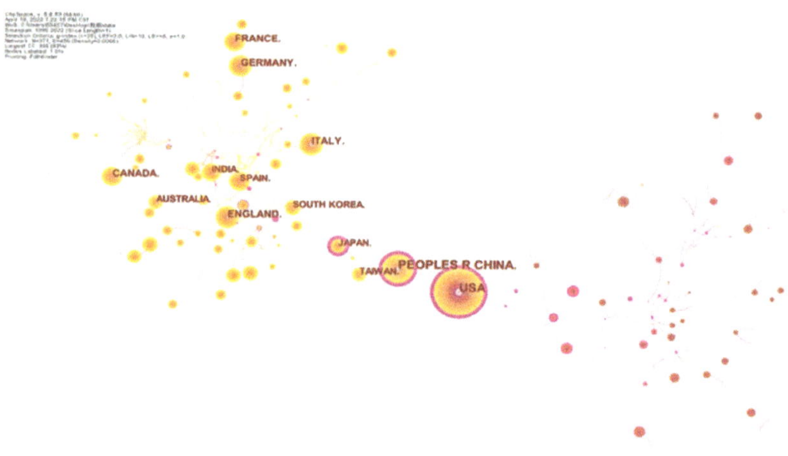

Fig. 8.3 Centrality analysis of papers in different countries or areas

Spain, Sweden, and the Netherlands) and Canada are the authorities in this field, while China's research is still in the developing stage.

8.4.2 *Analysis of Centrality and Citation Times of Review Articles*

Secondly, CiteSpace was employed to analyze 2903 review articles, obtain the articles with high centrality and high number of citations, and carry out burst detection. The

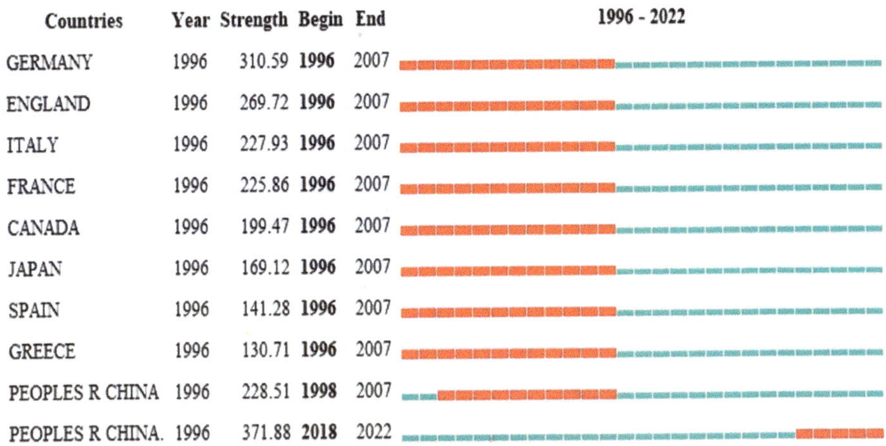

Fig. 8.4 Top 10 countries with the strongest citation bursts

Table 8.2 Centrality ranking statistics of papers in different countries or areas

No.	Country	Centrality	Output	Begin
1	USA	1.05	11,235	1996
2	ENGLAND	0.16	533	1996
3	GERMANY	0.1	597	1996
4	FRANCE	0.08	441	1996
5	ITALY	0.06	454	1996
6	SPAIN	0.06	271	1996
7	CANADA	0.05	397	1996
8	SWEDEN	0.05	247	1996
9	NETHERLANDS	0.05	237	1996

Table 8.3 Ranking statistics of papers in different countries or areas

No.	Country	Output	Centrality	Begin
1	USA	11,235	1.05	1996
2	PEOPLES R CHINA	8902	0	2003
3	ENGLAND	2088	0.02	1998
4	ITALY	1966	0.02	2003
5	GERMANY	1766	0.04	1997
6	CANADA	1539	0.03	2007
7	FRANCE	1507	0.04	2007
8	SPAIN	1382	0.03	2002
9	INDIA	1318	0.01	2007

ranking of the relevant literature is shown in Fig. 8.5, and the ranking of the top five burst articles is shown in Fig. 8.6 (Table 8.4).

The review article with the highest centrality (0.25) is Cohen et al. (2018), which mainly introduced the estimated number of global diseases caused by environmental air pollution, and believed $PM_{2.5}$ to be the fifth largest death risk factor after analyzing the trend within 25 years. Shah et al. (2013) is featured by the second highest centrality (0.2), which analyzed the relationship between global air pollution and heart failure using systematic review and meta-analysis. Another study with the same centrality of 0.2 is Song et al. (2014) which investigated the global contribution of outdoor air pollution to the incidence, prevalence, and mortality of chronic obstructive pulmonary disease. The centrality of Hoek et al. (2013) ranks third (0.19); it reviewed the association between long-term air pollution exposure and cardiopulmonary respiratory mortality. The top five burst review articles all appeared during

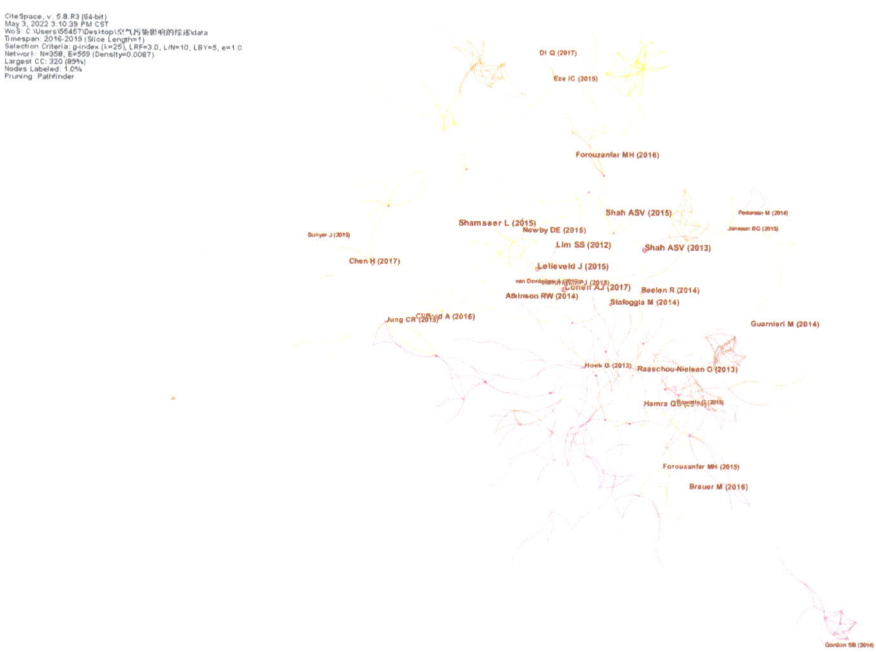

Fig. 8.5 High centrality review literatures on air pollution impacts

References	Year	Strength	Begin	End	2016–2019
Lim SS, 2012, LANCET, V380, P2224, DOI 10.1016/S0140-6736(12)61766-8, DOI	2012	9.68	**2016**	2017	
Bowatte G, 2015, ALLERGY, V70, P245, DOI 10.1111/all.12561, DOI	2015	2.98	**2016**	2017	
Becerra TA, 2013, ENVIRON HEALTH PERSP, V121, P380, DOI 10.1289/ehp.1205827, DOI	2013	2.38	**2016**	2017	
Brauer M, 2012, ENVIRON SCI TECHNOL, V46, P652, DOI 10.1021/es2025752, DOI	2012	2.08	**2016**	2017	
Song QK, 2014, INT J ENV RES PUB HE, V11, P11822, DOI 10.3390/ijerph111111822, DOI	2014	2.08	**2016**	2017	

Fig. 8.6 Top 5 references with the strongest citation bursts

Table 8.4 Ranking of high centrality review literatures on air pollution impacts

No.	The first author	Centrality	Cited times	Journal	Year
1	Cohen AJ	0.25	29	Lancet	2018
2	Shah ASV	0.2	23	Lancet	2013
3	Song QK	0.2	7	Int J Env Res Pub He	2014
4	Hoek G	0.19	13	Environ Health-Glob	2013
5	Burnett R	0.19	9	P Natl Acad Sci USA	2018
6	Gakidou E	0.19	5	Lancet	2017
7	Langrish JP	0.19	3	Environ Health Persp	2012
8	Stafoggia M	0.18	15	Environ Health Persp	2014
9	Bowatte G	0.18	10	Allergy	2015

2016–2017 and concerned the impact of air pollution on human health. Specifically, the burst intensity of Lim et al. (2012), which ranks fifth among the burst review articles above, holds a safe lead. It is followed by Bowatte et al. (2015), which used statistics and meta-analysis to examine the impact of traffic-related air pollution on the risk of asthma, allergy, and sensitization in children. From the above analysis, we could see that scholars have attached great importance to the study of the impact of air pollution on human health. Not only has the impact of air pollutants on human diseases and deaths been investigated, but the specific influence of certain pollutants on a certain population has also been focused. Nevertheless, the air pollution impact reviews to date are largely characterized by scattered research content and homogeneous research object.

8.4.3 Core Institution Analysis

The results of the core institution analysis of air pollution impact research are as follows. (1) From the perspective of institutional cooperation, the co-occurrence diagram of air pollution control research institutions generated by CiteSpace reveals that universities are the main institutions for air pollution impact research, but there is no close cooperation network between research institutions (Fig. 8.7). (2) With regard to institutions' literature quantity, the top ten are Chinese and American institutions. Chinese Science Academy is the leading institution in this regard, possessing 1842, followed successively by Harvard University (788), Peking University (786), and the US Environmental Protection Agency (743 articles); this is consistent with the ranking of countries in terms of the number of publications. These are all important institutions for air pollution impact research (Table 8.5). (3) Concerning the centrality of research institutions, University of Ulm in Germany (0.35) is far ahead of others, followed by NASA and National Center for Atmospheric Research. 60% of the institutions with high centrality are distributed in Europe (Table 8.6). (4) In terms of the publication burst intensity of institutions, the top three are all from the US. The centrality values of the US Environmental Protection Agency (92.42 from 1996 to 2009) and Harvard University (90.15 from 1998 to 2014) hold a safe lead, indicating the abundance of research of these two institutions in different periods. Among them, the research conducted by Harvard University exhibited the longest burst duration, which is up to 16 years (Fig. 8.8); the University of Southern California (with a centrality of 80.61 from 1999 to 2014) ranks third.

Fig. 8.7 Co-occurrence map of research institutions.

Table 8.5 The number of papers published by research institutions ranked in the top 5

No.	Organization Name	Output	Centrality	Year	Country
1	China Science Academy	1842	0.03	1999	China
2	Harvard University	788	0.1	1997	America
3	Peking University	786	0.04	2003	China
4	US Environmental Protection Agency	743	0.03	1996	America
5	Univ Chinese Acad Sci	543	0	2013	China

Table 8.6 Top 5 institutions with the highest centrality

No.	Organization Name	Centrality	Output	Year	Country
1	The Ulm University	0.35	12	2009	Genman
2	Danish Cancer Society Center research	0.27	21	2014	Denmark
3	National Aeronautics and Space Administration	0.26	441	1996	America
4	University Medical Center Utrecht	0.25	99	2011	Netherlands
5	National Center for Atmospheric Research	0.24	255	1998	America

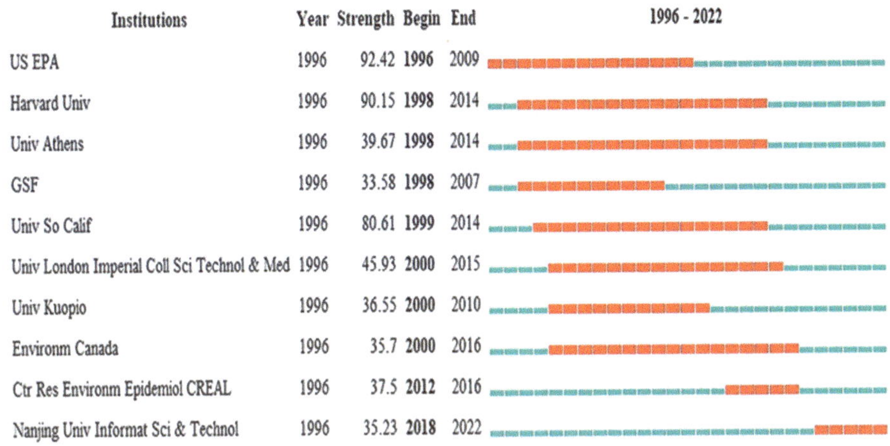

Fig. 8.8 Top 10 institutions with the strongest citation bursts

8.4.4 *Core Author Analysis*

CiteSpace was utilized to identify the highly prolific authors and high-centrality authors of air pollution impact research (Tables 8.7 and 8.8). There are 126,083 authors identified, eight of which have published 100 or more articles. Joel Schwartz is the most prolific author of this research topic (199 articles). In addition, Petros Koutrakis, Gerard Hoek, and Bert Brunekreef have published more than 150 articles.

In terms of author centrality, Min Hu maintains a strong lead with a centrality of 0.38, while D. R. Blake and Y. Kondo rank second and third respectively with a centrality of 0.26 and 0.19. Combining the number of publications and the authors,

Table 8.7 Statistics of the papers from productive authors

No.	The First Author	Output	Centrality	Year
1	Joel Schwartz	199	0.04	2006
2	Petros Koutrakis	155	0.02	2007
3	Gerard Hoek	154	0.06	2007
4	Bert Brunekreef	153	0.03	2006
5	Haidong Kan	147	0.02	2008
6	Brent A Coull	112	0.02	2007
7	Kees De Hoogh	103	0.01	2013
8	James J Schauer	103	0.15	2006
9	Francesco Forastiere	97	0.01	2007
10	Itai Kloog	96	0.03	2014

Table 8.8 Centrality ranking of authors

NO.	The First Author	Centrality	Output	Year
1	Min Hu	0.38	54	2006
2	D R Blake	0.26	19	2007
3	Y Kondo	0.19	9	2009
4	N Takegawa	0.17	4	2009
5	James J Schauer	0.15	103	2006
6	J F Burkhart	0.15	6	2011
7	G S Diskin	0.14	7	2011
8	A Alastuey	0.13	28	2002
9	J Pekkanen	0.12	31	1999

the top 15 highly prolific authors and the high-centrality authors exhibit a weak link. James J. Schauer is the only author with high output (103 articles) and high centrality (0.15).

8.4.5 Knowledge Base Analysis

Co-citation maps of literature drawn using CiteSpace can reveal the knowledge source of air pollution impact based on citation tracks. To arrive at a full picture of the literature that plays a key role in the research field of air pollution assessment, a high co-citation burst diagram was drawn (Fig. 8.9). Through the burst detection of highly cited articles, it was found that the difference in burst durations of the top ten articles is small, with the longest duration being four years and the shortest being three years. An article written by Brook R. D. (2011–2015, 237.67) exhibits the strongest burst intensity, followed by the research conducted by Cohen A. J. (2018–2022, 192.08) and then by Pope C. A. (2007–2011, 162.82). Another article written by Pope C. A. (2003–2007, 157.27) ranks fourth, so Pope C. A. has two articles with high burst intensity.

By analyzing the literature with high citation frequency (Table 8.9), we can get the hotspots in the field of air pollution impact research. Although the citation times of scientific research are also affected by other variables such as authors' social relationship, journal category, and impact factors, the analysis here is based on the citation frequency.

The following is a brief introduction to the top ten highly co-cited articles. The first one, Cohen et al. (2018), studied the trend of global disease burden caused by air pollution in the past 25 years, and estimated the relative risk of mortality from ischemic heart disease, cerebrovascular disease, chronic obstructive pulmonary

References	Year	Strength	Begin	End
Samet JM, 2000, NEW ENGL J MED, V343, P1742, DOI 10.1056/NEJM200012143432401, DOI	2000	92.62	**2001**	2005
Pope CA, 2002, JAMA-J AM MED ASSOC, V287, P1132, DOI 10.1001/jama.287.9.1132, DOI	2002	157.27	**2003**	2007
Pope CA, 2006, J AIR WASTE MANAGE, V56, P709, DOI 10.1080/10473289.2006.10464485, DOI	2006	162.82	**2007**	2011
Brook RD, 2010, CIRCULATION, V121, P2331, DOI 10.1161/CIR.0b013e3181dbece1, DOI	2010	237.67	**2011**	2015
Lim SS, 2012, LANCET, V380, P2224, DOI 10.1016/S0140-6736(12)61766-8, DOI	2012	146.4	**2013**	2017
Huang RJ, 2014, NATURE, V514, P218, DOI 10.1038/nature13774, DOI	2014	133.26	**2016**	2019
Lelieveld J, 2015, NATURE, V525, P367, DOI 10.1038/nature15371, DOI	2015	149.11	**2017**	2020
Stein AF, 2015, B AM METEOROL SOC, V96, P2059, DOI 10.1175/BAMS-D-14-00110.1, DOI	2015	100.45	**2017**	2020
Kim KH, 2015, ENVIRON INT, V74, P136, DOI 10.1016/j.envint.2014.10.005, DOI	2015	83.28	**2017**	2020
Cohen AJ, 2017, LANCET, V389, P1907	2017	192.08	**2018**	2022

Fig. 8.9 Top 10 references with the strongest citation bursts.

disease, lung cancer, and lower respiratory infections using non-linear exposure-response functions. It was also calculated that ambient $PM_{2.5}$ was the fifth-ranking mortality risk factor in 2015. The citation frequency of this paper outnumbers that of other papers substantially. The second one is Brook et al. (2010), a summary of which has been given in the literature review section above. The third one is Lelieveld et al. (2015), which studied the contribution of outdoor air pollution sources to world-wide premature death, and calculated that outdoor air pollution dominated by $PM_{2.5}$ caused 3.3% premature deaths in Asia every year. The study also discovered that the emissions from residential energy use (e.g., heating and cooking) prevalent in India and China had the greatest impact on global premature mortality, and predicted that the contribution of outdoor air pollution to premature death may double by 2050. The fourth one is Huang et al. (2014) which examined the secondary aerosol contribution to particulate pollution during haze events in China. The author investigated the chemical nature and sources of particulate matter in Beijing, Shanghai, Guangzhou and Xi'an during January 2013, and found that the severe haze pollution event was largely triggered by secondary aerosol formation. The fifth one is Lim et al. (2012), which is also the review article ranked fifth by burst intensity described above and the literature ranked first by co-citation burst. The sixth one is a study by Stein A. F. which introduced the atmospheric transport and deposition modelling, one of the most widely used atmospheric trajectory and dispersion calculation models. The latest applications of the system, including atmospheric tracer experiments and simulations of radioactive material, wildfire smoke, volcanic ash, mercury and wind-blown dust, were particularly explained, which provide a solid theoretical and technical basis for the study of the impact of atmospheric pollutants. The seventh and eighth articles, namely the two ranked third and fourth respectively by top 10 references with the Strongest Citation Bursts, are both written by Pope, which studied the influence of particulate matter on public health. In specific, the seventh one is about the impact of fine particles in air pollution on health (Pope & Dockery, 2006). The eighth one is about the relationship between lung cancer, cardiopulmonary mortality,

Table 8.9 Top 10 co-cited references paper

No.	First Author	Paper's name	Journal	Cited times	Year	Cluster NO.
1	Cohen AJ	Estimates and 25-year Trends of the Global Burden of Disease Attributable to Ambient Air Pollution: an Analysis of Data From the Global Burden of Diseases Study	Lancet	717	2017	1
2	Brook RD	Particulate Matter Air Pollution and Cardiovascular Disease An Update to the Scientific Statement from the American Heart Association	Circulation	479	2010	2
3	Lelieveld J	The Contribution of Outdoor Air Pollution Sources to Premature Mortality on a Global Scale	Nature	473	2015	3
4	Huang RJ	High Secondary Aerosol Contribution to Particulate Pollution During Haze Events in China	Nature	349	2014	4
5	Lim SS	A Comparative Risk Assessment of Burden of Disease and Injury Attributable to 67 Risk Factors and Risk Factor Clusters in 21 Regions, 1990–2010: a Systematic Analysis for the Global Burden of Disease Study 2010	Lancet	317	2012	5

(continued)

Table 8.9 (continued)

No.	First Author	Paper's name	Journal	Cited times	Year	Cluster NO.
6	Stein AF	Noaa's Hysplit Atmospheric Transport and Dispersion Modeling System	B Am Meteorl Soc	305	2015	6
7	Pope CA	Health Effects of Fine Particulate Air Pollution: Lines That Connect	J Air Waste Manage	297	2006	7
8	Pope CA	Lung Cancer, Cardiopulmonary Mortality, and Long-term Exposure to Fine Particulate Air Pollution	Jama-J Am Med Assoc	268	2002	8
9	Hegarty J	Evaluation of Lagrangian Particle Dispersion Models with Measurements from Controlled Tracer Releases	Hysplit Hybrid Sing	250	2013	9
10	Kim KH	A Review on the Human Health Impact of Airborne Particulate Matter	Environ Int	250	2015	10

and long-term exposure to fine particulate air pollution; it concluded that fine particulate matter and sulfur oxide-related pollution was related to all-cause mortality, lung cancer, and cardiopulmonary mortality (Pope et al., 2002). Hegarty et al. (2013) is the ninth one, which verified three widely applied Lagrangian models, and found that they had almost the same ability to simulate tracer plumes when inputting the same meteorological drive, indicating that the differences in the three models' frameworks played a secondary role in the study. The tenth one is Kim et al. (2015), which presented the progress of research into the impact of air particles on human health, and reviewed the basic evidence showing the impact of particles on human health.

Table 10 Top 6 keywords with the highest centrality.

No.	Keywords	Centrality	Frequency	Year
1	California	0.02	260	1999
2	hydrocarbon	0.02	201	1997
3	dioxide	0.02	171	1996
4	generation	0.02	167	1999
5	Asia	0.02	142	2000
6	diesel	0.02	124	2005

8.5 Research Focus Analysis

8.5.1 Keyword Analysis

The co-occurrence network analysis of keywords in academic papers can facilitate the exploration of the development trajectory and hotspots in a research field. To ensure the accuracy and rigour of the research, synonymous and congeneric keywords were merged based on the keyword frequency statistics of the original data. For example, NO_2, nitrogen oxides, and nitrogen dioxide were merged into NO_X; fine particulate matter, ultrafine particles, and nanoparticles were merged into PM. The analysis reveals that the centrality of keywords can be divided into three levels. There are six keywords with the highest centrality (0.2), which are California, hydrocarbon, dioxide, generation, Asia, and diesel (Table 10), implying that these keywords had been exhaustively discussed or strongly concerned at a certain stage, but their frequency is not high. There are 240 keywords with a centrality of 0.1, and the remaining keywords all have a centrality of 0. There are 17 keywords with a frequency of more than 1000 (Table 11), which are particulate matter, health, impact, aerosol, association, air quality, risk, transport, children, source application, disposition, polycyclic aromatic hydrocarbon, China, United States, long term exposure, area, and chemical composition. It can be found that the literature on particulate matter, health, impact, and aerosol has an absolute quantitative advantage; the countries with the largest number of research objects are China and the United States. Table 10 and Table 11 demonstrate that keyword frequency and centrality are weakly correlated. Combined with the keyword burst diagram (Fig. 8.10), it can be seen that the research on ozone, children, daily mortality, lung function, air, sulfur dioxide, sediment, symptoms, and $PM_{2.5}$ has a strong burst intensity.

8.5.2 Key Research Areas

CiteSpace was used to group the literature recorders of this paper into the clusters shown in Fig. 8.11. Among them, the top seven clusters are: daily mortality, long term

Table 11 Statistics of keywords with frequency over 1000.

No.	Keywords	Frequency	Centrality	Year
1	Particulate matter	7460	0.01	1999
2	Health	3654	0.01	1996
3	Impact	3070	0.01	1996
4	Aerosol	2800	0.01	1996
5	Association	2594	0.01	1996
6	Air quality	1966	0.01	1997
7	Risk	1825	0.01	1997
8	Transport	1556	0.01	1996
9	Children	1445	0.01	1996
10	Source apportionment	1408	0.01	2001
11	Deposition	1334	0.01	1996
12	Polycyclic aromatic hydrocarbon	1309	0.01	1996
13	China	1224	0.01	1996
14	United States	1057	0.01	1996
15	Long term exposure	1044	0.01	2000
16	Area	1026	0.01	1996
17	Chemical composition	1020	0.01	1998

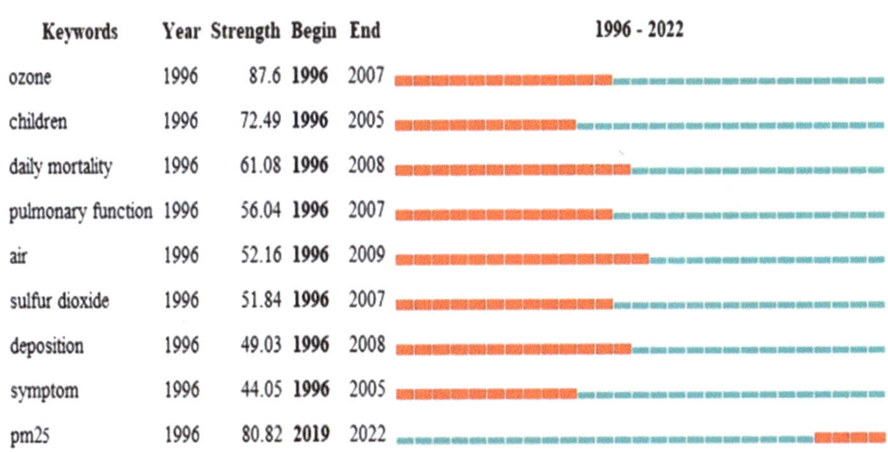

Fig. 8.10 Top 9 keywords with the strongest citation bursts

exposure, coronary heart disease, concentration, North China Plain, traffic-related air pollution, and air pollution (Table 8.12). The more keywords included in a cluster, the larger the cluster, and the richer the content. The description at the upper left corner of the cluster diagram shows that the mean silhouette (S for short) is 0.905, and the modularity (Q for short) is 0.7911. These values are much higher than the

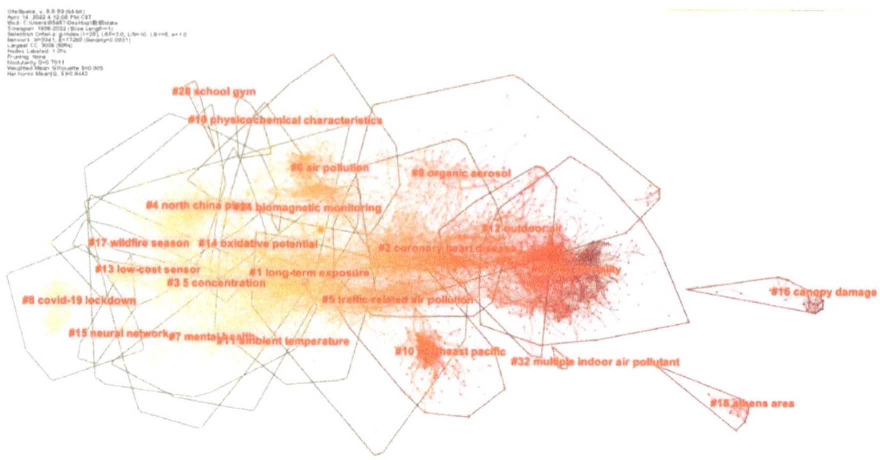

Fig. 8.11 Cluster map of literature

convincing clustering standard (S > 0.5 signifying reasonable clustering, S > 0.7 signifying convincing clustering; Q > 0.3 signifying significant clustering).

It can be seen from the cluster labels that the three most studied regions are North China, Northeast Pacific, and Athens area; the effects of air pollution on human diseases, mental health, and death (Coronary heart disease, Mental health, Oxidative potential, Daily mortality, Biometric monitoring) are mostly studied. At present, researchers have extended the impact of pollution to other fields, considering the effects not only on humans, but also on plants, such as the damage to tree crowns and lichens. The accumulation of airborne gases like CO, sulphur dioxide (SO_2), ozone (O_3), and NO_X has cascading effects on higher trophic levels, causing detrimental effect to various plantparts, including aerial structures and roots (Seaward & Richardson, 2024). Additionally, particulate matter (PM), comprising fine dust particles and non-gaseous pollutants fromanthropogenic and natural sources, disrupts plant reproduction by reducing pollen viabilityand hindering pollination processes, thereby threatening ecosystem stability (Das et al., 2024). Scholars have also found

Table 8.12 Top 7 cluster statistics

Cluster NO.	Cluster label	Relevant keywords quantity
0#	daily mortality	415
1#	long-term exposure	390
2#	coronary heart disease	374
3#	concentration	349
4#	north china plain	294
5#	traffic-related air pollution	190
6#	air pollution	161

that activities such as long-term exposure to polluted air, traffic emissions, and gym classes are most affected by air pollution (Long term exposure, Traffic air pollution, School gym). Bai et al. (2022) collected 97 dust samples from 10 indoor and 19 outdoor plastic playgrounds in Beijing, and found that the concentration of organophosphate esters was higher in indoor dust than in outdoor dust, and that in summer was higher than that in winter. Skin absorption was the main intake pathway for young people, while digestive tract absorption was the main intake pathway for children.

Furthermore, the most widely used method in relevant research is neural network; the research objects can be divided into indoor and outdoor (Indoor air & Multiple indoor air pollutants, Outdoor air); organic aerosol and particulate matter are the most studied air pollutants; because pollutant monitoring requires sensors, low-cost sensor is also a high-frequency keyword; in addition, the physical and chemical characteristics of air pollutants have also been investigated.

8.6 Research Status Analysis

Based on the keywords extracted above, synonymous and congeneric ones were combined to make a literature matrix (Support files Table 13). Then, theme clustering was carried out to obtain the high-frequency keywords network (Fig. 8.12). The centrality and density of the clusters were calculated according to the clustering results (Support files Table 14). Finally, a two-dimensional strategic diagram was drawn according to the coordinates of all clusters (Fig. 8.13). Clusters located in different quadrants have different implications: those in Quadrant I represent "core domains"; those in Quadrant II represent "potential domains"; those in Quadrant III represent "marginal domains"; those in Quadrant IV represent "basic domains".

It can be seen from the strategic diagram that 24 clusters were formed in this study (the cluster "air pollution" was excluded because it as the theme of this study has an extremely high centrality degree). In addition, the two keywords "epidemic" and COVID-19 are both in the cluster, indicating that the epidemic has a great impact on air pollution. Su et al. (2021) studied results show that COVID-19 cases cause unemployment for Germany, Italy, and the UK. The study's outcomes highlight that the pandemic increases the unemployment rate robustly in the mostly European economies. In turn, the impact of enterprises on the environment changes. So, the COVID-19 pandemic has directly or indirectly affected air pollution.

There is no cluster in Quadrant I, indicating that core clusters have not yet appeared at present. Quadrant II has 11 thematic clusters, the number of which is the largest among the four quadrants. They are Clusters A, B, D, F, G, E, C, H, K, I, and J. Clusters in this quadrant are featured by low density and high centrality. The centrality of Cluster A is the highest, demonstrating that it is closely related to all other clusters, appears frequently and actively in the research field, and is a research hotspot in this field. In addition, the low density of the clusters means that these clusters may evolve into core clusters with strong research potential. There are six

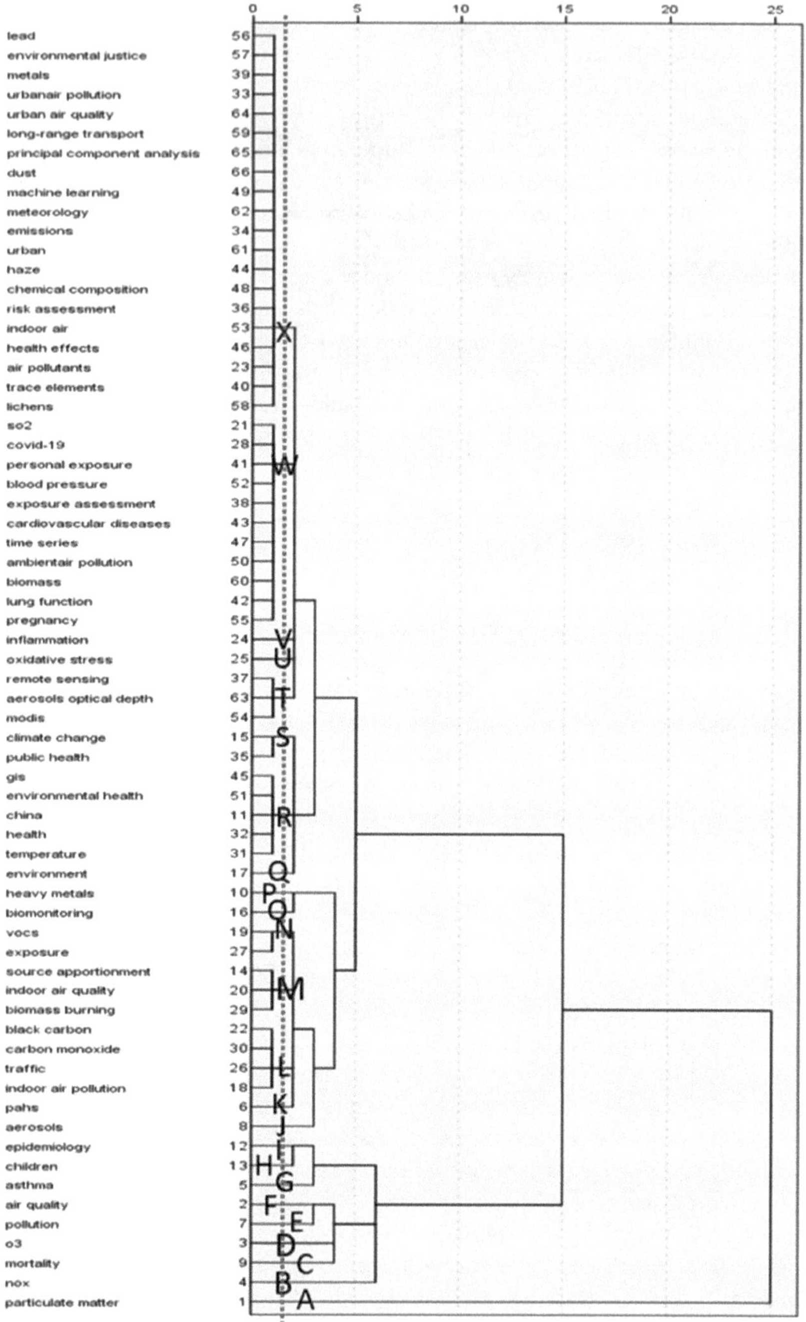

Fig. 8.12 Pedigree map of high frequency keywords

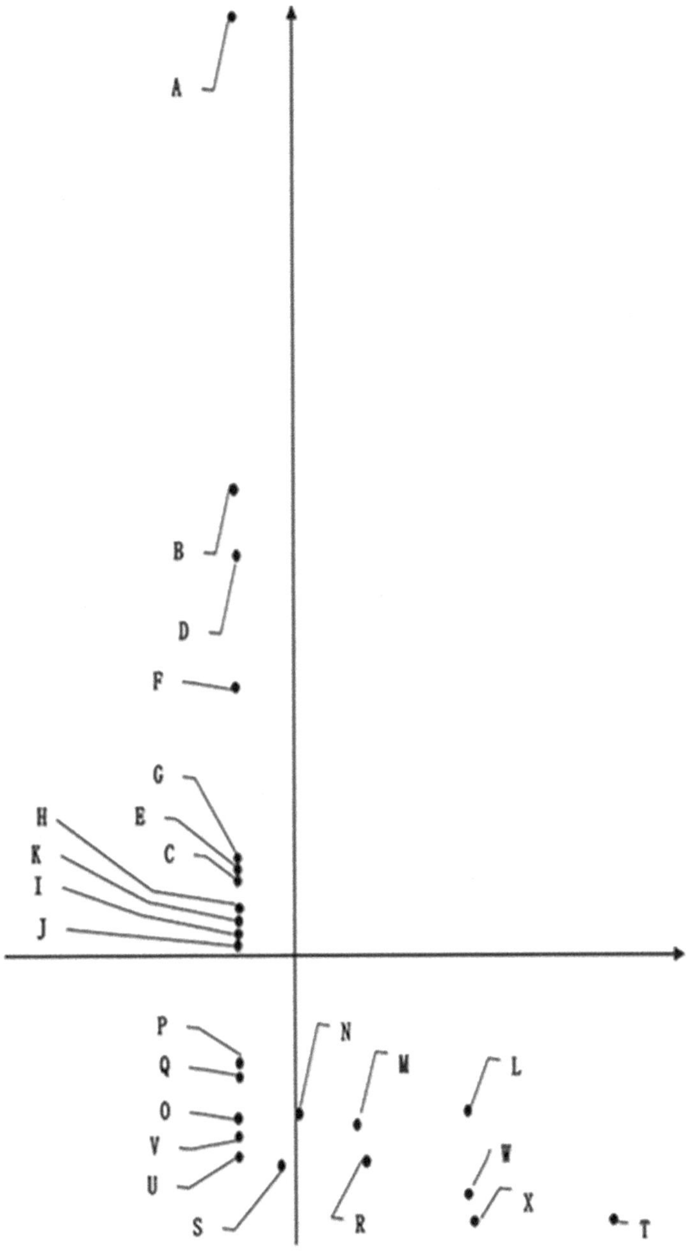

Fig. 8.13 Schematic diagram of strategy for studying the impacts of air pollution.

clusters in Quadrant III, namely Clusters P, Q, O, V, U and S. These clusters have low density and low centrality, indicating that they have the following characteristics: first, they are less related to other thematic clusters and more independent; second, they have attracted little scholarly attention. There are seven clusters in Quadrant IV, which are Clusters N, M, L, R, T, W, and X. These clusters are featured by low centrality but high density, capable of forming research systems by themselves. However, they are largely independent, loosely related to other clusters, and inactive. Therefore, after evolving into a certain stage, clusters in this quadrant may gradually fade out due to the lack of new research.

Support files Table 13 High frequency keywords matrix graph.

Support files Table 14 Clusters name and real value of strategic coordinates.

8.7 Conclusion

Taking air pollution impact research as the theme, this paper has visually analyzed the literature published between 1996 and 2022 sourced from WOS, and vividly presented the research status of this field. The main conclusions are as follows.

1. The quantity of literature has increased more than tenfold in the past twenty years; China and the US are the two significant leading countries by far in terms of the number of publications; the US started early in the field of air pollution assessment, and its academic leadership has lasted for a long term; the US, seven European countries, and Canada are the authorities in this research field, while China's research is currently in a vigorous development period; University of Ulm in Germany shows the highest institutional centrality, while the US Environmental Protection Agency and Harvard University have the highest publication burst intensity.
2. Many scholars have studied air pollution, but currently there lacks a cooperation network among highly prolific authors; colleges and universities are important institutions in this line of inquiry, but likewise no cooperation network has been developed between institutions.
3. Among the top ten articles ranked by the number of citations, those studying the impact of particulate matter on human health account for an overwhelming proportion.
4. California, hydrocarbon, dioxide, generation, Asia, and diesel are the six keywords with the highest centrality; keyword clustering has revealed that the six hottest clusters of air pollution impact research are: daily mortality, long term exposure, coronary heart disease, pollutant concentration, North China Plain, and traffic-related air pollution.
5. The keyword strategic diagram has demonstrated that no cluster has received considerable scholarly attention to date. Particulate matter, nitrogen oxide, mortality, ozone, pollution, air quality, asthma, children, epidemic disease,

aerosol, and polycyclic aromatic hydrocarbon are the 11 clusters that might become research hotspots.

Support files Table 13 High frequency keywords matrix graph of this paper.
Table 14 Clusters name and real value of strategic coordinates of this paper.

Acknowledgements Yanli Cao, Wenjun Han, and Jiaqi An also made great contributions to this manuscript. We express our heartfelt thanks to them. This research also was supported by: National Natural Science Foundation of China (72474127) and National Social Science Fund Major Bidding Project of China (18ZDA052). Part of this chapter was published: Cao Y, Wu X, Han W, An J. Visual analysis of global air pollution impact research: a bibliometric review (1996–2022). Environmental Science and Pollution Research 2024;31:40383–418. https://doi.org/10.1007/s11356-023-28468-y. The publication has been conducted with copyright permission.

References

Avnery, S., Mauzerall, D. L., Liu, J., & Horowitz, L. W. (2011). Global crop yield reductions due to surface ozone exposure: 2. Year 2030 potential crop production losses and economic damage under two scenarios of O_3 pollution. *Atmospheric Environment, 45*, 2297–2309. https://doi.org/10.1016/j.atmosenv.2011.01.002

Bai, L., Li, K., Shi, Y., Li, J., Wang, Y., & Jiang, G. (2022). Seasonal differences of organophosphate flame retardants(OPEs)in dust from plastic sports courts in Beijing, China, and implications for human exposure. *Scientific Bulletin, 67*, 2591–2604. https://doi.org/10.1360/TB-2022-0010

Bowatte, G., Lodge, C., & Lowe, A. J. (2015). The influence of childhood traffic-related air pollution exposure on asthma, allergy and sensitization: a systematic review and a meta-analysis of birth cohort studies. *Allergy, 70*(3), 245–256. https://doi.org/10.1111/all.12561

Brook, R. D., Rajagopalan, S., & Pope, C. A. (2010). Particulate matter air pollution and cardiovascular disease: An update to the scientific statement from the American Heart Association. *Circulation, 121*(21), 2331–2378. https://doi.org/10.1161/CIR.0b013e3181dbece1

Cheng, S. M., Zhang, J., & Wang, Y. (2019). Global research trends in health effects of volatile organic compounds during the last 16 years: a bibliometric analysis. *International Journal of Aerosol and Air Quality Research, 19*(8), 1834–1843. https://doi.org/10.4209/aaqr.2019.06.0327

Cohen, A. J., Brauer, M., & Burnett, R. (2018). Estimates and 25-year Trends of the global burden of disease attributable to ambient air pollution: an analysis of data from the global burden of diseases study. *Lancet, 391*, 1576–1576. https://doi.org/10.1016/S0140-6736(17)30505-6

Company, B. P. (2020). *Bp statistical review of world energy*. British Petroleum Company. https://www.gulfoilandgas.com/webpro1/main/mainnews.asp?id=901413

Crouse, D. L., & Peters, P. A. (2012). Risk of non accidental and cardiovascular mortality in relation to long-term exposure to low concentrations of fine particulate matter: A Canadian national-level cohort study. *Environmental Health, 120*(5), 708–714. https://doi.org/10.1289/ehp.1104049

Das, S., Barman, C., Roy, A., Mandal, M., Popek, R., Adit, A., & Sarkar, A. (2024). Air pollution may alter reproductive dynamics/efficacy in plants: correlative evidences from an urban sprawl in Eastern Indo-Gangetic plain. *Aerobiologia*, 1–19. https://doi.org/10.1007/s10453-024-09811-3

Han, M., Yang, F., & Sun, H. (2021). A bibliometric and visualized analysis of research progress and frontiers on health effects caused by $PM_{2.5}$. *Environmental Science and Pollution Research, 28*(24), 1–18. https://doi.org/10.1007/s11356-021-14086-z

Hegarty, J., Draxler, R. R., & Stein, A. F. (2013). Evaluation of Lagrangian particle dispersion models with measurements from controlled tracer releases. *Applied Mathematics and Computer Science, 40*(06), 171–178. https://doi.org/10.1175/JAMC-D-13-0125.1

Hoek, G., Krishnan, R. M., & Beelen, R. (2013). Long-term air pollution exposure and cardio—respiratory mortality: a review. *Environmental Health, 12*(1), 43–57. https://doi.org/10.1186/1476-069X-12-43

Huang, R., Zhang, Y., & Bozzetti, C. (2014). High secondary aerosol contribution to particulate pollution during haze events in China. *Nature, 514*(7521), 218–222. https://doi.org/10.1038/nature15371

Kim, K. H., Kabir, E., & Kabir, S. (2015). A review on the human health impact of airborne particulate matter. *Environment International, 74*, 136–143. https://doi.org/10.1016/j.envint.2014.10.005

Lelieveld, J., Evans, J. S., & Fnais, M. (2015). The contribution of outdoor air pollution sources to premature mortality on a global scale. *Nature, 525*(7569), 367–371. https://doi.org/10.1038/nature15371

Lim, S. S., Vos, T., & Flaxman, A. D. (2012). A comparative risk assessment of burden of disease and injury attributable to 67 risk factors and risk factor clusters in 21 regions, 1990-2010: A systematic analysis for the global burden of disease study 2010. *Lancet, 380*(9859), 2224–2260. https://doi.org/10.1016/S0140-6736(12)61766-8

Oliveira, M., Slezakova, K., & Delerue-Matos, C. (2019). Children environmental exposure to particulate matter and polycyclic aromatic hydrocarbons and biomonitoring in school environments: A review on indoor and outdoor exposure levels, major sources and health impacts. *Environment International, 124*, 180–204. https://doi.org/10.2307/25653851

Pope, C. A., Burnett, R. T., & Thun, M. J. (2002). Lung cancer, cardiopulmonary mortality, and long-term exposure to fine particulate air pollution. *Journal of the Air & Waste Management Association, 287*(9), 1132–1141. https://doi.org/10.1001/jama.287.9.1132

Pope, C. A., & Dockery, D. W. (2006). Health effects of fine particulate air pollution: lines that connect. *Journal of the Air & Waste Management Association, 56*, 709–742. https://doi.org/10.1080/10473289.2006.10464485

Saroj, P., Ghosh, C., Singh, S., & Soni, V. K. (2025). *Assessment of Crop Yield Loss Due, in: Select Proceedings of the 8th Indian International Conference on Air Quality Management (IICAQM 2023)* (p. 359). Springer Nature. https://doi.org/10.1007/978-981-96-2359-4_28

Seaward, M.R.D., Richardson, D.H.S., 2024. Atmospheric sources of metal pollution and effects on vegetation, in: Heavy metal tolerance in plants. CRC press, pp. 75–92.

Shah, A. S. V., Langish, J. P., & Nair, H. (2013). Global association of air pollution and heart failure: a systematic review and meta-analysis. *Lancet, 382*(9897), 1039–1048. https://doi.org/10.1016/S0140-6736(13)60898-3

Smith, D., & MedSc, M. P. H. (2009). A 30-year citation analysis of bibliometric trends at the archives of environmental health (1975–2004). *Archives of Environmental & Occupational Health, 64*, 43–54. https://doi.org/10.1080/19338240903293004

Song, M., Cui, L., & Zhou, Y. (2022). Management system and institution in China: current situation, problems and prospects. *Natural Resources, 37*(1), 1–16. https://doi.org/10.31497/zrzyxb.20220101

Song, M., Ma, X., Sha Ng, Y., & Zhao, X. (2020). Influences of land resource assets on economic growth and fluctuation in China. *Resources Policy, 68*, 101779. https://doi.org/10.1016/j.resourpol.2020.101779

Song, Q., Christiani, D. C., Wang, X., & Ren, J. (2014). The global contribution of outdoor air pollution to the incidence, prevalence, mortality and hospital admission for chronic obstructive pulmonary disease: A systematic review and Meta-analysis. *International Journal of Environmental Research and Public Health, 11*, 11822–11832. https://doi.org/10.3390/ijerph111111822

Stein, A. F., Draxler, R. R., & Rolph, G. (2015). Noaa's hysplit atmospheric transport and dispersion modeling system. *Bulletin of the American Meteorological Society, 96*(12), 2059–2077. https://doi.org/10.1175/BAMS-D-14-00110.1

Su, C., Dai, K., Ullah, S., & Andlib, Z. (2021). COVID-19 pandemic and unemployment dynamics in European economies. *Economic Research, 35*(1), 1752–1764. https://doi.org/10.1080/1331677X.2021.1912627

Sun, J. F., Zhou, Z., Huang, J., & Li, G. (2020). A bibliometric analysis of the impacts of air pollution on children. *International Journal of Environmental Research and Public Health, 17*(4), 1277–1287. https://doi.org/10.3390/ijerph17041277

Xue, J., & Zhao, L. (2017). *Study on joint prevention and control management mechanism of air pollution in China*. Beijing Science Press.

Part II
Economic Losses from Environmental Disasters in China

Chapter 9
Quantifying Aggregated Economic Damages from "Fat-Tail" Extremes High-Temperature Events in Climate Change

9.1 Introduction

With increasing climate change and more frequent heat events, assessing the economic damages associated with exacerbated risks from fat-tailed high temperatures is crucial. Fat-tailed high temperature risks refer to the increased frequency of extreme high-temperature events driven by cumulative carbon emissions and global warming, leading to heightened vulnerabilities and economic impacts.[1] This topic has garnered significant attention in recent years (Arribas et al., 2022). Recent studies have indicated that year 2022 was the hottest summer on record in Europe, resulting in approximately 62,000 deaths in Europe (Ballester et al., 2023). Additionally, data from the China Meteorological Administration shows that in 2023, China experienced the highest number of hot days (daily maximum temperature $\geq$ 35 °C) during the same period since 1961, with temperatures exceeding 40 °C in many regions. This leads to frequent tail high temperature events, severely impacting normal economic activities and daily lives of residents. With global warming, the likelihood of such tail high-temperature events and the "fat-tail" characteristics of climate risks are increasing (Coronese et al., 2019; Weitzman, 2013). However, while some scholars have employed classic Integrated Assessment Models (IAMs) to analyze the impacts of increasing climate risks (Cai & Lontzek, 2019; Van den Bremer & Van der Ploeg, 2021), there is currently limited research on the "fat-tail" characteristics of climate risks and quantification of economic damages from tail high-temperature. Such neglect in related IAMs gains the benefit of simplifying the analytical framework (such as DICE and RICE) and portraying multidimensional and complex associations (such as GCAM and IMAGE), but further refinement is urgently needed (see details in Table 9.1). Therefore, by constructing a dynamic stochastic equations-based climate

[1] (1) "Fat-tail" characteristics. Given the distribution of variables, "Fat-tail" is the frequent occurrence of events in the tails. Events in the tail refer to events that are far from the mean and have a very low probability of occurrence. (2) Tail high temperature. Given the distribution of temperature, it is the high temperature event in the tail. For example, the heatwaves discussed in Raei et al. (2018).

X. Wu, *Environmental Disaster in China*, https://doi.org/10.1007/978-981-92-0910-1_9

economic model (CEM), this paper aims to depict the characteristics of fat-tailed high temperatures and analyzes the aggregated economic damage under "fat-tail" characteristics of climate change. Our findings are crucial for the government in formulating climate policies to mitigate extreme high-temperature events.

Economic damage and the social cost of carbon (SCC) are the topic we focus on. SCC is defined as the present value of future economic damage caused by one ton of carbon emissions in the current period (Nordhaus, 2014, 2017). When evaluating economic damage, the results are influenced by multiple factors: (1) Climate risk factors, such as risks in the process of carbon stocks translating into temperature changes, risks in the process of temperature changes translating into damage rates, and risks in temperature changes leading to major disasters (also known as climate disruptions); (2) Economic risk factors, such as the economic growth fluctuations

Table 9.1 Comparison of various IAMs

Model	Main developer	Ranges	Types	Setting of economic damages
DICE	Nordhaus	Global	Optimization	A convex quadratic loss function
RICE	Nordhaus	Multi regions	Optimization	
GCAM	PNNL (Pacific northwest National Laboratory)	Multi regions	Comprehensive model	The interaction by running Hector with changes in policy, technique, and adaptation
IMAGE	PBL (Netherlands Environmental Protection Agency)	Multi regions	Simulation	Assessing direct economic and ecosystem losses with interventions of technology and policies
REMIND	PIK (Potsdam Institute for Climate Impact Research)	Multi regions	General equilibrium model	Setting economic impacts by focusing on the energy sector and related techniques
S-IAMs	C-R-2024 (Cruz and Rossi-Hansberg (2024))	$1^\circ \times 1^\circ$ resolution	Optimization and simulation	Production and utility loss by econometrics
DSGE-VV	V-V-2021 (Van den Bremer and Van der Ploeg (2021))	Global	Optimization	Setting stochastic process with multi risks
CEM	This paper	Global	Simulation	The role of "fat-tail" on economic damages

Note C-R-2023 is the work of Cruz and Rossi-Hansberg (2024). DSGE-VV refers to the DSGE model of V-V-2021 (Van den Bremer & Van der Ploeg, 2021)

and risks in trends (Cai & Lontzek, 2019; Gillingham et al., 2018; Van der Ploeg & de Zeeuw, 2016). As climate change intensifies and associated risks increase, the variance in temperature and damage processes expands, leading to a gradual flattening of the related distribution and a rising likelihood of tail high-temperature events. To account for the above risks, we construct a dynamic stochastic equation model, referred to as CEM, which captures the exacerbation of tail high-temperature risks and "fat tail" characteristics of climate risks. Subsequently, this study quantifies the response of tail characteristics to carbon emission increases and the related economic damages due to "fat-tail" trend.

To design the theoretical framework of the Climate Economic Model (CEM) with risk factors, this paper completes the following work: (1) **CEM-1 and CEM-2**: This introduces the accumulation of carbon emissions and carbon stocks, as well as the process of temperature change. Inspired by Golosov et al. (2014) and Rezai and Van der Ploeg (2016), a nonlinear carbon emissions process and a temperature change process incorporating temperature change risks are constructed. (2) **CEM-3**: This part focuses on the damage rate resulting from temperature change. The damage function is a crucial concept in climate economics, used to estimate the damage rates caused by temperature change (Nordhaus, 2014). In CEM-3, the disaster risks resulting from temperature change and damage risks are incorporated into the damage function. (3) **CEM-4 and CEM-5**: This part portrays the output process (GDP), the economic damages resulting from temperature change, and the calculation of SCC. Economic growth and economic trend risks are included in the economic output process, followed by the calculation of economic damages and SCC. After completing the model construction and calibrating related parameters, this paper quantified the "fat-tail" characteristics of climate change, the economic damages induced by tail high temperatures, and the response of these tail features to increasing carbon emissions under five Shared Socioeconomic Pathways (SSPs).

The major findings are as follows: (1) The "fat-tail" trends of climate change become more pronounced over time. By 2100, the temperature tail probability is projected to be about 3.45 times higher than in 2025, leading to an increase in the frequency of extreme temperatures and the associated economic damages. (2) Tail high-temperature events may lead to unknown risks. The economic damages of normal temperature paths are insufficient to provide a reliable reference under tail-high temperature scenarios, aligning with Weitzman (2015)'s discussion. This suggests that tail high temperatures may introduce some "unknown unknowns" beyond the traditional understanding of climate change. (3) The tail probability of climate change responds to increased carbon emissions, gradually increasing over time. Specifically, a 1 ppm increase in carbon stock leads to an average increase in the tail temperature probability of 0.15×10^{-4}.

The marginal contributions of this study are as follows: (1) Building on Cai and Lontzek (2019) and Van den Bremer and Van der Ploeg (2021), this study constructs a climate economic model (CEM) that incorporates multiple risk factors, noting as CEM. Applying the CEM, we analyze the trends in tail features of climate change, including the "fat-tail" characteristics of temperature change, damage rates and economic damages. This characterization is further supported by

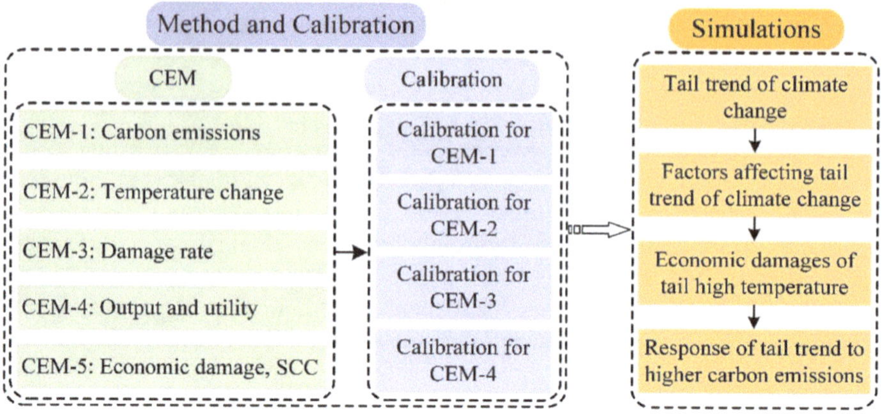

Fig. 9.1 Research framework (CEM is the constructed climate economic model in this paper)

Burke et al. (2018) and Coronese et al. (2019), ensuring our work supplemented the existing literature. (2) Inspired by Weitzman (2013), this study discusses the impacts of tail high-temperature risks, validating Weitzman (2015)'s analysis of "unknown unknowns" analysis. We also quantify the response of temperature tail probabilities to increasing carbon emissions, explaining the relationship between carbon emissions and the probability of high tail temperature occurrence. (3) Compared to other IAMs, the CEM presented in this paper is a global model based on a dynamic recursive and simulation approach, specifically focusing on the analysis of "fat-tail" temperature events and the associated damages. This approach complements the existing IAMs.

The rest part of this paper is organized as follows: Sect. 9.2 provides a literature review; Sect. 9.3 presents empirical evidence supporting our analysis; Sect. 9.4 outlines the research methods and calibration; Sect. 9.5 discusses the simulation analysis; and the final section presents the major conclusions. The research framework is illustrated in Fig. 9.1.

9.2 Literature Review

The research relevant to this paper focuses on economic damage associated with risk factors, fat-tailed high temperatures and the application of Integrated Assessment Models (IAMs). Quantifying economic damages came earlier from environmental issues (He, Wu, & Wang, 2025a; He, Wu, Wang, & Lu, 2025), and then received attention in climate economics. Based on the considered risk categories, the research of exacerbated risks and economic damage can be roughly divided into three types of literature. (1) **Economic risks**: Scholars have analyzed the relationship of economic risks and economic damage and the formulation of optimal climate policies, particularly the economic risks arising from climate governance measures

(Chen et al., 2023; Jensen & Traeger, 2014). For example, Gillingham et al. (2018) examined the impact of economic risks (mainly those related to population growth and technological progress) on economic damage and SCC in multiple economic models. They found that an exacerbation of economic risks leads to increased variance of temperature and economic output, thereby heightening economic damage and SCC. Danisman et al. (2025) examined the intensified economic risks leading to corporate-level climate exposure, particularly for companies with high climate vulnerability, stringent environmental regulations, and elevated risks of environmental litigation. (2) **Climate risks**: Research in this category addresses various climate risks, including climate sensitivity risks, major disaster risks, temperature change risks, and damage risks (Crost & Traeger, 2014; Dietz & Stern, 2015).[2] Cai et al. (2016) investigated the impact of disaster risks on the economic system and SCC, reporting that global SCC increases from $15/ton to $116/ton in the event of disaster shocks. Yin et al. (2025) studied the impacts of climate risk on property insurance demand from 2011–2020 in China. They found that intensified climate risks have increased demand for insurance, as more risk-averse expectations have emerged. Ha (2025) employed a panel vector autoregressive model to examine the relationship between digital public services and climate risk. The two exhibit a U-shaped correlation: digital public services decrease initially but increase later. This stems from subsequent adjustments and heightened attention to climate risk. (3) **Integrated economic and climate risks**: Relevant studies in this category simultaneously evaluate multiple risk factors (Barnett et al., 2020; Takakura et al., 2019). For example, Cai and Lontzek (2019) analyzed the impact of economic growth risks and major disaster risks on SCC, concluding that exacerbated risks contribute to increased SCC. Van den Bremer and Van der Ploeg (2021) explored the impacts of capital accumulation risks, temperature change risks, and damage risks on SCC. It is evident that extensive studies have been conducted on the impact of exacerbated risks on economic damage and SCC. Despite extensive research on the impact of exacerbated risks on economic damages and SCC, there remains a significant gap in studies addressing fat-tailed high-temperature risks and economic damages caused by tail high temperatures (Weitzman, 2013). This study integrates economic and climate risks, aligning with the third category of literature mentioned above, while focusing specifically on the "fat-tail" characteristics of climate risks and the economic damages caused by tail high temperatures, thereby expanding the scope of the existing literature.

Research on fat-tailed high temperatures has gained traction, particularly with Weitzman's early contribution. In 2013, Weitzman (2013) highlighted the significance of fat-tailed high-temperature risks, later arguing in 2015 of Weitzman (2015) that Nordhaus had overlooked the structural uncertainties of climate change.[3] The

---

[2] Climate sensitive risk is generally defined as the average global warming caused by a doubling of carbon emissions from pre-industrial levels.

[3] Weitzman emphasizes that the structural uncertainty of climate change consists of four main components: (1) Uncertainty in economic damage faced by different countries and regions; (2) Intergenerational uncertainty of economic damage; (3) Uncertainty in economic damage caused

economic damage and SCC at different warming levels represent a significant manifestation of this structural uncertainty. Weitzman (2015) posited that while moderate global warming aligns with the economic damages predicted by traditional Integrated Assessment Models (IAMs), once global warming exceeds a certain threshold, economic damages may experience an explosive growth, deviating from the damage rates trajectories established in conventional IAMs. This shift indicates the emergence of bad tail risks associated with global warming, where low-probability and high-temperature catastrophic events are more likely to occur. Consequently, the economic damage caused by fat-tailed climate events is expected to increase significantly. For example, Coronese et al. (2019) investigated the variation trends in economic damages from climate change by analyzing disaster event data. Their findings revealed that as climate change intensifies, extreme high temperatures lead to increased extreme economic damage, exhibiting fat-tail characteristics that hinder economic growth. Furthermore, Schmidt (2024) noted a significant spike in extreme high temperatures in 2023 that lacked a reasonable explanation with existing climate models, suggesting that we are in an unknown territory of climate change. Considering these insights, this study investigates the economic damage associated with tail high temperatures, thereby validating the view of Weitzman (2015)'s perspective and enriching the discourse surrounding tail high-temperature risks.

IAMs are the essential tools for climate impact assessment and climate policy design. Over the years, IAMs have evolved and are widely employed in climate science, economics and climate policy estimates (Van Benk et al., 2020). Several major IAMs have been developed (Table 9.1), each contributing to our understanding of climate-related issues. This section compares the difference of the major IAMs and our CEM about the setting of economic losses, including DICE (Dynamic Integrated Model of Climate and the Economy), RICE (Regional Integrated Model of Climate and the Economy), GCAM (Global Change Assessment Model), IMAGE (Integrated Model to Assess the Global Environment), REMIND (Regional Model of Investment and Development), and so on (Harfoot et al., 2014; Li et al., 2024). Of these, DICE and RICE, developed primarily by Nordhaus, reflected economic damage by setting a convex quadratic loss function and reducing productivity. GCAM used the carbon-cycle climate module-Hector for assessing the interactions and impacts of carbon emissions and climate change with policy constraints, technique change, and adaptation feedback. IMAGE focuses on economic impacts by assessing direct economic losses (agricultural economic losses, sea level rise, health cost, and so on) and ecosystem losses (reduced biodiversity and carbon sinks) with interventions of technology improvements and policies. REMIND, a general equilibrium model supported by Potsdam Institute for Climate Impact Research (PIK), assesses economic damage by focusing on the complete energy sector (multi energy resources and economic links) and related technologies improvement among multiple regions. Recently, new IAMs have appeared, such as the Spatial-IAMs (S-IAMs) by Cruz and Rossi-Hansberg (2024), which provides $1^\circ \times 1^\circ$ resolution spatial models for

by different levels of warming; (4) The cost uncertainty resulting from new climate actions and zero-carbon technologies.

assessing the climate impacts with production and utility damages. Another notable model is the Dynamic Stochastic General Equilibrium (DSGE) model by Van den Bremer and Van der Ploeg (2021), which emphasizes the impact of risk factors on economic damage by using a stochastic process. The CEM presented in this paper is a global model like DICE and DSGE by Van den Bremer & Van der Ploeg, 2021. The results of CEM are obtained by simulations based on dynamic recursive relationships and focus specifically on assessing tail-high temperature features and the associated economic damage. The comparison of the above IAMs is reported in Table 9.1.

9.3 Facts of Warming and "Fat-Tail"

This section provides empirical evidence to support our analysis in two parts. First, we examine the trends in global warming and heatwaves, using changes in the frequency of global heatwaves to illustrate the "fat-tail" risks associated with climate change. Second, we assess the economic impact of warming and heatwaves on GDP growth, highlighting the potential economic damage under "fat-tail" scenario of climate change.

Trend of warming and "fat-tail". Warming trend is measured by the temperature change data and the "fat-tail" characteristics of climate change are estimated by heatwaves data. The CRU-TS temperature data shows a significant global warming trend (Fig. 9.2-A). The frequency of heatwaves calculated using the Standardized Heat Index (SHI) method (15-day window) in the GHWR dataset has risen globally from 9.2 days in 1979 to 64.9 days in 2017 (Fig. 9.2-A to Fig. 9.2-C), indicating an increased likelihood of extreme temperatures in the "fat-tail" of climate change.

Impact of warming and heatwaves on GDP. Aligning Fig. 9.2 with *World Bank*, data provides a panel GDP versus heatwave frequency across 176 countries from 1979–2017. Regression results shown in Table 9.2 retain samples with non-zero heatwave frequency and include an interaction term between warming and heatwaves. The findings indicate that higher heatwave frequencies correlate with more substantial GDP losses due to warming, suggesting that "fat-tail" climate trends may significantly increase economic risks.

9.4 Method and Calibration

This section introduces the research method and the related calibration for matching the historical data. The "fat-tail" phenomenon and their dynamic characteristics of temperature change and damage rate are analyzed in our model and calibration.

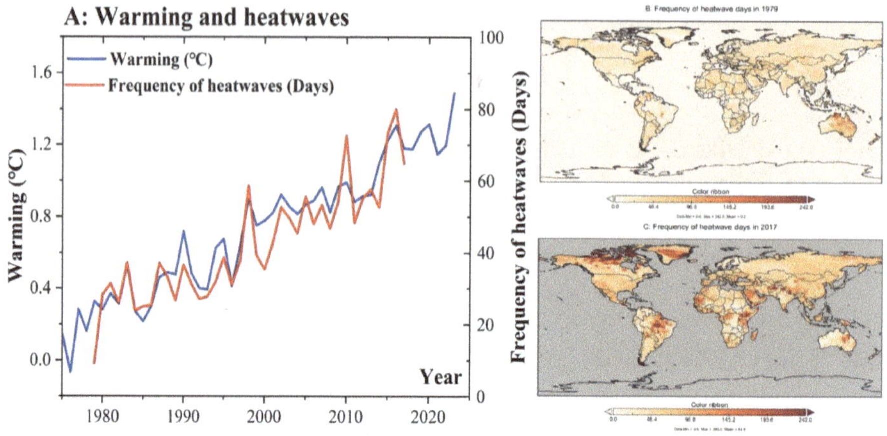

Fig. 9.2 Warming and heatwave trends. *Note* (1) Warming data comes from CRU-TS dataset (Harris et al., 2020). The heatwave data is provided by Raei et al. (2018). (2) The global map is developed by using software Panoply. (3) In Fig. 9.2-A, the base temperature is the mean temperature of 1901–1910

9.4.1 Research Method

This section describes the research methodology used in this paper, including a constructed climate-economic model (CEM), calculation of tail probability and model fitting criteria. CEM portrays the conversion process of fossil energy used in production to climate change and the damage rate.

Assumptions and descriptions of CEM. Our CEM assumes: (1) fossil fuel use leads to carbon emission, partially offset by natural and technological effect. (2) temperature changes depend on emissions, feedback, and risk factors, which increase disaster risks and damage rates. Calibration details: (1) carbon emissions and temperature trends were calibrated for real-world accuracy, particularly in capturing "fat-tail" temperature risks. (2) linear and quadratic damage rates of CEM align with findings from Nordhaus and Moffat (2017), Golosov et al. (2014) and Weitzman (2012), confirming reliability. (3) SSP1-SSP5 scenarios simulate future outcomes, with data from Burke et al. (2018).

CEM-1: carbon emissions and carbon stock. According to Rezai and Van der Ploeg (2016) and Van den Bremer and Van der Ploeg (2021), carbon emission mainly comes from the fossil fuel use F_t in production. The residual carbon emissions at period t from emissions at time k is represented as:

$$C_t = \left[\varphi_L + \varphi_0(1\text{-}\varphi_L)(1\text{-}\varphi)^{t\text{-}k+1}\right]\mu_1 F_k e^{\text{-}\varphi_I k} \tag{9.1}$$

Carbon emissions have a portion permanently retained in the natural environment, while the rest is absorbed and neutralized by nature. In Eq. (9.1), $\varphi_L + \varphi_0(1 - \varphi_L)(1 - \varphi)^{t-k+1}$, $(k \le t)$ is the residual share from period k to t. φ_L is the

Table 9.2 Impact of warming and heatwaves on GDP

Variables	GDP							
	(1)	(2)	(3)	(4)	(5)	(6)	(7)	(8)
L1.tc	−0.0439***	−0.0257***	−0.0351***	−0.0210**	−0.0439**	−0.0257*	−0.0351**	−0.0210*
	(−4.91)	(−3.40)	(−3.82)	(−2.72)	(−2.96)	(−2.43)	(−3.05)	(−2.51)
L2.tc × hw			−0.00725***	−0.00511**			−0.00725*	−0.00511*
			(−3.33)	(−2.79)			(−2.09)	(−2.05)
Control	No	Yes	No	Yes	No	Yes	No	Yes
FE of country	Yes	Yes	Yes	Yes	Yes	Yes	Yes	Yes
FE of year	Yes	Yes	Yes	Yes	Yes	Yes	Yes	Yes
Constants	23.99***	12.84***	24.02***	12.94***	23.99***	12.84***	24.02***	12.94***
	(3185.62)	(41.10)	(2715.84)	(40.26)	(2202.40)	(6.69)	(1359.49)	(6.71)
Observations	5726	5726	5561	5561	5726	5726	5561	5561

Note (1) T-statistics in parentheses. * $p < 0.05$, ** $p < 0.01$, *** $p < 0.001$. (2) The last four columns used the robust standard errors, clustering to country level. (3) Control variables include export scale and population. *L1. tc* is the temperature change with a lag period (using the mean temperature of 1901–1910 as the base period). *L2. tc* × hw is the interaction term of temperature change (*tc*) and heatwaves (hw). *GDP* is the output at constant 2015 price. *GDP*, hw and control variables take logarithms share of carbon emissions permanently retained in the atmosphere, and φ_0 is the reduction effect of $1-\varphi_L$ share carbon emissions. The reduction effect φ_0 gradually decreases by the rate of φ. F_t is defined as the fossil fuel use.[4] $e^{-\varphi_I k}$ represents the mitigating effect of technology on fossil fuel use.[5] φ_I denotes the exogenous improvement in energy efficiency. μ_1 is the conversion rate of energy consumption and carbon emissions. Then, carbon stock is:

$$AC_t = \mu_2\left\{AC_0 + \sum_k^t\left[\varphi_L + \varphi_0(1\text{-}\varphi_L)(1\text{-}\varphi)^{t\text{-}k+1}\right] \mu_1 F_k e^{\text{-}\varphi_I k}\right\}, k \le t \tag{9.2}$$

where AC_0 is the initial carbon stock. AC_t is composed of the initial carbon stock and the cumulative emissions from period k to t. μ_2 is the influence of other factors in the cumulative carbon emissions process.

9.4.1.1 CEM-2: Climate Change (Temperature Change)

Increased carbon stocks cause temperature changes. Based on Meinshausen et al. (2011) and Van den Bremer and Van der Ploeg (2021), we define the stochastic differential equation (SDE) of temperature change as follows:

$$dT_t = \left(\varepsilon\frac{AC_t}{AC_0} + m\frac{T_t}{T_0}\right)dt + \sigma_T T_t dW_t \tag{9.3}$$

T_t is the temperature level at period t. Noted that the modeling of temperature anomalies in previous research is influenced by different temperature baseline values. To avoid this issue, we directly model the temperature process T_t. AC_0 and T_0 are the initial carbon stock and temperature level, respectively. AC_t and T_t are the carbon stock and temperature level at period t. Parameters ε and m are the influence of carbon stock and temperature feedback factors on the temperature process. σ_T is the volatility of the temperature process. W_t means a standard Wiener process, reflecting the temperature change risks. From the Eq. (9.3), the change in temperature dT_t depends on three factors: (1) The impact of carbon stock $\varepsilon(AC_t/AC_0)dt$.

---

[4] A C-D production function ($Y_t = K_t^\alpha F_t^{1-\alpha}$) is used for obtaining fossil fuel use data. Y_t is output, K_t is capital, and α is the share of capital input. Solving the decision $\max\left(K_t^\alpha F_t^{1-\alpha} - rK_t - P_F F_t\right)$, we have $\alpha Y_t/r = K_t$. Output prices standardized to 1 as in Desmet and Rossi-Hansberg (2013). Interest rate r is set to 6.4% according to Van den Bremer and Van der Ploeg (2021). Bringing in $\alpha Y_t/r = K_t$, we have $F_t = Y_t(r/\alpha)^{\alpha/(1-\alpha)}$. We use the feasible generalized nonlinear least squares (FGNLS) and solve the problem $\min_\alpha\left[\Delta F_t - \Delta Y_t(r/\alpha)^{\alpha/(1-\alpha)}\right]$ for matching the fossil fuel use data from 1965 to 2020. Then, α is set to 0.55 and the share of energy input is 0.45.

[5] There is also literature on the negative impacts of climate change on productivity (Kumar & Maiti, 2024).

(2) The temperature feedback factor $m(T_t/T_0)dt$, such as radiative forcing. (3) The temperature change risk $\sigma_T T_t dW_t$. Then, the temperature process is calculated as:

$$\begin{aligned} T_t = T_0 \exp\left(\sigma_T W_t\text{-}\left(\tfrac{\sigma_T^2}{2}\text{-}\tfrac{m}{T_0}\right)t\right) \\ + \varepsilon \tfrac{AC_t}{AC_0} \int_0^t \exp\left[\sigma_T (W_t\text{-}W_s)\right. \\ \left.\text{-}\left(\tfrac{\sigma_T^2}{2}\text{-}\tfrac{m}{T_0}\right)(t\text{-}s)\right]ds \end{aligned} \tag{9.4}$$

hence, the expectation $E(T_t)$ and the variance $Var(T_t)$ of temperature process are obtained as $E(T_t) = T_0(1 - (\sigma_T^2/2 - m/T_0)t) + \varepsilon AC_t/AC_0(t - 0.5(\sigma_T^2/2 - m/T_0)t^2)$ and $Var(T_t) = T_0^2\sigma_T^2 + (\sigma_T^2/3)(\varepsilon AC_t/AC_0)^2 t^3$ by calculating the moment generating function (MGF) of temperature change.

9.4.1.2 CEM-3: Damages with Hazard Risks

Weitzman (2009) studied that significant temperature increases could lead to catastrophic events with unpredictable, low-probability, and massive economic damages. To characterize this potential catastrophe, we use the approach of Van Der Ploeg and De Zeeuw (2016) and introduce a hazard probability function $H(T_t - T_0)$. This function represents the probability of significant disaster risk caused by the temperature change $T_t - T_0$ (the temperature change of T_t and T_0). When $T_t - T_0 > 0$, $H(T_t - T_0)$ exhibits a monotonically increasing trend. Therefore, we define the damage rate after considering hazard risks as follows:

$$D_t = \frac{1}{1 - H(T_t - T_0)} \mu_T (T_t - T_0)^{1+\theta_T} {W_{Dt}}^{1+\theta_W} \tag{9.5}$$

where μ_T is the conversion ratio of temperature change $T_t - T_0$ and the damage rate D_t. When $\theta_T = 0$, it exhibits a linear relationship; when $\theta_T > 0$, this conversion follows a convex property. The quadratic function is commonly used in the studies of climate change economics. Referring to Cai and Lontzek (2019), $H(T_t - T_0)$ is defined as: $1 - \exp(-(1 + \lambda \max\{0, \mathbf{E}(T_t) - T_0\}))$. Parameter $\lambda (\lambda > 0)$ is the disaster parameter. As temperature change $\mathbf{E}(T_t) - T_0$ increases, the probability of no significant disaster occurring $\exp(-(1 + \lambda \max\{0, \mathbf{E}(T_t) - T_0\}))$ gradually declines. Then, the probability of a disaster occurring $H(T_t - T_0)$ increases. Drawing on Van den Bremer and Van der Ploeg (2021), the risk of damage rate W_{Dt} is introduced into Eq. (9.5). W_{Dt} is the result of a mean-reverting stochastic process.[6] Then, Eq. (9.5) can be transformed into:

[6] According to Van den Bremer and Van der Ploeg (2021), the risk of damage rate follows a normal distribution: $W_{Dt} \sim N(\mu_w, \Sigma_w^2)$. The mean and variance of ${W_{Dt}}^{1+\theta_W}$ are $\mu_w^{1+\theta_W}(1 + 0.5\theta_W(1 + \theta_W)(\Sigma_w^2/\mu_w^2))$ and $(1 + \theta_W)^2 \mu_W^{2(1+\theta_W)}(\Sigma_w^2/\mu_w^2)$, respectively. The setting of relevant parameters follows Van den Bremer and Van der Ploeg (2021). θ_W is set to 2.7.

$$D_t = \exp(1 + \lambda \max\{0, E(T_t)\text{-}T_O\}) \\ \mu_T(T_t\text{-}T_O)^{1+\theta_T} {W_{Dt}}^{1+\theta_W} \tag{9.6}$$

from Eq. (9.5) and Eq. (9.6), damage rate D_t depends on the temperature change $T_t - T_O$, the risk of damage rate W_{Dt}, and the hazard risk $H(T_t - T_O)$. We consider both the linear damage rate ($\theta_T = 0$) and the convex damage rate ($\theta_T = 1$).[7]

CEM-4: economic growth and utility. Following the study by Golosov et al. (2014) and Cai and Lontzek (2019), the dynamic process of economic output is defined as follows:

$$Y_{t+1} = \omega_Y Y_t + \chi_t + \zeta_Y W_{Yt} \tag{9.7}$$

where ω_Y is the contribution of economic output at period t to economic output at period $t + 1$. χ_t denotes the sustained trend of economic growth and meets: $\chi_{t+1} = \omega_\chi \chi_t + \zeta_\chi W_{\chi t}$. Here, $W_{\chi t}$ follows a normal distribution, reflecting the economic trend risk. ω_χ is the contribution of economic trend at period t to economic trend at period $t + 1$. ζ_Y and ζ_χ are the coefficients of economic growth risk W_{Yt} and economic trend risk $W_{\chi t}$ on Y_t and χ_t. W_{Yt} and $W_{\chi t}$ follow independent identically standard normal distributions. Solving Eq. (9.7), the expectation and variance of Y_t are obtained[8]:

$$E(Y_t) = {\omega_Y}^t Y_O, \ Var(Y_t) \\ = \left[\zeta_\chi^2 \frac{1-\omega_\chi^{2t}}{1-\omega_\chi^2} + \zeta_Y^2\right]\left[\frac{1-\omega_Y^t}{1-\omega_Y}\right]^2. \tag{9.8}$$

we adopt a logarithmic utility function as $U(C_t) = \ln(C_t)$. Residents consume the economic outputs to obtain resident utility: $C_t = Y_t$. Additionally, residents face the problem of utility maximization: $\sum_{t=0}^{n}(1 + r_t)^{-t}U(C_t)$. r_t is a Ramsey annual discounting rate that decreases over time, curbing the relationship between economic growth risk and time. Based on Weitzman (2012a) and Burke et al. (2018), $r_t = \rho + \eta g_{Yt} - 0.5\eta^2 \mathbf{Var}(g_{Yt})$. ρ is the time preference and the risk aversion coefficient

[7] We assume the temperature change and risk of damage rate are independent of each other. The expectations and variances of damage rate are $\boldsymbol{E}(D_t|_{\theta_T=0}) = \exp(1 + \lambda \max\{0, \boldsymbol{E}(T_t) - T_O\})\mu_T \boldsymbol{E}(T_t - T_O)\boldsymbol{E}\left(W_{Dt}^{1+\theta_W}\right)$, $\boldsymbol{E}(D_t|_{\theta_T=1})$

$$= \exp(1 + \lambda \max\{0, \boldsymbol{E}(T_t) - T_O\})\mu_T\left(\boldsymbol{E}\left(T_t^2\right) - 2T_O\boldsymbol{E}(T_t) + T_O^2\right)\boldsymbol{E}\left(W_{Dt}^{1+\theta_W}\right),\ \boldsymbol{Var}(D_t|_{\theta_T=0})$$

$$= \left[\exp(1 + \lambda \max\{0, \boldsymbol{E}(T_t) - T_O\})\mu_T\right]^2\Big[\boldsymbol{Var}(T_t - T_O)\boldsymbol{Var}\left(W_{Dt}^{1+\theta_W}\right)$$

$$+\boldsymbol{Var}(T_t - T_O)\boldsymbol{E}\left(W_{Dt}^{1+\theta_W}\right)^2 + \boldsymbol{Var}\left(W_{Dt}^{1+\theta_W}\right)\boldsymbol{E}(T_t - T_O)^2\Big],\ \boldsymbol{Var}(D_t|_{\theta_T=1})$$

$$= \left[\exp(1 + \lambda \max\{0, \boldsymbol{E}(T_t) - T_O\})\mu_T\right]^2\Big[(1 + \theta_T)^2\boldsymbol{Var}(T_t)\ \boldsymbol{Var}\left(W_{Dt}^{1+\theta_W}\right)$$

$$+ (1 + \theta_T)^2\boldsymbol{Var}(T_t)\boldsymbol{E}\left(W_{Dt}^{1+\theta_W}\right)^2 + \boldsymbol{Var}\left(W_{Dt}^{1+\theta_W}\right)\left(\boldsymbol{E}(T_t)^2 - 2T_O\boldsymbol{E}(T_t) + T_O^2\right)^2\Big]$$

[8] We set $\boldsymbol{Var}(Y_O)$ and $\boldsymbol{Var}(\chi_O)$ as 0.

η is 1 in the logarithmic utility function. g_Y is the economic growth rate and $\mathbf{Var}(g_{Yt})$ is the variance of g_{Yt}. If $\mathbf{Var}(g_{Yt})$ is 0, the Ramsey annual discounting rate declines to: $r_t = \rho + \eta g_{Yt}$.

CEM-5: economic damage and Social Cost of Carbon (SCC). Due to the negative externalities of climate change and based on Barrage (2020), the economic and utility impact of climate change is defined as follows:

$$EL_t = D_t Y_t + D_t \ln(Y_t) \tag{9.9}$$

where D_tY_t and $D_t\ln(Y_t)$ are production damages and utility damages due to climate change. It characterizes two pathways of damage caused by climate change: economic damage and utility damage (Barrage, 2020). In the traditional IAMs framework, previous research generally considers only production damage, leading to some underestimation.

Following Van Der Ploeg and De Zeeuw (2016), the SCC is defined as the present value of future cumulative damage caused by one ton of carbon emissions in the current period. The temperature change and damages rate caused by emitting one ton of carbon at period t are $\widetilde{T}_t$ and $\widetilde{D}_t$, respectively. Then, the SCC is expressed as:

$$SCC_t \equiv \sum\nolimits_{t=k}^{n} (1 + r_t)^{-t} \left[\widetilde{D}_t Y_t + \widetilde{D}_t \ln(Y_t) \right] \tag{9.10}$$

Based on Eq. (9.6) and Eq. (9.8), the expectation and variance of SCC are like Eq. (9.9).[9]

Calculation of tail probability. We use Chebyshev's inequality to compute the related tail probability of the temperature and damage rate variables for two reasons. First, the complexity of these variables makes their distribution features unknown, ruling out standard distribution-based approaches. Second, Chebyshev's inequality allows us to calculate the tail probability when only the mean and variance are known. Hence, it is shown as:

$$P[|X_t - \boldsymbol{E}(X_t)| \geq X_B] \leq \frac{\boldsymbol{Var}(X_t)}{{X_B}^2} \tag{9.11}$$

in Eq. (9.11), X_t is a random variable with expectation $\mathbf{E}(X_t)$ and variance $\mathbf{Var}(X_t)$. X_B means the deviation degree from $\mathbf{E}(X_t)$. An increase in Eq. (9.11) indicates an increase in the tail probability under condition X_B, reflecting the fat-tail process.

Fitting results criteria. The fit of the model result is measured by the mean square error (MSE) and the root mean square error (RMSE) as follows:

---

[9] The special formula for the expectation of SCC is shown as $\boldsymbol{E}(SCC_t) = \sum_{t=k}^{n}(1 + r_t)^{-t}\{(\omega_Y^t Y_0)\boldsymbol{E}(\widetilde{D}_t) + \ln(\omega_Y^t Y_0)\left[2\frac{1-\omega_Y^t}{1-\omega_Y}\boldsymbol{E}(\chi_t) - 0.5\left(\frac{1-\omega_Y^t}{1-\omega_Y}\right)^2(\boldsymbol{Var}(\chi_t) + \boldsymbol{E}(\chi_t)^2) - 1.5\right] \times \left[2\frac{1-\omega_Y^t}{1-\omega_Y}\zeta_Y \boldsymbol{E}(W_{Yt}) - 0.5\left(\frac{1-\omega_Y^t}{1-\omega_Y}\zeta_Y\right)^2(\boldsymbol{Var}(W_{Yt}) + \boldsymbol{E}(W_{Yt})^2) - 1.5\right] \times \boldsymbol{E}(\widetilde{D}_t)\}$.

$$MSE = \frac{1}{n}\sum_{i=1}^{n}(\hat{X}_i - X_i)^2,\ RMSE = \sqrt{MSE} \tag{9.12}$$

where n is the number of the sample. $\hat{X}_i$ and X_i are the results of our model and real data, respectively.

9.4.2 Calibration and Fitting Results

Calibration and fits for CEM-1. Referring to Golosov et al. (2014) and Rezai and Van der Ploeg (2016), we assume that carbon emissions at time *k* are reduced by 50% after 30 years, setting φ_L as 0.2 (reserving for two decimal places).[10] The changing rate φ meets: $(1-\varphi)^{299} = 0.5$, setting φ as 0.23%. The calibrated value for φ_0 is 0.4.[11] Then, μ_1 and μ_2 are determined by solving the following problems:

$$\begin{gathered}\min_{\mu_1}\sum\left[C_t\text{-}\left(C_t^{BP}+C_t^{IPCC}\right)/2\right]^2,\\ \min_{\mu_2}\sum\left[AC_t\text{-}\left(AC_t^{BP}+AC_t^{IPCC}\right)/2\right]^2\end{gathered} \tag{9.13}$$

where C_t^{BP} and AC_t^{BP} are real data from *BP Statistical Review of World Energy*. C_t^{IPCC} and AC_t^{IPCC} come from Lamb (2022). The solution yields μ_1 =5.94 and μ_2 =0.99. Figure 9.3 shows the simulated carbon emissions and carbon stock, with an MSE = 0.016 and RMSE = 0.126, indicating good fit. Figure 9.12 gives the fitted carbon emission and carbon stock, with fitness exceeding 98%.

Calibration and fits for CEM-2. According to Cai et al. (2023), the temperature feedback factor *m* is set to 0.01. Based on the result of Eq. (9.4), parameter ε and σ_T are determined by solving the following problem:

$$\begin{gathered}\min_{\varepsilon}\sum\left[E(T_t)\text{-}\left(T_t^{IPCC}+T_t^{NASA}\right)/2\right]^2\\ s.t.\sigma_T^2 = mean\left[VT_t^{CRU\text{-}TS}/\left(T_0^2+(\varepsilon AC_t/AC_0)^2t^3/3\right)\right]\end{gathered} \tag{9.14}$$

where T_t^{IPCC} and T_t^{NASA} are the temperature data come from IPCC and NASA. VT_t^{CRU-TS} is calculated using the CRU-TS dataset from Harris et al. (2020). The solution yields ε =0.012 and σ_T =0.037. Figure 9.4 shows the fitted results for the temperature process, with MSE = 0.005 and RMSE = 0.07, indicating a good fit. Appendix and Fig. 9.13 give other details for Eq. (9.14). Table 9.3 reported the results of our fitted temperature variance and tail probabilities. The MSE and RMSE results

[10] By applying impulse response functions, Joos et al. (2013) analyzed the decay process of carbon emissions in the atmosphere. They found that even after years of reduction efforts, 16%–24% of carbon emissions will remain in the atmosphere. The calibrated value of parameter φ_L is consistent with the conclusions of this study.

[11] The parameter φ_I is calibrated using the average energy consumed per unit of GDP released by the *World Bank* (sample period: 1970–2020). φ_I is calculated as 0.9%.

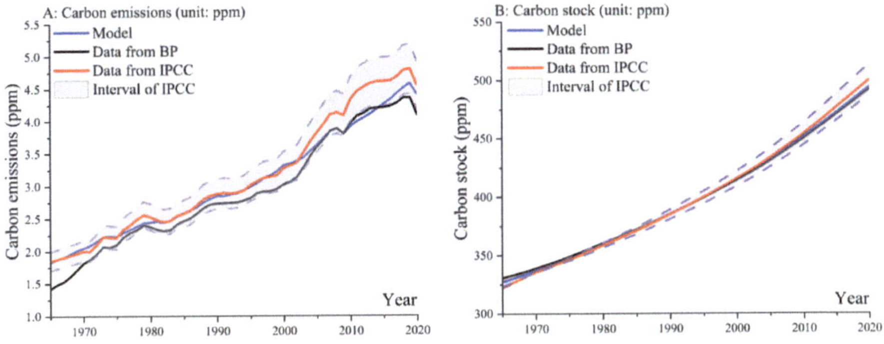

Fig. 9.3 Carbon emissions and carbon stock

of temperature variance are 0.012 and 0.11, indicating a better model fit. The MSE and RMSE results of tail temperature are also lower than 0.02 and 0.15, with a good fit.

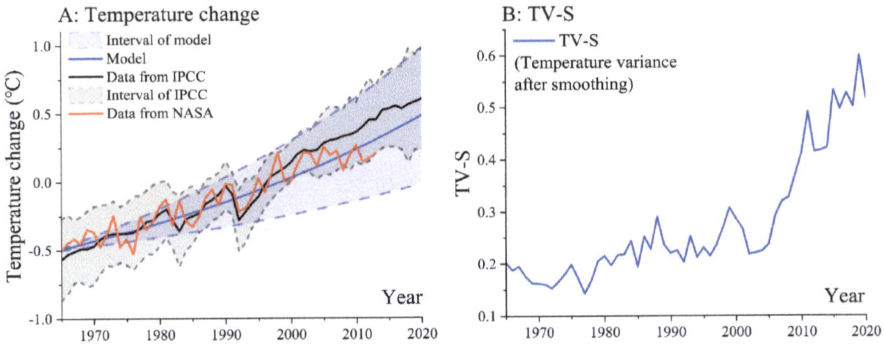

Fig. 9.4 Temperature change temperature variance. *Note* (1) IPCC data are from SROCC data in Figure SPM.1: past and future changes in the ocean and cryosphere in AR6. (2) Temperature change is calculated based on average temperatures from 1986–2005. (3) The interval of model is the results obtained by a 50% increase or decrease in carbon stock

Table 9.3 $Var(T_t)$ and tail feature of T_t

Year	Variance	Over 1 °C	Over 1.5 °C	Over 2 °C
Mean results from 1965 to 2020				
Data	0.276	0.276	0.122	0.069
Model	0.273	0.273	0.121	0.068
MSE	0.012	0.014	0.003	0.001
RMSE	0.11	0.12	0.05	0.03

Note Over 1 °C is obtained by using Eq. (9.11). This means the probability of deviating from the $E(T_t)$ by 1 °C

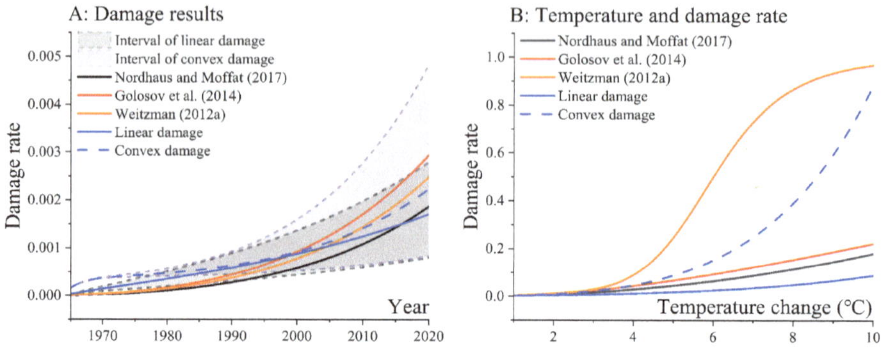

Fig. 9.5 Damage rate and damage trend. *Note* the interval of model is the results obtained by a 50% increase or decrease in carbon stock

Calibration and fits for CEM-3. To match the feature of catastrophic high-temperature events described by Weitzman (2012), parameter λ is calibrated to be 0.185.[12] When considering linear damage rate, θ_T is set to 0. θ_T is 1 if considering convex damage rate. Parameter μ_T is determined by solving the following problem:

$$\min_{\mu_T} \sum \left\{ \left[\mathbf{E}\left(D_t|_{\theta_T=0}\right) - D_t^B \right]^2 + \left[\mathbf{E}\left(D_t|_{\theta_T=1}\right) - D_t^B \right]^2 \right\} \tag{9.15}$$

where we consider both the linear damage $\mathbf{E}\left(D_t|_{\theta_T=0}\right)$ and convex damage $\mathbf{E}\left(D_t|_{\theta_T=1}\right)$. D_t^B is the fitted target, composed of $\left(D_t^{NM} + D_t^G + D_t^W\right)/3$. D_t^{NM} is the damage function in Nordhaus and Moffat (2017). D_t^G and D_t^W are the damage functions in Golosov et al. (2014) and Weitzman (2012).[13] The solution yields $\mu_T^* = 0.022$, with the results shown in Fig. 9.5. Our $\mathbf{E}(D_t)$ result (Fig. 9.5-A) aligns well with the previous results. The MSE of linear and convex damage are lower than 10^{-6}, while RMSE results are lower than 10^{-3}. When $\theta_T = 1$, our $\mathbf{E}\left(D_t|_{\theta_T=1}\right)$ result (Fig. 9.5-B) conforms to not only the estimated damage under normal warming but also the high-temperature catastrophe phenomenon described by Weitzman (2012). If the temperature change exceeds 2.7 °C, $\mathbf{E}\left(D_t|_{\theta_T=1}\right)$ surpasses the mean of Nordhaus and Moffat (2017) and Golosov et al. (2014), showing remarkable trends of damages. If the temperature change is close to 10 °C, the damage rate of our model and Weitzman (2012) approaches 1, a scenario known as deep structural uncertainty (Weitzman, 2015). In addition, Fig. 9.14 shows an increasing probability of tail damage rate with climate change.

[12] In addition, by varying λ from -10% to $+10\%$, the sensitivity tests are conducted and reported in the appendix.

[13] The specific calculations are as follows: $D_t^{NM} = d^{NM1} T_t^2$, d^{NM1} is the related damage parameter. $D_t^G = 1/\left(1 + d^{G1} T_t^2\right)$, d^{G1} is the related damage parameter. $D_t^W = 1/\left(1 + d^{W1} T_t^2 + d^{W2} T_t^{6.754}\right)$, d^{W1} and d^{W2} are the related damage parameters. Due to the use of convex functions for D_t^{NM} and D_t^G, the linear damage rate of this paper in Fig. 9.5 are lower.

Calibration and fits for CEM-4. Based on Cai and Lontzek (2019), we specify parameter ω_χ =0.775, economic growth volatility parameter ζ_Y =0.035, and economic trend volatility ζ_χ =0.008. We calibrate ω_Y by solving the following problem[14]:

$$\sum_{t=0}^{n}\Big[\big(E(Y_t)\text{-}Y_t^{SSP\text{-}1}\big)+\big(E(Y_t)\text{-}Y_t^{SSP\text{-}2}\big)+\big(E(Y_t)\text{-}Y_t^{SSP\text{-}3}\big) + \big(E(Y_t)\text{-}Y_t^{SSP\text{-}4}\big)+\big(E(Y_t)\text{-}Y_t^{SSP\text{-}5}\big)\Big]\leq \varepsilon_{stop} \tag{9.16}$$

where the Shared Socio-economic Pathways (SSPs) are chosen as the fitting objects for ω_Y. SSPs incorporate more complex calculations of population, economy, and technological progress for forecasting future economic growth. The details on SSPs are referred to O'Neill et al. (2014). Based on historical carbon emission data, our model obtains the corresponding carbon emission pathways by inputting SSPs. This is different from using a combination of SSPs and Representative Concentration Pathways (RCPs). The economic growth rate data for SSPs comes from Burke et al. (2018). We set the error as $\varepsilon_{stop} = 10^{-6}$. Then, the calibrated ω_Y satisfying Eq. (9.16) is 1.025. The five scenarios of SSPs are shown in Fig. 9.15. Based on SSPs, the result of Ramsey annual discounting rate is reported in Fig. 9.16.

9.5 Simulation Results

This section first analyzes the "fat-tail" trend of climate change. Then, we quantify the economic damage under the extreme tail high temperature and the response of tail feature to higher carbon emissions.

9.5.1 Estimates of Tail Trend

We first discuss the tail trends of climate change.

Distribution trend of climate change. Figure 9.6 reported the distribution trends of temperature, damage rate, and economic damages from 2020 (or 2025) to 2100. The temperature distribution (CEM-2) gradually flattens out and increases in mean value (Fig. 9.6-A and Fig. 9.6-B). Of these, SSP5 and 2100 are warmer and have thicker tails. The damage rates (CEM-3) also show an increasing mean and "fat-tail" trend, with the convex damage rate being more pronounced than the linear damage rate (Fig. 9.6-C and Fig. 9.6-D). Meanwhile, the economic damage shows the same trend of increasing mean and thickening tails (Fig. 9.6-E and Fig. 9.6-F). In particular, the thicker tails of economic damage are particularly evident over time

[14] The calculation interval [0, *n*] in Eq. (9.16) is consistent with the economic growth rate data provided by Burke et al. (2018).

(Fig. 9.6-F). Our results demonstrate an overall trend of increasing averages and thickening tails under climate change. The thicker tails of SSP5 and 2100 are more clearly characterized. In addition, Fig. 9.17 illustrates the temperature trends for the 5 SSPs scenarios.

Tail probability of climate change. How bad is the "fat-tail" phenomenon in Fig. 9.6? Using Eq. (9.11), we calculated the tail probabilities for temperature change, damage rates, and economic damages, as shown in Table 9.4. In Panel A, the temperature tail probability averaged among the 5 SSPs grows from 2.718% (2025) to 9.382% (2100), an increase of about 3.45 times. Of these, SSP5 has a greater temperature tail probability and increase. In Panel B, the mean tail probability of linear damage rate grows from 3.51% (2025) to 11.564% (2100), an increase of more than 3.29 times.

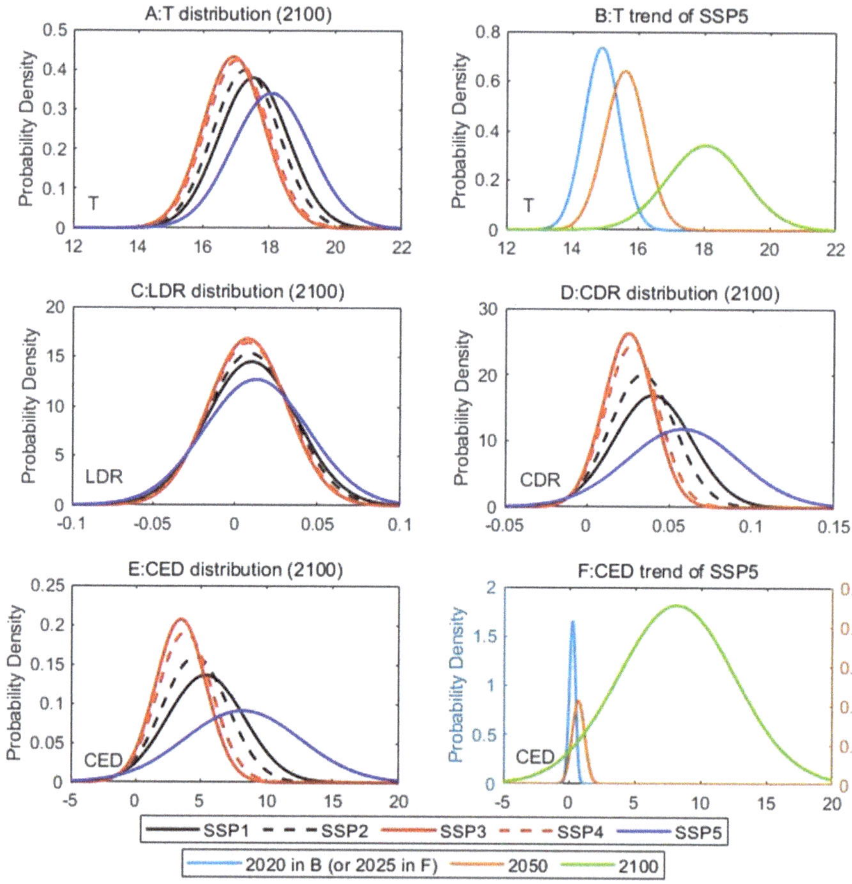

Fig. 9.6 "Fat-tail" trend of climate change. *Note* (1): T is temperature, LDR and CDR are the linear damage rate and convex damage rate, respectively. CED is the convex economic damages. (2) The lower left corner is labeled with horizontal label. In Fig. 9.6-F, the left and right axes use the same vertical label. (3) The above results are obtained under the normal distribution assumption

Meanwhile, the mean tail probability of convex damage rate grows from 0.12% (2025) to 8.382% (2100), an increase far beyond the linear damage ratio. In panel C, the mean tail probability of linear economic damage is 11.59% in 2100, much higher than the 1.67% in 2025. On the other hand, the mean tail probability of convex economic damages increased from 0.056% (2025) to 8.636% (2100). We find that the tail probabilities are higher at SSP5 and 2100. The tail probabilities under linear conditions are greater than the convex cases, but the increase is more pronounced under the convex conditions.

Factors affecting "fat-tail" trend. Another major question is what factors might change the tail probability under climate change? The related results are reported in Table 9.5. In panel A, higher economic risk raises the tail probability of climate change, but the effect is relatively small. Among them, the boost to SSP1 is more pronounced. In panel B, the tail probabilities are significantly higher when damage risks and hazard parameter increase. The higher the mean value of damage risks has a greater effect. In panel C, the higher the conversion of fossil fuel use and carbon emissions (or increasing the conversion of carbon emissions and temperature) both increase the tail probability of climate change. Of these, the impact of increasing the switch between carbon emissions and temperature on tail probability is more pronounced. We found that reducing damage risks and decreasing the efficiency of the conversion between carbon stock and temperature are important ways to suppress tail probabilities. The tail probabilities are more affected in SSP1 and SSP5. Table A1 also reported in detail the responses of tail probability (including temperature, damage rate, and economic damages) to changing influencing factors.

9.5.2 Economic Damages under Tail High-Temperature Events

Our results suggested that the tail probability of climate change will continue to rise. What will happen to global warming and economic damage if a tail high temperature event occurs?

Temperature and damage rate under tail high temperatures. To understand the potential unknown damages caused by extreme tail high temperatures discussed by Weitzman (2012), this subsection simulates scenarios where tail high temperatures occur. With a normal distribution, we assumed that occurs an extreme high-temperature event (right-hand side high-temperature event with a probability of 2.5%). Then, by using the results of $\boldsymbol{E}(T_t)$ and $\boldsymbol{Var}(T_t)$, we calculated the corresponding temperature and damage rate in Fig. 9.7.

Under this tail high temperature scenario, the mean global temperature change in year 2100 is 5.49 °C relative to the pre-industrial, higher than the base (without tail high temperature). The damage rates of tail high temperature in 2100 will reach 2.18% ($\theta_T = 0$) and 12.38% ($\theta_T = 1$). The damage rates of 12.38% ($\theta_T = 1$) even exceeds the results of worst-case climate shocks by Takakura et al. (2019). Without

Table 9.4 Trend of tail probability under climate change

	Year 2025	Year 2050	Year 2100	Change of 2025 to 2050	Change of 2050 to 2100
Panel A: Tail probability of temperature (CEM-1 and CEM-2, X_B is 3.34 °C)					
SSP1	2.71%	3.44%	9.93%	0.27	1.89
SSP2	2.72%	3.42%	8.99%	0.26	1.63
SSP3	2.72%	3.40%	7.66%	0.25	1.26
SSP4	2.72%	3.40%	7.99%	0.25	1.35
SSP5	2.72%	3.47%	12.34%	0.28	2.55
Panel B: Tail probability of damage rate (CEM-3, X_B is 7.86%)					
Linear damage rate ($\theta_T = 0$)					
SSP1	3.51%	4.70%	12.25%	0.34	1.61
SSP2	3.51%	4.66%	10.92%	0.33	1.34
SSP3	3.51%	4.62%	9.13%	0.32	0.98
SSP4	3.51%	4.63%	9.56%	0.32	1.07
SSP5	3.51%	4.77%	15.96%	0.36	2.35
Convex damage rate ($\theta_T = 1$)					
SSP1	0.12%	0.34%	8.97%	1.88	25.46
SSP2	0.12%	0.33%	6.42%	1.78	18.60
SSP3	0.12%	0.32%	3.73%	1.69	10.79
SSP4	0.12%	0.32%	4.31%	1.72	12.45
SSP5	0.12%	0.36%	18.48%	2.02	50.86
Panel C: Tail probability of economic damages (CEM-4 and CEM-5, X_B is 9.9 trillion dollars)					
Linear economic damage ($\theta_T = 0$)					
SSP1	1.59%	2.72%	11.68%	0.71	3.29
SSP2	1.63%	2.76%	10.65%	0.70	2.86
SSP3	1.67%	2.80%	9.12%	0.68	2.25
SSP4	1.71%	2.88%	9.78%	0.69	2.40
SSP5	1.75%	3.03%	16.71%	0.74	4.51
Convex economic damage ($\theta_T = 1$)					
SSP1	0.05%	0.20%	8.74%	2.69	43.41
SSP2	0.05%	0.19%	6.40%	2.56	31.87
SSP3	0.06%	0.19%	3.80%	2.44	18.75
SSP4	0.06%	0.20%	4.49%	2.48	21.52
SSP5	0.06%	0.23%	19.75%	2.87	85.90

Note (1) Growth degree is calculated by $(P_{final} - P_{start})/P_{start}$. (2) For easy comparison, X_B in each panel are unified. Based on the mean and variance in 2100, 1000 simulations are performed to obtain 95% quantile points. $0.5 \times X_B$ are the mean results of the 5 SSPs scenarios, linear damage and convex damage

Table 9.5 Tail probability under different factor conditions

	SSP1	SSP2	SSP3	SSP4	SSP5
Panel A: Changing economic factors (CEM-4)					
$1.1 \times \zeta_Y$					
Change in tail probability	0.16%	0.11%	0.10%	0.10%	0.012%
$1.1 \times \zeta_\chi$					
Change in tail probability	0.03%	0.03%	0.03%	0.03%	0.03%
Panel B: Changing damage rate factors (CEM-3)					
$1.1 \times \mu_W$					
Change in tail probability	60.55%	52.78%	52.57%	52.63%	53.10%
$1.1 \times \Sigma_W^2$					
Change in tail probability	22.51%	17.74%	17.93%	17.87%	17.46%
$1.1 \times \lambda$					
Change in tail probability	10.66%	8.40%	7.59%	7.80%	10.35%
Panel C: Changing carbon and temperature factors (CEM-1 and CEM-2)					
$1.1 \times \varepsilon$					
Change in tail probability	41.12%	29.61%	27.65%	28.17%	33.66%
$1.1 \times \mu_2$					
Change in tail probability	41.12%	29.61%	27.65%	28.17%	33.66%
$1.1 \times \mu_1$					
Change in tail probability	19.33%	13.21%	11.15%	11.70%	17.68%

Note The results in Table 9.5 are first averaged from 2025 to 2100 and then averaged over the 5 tail probabilities of Table 9.4 (5 SSP scenarios). See appendix for additional detailed results of Table 9.5

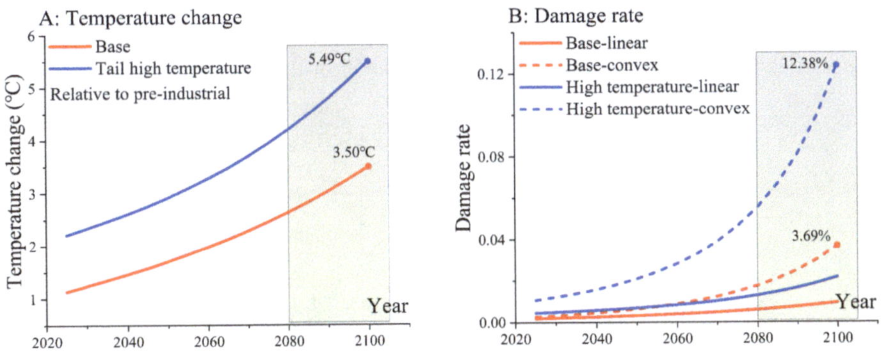

Fig. 9.7 Temperature change and damage rate under high temperature condition. *Note:* results in Fig. 9.7 are averaged from SSP1 to SSP5

climate measures, Takakura et al. (2019) suggested that the most pessimistic scenario of climate change could result in a damage rate of about [3.9%, 8.6%] by the end of the twenty-first century. Failing to effectively mitigate tail high temperature risks could mean a damage rate of 12.38% in 2100, resulting in huge production declines and economic damages. This conclusion is striking and emphasizes the urgent need to prevent high-temperature events as highlighted by Weitzman (2012). In addition, Fig. 9.18 reported the temperature and damage rate difference among 5 SSPs.

Economic damages under tail high temperatures. How much economic damage corresponds to high tail temperatures? Figure 9.8 reported the results. In Fig. 9.8-A ($\theta_T = 0$), although the mean economic damages from tail high temperatures are higher than those of the base, it still falls into the predicted range of the base. This indicates that the baseline results can still provide certain references for predicting damages from tail high temperatures. In Fig. 9.8-B ($\theta_T = 1$), the mean economic damages of tail high temperatures not only exceed the baseline results, but also surpass the predicted economic damage range of the baseline. When tail high temperatures occur, economic damages from regular warming pathways are not only underestimated but may not even provide reference. This goes beyond the traditional understanding of climate economics and aligns with the results of Weitzman (2015). Weitzman (2015) believed that extreme high temperatures could lead to "unknown unknowns". Meanwhile, under high tail temperature conditions, economic damages are more severe for SSP5 and SSP1. As discussed by Weitzman (2015), we now know less about the economic impact of tail high temperatures. This "unknown unknowns" could result in significant economic damages. This highlights the need for government to quantify the tail high temperature risks, avoiding the economic damages caused by tail high temperature.

Factors affecting economic damage under tail temperature. What factors affect economic damages under the tail high-temperature conditions? The results are reported in Table 9.6. In Panel A, increased economic volatility of growth and trend exacerbates the economic damages under the tail-high-temperature scenario, but the

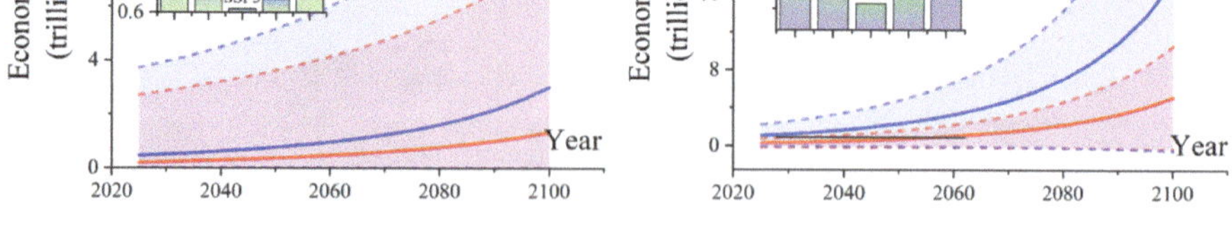

Fig. 9.8 Economic damages under high temperature condition. *Note* (1) The red and blue lines in Fig. 9.8 are the average results for the 5 SSPs. (2) The intervals are calculated by: $\boldsymbol{E}(EL_t) \pm 1.96 \times \boldsymbol{Var}(EL_t)$, EL_t is the economic damages. We only reported damage results greater than 0

effect is relatively small. Of these, the role of economic growth fluctuations and SSP5 is more pronounced. In Panel B, the boost to tail high-temperature economic damages is more pronounced with increased damage risks. Raising the hazard parameter is less effective than changing the damage risks. In Panel C, an increase in the conversion rate between carbon stock and temperature (or an increase in the conversion rate between carbon emissions and carbon stock) has a more pronounced effect on boosting tail high temperature economic damages. The role of higher conversion rates between fossil fuel use and carbon emissions is relatively small. In addition, the enhancement effect of changing factors on convex economic damages ($\theta_T = 1$) is more pronounced. The appendix further reported the impacts of higher tax and climate tipping points (CTP).

9.5.3 Response of Tail Probability and SCC

Further, we would like to explore the relationship between the tail probabilities of climate change and the social cost of carbon (SCC), and the implication from SCC calculation.

Response of tail probabilities to additional carbon emission. Drawing on the calculation of SCC, we first simulated the response of tail probability to additional unit carbon stock (1 ppm). The results are shown in Fig. 9.9. In Fig. 9.9-A, the temperature tail probability response to additional unit carbon stock is enhanced. Over time, the temperature tail probability due to additional unit carbon stock is higher. Of those, SSP5 and SSP1 are more responsive. In Fig. 9.9-B, the average responses for damage rate are higher, and the responses for temperature and economic damages are smaller. The average response for the temperature tail probability is 0.15×10^{-4}. The linear and convex damage rates are 0.20×10^{-4} and 0.22×10^{-4}, respectively. The linear and convex economic damages are 0.17×10^{-4} and 0.21×10^{-4}, respectively.

Note: (1) Fig. 9.9-A is the response of temperature tail probability to additional unit carbon stock (1 ppm in 2025). (2) Fig. 9.9-B is the mean response of additional unit carbon stock from 2025 to 2100. T is the response of temperature to additional carbon emissions, LD is the response of linear damage rate, CD is the response of convex damage rate, LED is the response of linear economic damages, and CED is the response of convex economic damages.

Relationship between tail probability and SCC. To assess the relationship between tail probability and SCC, we first calculated the SCC for the 5 SSPs. By applying Eq. (9.10), the results of SCC are calculated in Fig. 9.10. The SCC gradually increases over time, maintaining a positive proportional relationship with economic output, like the results of Golosov et al. (2014) and Rezai and Van der Ploeg (2016). SCC under the convex damage rate ($\theta_T = 1$) is higher than that of linear damage rate ($\theta_T = 0$). Under $\theta_T = 1$, SCC amounts to 34.54/ton and \$77.61/ton in year 2050 and 2100, respectively. This is significantly higher than the corresponding \$5.06/ton and \$9.05/ton in year 2050 and 2100 under $\theta_T = 0$. This finding is consistent

Table 9.6 Economic damages under different factor conditions

	SSP1	SSP2	SSP3	SSP4	SSP5
Panel A: Changing economic factors (CEM-4)					
$1.1 \times \zeta_Y$					
Economic damages ($\theta_T = 0$)	0.14%	0.14%	0.13%	0.13%	0.15%
Economic damages ($\theta_T = 1$)	0.16%	0.16%	0.15%	0.15%	0.17%
$1.1 \times \zeta_\chi$					
Economic damages ($\theta_T = 0$)	0.02%	0.02%	0.02%	0.02%	0.02%
Economic damages ($\theta_T = 1$)	0.02%	0.02%	0.02%	0.02%	0.02%
Panel B: Changing damage rate factors (CEM-3)					
$1.1 \times \mu_W$					
Economic damages ($\theta_T = 0$)	20.88%	20.88%	20.88%	20.88%	20.88%
Economic damages ($\theta_T = 1$)	20.88%	20.88%	20.88%	20.88%	20.88%
$1.1 \times \Sigma_W^2$					
Economic damages ($\theta_T = 0$)	18.20%	18.20%	18.20%	18.20%	18.20%
Economic damages ($\theta_T = 1$)	18.20%	18.20%	18.20%	18.20%	18.20%
$1.1 \times \lambda$					
Economic damages ($\theta_T = 0$)	8.24%	7.83%	7.28%	7.42%	9.19%
Economic damages ($\theta_T = 1$)	8.77%	8.27%	7.60%	7.77%	9.93%
Panel C: Changing carbon and temperature factors (CEM-1 and CEM-2)					
$1.1 \times \varepsilon$					
Economic damages ($\theta_T = 0$)	15.76%	15.14%	14.31%	14.53%	17.09%
Economic damages ($\theta_T = 1$)	26.31%	25.35%	23.99%	24.36%	28.38%
$1.1 \times \mu_2$					
Economic damages ($\theta_T = 0$)	15.76%	15.14%	14.31%	14.53%	17.09%
Economic damages ($\theta_T = 1$)	26.31%	25.35%	23.99%	24.36%	28.38%
$1.1 \times \mu_1$					
Economic damages ($\theta_T = 0$)	7.53%	6.88%	5.98%	6.22%	8.95%
Economic damages ($\theta_T = 1$)	12.63%	11.52%	9.95%	10.37%	15.03%

Note The results in Table 9.6 are averaged from 2025 to 2100

with the conclusions of Van den Bremer and Van der Ploeg (2021). It indicates that convex correlations are conducive to obtaining higher optimal carbon taxes, thereby reducing the risks of high emissions, low carbon taxes, and extreme high temperatures. Additionally, according to the *State and Trend of Carbon Pricing 2022* released by the *World Bank*, the current carbon price in many areas is lower than the SCC (convex link) obtained from this study, particularly in developing countries. As climate changes, the damage rates resulting from global warming escalate,

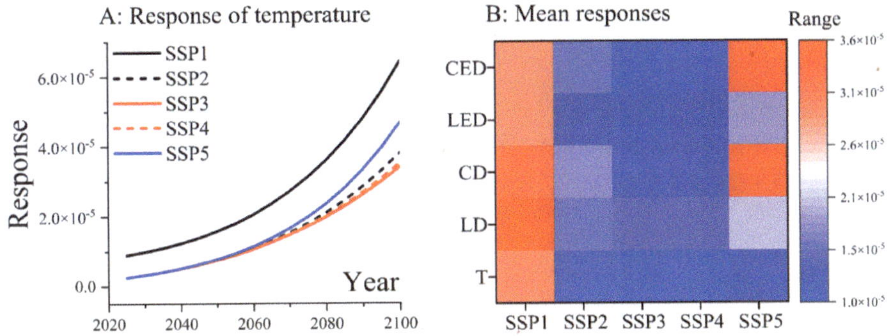

Fig. 9.9 Response of additional carbon emission

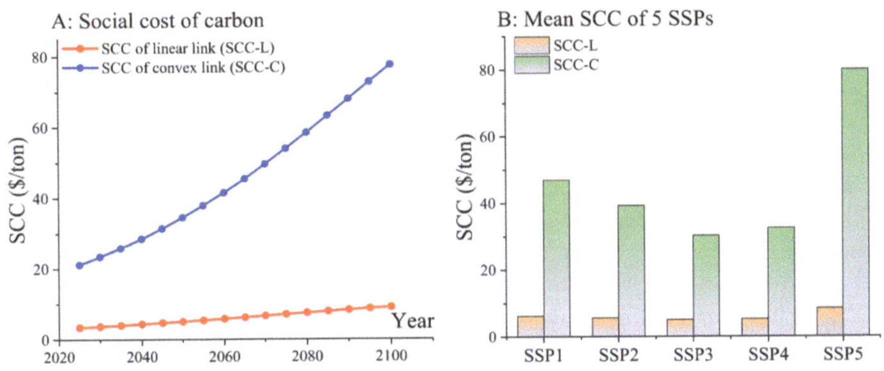

Fig. 9.10 Estimates of social cost of carbon (SCC)

leading to a gradual increase in SCC. This implies that governments should dynamically increase optimal carbon taxes rather than implement static, constant low carbon taxes.

Then, we discuss the relationship between tail temperature probability and SCC in Fig. 9.11. We fitted to obtain an interesting quadratic relationship. The tail temperature probability of economics will gradually increase at a faster and faster rate as the SCC is raised. Among them, the quadratic relationship of SSP5 and SSP1 is more obvious. This provides a warning. When increasing SCC and curbing climate change, one needs to be aware that tail temperature probabilities are increasing at a faster rate. If not handled well, extreme high temperature events could render many climate governance efforts worthless.

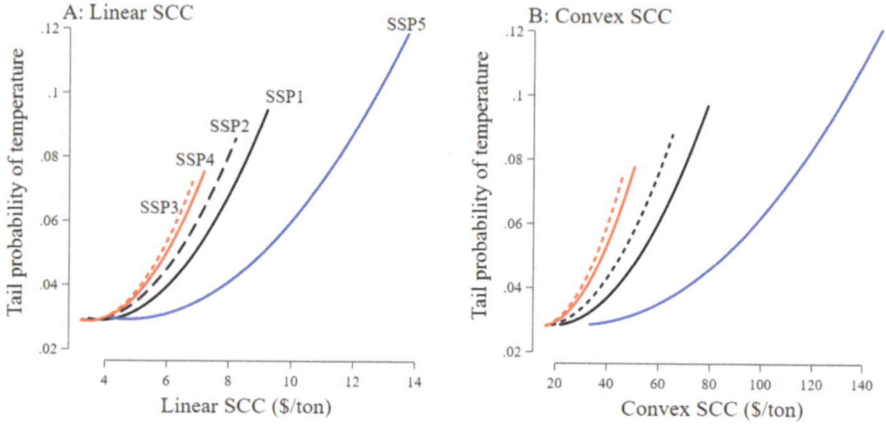

Fig. 9.11 Relationship between SCC and tail temperature probability

9.6 Conclusion and Implications

By constructing a climate economic model, this study analyzes the tail feature of climate change, the economic damage caused by tail high temperatures, and the response of tail features to increasing carbon emissions. The research findings indicate the following: (1) Over time, the tail trends of global climate change have gradually intensified, with the tail temperature probability increasing by about 3.45 times from 2025 to 2100, corresponding to a rise in extreme temperature events. (2) Escalating risks lead to an increase in the exposure of economic production activities, resulting in increased tail probability and a more pronounced fat-tailed trend. Among these, the impacts of damage risks and temperature change risks are more significant, while the impacts of economic risks are relatively minor. (3) Tail high temperatures can lead to substantial economic damage or even economic collapse. Traditional estimate of economic damages from climate shocks insufficient for assessing the economic impacts of tail high temperatures, corroborating Weitzman's (2015) assertion regarding the potential for these temperatures to "unknown unknowns". (4) The tail probability of climate change in response to increased carbon emissions gradually rises over time. Specifically, a 1 ppm increase in carbon stock leads to an average increase in the tail temperature probability of 0.15×10^{-4}.

Based on the above conclusions, the following policy recommendations are proposed: (1) Utilizing historical temperature data and frequency of heat events can aid in assessing regional tail heat risk and raising awareness of regional climate vulnerabilities. Quantifying regional climate risks will help determine exposure to climate risks and the likelihood of extreme high temperatures, thus mitigating potential economic damages. For example, effective daily heat monitoring and the formulation of climate policies aimed at curbing tail high temperatures are essential. (2) Further integrated assessments of economic damage resulting from climate change are necessary, particularly regarding tail-high temperature events explored in this paper. China's *National Climate Change Adaptation Strategy 2022* states that "the current analysis and assessment of climate change impacts and risks are still insufficient". Hence, it is recommended to further refine the economic damage assessments related to climate change and establish a climate risk monitoring mechanism to continuously evaluate the climate risks faced by different regions. (3) There is a need to enhance carbon mitigation measures and climate adaptation strategies to prevent increased tail probabilities of temperature and damage rates stemming from increased carbon emissions. Strengthening efforts to reduce carbon emissions, such as carrying out carbon pricing mechanisms, promoting low-carbon and zero-carbon technologies, and minimizing the occurrences of extreme tail high-temperature events is crucial. Additionally, it is important to bolster the adaptive capacity of firms and residents to climate change, thereby reducing exposure to and sensitivity to extreme heat. For example, increasing air conditioning and refrigeration units will be necessary, alongside a prompt decarbonization of the power sector.

This paper has several limitations: (1) The model simplifies the economic module for analytical convenience, omitting factors such as population growth and technological advancement. (2) A simplified logarithmic utility function is employed. Future work could explore alternative utility functions, including constant relative risk aversion (CRRA) utility functions and utility functions that satisfies Epstein-Zin preferences. (3) The model does not directly address the skewness and kurtosis of temperature changes and damage rates, leading to a lack of some higher-order moment information. However, empirical studies have confirmed that economic damage exhibits a thick and long right tail. These limitations can be further addressed in future research.

Acknowledgements Huai Deng, Hui Xu, You Wu, and Xin-Zhong Liang also made great contributions to this manuscript. We express our heartfelt thanks to them. This research also was supported by: National Natural Science Foundation of China (72474127) and National Social Science Fund Major Bidding Project of China (18ZDA052). Part of this chapter was published:Deng H, Wu X, Xu H, Wu Y, Liang X-Z. Quantifying aggregated economic damages from "fat-tail" extremes high-temperature events in climate change. Structural Change and Economic Dynamics 2025;75:108–21. https://doi.org/10.1016/j.strueco.2025.05.007. The publication has been conducted with copyright permission.

Appendix

Fitting carbon emissions and carbon stocks. For Eq. (9.13), the AC_0 is set to 329.2 ppm. The starting year for IPCC carbon emission data is 1970. The IPCC data of 1965 to 1969 are calculated by using the average annual growth rate from 1970 to 2020. We also reported the fit of carbon emissions and carbon stock to the real data from BP in Fig. 9.12. The results show that our model fits the realistic data well. The adjusted R^2 of simulated data to carbon emissions is more than 98%, and the adjusted R^2 of simulated data to carbon stocks is more than 99%.

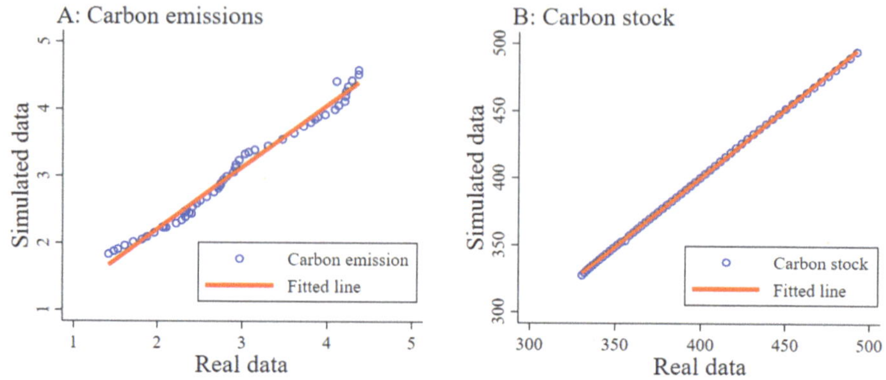

Fig. 9.12 Fit of carbon emissions and carbon stock

Fitting temperature process. For temperature variance in Eq. (9.14), we first collected annual average temperatures for each country (1901–2020) from CRU-TS. Using the average temperature from 1901–1964 as the baseline, the variance of the temperature change among countries is calculated from 1965–2020. The above data are then smoothed to obtain the results in Fig. 9.4-B. Figure 9.13 reported the simulated mean temperature and temperature tail probability. The results show that the model could fit the average temperature well (Fig. 9.13-A). As the carbon stock increases, the occurrence of tail temperatures is increasing (Fig. 9.12-B).

Note: (1) Tail probability is calculated by using Chebyshev's inequality. (2) The interval of model is the results obtained by a 50% increase or decrease in carbon stock.

Sensitivity test for parameter λ. We conducted a sensitivity analysis to test the plausibility of parameter λ. The related results are reported in Fig. 9.14. By varying the range of parameter values by 10%, our results show that the parameter λ is taken with some robustness.

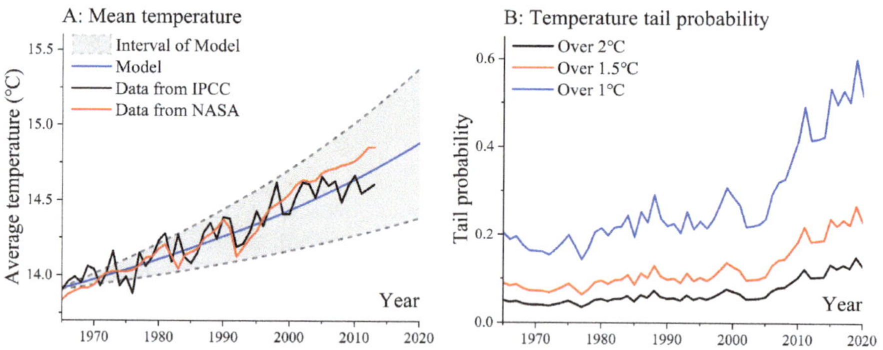

Fig. 9.13 Mean temperature and temperature tail probability

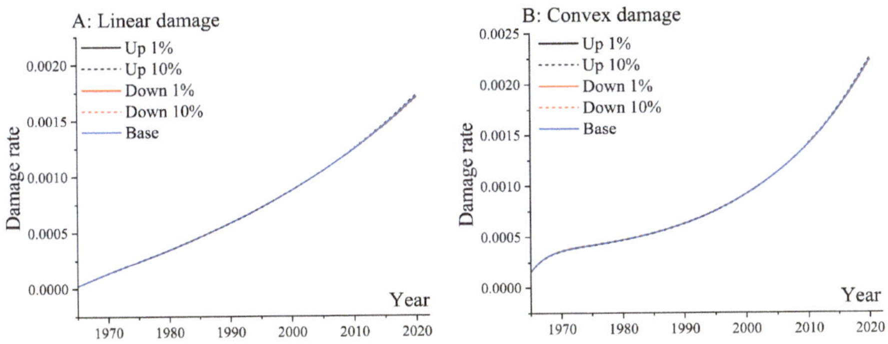

Fig. 9.14 Sensitivity test for parameter λ

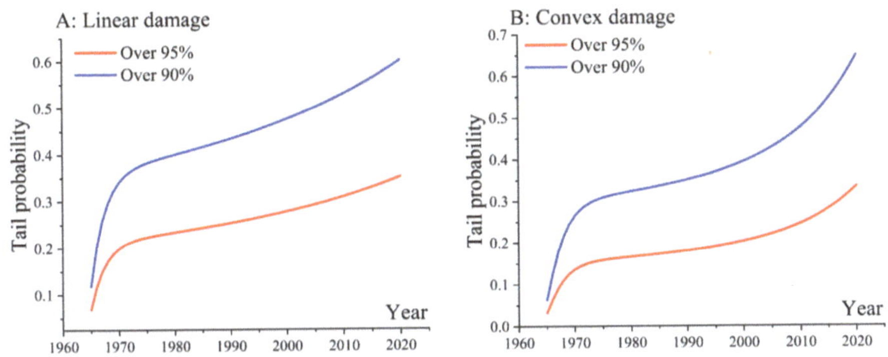

Fig. 9.15 Tail probability of damage rate

Tail probability of damage rates. We reported the tail probabilities under linear and convex damage conditions in Fig. 9.15. With global warming, the probability of tail damage rate is increasing, requiring caution from governments.

Note: (1) We assumed that the damage rate process follows a normal distribution. (2) Based on the mean and variance in 2020, 1000 simulations are performed to obtain quantile points.

Outputs of SSPs scenarios. We reported the future outputs for the five scenarios of SSPs in Fig. 9.16. SSP5 exhibits faster global economic growth rates, higher consumption of fossil fuels, and fewer implemented climate policies (O'Neill et al., 2014).

Ramsey annual discounting rate r_t. We estimated the Ramsey annual discount rate in Fig. 9.17. The Ramsey annual discounting rate adjusts downward over time. This implies that the results away from the present will receive a larger discounted value, reflecting the consideration of future risks.

Trend in temperature change. Figure 9.18 shows the global warming trend. An obvious global warming trend is observed. By 2100, global warming (relative to pre-industrial) will exceed 3 °C among 5 SSPs scenarios. This aligns closely with the

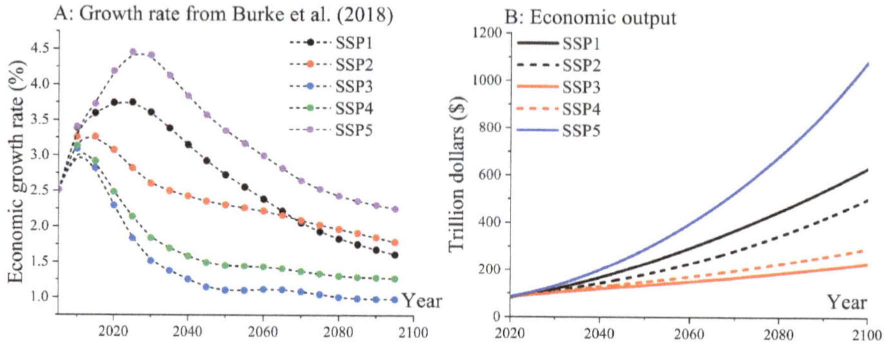

Fig. 9.16 Future scenarios under SSPs

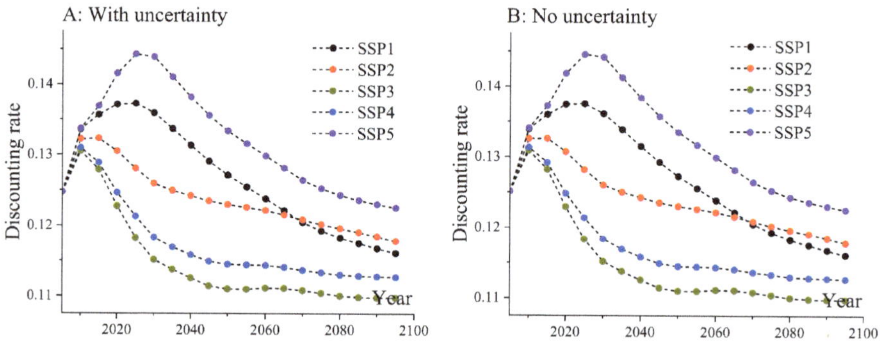

Fig. 9.17 Estimates of Ramsey annual discount rate

predictions of Cai and Lontzek (2019). If without proactive climate policies, global temperature rise is unlikely to be contained within the 2 °C and 1.5 °C temperature targets set by the Paris Agreement. As extreme tail high temperatures become more frequent, achieving the 2 °C and 1.5 °C temperature targets become increasingly challenging.

Detail results of tail probability. We reported detailed results on the factors affecting the tail probability of climate change in Table 9.7. The results show that changing factors have a greater role in enhancing the convex damage rates and convex economic damages. The damage risks play the largest role, the temperature and carbon emission factors play the second largest role, and the economic growth and trend factors play the smallest role.

Differences in temperature change and damage rate. Under the high tail temperature scenario, the corresponding temperature change and damage rates of the 5 SSPs are reported in Fig. 9.19. In Fig. 9.18-A, the warming of the tail high temperature scenario would be approximately 1.5 °C above the baseline. Of those, the warming is more pronounced for SSP5 and SSP1. In Fig. 9.18-B, the difference

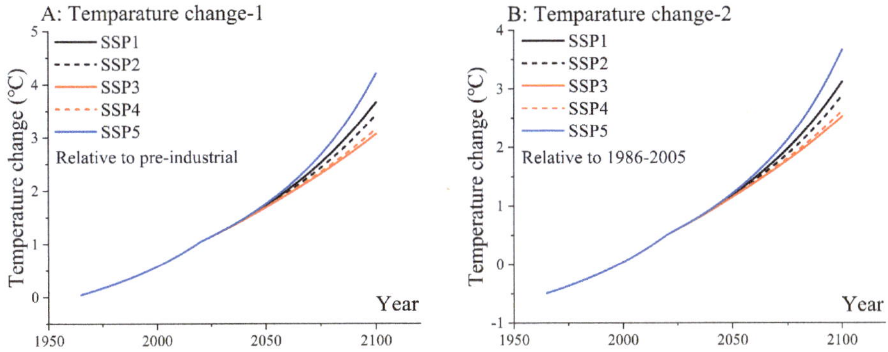

Fig. 9.18 Estimates of temperature change

of linear damage rates is more than 0.5% and the difference in the convex damage rates is more than 2%.

Impact of higher taxes and climate tipping point. In this section, we provide scenarios for increasing energy taxes and considering climate tipping points. (1) Case of higher energy taxes. Energy taxation is one of the most important measures to curb fossil fuel use and mitigate climate change (Deng et al., 2024). By incorporating energy taxes, we explore the role of higher taxes on tail-high temperature events. With the use of C-D production function, fossil fuel use is: $F_t = (1 - \alpha)Y_t/P_F$. Combined with $F_t = Y_t(r/\alpha)^{\alpha/(1-\alpha)}$, the price of fossil fuel use is: $P_F = (1 - \alpha)/(r/\alpha)^{\alpha/(1-\alpha)}$. Assuming a 10% increase in the capital share α, the prices P_F are raised by about 102.92%. This is the higher taxes scenario. The related results are shown in Fig. 9.19. Higher taxes could reduce the tail probability of high temperatures and damage rate, slowing the "fat-tail" trend of climate change (Fig. 9.20-A). Of these, the mitigation effect is more pronounced for SSP1, SSP4 and SSP5 (Fig. 9.20-B).

(2) Case of climate tipping points (CTP). CTP portrays the critical state of the climatic environment. If CTP is exceeded, climate change would cause severe damage. Dietz et al. (2021) explored in detail the role of the eight CTPs in assessing SCC (social cost of carbon), including Permafrost Carbon Feedback (PCF), Ocean Methane Hydrates (OMH), Arctic Sea ice/Surface Albedo Feedback (SAF), Amazon dieback (AMAZ), GIS disintegration, WAIS disintegration, Atlantic Meridional Overturning Circulation (AMOC) slowdown, and Indian summer monsoon (ISM) variability. Here, according to Heutel et al. (2016) and Dietz et al. (2021), we briefly introduce PTP_t to inscribe the probability of crossing the CTP. It is shown as:

$$\mathrm{PTP_t} = \begin{cases} 1-\exp\left[-b_{\mathrm{CTP}}\frac{\min(\mathrm{T_t},\overline{\mathrm{T}})}{\overline{\mathrm{T}}-\mathrm{T_t}}\right], \mathrm{T_t} < \overline{\mathrm{T}} \\ 1, \mathrm{T_t} \geq \overline{\mathrm{T}} \end{cases} \tag{9.17}$$

In Eq. (9.17), $\overline{\mathrm{T}}$ is the upper bound for the CTP and sets to 4.27 °C (Heutel et al., 2016). b_{CTP} is the hazard rate under OMH (mean result of Beta distribution) of Dietz et al. (2021) and sets to 0.22. Using forward differencing, we calculate the velocity of $\mathrm{PTP_t}$

Table 9.7 Detail changed results of tail probability under different factor conditions

	SSP1	SSP2	SSP3	SSP4	SSP5
Panel A: Changing economic factors (CEM-4)					
Increasing economic growth risks ($1.1 \times \zeta_Y$)					
Temperature	0	0	0	0	0
Linear damage rate	0	0	0	0	0
Convex damage rate	0	0	0	0	0
Linear economic damage	0.19%	0.12%	0.11%	0.11%	0.12%
Convex economic damage	0.53%	0.44%	0.40%	0.40%	0.46%
Increasing economic trend risks ($1.1 \times \zeta_\chi$)					
Temperature	0	0	0	0	0
Linear damage rate	0	0	0	0	0
Convex damage rate	0	0	0	0	0
Linear economic damage	0.04%	0.03%	0.02%	0.02%	0.03%
Convex economic damage	0.13%	0.11%	0.10%	0.10%	0.12%
Panel B: Changing damage rate factors (CEM-3)					
Increasing mean of damage risks ($1.1 \times \mu_W$)					
Temperature	0	0	0	0	0
Linear damage rate	76.45%	67.16%	67.17%	67.17%	67.14%
Convex damage rate	74.38%	64.86%	64.28%	64.44%	65.75%
Linear economic damage	77.09%	67.14%	67.16%	67.15%	67.11%
Convex economic damage	74.85%	64.74%	64.23%	64.39%	65.52%
Increasing variance of damage risks ($1.1 \times \Sigma_W^2$)					
Temperature	0	0	0	0	0
Linear damage rate	26.98%	21.13%	21.12%	21.12%	21.15%
Convex damage rate	28.80%	23.17%	23.68%	23.53%	22.38%
Linear economic damage	27.41%	21.15%	21.13%	21.14%	21.17%
Convex economic damage	29.37%	23.27%	23.72%	23.58%	22.58%
Increasing disaster parameter ($1.1 \times \lambda$)					
Temperature	0	0	0	0	0
Linear damage rate	11.77%	9.23%	8.53%	8.71%	11.01%
Convex damage rate	14.23%	11.30%	10.04%	10.36%	14.29%
Linear economic damage	12.59%	9.84%	9.03%	9.24%	11.83%
Convex economic damage	14.70%	11.62%	10.34%	10.67%	14.62%
Panel C: Changing carbon and temperature factors (CEM-1 and CEM-2)					
Increasing conversion of carbon stock and temperature ($1.1 \times \varepsilon$)					
Temperature	21.12%	10.33%	9.65%	9.83%	11.70%
Linear damage rate	21.33%	12.46%	11.52%	11.77%	14.29%
Convex damage rate	69.08%	55.40%	51.79%	52.77%	62.47%

(continued)

Table 9.7 (continued)

	SSP1	SSP2	SSP3	SSP4	SSP5
Linear economic damage	22.90%	13.29%	12.21%	12.49%	15.91%
Convex economic damage	71.08%	56.59%	53.07%	54.02%	63.40%
Increasing conversion of carbon emissions and carbon stock ($1.1 \times \mu_2$)					
Temperature	21.18%	10.33%	9.65%	9.83%	11.70%
Linear damage rate	21.33%	12.46%	11.52%	11.77%	14.29%
Convex damage rate	69.08%	55.40%	51.79%	52.77%	62.47%
Linear economic damage	22.90%	13.29%	12.21%	12.49%	15.91%
Convex economic damage	71.08%	56.59%	53.07%	54.02%	63.40%
Increasing conversion of fossil fuel use and carbon emissions ($1.1 \times \mu_1$)					
Temperature	10.51%	4.72%	4.06%	4.24%	6.16%
Linear damage rate	10.18%	5.57%	4.76%	4.97%	7.63%
Convex damage rate	31.89%	2.45%	20.59%	21.61%	32.74%
Linear economic damage	11.05%	6.06%	5.11%	5.36%	8.39%
Convex economic damage	33.02%	25.23%	21.25%	22.30%	33.50%

Note The results in Table 9.7 are averaged from 2025 to 2100

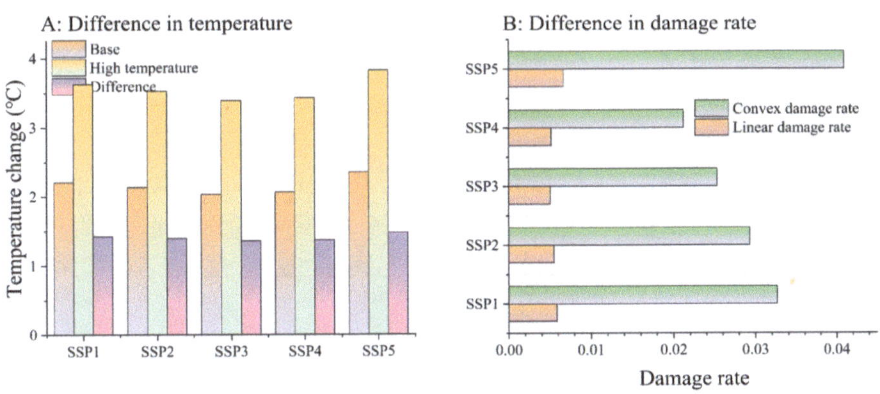

Fig. 9.19 Differences in temperature and damage rate

by $(\mathrm{PTP}_{t+\Delta t} - \mathrm{PTP}_t)/\Delta t$. If the next period exceeds the CTP, the velocity of PTP_t are not considered. The PTP_t are reported in Table 9.8. We found a gradual increase in the probability of crossing CTP PTP_t and a gradual increase in the velocity of PTP_t (Panel A of Table 9.8). The probability of crossing the CTP and the velocity of PTP_t are significantly higher in the tail high temperature scenario. This is especially true for the SSP5 and SSP1 scenarios (Panel B of Table 9.8). The increments of SSP1 and SSP5 scenarios are particularly noticeable. Of these, the earliest to reach 1 of PTP_t is the SSP5 scenario (2075) under tail high temperature case. The latest to reach 1 is the SSP3 scenario (2090) under tail high temperature case. This implies an

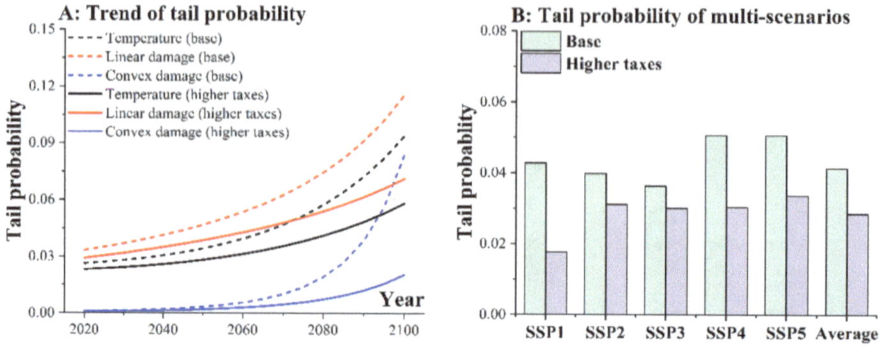

Fig. 9.20 Tail probability under higher energy taxes. *Note* (1) Fig. 9.19-A is the mean results of SSP1-SSP5. (2) Fig. 9.19-B is calculated over time (2020–2100) and over temperature and damage rates

Table 9.8 Results for the climate tipping points (CTP)

	Year	SSP1	SSP2	SSP3	SSP4	SSP5
Panel A: Mean results of base scenario (2020–2100)						
PTP_t	2020	0.067	0.067	0.067	0.067	0.067
	2100	0.724	0.581	0.420	0.456	0.999
The velocity of PTP_t	2020	0.002	0.002	0.002	0.002	0.002
	2095	0.031	0.022	0.011	0.013	0.031
Panel B: Mean results of tail high temperature (THT) scenario (2020–2100)						
PTP_t	THT	0.604	0.590	0.571	0.577	0.621
	Base	0.244	0.219	0.191	0.198	0.301
The velocity of PTP_t	THT	0.014	0.014	0.012	0.013	0.014
	Base	0.008	0.006	0.004	0.005	0.012
Panel C: Year of crossing CTP under THT scenario						
Year of crossing CTP		2080	2085	2090	2085	2075

Note: (1) Table 9.8 reports the probability of crossing the CTP. (2) Panel B is consistent with the setup of Fig. 9.7. (3) Due to the use of forward differencing, calculations can only be made up to 2095

accelerating trend of crossing the climate tipping points, needing to be treated with caution.

References

Arribas, A., Fairgrieve, R., Dhu, T., et al. (2022). Climate risk assessment needs urgent improvement. *Nature Communications, 13*(1). https://doi.org/10.1038/s41467-022-31979-w

Barrage, L. (2020). Optimal dynamic carbon tax in a climate-economy model with distortionary fiscal policy. *The Review of Economic Studies, 87*(1), 1–39. https://doi.org/10.1093/restud/rdz055

Burke, M., Davis, W. M., & Diffenbaugh, N. S. (2018). Large potential reduction in economic damages under UN mitigation targets. *Nature, 557*(7706), 549-+. https://doi.org/10.1038/s41586-018-0071-9

Ballester, J., Quijal-Zamorano, M., et al. (2023). Heat-related mortality in Europe during the summer of 2022. *Nature Medicine, 29*(7), 1857-+. https://doi.org/10.1038/s41591-023-02419-z

Barnett, M., Brock, W., & Hansen, L. P. (2020). Pricing uncertainty induced by climate change. *Rev. of Financial Stud., 33*(3), 1024–1066. https://doi.org/10.1093/rfs/hhz144

Cai, Y. Y., & Lontzek, T. S. (2019). The social cost of carbon with economic and climate risks. *Journal of Political Economy, 127*(6), 2684–2734. https://doi.org/10.1086/701890

Cai, Y. Y., Brock, W., & Xepapadeas, A. (2023). Climate change impact on economic growth: Regional climate policy under cooperation and noncooperation. *Journal of the Association of Environmental and Resource Economists, 10*(3). https://doi.org/10.1086/722274

Cai, Y. Y., Lenton, T. M., & Lontzek, T. S. (2016). Risk of multiple interacting tipping points should encourage rapid CO_2 emission reduction. *Nature Climate Change, 6*(5), 520-+. https://doi.org/10.1038/NCLIMATE2964

Coronese, M., Lamperti, F., Keller, K., et al. (2019). Evidence for sharp increase in the economic damages of extreme natural disasters. *Proceedings of the National Academy of Sciences, 116*(43), 21450–21455. https://doi.org/10.1073/pnas.1907826116

Chen, G. J., Chen, L. L., Jin, H., & Zhao, X. Q. (2023). Climate transition risk and macroeconomic policy regulation. *Economic Research Journal, 58*(5), 60–78.

Crost, B., & Traeger, C. P. (2014). Optimal CO_2 mitigation under damage risk valuation. *Nature Climate Chang, 4*(7), 631–636. https://doi.org/10.1038/NCLIMATE2249

Cruz, J. L., & Rossi-Hansberg, E. (2024). The economic geography of global warming. *Review of Economic Studies.* https://doi.org/10.1093/restud/rdad042

Desmet, K., & Rossi-Hansberg, E. (2013). Urban accounting and welfare. *The American Economic Review, 103*(6), 2296–2327. https://doi.org/10.1257/aer.103.6.2296

Dietz, S., & Stern, N. (2015). Endogenous growth, convexity of damage and climate risk: How Nordhaus' framework supports deep cuts in carbon emissions. *The Econometrics Journal, 125*(583), 574–620. https://doi.org/10.1111/ecoj.12188

Deng, H., Wu, X. H., Xu, H., & Zhang, D. W. (2024). Aggregated economic damages and the role of dynamic carbon tax under climate change: A dynamic general equilibrium model. *Economic Analysis and Policy, 84.* https://doi.org/10.1016/j.eap.2024.10.025

Dietz, S., Rising, J., Stoerk, T., & Wagner, G. (2021). Economic impacts of tipping points in the climate system. *Proceedings of the National Academy of Sciences of the United States of America, 118*(34). https://doi.org/10.1073/pnas.2103081118

Danisman, G. O., Bilyay-Erdogan, S., & Demir, E. (2025). Economic uncertainty and climate change exposure. *Journal of Environmental Management, 373.* https://doi.org/10.1016/j.jenvman.2024.123760

Golosov, M., Hassler, J., et al. (2014). Optimal taxes on fossil fuel in general equilibrium. *Econometrica, 82*(1), 41–88. https://doi.org/10.3982/ECTA10217

Gillingham, K., Nordhaus, W., Anthoff, D., et al. (2018). Modeling uncertainty in integrated assessment of climate change: A multi-model comparison. *Journal of the Association of Environmental and Resource Economists, 5*(4). https://doi.org/10.1086/698910

Harris, I., Osborn, T. J., Jones, P., & Lister, D. (2020). Version 4 of the CRU TS monthly high-resolution gridded multivariate climate dataset. *Scientific Data, 7*(1). https://doi.org/10.1038/s41597-020-0453-3

Heutel, G., Moreno-Cruz, J., & Shayegh, S. (2016). Climate tipping points and solar geoengineering. *Journal of Economic Behavior and Organization, 132*, 19–45. https://doi.org/10.1016/j.jebo.2016.07.002

Harfoot, M., Tittensor, D. P., et al. (2014). Integrated assessment models for ecologists: The present and the future. *Global Ecology and Biogeography, 23*(2), 124–143. https://doi.org/10.1111/geb.12100

He, S. L., Wu, X. H., & Wang, J. (2025a). Comprehensive economic impact assessment of regional shipping emissions: A case study of Shanghai, China. *Marine Pollution Bulletin, 211*. https://doi.org/10.1016/j.marpolbul.2024.117368

He, S. L., Wu, X. H., Wang, J., & Lu, Y. M. (2025b). Shipping emission inventory preparation with AIS data: A stratified random sampling method. *Transportation Research Part D-Transport and Environment, 138*. https://doi.org/10.1016/j.trd.2024.104520

Ha, L. (2025). Reconsider the role of digital revolution in public sector during the time of climate risk: Fresh insights from panel autoregression model. *Technological Forecasting and Social Change, 219*. https://doi.org/10.1016/j.techfore.2025.124243

Joos, F., Roth, R., Fuglestvedt, J. S., et al. (2013). Carbon dioxide and climate impulse response functions for the computation of greenhouse gas metrics: A multi-model analysis. *Atmospheric Chemistry and Physics, 13*(5), 2793–2825. https://doi.org/10.5194/acp-13-2793-2013

Jensen, S., & Traeger, C. P. (2014). Optimal climate change mitigation under long-term growth uncertainty: Stochastic integrated assessment and analytic findings. *European Economic Review, 69*, 104–125. https://doi.org/10.1016/j.euroecorev.2014.01.008

Kumar, N., & Maiti, D. (2024). Long-run macroeconomic impact of climate change on total factor productivity: Evidence from emerging economies. *Structural Change and Economic Dynamics, 68*, 204–223. https://doi.org/10.1016/j.strueco.2023.10.006

Lamb, W. F., 2022. Data for box TS.1 figure 1—technical summary of working group III contribution to the IPCC sixth assessment report, metadata W. https://doi.org/10.48490/hd2q-c531.

Li, H., Fan, P. R., Wang, Y. K., et al. (2024). Integrated assessment models for resource-environment-economy coordinated development. *Wiley Interdisciplinary Reviews: Energy and Environment, 13*(3). https://doi.org/10.1002/wene.514

Meinshausen, M., Raper, S. C. B., & Wigley, T. M. L. (2011). Emulating coupled atmosphere-ocean and carbon cycle models with a simpler model, MAGICC6-part 1: Model description and calibration. *Atmospheric Chemistry and Physics, 11*(4), 1417–1456. https://doi.org/10.5194/acp-11-1417-2011

Nordhaus, W. D., & Moffat, A. (2017). A survey of global impacts of climate change: Replication, survey methods, and a statistical analysis. *NBER working papers*. https://doi.org/10.3386/w23646

Nordhaus, W. D. (2017). Revisiting the social cost of carbon. *Proceedings of the National Academy of Sciences, 114*(7), 1518–1523. https://doi.org/10.1073/pnas.1609244114

Nordhaus, W. (2014). Estimates of the social cost of carbon: Concepts and results from the DICE-2013R model and alternative approaches. *Journal of the Association of Environmental and Resource Economists, 1*(1–2), 273–312. https://doi.org/10.1086/676035

O'Neill, B. C., Kriegler, E., Riahi, K., et al. (2014). A new scenario framework for climate change research: The concept of shared socio-economic pathways. *Climatic Change, 122*(3), 387–400. https://doi.org/10.1007/s10584-013-0905-2

Rezai, A., & Van der Ploeg, F. (2016). Intergenerational inequality aversion, growth, and the role of damages: Occam's rule for the global carbon tax. *Journal of the Association of Environmental and Resource Economists, 3*(2), 493–522. https://doi.org/10.1086/686294

Raei, E., Nikoo, M. R., et al. (2018). Data descriptor: GHWR, a multi-method global heatwave and warm spell record and toolbox. *Scientific Data, 5*. https://doi.org/10.1038/sdata.2018.206

Schmidt, G. (2024). Climate models can't explain 2023's huge heat anomaly-we could be in uncharted territory. *Nature, 627*(8004), 467. https://doi.org/10.1038/d41586-024-00816-z

Takakura, J., Fujimori, S., et al. (2019). Dependence of economic impacts of climate change on anthropogenically directed pathways. *Nature Climate Change, 9*(10), 737-+. https://doi.org/10.1038/s41558-019-0578-6

Van den Bremer, T. S., & Van der Ploeg, F. (2021). The risk-adjusted carbon price. *The American Economic Review, 111*(9), 2782–2810. https://doi.org/10.1257/aer.20180517

Van der Ploeg, F., & de Zeeuw, A. (2016). Non-cooperative and cooperative responses to climate catastrophes in the global economy: A north-south perspective. *Environmental and Resource Economics, 65*(3), 519–540. https://doi.org/10.1007/s10640-016-0037-z

Van Benk, L., Hajer, M., Pelzer, P., Van Vuuren, D., & Cassen, C. (2020). Anticipating futures through models: The rise of integrated assessment modelling in the climate science-policy interface since 1970. *Global Environmental Change-Human and Policy Dimensions, 65*. https://doi.org/10.1016/j.gloenvcha.2020.102191

Weitzman, M. L. (2015). The climate casino: Risk, uncertainty, and economics for a warming world. *Review of Environmental Economics and Policy, 9*(1), 145–156. https://doi.org/10.1093/reep/reu019

Weitzman, M. L. (2009). On modeling and interpreting the economic of catastrophic climate change. *The Review of Economics and Statistics, 91*(1), 1–19. https://doi.org/10.1162/rest.91.1.1

Weitzman, M. L. (2012). GHG targets as insurance against catastrophic climate damages. *Journal of Public Economic Theory, 14*(2), 221–244. https://doi.org/10.1111/j.1467-9779.2011.01539.x

Weitzman, M. L. (2012a). The Ramsey discounting formula for a hidden-state stochastic growth process. *Environmental and Resource Economics, 53*(3), 309–321. https://doi.org/10.1007/s10640-012-9594-y

Weitzman, M. L. (2013). A precautionary tale of uncertain tail fattening. *Environmental and Resource Economics, 55*(2), 159–173. https://doi.org/10.1007/s10640-013-9646-y

Yin, W. J., Liu, B., Yao, D. J., & Chen, G. (2025). Property insurance demand in response to climate risk: The role of green fiancé, economic policy uncertainty, and non-linearity. *Technology in Society, 83*. https://doi.org/10.1016/j.techsoc.2025.102957

Chapter 10
Comprehensive Economic Impact Assessment of Shipping Emissions in Port Cities: A Case Study of Shanghai, China

10.1 Introduction

Shipping activities are the third largest source of air pollution after industrial emissions and motor vehicle emissions. In particular, the negative impact of ship pollutants in coastal areas on the air quality of port cities is increasing day by day (Liu et al., 2017). According to the report of the International Maritime Organization (IMO) (Zincir et al., 2019), the annual consumption of fuel oil used by global shipping is 0.3 billion tons, and the resulting carbon dioxide accounts for 2.2% of the global total carbon dioxide emissions (Bilgili & Celebi, 2018). The emissions also contain a large amount of nitrogen oxides (NO_x), sulfur oxides (SO_x) and particulate matter (PM) (Kim & Seo, 2019; Sharafian et al., 2019). Shanghai is known as a major hub port of China. Its foreign trade throughput accounts for 20% of the total foreign trade throughput of major coastal ports. Its cargo turnover increased from 2000.52 billion tons/km in 2011 to 3,209,455 billion tons/km in 2020, with the proportion to the national total rising from 26.53%–30.33%. In 2021, the cargo throughput of Shanghai Port accounted for 5% of the national total cargo throughput. In 2022, its container throughput reached 47.3 million TEUs, ranking first in the world for 13 consecutive years. However, shipping emissions are increasing day by day, and the consequent negative environmental externalities have gained extensive attention of the government, academia and society.

The study on the impact caused by shipping emissions mainly covers the impact of pollutants on air quality, residents' health, industry development and social economy. With regard to the impact of exhaust emissions on air quality, this study focuses on the preparation of shipping emission inventory and exhaust diffusion. First, the exhaust emission inventory of ports, regions, river basins and other different scales was prepared using the "top-down" or "bottom-up" method (Dai et al., 2025; Yang Zhang Lv Zhang Yang, & Fu, 2022a; Yang Zhang Lv Zhang Yang Fu et al., 2022b).

X. Wu, *Environmental Disaster in China*,
https://doi.org/10.1007/978-981-92-0910-1_10

Second, the impact of shipping emissions on the air quality of navigation areas was estimated on the basis of measurement and simulation (He et al., 2023; He et al., 2022). Third, the impact of exhaust emissions on residents' health was calculated and the impact of shipping emissions on the health of local residents in port cities in terms of disease treatment expenditure, premature mortality, labor loss and other indexes was measured. Fourth, these scholars came to a conclusion regarding the impact of exhaust emissions on social economy. The monetary value of losses such as pollutant control, residents' health loss (Roy & Chakraborty, 2025; Xia et al., 2022), labor loss (Yin, 2022), industry loss (crop yield reduction) (Chen et al., 2022; Sharnil et al., 2022; Xu et al., 2022) and loss of fixed investment (building cleaning and material corrosion) (Fu et al., 2021) was selected as the quantitative basis to calculate the economic impact of shipping emissions (Gutenberg, 2014; Zhu & Wang, 2021). In this study, the treatment cost method and damage accounting method have been selected as the main methods to study the economic impact of shipping emissions, in constructing the CGE (Computable General Equilibrium) model to quantitatively analyze the direct economic losses caused by shipping emissions and simulate the impact of exhaust emissions on the regional economic system, with Shanghai as the study area in this paper. In this way, we not only measure the negative environmental externalities caused by shipping to port cities and then provide empirical reference for the promulgation of pollution control policies but also contribute a favorable supplement to the studies relating to economic losses caused by shipping emissions.

10.2 Literature Review

10.2.1 Study on the Impact of Shipping Emissions

Shipping emissions aggravate air pollution of port cities and areas nearby waterways. According to study of (Sheng et al., 2022), the concentration of exhaust pollutants in a 4-km area along the Yangtze River is 1.6 to 9.6 times the average concentration of the whole region in 2017. (Zhai et al., 2023) found that the impact of particulate matter and nitrogen oxides emitted by shipping in summer is as high as 21.4% and 39.5% respectively based on the real-time monitoring of shipping emissions. Long-term exposure to high concentrations of polluted air can easily lead to cardiovascular and respiratory diseases and increase premature mortality. (Natalie et al., 2023) predicted that shipping emissions led to 265,000 premature deaths in 2020, accounting for 0.5% of the global mortality. (Liu et al., 2020) estimated the premature deaths caused by exposure to $PM_{2.5}$ in the Yangtze River Delta region from 2001 to 2017 using the WRF/CMAQ air quality model, which reached 95,000. Mortalities and diseases cause labor losses and bring about additional economic burdens. (Sofiev et al., 2018) assessed the impact of low-sulfur fuel emissions in global shipping on healthy climate and discovered that 250,000 deaths and 6.4 million childhood asthma cases.

Shipping emissions have a greater impact on port cities and cities around waterways. The scope of such impact will be enlarged with the diffusion of exhaust gas. Moreover, exhaust gas produces an increasingly significant negative impact to other aspects. Emissions cause sectorial losses. For example, emissions affect the yield of crops sensitive to the atmospheric environment. These crops include corn, wheat, rice and so on (Chen et al., 2022; Sharnil et al., 2022; Xu et al., 2022). The reduction in the service life of buildings and the additional cleaning cost are defined as the economic loss value of fixed assets (Fu et al., 2021). Taking Baltic Sea and Kattegat Strait as the object, (Ytreberg et al., 2021) quantitatively analyzed the impact of shipping exhaust diffusion in three dimensions, including marine ecosystem (marine eutrophication and ecological toxicity), air quality (air pollution and climate change) and human welfare.

10.2.2 Study on the Economic Impact of Shipping Emissions

Scholars have tried to quantify the impact of shipping emissions on the regional economy, the extent of damage to the ecological environment and the health of residents in monetary terms. In light of measurement methods, exhaust treatment plays an important role. In other words, efforts are made to avoid the treatment expenditure against the loss caused by the negative externalities of exhaust gas, also the cost to be paid during the treatment process (Bindle et al., 2021; Irena et al., 2021; Progiou et al., 2021). The premature deaths caused by $PM_{2.5}$ emissions from shipping increased by 7.7% in the Iberian Peninsula, resulting in an economic loss of about € 910 million every year according to (Nunes et al., 2021), using the combined exposure-response function of $PM_{2.5}$ and NO_2. (Spengler & Tovar, 2022) calculated the cost of negative externalities caused by shipping emissions of four Spanish ports using impact path method, with a result up to € 115.4 million. (Yoo et al., 2022) estimated the social and environmental costs of shipping emissions from Busan Port using the international benefit transfer method, and discovered that the externality cost of emissions in the berthing period alone is about $ 31.55 million. With reference to the emission reduction cost method, Marie(Ing-Marie et al., 2021) estimated the total cost of negative externalities caused by shipping NO_x emissions to the ocean, which is approximately € 0.24 billion. However, there are few studies on the negative externalities of shipping emissions on the economic impact in China. Most of the relevant studies are subject to the foreign accounting methods. For example, (Joseph & Po-Hsing, 2012) estimated the combined environmental cost of emissions from ships and trucks in Kaohsiung Port, of Taiwan, (Song, 2014) estimated the shipping emission cost adding the ecological benefit to Yangshan Port in Shanghai and Liu et al., (2018) assessed the potential costs and benefits of DECA, etc. Utilizing a ship emission inventory model, Chen et al.(2025) assessed the feasibility and cost-effectiveness of implementing nitrogen oxide (NOx) Emission Control Areas (ECAs) in China's coastal regions. To cost-benefit analysis revealed benefit-cost ratios of 2.93 and 3.86

for expanding control zones to 50 and 100 nautical miles, respectively. Nunes et al. (2025) systematically assessed the impacts of shipping emissions on air quality, public health, and associated external costs across the Iberian Peninsula. Their analysis revealed that population exposure to shipping-related $PM_{2.5}$ during 2017–2022 caused approximately 1061 premature deaths and 3626 years of life lost (YLL), with estimated economic costs reaching €9.1 billion. Under projected 2050 emission scenarios, these impacts are expected to escalate dramatically to 4459 premature deaths and 68,081 YLL, translating to a total economic burden of €53.2 billion.

In summary, based on the degradation of air quality and residents' health damage caused by exhausts, the economic impact of negative externalities from shipping emissions rarely involves the comprehensive impact assessment of various pollutants. However, the direct loss calculation method leaves out the correlation between sectors of the economic system, fails to effectively calculate the comprehensive loss value caused by the input-output transmission between sectors and often underestimates the comprehensive economic impact caused by shipping emissions to port cities.

10.3 Modeling

The computable general equilibrium (CGE) model is a widely used tool (Li et al., 2020) to analyze the comprehensive impact of energy changes, carbon taxes and environmental policies on social economy (Lin et al., 2023; Rosa et al., 2018; Zhang et al., 2020). In order to study the impact of shipping emissions and diffusion on the economy and society of Shanghai, this paper complies with China's latest industry classification standard *Industrial classification for national economic activities (2017)* (GB/T4754). According to the 14th Five-Year Development Plan of Shanghai Municipality and the availability of data, the sectors in the input and output table are further merged or split. The output value of the financial service industry is used as the measurement index of the international financial center, the output value of shipping and air transportation is used as the variable of the international shipping center, and the output value of communication equipment, information services, research and experimental development is used as the index of technological innovation. The shipping CGE model of Shanghai is divided into 9 sectors (agriculture, forestry, animal husbandry and fishery, major manufacturing, shipping, air transportation, other transportation, financial services, technological innovation, real estate, and major service industries).

Using the standard LHR (Lofgren H., R. L. Harris, S. Robinson) system, we established a CGE model indicating the economic impact of shipping emissions in Shanghai. This model consists of a production module, a trade module, an economic module and an equilibrium closure module. The impact of shipping emissions on the economy and society is simulated by combining the shipping emission and diffusion scenario and applying different scenarios, such as economic loss, environmental protection investment and pollution tax.

In the actual import and export activities of Shanghai, the transfer-out of some sectors is greater than the production in the region, resulting in insufficient supply of products in the region, so that the model cannot be used for solving. According to the micro social accounting matrix, the supply value of sub-sectors in the region is negative, making it unable to effectively establish the flow share inside and outside the current region for the goods produced in the current region. For the above reason, we combine the interregional transfer-in and regional production into the regional total goods supply and combine the interregional transfer-out and regional sales into the regional total goods demand instead of establishing a separate interregional goods flow account in this paper. The above efforts are made to avoid the problem of calculating negative values due to transit trade.

Based on the classical CGE model, we monetized the total pollutants of shipping emissions, referred to the results of monetization as the economic loss value and substituted it into the original model as the exogenous parameters of this model. As a result, there will be changes in economic activity in terms of sector output, related factor prices, and income of residents and enterprises. Upon the application of CGE model to assess the economic impact of shipping emissions, it is necessary to describe the impact mechanism of such damage on the economic system. The exhaust loss module is introduced into the CGE model, in which the variable E is set as the economic loss impact parameter, Y´ as the output of the shipping sector calculated in shipping emissions and Y as the sector output without taking into account shipping emissions.

$$\mathrm{Y}' = \mathrm{E} \times \mathrm{Y} \tag{10.1}$$

In the scenario setup with simulated impact, the value of exhaust loss impact parameter is set between 0 and 1. The social economic cost of pollutants is taken as the direct economic loss value C´ while C indicates the total output value of the shipping sector. The proportion of direct economic loss to the total output value is taken as the impact of shipping emissions on the sector output. In other words, the impact of shipping emissions is ignored, all production sectors of Shanghai continue production according to the original equilibrium, the shipping sector is affected by the economic loss variable E, thereby causing changes to the output of each sector.

$$\mathrm{E} = 1 - \frac{\mathrm{C}'}{\mathrm{C}} \tag{10.2}$$

The total output equation of the shipping corresponding to the economic loss impact of shipping emissions is as follows:

$$\mathrm{QA}_a = \mathrm{E} * \alpha_{\mathrm{a}}^{\mathrm{t}}\left[\delta_a^t * \mathrm{QVA}_a^{\rho_a} + (1 - \delta_a^t)\mathrm{QINTA}_{\mathrm{a}}^{\rho_a}\right]^{\frac{1}{\rho_a}} \tag{10.3}$$

10.4 Empirical Analysis

10.4.1 Data Preprocessing

The SAM table is prepared on the basis of latest input-output tables of Shanghai in 2017 and other related tables, including the input-output tables of 42 sectors and the input-output tables of 149 sectors prepared for Shanghai in 2017, the input-output tables of 149 sectors nationwide in 2017, *the Statistical Yearbook of Shanghai 2018, China Statistical Yearbook 2018* and China's Balance of Payment Presentation 2018. With the macro social accounting matrix as the constraint, the data of the input-output tables prepared for Shanghai in 2017 is merged or split according to the sector division, the sectors in the corresponding input-output tables and the needs of 9 sectors, and then the micro social accounting matrix is formulated. Shanghai is known as a transit trade city. The taxes collected by its customs include all customs duties on imports and exports from the ports of Shanghai. According to the relevant statistics, the amount of tax levied by the customs of Shanghai is higher than the total amount of national customs duties. In consideration of the above facts, we split the customs duties by taking the total amount of national customs duties as the base value and referring to the import and export amount of different industries in the input-output tables of Shanghai (Table 10.1).

10.4.2 Estimation of Shipping Emissions

Based on ship AIS data, we estimated the emissions of CO_2, SO_2, NO_x, PM2.5, PM10, CO and VOC from shipping in Shanghai using the most widely used ship traffic emission model (STEAM). See the relevant literature for specific formulas (He et al., 2022; He et al., 2023; Yang, Zhang, Lv, Zhang, Yang, Fu, et al., 2022a).

$$E = E_m + E_a = \sum\nolimits_{m,a} P_{m,a} LF_{m,a} T_{m,a} LEF \tag{10.4}$$

E is the exhaust emissions of a single ship, and m and a represent the main engine and auxiliary engine respectively. P is the power of the ship power equipment; LF is the load factor; T is the operation (duration) time; LEF is the localized emission factor.

$$LF_m = \left(\frac{v_{actual}}{v_{max}}\right)^3 \tag{10.5}$$

v_{actual} represents the actual ship speed, also the ship speed corrected according to the change of wind speed and wind direction during navigation. v_{max} represents

Table 10.1 The Macro Social Accounting Matrix of Shanghai in the year 2017 (Unit: RMB million *yuan*)

	Activities	Commodities	Labor	Capital	Residents	Enterprises	Government	Foreign	Savings and Investment	Total
Activities		800,712,919								800,712,919
Commodities	499,617,432				129,701,000		45,808,700	134,934,445	256,516,668	1,066,578,245
Labor	136,072,392									136,072,392
Capital	117,103,074									117,103,074
Residents			136,072,392	35,755,382		224,361,149	1,534,000	6,921,477		404,644,400
Enterprises				78,793,675			104,000			78,897,675
Government	47,920,021	5,234,413			17,311,400	36,078,800		550,615		107,095,249
Foreign		260,630,913		2,554,017						263,184,930
Savings and investment					257,632,000	−181,542,274	59,648,549	120,778,393		256,516,668
Total	800,712,919	1,066,578,245	136,072,392	117,103,074	404,644,400	78,897,675	107,095,249	263,184,930	256,516,668	

the maximum ship speed during navigation, and Formula (10.5) is substituted into Formula (10.4).

The shipping emission of Shanghai in 2019 is estimated according to the ship traffic emission assessment model. First, we read the AIS data for the whole year and coordinated the number of ships sailing in the area of Shanghai in 2019. We used ship MMSI number to query the ships one by one at http://ship.chinaports.com and obtained the IMO number and name of these ships. Through IMO number and ship name as well as the static information of ships in the classification societies of various countries, a ship database of Shanghai with complete dynamic and static information was established at last. In 2019, the total amount of major types of shipping emissions in Shanghai is 154,977.147 tons, of which the amounts of CO_2, SO_2, NO_x, $PM_{2.5}$, PM_{10}, CO and VOC are 148,853.033 tons, 2288.37 tons, 2833.22 tons, 261.523 tons, 326.91 tons, 305.116 tons and 108.97 tons respectively, accounting for 96.5%, 1.48%, 1.83%, 0.17%, 0.21%, 0.2% and 0.07%.

10.4.3 *Social Economic Cost Calculation of Ship Exhaust Pollutants*

10.4.3.1 Unit Economic Cost of Exhaust Pollutants

See Table 10.2 for the external cost of shipping emissions. Exhaust pollutants mainly include sulfides, nitrides and suspended particulate matter. This paper takes the port area as the study object and takes the air treatment cost of the city as the shipping emission pollution treatment cost. The unit is mostly based on the marginal external cost of air pollution in Europe. Some indexes, such a population density, per capita GDP, economic development stage and technological progress, are utilized according to the changes in the study area and study time and the parameter calculation method of (Song, 2014) is selected to correct the unit economic cost value.

The external cost calculation methods of pollutants are well developed in foreign countries. However, they are still in the stage of start-up in China. The economic development stage, environmental capacity, environmental regulation capacity and population density are quite different in different regions and industries. The pollution cost factor in foreign countries is quite different from the current economic and social development and pollution treatment level in China. For the above reason, we corrected the unit social cost value of pollutants by combining the unit social cost of ports in Shanghai in the paper of (Song, 2014), using the parameter calculation method and referring to the per capita GDP and population density value of Shanghai.

$$C_{i,t2} = \frac{C_{i,t1} \times PGDP_{i,t2} \times D_{i,t2}}{PGDP_{i,t1} \times D_{i,t1}} \tag{10.6}$$

Table 10.2 Pollutant cost factors

Author	Region/time	External costs	Pollutant
(Ernestos, 2010)	Piraeus Port, Greece (2008–2009)	European Commission DG research extension project	NO_x SO_x $PM_{2.5}$
(Joseph & Po-Hsing, 2012)	Kaohsiung Port, Taiwan (2010)	Estimation of marginal external costs of air pollution in Europe(BeTa)	CO_2 N_2O PM_{10} $PM_{2.5}$ NO_x SO_x
(Castells et al., 2014)	Spanish (2009)	Marginal external costs of air pollution in Europe	NO_x SO_x $PM_{2.5}$ OC
(McArthur & Osland, 2013)	Bergen Norway (2010)	Marginal external costs of air pollution in Europe	NO_x SO_x $PM_{2.5}$ OC
(Song, 2014)	Yangshan Port, China (2009)	International average	CO_2 PM_{10} $PM_{2.5}$ NO_x SO_x
(Maragkogianni & Papaefthimiou, 2015)	Santorini Piraeus Port, Mykonos, Corfu and Katakolo Islands,Greece (2013)	CAFÉ and NEEDS	NO_x SO_x PM
(Miluse & Beatriz, 2015)	Puerto Las Palmas, Spain (2011)	Top-down approach	NO_x SO_x VOC $PM_{2.5}$ CO_2
(Dragovic et al., 2018)	Ports of Kotor and Dubrovnik (2012–2014)	CAFÉ and NEEDS	NO_x SO_x $PM_{2.5}$
(Tovara & Tichavska, 2019)	Las Palmas, St. Petersburg, Hong Kong	Marginal external costs of air pollution in Europe	NO_x SO_x $PM_{2.5}$ CO CO_2
(Progiou et al., 2021)	Piraeus Port (2017–2018)	Manual on external transportation costs	NO_x SO_x VOC PM

Where, $C_{i,\,t1}$ and $C_{i,\,t2}$ represent the unit social cost of ship pollutant i in the year t1 and t2 respectively (unit: RMB/t). $PGDP_{i,\,t1}$ and $PGDP_{i,\,t2}$ represent the regional GDP per capita in the year t1 and t2 respectively (unit: RMB 10,000). $D_{i,\,t1}$ and

Table 10.3 Social cost factors of ship emissions in Shanghai

Time	2014 year	2019 year
CO_2/(RMB/t)	178	266
SO_2/(RMB/t)	75,730	113,050
NO_x/(RMB/t)	65,644	97,993
$PM_{2.5}$/(RMB/t)	526,840	786,469
PM_{10}/(RMB/t)	472,148	704,824
CO/(RMB/t)	7039	10,508
VOC/(RMB/t)	8538	12,745
GDP per capita/ RMB	102,800	153,300
Population density (persons/km2)	3026	3830

Note GDP per capita and population density data are from the 2020 Shanghai Statistical Yearbook

$D_{i,\,t2}$ represent the regional population density in the year t1 and t2 respectively (unit: persons/km^2). According to the social cost factor of the year 2014, the unit economic cost of the year 2019 is calculated through per capita GDP and population density. See Table 10.3 for details.

10.4.3.2 Total Social Economic Cost of Exhaust Emissions

The social economic cost of shipping emissions is as follows:

$$C = \sum_{i} (c_i * Q_i) \tag{10.7}$$

Where, C indicates the total social economic cost of various pollutants, c_i indicates the unit social economic cost of different pollutants and Q_i indicates the total emissions of various pollutants.

The social economic cost caused by shipping emissions to Shanghai in 2019 is estimated according to the unit social economic cost c_i of pollutants and the total emissions Q_i of various pollutants. As can be seen from Table 10.4, the economic cost of shipping emissions in Shanghai in 2019 is RMB 1.017 billion, and the total economic cost caused by different pollutants corresponds to SO_2, NO_x, $PM_{2.5}$, PM_{10} and CO_2 in descending order. CO_2 emissions are much higher than the emissions other pollutants. However, the unit economic cost is lower, resulting in a lower total cost.

Table 10.4 Treatment costs by pollutant type

Pollutant	Total emissions/tons	Unit cost/(yuan/ton)	Total cost/million yuan
CO_2	148,853	266	3958.19
SO_2	2288	113,050	25,869.92
NO_x	2833	97,993	27,763.68
$PM_{2.5}$	261	786,469	20,568.36
PM_{10}	327	704,824	23,041.41
CO	305	10,508	320.62
VOC	109	12,745	138.89

Note The emissions from ships, calculated using stratified sampling emission modeling with ship AIS activity data and static data, are shown in the appendix

10.4.4 Scenario Setup

(1) Economic loss (Scenario 1)

Based on the empirical study of disaster loss assessment by (Tan et al., 2019), we assumed that the shipping emissions of Shanghai has caused a direct economic loss to the industry. Simulation and analysis are performed with shipping sector as the victim of direct economic loss according to the impact of exhaust gas and the Polluter Pays Principle (PPP). As an international trade center, Shanghai is significantly characterized by transit trade. In 2020, the proportion of water-water container transit was 51.6%, and the import and export volume of local enterprises through the ports of Shanghai accounted for about 40%. Moreover, the mooring, berthing and cruising status of the ships sampled for the calculation of shipping emissions is summarized and the number of transit ships accounts for about 50% of all sampled ships. Taking into account the factors of transit trade and ship transits, we assumed that S11 is the economic loss ratio of 50% for Shanghai to bear regional shipping emissions. A direct impact is caused to the region by the shipping emissions in Shanghai. Therefore, it bears the 100% negative damage of exhaust emissions. In consideration of coordinated exhaust treatment, the proportion of bearing the loss caused by exhaust emissions is not clarified between regions. Therefore, we assumed S12 that Shanghai bears the total economic loss caused by exhaust gas in the region at the ratio of 100%. The shadow price of environmental treatment elements fluctuates over time and shows an overall upward trend, resulting in an upward trend in the unit treatment unit of atmospheric environmental emissions (宋杰鲲 et al., 2023). According to the study of (张文静 & 马喜立, 2020) on the emission reduction cost in nine major regions nationwide, the average emission reduction cost in all these major regions has increased by more than 100% over the past decade. In order to review the impact of shipping emissions on Shanghai when the environmental treatment cost rises, we assumed that the unit cost of pollutants is doubled and S13 and S14 represent the loss of S11 and S12 under such impact.

(2) Environmental protection investment (Scenario 2)

The investment in innovative technologies and environmental infrastructure help improve energy efficiency and reduce shipping emissions. Increased environmental investment for a sector is a favorable factor in reducing the output of pollutants in the sector, while promoting the economic growth of this sector (Yang et al., 2021). The CO_2 emissions of high-income countries and low-income countries will be reduced by 0.04% and 0.02% as China's direct investment in the countries participating in the "Belt and Road Initiative" rises by 1% (Su et al., 2022). Investment in energy and innovative technologies has a positive effect on reducing carbon emissions (Huang & Wen, 2022). S21 converts the economic loss caused by shipping emissions into environmental investment by setting a growth rate of 2.67% in fixed assets. S22 takes the ratio of environmental protection investment to the total output of shipping industry as the amount of environmental protection investment, and the simulated fixed assets increase by 5.34%. S23 uses the planned amount of environmental protection investment as the fixed assets growth rate of 7.3%.

(3) Pollution tax (Scenario 3)

According to the price transmission mechanism, increased product prices will have an impact on the output and demand. Any change at producers and consumers will lead to the change of exhaust emissions, thus triggering economic fluctuations. Meanwhile, price changes will spread along product supply chains, causing an impact to the relevant sectors and regional economies. Therefore, the impact of price changes in the shipping sector on the regional economy, the sector economy and the formation of price levels is simulated by increasing prices. Price adjustment has an impact on the elasticity of shipping demand, especially in the sectors closely related to shipping. Changes in the output of shipping sector are reflected in the calculation of regional shipping emissions so as to explore the impact of sectoral price changes on pollutant emissions. In the scenario 3, the impact on the variables (S31, S32, S33 and S34) of shipping sector, including output, emissions and regional economy, are calculated when the pollution tax is levied to the shipping sector at the rate of 1%, 2%, 5% and 10%.

10.4.5 Analysis of the Simulation Results

The simulation results of economic loss caused by shipping emissions are shown in Table 10.5. Under the impact of economic loss, all the sectors, excluding transportation and technological innovation sectors, show a decline of total output in varying degrees. The shipping sector suffers the direct impact of economic loss. Therefore, more significant fluctuations occur to its output than other sectors and the sectoral industry goes down by 0.132%–0.498%. Shipping is one of the important industries in the transportation system of Shanghai. However, it has a lower output value than other industries. The fluctuation caused by the impact of economic loss is still slight even if the unit economic cost of pollutants is increased. The economic loss of shipping sector has a greater impact on the sector itself but has a slight impact on other industries. Under such circumstance, we can refer to the tax on the pollution exhaust gas of shipping sector as the unit cost in the simulation period.

When the output of shipping sector declines, the output of railway and highway transport as complementary means will increase, causing the total output of other transport types to fluctuate, with the output value growing by 0.018%–0.073%. The total output value of shipping sector accounts for 8.71% of the total output value of the transportation system, the economic impact of shipping has a slight impact on other transport types. In Shanghai, the total output value of air transportation accounts for 29.8% and the shipping industry has a great development potential. As for the change rates of total output in various industries under the scenario S12, agriculture, forestry, animal husbandry and fishery decrease by 0.046%, major manufacturing decreases by 0.013%, shipping decreases by 0.0259%, air transportation decreases by 0.005%, other transport types decrease by 0.036%, financial services decrease by 0.032%, technological innovation increases by 0.017%, real estate decreases by 0.03% and major service industries decrease by 0.023%. The technological innovation sector is different from traditional industries and less correlated with shipping. Therefore, it is hardly affected by the shipping sector.

See Table 10.6 for the simulation results of environmental protection investment. The total output of agriculture, forestry, animal husbandry and fishery, financial services, technological innovation, real estate industry and major service industries shows a downward trend. The total output of major manufacturing, shipping, air transportation and other types of transportation shows an upward trend. Direct investment in the shipping sector contributes to an increase in total output, resulting in a growth rate higher than that of other sectors, and the growth rate is 1.255%–6.873%. Total output in the transport sector has risen, with air transportation growing by 0.006% to 0.032% and other transport types growing by 0.122%–0.668%. Therefore, we believe that the investment in shipping sector has significantly increased the output of this sector. However, shipping only takes a small proportion in the economic system, based on which the environmental protection investment has a weak impact on other industries.

See Table 10.7 for the simulation results of pollution tax collection. All the sectors, excluding the technological sector, see a downward trend in the total output. The total output of shipping sector declines by 0.08%–0.861%. The collection of pollution tax causes the price of shipped products to rise, thus inhibiting consumer demand and reducing production. The shipping tax rate of 1% to 10% causes not only changes to the output of shipping sector but also a great impact to other sectors. Other types of transport suffer the greatest impact with the output down by 0.029%–0.327%.

Table 10.5 Changes in sectoral output after economic loss shocks

Sector/scenario		Value added	Intermediate inputs	Total output	Imports	Exports
Agriculture, forestry, animal husbandry and fishery	S11	−0.015	−0.024	−0.023	−0.022	−0.028
	S12	−0.029	−0.049	−0.046	−0.044	−0.057
	S13	−0.045	−0.075	−0.069	−0.066	−0.087
	S14	−0.06	−0.1	−0.093	−0.088	−0.117
Major manufacturing	S11	0.002	−0.008	−0.007	−0.005	−0.012
	S12	0.003	−0.015	−0.013	−0.01	−0.024
	S13	0.005	−0.024	−0.02	−0.015	−0.037
	S14	0.007	−0.032	−0.028	−0.02	−0.051
Shipping	S11	0.472	0.374	−0.132	0.905	−2.048
	S12	0.955	0.756	−0.259	1.824	−4.064
	S13	1.449	1.147	−0.381	2.76	−6.047
	S14	1.954	1.547	−0.498	3.711	−7.999
Air transportation	S11	0.006	−0.003	−0.003	−0.00053	−0.007
	S12	0.012	−0.006	−0.005	−0.001	−0.015
	S13	0.019	−0.009	−0.008	−0.002	−0.023
	S14	0.025	−0.012	−0.011	−0.002	−0.031
Other transportation	S11	0.037	0.014	0.018	0.044	−0.039
	S12	0.074	0.028	0.036	0.09	−0.078
	S13	0.113	0.042	0.054	0.136	−0.118
	S14	0.152	0.057	0.073	0.183	−0.158
Financial services	S11	−0.013	−0.019	−0.016	−0.027	0.008
	S12	−0.026	−0.038	−0.032	−0.054	0.017
	S13	−0.039	−0.057	−0.049	−0.082	0.026
	S14	−0.052	−0.077	−0.065	−0.11	0.034
Technological innovation	S11	0.014	0.008	0.009	−0.005	0.022
	S12	0.028	0.015	0.017	−0.01	0.044
	S13	0.042	0.023	0.026	−0.016	0.066
	S14	0.056	0.03	0.035	−0.021	0.089
Real estate	S11	−0.011	−0.016	−0.015	0	0
	S12	−0.023	−0.033	−0.03	0	0
	S13	−0.034	−0.05	−0.046	0	0
	S14	−0.046	−0.067	−0.062	0	0
Major service industries	S11	−0.005	−0.014	−0.011	−0.016	−0.003
	S12	−0.01	−0.029	−0.023	−0.031	−0.006
	S13	−0.016	−0.043	−0.034	−0.048	−0.01
	S14	−0.021	−0.058	−0.046	−0.064	−0.013

Table 10.6 Changes in sectoral output after increased environmental investment

Sector/scenario		Value added	Intermediate inputs	Total output	Imports	Exports
Agriculture, forestry, animal husbandry and fishery	S21	−0.031	−0.031	−0.031	−0.032	−0.03
	S22	−0.063	−0.063	−0.063	−0.063	−0.059
	S23	−0.086	−0.086	−0.086	−0.087	−0.081
Major manufacturing	S21	0.007	0.007	0.007	0.006	0.009
	S22	0.014	0.014	0.014	0.012	0.018
	S23	0.019	0.019	0.019	0.017	0.025
Shipping	S21	1.255	1.255	1.255	1.255	1.256
	S22	2.511	2.511	2.511	2.51	2.512
	S23	3.437	3.437	3.437	3.435	3.439
Air transportation	S21	0.006	0.006	0.006	0.006	0.007
	S22	0.012	0.012	0.012	0.011	0.013
	S23	0.016	0.016	0.016	0.015	0.018
Other transportation	S21	0.122	0.122	0.122	0.122	0.123
	S22	0.244	0.244	0.244	0.243	0.246
	S23	0.334	0.334	0.334	0.333	0.336
Financial services	S21	−0.024	−0.025	−0.024	−0.025	−0.023
	S22	−0.049	−0.049	−0.049	−0.05	−0.046
	S23	−0.067	−0.067	−0.067	−0.069	−0.063
Technological innovation	S21	−0.006	−0.006	−0.006	−0.007	−0.005
	S22	−0.012	−0.012	−0.012	−0.014	−0.01
	S23	−0.016	−0.016	−0.016	−0.019	−0.013
Real estate	S21	−0.024	−0.024	−0.024	0	0
	S22	−0.047	−0.048	−0.048	0	0
	S23	−0.065	−0.065	−0.065	0	0
Major service industries	S21	−0.012	−0.012	−0.012	−0.013	−0.012
	S22	−0.025	−0.025	−0.025	−0.025	−0.023
	S23	−0.034	−0.034	−0.034	−0.035	−0.032

Major manufacturing is subjected to the second greatest impact with the maximum output decrease of 0.142%. Shanghai is one of China's transit trade centers. The construction of international shipping centers promotes the shipping development of Shanghai. Driven by the double goals of ecological construction and ecological development, the multimodal transport is a new mode of transportation system construction to promote the coordinated development of shipping, railway transportation and highway transportation. Changes in the simulation results unveil the strong correlation between different modes of transport.

Table 10.7 Changes in sectoral output after the imposition of pollution taxes

Sector/scenario		Value added	Intermediate inputs	Total output	Imports	Exports
Agriculture, forestry, animal husbandry and fishery	S31	−0.0004	−0.011	−0.009	−0.006	−0.023
	S32	−0.00084	−0.021	−0.018	−0.012	−0.046
	S33	−0.002	−0.056	−0.047	−0.031	−0.121
	S34	−0.006	−0.121	−0.101	−0.067	−0.262
Major manufacturing	S31	−0.004	−0.014	−0.012	−0.008	−0.027
	S32	−0.008	−0.027	−0.025	−0.016	−0.056
	S33	−0.021	−0.071	−0.066	−0.042	−0.145
	S34	−0.044	−0.154	−0.142	−0.089	−0.313
Shipping	S31	0.016	−0.088	−0.08	0.155	−0.518
	S32	0.033	−0.178	−0.161	0.314	−1.046
	S33	0.09	−0.456	−0.413	0.82	−2.686
	S34	0.211	−0.952	−0.861	1.765	−5.628
Air transportation	S31	0.004	−0.006	−0.005	−0.002	−0.013
	S32	0.007	−0.011	−0.011	−0.003	−0.027
	S33	0.019	−0.03	−0.028	−0.009	−0.071
	S34	0.042	−0.063	−0.061	−0.018	−0.153
Other transportation	S31	−0.009	−0.034	−0.029	0.001	−0.093
	S32	−0.018	−0.068	−0.059	0.002	−0.188
	S33	−0.047	−0.176	−0.154	0.007	−0.489
	S34	−0.098	−0.375	−0.327	0.017	−1.047
Financial services	S31	−0.001	−0.007	−0.004	−0.014	0.017
	S32	−0.002	−0.014	−0.009	−0.028	0.035
	S33	−0.006	−0.038	−0.022	−0.073	0.091
	S34	−0.012	−0.082	−0.048	−0.157	0.195
Technological innovation	S31	0.013	0.007	0.008	−0.003	0.018
	S32	0.027	0.013	0.016	−0.005	0.036
	S33	0.069	0.035	0.04	−0.014	0.094
	S34	0.148	0.075	0.087	−0.031	0.202
Real estate	S31	0.000614	−0.005	−0.003	0	0
	S32	0.001	−0.009	−0.007	0	0
	S33	0.003	−0.024	−0.018	0	0
	S34	0.006	−0.052	−0.039	0	0
Major service industries	S31	0.000509	−0.009	−0.006	−0.008	−0.001
	S32	0.001	−0.018	−0.012	−0.017	−0.002
	S33	0.003	−0.047	−0.031	−0.045	−0.006
	S34	0.005	−0.101	−0.068	−0.096	−0.013

By analyzing the simulation of impacts caused under the scenarios of economic loss, environmental protection investment and pollution tax and referring to the results of impact simulation in the shipping industry, we can learn that the industry suffers a significant impact. In addition, shipping emissions have a certain impact on other transport sectors with strong correlation. However, the output share of shipping industry in the transportation system and the proportion of this industry in the whole economic system are small. Therefore, it has a slight impact on other industries. Both economic losses and pollution tax methods reduce the total output of shipping industry. However, environmental protection investment has increased the total output of shipping industry, causing a positive effect to the shipping development. See Table 10.8 for the output of shipping industry based on the shipping emissions and unit economic cost in the base period and the different impacts. We can learn that economic loss and pollution tax can reduce shipping emissions. Environmental protection investment contributes to a rising output but results in increased shipping emissions. Reduction of shipping emissions based on the reduction of production is not a favorable factor for the development of industry and economy. A reduction of production in a small scale cannot achieve the reduction of shipping emissions. In case of a drastic reduction of production, the industry development will be affected.

See Table 10.9 for the results of three types of impact. The impact of economic loss leads to lower demand for shipping, affecting input factor prices, sectoral output and prices. Furthermore, it affects the production, trade and entity modules, and social welfare and causes changes to the income consumption and savings of each entity. The exhaust gas from shipping emissions has a negative impact on economic development, leading to a downturn in the regional economy. When an impact equivalent to 100% of the direct economic loss occurs, GDP declines by 0.035% eventually.

Table 10.8 Changes in emissions and economic losses

Scenario		Emissions /tons	Change in Emissions /tons	Economic losses / billion yuan
Scenario 1 (economic loss)	S11	154772.97	−204.17	10.16
	S12	154576.44	−400.71	10.14
	S13	154387.38	−589.77	10.13
	S14	154205.49	−771.65	10.12
Scenario 2 (environmental protection investment)	S21	156922.57	1945.42	10.30
	S22	158868.15	3891.00	10.43
	S23	160303.13	5325.99	10.52
Scenario 3 (pollution tax)	S31	154853.24	−123.91	10.16
	S32	154727.46	−249.68	10.15
	S33	154337.66	−639.49	10.13
	S34	153642.85	−1334.30	10.08

Note Change in emissions = emissions in each scenario - emissions in the base period

Income decreases, resulting in a consequent reduction in consumption and savings. Government revenues decline by 0.057%, while consumption and savings go down by 0.036% and 0.052%, respectively. Residents' income decreases by 0.076% and consumption by 0.077%. Regional economic indexes show a downward trend, but compared with the impact caused by direct economic loss, the change of various indexes has slowed down, and the fixed capital indexes rise by 0.04%–0.218%. When the same amount of economic loss is used for fixed investment, the GDP loss is effectively mitigated. In the scenario S12 and S21, the change in GDP is 0.035% and – 0.0006% compared to the change in the scenario free of impact. The comparison shows that environmental protection investment is more conducive to the development of industry and economy. If pollution tax is levied to the products of shipping sector, the government revenue rises. When the tax rate is 10%, the government revenue rises by 0.848%, savings rise by 0.818% and consumptions rise by 0.221%. However, taxation inhibits industry development and reduces output. Meanwhile, the industry sees a decline in both income index and consumption index.

10.4.6 Robustness Test of the Results

The selection of key parameters (different elastic parameters) may affect the simulation results during the analysis of economic loss impact caused by shipping emissions to the economic and product sectors of Shanghai. Therefore, a robustness test is necessary for the parameters corresponding to the CGE model. (Mahmood & Marpaung, 2014) set the elastic parameters relating to the impact with high elasticity and low elasticity and simulated the impact using the new elastic parameters. Based on this method, they updated the elasticity values in the production function using the high elasticity value (increase by 20%) and low elasticity value (decrease by 20%), performed a sensitivity test to the new elasticity values and reanalyzed the value of economic loss caused by shipping pollutants. Table 10.10 shows the changes of each industry to the initial values.

As reflected by the sensitivity simulation analysis results of elastic parameters in the production function, the new elastic parameters in the total output function of each industry have a slight impact on each index of sectoral output. The total output of subsector is taken as an example and the value of alternative elastic parameter in the CES production function is increased by 20%. Then the variable ratio between the total output and the benchmark amount is −0.069%. The elastic parameters are decreased by 20% in the same way and the variable ratio between the total output and the benchmark amount is 0.103%. As can be seen from the changes in sectoral output, the CGE model constructed in this section uses alternative elastic parameter values to change the simulation results within an acceptable range, and its sensitivity is low, indicating that the model results are robust.

Table 10.9 Impact of regional socio-economic variables (%)

Scenario		Nominal GDP	Real GDP	Government Consumption	Government Savings	Government Revenue	Residents' Consumption	Resident's Income	Enterprise Income	Labor Income	Capital Revenue	Fixed Investment
Scenario 1 (economic loss)	S11	−0.006	−0.017	0.018	−0.026	−0.028	−0.038	−0.038	−0.047	−0.024	−0.026	0
	S12	−0.013	−0.035	0.036	−0.052	−0.057	−0.077	−0.076	−0.096	−0.048	−0.052	0
	S13	−0.019	−0.052	0.055	−0.079	−0.086	−0.116	−0.115	−0.145	−0.072	−0.078	0
	S14	−0.025	−0.07	0.074	−0.106	−0.115	−0.156	−0.155	−0.195	−0.097	−0.105	0
Scenario 2 (environmental protection investment)	S21	−0.00079	−0.00057	−0.0008	−0.003	−0.003	−0.07	−0.002	−0.003	0.00064	−0.001	0.04
	S22	−0.002	−0.001	−0.002	−0.006	−0.006	−0.141	−0.004	−0.005	0.001	−0.003	0.08
	S23	−0.004	−0.002	−0.002	−0.008	−0.008	−0.193	−0.005	−0.007	0.002	−0.004	0.109
Scenario 3 (pollution tax)	S31	0.011	−0.001	0.019	0.072	0.075	−0.003	−0.038	−0.047	−0.025	−0.026	0
	S32	0.023	−0.002	0.039	0.146	0.152	−0.006	−0.077	−0.096	−0.05	−0.052	0
	S33	0.06	−0.007	0.103	0.381	0.394	−0.017	−0.2	−0.251	−0.13	−0.135	0
	S34	0.129	−0.014	0.221	0.818	0.848	−0.038	−0.431	−0.539	−0.28	−0.291	0

Table 10.10 Sensitivity test of elasticity parameters of production function

Sector	Production function elasticity parameter	
Rate of change in output	High elasticity (20% increase)/%	Low elasticity (20% reduction)/%
Agriculture, forestry, animal husbandry and fishery	−0.050	0.074
Major manufacturing	−0.076	0.111
Shipping	−0.052	0.078
Air transportation	−0.292	0.044
Other transportation	−0.110	0.163
Financial services	−0.060	0.090
Technological innovation	−0.742	0.111
Real estate	−0.050	0.075
Major service industries	−0.057	0.086
Total output	−0.069	0.103

10.5 Conclusion and Prospect

Based on the LHR system, this paper simulates the impact of shipping emissions in Shanghai on sectoral output and regional economic variables. It also introduces the basic framework structure, core modules, data source, data processing, parameter solving. This paper also sets the simulated scenarios of regional economic model based on the monetized regional shipping emissions. A comparative analysis is performed to measure the output of different sectors and the impact of economic variables. The main conclusions are as follows:

1. The resulting social economic cost caused by the shipping emissions of Shanghai in 2019 is RMB 1.017 billion. The direct economic losses caused by SO_2, NO_x, $PM_{2.5}$ and PM_{10} are RMB 0.2587 billion, RMB 0.2776 billion, RMB 0.2057 billion and RMB 0.2304 billion respectively. At present, the emission control area of Shanghai only restrict the sulfur content of fuel rather than nitrogen content. The hazard caused by the climate change due to increased concentration of CO_2 is much higher than the impact on health, industry and buildings in the area. As a result, the social economic cost is underestimated.
2. Changes of sectoral and regional economy under different impacts Shipping suffers the greatest impact, followed by other transport sectors in the transportation system. Economic losses and pollution taxes lead to higher product prices, less sectoral output and lower shipping emissions. The impact of environmental protection investment increases sectoral input and output, resulting in an increase in shipping emissions.

3. A new equilibrium is established when the regional economic system is subjected to external impacts. In the new situation of equilibrium, more stable shipping emission factor and unit economic cost lead to smaller changes in shipping emissions and economic loss values. Therefore, the use of technological innovation, energy structure optimization and technological development supplemented by pollution taxes helps achieve a better emission reduction effect than the reduction of production.

This paper analyzes the impact caused by the negative externalities of shipping emissions on the regional economy, combines the microscopic shipping emission model with the economic model, and realizes an effective correlation between the two models. However, no in-depth studies are conducted on the parameters of technological innovation and the optimization of energy structure. In consideration of the above facts, the changes in shipping emission factor and unit economic loss will be taken into account in the follow-up studies to study the changes of regional economic indexes, thus helping work out the best solution for shipping emission reduction.

Acknowledgements Shaoli He and Jun Wang also made great contributions to this manuscript. We express our heartfelt thanks to them. This research also was supported by: National Natural Science Foundation of China (72474127) and National Social Science Fund Major Bidding Project of China (18ZDA052). Part of this chapter was published:He S, Wu X, Wang J. Comprehensive economic impact assessment of regional shipping emissions: A case study of Shanghai, China. Marine Pollution Bulletin 2025;211:117368. https://doi.org/10.1016/j.marpolbul.2024.117368. The publication has been conducted with copyright permission.

Appendix

The stratified random sampling method consists of 4 steps, including data collection and processing, stratified sampling model (sample ship selection), STEAM model (exhaust emission calculation) and emission inventory preparation.

Step 1: Data collection and processing. Coordinate the AIS data of the study area, test and handle abnormal values and add the missing data through interpolation. Use web crawlers to collect static information from the official websites of classification societies of different countries. Check the integrity of the collected ship information and supplement the missing values. At last, classify and match the AIS data and static ship information.

Step 2: Select sample ships and build a stratified sampling model. Perform three-layer sampling according to waterways, ship types and engine power characteristics in the study area. Finish sampling according to the density of ships in each route section of the study area, types of ships in each density layer and power clustering results from each type of ship. At last, determine the sample size of ships for the regional exhaust emission calculation.

Step 3: Build the STEAM model and calculate the exhaust emissions of every single ship. Determine the parameters required for the calculation according to the STEAM model. Determine the design ship speed, power of main engine and auxiliary engine, emission factor and load factor according to the area of study object, characteristic data of ships and the existing literature. Calculate the exhaust emissions of every single ship based on the above efforts.

Step 4: Prepare the shipping emission inventory. Use the sample estimation method to calculate the total shipping emissions in the area according to the sample ships in Step 2, and the emissions of every single ship in Step 3. Finally, prepare the emission inventories according to specific main engines, auxiliary engines, different ship types and months, and analyze the emission characteristics (Fig. 10.1).

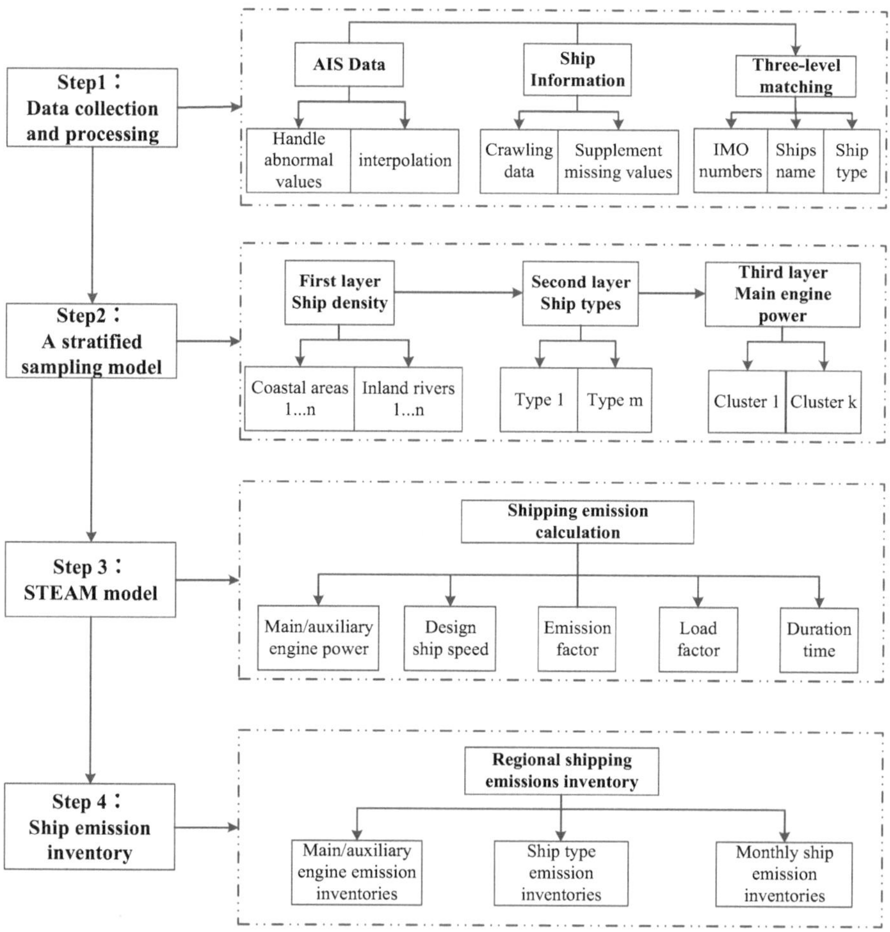

Fig. 10.1 Stratified sampling emission modeling

The emission of each type of pollutant from the ship's main engine and auxiliary engine were obtained separately based on the stratified sampling model calculations, as shown in Appendix Tables 10.11,10.12,10.13,10.14.

Table 10.11 Emission inventories by ship type (main engine emissions/t)

Ship type	CO_2	SO_2	NO_x	$PM_{2.5}$	PM_{10}	CO	VOC
Cargo ships	98265.84	1510.68	1870.36	172.65	215.81	201.42	71.94
Other ships	24045.42	369.66	457.67	42.25	52.81	49.29	17.60
Oil tankers	18468.14	283.92	351.52	32.45	40.56	37.86	13.52
Tugboats	739.99	11.38	14.09	1.30	1.63	1.52	0.54
Passenger ships	1899.82	29.21	36.16	3.34	4.17	3.89	1.39
Total	143419.20	2204.83	2729.79	251.99	314.98	293.98	104.99

Table 10.12 Emission inventories by month (main engine emissions/t))

Month	CO_2	SO_2	NO_x	$PM_{2.5}$	PM_{10}	CO	VOC
January	9273.34	142.56	176.51	16.29	20.37	19.01	6.79
February	6720.10	103.31	127.91	11.81	14.76	13.78	4.92
March	8364.68	128.59	159.21	14.70	18.37	17.15	6.12
April	6995.93	107.55	133.16	12.29	15.36	14.34	5.12
May	6906.19	106.17	131.45	12.13	15.17	14.16	5.06
June	7526.34	115.71	143.25	13.22	16.53	15.43	5.51
July	8350.13	128.37	158.93	14.67	18.34	17.12	6.11
August	18291.69	281.21	348.16	32.14	40.17	37.49	13.39
September	16037.56	246.55	305.25	28.18	35.22	32.87	11.74
October	18554.66	285.25	353.16	32.60	40.75	38.03	13.58
November	18407.12	282.98	350.36	32.34	40.43	37.73	13.48
December	17991.47	276.59	342.44	31.61	39.51	36.88	13.17
Total	143419.20	2204.83	2729.79	251.98	314.98	293.98	104.99

Table 10.13 Emission inventories by ship type (Auxiliary engine emissions/t)

Ship type	CO_2	SO_2	NO_x	$PM_{2.5}$	PM_{10}	CO	VOC
Cargo ships	3871.38	59.52	73.69	6.80	8.50	7.94	2.83
Other ships	937.31	14.41	17.84	1.65	2.06	1.92	0.69
Oil tankers	501.34	7.71	9.54	0.88	1.10	1.03	0.37
Tugboats	14.90	0.23	0.28	0.03	0.03	0.03	0.01
Passenger ships	108.89	1.67	2.07	0.19	0.24	0.22	0.08
Total	5433.83	83.54	103.43	9.55	11.93	11.14	3.98

Table 10.14 Emission inventories by month (auxiliary engine emissions/t)

Month	CO_2	SO_2	NO_x	$PM_{2.5}$	PM_{10}	CO	VOC
January	295.93	4.55	5.63	0.52	0.65	0.61	0.22
February	214.66	3.30	4.09	0.38	0.47	0.44	0.16
March	282.03	4.34	5.37	0.50	0.62	0.58	0.21
April	215.21	3.31	4.10	0.38	0.47	0.44	0.16
May	228.66	3.52	4.35	0.40	0.50	0.47	0.17
June	241.07	3.71	4.59	0.42	0.53	0.49	0.18
July	280.85	4.32	5.35	0.49	0.62	0.58	0.21
August	754.90	11.61	14.37	1.33	1.66	1.55	0.55
September	697.28	10.72	13.27	1.23	1.53	1.43	0.51
October	733.20	11.27	13.96	1.29	1.61	1.50	0.54
November	737.08	11.33	14.03	1.30	1.62	1.51	0.54
December	752.96	11.58	14.33	1.32	1.65	1.54	0.55
Total	5433.83	83.54	103.43	9.55	11.93	11.14	3.98

References

Bilgili, L., & Celebi, U. B. (2018). Developing a new green ship approach for flue gas emission estimation of bulk carriers. *Measurement, 120*, 121–127. https://doi.org/10.1016/j.measurement.2018.02.002

Bindle, L., Martin, R. V., Cooper, M. J., Lundgren, E. W., Eastham, S. D., Auer, B., & M.,…Weng, H. (2021). Grid-stretching capability for the GEOS-Chem 13.0.0 atmospheric chemistry model. *Geoscientific Model Development, 14*(10), 5977–5997. https://doi.org/10.5194/gmd-14-5977-2021

Castells, S. M., Santamaria, J. J. U., & Oses, F. X. M. D. (2014). Manoeuvring and hotelling external costs: Enough for alternative energy sources? *Maritime Policy & Management, 41*(1), 42–60. https://doi.org/10.1080/03088839.2013.782441

Chen, H., Zeng, W., Li, J., Ma, T., Liu, S., & Lei, G.,…Amit, K. S. (2022). Impact of air pollution on maize and wheat production. *Ecological Chemistry and Engineering S-Chemia I Inzynieria Ekologiczna S, 29*(2), 237–256. https://doi.org/10.2478/eces-2022-0018

Chen, J., Zhang, Y., & Shi, J. (2025). Cost-benefit analysis of ships NOx emission control areas (NECA) along China's coastal area. *Marine Pollution Bulletin, 215*, 117830. https://doi.org/10.1016/j.marpolbul.2025.117830

Dai, Y., Shi, X., & Deng, Q. (2025). Synergistic effects of CO2 and air pollutants from ship emissions in Shanghai, China: Spatial-temporal characteristics, prediction assessment, policy implications. *Journal of Environmental Management, 376*. https://doi.org/10.1016/j.jenvman.2025.124417

Dragovic, B., Tzannatos, E., Tselentis, V., Mestrovic, R., & Skuric, M. (2018). Ship emissions and their externalities in cruise ports. *Transport Research Part D Transport and Enivronment, 61*, 289–300. https://doi.org/10.1016/j.trd.2015.11.007

Ernestos, T. (2010). Ship emissions and their externalities for Greece. *Atmospheric Environment, 44*(18), 2194–2202. https://doi.org/10.1016/j.atmosenv.2010.03.018

Fu, S., Brian, V. V., & Zhang, P. (2021). Air pollution and manufacturing firm productivity: Nationwide estimates for China. *Economic Journal, 131*(640), 3241–3273. https://doi.org/10.1093/ej/ueab033

Gutenberg, P. (2014). Directorate-general for mobility and transport. *Handbook on the external costs of transport, 1-202.* https://doi.org/10.2832/27212

He, S., Wu, X., & Wang, J. (2023). A calculation algorithm for ship pollutant gas emissions and diffusions based on real-time meteorological conditions and its application. *Ocean Engineering, 287*, 1–12. https://doi.org/10.1016/j.oceaneng.2023.115825

He, S., Wu, X., Wang, J., & Guo, J. (2022). A ship emission diffusion model based on translation calculation and its application on Huangpu River in Shanghai. *Computers & Industrial Engineering, 172*, 1–15. https://doi.org/10.1016/j.cie.2022.108569

Huang, J., & Wen, C. (2022). The impact of private sector energy investment, innovation and energy consumption on China's carbon emissions. *Renewable Energy, 195*, 1291–1299. https://doi.org/10.1016/j.renene.2022.06.131

Ing-Marie, G., Andreas, B., & Annika, J. (2021). Air pollutants from shipping: Costs of NOx emissions to the Baltic Sea. *Journal of Environmental Management, 300*, 1–10. https://doi.org/10.1016/j.jenvman.2021.113824

Irena, K., Ernst, W., & Alexandros, C. G. (2021). The cost-effectiveness of CO2 mitigation measures for the decarbonisation of shipping. The case study of a globally operating ship-management company. *Journal of Cleaner Production, 316*, 1–20. https://doi.org/10.1016/j.jclepro.2021.128094

Joseph, B., & Po-Hsing, T. (2012). Estimating the environmental costs of port related emissions: The case of Kaohsiung. *Transportation Research Part D, 17*, 35–38.

Kim, A.-R., & Seo, Y.-J. (2019). The reduction of SOx emissions in the shipping industry: The case of Korean companies. *Marine Policy*, 98–106.

Li, X., Yao, X., Guo, Z., & Li, J. (2020). Employing the CGE model to analyze the impact of carbon tax revenue recycling schemes on employment in coal resource-based areas: Evidence from Shanxi [employing the CGE model to analyze the impact of carbon tax revenue recycling schemes on employment in coal resource-based areas: Evidence from Shanxi]. *The Science of the Total Environment, 720*(10), 1–14. https://doi.org/10.1016/j.scitotenv.2020.137192

Li, Y., Jia, P., Jiang, S., Li, H., Kuang, H., Hong, Y., et al. (2023). The climate impact of high seas shipping. *National Science Review, 10*(3), 1–15. https://doi.org/10.1093/nsr/nwac279

Li, Y., Xie, W., Yang, Y., Mei, Q., Wang, Z., Li, Z., & Wang, P. (2023). Research on the carbon emissions traceability inventory and multi-horizon prediction of ship carbon emissions: A case study of Tianjin port. *Frontiers in Marine Science, 10*, 1–19. https://doi.org/10.3389/fmars.2023.1174411

Lin, Z., Wang, P., Ren, S., & Zhao, D. (2023). Economic and environmental impacts of EVs promotion under the 2060 carbon neutrality target—A CGE based study in Shaanxi Province of China. *Applied Energy, 332*, 1–12. https://doi.org/10.1016/j.apenergy.2022.120501

Liu, H., Meng, Z., Shang, Y., Lv, Z., Jin, X., Fu, M., & He, K. (2018). Shipping emission forecasts and cost-benefit analysis of China ports and key regions' control. *Environmental Pollution, 236*, 49–59. https://doi.org/10.1016/j.envpol.2018.01.018

Liu, T., Wang, C., Wang, Y., Huang, L., Li, J., & Xie, F.,...Hu, J. (2020). Impacts of model resolution on predictions of air quality and associated health exposure in Nanjing, China. *Chemosphere, 249*, 1–11. https://doi.org/10.1016/j.chemosphere.2020.126515

Liu, Z., Lu, X., Feng, J., Fan, Q., Zhang, Y., & Yang, X. (2017). Influence of ship emissions on urban air quality: A comprehensive study using highly time-resolved online measurements and numerical simulation in Shanghai. *Environmental Science & Technology, 51*(1), 202–211. https://doi.org/10.1021/acs.est.6b03834

Mahmood, A., & Marpaung, C. O. P. (2014). Carbon pricing and energy efficiency improvement--why to miss the interaction for developing economies? An illustrative CGE based application to the Pakistan case. *Energy Policy, 67*, 87–103. https://doi.org/10.1016/j.enpol.2013.09.072.

Maragkogianni, A., & Papaefthimiou, S. (2015). Evaluating the social cost of cruise ships air emissions in major ports of Greece. *Transportation Research Part D Transport and Environment, 36*, 10–17. https://doi.org/10.1016/j.trd.2015.02.014

McArthur, D. P., & Osland, L. (2013). Ships in a city harbour: An economic valuation of atmospheric emissions. *Transportation Research Part D Transport and Environment, 21*, 47–52. https://doi.org/10.1016/j.trd.2013.02.004

Miluse, T., & Beatriz, T. (2015). Environmental cost and eco-efficiency from vessel emissions in Las Palmas port. *Transportation Research Part E, 83*, 126–140.

Natalie, M., Marie, W., & Mark, N. (2023). Health impact assessments of shipping and port-sourced air pollution on a global scale: A scoping literature review. *Environmental Research, 216*, 1–24. https://doi.org/10.1016/j.envres.2022.114460

Nunes, R. A. O., Alvim-Ferraz, M. C. M., Jalkanen, J.-P., Sousa, S. I. V., & Martins, F. G. (2025). Shipping emissions forecast in the Iberian Peninsula for 2050 impact on air quality, mortality and related costs. *Journal of Hazardous Materials, 497*, 1–13. https://doi.org/10.1016/j.jhazmat.2025.139622

Nunes, R. A. O., Alvim-Ferraz, M. C. M., Martins, F. G., Penuelas, A. L., Duran-Grados, V., & Moreno-Gutierrez, J.,…Sousa, S. I. V. (2021). Estimating the health and economic burden of shipping related air pollution in the Iberian Peninsula. *Environment International, 156*, 1–11. https://doi.org/10.1016/j.envint.2021.106763

Progiou, A. G., Bakeas, E., Evangelidou, E., Kontogiorgi, C., Lagkadinou, E., & Sebos, I. (2021). Air pollutant emissions from Piraeus port: External costs and air quality levels [air pollutant emissions from Piraeus port: External costs and air quality levels]. *Transportation Research Part D Transport and Environment, 91*(2017), 1–14. https://doi.org/10.1016/j.trd.2020.102586

Rosa, D., Julio, S.-C., & Cristina, S. (2018). Consumer-side actions in a low-carbon economy: A dynamic CGE analysis for Spain. *Energy Policy, 118*, 199–210. https://doi.org/10.1016/j.enpol.2018.03.065

Roy, A., & Chakraborty, M. (2025). A review of ship emissions impacts on environmental, health, societal impacts and IMO's mitigation policies. *Regional Studies in Marine Science, 81*, 1–18. https://doi.org/10.1016/j.rsma.2024.103964

Sharafian, A., Blomerus, P., & Mérida, W. (2019). Natural gas as a ship fuel: Assessment of greenhouse gas and air pollutant reduction potential. *Energy Policy, 131*, 332–346. https://doi.org/10.1016/j.enpol.2019.05.015

Sharnil, P., Reddy, G. T., Reddy, M. P. K., & Rohit, S. (2022). A study of the impacts of air pollution on the agricultural community and yield crops (Indian context). *Sustainability, 14*(20), 1–17. https://doi.org/10.3390/su142013098

Sheng, L., Qin, M., Li, L., Wang, C., Gong, K., & Liu, T.,…Hu, J. (2022). Impacts of emissions along the lower Yangtze River on air quality and public health in the Yangtze River delta, China. *Atmospheric Pollution Research, 13*, 1–9. https://doi.org/10.1016/j.apr.2022.101420

Sofiev, M., Winebrake, J. J., Johansson, L., Carr, E. W., Prank, M., & Soares, J.,…Corbett, J. J. (2018). Cleaner fuels for ships provide public health benefits with climate tradeoffs. *Nature Communications, 406*, 1–12. https://doi.org/10.1038/s41467-017-02774-9

Song, J., Liu, Z., Chen, R., & Leng, X. L. (2023). Cost calculation and allocation of regional atmospheric environment governance: A case of the Yangtze River Economic Belt. *Journal of Arid Land Resources and Environment, 37*(8),3–22.

Song, S. (2014). Ship emissions inventory, social cost and eco-efficiency in Shanghai Yangshan port. *Atmospheric Environment, 82*, 288–297.

Spengler, T., & Tovar, B. (2022). Environmental valuation of in-port shipping emissions per shipping sector on four Spanish ports. *Marine Pollution Bulletin, 178*, 1–12. https://doi.org/10.1016/j.marpolbul.2022.113589

Su, X., Li, Y., Fang, K., & Long, Y. (2022). Does China's direct investment in "belt and road initiative" countries decrease their carbon dioxide emissions? *Journal of Cleaner Production, 339*, 1–12. https://doi.org/10.1016/j.jclepro.2022.130543

Tan, L., Wu, X., Xu, Z., & Lic, L. (2019). Comprehensive economic loss assessment of disaster based on CGE model and IO model—A case study on Beijing "7.21 rainstorm". *International Journal of Disaster Risk Reduction, 39*, 1–11. https://doi.org/10.1016/j.ijdrr.2019.101246

Tovara, B., & Tichavska, M. (2019). Environmental cost and eco-efficiency from vessel emissions under diverse SOx regulatory frameworks: A special focus on passenger port hubs [environmental cost and eco-efficiency from vessel emissions under diverse SOx regulatory frameworks: A special focus on passenger port hubs]. *Transportation Research Part D, 69*, 1–12. https://doi.org/10.1016/j.trd.2019.01.019

Xia, F., Xing, J., Xu, J., & Pan, X. (2022). The short-term impact of air pollution on medical expenditures: Evidence from Beijing. *Journal of Environmental Economics and, Management(114)*, 2194–2202. https://doi.org/10.1016/j.jeem.2022.102680

Xu, J., Dong, X., Zhang, T., Liu, J., & Tao, S. (2022). Mitigation of air pollutant impacts on rice yields in China by sector. *Environmental Research Letters, 17*(5), 1–12. https://doi.org/10.1088/1748-9326/ac681c

Yang, L., Zhang, Q., Lv, Z., Zhang, Y., Yang, Z., & Fu, F. (2022a). Efficiency of DECA on ship emission and urban air quality: A case study of China port. *Journalj of Cleaner Production, 362*, 1–13.

Yang, L., Zhang, Q., Lv, Z., Zhang, Y., Yang, Z., & Fu, F.,...Mao, H. (2022b). Efficiency of DECA on ship emission and urban air quality: A case study of China port. *Journal of Cleaner Production, 362*, 1–13. https://doi.org/10.1016/j.jclepro.2022.132556

Yang, Q., Hu, X., Wang, Y., Liu, Y., & Liu, J. (2021). Comparison of the impact of China's railway investment and road investment on the economy and air pollution emissions. *Journal of Cleaner Production, 293*, 1–11. https://doi.org/10.1016/j.jclepro.2021.126100

Yin, S. (2022). Decadal changes in PM2.5-related health impacts in China from 1990 to 2019 and implications for current and future emission controls [decadal changes in PM2.5-related health impacts in China from 1990 to 2019 and implications for current and future emission controls]. *Science of the Total Environment, 834*, 1–13. https://doi.org/10.1016/j.scitotenv.2022.155334

Yoo, Y., Moon, B., & Kim, T.-G. (2022). Estimation of pollutant emissions and environmental costs caused by ships at port: A case study of Busan port. *Marine Science and Engineering, 10*(5), 1–21. https://doi.org/10.3390/jmse10050648

Ytreberg, E., Astrom, S., & Fridell, E. (2021). Valuating environmental impacts from ship emissions - the marine perspective. *Journal of Environmental Management, 282*, 1–10. https://doi.org/10.1016/j.jenvman.2021.111958

Zhai, J., Yu, G., Zhang, J., Shi, S., & Yuan, Y. (2023). Impact of ship emissions on air quality in the Greater Bay Area in China under the latest global marine fuel regulation. *Environmental Science & Technology, 57*(33), 12341–12350. https://doi.org/10.1021/acs.est.3c03950

Zhang, W., & Ma, X. (2020). Emission reduction effect and emission reduction cost of haze by environmental tax. *Journal of Beijing Institute of Technology (Social Sciences Edition), 22*(3), 36–47.

Zhang, W., Zhao, B., Gu, Y., Sharp, B., Xu, S., & Liou, K. (2020). Environmental impact of national and subnational carbon policies in China based on a multi-regional dynamic CGE model. *Journal of Environmental Management, 270*, 1–9. https://doi.org/10.1016/j.jenvman.2020.110901

Zhu, J., & Wang, J. (2021). The effects of fuel content regulation at ports on regional pollution and shipping industry. *Journal of Environmental Economics and Management, 106*, 1–28. https://doi.org/10.1016/j.jeem.2021.102424

Zincir, B., Deniz, C., & Tunér, M. (2019). Investigation of environmental, operational and economic performance of methanol partially premixed combustion atslowspeed operation of a marine engine. *Journal of Cleaner Production, 235*, 1006–1019.

Chapter 11
Aggregated Economic Damages and the Role of Dynamic Carbon Tax Under Climate Change: A Dynamic General Equilibrium Model

11.1 Introduction

The whole world is faced with the challenge of climate change.[1] It is necessary to conduct an in-depth study on how to assess economic damages from climate change reasonably and apply climate policies, such as carbon tax. Meanwhile, there are significant regional differences between countries, making it more difficult to assess economic damages and use climate policies. Considering IPCC (Intergovernmental Panel on Climate Change) Working Group II's report *Climate Change 2022: Impacts, Adaptation and Vulnerability*, the climate shocks are significantly heterogeneous on a global scale due to the differences in climate adaptation capabilities among regions. Specifically, impoverished, and hot countries are more vulnerable to climate change (Burke et al., 2015; Tol, 2018). According to the report *Economic Losses, Poverty and Disasters* (1998–2017) released by the *United Nations*, developing economies are about six times more likely to suffer losses from climate-related disasters than developed economies. How could regional economic damages from climate change be rationally analyzed? Moreover, carbon tax is one of the crucial measures to address climate change. There are now more than 30 carbon taxes around the world, and more countries are preparing to implement carbon tax. However, there are still many problems in the current global carbon pricing (carbon tax and emission trading system). According to the report *Global Carbon Pricing Development Trend 2022* released by the *World Bank*, the global carbon pricing only covered 23% of total carbon emissions in 2022 and was concentrated in developed regions (carbon prices were also too low). What are the emissions reductions and economic impacts of using carbon tax? Should the application of carbon taxes be expanded? Hence, a reasonable assessment of the regional economic damages from climate change and recognition of the role of carbon taxes are essential for climate policymakers. To achieve the above

[1] According to the *United Nations Framework Convention on Climate Change*, climate change here mainly refers to temperature changes caused by anthropogenic carbon emissions (also called temperature anomaly).

X. Wu, *Environmental Disaster in China*,
https://doi.org/10.1007/978-981-92-0910-1_11

research goals, this work constructed a multi-regional dynamic general equilibrium model to re-assess the regional economic damages (including the often-overlooked utility loss) and analyze the role of the carbon tax in each region. Our model will incorporate both production and utility damages from climate change and calibrate the parameters separately for each region to portray regional differences. The above considerations are still relatively rare in the existing literature (Barrage, 2020; Cruz & Rossi-Hansberg, 2023).

There are various studies assessing the regional economic damages of climate change (Burke et al., 2015; Cai et al., 2023; Nordhaus, 2013). When estimating the external impact of climate change on different regions, some research concerns the GDP losses, the endogenous correlation between energy consumption and climate change (Acemoglu et al., 2016; Golosov et al., 2014) and the uncertainty in the conversion of carbon stocks to climate change (Cai & Lontzek, 2019; Jensen & Traeger, 2021). However, the utility losses of households from climate change are often overlooked (Barrage, 2020; Cruz & Rossi-Hansberg, 2023), resulting in some estimate bias of regional economic damages. Drawing on the above literature and for measuring the regional economic damage differences in climate change, we constructed a dynamic general equilibrium model with multiple regions for quantifying regional climate shocks. Unlike the only consideration of GDP losses in Cai et al. (2023) and Golosov et al. (2014), the economic damages in our model consisted of GDP losses in production and utility losses of households. Besides, an attempt is made to calibrate the parameters separately for each region for charactering the regional economic differences.

Carbon tax is a major component of the carbon price mechanism and can encourage decision makers to promote the "zero carbon emission" of production activity, ultimately helping the achievement of "decarbonization". The carbon tax could be traced back to the Pigouvian tax, used to mitigate the negative externalities of climate change. Due to issues such as lower application and prices of carbon tax, it is an urgent need to theoretically analyze the emission reduction and economic impacts of carbon tax for aiding in promoting carbon tax and low-carbon development. Based on the multi-regional dynamic general equilibrium model in this paper, we discuss in detail the role of static carbon tax, dynamic carbon tax, and optimal dynamic carbon tax. We calculate the regional optimal dynamic carbon tax based on the social cost of carbon (considering both production and utility damage). This has assisted countries in recognizing the emission reduction and economic impacts of carbon tax and actively implementing carbon tax.

To achieve the research objectives, this paper first constructs a multi-regional dynamic general equilibrium model. Production damage and utility damage due to climate change are modeled. Drawing on Conte et al. (2021) and Xu and Deng (2024), the regionally differentiated impacts of climate change are determined by the share of carbon emissions, the size of labor force, and the temperature discount factor. Then, the model is calibrated by selecting the top 10 economies (based on GDP) of the world and using the *World Bank database*, the *IMF database*, the *BP Energy Statistics Yearbook* and the *CRU-TS database*. After calibration, we simulated future scenarios for 2020–2100 and analyzed the corresponding regional economic damages from climate change (both production and utility damage). The emission reduction and economic impacts of using static carbon tax, dynamic carbon tax and optimal regional carbon tax (based on the SCC, social cost of carbon) are analyzed. We also discussed the impact of carbon tax policies on temperature control goals.

The marginal contribution of this paper is as follows: (1) Different from the current literature discussing the production and utility damage caused by climate change respectively, this paper re-estimates the aggregated economic damages in 10 major economics by drawing on Barrage (2020) and considering both production and utility damage, thereby modifying the estimation bias of climate shocks in the existing literature. (2) This paper discusses in detail the different impacts of carbon taxes: static carbon tax, dynamic carbon tax pathway, and regional optimal dynamic carbon tax (based on regional SCC). This adds to the existing literature and improves understanding of carbon tax pathways, carbon tax implementation regions, and the economic impacts of carbon taxes.

The remaining parts are structured as follows: The second part is a literature review, the third part introduces the theoretical model framework, the fourth part introduces parameter calibration and variable estimation, the fifth part is simulation analysis, and a summary is provided in the last part.

11.2 Literature Review

This section presents the literature review, consisting of two main strands of research: the assessment of economic damages from climate change and the studies on carbon taxes.

The literature closely related to this study provides the assessment of economic damages from climate change. There are two main types of assessment: production damage and utility damage. For the production damage pathway, climate change directly causes a proportionate loss of economic output (Cai et al., 2023; Cai & Lontzek, 2019) or indirectly has a negative impact on production by reducing the production efficiency (Jensen & Traeger, 2021). This path highlights the impact on the production process. For example, Cruz and Rossi-Hansberg (2023) constructed a global spatial growth model at $1^\circ \times 1^\circ$ resolution to analyze the global losses caused by climate change. The results show that climate change causes significant global welfare losses. Waidelich et al. (2024) estimated the combined empirical dose-response damage functions among temperature change, rainfall, and extreme precipitation. The results indicated that a 3 °C temperature rise would cause a 10% loss in global GDP, with low-latitude regions suffering greater losses. van der Wijst et al. (2023) combined multiple IAMs (Integrated Assessment Models) to construct various composite damage functions. Results indicate that the aggregate economic losses from 3 °C temperature rise are significantly higher than those from 2 °C rise. When factoring in discount rates, biodiversity, and other considerations, the damage could be even greater. Regarding the path of utility damage, a temperature variable (or the environmental quality related to temperature change and carbon stock) is incorporated into the utility function for estimating economic damages on utility, emphasizing such factors as consumption, amenities, and overall welfare (Acemoglu et al., 2012; Zheng et al., 2016). For example, Barrage (2020) introduced a temperature variable in the utility function and analyzed the social cost of carbon (SCC) incorporating the utility damage from climate change. In most cases, the IAM (Integrated Assessment Model) and dynamic stochastic general equilibrium (DSGE) models under risk uncertainty are used to describe the above two impact paths. However, most of the studies have separated production and utility damages. This may underestimate the aggregated economic damages caused by climate change (Barrage, 2020). The multi-regional dynamic general equilibrium model constructed in this paper considers both the production and utility damages, enabling us to explore the compound regional impact of climate change.

Related studies on carbon tax could be divided into two categories: how to design optimal carbon tax and the economic impacts of carbon tax. For literature on designing carbon tax, Golosov et al. (2014) provided an important analysis. They provide an easy expression for calculating the optimal carbon tax (based on marginal damages from climate change). Their results show that the optimal carbon tax is positively related to GDP. Further, Cai and Lontzek (2019) and Van den Bremer and Van der Ploeg (2021) analyzed the social cost of carbon (SCC, an essential part for calculating optimal carbon tax) under economic and climate risks. Their results indicate that risk factors would enhance the value of SCC, thereby necessitating higher levels of carbon tax. Fikru et al. (2025) designed the optimal carbon tax via comparing tax and subsidy. They proposed linking the carbon tax to pollution intensity and the extent of damage. For the research on the impacts of carbon tax, Acemoglu et al. (2016) discussed the role of carbon tax and research subsidies. Their conclusions show that carbon taxes can promote innovation in clean technologies and low-carbon development. Desmet and Rossi-Hansberg (2015) analyzed the impact of energy tax (or carbon tax) in a multi-regional model. Their conclusions shown that energy tax would reduce carbon emissions, curb climate change, and enhance the total welfare. Timilsina et al. (2024) also concluded similar results with reducing carbon emission and rising welfare. Such literature mainly uses the DICE (Dynamic Integrated Model of Climate and Economy), CGE (Computable General Equilibrium) model, and dynamic general equilibrium (DGE) model to assess the economic impacts of carbon tax. Different from the previous literature, this paper provides a detailed discussion of the different scenes for the use of carbon taxes: static carbon tax, dynamic carbon tax pathways, and the role of regional optimal dynamic carbon tax (based on the social cost of carbon, considering both production and utility damages). This is essential for recognizing the role of carbon tax and promoting its implementation. See Fig. 11.1 for the study framework.

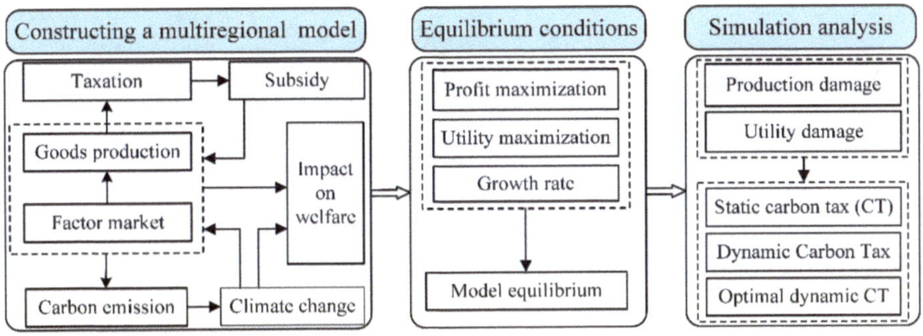

Fig. 11.1 Study framework

11.3 Theoretical Model

To reflect the regional heterogeneity of climate change, this part introduces a multi-regional general equilibrium model. The model has the following characteristics: (1) It contains two types of economies: developed economies and developing economies.[2] A single economy is represented by i and N indicates the total number of economies. t indicates a period; (2) According to Cruz and Rossi-Hansberg (2023), each economy needs to input traditional fossil energy (mainly including coal consumption) for production; (3) Production would increase the carbon emissions, thus resulting in climate change and causing negative externalities to the economic system.

For the regional variation in economies and climate shocks, we made the following considerations: (1) Drawing on the RICE (Regional Integrated Model of Climate and Economy) model of Nordhaus (2010), regional differences are reflected in differential economic variables, such as capital, labor, energy, and productivity. Moreover, the differential model parameters in each region reflect the heterogeneity of economies, such as differentiated factor input shares and consumption preferences. (2) Referring to Conte et al. (2021) and Xu and Deng (2024), we estimated regional climate shock factor to reflect the regionally differentiated impacts of climate change. It consists of the carbon emissions share of each region, the labor force size (Solomon et al., 2017), and the regional temperature discount factors (gap between the regional real temperature and the optimal production temperature). See Fig. 11.2 for the framework of the economy i in the model.

[2] It is classified by *World Economic Outlook update* released by IMF (International Monetary Fund) in 2024.

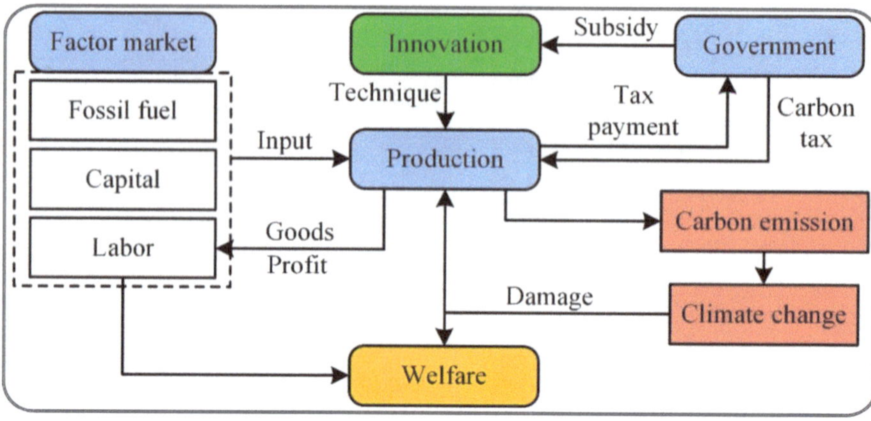

Fig. 11.2 Theoretical model framework

11.3.1 Technology and Production

Production technology. With reference to Acemoglu and Restrepo (2022), the final output production consists of a CES (Constant Elasticity of Substitution) type, which requires a variety of intermediate products. The technical characteristics are as follows:

$$Y_t = \left(\sum\nolimits_{i=1}^{N} Y_{it}^{\frac{\sigma-1}{\sigma}}\right)^{\frac{\sigma}{\sigma-1}}, \sigma > 1 \tag{11.1}$$

In Eq. (11.1), Y_{it} indicates the output of the economy i. σ indicates the elasticity of substitution. According to Cruz and Rossi-Hansberg (2023), output Y_{it} of each economy is a C-D technology that introduces energy consumption. Its expression is as follows:

$$Y_{it} = A_{it}\left[A_{it}^{E}E_{it}\right]^{\alpha_i}\left[\Psi_i L_{it}\right]^{\beta_i} K_{it}{}^{1-\alpha_i-\beta_i}, 0 < \alpha_i < 1, 0 < \beta_i < 1 \tag{11.2}$$

In Eq. (11.2), A_{it} indicates the technical level of the economy i at period t. A_{it}^{E} means the efficiency of traditional fossil energy.[3] E_{it} is the use of traditional fossil energy. K_{it} indicates the capital input. L_{it} is the labor input. α_i and β_i are the input share of traditional fossil energy and labor, respectively. Ψ_i is the rate of labor in production.

Production decisions. Based on Eq. (11.1), the profit maximization decision of the total output is: $\max P_t Y_t - \sum_{i=1}^{N} P_{Y,it} Y_{it}$. The demand function of the above problem is shown as: $Y_{it} = (P_{Y,it}/P_t)^{-\sigma} Y_t$. Where, P_t and $P_{Y,it}$ represent the price of the total output and the economy i. The total price index is shown as: $P_t =$

[3] Grossman and Helpman (1991) integrated two innovative modes in the dynamic equilibrium model, including "increasing product variety" and "improving product quality". Here, "improving product quality" is used to indicate the technological innovation mechanism of each economy.

$\left(\sum_{i=1}^{N} P_{Y,it}{}^{1-\sigma}\right)^{1/(1-\sigma)}$. Capital inputs require the payment of interest rate r, wage of labor input is w, and the factor cost of traditional fossil energy is the per unit $\eta(0 < \eta < 1)$ of Y_{it} (Bretschger & Valente, 2012). The optimization problem of Y_{it} is: $maxP_{Y,it}Y_{it} - rK_{it} - wL_{it} - \eta P_{Y,it}E_{it} - f$. f is the fixed innovation cost. For the Y_{it}, the demand of total output is competitive and there is no threshold cost on entering related market. The production market is purely competitive. Combining with Eq. (11.2), the first order condition for maximizing the profit of Y_{it} is: $\alpha_i Y_{it} = \eta E_{it}, \beta_i P_{Y,it}Y_{it} = wL_{it}, (1 - \alpha_i - \beta_i)P_{Y,it}Y_{it} = rK_{it}$.

11.3.2 R&D and Climate Change

R&D of each economy. Two types of efficiency $\{A_{it}, A_{it}^E\}$ are involved in each economy. New efficiency R&D depend on government support and the R&D capability of the domestic workforce. Referring to Pan et al. (2019) and Van den Bremer and Van der Ploeg (2021), the production efficiency is defined as follows:

$$\mathrm{A_{it}} = \mathrm{A_{it}^*}(1 - \mathrm{D_{it}}), \mathrm{A_{it}^E} = \mathrm{A_{it}^{E*}}(1 - \mathrm{D_{it}}) \tag{11.3}$$

In Eq. (11.3), $\{A_{it}^*, A_{it}^{E*}\}$ represents the efficiency decoupled from climate change. It means the technical level without the impact of carbon emissions and climate change. D_{it} is the ratio of efficiency losses due to climate change. This is generally referred to as production damage (Golosov et al., 2014). We define the following integral to denote the efficiency increment of A_{it}^*:

$$\dot{\mathrm{A}}_{\mathrm{mt}}^* = \int \Phi(\mathrm{G_{mt}})\mathrm{A_{mt}^*}(1 - \Psi_{\mathrm{m}})\mathrm{L_{mt}}\mathrm{f(a)da} \tag{11.4}$$

$$\dot{\mathrm{A}}_{\mathrm{nt}}^* = \int \Phi(\mathrm{G_{nt}})\left[(1 - \theta)\mathrm{A_{mt}^*} + \theta\mathrm{A_{nt}^*}\right](1 - \Psi_{\mathrm{n}})\mathrm{L_{nt}}\mathrm{f(a)da} \tag{11.5}$$

Equations (11.4) and (11.5) are the current R&D increment of developed economies $\dot{A}_{mt}^*$ and developing economies $\dot{A}_{nt}^*$, respectively. For developing economics, the technological innovation depends on A_{mt}^* and their own technologies A_{nt}^*. $\theta(0<\theta<1)$ is the rate of independent innovation in developing economies. In other words, developing economies implemented $1 - \theta$ ratio of simulated innovation form A_{mt}^* of developed economies and θ ratio of independent innovation from A_{nt}^*. With the technological superiority, developed economies depend on their own technical level A_{mt}^* for independent innovation. a is the innovation capability or efficiency of workforce. $f(a)$ indicates the probability density function of innovation capability, following Pareto distribution. $(1 - \Psi_m)L_{mt}$ and $(1 - \Psi_n)L_{nt}$ are the labor force in the economy for R&D activities. The R&D capability of workforce should be higher than the lower limit of innovation capability.[4] It is expressed as a_i^*. $\Phi(\cdot)$ is the contribution of government subsidies G_{it} to R&D.

The change of traditional fossil energy efficiency A_{it}^E is expressed as efficiency improvement. With reference to Acemoglu et al. (2016) on the transformation of clean production technologies, the efficiency of traditional fossil energy decoupled from climate change is dynamically set as: ${A_{it+1}^E}^* = (1+\mu_i)A_{it}^{E*}$. Where, $\mu_i(\mu_i > 0)$ is the energy efficiency improvement ratio of economy i in each period.

Production damage of climate change. It is also called as productivity damage rate (or GDP damage). With reference to the study of Golosov et al. (2014) and Van den Bremer and Van der Ploeg (2021), D_{it} in the production damage is expressed as:

$$\mathrm{D_{it}} = \rho_{\mathrm{it}} \log\left(\frac{\mathrm{O_t}}{\mathrm{S_P}}\right)\lambda_t {\chi_t}^{1+\theta_{\chi T}} \tag{11.6}$$

In Eq. (11.6), ρ_{it} indicates the differential impact on each economy. Based on the study of Tol (2018) and Xu and Deng (2024), the heterogeneous impact composed of regional carbon shares, labor force size and special temperature discount factor: $\rho_{it} = (1+\varpi_{it})^o x_{it} g(T_{it})$. ϖ_{it} and x_{it} are the heterogeneous labor scale and carbon shares of each economy. o is the elasticity of governance inputs. $g(T_{it})$ shows the difference between realistic and optimal temperatures for production (Conte et al., 2021). T_{it} is the temperature. $O_t = S_t - S_P$. O_t are the cumulative anthropogenic carbon emissions from pre-industrial period carbon stocks S_P to the moment t carbon stocks S_t. The current carbon emission increments for the economy i is: $\varphi_f E_{it} + \varphi_0(1-\varphi_f)E_{it}$. As supplementary information, φ_f is the share of permanent carbon emissions in the atmosphere, while $\varphi_0(1-\varphi_f)$ means the ratio of carbon emissions that will gradually disappear through natural attenuations. The dynamics of carbon emissions in economy i are: $O_{it+1} = O_{it} + \varphi_f E_{it+1} + \varphi_0(1-\varphi_f)E_{it+1} - \varphi O_{it}$ The total carbon emissions are: $O_t = \sum_{i=1}^N O_{it}$. The attenuation effect φ does not vary with natural conditions (Millar et al., 2017). χ_t is the uncertainty of climate sensitivity at a given level of carbon stock.[5] $\theta_{\chi T}$ is the combined effect of the above uncertainty and temperature changes on D_{it}. λ_t is the uncertainty of production damage at a certain temperature level.

11.3.3 Preferences and Welfare

Utility of residents. CRRA (Constant of Relative Risk Aversion) function is widely used in the preference of residents under climate change. The preference of residents

[4] The workforce can enter the R&D market only when the R&D capability is higher than a_i^*.

[5] According to the hypothesis of Van den Bremer and Van der Ploeg (2021), we can obtain the calculation results of χ_t and λ_t. It subjects to a normal distribution with mean value of $\{E_\chi, E_\lambda\}$ and standard deviation of $\{V_\chi, V_\lambda\}$. Where, $E_i = i_0 e^{-\xi_i t} + \bar{i}(1-e^{-\xi_i t})$, $V_i = \sigma_i\sqrt{(1-e^{-2\xi_i t})/2\xi_i}$, $i = \{\chi, \lambda\}$. Refer to the study of Van den Bremer and Van der Ploeg (2021) for the relevant parameters.

in the economy i is described as:

$$\mathrm{U_{it}} = \frac{\left[\prod_{\mathrm{j=1}}^{\mathrm{N}}\left(\mathrm{C_{it}^{j}}\right)^{\gamma_{\mathrm{j}}}\right]^{1-\phi_{\mathrm{i}}}}{1-\phi_{\mathrm{i}}}, \ \phi_{\mathrm{i}} \neq 1 \text{ and } 0 < \gamma_{\mathrm{j}} < 1 \tag{11.7}$$

In Eq. (11.7), ϕ_i is the CRRA of economy i. A greater value here corresponds to a higher degree of risk aversion. γ_j means the consumption preference for the goods produced by economy j. This constitutes a C-D preferences and meets $\sum_{j=1}^{N}\gamma_j = 1$. C_{it}^{j} is the consumption of i produced from economy j.

Resident decisions maximize the lifetime utility: $\max\int_0^{\infty} e^{-\rho t} U_{it} dt$ and the constraint of decision is: $\sum_j C_{it}^{j} + \dot{F}_{it} = rF_{it} + wL_{it} + \pi_{it}$. ρ is the discount factor. Output price $P_{Y,\,it}$ standardized to 1. F_{it} represents the final assets held by residents, r is the fully competitive market interest rate. wL_{it} and π_{it} are the total wage and corporate profits earned by residents. We constructed the current-value Hamiltonian function and solved the first order condition of consumption. Then, we obtained the consumption dynamic constraint for residents:

$$\mathrm{g_{C_{it}}^{j}} = \dot{\mathrm{C}}_{\mathrm{it}}^{\mathrm{j}}/\mathrm{C_{it}^{j}} = (\mathrm{r}-\rho)/\phi_{\mathrm{i}} \tag{11.8}$$

Where, $g_{C_{it}}^{j}$ is the consumption change rate of the economy i. According to Eq. (11.8), we obtained that: $\sum_j \gamma_j\left(\dot{C}_{it}^{j}/C_{it}^{j}\right) = (r-\rho)/\phi_i$.

Utility damage of climate change. Climate change impacts are reflected not only in production, but also in direct welfare of residents (Barrage, 2020). For example, adverse impacts on household goods, human capitals, and natural capitals. They are collectively referred to as marginal welfare damage. The key to this path is the setting of damage function. Here we define the damage function as: $\Lambda(\Delta T_t)$, satisfying $\Lambda(\Delta T_t)' > 0$. ΔT_t is the temperature change due to anthropogenic carbon emissions of Eq. (11.6). Then, we have the adjusted resident preference into the following form: $U_{it}^{'} = U_{it}(1-\Lambda(\Delta T_t))$. When $\Delta T_t = 0$, the preference is the initial scenario. Due to the absence of temperature change under such circumstance, $D_{it} = 0$. According to the production damage set by Eq. (11.6), there is no production damage. When $\Delta T_t \neq 0$, the resident utility damage is expressed as:

$$\Delta\mathrm{U}^{\Delta\mathrm{T_t}} = \mathrm{U_{it}} - \mathrm{U_{it}^{'}} = \mathrm{U_{it}}\Lambda(\Delta\mathrm{T_t}) \tag{11.9}$$

Eq. (11.9) describes the marginal welfare damage brought by climate change. According to Barrage (2020), damage function of utility defined as: $\Lambda(\Delta T_t) = \rho_{it}\left[\alpha_1(\Delta T_t)^2\right]^{(1-\phi_i)}/(1-\phi_i)$. It satisfies the classical quadratic damage function and α_1 is the marginal damage parameter inspired by Barrage (2020).

11.3.4 Government and Model Equilibrium

Government. The government of each economy mainly makes tax decisions and uses the tax for innovative R&D, etc. We suppose that the government imposes tax $\tau_i(0 < \tau_i < 1)$ on the production, the final manufacturer's optimization decision is: $max(1 - \tau_i)P_{Y,it}Y_{it} - rK_{it} - wL_{it} - \eta P_{Y,it}E_{it} - f$. The government tax revenue is: $Tax_{it} = \tau_i P_{Y,it}Y_{it}$. Under the above conditions, the demand for input factors is: $\alpha_i(1 - \tau_i)Y_{it} = \eta E_{it}, \beta_i(1 - \tau_i)P_{Y,it}Y_{it} = wL_{it}, (1 - \alpha_i - \beta_i)(1 - \tau_i)P_{Y,it}Y_{it} = rK_{it}$. Assuming that all the government tax revenues are used as the R&D subsidies G_{it+1} of the next year. It means that $G_{it+1} = Tax_{it}$.

Equilibrium analysis. When the model reaches equilibrium, the economy meets the following conditions: $\sum_{i=1}^{N} K_{it} + \sum_{i=1}^{N}\sum_{j} C_{it}^{j} = \sum_{i=1}^{N} Y_{it}, \sum_{i=1}^{N} \Psi_i L_{it} + (1 - \Psi_i)L_{it} = L_t, \sum_{i=1}^{N} K_{it} = \sum_{i=1}^{N} F_{it}$. The cumulative global capital and the cumulative consumption of residents are equal to total output in the current period. The equilibrium condition of financial markets means that the financial assets held by residents are equal to the total capital of financial markets. The sum of labor in the innovation and production sectors equals the total labor force. The equilibrium stability condition is shown in the Appendix.

After obtaining the equilibrium stability condition, the growth rate of each economy in the equilibrium state is solved. According to Eq. (11.8), the consumption growth rate is: $g_{C_{it}^{j}} = (r - \rho)/\phi_i$. As per Eq. (11.1), the growth rate of the total economy is: $g_t = \sum_{i=1}^{N} \Delta_{it} g_{it}$. $\Delta_{it} = Y_{it}^{\frac{\sigma-1}{\sigma}} / \sum_{i=1}^{N} Y_{it}^{\frac{\sigma-1}{\sigma}}$. It depends on the growth rate of each economy g_{it} and output share Δ_{it}. According to Eq. (11.2) and the FOC conditions of each economy, we obtained the equilibrium growth rate of each economy: $g_{it} = \dot{A}_{it}/A_{it} + \alpha_i\mu_i - \alpha_i[D_{it}/(1 - D_{it})]g_{D_{it}} + \alpha_i g_{E_{it}} + \beta_i g_{L_{it}} + (1 - \alpha_i - \beta_i)g_{K_{it}}$. It depends on the growth rate of productivity, efficiency of the energy μ_i, the growth rate of energy $g_{E_{it}}$, labor $g_{L_{it}}$, capital $g_{K_{it}}$, and the change rate of production damage $g_{D_{it}}$. $[D_{it}/(1 - D_{it})]g_{D_{it}}$ is the negative effect of climate change on the growth rate through the production damage. We can discover that climate change has a negative effect on the equilibrium growth rate g_{it} of each economy by inhibiting the growth rate of energy efficiency. The calculation of productivity growth rate $\dot{A}_{it}/A_{it}$ is shown in the Appendix.

Based on the above analysis, the equilibrium growth rate of developed and developing economies is:

$$\begin{aligned} g_{mt} =& \Phi(G_{mt})(1 - \Psi_m)L_{mt}\left(a_m^*/a_{\min}\right)^{-\delta} + \alpha_{\mathrm{m}}\mu_{\mathrm{m}} - (1 + \alpha_m)[D_{mt}/(1 - D_{mt})]g_{D_{mt}} \\ &+ \alpha_m g_{E_{mt}} + \beta_m g_{L_{mt}} + (1 - \alpha_m - \beta_m)g_{K_{mt}} \end{aligned} \tag{11.10}$$

$$\begin{aligned} nt =& \Phi(G_{nt})[\theta + (1 - \theta)\nabla^*](1 - \Psi_n)L_{nt}(a_n^*/a_{\min})^{-\delta} + \alpha_n\mu_n - \alpha_n[D_{nt}/(1 - D_{nt})]g_{D_{nt}} \\ &+ \alpha_n g_{E_{nt}} + \beta_n g_{L_{nt}} + (1 - \alpha_n - \beta_n)g_{K_{nt}} \end{aligned} \tag{11.11}$$

The difference between Eqs. (11.10) and (11.11) is reflected by the innovation and the heterogeneity between regions, including the share of inputs $\{\alpha_i, \beta_i\}$, differences in energy efficiency μ_i, differences in input growth rate $\{g_{E_{it}}, g_{L_{it}}, g_{K_{it}}\}$, differences in the lower limit a_i^* of R&D capabilities, and differences in production damage D_{it}.

Based on the dynamic of capital, the capital meets: $\dot{K}_{it} + [\delta^K - r/(1-\alpha_i-\beta_i)]K_{it} = -C_{it}$. Consumption meets: $\dot{C}_{it} = (r-\rho)C_{it}/\phi_i$. Here, $C_{it} = \sum_j C_{it}^j$. Solving this system of first-order ordinary differential linear equations and we have:

$$K_i^* - K_{it} = K_{i0}\exp\left[\left(\frac{r}{1-\alpha_i-\beta_i} - \delta^K\right)t\right] + \frac{C_{i0}}{\frac{r}{1-\alpha_i-\beta_i} - \delta^K - \frac{r-\rho}{\phi_i}}\exp\left(\frac{r-\rho}{\phi_i}t\right)$$

$$C_i^* - C_{it} = C_{i0}\exp\left(\frac{r-\rho}{\phi_i}t\right) \tag{11.12}$$

K_{i0} and C_{i0} is the initial value. Then. we obtained the consumption path for each economy: $C_{it}^j = C_{i0}^j\exp[(r-\rho)t/\phi_i]$. C_{i0}^j is the initial consumption state. Based on the above calculations, we could obtain: $\prod_{j=1}^N\left(C_{it}^j\right)^{\gamma_j}$. Considering the above conditions, the resident equilibrium utility in the current period could be determined: U_{it}^*.

When solving the equilibrium utility $U_{it}^{'}$, we need to calculate the temperature change ΔT_t caused by carbon stock. The temperature change is expressed as: $\Delta T_t = \log(O_t/S_P)\chi_t^{1+\theta_{\chi T}}$. ΔT_t means the temperature change level. When reviewing ΔT_t, we used the cumulative temperature change to indicate the long-term marginal welfare loss caused by climate change (Golosov et al., 2014). It is also noted as ΔT_t^{acc} to reflect the long-term marginal welfare loss with the temperature change $T_t - T_0$. Based on the resident equilibrium utility $U_{it}^{'}$, we can obtain the resident welfare of the economic system. It means $\partial\left(U_{it}^{'}\right)^*/\partial\Delta T_t < 0$, indicating the negative marginal utility damages from climate change.

11.4 Calibration and Variable Estimation

11.4.1 Parameter and Sample

In this section, we describe the calibration process for the main parameters in the model and the selected study sample.

Calibration of key parameters. The relevant parameters of the model are set as follows: (1) Based on Jaimovich and Rebelo (2017), parameter δ of the R&D capabilities following the Pareto distribution is set to 1.5. (2) According to Van den Bremer and Van der Ploeg (2021) for climate change damages, the temperature change uncertainty $\theta_{\chi T}$ is set to 3 for matching temperature anomalies. (3) The rate of simulation-based innovation in developing economies θ is calibrated to 0.5 (Pan

Table 11.1 Model parameter settings

Parameter	Value	Implication	Sources
o	0.74	Governance elasticity	Ye and An (2017)
$\theta_{\chi T}$	3	Temperature uncertainty impacts	Van den Bremer and Van der Ploeg (2021)
δ	1.5	R&D efficiency	Jaimovich and Rebelo (2017)
θ	0.5	Simulation innovation rate	Pan et al. (2019)
δ^K	0.1	Depreciation rate	Cai and Lontzek (2019)
ρ	0.1	Discounting rate	Burke et al. (2018)
σ	1.5	Elasticity of substitution	Fan et al. (2016)

et al., 2019). (4) According to Cai and Lontzek (2019) and Burke et al. (2018), the depreciation rate δ^K and the discounting rate ρ are set to 0.1. (5) Based on the study of Ye and An (2017), the governance elasticity o is set to 0.74. With reference to Fan et al. (2016), the elasticity of substitution σ is set to 1.5. See Table 11.1 for the summary of parameters.

Sample description. Based on the GDP of the world's major economies, we selected 10 economies, including the United States (US), China, Japan, Germany, India, the United Kingdom (UK), France, Italy, Canada, and South Korea. Other economies are recorded as "others". The coal consumption by the study objects accounted for 78.95% of the global and the economic output accounted for 67.79% in 2022. In this view, the study objects are typical. According to the *World Economic Outlook update* released by IMF in 2024, the developing economies include: China and India. Other economies are developed countries.

Calibration of tax rate τ_i. Based on the corporate income tax rate provided by *Trading Economics*, the tax rates paid by producers in our model are respectively: 0.21 of the United States, 0.25 of China, 0.3062 of Japan, 0.3 of Germany, 0.3494 of India, 0.25 of the United Kingdom, 0.25 of France, 0.24 of Italy, 0.265 of Canada, and 0.275 of South Korea.

Calibration of $\alpha_i, \beta_i, \gamma_j$ and ϕ_i. Based on the production decisions, the labor share β_i is obtained as: $\beta_i = average(\sum_t wL_{it}/Y_{it})$. Here, we take the average of 1960–2020 from the *World Bank database*. Further, according to the production decisions, the capital share $1 - \alpha_i - \beta_i$ is shown as: $1 - \alpha_i - \beta_i = average[\sum_t(\beta_i rK_{it})/(wL_{it})]$. Also, we take the annual mean (1960–2020). For the capital K_{it}, it is calculated by: $K_{it} = IK_{it}/(g_{IK} + \delta^K)$. IK_{it} is the formed capital and total saving, and g_{IK} means the growth rate of formed capital.[6] After above calculation, the energy share α_i could be determined. The results show that the average labor share β_i is 0.35 and the average capital share $1 - \alpha_i - \beta_i$ is 0.26. The estimation process of γ_j and ϕ_i are shown in the Appendix. It is shown that residents prefer the goods of developed economies and the average results of ϕ_i is 0.57.

[6] Here, r is set to 0.072 to match the setting of Van den Bremer and Van der Ploeg (2021) and reduce the underestimation of unformed capital.

11.4.2 Variable Estimation

In this section we describe the process of estimating the core variables $\{D_{it}, \Lambda(\Delta T_t), a_i^*\}$ in our model, as well as illustrating the essentials other estimates for quantitative analysis.

Estimation of production damage D_{it} and utility damage $\Lambda(\Delta T_t)$. For production damage, Eq. (11.6) defines the temperature uncertainty (χ_t) and damage uncertainty (λ_t). Where, χ_t and λ_t are of normal distribution and they are assumed to be independent of each other. With reference to Van den Bremer and Van der Ploeg (2021), we obtained the approximate expectation and variance of ${\chi_t}^{1+\theta_{\chi T}}$: $E\left({\chi_t}^{1+\theta_{\chi T}}\right) = E_\chi^{1+\theta_{\chi T}}\left(1 + 0.5 * \theta_{\chi T}\left(1 + \theta_{\chi T}\right)\left(\frac{\sqrt{V_\chi}}{E_\chi}\right)^2\right)$ and $Var\left({\chi_t}^{1+\theta_{\chi T}}\right) = \left(1 + \theta_{\chi T}\right)^2 E_\chi^{2\left(1+\theta_{\chi T}\right)}\left(\frac{\sqrt{V_\chi}}{E_\chi}\right)^2$. we ignored the infinitesimal of higher order in the results. The parameter setting of D_{it} and results of ρ_{it} are shown in Appendix. See Fig. 11.3 for the estimation results of D_{it}. Production damage (Figs. 11.3a and 11.3b) is gradually increasing, and all the countries are facing severe climate challenges. China suffered the most significant production damage, about 0.76% from 1990 to 2020 (Fig. 11.3b). Developing economies, such as China and India, are more vulnerable to climate change and have weaker climate risk resistance and post-sdisaster resilience.

For utility damage, parameter α_1 is set to 0.026% based on the mean result of Barrage (2020). Then, the calculation results of $\Lambda(\Delta T_t)$ is shown in Figs. 11.3 (Fig. 11.3c and 11.3d). The results are generally consistent with the production damage. In addition, to calculate T_{it} of each economy, we define that $T_{it} - T_{i1901} = ft_i(\Delta T_t) + \varepsilon_{it}$. ε_{it} is the error term and ft_i is the conversion parameter. Estimating the following problem $argmin\left[\sum_{t=1990}^{2020}\left(ft_i(\Delta T_t) - (T_{it} - T_{i1901})^2\right)\right]$, we obtain the results of ft_i that 1.81 of the United States, 1.76 of China, 2.83 of Japan, 3.14 of Germany, 1.04 of India, 1.89 of the United Kingdom, 3.45 of France, 2.93 of Italy, 1.86 of Canada, and 2.96 of South Korea. Data for the above estimation comes from the *CRU-TS database* and the results of D_{it}.

Estimation of a_m^* and a_n^*. First, the R&D subsidy function $\Phi(\bullet)$ is defined as: $\Phi(G_{it}) = \left(1 + \tau_i P_{Y,it-1} Y_{it-1}\right)^{1/\sigma} / Y_{it-1}$ according to Pan et al. (2019). Based on the condition of profit maximization, Growth rate of E_{it} meets $g_{E_{mt}} = g_{K_{mt}}$. According to the growth rate of productivity, the lower limit of innovation capacity a_i^* is:

$$a_m^* = a_{min}\left[\frac{g_{A_{mt}} + \left(\frac{D_{mt}}{1-D_{mt}}\right) g_{D_{mt}}}{\Phi(G_{mt})(1 - \Psi_m)L_{mt}}\right]^{-\frac{1}{\delta}} \tag{11.13}$$

Eq. (11.13) is the a_m^* of developed economies. Similarly, it is possible to calculate a_n^* for developing economies. Based on the data from the *World bank database*, the

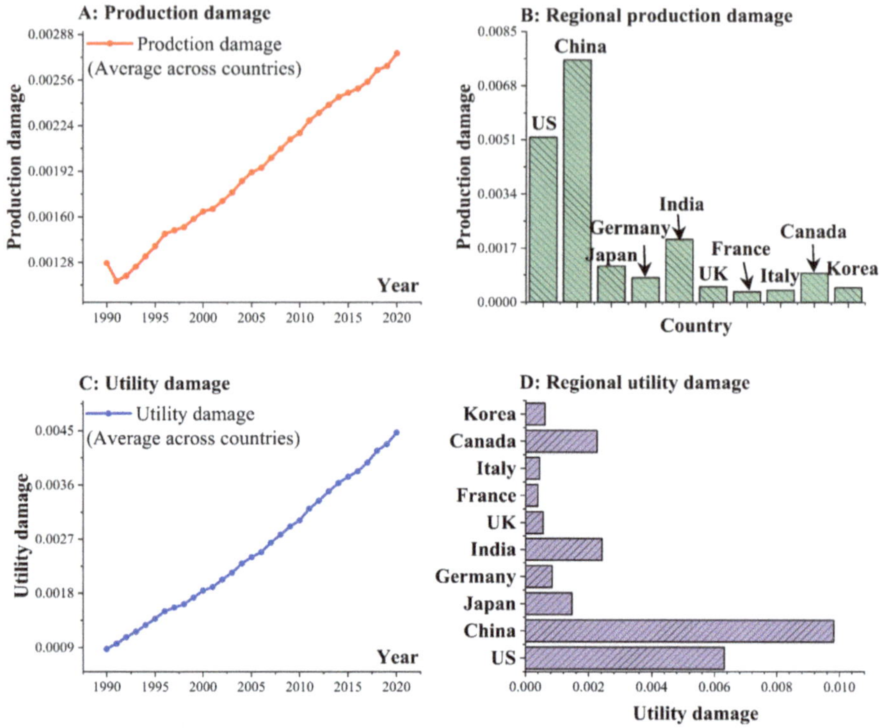

Fig. 11.3 Production damage and utility damage

mean results of $\{\Psi_i, L_{it}, g_{A_{it}}, Y_{it}\}$ from 1990 to 2020 are obtained. $g_{A_{mt}}$ is replaced by the growth rate of the scientific research population. The lowest innovation capability a_{min} is set to 1 (Jaimovich & Rebelo, 2017), the a_i^* of each economy is shown in Fig. 11.4. After considering the impact of climate change, the lower limit of workforces' R&D capabilities in developing economies is higher: $a_m^* < a_n^*$ (Fig. 11.4b). This may be due to the weak research base and high technological barriers in developing economies. Lower a_i^* in the US and many European countries, such as Germany, UK, France, Italy, and so on.

Description of other estimates. Before quantitative analysis, this section provides a description of other estimates. Using growth rates of energy consumption per output to characterize changes in energy efficiency μ_i. The relevant results are 3.87% of the United States, 6.09% of China, 2.97% of Japan, 4.1% of Germany, 3.42% of India, 5.28% of the United Kingdom, 3.45% of France, 2.83% of Italy, 3.11% of Canada, and 2.32% of South Korea. The growth rate of labor is exogenous, and the relevant results are 0.77% of the United States, 0.32% of China, 0.69% of Japan, 0.8% of Germany, 1.01% of India, 0.65% of the United Kingdom, 0.77% of France, 0.53% of Italy, 1.25% of Canada, and 1.67% of South Korea. The above data comes from the *World Bank database* (1990–2020). To get the total energy consumption, the optimal

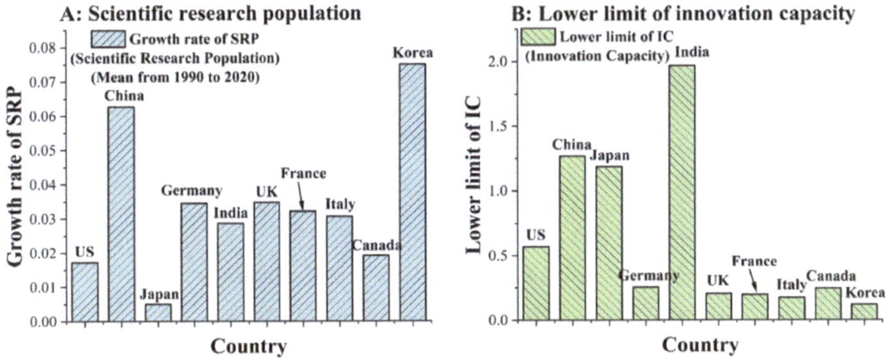

Fig. 11.4 Lower limit of innovation capacity a_i^*

solution for $\eta = argmin\sum_t\left[\sum_i^N \alpha_i Y_{it}/\eta - \sum_i^N E_{it}\right]^2$ is computed. The result of η is 0.11. Data of estimating η is supported by the *World Bank database* from 1970 to 2020. Interest rate r is set to 0.01 for meeting conditions $r<\rho$ and $r/(1-\alpha_i-\beta_i)<\delta^K$. The model sets a simulation period of two years. The model data and variables are described in Table 11.2 below.

11.5 Simulation Analysis

This section will discuss the economic damages of climate change on major global economies and analyze the emission reduction and economic effects of carbon tax policies.

11.5.1 Base Scenario

Carbon stock and damage from climate change. The benchmark scenario under climate change is simulated at first. See Fig. 11.5 for the simulation results. Global carbon emissions and carbon stocks continue to grow because of the failure to implement climate policies (Fig. 11.5a). By 2100, carbon emissions reach 33.01GtC and the carbon stocks reach 1531.14GtC. With the accumulation of carbon stocks, production damage, utility damage, and temperature anomalies due to climate change progressively increase (Figs. 11.5b and 11.5c). By 2100, the temperature anomaly reaches 4.38 °C, with production damage of 8.62% and utility damage of 1.25%. Based on Figs. 11.5d to 11.5f, the mean production damage due to climate change grows from \$0.032 trillion in 2020 to \$0.19 trillion in 2100. The mean utility damage grows from \$0.004 trillion in 2020 to \$0.014 trillion in 2100. If no actions are taken to address

Table 11.2 Parameters and Variables

	Implication/description
Panel A: Parameter	
δ is 1.5	R&D efficiency from Jaimovich and Rebelo (2017)
θ is 0.5	Simulation innovation rate of developing countries from Pan et al. (2019)
δ^K is 0.1	Depreciation rate of capital from Cai and Lontzek (2019)
ρ is 0.1	Discounting rate of utility from Burke et al. (2018)
σ is 1.5	Elasticity of substitution of production from Fan et al. (2016)
$\theta_{\chi T}$ is 3	Temperature uncertainty impacts from Van den Bremer and Van der Ploeg (2021)
o is 0.74	Governance elasticity from Ye and An (2017)
r is 0.01	Interest rate for matching equilibrium conditions
α_1	Impact of temperature anomalies on utility from Barrage (2020)
τ_i	Corporate income tax rate of each economy from the *trading economics*
α_i and β_i	Energy share and labor share of production from the *World Bank database*
γ_j and ϕ_i	Consumption preferences and CRRA from the *IMF database*
Panel B: Variable and description	
φ_f and φ_0	Parameters of carbon emissions from Golosov et al. (2014) and the *BP Energy Statistics Yearbook 2021*
φ is 0.0037	Parameter of carbon stock for matching carbon stock in 2015 (854GtC)
ϖ_{it} and x_{it}	Variables of ρ_{it} from the *World Bank database* and the *BP Energy Statistics Yearbook 2021*
T_{it} and ς_{it}	Variables for calculating ρ_{it} from the *CRU-TS database* and the *World Bank* database
D_{it}	Production damage calculated by authors
$\Lambda(\Delta T_t)$	Utility damage calculated by authors
ft_i	Conversion parameter for temperature anomaly of each economy from the *CRU-TS database*
K_{i0}	Initial value of capital from the *World Bank* database
C_{i0}	Initial value of consumption from the *IMF database* (trade expenditures)
a_i^*	Lower limit of innovation capacity calculated by authors from the *World Bank* database
μ_i	Grow rate of energy efficiency from the *World Bank* database
$g_{L_{it}}$	Exogenous population growth rate from the *World Bank* database
η	Conversion parameter for energy consumption from *the World Bank database*
t	A simulation period of two years (initial period 2019, simulation period 2020–2100)
Panel C: Classification	
Developed	Divided by the *world economic outlook update* released by IMF in 2024
Developing	
Sample	US, China, Japan, Germany, India, the UK, France, Italy, Canada, and south Korea

climate change, all economies around the world will suffer serious economic and welfare losses.

For regional differences, the largest production and utility damages are in the US and China (Figs. 11.5d and 11.5f). The possible reasons for this are as follows: (1) China and the US have high economic output and large populations. The economic activities and labor force are more easily exposed to the impacts of climate change. (2) China and the US are vast and have significant differences in domestic development. This exacerbates the difficulty of climate governance. (3) China and the US are large carbon emitters, with both countries accounting for more than 40% of global carbon emissions in 2020 (*BP Energy Statistics Yearbook 2021*), corresponding to higher economic and utility damages (Rickels et al., 2023). The enormous utility damage in China may lead to a widening relative gap in welfare between developing and developed economies, exacerbating global inequalities (Burke et al., 2015). The possible reasons are as follows: (1) Developing economies is limited in the scale of infrastructure, the utilization rate, and benefit rate per capita. Then, the impact of climate change is more significant; (2) Developing economies have poor climate adaptability, insufficient infrastructure "resilience", and large fluctuations due to climate change.

Turning point of GDP growth rate. The negative effect of climate change on economy is analyzed in the previous section. Does climate change lead to negative GDP growth rates? What are the turning conditions for negative GDP growth rates? We discuss this part. Letting g_{mt} and g_{nt} of Eqs. (11.10) and (11.11) equal to 0, the turning conditions of $g_{mt} = g_{nt} = 0$ is obtained (a combination of D_{it} and $g_{D_{it}}$). Using the data of year 2019 as an example, the turning point conditions of production damage D_{it} and the change rate of D_{it} is reported in Fig. 11.6. The results showed a negative relationship. As D_{it} or $g_{D_{it}}$ increases, the value of $g_{D_{it}}$ or D_{it} gradually decreases. According to Fig. 11.6-A, when production damage is 0.1, the turning point condition needs to be satisfied that $g_{D_{it}}$ is 16.8%. This is much higher than the average growth rate 1.61% of $g_{D_{it}}$ from 2020 to 2100. Based on the Fig. 11.6-B, when $g_{D_{it}}$ is 0.1, the corresponding turning point condition meets that D_{it} is 0.156. This is much higher than the production damage rate of 8.62% in 2100. Despite the existence of a turning point of $g_{it} = 0$, the damage rate D_{it} and change rate of D_{it} required to achieve this condition are demanding. The above conclusion suggests that all economies should actively implement policies to curb climate change and reduce the negative impact caused by climate change. Therefore, how to effectively mitigate climate change and reduce the production and utility damage is a problem that needs urgent solutions.

11.5.2 Application of Static Carbon Tax

Simulation of static carbon tax. This section first discusses the policy effect of static carbon tax. A static carbon tax means that all economies have the same carbon tax, and it does not change over time. After using a static carbon tax, the price of

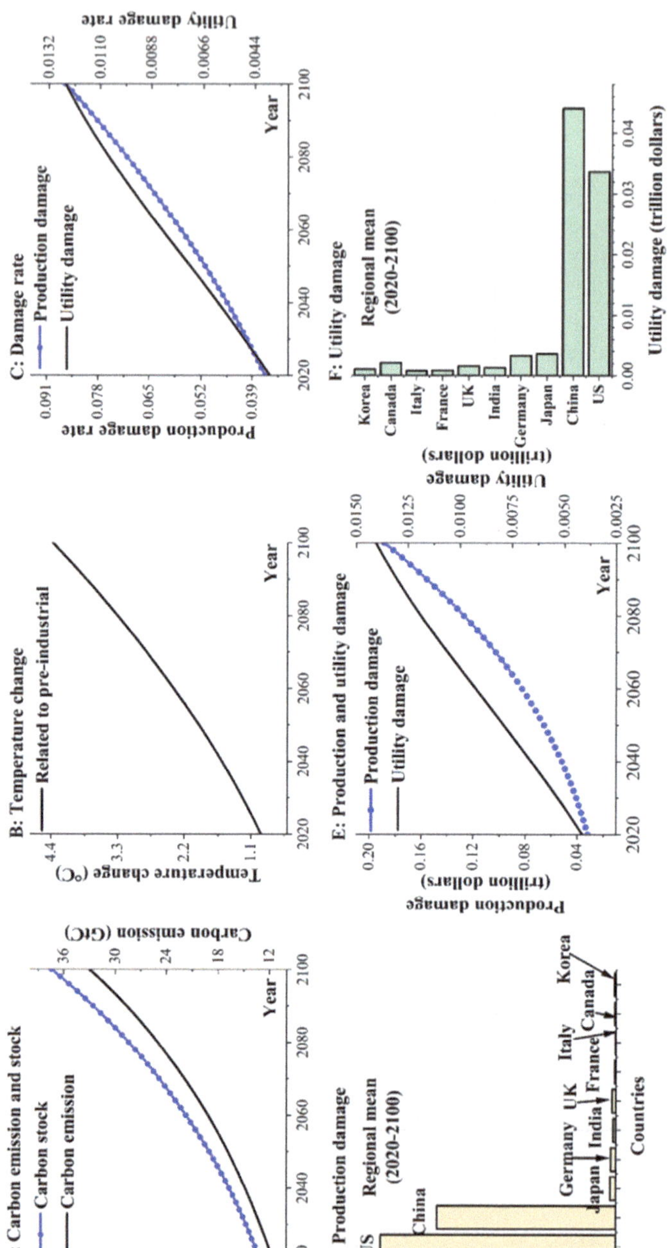

Fig. 11.5 Benchmark scenario from 2020 to 2100 *Note* All image data in this part are obtained based on model calculations and all images are drawn by using Origin software

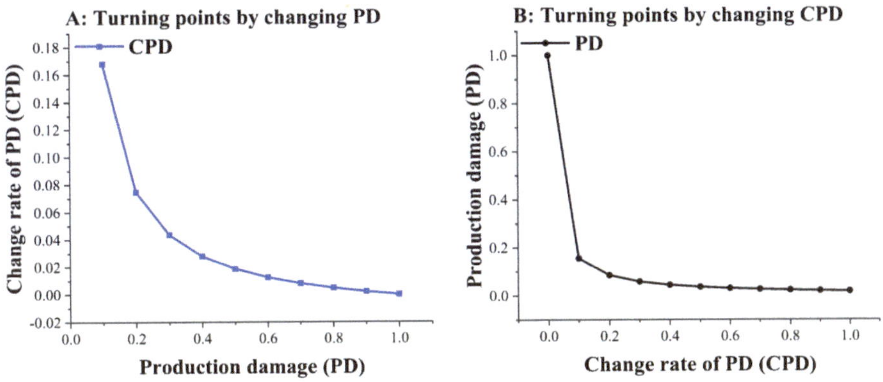

Fig. 11.6 Turning point of GDP growth rate (year of 2019)

coal consumption is: $(1 + \tau^E)\eta P_{Y,\,it}$. τ^E is the rate of static carbon tax. Then, the new condition of coal consumption is: $\alpha_i Y_{it} = (1 + \tau^E)\eta E_{it}$. Considering three scenarios: (1) τ^E is set to 10%; (2) τ^E is 100%; (3) τ^E is 100% only in developed economies. Related results are reported in Fig. 11.7. According to Fig. 11.7a, the total carbon emission gradually decreases. The emission reduction effect of implementing static carbon tax τ^E gradually increases over time. In addition, when static carbon tax is implemented only in developed economies, it does not have the desired effect of reducing emissions. Due to the public goods nature of climate change, the regions without static carbon tax may turn into a "free rider" (Nordhaus, 2015). It means that climate change is a global crisis, and all economies need to shoulder the corresponding responsibility. As the carbon emissions of using static carbon tax decline, the corresponding production and utility damage are significantly lower (Figs. 11.7b and 11.7c).

Figure 11.7d present the impact of using static carbon tax on utility (scene of τ^E is 100%). The results show that the use of static carbon tax resulted in utility gains for all economies, but increased utility gains vary significantly across regions. Of these, China and the US gained the most utility gain. The possible reasons are as follows: (1) Earnings factor. The static carbon tax drives the production to reduce the use of traditional fossil energy. Further, carbon emissions and carbon stocks are reduced, the rate of production and utility damage decreases. Then, the utility gains are realized. (2) Loss factor. Taking the sector-specific temperature discount factor $g(T_{it})$ as an example. Due to regional differences, warming may be beneficial to productive activities in some areas (Cruz & Rossi-Hansberg, 2023). When using a static carbon tax, warming is suppressed. This may result in an increase in the deviation of regional temperatures from optimal production temperatures, leading to an increase in utility damage. Despite these factors, the total benefits of applying a static carbon tax are clear. The Appendix also reported the economic impacts of using carbon tax, such as economic growth and income distribution.

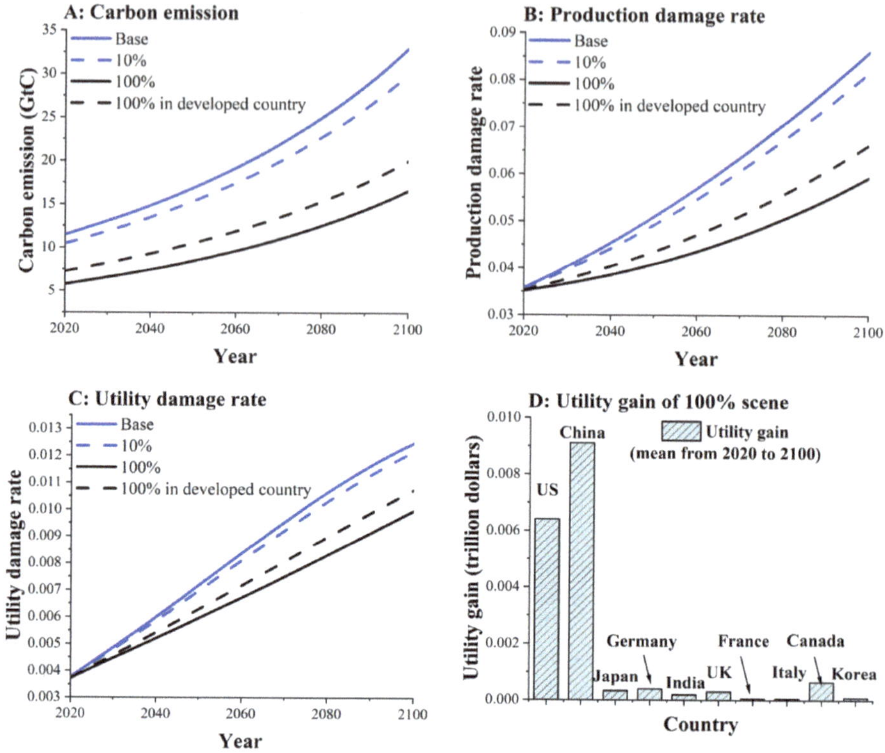

Fig. 11.7 Impact of static carbon tax

11.5.3 Application of Dynamic Carbon Tax

Setting of dynamic carbon tax. According to the *State and Trends of Carbon Pricing 2021*, carbon prices are currently low in most economies and should be increased in the future. Therefore, the dynamic carbon tax is considered in this section. A dynamic carbon tax indicates that the carbon tax increases over time but is consistent across regions. Based on Wu et al. (2021), three dynamic paths are assumed, including the environmental path (fast first and slow then, EP), neutral path (constant growth, NP), and development path (slow first and fast then, DP). Dynamic carbon tax τ_t^E grows from 0 to 100% under the three paths described above. For the above three paths, $\tau_t^E = \ln\left(a_1^E t + 1\right)$ for EP, $\tau_t^E = a_2^E t$ for NP, and $\tau_t^E = \exp\left(a_3^E t\right) - 1$ for DP. Parameters $\left\{a_1^E, a_2^E, a_3^E\right\}$ is matched to the dynamic carbon tax value.

Simulation of dynamic carbon tax. The above simulation results are reported in Fig. 11.8. Observing Fig. 11.8a, the difference between the use of dynamic carbon tax and static carbon tax is very clear. The dynamic carbon tax starts from the base case and gradually grows to the result of using a 100% static carbon tax. Such changes give economic agents the time and space to adjust their decisions, with correspondingly

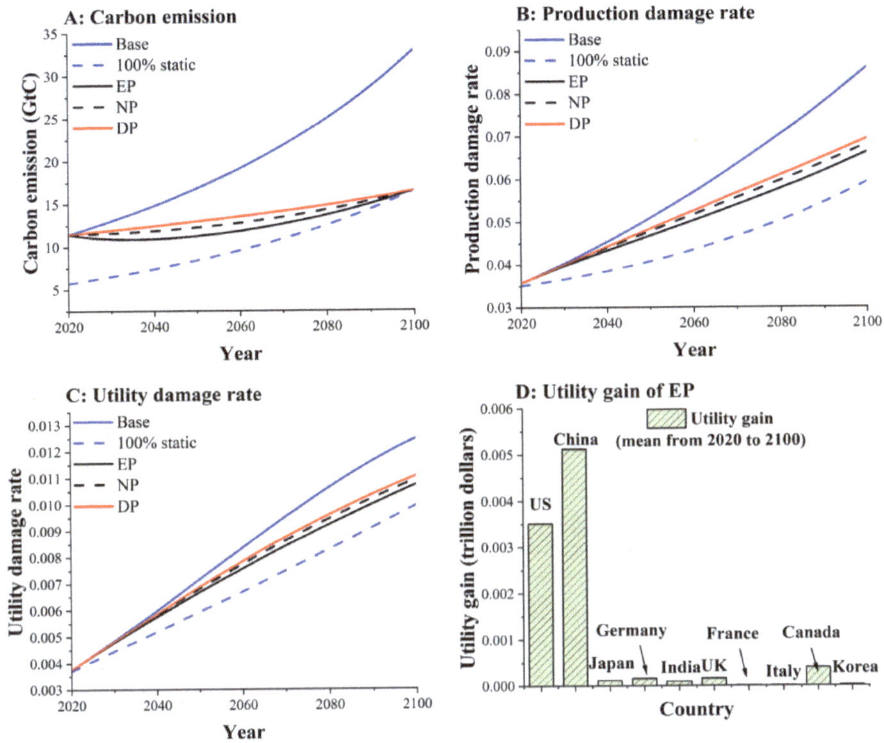

Fig. 11.8 Impact of dynamic carbon tax
Note 100% static is the scene (2) in static carbon tax. EP is environmental path, NP is neutral path, DP is development path. EP, NP, and DP increase from 0 to 100%

less economic volatility. Comparing the above three paths, the EP contributes the most significant carbon reduction and damage decline effect. This situation inspires government departments not to persevere in the principle of "treatment after development". Government should pay attention to environmental issues in the early stage of development and release the expected signals to decision makers, forcing them to accelerate technological innovation for carbon emission reduction. Results of Fig. 11.8d show that the dynamic carbon tax meeting EP results in utility gains for all economies. This is consistent with the use of 100% static carbon tax. More effectively, dynamic carbon tax enables economic agents to adjust their production decisions in a timely manner, reducing the impact of loss factor.

11.5.4 Application of Optimal Dynamic Carbon Tax

Calculation of optimal dynamic carbon tax. In the previous analysis, the use of carbon tax ignores the regional differences. In this section, the optimal dynamic carbon tax of each economy would be calculated. Our model formulates an optimal dynamic carbon tax based on the social cost of carbon (SCC) faced by each economy. Inspired by Burrage (2020), both production and utility damage are considered. The corresponding SCC is expressed as follows according to Rezai and Van der Ploeg (2016):

$$\mathrm{SCC_{it}} = \sum\nolimits_{t=0}^{\infty} (1+\rho)^{-t} \Upsilon \xi_t \left[\underbrace{\mathrm{Y_{it}^{\varrho} Y_{i0}^{1-\varrho}}}_{\text{Production damage}} + \underbrace{\mathrm{U_{it}^{\varrho} U_{i0}^{1-\varrho}}}_{\text{Utility damage}} \right] \tag{11.14}$$

Eq. (11.14) means that the present value of damage per unit of carbon emissions in period t. Parameter ρ is the discounting rate. Υ is the marginal production and utility damage caused by carbon emissions per unit. It is calculated by the ratio of carbon stocks to total damage (production and utility). Υ is set to 0.187% based on the simulation from 2020 to 2100. ξ_t is the remainder factor of carbon emissions. It is shown as: $\xi_t = \sum_s^t \phi^C \left[\varphi_f + \varphi_0 \left(1 - \varphi_f \right) \right] (1 - \varphi)^{t-s}$. This represents the remainder share of carbon emissions in period t caused by the carbon emissions in period s. ϕ^C is the conversion parameter between GDP and carbon emissions. It is calculated by $\sum_t \sum_i Y_{it} / \sum_t \sum_i EL_{it}$ and set to 1.993 according to the simulation from 2020 to 2100. ϱ denotes the trade-off between initial GDP and current GDP. The results of Golosov et al. (2014) indicated that SCC is increasing gradually with GDP. Therefore, we set ϱ to 1. Considering three scenarios: (1) ϱ is 1, production and utility damage are both considered; (2) ϱ is 1, only production damage is considered; (3) ϱ is 0.5, production and utility damage are both considered. See Appendix for SCC calibration details.

The results are reported in Fig. 11.9. The results showed a gradual increase in SCC. Neglecting utility damage and biasing initial GDP may lead to underestimation of SCC (Fig. 11.9a). By 2100, the average SCC for the scene (1) reaches 436.6 \$/ton. Although low, there are significant regional variations (Fig. 11.9b). Of these, the US and China face greater SCC. What would be the effect of using an optimal dynamic carbon tax based on SCC, and would countries with low SCC exhibit the "free rider" behavior that Nordhaus (2019) worries about?

Simulation of optimal dynamic carbon tax. Using the coal price data of initial year 2019 provided by the *BP Energy Statistics Yearbook 2021*, the global average coal prices are obtained. On this basis, the ratio between the scene (1) result of SCC and the average coal price is computed. This ratio is defined as the optimal carbon tax τ_{it}^E based on the SCC (a composite of production and utility damage). Energy tax rates τ_{it}^E based on SCC vary both over time and regionally. The related results are reported in Fig. 11.10.

The optimal dynamic carbon tax significantly dampens the growth trend in carbon emissions, only 1.39GtC from 2020–2100 (Fig. 11.10a). Correspondingly, the optimal dynamic carbon tax has lower production and utility damage ratios (Fig. 11.10b and 11.10c). This contributes significantly to global low-carbon development. Figure 11.10d compares the utility gain from the optimal dynamic carbon tax τ_{it}^E with two other scenarios. The results show the highest utility gains from the

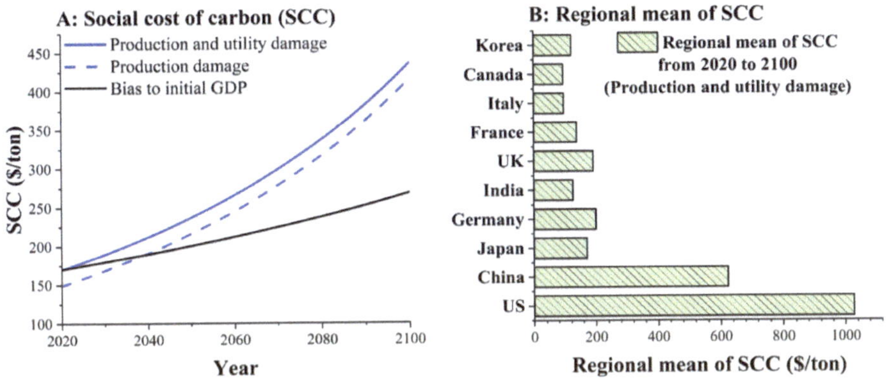

Fig. 11.9 Social cost of carbon (both production and utility damage)

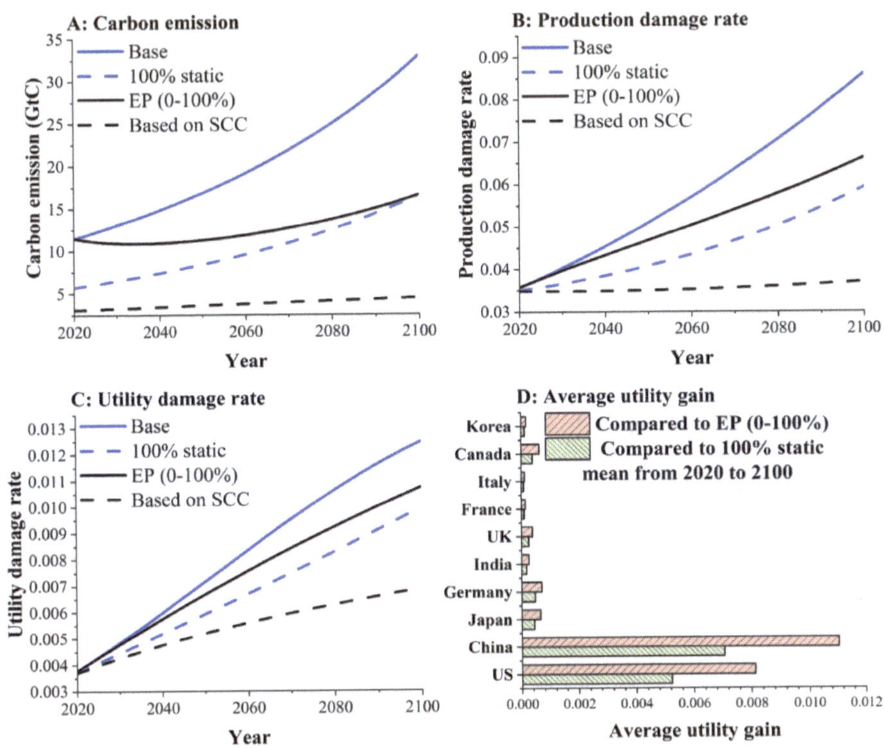

Fig. 11.10 Effect of regional optimal dynamic carbon tax
Note "Base" is baseline model, "100% static" is scene (2) of using static carbon tax, "EP (0–100%)" is scene (1) of using dynamic carbon tax, "Based on SCC" is the scene of optimal dynamic carbon tax

Table 11.3 Temperature anomaly (TA) of simulation

	TA of 2050	TA of 2100	2020 to 2050	2050 to 2100
Base	1.96 °C	4.38 °C	1.03 °C	2.41 °C
Static carbon tax (100%)	1.34 °C	3.01 °C	0.65 °C	1.44 °C
Dynamic CT with EP	1.80 °C	3.36 °C	0.87 °C	1.56 °C
Optimal dynamic CT based on SCC	1.35 °C	1.88 °C	0.44 °C	0.53 °C

Note CT is carbon tax; EP is the environmental path; SCC is the social cost of carbon

optimal dynamic carbon tax. The utility gains from using a regionally differentiated optimal dynamic carbon tax are clear, despite the risk of free riding by economies with low carbon prices. This may suggest that it is worthwhile to take this risk when differences in economic factors across countries and higher utility gains are considered. Meanwhile, the possibility of low-carbon priced regions becoming "carbon havens" is reduced if regions use a regional dynamic optimal carbon tax. This is because the influx of new carbon-emitting firms will increase the regional social cost of carbon (SCC) and the optimal carbon tax will be adjusted upwards. New carbon emitters would not receive additional benefits. Therefore, the risk of "free riding" on the implementation of a regional dynamic optimal carbon tax may be greatly reduced. The Appendix also discussed the use of carbon taxes and other economic impacts of carbon taxes.

Achieving of temperature control goal. In addition, we are concerned about whether the use of the dynamic optimal carbon tax is conducive to the achievement of temperature control goal. The relevant results are summarized in Table 11.3. None of the first three scenarios will achieve the Paris Agreement's 1.5 °C or 2 °C temperature control targets. For the baseline scenario, temperature anomaly reaches 4.38 °C in 2100. This far exceeds the requirements of the temperature control target. However, when the optimal dynamic carbon tax based on SCC is implemented, global temperature control goals are largely achieved. This is essential for the realization of the Paris Agreement. By 2050, the temperature anomaly is only 1.35 °C. By 2100, it will be 1.88 °C, an increase of only 0.53 °C.

11.6 Conclusions and Revelations

By constructing a multi-regional dynamic general equilibrium model, this paper explores the heterogeneous production and utility damages of climate change on 10 major economies and evaluates the effects of static carbon tax, dynamic carbon tax, and regional optimal dynamic carbon tax (calculated by SCC). It is found that: (1) Climate change has a negative impact on production and utility with regional differences. The US and China will suffer more severe production and utility damages. (2) By the end of the twenty-first century, climate change will cause more than 8% of production damage and more than 1% of utility damage. Applying a carbon tax faces the risk of increasing global inequality. (3) An effective carbon tax policy should consider the regionality of policy implementation, characteristics of policy path and features of regional economy. The regional optimal dynamic carbon tax (based on SCC) could significantly reduce global carbon emissions, and the production and utility damage faced by each economy. Meanwhile, the implementation of a regional dynamic optimal carbon tax would also effectively contribute to the achievement of global temperature control goals. Since countries have already used the regional optimal carbon tax based on SCC, this paper argues that the risk of "free riding" in low carbon priced regions is not high.

The above conclusions also bring some revelations for the climate governance and carbon tax implementation. (1) Due to the heterogeneity of climate shocks, the use of carbon tax policies should consider both international and domestic variability. For example, developed countries could actively implement carbon tax. This can help accelerate economic transformation (developing service industries, such as finance) and improve climate resilience, reducing the exposure and sensitivity of economic activities to climate change. On the other hand, developing countries should implement carbon taxes as appropriate, with more focus on poverty reduction and economic development. Domestically, the carbon tax can be implemented first in high-carbon emitting industries and regions, forcing the optimization of economic structure. The level of carbon tax should also be determined by the development situation of each region, and attempts should be made to design a top-down carbon tax regulatory mode. (2) The use of carbon taxes could be an important complement to the Emission Trading System (ETS). At present, many regions have only implemented carbon pricing like the ETS. However, the ETS still has some defects. For example, the market has a high industry threshold, insufficient industry coverage and free quotas in its early development. As another carbon price, the carbon tax can make up for the above defects and become an important supplement to the ETS. Carbon taxes have no industry thresholds and are cheaper to run. The revenues of carbon taxes can also be used to subsidize businesses through fiscal means. Therefore, it is recommended that carbon tax policies be vigorously promoted to supplement the ETS and help achieve global temperature control goals. To match the practice of carbon tax, a reasonable corporate carbon accounting mechanism should be constructed. (3) The price of carbon taxes implemented and planned needs to be further increased for clarifying the carbon emissions cost and boosting emission reductions technology. The current carbon prices are still very low in many parts of the world, especially in some developing countries. Based on the quantitative results of our model, a reasonable increase in the carbon price is needed. In the process of upgrading carbon taxes, other tax and fiscal subsidies can be given to help firm's transition. The risk of "free riding" in low carbon-priced regions may be avoidable if an optimal carbon price (approaching the marginal production and utility damage of climate change) is implemented.

Admittedly, this paper still has many limitations: (1) Our model has taken into account the uncertainty of climate sensitivity and loss proportion, but it is limited in considering economic uncertainty; (2) In the modes of technological innovation, our model only describes the production damage caused by climate change from a macro perspective and ignores the impact of climate change on the innovation R&D capabilities of the workforce from a micro perspective; (3) Our model sets both corporate income tax and carbon tax, but ignores the issue of double taxation. A future careful study is thus needed to eliminate the above limitations.

Acknowledgements Huai Deng, Hui Xu, and Dawei Zhang also made great contributions to this manuscript. We express our heartfelt thanks to them. This research also was supported by: National Natural Science Foundation of China (72474127) and National Social Science Fund Major Bidding Project of China (18ZDA052). Part of this chapter was published:Deng H, Wu X, Xu H, Zhang D. Aggregated economic damages and the role of dynamic carbon tax under climate change: A dynamic general equilibrium model. Economic Analysis and Policy 2024;84:1099–119. https://doi.org/10.1016/j.eap.2024.10.025. The publication has been conducted with copyright permission.

Appendix

This section provides Appendices to this paper, including a model Appendix, a calibration Appendix, and a simulation results appendix.

Appendix A for Model

Description for part "Technology and production". The production setting of Eq. (11.2) reflect the differences between economies in the following aspects: Parameter differences $\{\alpha_i, \beta_i, \Psi_i\}$, technical and endowment differences $\left\{A_{it}, A_{it}^E, E_{it}, L_{it}, K_{it}\right\}$. Parameter differences elaborate on the heterogeneity in the shares of traditional fossil energy and labor force for different economies. They also reflect the varying economic structure and production technology characteristics. Endowment differences indicated the heterogeneity of factor inputs that can be utilized by different economies.

After obtaining the first best condition of each economy's production, we have the ratio of inputs: $E_{it}/K_{it} = \alpha_i r/(1-\alpha_i-\beta_i)\eta P_{Y,\,it}$, $L_{it}/K_{it} = \beta_i r/(1-\alpha_i-\beta_i)w$. The ratio of inputs in the production is determined by the relative payment share of input factors and the relative price of input factors. Taking energy and capital as an example, when the share of energy payments is higher and other conditions remain unchanged, energy inputs increase relatively; when energy factor prices are higher and other conditions remain unchanged, capital inputs increase relatively. In actual productions, the actual factor inputs in the production activity are jointly determined by the above two paths. The capital dynamics are set as: $\dot{K}_{it} = Y_{it} - \sum_{j=1}^{N} C_{it}^j - \delta^K K_{it}$. Where, $\sum_{j=1}^{N} C_{it}^j$ is the total consumption of the economy i, and C_{it}^j is the consumption of i produced from economy j. $\delta^K K_{it}$ means the depreciation of capital.

Description for part "R&D and climate change". In Eq. (11.3), climate change does not directly affect technological innovation. In other words, it does not affect the $\{A_{it}^{*}, A_{it}^{E*}\}$, but rather has a negative effect on the economic system by reducing the production efficiency of R&D.

In Eq. (11.6), $\log\left(\frac{O_L}{S_P}\right)\chi_t^{1+\theta_{\chi T}}$ indicates the level of temperature change corresponding to climate change. Eq. (11.6) describes how the negative impacts of climate change D_{it} are transformed from traditional fossil fuels and carbon stocks.

Description for part "Preferences and welfare". In Eq. (11.7), since the residents consumes the outputs of all regions simultaneously, the final goods market is purely competitive.

Description for part "Government and model equilibrium". Before solving the economic growth rate of Eqs. (11.10) and (11.11), we discussed the dynamic equilibrium stability condition here. The sum of the dynamic of consumption and capital is transformed into the following form: $\dot{C}_{it} = (r - \rho)C_{it}/\phi_i$ and $\dot{K}_{it} = Y_{it} - C_{it} - \delta^K K_{it}$. Here, $C_{it} = \sum_j C_{it}^j$ and $(1 - \alpha_i - \beta_i)P_{Y,it}Y_{it} = rK_{it}$ based on the condition of base firm's problem. The Taylor expansion of $\dot{C}_{it}(C_{it}, K_{it})$ and $\dot{K}_{it}(C_{it}, K_{it})$ is performed to the consumption and capital dynamics. Then, the dynamics are converted into the matrix expression:

$$\begin{bmatrix} (r-\rho)/\phi_i & 0 \\ -1 & \frac{r}{1-\alpha_i-\beta_i} - \delta^K \end{bmatrix}\begin{bmatrix} C_i^* - C_{it} \\ K_i^* - K_{it} \end{bmatrix} \tag{11.15}$$

In Eq. (11.15), C_i^* and K_i^* is the equalized value. The eigenvalue of the coefficient matrix is solved, and the equilibrium stability condition of the system is obtained as follows: $r < \rho$, $r/(1 - \alpha_i - \beta_i) < \delta^K$.

For calculating the growth rate of each economy, we solved the technological growth rate here. Technological progress is identified as a source of economic growth. According to Eq. (11.3), the overall technology growth rate is: $\dot{A}_{it}/A_{it} = \dot{A}_{it}^*/A_{it}^* - [D_{it}/(1 - D_{it})]g_{D_{it}}$. Here, $\dot{A}_{it}^*/A_{it}^*$ is the growth rate of "decoupling" technology. The "decoupling" technology equilibrium growth rate of each economy is solved separately by using Eqs. (4) and (5). a_i^* indicates the lower limit of innovation capability owned by workforce to carry out technological innovation. Thus, the set of innovation capabilities of workforce can be expressed as $[a_i^*, +\infty]$. Assuming that the innovation capability of each economy follows the Pareto distribution, we obtained $g_{A_{mt}^*} = \dot{A}_{mt}^*/A_{mt}^* = \Phi(G_{mt})(1 - \Psi_m)L_{mt}\left(a_m^*/a_{min}\right)^{-\delta}$ and $g_{A_{nt}^*} = \dot{A}_{nt}^*/A_{nt}^* = \Phi(G_{nt})[\theta + (1 - \theta)\nabla^*](1 - \Psi_n)L_{nt}\left(a_n^*/a_{min}\right)^{-\delta}$. Here, $\nabla^* = A_{mt}^*/A_{nt}^*$ is the average technology gap between the developed and developing economies in equilibrium (∇^* is calculated by the average gap). When the technology gap narrows (the value of ∇^* decreases), assuming other conditions remain unchanged, the equilibrium technology growth rate of developing economies driven by the technology gap will gradually decrease and the economic growth rate will decline.

In addition, the production damage is $[D_{it}/(1 - D_{it})]g_{D_{it}}$. It depends on the regional impact heterogeneity ρ_{it} of carbon emissions, climate sensitivity uncertainty and damage uncertainty. Meanwhile, $[D_{it}/(1 - D_{it})]$ indicates the heterogeneous impact of climate change due to endowments, environment, and economic features.

Appendix B for Calibration

Databases used for calibration. In part "**4. Calibration and variable estimation**", we mainly used four databases, including the *World Bank database*, the *IMF database*, the *BP Energy Statistics Yearbook* and the *CRU-TS database.*

The *World Bank database* provides free access to global country-level multidimensional data. For example, health, economic development, poverty, environment, energy and so on. The data website is https://data.worldbank.org.cn/. When using this database, we removed data from outside the sample countries and chose data from a uniform calibrator for parameter estimation.

The *IMF database* a dataset is provided by the International Monetary Fund (IMF) and consists of four main components, including Financial Statistics, Balance of Payments Statistics (BOP), Direction of Trade Statistics (DOT), and Government Finance Statistics (GFS). This paper mainly used the bilateral trade dataset from DOTs (1948–2021). When using this database, we only retain data for countries where both sides of the trade are in the sample. Trade gap is recognized in national consumption.

The *BP Energy Statistics Yearbook* is provided by the Royal Dutch Shell Group of Companies, reporting the energy production, energy consumption, energy price and carbon emissions data for a wide range of energy sources of the major economies in the world. In our model, the *BP Energy Statistics Yearbook 2021* is mainly used.

The *CRU-TS database*, noting as the most widely used climate datasets, is produced by the UK's National Center for Atmospheric Sciences (NCAS). This database provides monthly data at 0.5° resolution covering the land surface globally for the years 1901 to 2020, including temperature, precipitation, humidity, number of frosty days, cloudiness and potential transpiration. In our model, we mainly used the temperature data matched to the country level from the *CRU-TS database*. This database could be downloaded at: https://crudata.uea.ac.uk/cru/data/hrg/.

Calculation of γ_j and ϕ_i. For parameter γ_j, consumption preferences of residents satisfy: $\sum_j \gamma_j = 1$. Assuming that the trade correlation between economies in our model is closed, consumer preference γ_j of each economy is defined as: $\gamma_j = \sum_t Import_{jt} / \sum_t \sum_j Import_{jt}$. $Import_{jt}$ means the output in period t imported from economy j. Based on the bilateral trade data in the *IMF database* from 1948 to 2021 (Data source: IMF Data), we calculated the γ_j of each economy in our model. Then, we computed the average value for each economy. The results shown that representative residents prefer to the goods of developed economies, with an average difference of about 0.1%. This fact implies that the goods of developed economies are better in terms of quality, added value, and market brand effect. For parameter ϕ_i, based on the utility problem, consumption C_{it}^j meets: $\left[\prod_j^N \left(C_{it}^j\right)^{\gamma_j}\right]^{1-\phi_i} \left(\gamma_j / C_{it}^j\right) = \exp[(\rho - r)t]$ and we could obtain the following conditions: $\sum_j^N \gamma_j \ln\left(C_{it}^j\right) = \left[\sum_j^N \gamma_j \ln(\gamma_j) + (r - \rho)t\right] / \phi_i$. Meanwhile, based on the consumption path: $C_{it}^j = \gamma_j C_{i0}^j \exp[(r - \rho)t / \phi_i]$, we calculated the following condition: $\sum_j^N \gamma_j \ln\left(C_{it}^j\right) = (r - \rho)t / \phi_i + \sum_j^N \gamma_j \ln\left(\gamma_j C_{i0}^j\right)$. Combined the conditions of $\sum_j^N \gamma_j \ln\left(C_{it}^j\right)$, we can realize the estimation of ϕ_i for each economy as: $\phi_i = \sum_j^N \gamma_j \ln(\gamma_j) / \sum_j^N \gamma_j \ln\left(\gamma_j C_{i0}^j\right)$. The setting of C_{i0}^j comes from the *IMF database* (Normalizing based on national consumption and using sample means of year 1948–2021) and the data of γ_j from the above estimation. The average results of ϕ_i is 0.57.

Calibration for production damage D_{it}. To calculate production damage, we firstly calibrated the parameter of carbon emissions. φ_f is set to 0.2 from Golosov et al. (2014). For parameter φ_0, we solve the following optimization problem: $argmin\left[\sum_{t=1965}^{2020} \left(\varphi_f \sum_{i=1}^N E_{it} + \varphi_0 (1 - \varphi_f) \sum_{i=1}^N E_{it} - CE_t\right)^2\right]$. Here, CE_t is the carbon emission in period t. Data of E_{it} and CE_t comes from the *BP Energy Statistics Yearbook 2021*. We have that φ_0 is 0.026. For parameter φ, we match the carbon stock in 2015 according to the data of Van den Bremer and Van der Ploeg (2021). Let $\varphi_f \sum_{i=1}^N E_{it} + \varphi_0 (1 - \varphi_f) \sum_{i=1}^N E_{it}$ as EL_t, we have the carbon stock as: $O_t = (1 - \varphi)^{t-1} O_1 + (1 - \varphi)^{t-2} EL_1 + \ldots + (1 - \varphi)^{t-t} EL_{t-1}$. By solving this equation (the data interval is 1965–2020), the result of φ is 0.0037. The fitting results are shown in Fig. 11.11.

Then, the calculation of ρ_{it} for production damage is as follows. ϖ_{it} is defined as: $L_{it} / \sum_{i=1}^N L_{it}$ and data from the *World Bank database*. x_{it} is the share of total carbon

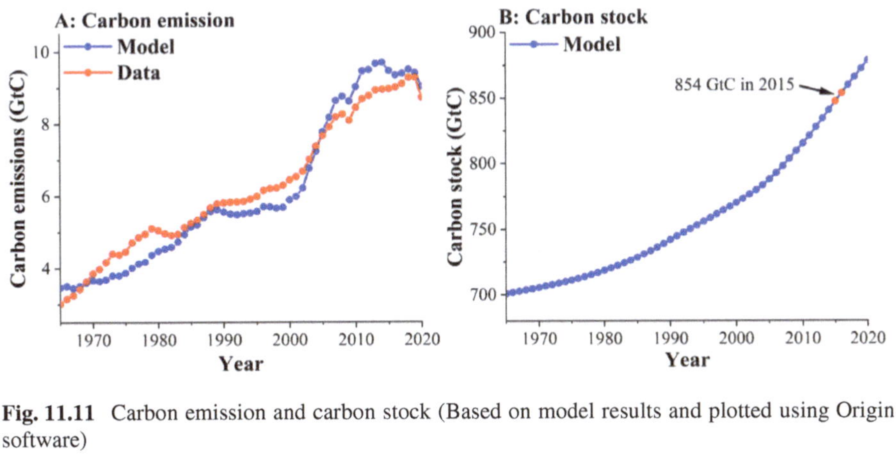

Fig. 11.11 Carbon emission and carbon stock (Based on model results and plotted using Origin software)

emissions: $O_{it}/\sum_{i=1}^{N} O_{it}$ and data from the *BP Energy Statistics*. Inspired by Conte et al. (2021), The sector-specific temperature discount factor $g(T_{it})$ is defined as: $g(T_{it}) = \varsigma_{it}\exp\left[-0.5\left(\frac{T_{it}-T^{aopt}}{T^{avar}}\right)^2\right] + (1-\varsigma_{it})\exp\left[-0.5\left(\frac{T_{it}-T^{mopt}}{T^{mvar}}\right)^2\right]$. Here, $\{T^{aopt}, T^{avar}\}$ are the optimal temperature and temperature variance for agricultural production. $\{T^{mopt}, T^{mvar}\}$ for non-agricultural production. ς_{it} is the share of agricultural value added in GDP. When modeling future scenarios, ς_{it} and x_{it} is averaged from 1990 to 2020 (based on data availability), noting as ς_i and x_i. Data comes from the *CRU-TS database* and the *World Bank* database. The *CRU-TS database* is supported by Harris et al. (2020).

Calibration for social cost of carbon. To calculate SCC of Eq. (11.14), we provided the following details. For the remainder factor ξ_t, it is shown as:

$$\xi_t = \sum_{s=0}^{t} \phi^C[\varphi_f + \varphi_0(1-\varphi_f)](1-\varphi)^{t-s} = \phi^C[\varphi_f + \varphi_0(1-\varphi_f)]\sum_{s}^{t}(1-\varphi)^{t-s} \tag{11.16}$$

Due to the isoperimetric series property, we have $\sum_s^t (1-\varphi)^{t-s} = \left[1-(1-\varphi)^{t+1}\right]/\varphi$. Then, we could solve the factor $\sum_{t=0}^{\infty}(1+\rho)^{-t}\Upsilon\xi_t$ of regional SCC as follows:

$$\sum_{t=0}^{\infty}(1+\rho)^{-t}\Upsilon\xi_t = \frac{\Upsilon\phi^C}{\varphi}[\varphi_f + \varphi_0(1-\varphi_f)] \sum_{t=0}^{\infty}\left[\left(\frac{1}{1+\rho}\right)^t - (1-\varphi)\left(\frac{1-\varphi}{1+\rho}\right)^t\right] \tag{11.17}$$

Using the infinite series property and $0 < 1/(1+\rho) < 1$, $0 < (1-\varphi)/(1+\rho) < 1$, we obtained:

$$\frac{\Upsilon\phi^C}{\varphi}[\varphi_f + \varphi_0(1-\varphi_f)]\left[\frac{1+\rho}{\rho} - \frac{(1-\varphi)(1+\rho)}{\varphi+\rho}\right] \tag{11.18}$$

For Υ and ϕ^C, we calibrated by using simulation results from the benchmark model 2020–2100 (Base scenario in Part 5.1). Υ is calculated by the ratio of total production and utility damage to carbon stocks (all sample countries from 2020 to 2100). The result of Υ is 0.187%. ϕ^C is calculated by the ratio of total GDP to carbon emissions (all sample countries from 2020 to 2100) and the related result is 1.993. The setting of $\{\varphi, \varphi_f, \varphi_0, \rho\}$ are like the results of Table 11.2.

Appendix C for Simulation

Trend of global development. we reported the global economic trend in Fig. 11.A2. In Fig. 11.12a, the increase in the average technology growth rate stems from the role of government subsidies and population growth. The technological gap between developing and developed countries is still widening because of the higher lower limit of innovation a_i^* in developing countries. This factor constraints on technological progress. From 2020 to 2100, global GDP grew by \$149.82 trillion. The average annual growth rate is 2.54% (Fig. 11.12b). Global GDP to reach \$236.79 trillion by 2100. In terms of regional differences, the US has the highest GDP, followed by China. Meanwhile, there are large differences between China and the US with other economies (Fig. 11.12c). According to our model, international free trade (Caliendo & Parro, 2015) and higher technological growth rates may contribute to the rapid growth of China. For GDP per capita, developing countries (China and India) have always lower GDP per capita than other developed countries due to the size of populations.

Robustness test of simulation. In this section, we first conducted a robustness test on the baseline model. Then, using the implementation of dynamic carbon tax as an example, the robustness tests are carried out. The results are shown in Table 11.4. The results of Panel A (baseline model) show that enhancing the offsetting effects of the natural conditions φ can mitigate the negative impacts of climate change and reduce production and utility damages. Lowering energy prices η increases energy consumption, resulting in higher carbon emissions and climate shocks. Increasing the ratio of carbon emissions retained in the natural environment φ_f also exacerbates climate shocks. In Panel B, we test for robustness by varying the carbon tax growth rate and initial values. The results show that increasing both the growth rate and the initial value of carbon tax can help curb climate change and reduce production (GDP) and utility damages.

Discussion on the use of carbon tax. In this section, we briefly discuss two issues in the use of carbon taxes: how carbon taxes fit into existing climate policies; the impact of different economic and political environments on the practice of carbon taxes. For the first issue, a closely related climate policy to carbon taxes is the emissions trading system (ETS). ETS is a widely used carbon pricing mechanism, and the carbon taxes should complement and improve the ETS. The ETS is a quantitative measure with carbon emissions and carbon intensity as the emission reduction targets, helping countries to reduce emissions. However, ETS also has many shortcomings, such as not covering a wide range of industries (for example, the current China's national carbon market), high operating costs of ETS, and difficulties in the design of the quota mechanism. As tax measures, using carbon tax has no industry thresholds and low operating costs. This complements the ETS and promotes carbon emission reduction and low-carbon development.

For the second issue, the introduction of carbon tax depends on the economic features and institutional mode of each country and requires consideration of many factors. Foreign experience in the practice of carbon tax provides the following

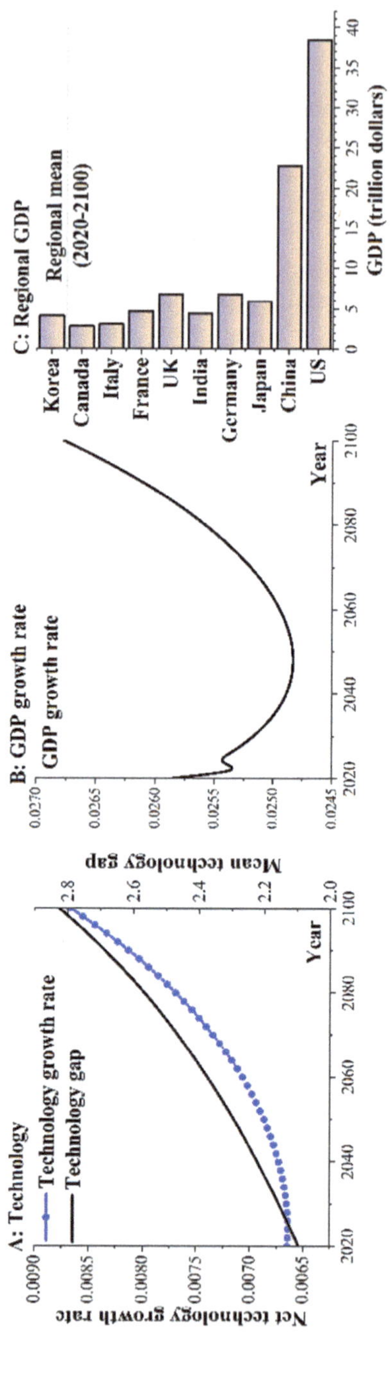

Fig. 11.12 Additional benchmark results from 2020 to 2100

Table 11.4 Robustness tests of our model

Changes	Carbon emissions	D_{it}	$\Lambda(\Delta T_t)$	Production damage	Utility damage
Panel A: Robustness of baseline model					
$\varphi \times 2$	+0.024	−0.56	−0.06	−0.008	−0.001
$\eta \times 0.5$	+20.09	+2.11	+0.17	+0.024	+0.002
$\varphi_f \times 2$	+17.71	+1.89	+0.16	+0.022	+0.002
Panel B: Robustness of using dynamic carbon tax					
$a_1^E \times 2$	−9.816	−1.08	−0.12	−0.017	−0.001
$a_2^E \times 2$	−10.088	−1.03	−0.11	−0.016	−0.001
$a_3^E \times 2$	−10.526	−1.02	−0.11	−0.016	−0.001
b_1^E is 1.105	−8.145	−0.88	−0.09	−0.014	−0.001

Note (1) Production and utility damages retained to three decimal places. (2) In the fourth row of Panel B, the dynamic carbon tax is $\tau_t^E = \ln(a_1^E t + b_1^E)$ of the EP case (fast first and slow then). We solved b_1^E for meeting the new setting of initial carbon tax value 0.1. (3) The unit of carbon emissions GtC. The unit of D_{it} and $\Lambda(\Delta T_t)$ is %. The unit of production and utility damage is trillion dollars

insights into whether to use carbon tax: (1) The mode of carbon tax collection. An independent carbon tax (such as a carbon emissions tax or a climate change tax) can be set up, or the original related taxes can be reformed (such as an energy tax or a fuel tax). (2) Carbon tax price setting. Progressive carbon tax price setting (allows firms and residents to gradually adapt), or high carbon taxes matching with other incentives (protect national business competition, but the incentives would gradually decline) could be used. (3) Calculation of carbon tax base (construct a carbon emissions accounting mechanism). The actual carbon emissions caused in the process of energy consumption can be used as the tax base. The estimated carbon emissions from energy consumption can also be used as the tax base. In addition, factors such as the scope of taxpayers and the use of carbon tax incomes should further be considered for the implementation of carbon tax.

Other economic impacts of carbon tax. We mainly focused on three economic impacts of carbon tax, including economic growth, labor employment, and GDP distribution (GDP is treated as an indication of income in each economy). For economic growth, Fig. 11.13 reported the GDP gain and GDP growth rate gain from using carbon tax in three cases. The results show that the use of carbon taxes would curb climate change, reduce GDP damages (Fig. 11.13a), and boost the GDP growth rate of economies (Fig. 11.13b). This is consistent with the fact that using carbon tax would improve residents' welfare and output in the long run (Cruz & Rossi-Hansberg, 2023).

For labor employment, in the face of climate change, the labor force will generally suffer the following impacts: the negative impact on the labor marginal productivity of climate change (Xu & Deng, 2024), the trade-off between work and leisure (Somanathan et al., 2021), and the climate change induced migration (Cruz &

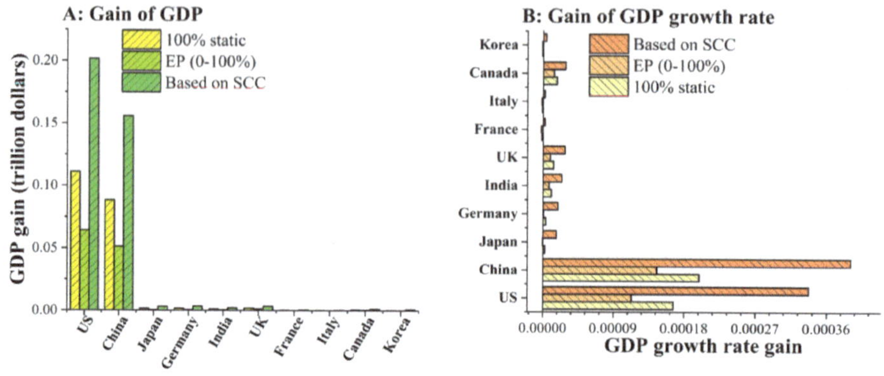

Fig. 11.13 Impact of carbon tax on economic growth

Rossi-Hansberg, 2023; Xu & Deng, 2024). The results of the above impacts are lower labor wages, higher leisure, and mitigation to place with comfortable climate conditions and non-farm sectors. Especially, the cleaner production sector of non-agricultural sectors is more attracted (Conte et al., 2021; Xu & Deng, 2024). Thus, after reducing carbon emissions and curbing climate change, using carbon tax will help improve labor wages and reduce idleness (boosting labor supply). The scale of climate change-induced migration may shrink.

For GDP distribution, we calculated the Gini coefficient of GDP using the population of each economy as a benchmark. The results show a gradual increase in the global Gini coefficient, indicating an increase in global income inequality. Meanwhile, the implementation of carbon tax would raise the global Gini coefficient and increase global income inequality. This could be regarded as the potential risk of using carbon taxes (Table 11.5).

Table 11.5 Impact of carbon tax on income distribution (Gini coefficient)

	Year 2020	Year 2100	Changes 2020–2100	Difference from base
Base	0.161	0.286	0.125	–
Static carbon tax (100%)	0.161	0.288	0.127	0.002
Dynamic CT with EP	0.161	0.287	0.126	0.001
Optimal dynamic CT (SCC)	0.161	0.290	0.128	0.003

Note (1) CT is carbon tax; EP is the environmental path; SCC is the social cost of carbon. (2) Difference from base is results in 2100

References

Acemoglu, D., Aghion, P., Bursztyn, L., & Hemous, D. (2012). The environment and directed technical change. *American Economic Review, 102*(1), 131–166. https://doi.org/10.1257/aer.102.1.131

Acemoglu, D., Akcigit, U., Hanley, D., & Kerr, W. (2016). Transition to clean technology. *Journal of Political Economy, 124*(1), 52–104. https://doi.org/10.1086/684511

Acemoglu, D., & Restrepo, p. (2022). Demographics and automation. *Review of Economic Studies, 89*(1), 1–44. https://doi.org/10.1093/restud/rdab031

Barrage, L. (2020). Optimal Dynamic Carbon Taxes in a Climate–Economy Model with Distortionary Fiscal Policy. *Review of Economic Studies, 87*(1):1–39. https://doi.org/10.1093/restud/rdz055.

Bretschger, L., & Valente, S. (2012). Endogenous growth, asymmetric trade and resource dependence. *Journal of Environmental Economics and Management, 64*(3), 301–311. https://doi.org/10.1016/j.jeem.2012.01.008

Burke, M., Davis, W. M., & Diffenbaugh, N. S. (2018). Large potential reduction in economic damages under UN mitigation targets. *Nature, 557*(7706), 549-+. https://doi.org/10.1038/s41586-018-0071-9

Burke, M., et al. (2015). Global non-linear effect of temperature on economic production. *Nature, 527*(7577), 235–239. https://doi.org/10.1038/nature15725

Cai, Y. Y., Borck, W., & Xepapadeas, A. (2023). Climate change impact on economic growth: Regional climate policy under cooperation and noncooperation. *Journal of the Association of Environmental and Resource Economists, 10*(3), 569–605. https://doi.org/10.1086/722274

Cai, Y. Y., & Lontzek, T. S. (2019). The social cost of carbon with economic and climate risks. *Journal of Political Economy, 127*(6), 2684–1734. https://doi.org/10.1086/701890

Caliendo, L., & Parr, F. (2015). Estimates of the trade and welfare effects of NAFTA. *Reviews of Economic Studies, 82*(1), 1–44. https://doi.org/10.1093/restud/rdu035

Conte, B., Desmet, K., Nagy, D. K., & Rossi-Hansberg, E. (2021). Local sectoral specialization in a warming world. *Journal of Economic Geography, 21*(4), 493–530. https://doi.org/10.1093/jeg/lbab008

Cruz, J. L., & Rossi-Hansberg, E. (2023). The economic geography of global warming. *Review of Economic Studies.* https://doi.org/10.1093/restud/rdad042

Desmet, K., & Rossi-Hansberg, E. (2015). On the spatial economic impact of global warming. *Journal of Urban Economics, 88*, 16–37. https://doi.org/10.1016/j.jue.2015.04.004

Fan, Q. Q., Zhou, X. H., & Zhang, T. B. (2016). Externalities of dynamic environmental taxation, paths of accumulative pollution and long-term economic growth. *Economic Research Journal, 8*, 116–128.

Fikru, M. G., Ahmed, B., & Daher, W. (2025). Optimal policies for decarbonization via carbon capture: Tax, subsidize, or both? *Economic Modelling, 153.* https://doi.org/10.1016/j.econmod.2025.107319

Golosov, M., Hassler, J., Krusell, P., & Tsyvinski, A. (2014). Optimal taxes on fossil fuel in general equilibrium. *Econometrica, 82*(1), 48–88. https://doi.org/10.3982/ECTA10217

Grossman, G. M., & Helpman, E. (1991). Quality ladders and product cycles. *Quarterly Journal of Economics, 106*(2), 557–586. https://doi.org/10.3386/w3201

Harris, I., Osborn, T. J., Jones, P., & Lister, D. (2020). Version 4 of the CRU TS monthly high-resolution gridded multivariate climate dataset. *Scientific Data, 7*(1). https://doi.org/10.1038/s41597-020-0453-3

Jaimovich, N., & Rebelo, S. (2017). Nonlinear effects of taxation on growth. *Journal of Political Economy, 125*(1), 265–291. https://doi.org/10.1086/689607

Jensen, S., & Traeger, C. (2021). *Pricing climate risk.* CESifo Working Paper.

Millar, R. J., Zebedee, R., et al. (2017). A modified impulse-response representation of the global near-surface air temperature and atmospheric concentration response to carbon dioxide emissions. *Atmospheric Chemistry and Physics, 17*(11), 7213–7228. https://doi.org/10.5194/acp-17-7213-2017

Nordhaus, D. (2015). Climate clubs: Overcoming free-riding in international climate policy. *American Economic Review, 105*(4), 1339–1370. https://doi.org/10.1257/aer.15000001

Nordhaus, W. (2019). Climate change: The ultimate challenge for economics. *American Economic Review, 109*(6), 1991–2014. https://doi.org/10.1257/aer.109.6.1991

Nordhaus, W. D. (2010). Economic aspects of global warming in a post-Copenhagen environment. *Proceedings of the National Academy of Sciences of the United States of America, 107*(26), 11721–11726. https://doi.org/10.1073/pnas.1005985107

Nordhaus, W. D. (2013). *The climate casino: Risk, uncertainty, and economics for a warming world.* Yale University Press.

Pan, S. Y., Zhu, D. D., & He, Y. Y. (2019). China's response to tax cuts in the U. S. from a brain drain perspective. *Economic Research Journal, 10*, 183–198.

Rezai, A., & Van der Ploeg, F. (2016). Intergenerational inequality aversion, growth, and the role of damages: Occam's rule for the global carbon tax. *Journal of the Association of Environmental and Resource Economists, 3*(2), 493–522. https://doi.org/10.1086/686294

Rickels, W., Meier, F., & Quaas, M. (2023). The historical social cost of fossil and industrial CO2 emissions. *Nature Climate Change.* https://doi.org/10.1038/s41558-023-01709-1

Solomon, H., Robert, K., Amir, J., et al. (2017). Estimating economic damage from climate change in the United States. *Science, 356*(6345), 1362–1368. https://doi.org/10.1126/science.aal4369

Somanathan, E., Somanathan, R., Sudarshan, A., et al. (2021). The impact of temperature on productivity and labor supply: Evidence from Indian manufacturing. *Journal of Political Economy, 129*(6), 1797–1827. https://doi.org/10.1086/713733

Timilsina, G. R., Dissou, Y., Toman, M., & Heine, D. (2024). How can a carbon tax benefit developing economies with informality? A CGE analysis for Cote d'Ivoire. *Climate Policy, 24*(1), 71–86. https://doi.org/10.1080/14693062.2023.2223530

Tol, R. S. J. (2018). The economic impacts of climate change. *Review of Environmental Economics and Policy, 12*(1), 4–25. https://doi.org/10.1093/reep/rex027

Van den Bremer, T. S., & Van der Ploeg, F. (2021). The risk-adjusted carbon price. *American Economic Review, 111*(9), 2782–2810. https://doi.org/10.1257/aer.20180517

van der Wijst, K. I., Bosello, F., Dasgupta, S., Drouet, L., et al. (2023). New damage curves and multi model analysis suggest lower optimal temperature. Nature. *Climate Change, 13*(5), 434-+. https://doi.org/10.1038/s41558-023-01636-1

Waidelich, P., Batibeniz, F., Rising, J., Kikstra, J. S., & Seneviratne, S. I. (2024). Climate damage projections beyond annual temperature. Nature. *Climate Change, 14*(6). https://doi.org/10.1038/s41558-024-01990-8

Wu, X. H., Deng, H., et al. (2021). The impact of energy structure adjustment and environmental regulation on air pollution in China: Simulation and measurement research by the dynamic general equilibrium model. *Technological Forecasting and Social Change, 172*. https://doi.org/10.1016/j.techfore.2021.121010

Xu, H., & Deng, H. (2024). Short-term impact of climate change on labor market in China: Quantitative spatial analysis based on an oligopsony model. *China Economic Review.* https://doi.org/10.1016/j.chieco.2024.102142

Ye, J. Z., & An, H. S. (2017). Can environmental taxes effectively control air pollution? *China Industrial Economics, 5*, 54–74. https://doi.org/10.19581/j.cnki.ciejournal.2017.05.005

Zheng, Y., Pan, J. H., Xie, X. L., Zhou, Y. M., & Liu, C. Y. (2016). From climate change vulnerability to adaptation planning: A perspective of welfare economics. *Economic Research Journal, 2*, 140–153.

Chapter 12
Air Pollution, Migration Costs, and Urban Residents' Welfare: A Spatial General Equilibrium Analysis from China

12.1 Research Question

The industrialization has brought air pollution to developing countries, and increased the difficulty of sustainable development in developing countries. With the warming of the climate and the increasing of extreme weather events, air pollution such as haze has become an important problem faced by many developing countries, including China. In 2018, the annual average $PM_{2.5}$ concentration of 338 cities in China exceeded the secondary-level limit of national standards by about 11.4%, and the number of cities meeting air quality standards accounted for less than 50%, according to the *China's 2019 Report on Atmosphere: China's Process in Air Pollution Prevention and Control*. Severe air pollution not only deteriorates the urban environment, but also has obvious negative effects on the welfare of urban residents (Shi & Yu, 2020). For example, air pollution will increase the prevalence of respiratory diseases and cancer among urban residents (Thurston et al., 2017), and endanger the health of residents, thereby reducing the residents' welfare. Moreover, air pollution will also worsen the urban environment, making cities less attractive (Hanlon, 2016), leading to a decline in economic growth and in the welfare of urban residents. Therefore, studying the impact of air pollution on the welfare of Chinese urban residents is of great significance for China to control air pollution and improve urban residents' welfare.

Air pollution has a significant negative effect on urban residents' welfare, which makes residents pay more attention to urban air quality when choosing a city to live in. For example, Liu and Yu (2020) studied the importance of air pollution in the migration decision of urban residents, and found that the more serious the air pollution in a city, the stronger the residents' willingness to move out of the city. Especially in areas with severe weather pollution, some residents move out due to air quality problems (Hong et al., 2016). The increasing air pollution in China has become an important reason for the migration of middle- and high- income classes, according to the *2014 China International Migration Report* jointly released by the Center

X. Wu, *Environmental Disaster in China*,
https://doi.org/10.1007/978-981-92-0910-1_12

for China and Globalization (CCG) and the Chinese Academy of Social Sciences. Residents moving out of cities with severe air pollution can not only improve their own welfare, but also improve the welfare of residents in the places where they move to by expanding the market size (Aubry et al., 2016). However, due to the migration costs, including the China's unique household registration control (HRC) system, it is difficult for urban residents to move freely in China. The higher the cost of migration, the higher the cost of moving away from the city with severe air pollution, and the lower the benefits residents can get after migration. Nevertheless, few studies on the impact of air pollution on urban residents' welfare in China (Ferreira et al., 2013; Nam et al., 2010; Shi & Yu, 2020) have considered the moderating role of migration costs despite its importance.

In order to answer the above questions, this paper mainly carried out the following work. (1) A theoretical spatial general equilibrium model including air pollution variables and migration costs is constructed based on the Rosen-Roback model. (2) Based on the equilibrium solution of the above model, the resident's welfare and the impact of the change of air pollution and migration costs on the average resident's welfare in China from 2007 to 2017 are estimated. (3) Panel data of 274 cities at the prefecture level and above in China from 2007 to 2017 are collected to examine the impact of air pollution on urban residents' welfare, and to discuss the moderating effect of migration costs. The possible innovations of this paper are: (1) The influence mechanism of air pollution on urban residents' welfare is analyzed from the perspective of spatial equilibrium. (2) This paper constructs a spatial general equilibrium model of air pollution, residents' migration costs and residents' welfare, thereby providing a systematic framework for studying the relationship among the three. (3) The results of the theoretical analysis are tested by collected data, and the adjustment effect of the migration costs are discussed, from which a series of enlightening conclusions are drawn. The research framework is shown in Fig. 12.1.

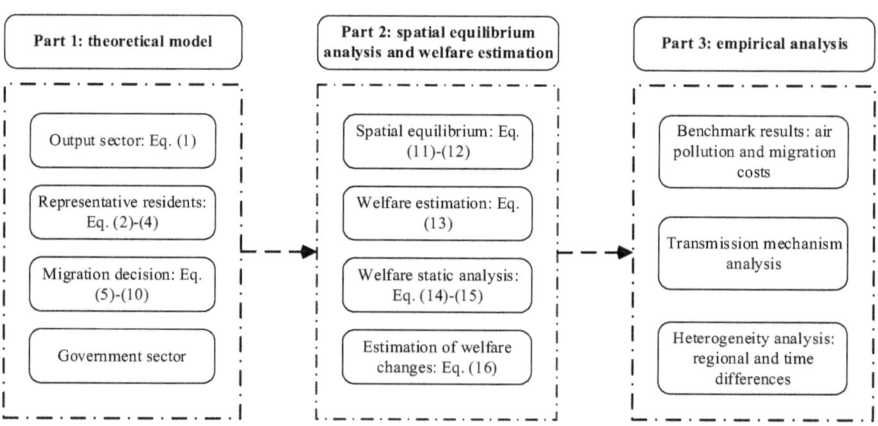

Fig. 12.1 Research framework

The rest of this paper is arranged as follows: The relevant studies review is presented in the second part, while the third part introduces the spatial general equilibrium model. The fourth part presents model parameters setting, feature variables estimation and simulation results. The empirical analysis results are shown in the fifth part and the conclusions of this paper are in the final section.

12.2 Literature Review

12.2.1 Research on the Impact of Air Pollution on Residents' Welfare and Migration

There are extensive studies on the effects of air pollution on migration and residents' welfare. As far as residents' welfare is concerned, a lot of studies had shown that air pollution would cause losses of resident's welfare (Chen & He, 2014; Guo et al., 2010; Matus et al., 2012; Nam et al., 2010; Yaduma et al., 2013; Zhang et al., 2017). For example, Shi and Yu (2020) evaluated the impact of air pollution on residents' subjective welfare loss by using micro data at the individual level, and found that residents' welfare loss caused by an increase of 1ug/m^3 in urban average annual $PM_{2.5}$ concentration is equivalent to a 7.7% reduction in the disposable income of urban residents. Zhang and Mu (2018) explored the relationship between air pollution and residents' welfare based on the data of residents purchasing particulate filtering face-masks. Their research showed that a 10% reduction in the number of heavy pollution days (AQI $\geq$ 201) in a city each year would reduce residents' spending about $187 million on particulate filter masks. This meant that the government's reduction of the possibility of the "Airpocalypse" would directly increase the residents' disposable income and improve residents' welfare. Xie et al. (2024) estimated the health and welfare gains from air pollution regulations in China. Using the China's stringent "2 + 26" and spatial regression discontinuity model, they found lower medical spendings and reduced disease prevalence, thereby leading to more subjective well-being and welfare. Borgschulte et al. (2024) used satellite-based smoke plume data and the labor market data and studied the impact of air pollution on income and welfare in US. The exposure to smoke plume leads to lower income and reduced welfare. Jin and Li (2024) focused on the inequality of healthcare services and potential welfare losses due to air pollution in China. They found that air pollution results in more inpatient care use of 60 years of age and older, leading to corresponding inequality and welfare damage.

Another research perspective was that air pollution also had a negative impact on resident's welfare by reducing their mental and subjective well-being (Ferreira et al., 2013; Liu et al., 2019a; 2019b; Menz, 2011; Zhang et al., 2017; Zhou & Liu, 2020). Levinson (2012) established a function of residents' happiness, including individual characteristics, income, degree of air pollution, location, date, and weather of residence to study the relationship between air pollution and residents' happiness.

The results showed that in locations with severe air pollution, residents have a lower sense of well-being. Li et al. (2014) constructed a structural equation model for residents' well-being to study the effects of residents' perceived risk (of exposure to polluted environment and being harmed by pollutants), and objective risk (of approaching pollution sources) on their happiness. The results showed that both risks have a significant negative impact on residents' well-being. And increasing the information disclosure of regional air pollution could improve the resident's awareness of environmental quality, reduce their perceived and objective risks, and improve their well-being.

In addition to some qualitative and quantitative studies on welfare losses caused by air pollution, some scholars had calculated resident's willingness to pay to improve their welfare status by treating air pollution and researched the positive impact of reducing air pollution on resident's welfare (Chu et al., 2017). For example, Gonzalez et al. (2013) measured residents' marginal willingness to pay for air pollution (PM_{10}) in three cities with the largest population in Mexico, and found that the residents of Mexico City, Guadalajara, and Monterrey were willing to pay 41.73, 36.34, and 43.47 US dollars to reduce a unit of PM_{10}, respectively. Moreover, based on historical experience, some scholars emphasized that future air pollution control should pay attention to the impact of air pollution on residents' welfare, rather than only focusing on the regional economic growth rate. It was meaningful that the transition from an economic output-oriented growth model to a national welfare-oriented development model was implemented (Zhang & Liang, 2014).

On the other hand, in terms of resident's migration, the popular view is that the motivation of migration mainly comes from higher incomes and better career development. However, with the aggravation of air pollution, the kind of literatures on the impact of air pollution on the resident's migration had gradually increased. Usually, scholars used micro-survey data for empirical analysis to study the migration of residents brought by air pollution (Chen et al., 2017; Jia & Chen, 2021; Liu & Yu, 2020; Wang et al., 2021). Germani et al. (2021) paid attention to the role of air pollution in the movement of the population in Italy. They found that after considering the other important factors, it is still a significant relationship between air pollution and migration choices. In the process of migration, the migration costs played an important role to determine whether migration took place (Falck et al., 2018; Tombe & Zhu, 2019). In China, migration costs brought by household registration control (HRC) and regional discrimination often had a significant impact on residents' migration decisions. For example, Liu and Li (2017) considered the impact of migration costs when studying the urban size spatial distribution in China. They found that reducing migration costs led to urban population redistribution and significant welfare enhancement effects. Nevertheless, from the perspective of spatial equilibrium, few literature had included the migration costs into the research framework of the impact of air pollution on residents' welfare in China. Few articles provided a systematic analysis of all three.

12.2.2 Application Research of Spatial General Equilibrium Model

The theoretical model adopted in this paper is based on the spatial general equilibrium model (known as Rosen-Roback model) proposed by Roback (1982). In the model, the labor made the migration decision according to wages, housing rents, and the level of exogenous urban amenities in each region and chose a residential city that maximizes their utility. Later, many scholars expanded Rosen-Roback model from the urban agglomeration effects, heterogeneous labor supply, endogenization of amenities (Diomand, 2016; Duan et al., 2020), and urban commuting costs (Chen et al., 2019; Desmet & Rossi-Hansberg, 2013). For example, Desmet et al. (2018) considered the differences in economic development levels between cities and found that the economic agglomeration effect promoted the concentration of labor and industries in cities with a dense population and high degree of economic agglomeration. Glaeser (2008), from the perspective of heterogeneous labor, divided labor into skilled and unskilled labor, and found that skilled labor tends to cluster in cities with higher levels of convenience facilities and greater housing supply. Similarly, Zong and Zhou (2015) divided labor into high-skilled and low-skilled labor from the perspective of heterogeneous labor.

Generally and historically speaking, scholars mostly used the computable general equilibrium model, willingness-to-pay approach, and micro survey data to study the correlation between air pollution and residents' welfare. Few studies have simultaneously considered the residents' migration behavior that may be caused by increased air pollution, explored the theoretical mechanism of air pollution affecting the welfare of urban residents from the perspective of spatial equilibrium, and investigated the adjustment effect of migration costs. As mentioned above, with the urban air pollution increases, residents are increasingly concerned about urban air quality when they choose where to settle. In view of this, this study introduces air pollution variable and migration costs variable into the Rosen-Roback model, and adds residents' health levels to the resident utility function of the model, so as to investigate the impact of air pollution on urban residents' welfare from the perspective of spatial equilibrium. In addition, in the empirical analysis, the two indexes of the measured urban migration costs and the urban household registration control (HRC) are used as proxy variables of residents' migration costs, and the moderating effect of migration costs on the impact of air pollution on urban residents' welfare is systematically studied.

12.3 Theoretical Model

12.3.1 Output Sector Analysis

The model assumes that there are n cities in the economy, each of which is represented by i and t is the year. Residents migrate between cities and choose the most suitable cities to live in so as to maximize their utility. At the same time, residents need to bear certain migration costs. Herein, residents' transfer between urban and rural areas and transnational residents' migration are not considered, and only residents' transfer between cities is considered out of simplification.

The input elements of the output sector include labor, land, and capital. The urban land supply L_{it} has two uses, economic production and construction of residential housing. Referring to Duan et al. (2020), the output function is:

$$Y_{it} = (\varepsilon L_{it})^{\alpha}\left[(1-\gamma)\mu A_{it}N_{it}\right]^{\beta}K_{it}^{1-\alpha-\beta}(0<\alpha<1,0<\beta<1). \quad (12.1)$$

where Y_{it} is the urban output, A_{it} is urban productivity level, L_{it} is urban land supply, ε is the proportion of land supply for economic production ($0 < \varepsilon < 1$), N_{it} is the urban population, μ is the proportion of adult labor in urban population, α and β are the output elasticity of land and labor, respectively. In addition, because urban air pollution can damage the physical condition of labor (Kan et al., 2012), and health condition is an important factor affecting labor productivity, γ is introduced to represent the loss degree of labor efficiency caused by air pollution, $0 < \gamma < 1$. K_{it} is capital level, and the capital market is free-flowing.

The output sector will also cause air pollution during production, and the concentration of air pollution is denoted by P_{it}. Referring to Nordhaus (2010), the expression of air pollution concentration is:

$$P_{it} = (1-V_t)\alpha'' \frac{\alpha' Y_{it}}{A_{it}^{\beta'}}. \quad (12.2)$$

Here, α'' is the natural purification capacity, α' is the influence of economic production process on environment, β' reflects the air pollution abatement elasticity of productivity. V_t ($0 \le V_t < 1$) is the air pollution control intensity of the government.

12.3.2 Analysis of Representative Residents

With reference to the research of Monte et al. (2018) and Wu et al. (2015), the utility function of representative urban residents is expressed as:

$$U_{it} = q_{it}\left[C_{it}^{\alpha_1}H_{it}^{1-\alpha_1}\right]^{\beta_1}F_{it}^{1-\beta_1}(0<\alpha_1<1,0<\beta_1<1). \quad (12.3)$$

In Eq. (12.3), U_{it} is the utility of representative urban residents, q_{it} is the city's livability, C_{it} is commodity consumption, H_{it} is housing consumption, and α_1 is the contribution of commodity consumption to the utility of representative residents ($0 < \alpha_1 < 1$). In addition to commodity and housing consumption, representative residents are also concerned about residents' health. F_{it} is the health level of the representative residents, and $1 - \beta_1$ is the contribution of health to the representative resident's welfare. Wu et al. (2015) believed that the function of residents' health level is a convex function of environmental pollution and a concave function of health investment expenditure. Based on this, the function of residents' health level is written as:

$$F_{it} = \frac{I_{it}^{\zeta}}{P_{it}^{\kappa}}(0 < \zeta < 1, 0 < \kappa < 1). \tag{12.4}$$

In Eq. (12.4), the residents' health level F_{it} is determined by the health investment expenditure I_{it} and the urban air pollution level P_{it}. ζ, and κ are respectively the efficiencies of health investment expenditure and the negative impact of air pollution on residents' health. The health investment expenditure herein mainly refers to the protective expenditure of residents to improve their health. Since the urban environment is a public good, if the health investment expenditure in the model set can improve the overall air quality of the city, some urban residents who are insensitive to environmental quality may have "free-riding" behavior and benefit from this as well. That is, some urban residents have enjoyed the benefits of improved urban air quality without spending more on health investments. To avoid this inefficient situation, health investment expenditures here can only improve the health of residents themselves.

Duan et al. (2020) believed that urban livability is composed of the city's natural endowment conditions and fiscal expenditure of the city's government on public services. On this basis, this paper introduces air pollution variables to the expression of urban livability:

$$\mathrm{q_{it}} = \mathrm{B_i}\left(\frac{\mathrm{G_{it}}}{\mathrm{P_{it}}}\right)^{\delta}(0 < \delta < 1). \tag{12.5}$$

In Eq. (12.5), B_i is the exogenous natural endowment condition of city i, which does not change over time. G_{it} is government expenditure on public service, δ is the efficiency of public service expenditure, and P_{it} is the degree of urban air pollution. Urban air pollution is negatively correlated with the degree of urban livability. That is, the more serious the air pollution, the lower the degree of urban livability. The urban housing land is denoted by $(1 - \varepsilon)L_{it}$. In the equilibrium status, the housing consumption of the representative residents refers to the ratio of housing land to the equilibrium urban population.

12.3.3 Migration Decisions of Representative Residents

Depending on the expected utility, representative residents firstly decide whether to migrate or stay. And then maximize the utility through consumption and investment decisions. The following two sides are considered to describe the trade-off of migration, the welfare loss due to the migration costs and the welfare increment of migration incentive. To calculate the expected utility, we refer to the research of Eaton and Kortum (2002) and Tombe and Zhu (2019). Modeling the migration costs as proportional to utility. It is assumed that the migration costs cause a certain loss to resident's welfare during migration, which is measured as T_{it} $(0 \leq T_{it} \leq 1; i \in n)$. Thus, the migration costs are $T_{it} U_{it}^{New}$, in which U_{it}^{New} is the utility of the destination city where residents move to. It is assumed that the migration behavior completes in the current period.

It is also an assumption that th_i means an increment of welfare due to migration. By giving the distribution function of th_i, we could solve the migration costs of each city. Tombe and Zhu (2019) and Duan et al. (2020) introduced Frechet distribution into the analysis of migration for implying welfare change (welfare loss or improvement) of migration. It is useful in relating the migration to the underlying characteristic of different regions, and solving the migration proportion. Initially, Frechet distribution is commonly applied to reflect the different distribution of technology and productivity in a quantitative model (Eaton & Kortum, 2002; Tombe & Zhu, 2019) and then widely used in other aspects. Hence, the th_i followed a Frechet distribution and could be considered as the heterogeneous incentive factors of migration, such as more productivity, diverse working opportunity, better living environment. The distribution formula is $F(th) = e^{-th^{-\theta}}$, where θ means the dispersion situation of th_i. The higher the value of θ, the denser the distribution. Hence, the migration probability is shown as below:

$$M_{it} = Pr\left[(1 - T_{it})\text{th}_i U_{it}^{New} \geqslant \text{th}_j U_{jt,j\neq i}^{Pre}\right] \tag{12.6}$$

Then, based on the solving logic of Tombe and Zhu (2019), the solution equation for the migration probability is $(1 - T_{it})^{\theta} U_{it}^{New\theta} / \left[(1 - T_{it})^{\theta} U_{it}^{New\theta} + U_{jt,j\neq i}^{Pre\ \ \theta}\right]$. Then, the new expected utility function is measured as $U_{it}^{E} = U_{it}^{New} M_{it}^{\frac{1}{\theta}}$.

In Eq. (12.6), $U_{jt,j\neq i}^{Pre}$ is the utility of the initial city where residents move from. Only when the net welfare after deducting migration costs is higher than initial welfare, the migration behavior of representative residents could happen. The calculated migration condition means that if $(1 - T_{it})\frac{U_{it}^{New}}{U_{jt,j\neq i}^{Pre}} < \frac{\text{th}_j}{\text{th}_i}$, it implies that migration can't improve the representative resident's utility and the initial urban size distribution is an equilibrium status. Here, we could find that the increment of welfare is of great significance. If the potential welfare increment of migration exceeded migration costs, it may be beneficial for residents.

Given the first-stage decision, it means that the possible migration is complete. After that, representative residents make the consumption and investment decisions to maximize utility. This problem is:

$$\max_{C_{it}, H_{it}, I_{it}} U_{it} \tag{12.7}$$

$$s.t: p_{it}^{C} \mathrm{C}_{it} + p_{it}^{H} \mathrm{H}_{it} + (1 - \mathrm{tax})\mathrm{w}_{it}. \tag{12.8}$$

Residents' decision variables include commodity consumption, housing consumption and health investment expenditures. The total expenditure of a resident shall not exceed the disposable income of the resident in the current period $(1 - tax)w_{it}$. Here, w_{it} is the labor wage, and tax is the tax rate. Given the C-D type technologies, disposable income could be expressed as $(1 - tax)\ w_{it}\ (1 - \gamma)\ \mu A_{it}\ N_{it} = \beta Y_{it}$. The optimal utility makes the model in the equilibrium state, and the representative resident's utility of each city will be equal. Rearranging, we obtained that $U_{jt,j\neq i}^{Pre} = U_{it}^{New} = \overline{U}_{t}$.

For simplification, the housing price p_{it}^{H} is normalized to 1. Lagrange multiplier method is used to solve the above $max\{U_{it}\}$ problem, and the equilibrium condition of representative residents is obtained:

$$\frac{H_{it}}{C_{it}} = \frac{(1 - \alpha_1)\beta_1}{\alpha_1 \beta_1} \times p_{it}^{C} \tag{12.9}$$

$$\frac{I_{it}}{C_{it}} = \frac{(1 - \beta_1)\zeta}{\alpha_1 \beta_1} \times p_{it}^{C}. \tag{12.10}$$

In addition, the government revenue in the model is $tax \times w_{it}$. Government expenditures are mainly expenditures for public service construction G_{it}. The government is in a state of balance of payments in the current period, without considering the inter-period lending, issuing government bonds, and other financing and investment activities. It means $G_{it} = tax \times w_{it}$.

12.3.4 Spatial Equilibrium and Static Analysis

When the model reaches a spatial equilibrium, residents move between cities until the utility level of the representative residents in each city is equal, and the equilibrium utility is expressed as $\overline{U}_{t}$. Meanwhile, the housing consumption of representative residents in equilibrium is $H_{it}^{*} = (1 - \varepsilon)L_{it}^{*}/N_{it}^{*}$ and commodity's consumption is $C_{it}^{*} = Y_{it}^{*}/N_{it}^{*}$. For the labor market, the output sector employs labor for production after the size of the city is determined. And labor obtains wages according to their marginal output. Substituting housing consumption and commodity consumption conditions into Eq. (12.3), the population of each city in equilibrium can be obtained and expressed as:

$$N_{it}^{*} = [\frac{M_0 P_{it}^{-\delta-\kappa(1-\beta_1)} G_{it}^{\delta} Y_{it}^{\alpha_1\beta_1} L_{it}^{(1-\alpha_1)\beta_1+\zeta(1-\beta_1)}}{\bar{U}_t}]^{\frac{1}{\alpha_1\beta_1+(1-\alpha_1)\beta_1+\zeta(1-\beta_1)}} \quad (12.11)$$

In Eq. (12.11), $M_0 = \left[\frac{\zeta(1-\beta_1)}{\alpha_1\beta_1}\right]^{\zeta(1-\beta_1)} \left[\frac{(1-\varepsilon)\alpha_1\beta_1}{(1-\alpha_1)\beta_1}\right]^{\zeta(1-\beta_1)} (1-\varepsilon)^{\beta_1(1-\alpha_1)} B_i$. For the commodity price in Eq. (12.11), it can be obtained by Eq. (12.9), $p_{it}^{c} = \frac{\alpha_1\beta_1}{(1-\alpha_1)\beta_1} \frac{(1-\varepsilon)L_{it}^{*}}{Y_{it}^{*}}$. When the model is in spatial equilibrium, the equilibrium conditions Eqs. (12.9) and (12.10) of representative residents are substituted into the budget constraints Eqs. (12.8) and (12.3). Then the equilibrium expected utility of representative urban residents is calculated as indirect utility function by:

$$\bar{U}_{it}^{E} = M_3 B_i P_{it}^{-\delta-\kappa(1-\beta_1)} w_{it}^{M_2} (p_{it}^{c})^{-\alpha_1\beta_1} M_{it}^{\frac{1}{\theta}}. \quad (12.12)$$

Where $M_1 = 1 + \frac{(1-\alpha_1)\beta_1}{\alpha_1\beta_1} + \frac{(1-\beta_1)\zeta}{\alpha_1\beta_1}$, $M_2 = \alpha_1\beta_1 + (1-\alpha_1)\beta_1 + (1-\beta_1)\zeta$, $M_3 = tax^{\delta}\left(\frac{1-tax}{M_1}\right)^{M_2}\left(\frac{(1-\alpha_1)\beta_1}{\alpha_1\beta_1}\right)^{(1-\alpha_1)\beta_1}\left(\frac{(1-\beta_1)\zeta}{\alpha_1\beta_1}\right)^{(1-\beta_1)\zeta}$.

The overall welfare of urban residents[1] is expressed by the product of the equilibrium urban population and the expected utility of urban representative residents. Put in the equilibrium conditions of the labor market, multiplying Eqs. (12.11) and (12.12) to obtain the overall welfare of urban residents:

$$U_t^{*} = \sum_{i=1}^{n} f_{it} \bar{U}_{it}^{E} N_{it}^{*M_4} = \sum_{i=1}^{n} f_{it} M_0 P_{it}^{-\delta-\kappa(1-\beta_1)} G_{it}^{\delta} Y_{it}^{\alpha_1\beta_1} L_{it}^{(1-\alpha_1)\beta_1+\zeta(1-\beta_1)} M_{it}^{\frac{1}{\theta}} \quad (12.13)$$

In the above formula, $N_{it}^{*M_4}$ is added to simplify the calculation, $M_4 = M_2 + \delta$, $\phi_{it} = N_{it}^{o}/\sum_{k=1}^{n} N_{kt}^{o}$, N_{it}^{o} means the initial population in every city. Then U_t^{*} is the overall welfare of urban residents. To simplify the analysis, it is assumed that U_{it}^{*} is equal to $\phi_{it}\overline{U}_{it}^{E} N_{it}^{*M_4}$, and U_t^{*} is equal to $\sum_{i=1}^{n} U_{it}^{*}$. By differentiating $U_{it}^{*} = \phi_{it}\overline{U}_{it}^{E} N_{it}^{*M_4}$, the derivative of concerned variables can be got:

$$\frac{dU_{it}^{*}}{dP_{it}}<0, \frac{dU_{it}^{*}}{dT_{it}}<0, \frac{dU_{it}^{*}}{dY_{it}}>0, \frac{dU_{it}^{*}}{dL_{it}}>0, \frac{dU_{it}^{*}}{dG_{it}}>0 \quad (12.14)$$

It can be seen that the increase in urban air pollution and in $T_{i,t}$ will reduce the overall welfare of urban residents. Increased economic output, increased public service construction expenditure and increased land supply will improve the overall

[1] By calculating the partial derivative of the air pollution variable from equation (12.11): $\frac{dN_{it}^{*}}{dP_{it}} < 0$, it can be found that the worse the air quality in a city, the smaller its population size. The decline in the size of urban population makes it difficult for cities to exert agglomeration effects of population and economy, and damages the overall welfare of urban residents. Therefore, this paper uses the product of the equilibrium population size and the welfare of representative residents to represent the overall welfare of urban residents.

welfare of urban residents. Further, by taking the second derivatives of air pollution and migration costs, these can be got:

$$\frac{d^2U_{it}^*}{d(P_{it})^2} < 0, \frac{d^2U_{it}^*}{dp_{it}dT_{it}} > 0. \tag{12.15}$$

The above formula shows that the negative impact of air pollution on the overall welfare of urban residents is increasing. The more serious the urban air pollution, the more significant the negative impact of air pollution on the overall welfare of urban residents will be. This means that if air pollution is not treated in time, the cumulative effect of air pollution will increase, as will the negative impact on the welfare of urban residents. From the perspective of the transmission mechanism, it can be known that the higher the fraction loss of resident's welfare caused by migration costs, the less the migration benefits. As the migration costs increase, the negative effect of air pollution on the overall welfare of urban residents gradually increases. Despite the derivative and Hessian matrixes of each variable could be obtained through differentiating Eq. (12.13) directly, but the calculation process is more complex than before.

In the previous analysis, it is assumed that migration would be completed in one period. Here, considering the complexity of migration, such as finding a place to live, long-distance migration, and the uncertainty in the migration process, it is assumed that migration is an inter-temporal behavior. Representative residents still make migration decisions in period t, we obtain the new expected utility, $\left(U_{it}^E\right)^{'} = e^{-\overline{r}|t^{'}-t|}\left(U_{it}^{New}{}^{'}\right)^E M_{it}^{\frac{1}{\theta}}, \left(U_{it}^{New}{}^{'}\right)^E$ represents the expected utility in period $t^{'}$ calculated based on the utility of the place moved to in period t. To solve the above formula, it is needed to define the inter-period utility function. It means the traditional independence assumption of solving infinite period utility need to be relaxed. The new condition of migration is $\left(1 - T_{it,t^{'}}\right)\frac{e^{-\overline{r}|t^{'}-t|}\left(U_{it}^{New}{}^{'}\right)^E}{U_{jt,j\neq i}^{Pre}} \geq \frac{th_{j,t}}{th_{i,t^{'}}}$. The hypothesis of inter-temporal migration changes the expected utility function of representative residents and increases the complexity of calculation, but no new findings appear.

Based on the above analysis, this paper puts forward the following hypotheses:

Hypothesis 1: Air pollution has a significant negative impact on residents' welfare.

Hypothesis 2: A reduction in residents' migration costs will contribute to residents transferring between cities and moving away from cities with severe air pollution, thereby improving the welfare of urban residents.

12.4 Estimated Results

12.4.1 Parameter Setting and Characteristic Variable Estimation

To make the model matching reality as closely as possible, it is necessary that calibrate model parameters and estimate model characteristic variables here. Before the estimation, the relevant parameters in the model are set first, and the main urban characteristic variables in the model are estimated.

The parameter settings in this paper are as follows.

1. According to the research on China's return on capital investment by Bai and Qian (2006), the equilibrium interest rate $\overline{r}$ is set to 0.2008. Pan et al. (2018) measured the capital output elasticity of the output function after including the land factors, with 1-α-β set to 0.425.
2. Regarding the setting of labor output elasticity, Wu et al. (2015) believed that with the development of the economy, the output elasticity of capital would dwindle, while the output elasticity of labor will gradually increase. Hence, according to the research conclusion of Bai and Qian (2006), the labor output elasticity is calculated, and β is set to 0.4779, higher than the 0.4 set by Wang and Fan (2000) and the 0.4687 set by Duan et al. (2020). The land output elasticity α is calculated to be 0.0971.
3. With reference to the results of China's sixth census in 2010, μ is set to 0.75. According to the research on environment, health, and economic growth by Chen and He (2017), the negative impact of air pollution on labor efficiency γ is set to 0.045. For the estimation of air pollution, referring to the research of Ye and An (2017), α'' set to 0.9975, α' set to 0.0052 and β' set to 0.4.
4. In this paper, the industrial land, transportation and storage land, and commercial and public construction land in the *China Urban Construction Statistical Yearbook* are regarded as production land,[2] and residential land is regarded as housing construction land. The sum of the two land-use types is the total available urban land. The average proportion of the production land to the total usable land from 2002 to 2017 is calculated to be 0.6445. Herein ε is set to 0.6, given that some land supply is not included in the calculation.
5. With reference to Wang's (2018) study on the correlation between the welfare and health of Chinese urban residents, the contribution of residents' health to utility $1-\beta_1$ is set to 0.22. With reference to Duan et al.'s (2020) estimation of Chinese urban residents' welfare based on the Rosen-Roback model, $\alpha_1\beta_1$ is set to 0.667, and the efficiency of public service expenditure δ is set to 0.3. Then the contribution of housing consumption to residents' utility $(1-\alpha_1)\beta_1$ is calculated

[2] Production land specifically includes land for public administration and public services, land for commercial service facilities, industrial land, land for logistics and storage, and land for transportation and public facilities prescribed in *China Urban Construction Statistical Yearbook*.

to be 0.113. Based on the estimation of Duan et al. (2020), the dispersion situation θ is set to the estimated average value of 1.92.

6. According to Qi and Lu's (2015) research on how to overcome "the environmental health poverty" trap, the health investment expenditure efficiency herein ζ is set to 0.12. Pautrel (2009) studied the relationship between pollution and residents' health. This paper follows his parameter settings of the impact of pollution on health and sets κ to 0.025. The tax rate *tax* is set to the median 0.25 of the excess progressive tax rates of personal income at each stage in China. The parameter settings in this article are shown in Table 12.1.

Table 12.1 Parameter setting

Parameter	Calibration value	Implication	Parameter's source
β	0.4779	The labor output elasticity	Bai and Qian (2006)
$\overline{r}$	0.2008	The equilibrium interest rate	Bai and Qian (2006)
α	0.0971	The land output elasticity	Calculated by the authors
μ	0.75	The rate of the urban population in the labor	China's sixth census
γ	0.045	The impact of air pollution on labor efficiency	Chen and He (2017)
$\alpha_1\beta_1$	0.667	The contribution of commodity consumption	Duan et al. (2020)
ζ	0.12	The health investment expenditure efficiency	Qi and Lu (2015)
ε	0.6	The proportion of land supply for production	Calculated by the authors
δ	0.3	The efficiency of public service expenditure	Duan et al. (2020)
θ	1.92	The dispersion situation of *th*	Duan et al. (2020)
$1-\beta_1$	0.22	The contribution of residents' health to utility	Wang (2018)
$(1-\alpha_1)\beta_1$	0.113	The contribution of housing consumption	Calculated by the authors
κ	0.025	The impact of pollution on health	Pautrel (2009)
α''	0.9975	The natural purification capacity	Ye and An (2017)
α'	0.0052	The impact of production on environment	Ye and An (2017)
β'	0.4	Abatement elasticity of productivity	Ye and An (2017)
tax	0.25	Tax rate (tax rate of personal income)	Calculated by the authors

Data source According to the related literatures reviewed by the authors

The relevant characteristic variables are estimated as follows:

1. Referring to Liu and Li (2017), Eq. (12.1) is converted to obtain the estimated expression of the overall urban productivity A_{it}. Since it is difficult to obtain the urban capital stock, the partial derivative of Eq. (12.1) is taken to obtain the correlation between capital input and urban population: $K_{it} = \frac{w_{it}}{\overline{r}}(1-\alpha-\beta)N_{it}$, where $\overline{r}$ is the capital equilibrium interest rate. By substituting this formula into the expression of A_{it}, the estimated A_{it} can be obtained.
2. To estimate the natural endowment condition of a city B_i, this paper refers to Desmet and Rossi-Hansberg (2013) and sets the equilibrium utility of representative residents in 2007 $\overline{U}_{2007}$ to 1. Substituting this value into Eq. (12.11), then the natural endowment conditions B_i of each city can be solved.
3. Due to the air pollution is mainly caused by economic activities, we do not simply used the observed air pollution concentration but measured the air pollution by using the Eq. (12.2). Using Eq. (12.2), simulation values of air pollution are measured. To compare the accuracy of simulation values, the relationship test result is shown in Appendix Fig. 12.3. It showed a significant positive correlation between simulated values and real values, and meant that the simulation values of air pollution are pretty valid.
4. Due to the absence of more detailed bilateral migration population data, we were inspired by Cai and Wang (2009) research method to measure the migration probability. It is expressed as: $M_{it} = \frac{(\textit{resident population} - \textit{registered population})}{\textit{registered population}} - \textit{natural population growth rate}$. It implied the capacity of a city to absorb the residential population after deducting the natural growth. This rate is regarded as the proxy variable of migration probability. Substituting the migration probability into the migration probability expression result from Eq. (12.6), then the T_{it} of each city can be obtained. The average value of migration costs in the sample period is 0.595 (please see Appendix Table 12.7). Also, Table 12.1 in the Appendix shows that migration costs shrank by 42.5% from the initial year of 2007. It is constant with the policies of the Chinese government promoting the free movement of labor.
5. In addition, referring to Chen and Chen et al. (2018), the proportion of environment-related words in central government work reports is seen as the proxy variable of V_t. It is only used the central government work report as the sample to alleviate the possible endogenous problems.

12.4.2 Welfare Estimation and Inter-Period Welfare Changes

To analyze the effect of the changes of air pollution and migration costs on the resident welfare, let x denote variables in the model and $x^{'}$ denote the value in the counterfactual situation. Then, $\hat{x} = \frac{x'}{x}$ means the proportional change under the counterfactual situation. $\hat{x} > 1$ implies that x increased by $x(1-\hat{x})$. This is a research methodology proposed by Dekle et al. (2008). First, to calculate the counterfactual population, we originally have $\hat{P}_{it} = \frac{(1-\hat{V}_t V_t)}{(1-V_t)} \hat{Y}_{it}/\hat{A}_{it}^{\beta'}$, $\hat{G}_{it} = \hat{w}_{it}$ and $\hat{w}_{it} = \hat{Y}_{it}/(\hat{A}_{it}\hat{N}_{it}^*)$. From Eq. (12.11), it implies counterfactual population is:

$$\hat{N}_{it}^* = [\frac{\hat{B}_i \hat{P}_{it}^{-\delta-\kappa(1-\beta_1)} \hat{G}_{it}^{\delta} \hat{Y}_{it}^{\alpha_1\beta_1} \hat{L}_{it}^{(1-\alpha_1)\beta_1+\zeta(1-\beta_1)}}{\widehat{U}_t}]^{\frac{1}{M_2}} \tag{12.16}$$

Then, we need to solve the counterfactual overall welfare of urban residents. The counterfactual migration probability is $\hat{M}_{it} = (1-\hat{T}_{it}T_{it})^{\theta}/M_{it}\left[(1-\hat{T}_{it}T_{it})^{\theta}+1\right]$. Therefore, the counterfactual expected utility is $\widehat{U}_{it}^{E} = \hat{B}_i \hat{P}_{it}^{-\delta-\kappa(1-\beta_1)} \hat{w}_{it}^{M_2} \left(\hat{p}_{it}^{c}\right)^{-\alpha_1\beta_1} \hat{M}_{it}^{\frac{1}{\theta}}$, where $\hat{p}_{it}^{c} = \hat{L}_{it}/\hat{Y}_{it}$. Rewritten the overall welfare of urban residents by the above method, Eq. (12.13) becomes:

$$\widehat{U}_t^* = \sum\nolimits_{i=1}^{n} \chi_{it} \widehat{U}_{it}^{E} \widehat{N}_{it}^{*M_4}. \tag{12.17}$$

Here, $\hat{U}_t^* = \sum_{i=1}^{n} \chi_{it} \widehat{U}_{it}^{E} \hat{N}_{it}^{*M_4}$.

This paper uses the data from 274 cities at the prefecture-level and above in China to estimate the welfare of urban residents. Specifically, the urban economic output Y_{it}, urban construction land L_{it}, and urban population data N_{it}^{o} in 2007–2017 are from *China Statistical Yearbook*, *China Urban Construction Statistical Yearbook*, and *China City Statistical Yearbook*. The economic output Y_{it} is all deflated to 2007 using the consumer price index. Before calculating the resident's welfare change, and the effect of air pollution and migration costs on resident's welfare, it is necessary to test the efficiency of the initial equilibrium of our model. Using Eq. (12.13), we measured the overall resident's welfare of each city in China.

First, from Eq. (12.13), the size of the equilibrium welfare estimation results of the model are shown in Fig. 12.2. It is observed that the average welfare of urban residents increases gradually over time. The welfare of urban residents in China increased by 169.32% from 2007 to 2017, 60.39% from 2007 to 2012, and 106.91% from 2012 to 2017. The growth rate of residents' welfare has increased, which is not in line with the fact that China's industrial structure was upgrading, China has entered a new normal in economic development, and its economic growth has declined. This means that the improvement of resident's welfare is continuous and cumulative. The higher the economic level, the more obvious the improvement of residents' welfare. Moreover, after migration costs and air pollution are considered in the model, the estimation results show that the average growth rate per annum of welfare is 9.67%. Compared with the Chinese welfare calculation results of Duan et al. (2020) (without considering migration costs and pollution factors), it is less than the results of Duan et al. (2020) by 10.94%. Therefore, promoting free movement of labor and increasing air pollution abatement is of great importance in boosting the resident's welfare.

Next, we consider how the importance of air pollution abatement and reducing migration costs on resident's welfare? Air pollution change is concerned: as we need

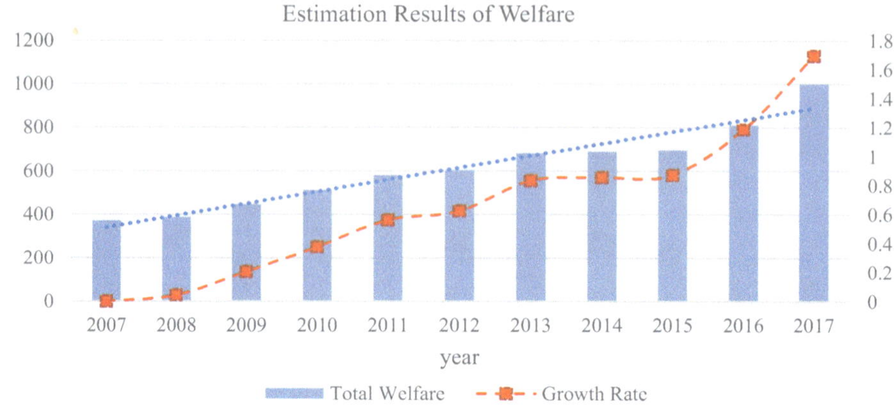

Fig. 12.2 Estimation results of welfare. *Notice* (1) Total welfare is calculated by Eq. (12.13); (2) The growth rate of welfare is calculated by $\frac{U_t^* - U_{2007}^*}{U_{2007}^*}$, and the unit is %; (3) The value in the left axis of the above Figure represents the value of welfare. The larger the value, the higher the welfare

to reflect the influence of the change of air pollution on welfare, we adjusted the air pollution control intensity of government firstly and then introduced the counterfactual air pollution into Eq. (12.17). Specific as follows: (1) If $\hat{V}_t = 2$ (it means the regulation intensity improved 100%), the improvement of welfare is 2.89%; (2) If $\hat{V}_t = 3$, the gain of welfare is 6.2%; (3) If $\hat{V}_t = 4$, total welfare of residents would increase by 10.04%; (4) we also set the $\hat{V}_t = 1.032$ based on the measurement results change in the estimation (5) of characteristic variables in Sect. 12.4.1 from 2007–2017, and the welfare gain of the change is only 0.087%. It implied that the improvement of air pollution abatement by the government could enhance resident's welfare and the government can still significantly increase the intensity of environmental regulation for improving air quality to improve residents' welfare. That is consistent with Hypothesis 1. Besides, the effect of air pollution control is increasing return. As the intensity of regulation increases, the improvement of residents' welfare increases.

On the other side, Lower migration costs is concerned: (1) Set a perfect free movement of labor state $\hat{T}_{it} = 0$, it means that residents have the obvious free in choosing a city to settle down. The $\hat{T}_{it} = 0$ has the total welfare gains of 78.7%; (2) To show the effect of the changes of migration costs on welfare, we set the $\hat{T}_{it} = 0.896$ (Change of migration costs from 2007–2012) from the Appendix Table 12.8. The welfare results could increase by 10.9%; (3) Furthermore, from the Appendix Table 12.8, it also sets the $\hat{T}_{it} = 0.575$ (Change of migration costs from 2007–2017). The increase of resident's welfare is 41.59%. It is in line with Hypothesis 2.

Another important question is how the effect of comprehensive change of both air pollution abatement and migration costs on welfare? we choose the set (4) of the air pollution change and the set (3) of the lower migration costs as the comprehensive situation. It is found the welfare's improvement is 41.74%. Hence, from a migration and space equilibrium perspective, the decline of migration costs and the improvement of air quality accounts for roughly 24.65% improvement of welfare from 2007–2017 in China. Despite the past change of air pollution and migration costs, the potential welfare gain of reform about lower migration costs and higher air quality is still large. Further Policies, such as Hukou registration reform, more accurate air pollution regulation, are still very meaningful.

12.5 Empirical Analysis

12.5.1 Measurement Model Design and Data Description

Taking the logarithm of Eq. (12.12), and taking the logarithm of urban residents' welfare as the explanatory variable, and the logarithm of air pollution as the explanatory variable, the regression model can be expressed by:

$$\ln \mathrm{U_{it}} = \alpha_0 + \alpha_2 \ln \mathrm{P_{it}} + \sum \theta \mathrm{X_{it}} + \mathrm{y_t} + \mathrm{c_i} + \varepsilon_{\mathrm{it}}. \tag{12.18}$$

In Eq. (12.18), U_{it} is the residents' welfare of city i in year t, P_{it} is the concentration of air pollution in every city. α_2 is predicted to be negative, indicating that air pollution has a negative impact on urban residents' welfare, and X_{it} is control variables. The last three terms of the formula are the year fixed effect, the city fixed effect and the error term, respectively.

The data description is as follows. First, the measurement of welfare. (1) An important method of estimating welfare is index construction. A well-known resident's welfare index (Human Development Index (HDI) reported by the United Nations) includes not only the GDP per capita, but also the life expectancy and the adult literacy. It reflects the economic level, education situation and medical and environmental conditions of a region. Therefore, according to the research of Wang (2018), the welfare index in this paper of urban residents is divided into living welfare, basic welfare, continuous welfare, security welfare and environmental welfare. And they are respectively represented by the city's per capita GDP, the number of public transport vehicles per 10,000 people, the number of college students, the number of beds per 10,000 people, and the urban green coverage. The former reflects economic conditions, the latter two represent the levels of education and infrastructure, and the last two are proxies for the environment and health. The year 2007 was set as the base year, and the weight of each component was obtained by factor analysis. By calculating the weight of the above five variables, the total welfare index of urban residents is calculated by the weighted expressions of the above five indicators. The above total welfare index is a proxy variable of resident's welfare (welfare index). In addition, KMO tests of the factor analysis are made to verify the reasonability of the above analysis. The KMO test results are valid with an average value 0.6609.

(2) However, the above method is in the debate because of lacking theoretical background. Further, by constructing a macroeconomic model, welfare is discomposed to life expectancy, consumption share, leisure, consumption inequality and leisure inequality in the research Jones and Klenow (2016), Chen and Li (2018). Despite other factors, it still shows a high positive correlation between GDP per capita and resident's welfare. Hence, subjected to data availability, GDP per capita is used as a proxy variable of resident's welfare (GDP per capita) with a strong theoretical basis (the related regression results shown in the Appendix).

Second, migration costs are replaced by two indicators to investigate the regulating effect of migration costs. (1) The migration costs calculation results in Sect. 12.4.1 are used as a proxy variable of migration costs. (2) Another important barrier to migration in China is hukou control. With reference to the research of Deng and Ding (2010), this paper divides the number of new mobile phone users in the city by the number of newly registered population, and uses this index to reflect the urban household registration control (HRC). This index reflects that the ratio of the number of people who are willing to settle in the local area (new mobile communication equipment users) to the actual number of people who settled in that year (newly registered population). This ratio could reflect the HRC according to He (2004). In reality, the gap between the above two indicators is not only determined by the intensity of household registration control (HRC), but also by economic factors and individual heterogeneity. This may be an overestimate, but it basically covers the effect of HRC. The larger the value, the stricter the household registration control. It is used as a substitute scalar for household registration control and is processed logarithmically.

Finally, urban $PM_{2.5}$ is selected to represent the concentration of air pollution. The control variables adopted in this paper include urban GDP, urban road area, the proportion of the secondary industry, the number of Internet users, and urban residents' consumption of water, gas and electricity. Some missing data are filled with the mean value of data of adjacent cities or adjacent years. The data sources of this paper are *China Statistical Yearbook*, *China City Statistical Yearbook*, and *China Statistical Yearbook for Regional Economy*. $PM_{2.5}$ data from 2007 to 2016 are from SEDAC,[3] and $PM_{2.5}$ data in 2017 are from *China City Statistical Yearbook*. Table 12.2 lists the data sources, and Table 12.3 reports the statistical descriptions of data.

12.5.2 Benchmark Results: The Impact of Air Pollution on Residents' Welfare, and Migration Costs

Table 12.4 reports the benchmark regression results. The first two columns are the estimated results of the cross-sectional data model, and it shows that air pollution has a significant negative impact on the welfare of urban residents. The estimates in

[3] *Sources* NASA Socioeconomic Data and Applications Center. Address: https://doi.org/10.7927/H4ZK5DQS.

Table 12.2 The data source's summary of this article

Variables	Data source
Urban population, urban construction land	2007 to 2017 from *China urban construction statistical yearbook*
Urban economic output	2007 to 2017 from *China City statistical yearbook*
$PM_{2.5}$	2007 to 2016 from SEDAC
	2017 from *China City statistical yearbook*
The control variables in this article	2007 to 2017 from *China City statistical yearbook*
Urban resident's average urban deaths	2007 to 2012 from *China statistical yearbook for regional economy*

Table 12.3 The statistical descriptions of related data from 2007–2017

Variables	Mean	SE	Min	Max
Resident's welfare (welfare index)	38.79	18.48	0.40	206.66
$PM_{2.5}$ (Ug/m^3)	38.08	16.72	4.68	116
MC	0.59	0.26	−2.5	0.99
Ln (HRC)	2.21	1.49	−5.29	8.32
GDP (100 million yuan)	1470.71	2082.08	25.82	22626.39
URA (million square meters)	1761.22	4304.85	1	175,500
PSI (%)	48.78	10.53	13.57	85.08
NIU (ten thousand people)	69.59	135.53	0.02	5174
UCW (ten thousand tons)	5575.98	11484.55	70	105,493
UCE (million kilowatt/hour)	1091.45	2022.25	11.84	22528.91
UCG (thousands of cubic meters)	8023.39	38178.38	1	1,622,393

Notice (1) GDP was deflated to 2007 using the consumer price index; (2) In the above Table, "MC" means migration costs, "HRC" is Household registration control, "URA" means urban road area, "PSI" means the proportion of the secondary industry, "NIU" means the number of internet users, "UCW" means the urban resident's consumption of water, "UCE" means the urban resident's consumption of electricity, "UCG" means the urban resident's consumption of gas; (3) Variables of transmission mechanism analysis are introduced in the section "analysis of transmission mechanism"

Column (2) show that an average of 1% increase in $PM_{2.5}$ reduces urban residents' welfare by 0.137% on average, which verifies Hypothesis (1).

Taking into account the differences between cities and the time trend, Columns (3) to (6) are the estimation results of the static panel model. It can be observed that air pollution still has a negative impact on the welfare of urban residents. Estimation results in Column (4) show that the negative impact of air pollution on urban residents' welfare is still significant even after considering the possible lag effect of air pollution. Hypothesis (1) is again proved by the estimation in Column (5), which has taken into account the possible estimation bias caused by the omitted variables. Chen and Chen et al. (2018) believed that aggravated air pollution is beneficial to economic growth in

Table 12.4 The benchmark results of regression

Variables	Resident's welfare (welfare index)						
	The cross-sectional data model		The static panel model				The dynamic panel model
	(1)	(2)	(3)	(4)	(5)	(6)	(7)
	Year 2007	The annual average					Systematic GMM method
Lag. (resident's welfare)							0.0572***
							(0.0197)
$PM_{2.5}$	−0.310***	−0.137***	−0.26***		−0.239***	1.157***	−0.104***
	(0.0537)	(0.0378)	(0.0503)		(0.052)	(0.365)	(0.0319)
$(PM_{2.5})^2$						−0.209***	
						(0.053)	
Lag. $PM_{2.5}$				−0.341***			
				(0.0688)			
Control variables	Yes	Yes	No	No	Yes	Yes	Yes
FE of cities			Yes	Yes	Yes	Yes	
FE of years			Yes	Yes	Yes	Yes	Yes
Test of AR (1)							0.000
Test of AR (2)							0.138
Constant	3.452***	3.266***	4.288***	4.622***	3.969***	1.671***	3.11***
	(0.213)	(0.177)	(0.202)	(0.263)	(0.229)	(0.644)	(0.147)
Observations	274	274	3008	2734	3007	3007	2732
Adjusted R^2	0.383	0.421	0.632	0.625	0.635	0.637	

Notice (1) robust standard error in parentheses; (2) ***, **, * respectively express significant at the significance level of 1%, 5% and 10%; (3) "The annual average" in the second column is the annual average from 2007 to 2017 the early stage of economic development. In order to verify whether this view is also valid between air pollution and resident's welfare, the second term of air pollution is added to the estimation results in Column (6), and it is found that $PM_{2.5}$ has an inverted U-shaped relationship with urban residents' welfare. This means that in the early stages of economic development, urban air pollution can improve the welfare of urban residents. However, as air pollution continues to increase, the daily life and health of urban residents are affected, and air pollution has a significant negative effect on the welfare of urban residents.

Column (7) is the estimation result of the dynamic panel model. Considering the inverse causality between variables, the estimation in Column (7) uses the systematic GMM estimation method of the dynamic panel data model. The results show that the negative impact of $PM_{2.5}$ on the welfare of urban residents is still significant, which is consistent with the estimated results in the first six columns. The result of the AR test shows that the GMM estimation result is effective, which better solves the endogenous problem of the model. At the same time, the estimates in Column (7) also indicate that there is a cumulative effect on urban residents' welfare over time. That is, the higher the welfare of urban residents in the previous period, the larger the increase in urban residents' welfare in the next period. It is in line with the conclusion in Sect. 12.4.2.

Table 12.5 presents the estimates when migration costs are taken into account. The first three columns are the estimated results of the indicator migration cost (MC). The results show that both $PM_{2.5}$ and migration costs have a significant negative impact on the welfare of urban residents. Comparing Column (1) and (2), a simple conclusion is that the negative effect of migration costs (MC) is reducing with time passed. Estimation in Column (3) includes the interaction term between $PM_{2.5}$ and migration costs, and it is found that the higher the migration costs (MC), the greater the negative impact of $PM_{2.5}$ on urban residents' welfare. The higher the migration costs (MC), the higher the cost for urban residents to reduce the negative impact of air pollution on their welfare through migration, thus amplifying the negative impact of air pollution on urban residents' welfare, which verifies the view of Hypothesis (2). That is, reducing the migration costs (MC) of urban residents and strengthening the mobility of residents between cities can improve the welfare of residents.

The last three columns re-estimate the impact of migration costs through the indicator household registration control (HRC). The substitution variable of the migration costs in the last three columns is the household registration control (HRC) of each city, which is processed in logarithm. This is a special kind of migration limitation in China. From the estimated results in the last three columns, it is found that the household registration control (HRC) has an overall negative yet insignificant impact on the welfare of urban residents. However, from 2007 to 2012, the household registration control (HRC) has a significant negative effect on urban residents' welfare. The regression results in Column (6) show that the higher the household registration control (HRC), the greater the negative impact of $PM_{2.5}$ on urban residents' welfare, which is consistent with the estimated results of the first three columns. To be more specific, the negative impact of the household registration control (HRC) on urban residents' welfare is greater from 2007 to 2012 and not significant from 2007–2017.

Table 12.5 The estimation results with migration costs

Variables	Resident's welfare (welfare index)					
	Migration cost (MC)			Household registration control (HRC)		
	(1)	(2)	(3)	(4)	(5)	(6)
	2007–2017	2007–2015	2007–2015	2007–2017	2007–2012	2007–2012
$PM_{2.5}$	−0.264***	−0.510***	−0.425***	−0.220***	−0.460***	−0.445**
	(0.0554)	(0.0984)	(0.109)	(0.0676)	(0.176)	(0.177)
MC	−0.0321	−0.0669*				
	(0.0242)	(0.0352)				
MC × $PM_{2.5}$			−0.0206**			
			(0.00998)			
HRC				−0.00732	−0.0263**	
				(0.00692)	(0.0113)	
HRC × $PM_{2.5}$						−0.00705**
						(0.00321)
Constant	4.213***	5.230***	4.951***	4.036***	5.758***	5.692***
	(0.264)	(0.411)	(0.393)	(0.279)	(0.653)	(0.653)
Observations	2927	2425	2425	2024	1457	1457
Adjusted R^2	0.631	0.622	0.622	0.677	0.656	0.656

Notice (1) robust standard error in parentheses; (2) ***, **, * respectively express significant at the significance level of 1%, 5% and 10%

The reason is that with the reform of China's household registration system, the impact of migration costs brought by household registration control (HRC) on the welfare of urban residents has gradually decreased. Hence, for the entire sample interval, although the estimated result is still negative, it is insignificant. Another finding is that the decline of migration costs (significant from 2007–2015) in the first three column lags behind the relaxation of household registration control (significant from 2007–2012). It implies that policymakers should pay more attentions on the other factors related migration rather than only on household registration control (HRC) in the reform for eliminating labor migration barriers in China. By the way, the results of resident's welfare (GDP per capita indicator) are shown in Appendix Table 12.9. If we compared the results of resident's welfare (GDP per capita) with the results of resident's welfare (welfare index) in the situation of considering the migration costs (MC), it is shown that MC has a significant negative impact on resident's welfare (GDP per capita) in the period 2007–2017 but only has the significant adverse impact on resident's welfare (welfare index) in the period 2007–2015. It implied that if we only consider the economic influence in the resident's welfare, the reduction effect of migration costs will be amplified and lasts longer.

12.5.3 Analysis of Transmission Mechanism

Benchmark regression analysis has confirmed the negative effects of air pollution on the welfare of urban residents, but what is the transmission mechanism of this effect? Herein, residents' health is considered to be the transmission variable of air pollution affecting urban residents' welfare.

On the one hand, air pollution will increase the health risks of residents during outdoor activities and their prevalence rate of respiratory diseases and other diseases (Chen et al., 2017), which further reduces the labor time and outdoor activities time of the residents, and increases the medical expenditure. Moreover, reduced working hours caused by diseases will decrease the income of urban residents; the decline in outdoor activity time will reduce residents' outdoor recreational activities as well as their happiness; the increase in medical expenditure will lower the residents' disposable income in other aspects. All three will have a negative impact on the welfare of residents. On the other hand, air pollution can also cause a decrease in air visibility, cause frequent traffic accidents (Wan et al., 2020) and more urban death toll, which also have a negative effect on the welfare of urban residents.

In order to test the transmission mechanism of residents' health, this paper chooses two substitution variables to represent residents' health. (1) The risk-based approach is applied to estimate health risks from air pollution. With reference to the work of Pandey et al. (2005), the specific estimation formula is

$$\mathrm{HR} = \mathrm{Pop} \times \ (\mathrm{DR/LOAEL}) \tag{12.19}$$

$$\mathrm{DR} = (\mathrm{BR/BW}) \int_0^{24} P(t)\mathrm{OF}(t)dt \tag{12.20}$$

In Eq. (12.19), *HR* is the overall health risk of urban residents, which is determined by the urban labor population *Pop* and the ratio of the dosage rate *DR* to the lowest observed dose *LOAEL*. Eq. (12.20) is the expression of the dosage rate, where *BR* is the respiratory rate, *BW* is the body weight, $P(t)$ is the air pollution concentration, and $OF(t)$ is the retention factor of the pollutant in the area, which is assumed to be 1. Related parameter values refer to the parameter settings of Pandey et al. (2005). The calculated residents' health (health risk) is a substitution variable for residents' health. A larger value indicates a higher health risk of urban residents. (2) The urban death toll is used as another substitution variable for residents' health[4] (the number of deaths). The higher the value of resident's health (the number of deaths), the higher the risk to residents' health.

Table 12.6 reports the results of the mechanism test. The estimated results in Columns (1) and (2) of Table 12.6 showed that $PM_{2.5}$ significantly increases the number of urban deaths and damages the health of urban residents. As is shown in

[4] Because the *China Statistical Yearbook for Regional Economy* only releases data by 2013, the data interval of the residents' health (death toll) data in the transmission mechanism analysis is from 2007 to 2012.

Columns (3) and (4), the increase in residents' health risks remarkably reduces the welfare of urban residents. The values in Column (4) show that an average increase of 10,000 in urban death toll leads to a 10.6% decrease in the overall welfare of residents on average. The estimates in the last four columns are the results of the static panel data model, which shows that the mechanism of residents' health (health risk) still holds. It can be observed from Column (8) that an average increase of 1% in urban health risks results in an average reduction of 0.214% in urban residents' welfare. It means that air pollution has a significant negative impact on the welfare of urban residents by increasing their health risks.

12.5.4 Heterogeneity Analysis

To investigate the regional, temporal and city size heterogeneity, the effects of urban $PM_{2.5}$ on urban residents' welfare in different regions, years and city size are studied. The estimation results are shown in Table 12.7. The first two columns of Table 7 list the estimation results of different regions. It is found that $PM_{2.5}$ has a greater negative impact on urban residents' welfare in the economically developed eastern and north-eastern regions. This indicates that with the economic growth, urban residents pay more attention to the environmental quality that affects their health, and economically developed areas should attach greater importance to health, leisure, and education while improving residents' welfare. The last three columns are the cross-sectional data of 2007, 2011 and 2017 respectively for regression. It is found that the negative effect of air pollution on urban residents' welfare has decreased over time by comparing year 2007 and 2017, but it is still significantly negative, which is in line with the gradually improved yet still prominent air pollution in China. This indicates that although air pollution has a significant negative impact on the welfare of urban residents on the whole, this negative influence is dwindling because China has achieved remarkable results in air pollution control in recent years. The estimation of the final three columns are the results by dividing the sample according to city size. Small cities with a population of less than one million, medium cities with a population of one million to five million, and large cities with more than five million. It is still robust with the benchmark results, yet medium cities and big cities are more affected by air pollution.

12.6 Conclusion

In this paper, the effects of air pollution on urban residents' welfare are studied theoretically and empirically. The theoretical analysis shows that: (1) Between 2007 and 2017. the reduction of migration costs and the improvement of air quality accounts for roughly 24.65% improvement of resident's welfare from 2007–2017 in China. (2) In addition, the higher the costs of migration, the more obvious the negative

Table 12.6 The estimation results of the transmission mechanism

Variables	The cross-sectional data model				The panel data model			
	RH (Death toll)		Resident's welfare (WI)		RH (Health risk)			Resident's welfare (WI)
	(1)	(2)	(3)	(4)	(5)	(6)	(7)	(8)
	Year 2007	The annual average	Year 2007	The annual average	2007–2017	2007–2012	2013–2017	2007–2012
$PM_{2.5}$	0.927***	1.233***			0.875***	0.990***	0.961***	
	(0.151)	(0.142)			(0.0242)	(0.0271)	(0.0192)	
RH (death toll)			−0.128***	−0.106***				
			(0.0331)	(0.0194)				
RH (health risk)								−0.214*
								(0.111)
Control variables	Yes	Yes	Yes	Yes	Yes	Yes	Yes	Yes
FE of cities					Yes	Yes	Yes	Yes
FE of years					Yes	Yes	Yes	Yes
Constant	−0.0318	−0.156	2.829***	2.714***	−2.712***	−2.898***	−2.984***	4.581***
	(0.517)	(0.554)	(0.157)	(0.161)	(0.12)	(0.131)	(0.103)	(0.384)
Observation	274	274	274	274	3012	1643	1369	1643
Adjusted R^2	0.457	0.541	0.407	0.455	0.981	0.990	0.994	0.599

Notice (1) robust standard error in parentheses; (2) ***, **, * respectively express significant at the significance level of 1%, 5% and 10%; (3) "RH" means to the Resident's Health, "WI" is welfare index; (4) "The annual average" in the Column (2) and (4) are the annual average from 2007 to 2012. 12.5.4 Heterogeneity analysis.

Table 12.7 The estimation results of regional, city size and time differences

Variables	Resident's welfare (welfare index)							
	The eastern and northeastern regions	The central and western regions	Year 2007	Year 2011	Year 2017	Small cites	Medium cities	Big cities
	(1)	(2)	(3)	(4)	(5)	(6)	(7)	(8)
$PM_{2.5}$	−0.187***	−0.176***	−0.310***	−0.0985*	−0.151***	−0.0833	−0.281***	−0.291***
	(0.0674)	(0.0635)	(0.0537)	(0.0575)	(0.0363)	(0.263)	(0.0660)	(0.101)
Control variables	Yes	Yes	Yes	Yes	Yes	Yes	Yes	Yes
FE of cities	Yes	Yes	No	No	No	Yes	Yes	Yes
FE of years	Yes	Yes	No	No	No	Yes	Yes	Yes
Constant	3.651***	3.357***	3.452***	3.881***	3.417***	3.426***	3.928***	7.147***
	(0.255)	(0.296)	(0.213)	(0.215)	(0.153)	(1.014)	(0.267)	(1.038)
Observation	1269	1738	274	269	274	107	1843	1057
Adjusted R^2	0.710	0.678	0.383	0.282	0.314	0.615	0.626	0.683

Notice (1) robust standard error in parentheses; (2) ***, **, * respectively express significant at the significance level of 1%, 5% and 10%

effect of air pollution on urban residents' welfare. The empirical study shows that: (1) Air pollution has a significant negative effect on urban residents' welfare, and that there is an inverted U-shaped relationship between them. (2) This negative impact of air pollution on urban welfare is amplified by migration costs. Nevertheless, with the deepening of HRC reform and migration barriers reduction change in China, the negative moderating effect of migration costs (MC) and household registration control (HRC) is decreasing. (3) It also shows that the health risk of urban residents is the transmission mechanism through which air pollution affects the welfare of urban residents. Specifically, air pollution firstly significantly increases the death toll and the health risk of urban residents, and further negatively impacts urban residents' welfare. (4) Although the negative effects of air pollution on urban residents' welfare will exist for a long time, it will gradually decrease on the whole. From a regional perspective, the negative effects of air pollution on the welfare of urban residents exist in different regions.

The policy implications of this paper are as follows: (1) In the process of sustainable development, weigh residents' welfare against air pollution according to different stages of economic development. It is found that air pollution has an inverted U-shaped relationship with residents' welfare, and that at the initial stage of economic development, a moderate amount of air pollution can promote residents' income and improve residents' welfare. This suggests that in the early stage of development, when the environmental carrying capacity is large and residents are more concerned about income growth, local governments can moderately sacrifice part of environmental quality in exchange for rapid economic growth and better residents' welfare. As the economy further develops, the environmental carrying capacity decreases and residents pay more attention to the living environment, if local governments continue to develop the regional economy by expensing the environment, it will have a significant negative impact on residents' welfare. And this can even lead to aggravated air pollution, the outflow of labor and talent, economic recession, and further outflow of population. Therefore, in the later stage of development, local governments need to formulate a more reasonable policy to promote the high-quality growth of the local economy and industrial upgrading while taking into account air pollution control and improvement of residents' welfare.

(2) From the perspective of urban residents, China should further deepen the reform of the household registration system to reduce the costs of migration between cities. On the one hand, the removal of restrictions (such as easier access to health insurance reimbursement, and pension payments, etc.) in heavily polluted cities will enable urban residents to freely move out of the polluted cities and improve their overall welfare, which in turn will help urban residents to "feet voting" and force the city's government to reduce the pollution. On the other hand, in cities with a good air environment, the restriction on immigration should be relaxed, for example, implementing less strict household registration controls, and providing foreign students with opportunities for enrollment and college entrance examinations.

(3) Reducing the costs of inter-city migration can not only improve residents' welfare in the short term, but also help break the boundaries between cities and regions, further promote the free flow of labor, give full play to the role of the

market in labor allocation, and realize integrated regional development in the long run. *The 14th Five-Year Plan* released by the Chinese government clearly states that it is necessary to remove the system and mechanism barriers that hinder the market-oriented allocation of production factors and the circulation of goods and services, so as to build a stronger domestic market with smooth circulation service. In addition, a lower cost of migration between cities will effectively reduce regional restrictions on labor supply as well as labor employment restrictions, and form a more unified and complete national labor factor market, thereby providing conditions for sustainable economic development and improvement of residents' welfare.

The possible limitations and future research of this paper are as follows: (1) In the theoretical model, the government was assumed to be in a state of balance of fiscal revenue and expenditure, which was not true to most local governments in reality. Hence, the theoretical assumption was not completely consistent with the practical situation. (2) The model failed to consider the heterogeneous characteristics of residents, such as age and gender, and also did not describe the spatial distribution and variation of air pollution (Wu et al., 2021). Therefore, the analysis was not in-depth enough when estimating the impact of air pollution on residents' welfare because the abatement capacities of air pollution which is caused by the change of economic structure and industrial structure is not considered. (3) Many parameter settings in this paper were based on existing research, some of which might not conform to reality. Some uncertainties (Wu et al., 2019) and policy factors (Wu et al., 2020) are not taken into account. (4) Due to the difficulty in data acquisition, lung diseases and respiratory diseases directly related to air pollution were not included as health risk variables in this paper, which may leave a certain deviation in the mechanism test. (5) Due to the data's availability, factors such as housing price and transportation costs were not included in the migration costs between cities. All of these are worth studying in the future.

Acknowledgements Huai Deng, Hui Xu, and Yuxiang Huang also made great contributions to this manuscript. We express our heartfelt thanks to them. This research also was supported by: National Natural Science Foundation of China (72474127) and National Social Science Fund Major Bidding Project of China (18ZDA052). Part of this chapter was published:Wu X, Deng H, Huang Y, Guo J. Air pollution, migration costs, and urban residents' welfare: A spatial general equilibrium analysis from China. Structural Change and Economic Dynamics 2022a; 63: 396–409. https://doi.org/10.1016/j.strueco.2022.05.010. The publication has been conducted with copyright permission.

Appendix

See Fig. 12.3, and Tables 12.8 and 12.9.

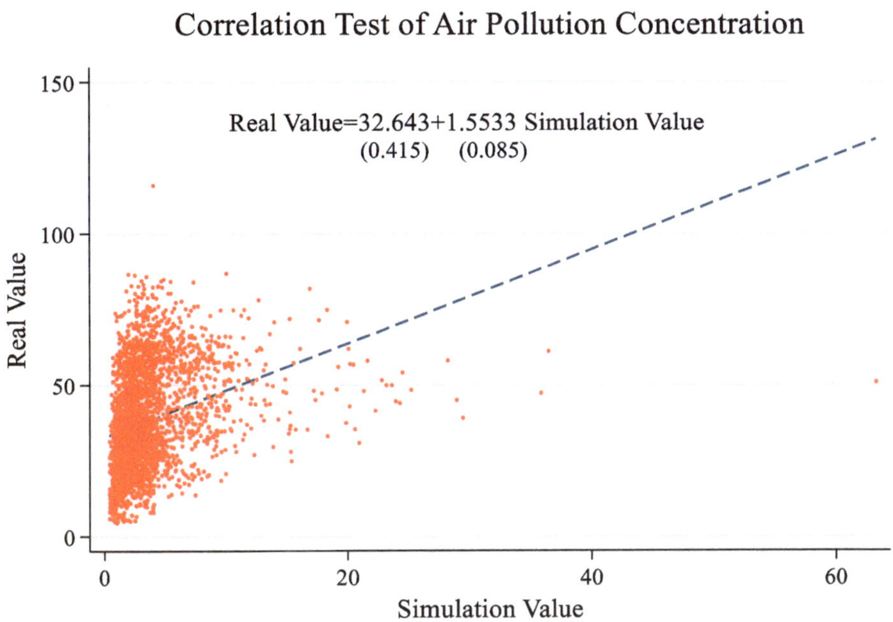

Fig. 12.3 Correlation test of air pollution concentration (simulation value and real value)

Table 12.8 Migration rate and migration costs (Unit: %)

		Migration probability (M_{it})	Change of M_{it}	Migration cost (T_{it})	Change of T_{it}
Average value		19.6		59.5	
Time trend	Initial (2007)	12.5		68.3	
	Year 2012	15.4	23.2	61.2	−10.4
	Year 2017	28.5	128	39.3	−42.5

Table 12.9 The benchmark results of GDP per capita indicator

Variables	Resident's welfare (GDP per capita)					
	Benchmark results			Dynamic panel	Migration cost (MC) results	
	(1)	(2)	(3)	(4)	(5)	(6)
				Systematic GMM	2007–2017	2007–2017
Lag. (resident's welfare)				0.579***		
				(0.122)		
$PM_{2.5}$	−0.143***		−0.0635**	−0.0425*	−0.0619***	0.0228
	(0.0296)		(0.0258)	(0.0236)	(0.0220)	(0.0299)
Lag. $PM_{2.5}$		−0.263***				
		(0.0369)				
MC					−0.0701***	
					(0.0175)	
MC × $PM_{2.5}$						−0.0204***
						(0.00474)
Control variables	No	No	Yes	Yes	Yes	Yes
FE of cities	Yes	Yes	Yes		Yes	Yes
FE of years	Yes	Yes	Yes	Yes	Yes	Yes
Test of AR (1)				0.087		
Test of AR (2)				0.167		
Constant	0.793***	1.253***	−0.383***	0.0651	−0.0658	−0.358***
	(0.129)	(0.152)	(0.126)	(0.0844)	(0.141)	(0.110)
Observations	3007	2733	3006	2729	2926	2926
Adjusted R^2	0.918	0.913	0.931		0.937	0.937

Notice (1) robust standard error in parentheses; (2) ***, **, * respectively express significant at the significance level of 1%, 5% and 10%

References

Aubry, A., Burzynski, M., & Docquier, F. (2016). The welfare impact of global migration in OECD countries. *Journal of International Economics, 101*, 1–21. https://doi.org/10.1016/j.jinteco.2016.03.005

Bai, C. E., & Qian, Y. (2006). The return to Capital in China. *Brookings Papers on Economic Activity, 74*(2), 61–102.

Borgschulte, M., Molitor, D., & Zou, E. Y. (2024). Air pollution and the labor market: Evidence from wildfire smoke. *Review of Economics and Statistics, 106*(6), 1558–1575. https://doi.org/10.1162/rest_a_01243

Cai, F., & Wang, M. Y. (2009). Why isn't labor mobility narrowed the urban-rural income gap? *Economic Perspectives, 8*, 4–10.

Chen, J., Hao, Q. J., & Yoon, C. (2018). Measuring the welfare cost of air pollution in Shanghai: Evidence from the housing market. *Journal of Environmental Planning and Management, 61*(10), 1744–1757. https://doi.org/10.1080/09640568.2017.1371581

Chen, S. M., & He, L. Y. (2014). Welfare loss of China's air pollution: How to make personal vehicle transportation policy. *China Economic Review, 31*, 106–118. https://doi.org/10.1016/j.chieco.2014.08.009

Chen, S. M., & He, L. Y. (2017). Environment, health, and economic growth: The optimal allocation of energy tax revenue. *Economic Research Journal, 52*(4), 120–134.

Chen, S. Y., & Chen, D. K. (2018). Air pollution, government regulations and high-quality economic development. *Economic Research Journal, 53*(2), 20–34.

Chen, S. Y., Liu, C. L., & Feng, B. (2019). Capital allocation efficiency, City-size distribution and social welfare analysis. *Economic Research Journal, 54*(2), 133–147.

Chen, S., Oliva, P., & Zhang, P. (2017). *The effect of air pollution on evidence from China.* (NBER Working Paper) https://doi.org/10.2139/ssrn.3070762

Chen, X. Y., Shao, S. A., Tian, Z. H., Xie, Z., & Yin, P. (2017). Impacts of air pollution and its spatial spillover effect on public health based on China's big data sample. *Journal of Cleaner Production, 142*, 915–925. https://doi.org/10.1016/j.jclepro.2016.02.119

Chen, Z. H., & Li, Y. (2018). A measurement of Chinese urban Resident's welfare by region. *Journal of Finance and Economics, 44*(10), 111–124. https://doi.org/10.16538/j.cnki.jfe.2018.10.008

Chu, D. Y., He, P. F., & Liang, R. B. (2017). Subjective air pollution and resident's Well-being: A micro-data verification based on breakpoint regression design. *Economic Perspectives*, 88–101.

Dekle, R., Eaton, J., & Kortum, S. (2008). Global rebalancing with gravity: Measuring the burden of adjustment. *IMF Staff Papers, 55*(3), 511–540. https://doi.org/10.1057/imfsp.2008.17

Deng, K. B., & Ding, J. H. (2010). *Household registration.* Economy Growth and Income Disparity.

Desmet, K., & Rossi-Hansberg, E. (2013). Urban accounting and welfare. *American Economic Review, 103*(6), 2296–2327. https://doi.org/10.1257/aer.103.6.2296

Desmet, K., Nagy, D. K., & Rossi-Hansberg, E. (2018). The geography of development. *Journal of Political Economy, 126*(3), 903–982. https://doi.org/10.1086/697084

Diomand, R. (2016). The determinants and welfare implications of US-Workers' diverging location choices by skills: 1980-2000. *American Economic Review, 106*(3), 479–524. https://doi.org/10.1257/aer.20131706

Duan, W., Wang, M., & Wu, F. X. (2020). Welfare accounting of Chinese urbanization from 2000 to 2017: A structural estimation using the quantitative spatial model. *Economic Research Journal, 55*(5), 166–182.

Eaton, J., & Kortum, S. (2002). Technology, geography and trade. *Econometrica, 70*(5), 1741–1779. https://doi.org/10.1111/1468-0262.00352

Falck, O., Lameli, A., & Ruhose, J. (2018). Cultural biases in migration: Estimating non-monetary migration costs. *Papers in Regional Science, 97*(2), 411-+. https://doi.org/10.1111/pirs.12243

Ferreira, S., Akay, A., Brereton, F., et al. (2013). Life satisfaction and air quality in Europe. *Ecological Economics, 88*, 1–10. https://doi.org/10.1016/j.ecolecon.2012.12.027

Germani, A. R., Scaramozzino, P., Castaldo, A., & Talamo, G. (2021). Does air pollution influence internal migration? An empirical investigation on Italian provinces. *Environmental Science and Policy, 120*, 11–20. https://doi.org/10.1016/j.envsci.2021.02.005

Glaeser, E. L. (2008). *Cities, agglomeration and spatial equilibrium.* Oxford University Press.

Gonzalez, F., Leipnik, M., & Mazumder, D. (2013). How much are urban residents in Mexico willing to pay for cleaner air? *Environment and Development Economics, 18*(3), 354–379. https://doi.org/10.1017/S1355770X13000077

Guo, X. R., Cheng, S. Y., & Chen, D. S. (2010). Estimation of economic costs of particulate air pollution from road transport in China. *Atmospheric Environment, 44*(28), 3369–3377. https://doi.org/10.1016/j.atmosenv.2010.06.018

Hanlon, W. W. (2016). Coal smoke, City growth, and the costs of the industrial revolution. *National Bureau of Economic Research, 130*(626), 462–488. https://doi.org/10.1093/ej/uez055

Hong, D. Y., Fan, Y. C., & Li, P. F. (2016). Socio-economic status, Behavioral adaptation, and expectation for government performance: An analysis of the differentiation of resident's smog-related out-migration intention. *Sociological Studies, 31*(3), 1–24. https://doi.org/10.19934/j.cnki.shxyj.2016.03.001

Jia, K., & Chen, S. W. (2021). Escaping from pollution: Air pollution and the settlement intentions of floating migrants in Chinese cities. *Migration Studies*. https://doi.org/10.1093/migration/mnab029

Jin, B. H., & Li, Z. (2024). Air pollution, healthcare use, and inequality: Evidence from China. *Economic Modelling, 141*. https://doi.org/10.1016/j.econmod.2024.106905

Jones, C. I., & Klenow, P. J. (2016). Beyond GDP? Welfare across countries and time. *American Economic Review, 106*(9), 2426–2457. https://doi.org/10.1257/aer.20110236

Kan, H. D., Chen, R. J., & Tong, S. L. (2012). Ambient air pollution, climate change, and population health in China. *Environmental International, 42*, 10–19. https://doi.org/10.1016/j.envint.2011.03.003

Levinson, A. (2012). Valuing public goods using happiness data: The case of air quality. *Journal of Public Economics, 96*(9), 869–880. https://doi.org/10.1016/j.jpubeco.2012.06.007

Li, Z. T., Folmer, H., & Xue, J. H. (2014). To what extent does air pollution affect happiness? The case of the Jinchuan mining area. *China. Ecological Economics, 99*, 88–99. https://doi.org/10.1016/j.ecolecon.2013.12.014

Liu, K. H., Tong, J. D., & Liu, R. J. (2019). The health cost of China's export expansion: Evidence from adult morbidity. *China Industrial Economics*, 118–135. https://doi.org/10.19581/j.cnki.ciejournal.2019.08.007

Liu, W. H., Pan, W. C., & Yi, C. C. (2019). *"Happiness in the air?" the effects of air pollution on adolescent happiness*. BMC Public Health. https://doi.org/10.1186/s12889-019-7119-0

Liu, X. Y., & Li, S. L. (2017). Housing Price, migration friction, and City-size distribution in China: Theory model and structural estimation. *Economic Research Journal, 52*(7), 65–78.

Liu, Z. M., & Yu, L. (2020). Stay or leave? The role of air pollution in urban migration choices. *Ecological Economics*. https://doi.org/10.1016/j.ecolecon.2020.106780

Matus, K., Nam, K. M., Selin, N. E., et al. (2012). Health damages from air pollution in China. *Global Environmental Change-Human and Policy Dimensions, 22*(1), 55–66. https://doi.org/10.1016/j.gloenvcha.2011.08.006

Menz, T. (2011). Do people habituate to air pollution? Evidence from international life satisfaction data. *Ecological Economics, 71*, 211–219. https://doi.org/10.1016/j.ecolecon.2011.09.012

Monte, F., Redding, S. J., & Rossi-Hansberg, E. (2018). Commuting, migration and local employment elasticities. *American Economic Review, 108*(12), 3855–3890. https://doi.org/10.1257/aer.20151507

Nam, K. M., Selin, N. E., Reilly, J. M., & Paltsev, S. (2010). Measuring welfare loss caused by air pollution in Europe: A CGE analysis. *Energy Policy, 38*(9), 5059–5071. https://doi.org/10.1016/j.enpol.2010.04.034

Nordhaus, W. D. (2010). Economic aspects of global warming in a post-Copenhagen environment. *Proceedings of the National Academy of Sciences of the United States of America, 107*(26), 11721–11726. https://doi.org/10.1073/pnas.1005985107

Pandey, J. S., Kumar, R., & Devotta, S. (2005). Health risks of NO2, SPM and SO2 in Delhi (India). *Atmospheric Environment, 39*(36), 6868–6874. https://doi.org/10.1016/j.atmosenv.2005.08.004

Pautrel, X. (2009). Pollution and life expectancy: How environmental policy can promote growth. *Ecological Economics, 68*(4), 1040–1051. https://doi.org/10.1016/j.ecolecon.2008.07.011

Qi, S., & Lu, H. Y. (2015). Pollution, health and inequality: Crossing the "Environmental Health Poverty" Trap. *Management World*, 32–51. https://doi.org/10.19744/j.cnki.11-1235/f.2015.09.004

Roback, J. (1982). Wages, rents, and the quality of life. *Journal of Political Economy, 90*(6), 1257–1278.

Shi, D. Q., & Yu, H. W. (2020). Reevaluating the subjective welfare loss of air pollution. *Journal of Cleaner Production*. https://doi.org/10.1016/j.jclepro.2020.120445

Thurston, G. D., Kipen, H., Annesi-Maesano, I., et al. (2017). A joint ERS/ATS policy statement: What constitutes an adverse health effect of air pollution? *European Respiratory Journal*. https://doi.org/10.1183/13993003.00419-2016

Tombe, T., & Zhu, X. D. (2019). Trade, migration, and productivity: A quantitative analysis of China. *American Economic Review, 109*(5), 1843–1872. https://doi.org/10.1257/aer.20150811

Wan, W., Zhang, W. H., Cheng, H. H., & Li, Q. W. (2019). *Air China 2019 report: Progress in air pollution control in China*. Asia Clean Air Center.

Wan, Y., Li, Y. H., Liu, C. H., & Li, Z. Q. (2020). Is traffic accident related to air pollution? A case report from an island of Taihu Lake. *China. Atmospheric Pollution Research, 11*(5), 1028–1033. https://doi.org/10.1016/j.apr.2020.02.018

Wang, H. Y., & Liu, G. F. (2014). *China international migration report*. Social Sciences Academic Press.

Wang, X. L., & Fan, G. (2000). *Sustainability of China's economic growth: Review and Prospect of the cross century*. National Economic Research Institute.

Wang, Z. H., Ma, J. H., Zhang, B., & Wang, B. (2021). Air pollution and residential migration: Empirical evidence from smart meter data. *Management World, 3*, 19–33. https://doi.org/10.19744/j.cnki.11-1235/f.2021.0033

Wang, Z. S. (2018). Urban Resident's welfare: Measurement, health relevancy and improvement path. *Journal of Social Science of Hunan Normal University, 47*(2), 84–91. https://doi.org/10.19503/j.cnki.1000-2529.2018.02.013

Wu, G. P., Wang, W. B., & Zhang, X. N. (2015). Analysis of economic potential growth rate of China under the new Normal. *Journal of Financial Research*, 46–63.

Wu, K. P., Tong, J., & Chu, C. J. (2015). Impact of environment quality on people's happiness. *Journal of Technology Economics, 34*(6), 95–105.

Wu, X. H., Deng, H., Li, H., & Guo, Y. M. (2021). Impact of energy structure adjustment and environmental regulation on air pollution in China: Simulation and measurement research by the dynamic general equilibrium model. *Technological Forecasting & Social Change*. http://doi.org/j.techfore.2021.121010

Wu, X. H., Wang, Z. J., Guo, J., & Xue, P. P. (2020). Disaster probability, optimal government expenditure for disaster prevention and mitigation, and expected economic growth. *Science of the Total Environment*. https://doi.org/10.1016/j.scitotenv.2019.135888

Wu, X. H., Xu, Z., Liu, H., Guo, J., & Zhou, L. (2019). What are the impacts of tropical cyclones on employment? *Weather, Climate, and Society*. https://doi.org/10.1175/WCAS-D-18-0052.1

Xie, T. T., Wang, Y., & Yuan, Y. (2024). Health benefits from improved air quality: Evidence from pollution regulations in China's "2+26" cities. *Environmental & Resource Economics, 87*(5), 1175–1221. https://doi.org/10.1007/s10640-024-00860-3

Yaduma, N., Kortelainen, M., & Wossink, A. (2013). Estimating mortality and economic costs of particulate air pollution in developing countries: The case of Nigeria. *Environmental & Resource Economics, 54*(3), 361–387. https://doi.org/10.1007/s10640-012-9598-7

Ye, J. Z., & An, H. S. (2017). *Can environmental taxes effectively control air pollution?* China Industrial Economics. https://doi.org/10.19581/j.cnki.ciejournal.2017.05.005

Zhang, J. J., & Mu, Q. (2018). Air pollution and defensive expenditures: Evidence from particulate-filtering facemasks. *Journal of Environmental Economics and Management, 92*, 517–536. https://doi.org/10.1016/j.jeem.2017.07.006

Zhang, X. D., & Liang, J. (2014). From London to Beijing: A comparison and reflection on smog controls in China and Britain. *Frontiers*, 51–63. https://doi.org/10.16619/j.cnki.rmltxsqy.2014.03.001

Zhang, X., Ou, X. M., Yang, X., et al. (2017). Socioeconomic burden of air pollution in China: Province-level analysis based on energy economic model. *Energy Economics, 68*, 478–489. https://doi.org/10.1016/j.eneco.2017.10.013

Zhang, X., Zhang, X. B., & Chen, X. (2017). Happiness in the air: How does a dirty sky affect mental health and subjective Well-being? *Journal of Environmental Economics and Management, 85*, 81–94. https://doi.org/10.1016/j.jeem.2017.04.001

Zhou, Y. S., & Liu, J. Y. (2020). Air pollution and mental health of older adults in China. *Sustainability.* https://doi.org/10.3390/su12030950

Zong, J. F., & Zhou, L. (2015). Do big cities pay higher wages? *China Economic Quarterly, 14*(4), 1467–1496.https://doi.org/10.13821/j.cnki.ceq.2014.04.12

Zou, Y. N., & Li, A. M. (2013). Household registration control, City size and City development. *Contemporary Economic Research*, 53–60.

Chapter 13
Assessment of the Impact of Fukushima Nuclear Wastewater Discharge on the Global Economy Based on GTAP

13.1 Introduction

Nuclear pollution has always been an issue of global concern. There have been many nuclear pollution accidents in history, such as the Three Mile Island accident in the United States in March 1979, the Chernobyl nuclear power plant accident in the Soviet Union in April 1986, and the Fukushima nuclear accident in Japan in March 2011. Nuclear radiation will not only induce severe radiation diseases and even cancer, but also lead to biological deformities and mutations. Along with the dangerous health effects caused by the release of nuclear pollution, these accidents have consistently resulted in serious damage to the environment and continue to be a significant concern among governments, academics, and society. When these disasters happen near the coast, the marine environment becomes a focus of attention, mainly because of the vital role that oceans play in the global carbon cycle, regulating climate, ensuring food security, and supporting the livelihoods of billions of people. (Suarez et al., 2014). Increasing ocean pollution has threatened the ocean system in recent years, and public awareness of marine environmental protection is growing worldwide. Lotze et al. (2018) compared public perceptions of ocean threats and protection from more than 32,000 respondents in 21 countries, and the results showed that 70% of respondents believe that human activities threaten the marine environment. Wright et al. (2006) have demonstrated that consultative arrangements should be strengthened to improve environmental governance and deal with increasing environmental threats.

Since 2011, TEPCO (Tokyo Electric Power Company) has been injecting seawater into the Fukushima Daiichi Nuclear Power Plant to cool the reactors in order to avoid a meltdown after the Fukushima nuclear accident in Japan on March 11, 2011. As groundwater and rainwater seep into the Fukushima Nuclear Power Plant, a large amount of nuclear waste has been stored in the water storage tower. According to TEPCO, the tower has a total capacity of 1.37 million tons and will reach its maximum in summer 2022. On April 13, 2021, the Japanese government announced

X. Wu, *Environmental Disaster in China*,
https://doi.org/10.1007/978-981-92-0910-1_13

that it plans to discharge nearly 1.26 million tons of nuclear wastewater from the Fukushima Daiichi Nuclear Power Plant into the Pacific Ocean in 40 years, and the discharge will begin at a distance of 1 km offshore in 2023 (Normile, 2021; Yang et al. 2021).[1] This decision has already provoked opposition from neighboring countries and residents of Japan. And more, the discharge of nuclear wastewater into the ocean is constantly opposed by the citizens of Fukushima, most of the municipal assemblies of Fukushima Prefecture, the wider Japanese society, and Japan's national federation of fisheries cooperatives, etc. (Shaun, 2020). Internationally, the decision has also drawn condemnation from neighboring countries and international environmental groups. Shaun Burnie, an expert at global environmental group Greenpeace, believes the discharge of nuclear wastewater into the ocean is "unacceptable". The Pacific Ocean has an enormous powerful current, and the most directly affected would be the Japanese people living on the Pacific coast and the fishing industry, he said in an interview.[2] China and South Korea have expressed concern and dissatisfaction with this proposal. For example, the Chinese government demands that Japan take a prudent approach to the disposal of nuclear wastewater at Fukushima.[3] The South Korean government said the move would pose risks to neighboring countries' security and marine environment. The South Korean side expresses its opposition to the decision of the Japanese government.[4] The British "Daily Mail" also published an article indicating that the discharge of such a large amount of polluted nuclear wastewater into the sea is undoubtedly a disaster for human beings (Liu et al. 2021).

Oceanographers, environmental groups, and fishing groups have also raised concerns about the discharge's impact on marine life and fisheries. According to Buesseler (2020), no purification system can remove 100% of all radioactive pollutants. In addition to tritium, the treated Fukushima nuclear wastewater also contains isotopes such as carbon 14, cobalt 60 and strontium 90. Among these isotopes, tritium not only has a short half-life of a little over 12 years, but as a helium isotope it is not readily absorbed by living things and as a primarily inert gas, generally does not settle in sediments. Carbon, however, has an extremely long half-life in human terms, at 5700 years, and is easily absorbed into living things and entrained in sea-floor sediments. These isotopes are potentially harmful to human genes. At certain levels in the ocean, radioactive elements can accumulate in marine organisms through the food chain. The longer the food chain, the greater the aggregation effect, which will eventually affect humans (Lu et al., 2021). Experiments have shown that long-term and large consumption of radioactive seafood will cause the radioactive substances accumulated in the body to exceed the allowable amount, which will cause various diseases. This shows that the discharge of nuclear wastewater into the sea would affect not only Japan itself, but also the countries along the Pacific Ocean as the pollutants spread with ocean currents, and in a few years, other waters as well.

---

[1] https://www.sciencemag.org/news/2021/04/japan-plans-release-fukushima-S-contaminated-water-ocean.

[2] http://www.xinhuanet.com/english/asiapacific/2021-04/22/c_139899269.htm.

[3] https://www.fmprc.gov.cn/fyrbt_673021/dhdw_673027/202104/t20210412_9171341.shtml.

[4] https://www.mofa.go.kr/www/brd/m_4080/view.do?seq=371085.

The discharge of nuclear waste water into the Pacific Ocean will impact more than ten countries or regions along the Pacific coast, and the total population of these regions exceeds 2 billion, and their GDP accounts for 55.8% of the world's total.[5] Yet few quantitative studies have been carried out on the economic impact of discharging nuclear wastewater from the Fukushima Nuclear Power Plant into the Pacific Ocean on Japan's neighboring countries. The authors intended to use the GTAP model to quantitatively evaluate the impact of nuclear wastewater on countries and industries under different scenarios. The results obtained can provide an empirical reference for the pre-disaster prevention and post-disaster response of governments and industries in many countries, including Japan.

The rest of this paper is arranged as follows. Section 13.2 introduces related research on nuclear leakage accidents and GTAP model applications. Section 13.3 analyzes the hazards of nuclear wastewater discharge. The model introduction and data processing are presented in Sect. 13.4. Section 13.5 gives the simulation scenarios. Section 13.6 is the result analysis. Conclusions and suggestions are given in Sect. 13.7.

13.2 Literature Review

13.2.1 Research on Impact Assessment of Nuclear Leakage Accidents

After the Fukushima nuclear leakage, many studies have analyzed the accident's impact on ecology and human health. For example, Tateda et al. (2016) used a dynamic food chain transfer model to simulate 137Cs pollution in the marine biota on the eastern Pacific coast of Japan two years after the Fukushima accident. Merz et al. (2013) pointed out that radionuclides released by nuclear reactors may affect human health through external exposure and intake of contaminated air, food or water. Fisher et al. (2013) assessed the risk associated with radiation doses that may be absorbed by humans after consuming contaminated seafood. Wada et al. (2016) described the data of radioactive pollutants five years after the accident and analyzed the trend of chemical substance reduction around the Fukushima Nuclear Power Plant. Research has also been conducted on the distribution and pollution of radioactive substances in rivers (Matsuo et al., 2021) and soil (Iwagami et al., 2017), their temporal changes in diffusion into the atmosphere (Niisoe, 2021), their impact on physical and mental health (Hayashi et al., 2019; Moriyama et al., 2020).

Nuclear wastewater discharge seriously threatens human health, marine environment and ecological safety. Scholars have studied the Fukushima nuclear wastewater discharge from multiple perspectives. First, in terms of the issue of Japan's nuclear

[5] Data source: calculated according to the data from the French Economic Research Center (CEPII).

wastewater discharge Decision, Liu et al. (2021a, 2021b, 2021c) established a tripartite evolutionary game model for Japan's nuclear wastewater discharge decision-making, discussed the evolutionary stabilization strategies (ESSs) of Japan, other countries, and the International Environmental Protection Organization (IEPO), and put forward policy recommendations on promoting Japan's no-emission strategy. Zhang et al. (2022) established a game model of the optimal strategic behavior of the Japanese government, the domestic public and stakeholder countries in the nuclear wastewater discharge plan by using evolutionary game theory and found a strategy to prevent the Japanese government from discharging nuclear waste water through the analysis of the interests of various decision-makers. Pu et al. (2022) taking Weibo comments as a case to analyze the Chinese public's views on Japan's nuclear wastewater, the research showed the public's main concerns were nuclear pollution and marine ecology, public ethics, and international responsibility as seafood imports and food safety. Wu (2022) analyzes the issue of Japan's nuclear wastewater discharge into the ocean from a legal perspective. Yang et al. (2021) proposed a novel grey and unknown preference framework of GMCR (the graph model for conflict resolution), and the method could help resolve conflicts about the discharge of nuclear wastewater into the sea. Secondly, in terms of the impact of nuclear wastewater discharge, Men (2021) briefly reviewed the discharge of radionuclides derived from the Fukushima Daiichi Nuclear Power Plant Accident (FDNPPA) in the Pacific and the resulting impact on the marine organism to assess the possible impact of the problem. Lu et al. (2021) believed that potential risks would last for hundreds of years once nuclear waste water is discharged into the Pacific Ocean. Zhao et al. (2021) studied the diffusion and transfer of radioactive tritium water after the discharge of Fukushima nuclear waste water, and evaluated the impact of the plan. Periáez et al. (2021) used a numerical model to simulate the diffusion of radioactive 137Cs in the cooling water of the Fukushima nuclear reactor and estimated the concentration of 137Cs in the marine biota resulting from the discharge. Liu et al. (2021a, 2021b, 2021c) analyzed the macroscopic and microscopic diffusions of nuclear pollutants in the ocean. In terms of socio-economic impacts, Guo (2022) predicted the impact of nuclear wastewater discharge on aquatic products trade based on the current status of China's aquatic products industry and the competitiveness of foreign trade. Scholars (such as Dai & Yang, 2021; Jin, 2021; Wang, 2021; Xiang, 2021), through qualitative analysis, pointed out that nuclear wastewater discharge will affect the global fisheries, change the import and export trade in related countries or regions dramatically, hit the aquaculture industry badly, and inflict certain losses on the global economic system. Unfortunately, there has been little quantitative analysis of the effects of nuclear wastewater discharge.

13.2.2 Application Research of the GTAP Model

The Global Trade Analysis Project (GTAP) can link the industrial systems of various countries. With the impact of a certain policy or event being regarded as an external shock, GTAP can be used to evaluate the economic effects of various sectors in each country by analyzing the interaction between various entities in the economic system (Chen & Bian, 2015). The model began to develop in 1992 (Aguiar et al., 2019a), and Professor Hurt published the first article on the model in 1997 (Hertel, 1997), followed by Huff and Hertel (2001), McDougall (2003), etc. Hertel et al. (2000) added and modified the welfare and final demand of the model. Ianchovichina et al. (1999) developed a recursive dynamic GTAP Model. In recent years, the GTAP model has also been continuously improved in the application of various issues, such as the GTAP-HET model, which introduces corporate heterogeneity (Akgul et al., 2016), and the GTAP-HS model that decomposes GTAP commodities according to HS6 levels (Aguiar et al., 2019b; Grant et al., 2009; Narayanan et al., 2010) and the GTAP-E model (Burniaux & Truong, 2002) for extended applications in the environmental field, etc.

In terms of applications, the GTAP model was mainly used to assess various international trade issues (Stenberg & Siriwardana, 2015), energy-economic issues (Qian et al., 2025; Xu et al., 2025; Zahra et al., 2019), climate change (Wang et al., 2025; Zer et al., 2020), agricultural economic issues (Ko & Ito, 2017; Yang et al., 2017), and economic assessment of disasters or negative events, etc. In 2012, Boisvert et al. (2012) used the GTAP model to analyze the impact on the economy after the outbreak of foot-and-mouth disease (FMD). Maliszewska et al. (2020) simulated the potential impact of the Covid-19 pandemic on GDP and trade from the pandemic-induced increase in trade costs, and the decline in employment, tourism, and household demand. Zhu et al. (2020), Tang and Liu (2020) and Zhou and Zhang (2020a) applied the GTAP to analyze the impact of the Covid-19 pandemic on China's economy and trade. Therefore, the authors adopted the GTAP model to quantitatively analyze the impact of Japan's nuclear wastewater discharge on the output of industries, trade and macroeconomics of relevant regions by considering the changes in fishery output value and import and export trade among countries. The research in this paper is not only a further application of the GTAP model in the field of disaster assessment, but also provides more references for the methods of economic loss assessment of disasters and adverse events.

13.3 Impact Analysis of Nuclear Wastewater Discharge

The authors considered the possible impact of nuclear wastewater discharge on fisheries development from the perspective of supply and demand. First of all, from the perspective of demand, people of various countries are afraid to eat aquatic products in contaminated areas due to fear of nuclear pollution, and the demand for aquatic products in various countries has declined, resulting in a reduction in the trade of aquatic products in contaminated areas. Secondly, from the supply point of view, pollution caused by nuclear wastewater discharge may lead to the variation and death of marine life, and the decline of fishery products such as mutation.

13.3.1 Impact on Trade of Related Products

After the Fukushima nuclear power plant leakage accident, 54 countries and regions, including China, the United States, Indonesia, Malaysia and South Korea, imposed restrictions on agricultural products imported from Japan (Bachev, 2021). As shown in Fig. 13.1, affected by import and export restrictions, the growth rate of Japan's total aquatic product exports declined in 2011. Except for a slight increase in the United States, Japan's export of aquatic products to other major trading countries fell in 2011. United States' import of aquatic products from Japan decreased from 16.27% to 15.54% of its total import. Inevitably, Japan's agri-food sector has also been severely impacted. Demand for and prices of many products, such as rice, fruits, vegetables, mushrooms, milk, butter and beef, have fallen sharply.

On March 24, 2011, the Administration of Quality Supervision, Inspection and Quarantine of China (AQSIQ) embargoed dairy products, vegetables and their products, fruits, aquatic animals and aquatic products from related counties in Japan due to concerns about the contamination caused by the Fukushima nuclear leakage. China is not alone. After the nuclear accident, the U.S. Food and Drug Administration (FDA) issued import alert No. 99–33 for Japanese food, which was updated in March 2021 to cover fish, sea urchins, clams and other marine biological products, as well as milk, butter, and dry dairy products. A range of other Japanese foods like meat, meat products, poultry, vegetables, citrus fruits and kiwifruit are banned from

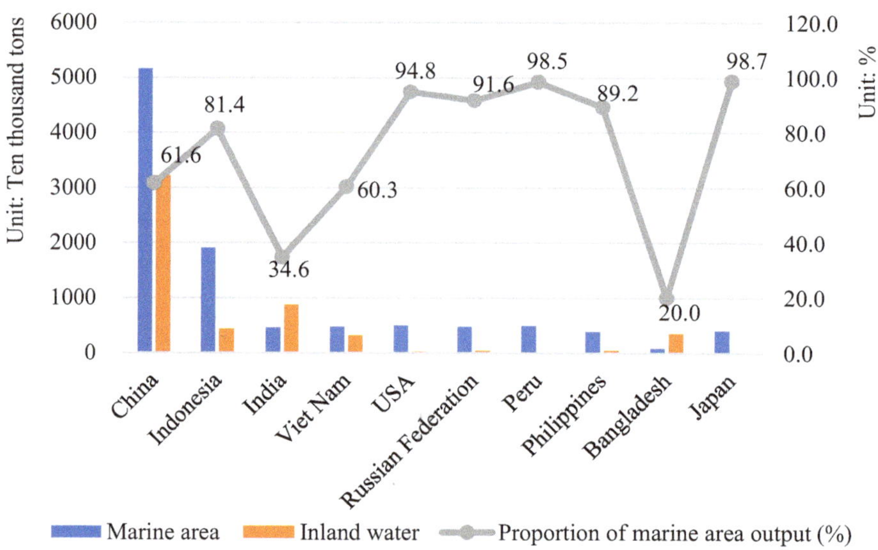

Fig. 13.1 Growth rate of Japan's total aquatic product exports to the world and major trading countries. *Data Source* U.N. Comtrade Database (https://comtrade.un.org/)

entering the United States. Referring to the impact of the Fukushima nuclear accident on the local fishery, the discharge of the Fukushima nuclear wastewater into the sea would have a blow to the local fishery and other related affected countries. Some people would be afraid to eat seafood because they are worried about nuclear pollution, which would decrease the demand for seafood products, and the fishery exports of various countries would suffer a serious blow. The nuclear pollution caused by Japan's discharge of nuclear wastewater will affect the fishery trade and the related food industry. At the same time, other countries will also be affected by the spread of nuclear waste.

13.3.2 Impact on the Fishery Industry

According to the research of Buesseler et al., the residual radioactive elements in nuclear wastewater may cause harm to marine organisms, and even produce mutations, which will reduce fishery production. A study on low-dose radiation effects on animals and ecosystems showed that since the FNPP (Fukushima Nuclear Power Plant) accident, the catch amount of the trial fishing operation in Fukushima Prefecture in 2017 was only 12.7% before the accident. The reason is that the FNPP accident has caused consumers and distributors to lose confidence in aquatic products in nuclear-contaminated areas, and people will reduce or not buy aquatic products in polluted areas, and instead buy aquatic products produced in other uncontaminated areas (Morita et al., 2020). Therefore, pollution caused by nuclear wastewater discharge may reduce fishery output. At the same time, as countries lose confidence in aquatic products in polluted areas and demand decreases, the fishery market in various countries will shrink, a large number of fishing grounds will be closed, and fishery output will be further reduced. The total fishery output is divided into inland water output and marine output. According to the *Report on the State of the World's Fisheries and Aquaculture*, the total output of global fisheries and aquaculture in 2020 was 178.5 million tons, of which 84.4 and 30.8 million tons were caught and raised respectively in marine areas, accounting for 64.5% of the total. Figure 13.2 shows the proportions of marine area production, inland water production, and ocean production of the top ten countries in fishery production in 2019. It can be seen that marine production accounts for more than 60% in most countries. The total fishery output mainly comes from marine areas. Due to the discharge of nuclear waste water, mutated marine organisms may appear in the offshore area, which will impact marine aquaculture and fishing, and further processed seafood and related food industries.

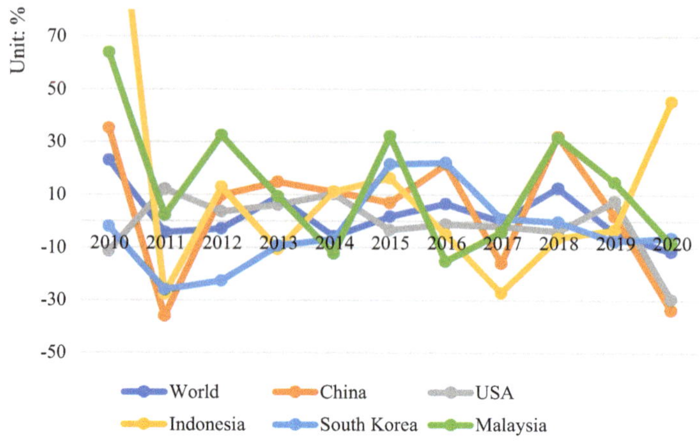

Fig. 13.2 Inland water production, ocean production, and the proportion of marine area production of the top ten countries in global fishery production in 2019. *Data Source* Food and Agriculture Organization of the United Nations (FAO) (https://www.fao.org/fishery/en/statistics/zh)

13.4 Model Introduction and Data Processing

13.4.1 Model introduction

GTAP is a computable general equilibrium model developed by Hertel (1997) from Purdue University. Based on neoclassical economic theory, the model assumes a perfect competition market and constant return to scale of the production function, and follows the principles of consumer utility maximization and producer cost minimization. The model assumes that there are five production factors (land, capital, skilled labor, unskilled labor, and natural resources), as well as three representative entities (private, government, and manufacturers).

Among them, manufacturers adopt the CES (Constant Elasticity of Substitution) production function to minimize costs:

$$\min C = \sum_i P_i X_i \tag{13.1}$$

And which is bound by the CES production function:

$$X = \left(\sum_i \delta_i X_i^{-\rho}\right)^{-\frac{1}{\rho}} \tag{13.2}$$

where P_i and X_i respectively are the price and quantity of a certain input factor i, X is the total output, δ_i is the contribution share of this factor, $\sum_i \delta_i = 1$, and ρ is a substitution parameter. The Lagrange function is constructed and the optimal solution of the optimal factor input X_i is obtained. After linearization, the optimal solution is expressed as

$$x_i = q - \sigma(p_i - p) \tag{13.3}$$

x_i is the he change ratio of the factor input i. Eq. (13.3) shows that the factor input is related to the total output, the elasticity of substitution, the factor price and the product price.

All the income (mainly including the government tax and the factor income of private consumers) in the model is collected into the virtual regional sector. All the income collected in this sector is allocated to private consumption, government consumption and savings in a fixed proportion in the form of Cobb-Douglas function. Assuming that the regional income is Y, the utility of each regional sector i is U_i (where $i = P, G$, and S, representing private, government, and savings, respectively), P_i is the consumer price of each sector, B_i is the share of different subjects i, then the total utility function U is expressed as[6]:

$$U = AU_P^{B_P} U_G^{B_G} U_S^{B_S} \tag{13.4}$$

The budget constraint is

$$\sum P_i U_i = Y \tag{13.5}$$

The optimal solution for the expenditure proportion of each sector is obtained:

$$\sum P_i U_i = Y \tag{13.6}$$

Therefore, in the standard model closure, the proportions of government, private, and savings in total income are constant. Consumer demand behavior adopts the CDE (Constant Differences of Elasticities) consumption function constructed by Hanoch (1975). Unlike government consumption, private consumption is non-homothetic, so a more realistic CDE function is adopted. The CDE function can provide flexible free parameters to match its own price elasticity, cross-price elasticity, and income elasticity. It is more flexible than the CES function and can reflect consumers' preferences for different types of commodities by adjusting parameters, so it is more suitable for the application research of the CGE (Zhou & Zhang, 2020b). The CDE function is expressed as follows

$$\sum_i \delta_i U^{\sigma_i B_i} \left(\frac{P_i}{Y}\right)^{\sigma_i} = 1 \tag{13.7}$$

[6] http://www.econmod.cn.

where i is the type of goods consumed, δ_i is allocation parameter, B_i is the expansion parameter of income, which is expressed as SUBPAR (CDE Substitute Parameters) in GTAP, σ_i is the substitution parameter which is expressed as INCPAR (CDE Expansion Parameters) in GTAP, Y is the consumption per person and P_i is price.

Private and government consumption of imported and domestic goods, and the substitution between imported and domestic goods follow the Armington hypothesis (Hillberry et al., 2013). The aggregate savings flow to the World Bank, which is then transformed into investment in various countries. When the product and input factor markets clear, the model reaches equilibrium.

13.4.2 Data Processing

The results of this quantitative study are based on the newly published standard GTAP10 database. The database is an independent database based on the I-O tables of various countries, covering 141 regions and 65 products, as well as various macroeconomic data, fiscal trade data, and commodity tariff and energy data.

First of all, for the research of this paper, it is necessary to aggregate the database, which is mainly divided into the following two aspects:

1. Country classification. According to Zhao et al. (2020), nuclear wastewater discharge not only affects Japan, but also Pacific rim countries. Thus, the original countries in the GTAP database are reclassified after considering the ranking of GDP, total fishery output value,and total aquatic product export of Pacific rim countries.[7] Countries and regions were divided into two categories: Pacific Rim areas with high pollution and high economic impact, and the other is areas that are not affected, such as India, Peru and Norway, which have large fisheries output and trade. Based on this, the original 141 countries and regions in GTAP have been reclassified into 13 countries or regions: the United States, China, Japan, Canada, Korea Rep, Russia, Mexico, Indonesia, Vietnam, India, Peru, Norway and other regions of the world.
2. Industry classification. After the discharge of nuclear waste water, the import and export of aquatic products and all kinds of related food will be significantly impacted. According to the research needs and the closing conditions required for the GTAP operation, the original 65 sectors in the GTAP database are further subdivided into the following 20 industries: cereals, fruits, vegetables, nuts, fisheries, dairy products, meat, animal products, oil and sugar, processed foods, other agricultural products, textiles and clothing, light industry, machinery and equipment, mining industry, chemical industry, transportation equipment, electronic products, steel and metal, public service and construction, other manufacturing, and other services (see Appendix A Table 13.7 for specific classification).

[7] See the appendix 1 for specific data.

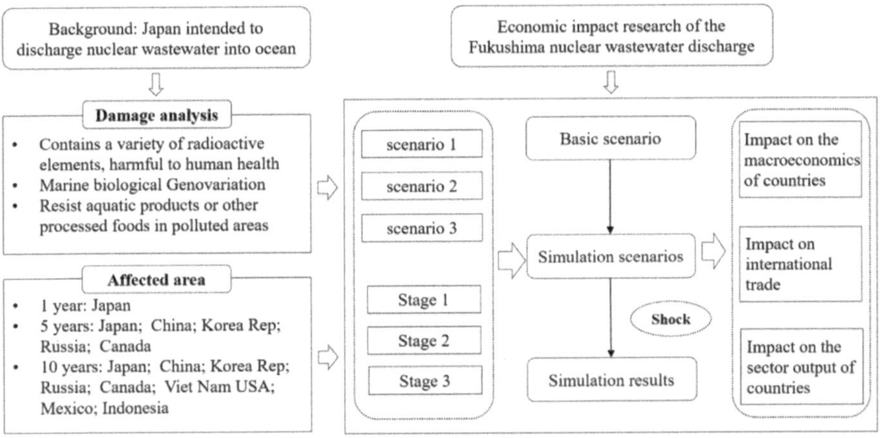

Fig. 13.3 The research idea of the comprehensive economic loss of nuclear wastewater discharge

13.5 Scenario Setting

13.5.1 Scenario Setting

Figure 13.3 shows the research idea of the comprehensive economic loss of nuclear wastewater discharge.

The authors hold that the Fukushima nuclear wastewater discharge harms the world mainly through two channels: (1) Many countries embargo aquatic products from Japan; (2) the impact of nuclear wastewater discharge on local fisheries output. (Including aquaculture production and Capture production).

Thus, the following three simulation scenarios are set up in this paper.

Scenario 1 (S1): We assume that nuclear wastewater discharge impacts aquatic product exports from Japan and other affected countries and Japan is the most affected by nuclear wastewater discharge. Let's assume that Japan's exports of aquatic products will plunge by 80%; With reference to the impact of the nuclear accident on the Japanese food industry and agriculture, it is assumed that Japan's exports of dairy products, fruits and vegetables, meat, rice and other grains and processed foods to various countries will be reduced by 50%. Less affected than Japan, other countries will decline by 50%.[8]

Scenario 2 (S2): Japan's nuclear wastewater discharge would have a certain impact on the fisheries of Japan and the Pacific Rim countries. We assume that the nuclear wastewater discharge will only affect the fishery production of various countries in

[8] As the disaster event has not yet occurred, it is difficult to estimate the loss caused by the nuclear wastewater discharge exactly. Thus, only the loss from one of the assumptions is analyzed here. The simulation results of other assumptions, such as enlargement (please see the result in appendix B) and reduction (please see the result in appendix C) of the damaged scale are similar to the original simulation result in this paper.

Table 13.1 The ratio of the fishery output in the North Pacific waters to the total fishery output of each country and the shock ratio of the scenario simulation (unit: %)

Country	Pacific Ocean output ratio	Pacific output loss ratio	Total output loss ratio
Japan	97.32	60	58.39
USA	57.64	30	17.29
China	99.47	30	29.84
Canada	29.79	30	8.94
Russia	68.02	30	20.41
Korea Rep	96.17	30	28.85
Mexico	73.91	30	22.17
Indonesia	74.21	30	22.26
Viet Nam	61.6	30	18.48

Data Source It is calculated according to the FAO (Food and Agriculture Organization of the United Nations) "Global production by production source" database.[10]

the North Pacific waters[9] and that Japan is the most affected area. Therefore, we assume that Japan's fisheries in the Pacific Ocean will drop by 60% after nuclear wastewater discharge, while other affected countries' fishery output will drop by 30%. The following table shows the ratio of the fishery output in the North Pacific waters of each country to the total fishery output of each country and the ratio of the total output loss of each country in the scenario simulation:

Scenario 3 (S3): Scenario 3 is set to comprehensively investigate the economic impact of nuclear wastewater discharge on various countries through the common impact of the above two aspects (Table 13.1).

13.5.2 Stage Division

Zhao et al. (2020) simulated the diffusion of radioactive substance tritium in nuclear waste water. It shows that one year after the discharge, the tritium plume from Japan will soon spread to the Pacific Ocean. And the polluted water will gradually spread to the Northeast Pacific in 4 to 5 years along with the North Pacific Ocean current. After ten years of advection and spread, almost the entire North Pacific Ocean will be contaminated. According to the third and fourth simulations in the study, the tritium core contamination area (tritium concentration greater than 0.1 Bq/m^3) is confined to the east coast of Japan, assuming that Japan discharges all the nuclear water within 5

[9] The definition of this area refers to the Major fishing areas classified by FAO (Food and Agriculture Organization of the United Nations). https://www.fao.org/cwp-on-fishery-statistics/handbook/general-concepts/main-water-areas/en/.

[10] https://www.fao.org/fishery/zh/topic/16003.

Table 13.2 Simulation scenarios of the impact of nuclear wastewater discharge on the economy

S1	Stage 1	Japan's exports of aquatic products will plunge by 80% Due to nuclear waste water discharge; Exports of meat, milk, and processed food will plunge by 50%
	Stage 2	On the basis of stage 1, Exports of aquatic products from China, Korea Rep, Russia and Canada will plunge by 50%
	Stage 3	On the basis of stage 2, Exports of aquatic products from the United States, Mexico, Indonesia and Viet Nam will plunge by 50%
S2	Stage 1	The fishery output value of Japan's marine areas will drop by 60% because of nuclear wastewater discharge
	Stage 2	On the basis of stage 1, the fishery output value of China, Korea Rep, Russia and Canada's marine areas will drop by 30%
	Stage 3	On the basis of stage 1, the fishery output value of the United States, Mexico, Indonesia and Viet Nam's marine areas will drop by 30%
S3	Stage 1	In the 1 stage, the comprehensive impact of Scenario 1 and Scenario 2
	Stage 2	In the 2 stage, the comprehensive impact of Scenario 1 and Scenario 2
	Stage 3	In the 3 stage, the comprehensive impact of Scenario 1 and Scenario 2

to 10 years. The simulation can be divided into three stages according to the influence ranges in different periods after the discharge.

The first stage (within one year of nuclear wastewater discharge): Fukushima nuclear wastewater discharge only affects Japan.

The second stage (5 years after nuclear wastewater discharge): The impact spreads to the North Pacific, affecting countries including Japan, China, South Korea, Russia, and Canada. The import and export of aquatic products and food, as well as fish farming in relevant countries, will be impaired.

The third stage (10 years after the discharge): The affected countries are expanded to the entire North Pacific, and the affected countries include Japan, China, South Korea, Russia, Vietnam, the United States, Canada, Mexico, and Indonesia. It should be noted that since the tritium core pollution area is limited to the east coast of Japan, the impact of Japan in each scenario simulation is greater than that of other affected countries. The specific scenario settings are shown in Table 13.2.

13.6 Simulation Result Analysis

RunGTAP is used to simulate the three stages of the three scenarios and investigate the changes in the economic effects of the relevant countries and regions under each scenario. The changes in the trade scale and industry level are further analyzed.

13.6.1 Impact of Nuclear Wastewater Discharge on Macroeconomics

1. Changes in GDP

In general, once nuclear waste water is discharged, the GDP of most affected countries will decline under various scenarios, and the decline will become greater with the expansion of the affected area.

First, in Scenario 1, from a vertical perspective, when the exports of aquatic products and other related food products from affected countries to other countries in the world are reduced, the GDP of all countries except Norway will be slightly negatively affected. But considering that the overall trade volume of the world is relatively high, fisheries only account for a small share of all trade in goods, so the overall impact on GDP is small. From a horizontal perspective, Norway's GDP has increased slightly in all three stages. The reason may be that when Japan and other countries suffer losses in fishery exports, part of the trade volume will be transferred to Norway, which has fishery export advantage, thus contributing to Norway's economic growth.

Second, in Scenario 2 (production reduction), it can be seen that when the impact reaches the third stage, GDP falls in most countries except Norway, especially Vietnam and China, with the highest declines of 1.9% and 1.3%, respectively. Therefore, in Scenario 2, China and Vietnam are the countries that lose the most. From Appendix A Table 13.9, it can be seen that the fishery output value of these two countries is higher, indicating that nuclear wastewater discharge will have a greater impact on countries with rich fishery resources. With the deepening of international fishery cooperation, India and Peru, which are also rich in fishery resources but not affected by nuclear waste water, are also negatively affected to a certain extent. Only Norway still maintains positive growth in the scenario. The possible reason is that the aquatic products of Norway and Vietnam have strong competitive advantages in the international market and are larger than India, Peru and other countries (He & Zhou, 2021). Therefore, when the fishing industry is hit (S2), Viet Nam will be greatly negatively affected. When the fisheries of other countries with fishery resource advantages suffer losses, Norway will benefit by highlighting its own competitive advantages. Finally, when the two effects are added together, South Korea will be the most affected country, with GDP falling by as much as 2.8% (Table 13.3).

2. Changes in social welfare

The E.V. (Equivalent Variation) variable in GTAP represents the change in the welfare of national residents. When the domestic price remains unchanged, the E.V. variable represents the monetary compensation required from the initial to the final utility level. The larger the E.V., the bigger improvement in welfare (Zhu, 2020). There are four significant welfare source changes: allocative efficiency effect, endowment effect, technology effect, and terms of trade effect.

The welfare changes under the three scenarios are shown in Table 13.4: Generally speaking, the discharge of nuclear wastewater will cause welfare losses on a global

Table 13.3 Changes in GDP of global countries and regions (unit: %)

	Scenario 1			Scenario 2			Scenario 3		
	Stage 1	Stage 2	Stage 3	Stage 1	Stage 2	Stage 3	Stage 1	Stage 2	Stage 3
China	−0.0066	−0.0091	−0.0145	−0.0009	−1.2528	−1.2808	0.0071	−1.3294	−1.3959
Japan	−0.0023	−0.0144	−0.0246	−0.2329	−0.2834	−0.2930	−0.2637	−0.4118	−0.1153
Korea Rep	−0.0073	−0.2257	−0.3700	0.0003	−0.3766	−0.3856	0.0127	2.9260	−2.7783
Russia	−0.0005	−0.0015	−0.0014	0.0011	−0.1886	−0.1983	0.0013	−0.3491	−0.3776
USA	−0.0013	−0.0118	−0.0261	0.0002	0.0019	−0.0147	0.0016	−0.0081	−0.0089
Canada	−0.0016	−0.0057	0.1112	0.0009	−0.0340	−0.0490	0.0023	0.0281	−0.0465
Mexico	−0.0003	−0.0001	0.0008	0.0003	−0.0029	−0.0794	0.0003	−0.0073	−0.0638
Indonesia	−0.0010	−0.0169	−0.0171	0.0016	−0.0016	−1.0594	0.0013	0.0491	−1.0433
Viet Nam	−0.0683	−0.0781	−0.1069	−0.0079	0.0212	−1.9055	0.1100	0.1830	−1.7025
India	−0.0008	−0.0018	−0.0020	−0.0007	−0.0073	−0.0064	−0.0001	−0.0022	−0.0024
Peru	−0.0003	−0.0003	−0.0003	0.0000	−0.0014	−0.0016	−0.0003	−0.0022	−0.0014
Norway	0.0004	0.0060	0.0101	0.0056	0.0361	0.0422	0.0039	0.0184	0.0568
Rest of World	−0.0011	−0.0024	−0.0033	0.0001	0.0025	0.0022	0.0007	0.0084	0.0108

Data Source Sorted out according to GTAP simulation results

scale, and the loss will increase with the enlarging scope of nuclear wastewater pollution.

Scenario 1: In the third stage, when the export of aquatic products of all affected countries decreases, the welfare of these countries declines with the exception of Canada, among which South Korea and the United States have been hit the hardest. As can be seen from Tables 13.10 of the appendix, both these two countries have trade deficits in aquatic products and the consumption of aquatic products mainly comes from imports. Therefore, welfare losses to these two countries might be due to a reduction in fisheries exports from all affected regions, which would significantly affect fish imports in these two countries. In addition, India, Peru and Norway, which are outside the areas of influence of nuclear wastewater, have seen increases in welfare. Despite an increase in other countries or regions, this would still result in a loss of $12.015 billion in welfare worldwide. This indicates that the occurrence of negative events hinders the development of free trade leading to global welfare losses.

Scenario 2: When the fishery output value of various countries declines owing to the impact of nuclear wastewater discharge, the world's welfare would be $182.65 billion. China and Japan are the countries with the most significant losses, reaching $14.526 billion and $1.696 billion, respectively. For China, the main reason might be that China is the country with the highest fishery output in both capture and aquaculture.[11] Second, China's fishery output has a larger proportion within the range polluted by nuclear wastewater. Japan may have higher welfare losses than other countries because nuclear wastewater would greatly impact its own fisheries output. Similarly, although welfare rises in some regions, it will remain negative globally.

Scenario 3: When combined, the two negative effects of nuclear wastewater discharge could cause a global welfare loss of up to $219.80 billion. Specifically, in the first stage, when nuclear wastewater only affects Japan, Japan's welfare would be reduced by $12.62billion. In the second stage, when the nuclear wastewater's impact expands to China, South Korea, Russia, and Canada, their welfare begins to decline except in South Korea. And China's largest welfare loss is up to $150.09 billion. In the third phase, all affected countries except the U.S. would suffer a welfare loss, and welfare has slightly increased in unaffected areas. In the unaffected regions, Norwegian welfare increased more than the other two countries, up to $2167 million. The result indicates that while the effects of nuclear wastewater discharge may positively impact a few countries, they are detrimental to the efficient allocation of resources around the world, and the total global welfare would be lost.

[11] Refer to Appendix Table 13.9.

Table 13.4 Changes in the welfare of global countries and regions (unit: million dollars)

	Scenario 1			Scenario 2			Scenario 3		
	Stage 1	Stage 2	Stage 3	Stage 1	Stage 2	Stage 3	Stage 1	Stage 2	Stage 3
China	−644.4	−988.7	−1504.4	−318.9	−141728.4	−145264.9	522.0	−150090.4	−157158.9
Japan	−608.9	−1329.3	−1884.3	−11335.9	−16048.1	−16959.8	−12619.3	−19853.5	−10890.8
Korea Rep	−115.4	−3340.9	−5462.8	−92.5	−7310.6	−7555.5	93.7	40096.6	−42539.8
Russia	−8.7	−65.2	−53.1	85.4	−3535.1	−3778.5	112.0	−6159.7	−6495.3
USA	−153.3	−1593.8	−3490.6	77.5	2480.3	−561.6	361.2	1149.4	1355.6
Canada	21.8	−774.2	537.0	109.8	309.3	70.1	56.4	−383.1	−610.2
Mexico	10.4	69.5	−1.1	52.9	266.3	−567.8	48.3	139.3	−380.9
Indonesia	9.4	−70.5	−182.6	120.2	221.6	−8321.6	124.9	577.8	−8157.0
Viet Nam	−98.9	−105.5	−205.9	−78.5	209.6	−3236.9	101.1	195.6	−2640.5
India	4.9	16.7	18.2	39.4	482.8	587.1	47.4	475.7	836.3
Peru	5.1	14.0	18.3	15.7	102.1	115.9	16.6	68.1	134.2
Norway	41.4	328.5	539.2	242.5	1419.5	1689.2	154.9	537.6	2167.8
Rest of World	−220.8	−344.0	−343.4	401.0	1580.6	1136.8	653.9	2750.3	4578.0
Total	−1757.3	−8183.5	−12015.4	−10681.5	−161550.0	−182647.5	−10326.8	−130496.4	−219801.5

Data Source Sorted out according to GTAP simulation results

13.6.2 *Impact of Nuclear Wastewater Discharge on International Trade*

In the third stage of nuclear wastewater pollution, the changes in terms of trade of relevant countries and regions under each scenario are as follows.

Terms of Trade (TOT) refers to the ratio of how many units of imported goods a country can exchange for each unit of exported goods in a certain period, which can reflect a country's macroscopic trade. Analysis from the perspective of TOT: In Scenario 1, in the affected areas, only China, the United States, and Viet Nam have a slight improvement. The United States was the country with the most significant increase in terms of trade, increasing by 0.04%. In recent years, the aquatic products of the United States have been at a disadvantage in international trade competition, and the market share has also declined year by year. It is also a country that imports significantly more than exports (see Table A4). Therefore, when nuclear wastewater affects the export of aquatic products of various countries, the United States is not only less affected than other countries but may improve its own TOT. The TOT of other countries has deteriorated. Among them, Canada is the most affected. In Scenario 2 and Scenario 3, only China, Japan, and South Korea's terms of trade deteriorate, and all other countries' terms of trade improve. It shows that the decline of fishery output caused by nuclear wastewater discharge will bring negative economic benefits to the overall foreign trade of these three countries (Table 13.5).

Finally, the TOT of countries not affected by nuclear wastewater discharge has improved, such as India, Peru, and Norway. Norway is the country with the largest increase in terms of trade. A possible explanation is that Norway's aquatic product trade has a strong international competitive advantage, and it is a major exporter of aquatic products. Therefore, the aquatic product trade of other countries will

Table 13.5 Changes in terms of trade of global countries and regions in the third stage (unit: %)

	Scenario 1	Scenario 2	Scenario 3
China	0.0033	−0.1212	−0.0776
Japan	−0.0809	−0.3637	−0.5875
Korea Rep	−0.0423	−0.3372	−0.5625
Russia	−0.0071	0.0049	0.1368
USA	0.0484	0.1080	0.1381
Canada	−0.2925	0.2022	0.0602
Mexico	−0.0030	0.1178	0.1152
Indonesia	−0.0152	0.5577	0.5770
Viet Nam	0.0066	0.2934	0.3902
India	0.0119	0.1641	0.2031
Peru	0.0419	0.2796	0.3236
Norway	0.2757	0.8603	1.1000
Rest of World	0.0063	0.0114	0.0177

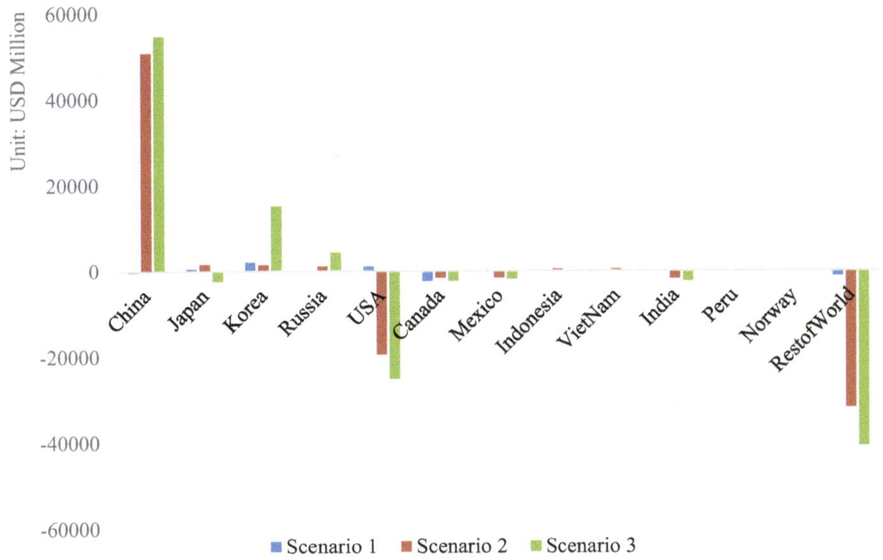

Fig. 13.4 Changes in the trade balance of global countries and regions in the third stage. *Data Source* Sorted out according to GTAP simulation results

be more transferred to Norway, which will greatly promote the economic situation of Norway's foreign trade. Combined with the previous analysis, except for a slight decline in the GDP of India and Peru, the economies of countries outside the scope of nuclear wastewater are almost unaffected or benefited from it. However, the development of global economic integration and the expansion of the impact of wastewater makes it difficult to ensure that these countries remain immune to the impact of nuclear wastewater discharges.

In the third stage of nuclear wastewater pollution, the changes in the trade balance of countries and regions in the world under each scenario are shown in the Fig. 13.4 below. It can be seen that the discharge of nuclear wastewater will enlarge several countries' trade surplus. China would reach a maximum of $54.75 billion, while the trade deficit of the United States would further increase to $25.02 billion.

13.6.3 Impact of Nuclear Wastewater Discharge on Sectoral Output

From the previous analysis, it can be seen that nuclear wastewater discharge negatively affects the GDP, welfare, and trade of most countries. The following will analyze the possible impacts of nuclear wastewater discharge from the sectoral output.

This subsection considers the growth ratio of the total output value of each country when the pollution range of nuclear wastewater discharge is the maximum (that is, the third stage). The growth ratios of the total output value of global countries or regions are shown in Fig. 13.1. It can be seen that the growth ratios of the total output value of most regions in the world decline to varying degrees. The growth ratio of South Korea's total output value increases significantly in Scenario 1 and Scenario 3. Combined with the analysis in Table 13.6, the reduction of South Korea's fishery and related food industries will promote the development of high value-added industries such as electronic equipment, thereby increasing the total output. Therefore, from the perspective of output, South Korea will benefit from the discharge of nuclear wastewater in various countries. Among them, Vietnam's total output value growth ratio decreases by 1.22%, mainly in the fishing and food industries. Vietnam has a long coastline, densely covered rivers, and rich fishery resources. Therefore, when the discharge of nuclear wastewater reduces fishery output, Vietnam's related food industry output value would also be shocked. The total output value of Vietnam will be greatly affected and decreased. It is followed by China and Indonesia at 0.57% and 0.73%, respectively, and the rest of the world will also see a slight decline in aggregate output.

Table 13.6 presents the output change ratios of various sectors in the third stage of Scenario 3, under the comprehensive impact of nuclear wastewater discharge. It can be seen that under the impact of nuclear wastewater discharge, the output ratio of various countries' sectors fluctuates to varying degrees due to the import and export volume, product prices, and the competitiveness of products. Hit by trade barriers and a decline in the fisheries in various countries,' the output of China's grain and processed food sectors dropped by 4.38% and 5.46%, respectively, followed by meat and milk, which fell by 2.88% and 2.79%. When the United States, Canada, and Mexico are affected, the output changes of other industries are not particularly obvious. Therefore, combined with Fig. 13.5, the changes in the total output value of these three countries are also relatively small. In addition, the simulation results show that when the fishery output in the areas affected by nuclear wastewater discharge decreases, the fishery output value of the unaffected countries increases. Among them, Norway has the most obvious increase. Combined with the previous analysis, Norway is also the only country to see an increase in GDP and a more pronounced improvement in terms of trade when all countries are affected. An increase in the output of the Norwegian fishery is associated with a corresponding decrease in the output of other industries. This indicates that, under the influence of nuclear wastewater discharge, the resources of each country will be changed. Countries affected by nuclear wastewater discharge would transfer from the sectors with reduced output value to others with comparative advantages, and countries that are not affected by nuclear wastewater discharge would suffer losses to other industries due to the transfer of resources to the aquaculture industry and the vigorous development of fisheries. But the redistribution of resources after the shock will reduce the gross output value, and thus the country will suffer losses.

Table 13.6 The output change ratios of various sectors in the third stage of Scenario 3

	China	Japan	Korea Rep	Russia	USA	Canada	Mexico
Cereals	−4.38	−0.52	−3.97	1.35	0.80	0.97	0.41
Fruits vegetables nuts	−0.17	−1.63	−0.76	0.09	−0.11	0.03	0.12
Milk products	−29.84	−58.39	−28.85	−20.41	−17.29	−8.94	−22.17
Meat products	−2.79	−0.73	−10.00	−0.91	0.13	−0.17	−0.08
Animal products	−2.88	−2.28	−10.16	−1.17	0.25	0.01	0.10
Oils and sugars	−2.44	−2.61	−10.78	−1.20	0.38	−0.31	0.09
Food products	−1.79	−2.24	−5.77	−0.36	−0.53	−0.75	0.02
Other agricultural products	−5.46	−3.37	−28.99	−0.85	1.40	−0.44	0.92
Textile and apparel	−1.36	−0.09	−4.73	−0.32	0.07	−0.10	0.09
Light industry	0.62	−0.05	3.43	0.25	−0.54	−0.76	−0.50
Mechanical equipment	−0.25	0.08	1.43	0.00	0.06	−0.25	−0.05
Mining industry	0.45	0.14	4.13	0.56	−0.34	−0.50	−0.36
Chemical industry	0.25	0.27	2.70	0.55	−0.17	−0.14	−0.15
Transportation Equipment	0.12	0.36	2.95	0.66	−0.37	−0.39	−0.13
Electronic product	−0.32	0.67	4.34	0.13	−0.28	−0.25	−0.21
Steel and metal	1.26	−0.35	5.50	0.60	−1.25	−1.41	−1.42
Public service and construction industry	0.41	0.14	3.75	1.11	−0.58	−0.68	−0.64
Manufactures	−1.06	0.13	−2.06	−0.29	0.40	0.35	0.53
Other service industries	0.48	−0.59	2.60	−0.56	−0.24	0.03	−0.16
Cereals	−0.80	−0.02	−1.73	−0.44	0.01	0.03	0.00

	Indonesia	Viet Nam	India	Peru	Norway	Rest of World
Cereals	−0.92	0.19	0.50	0.72	−2.56	0.59
Fruits vegetables nuts	−0.25	−0.43	0.20	−0.08	0.30	0.06
Milk products	−22.26	−18.48	1.07	2.94	14.97	5.98
Meat products	−1.10	−5.34	0.00	0.62	−5.63	0.21
Animal products	−0.89	−7.55	−1.12	0.42	−2.55	0.35
Oils and sugars	−1.27	−1.86	−0.10	0.20	−2.16	0.47
Food products	0.35	−0.71	−0.03	0.14	−8.39	0.00
Other agricultural products	−4.52	−21.93	1.71	5.22	−6.12	2.17

(continued)

Table 13.6 (continued)

	Indonesia	Viet Nam	India	Peru	Norway	Rest of World
Textile and Apparel	−0.24	0.00	0.11	0.28	−0.58	0.16
Light Industry	−0.90	1.39	−0.64	−0.59	−1.12	−0.83
Mechanical equipment	−0.30	1.30	−0.22	−0.09	−0.57	−0.11
Mining industry	−0.15	1.19	−0.30	−0.65	−2.64	−0.78
Chemical industry	0.12	0.33	−0.21	−0.83	−0.49	−0.24
Transportation Equipment	−0.25	0.40	−0.35	−0.55	−1.16	−0.36
Electronic product	−0.38	0.68	−0.15	−0.48	−0.74	−0.44
Steel and metal	−0.95	1.30	−0.94	−1.19	−3.06	−1.67
Public service and construction industry	−0.14	1.34	−0.44	−1.81	−2.01	−0.85
Manufactures	0.07	−0.04	0.21	0.30	0.34	0.29
Other service industries	−0.55	0.19	−0.17	−0.25	−1.84	−0.14
Cereals	−0.69	−0.60	−0.02	0.02	0.23	0.03

Data Source Sorted out according to GTAP simulation results
Note The change in fishery output value is set by scenario simulation, so it is not reflected in the table

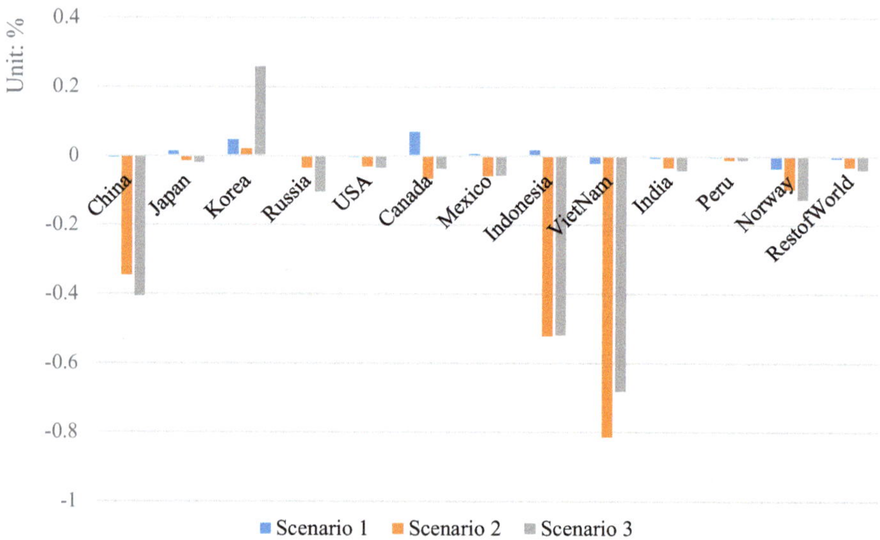

Fig. 13.5 The growth ratios of the total output value of global countries and regions. *Note* Output changes for the nine countries affected by nuclear wastewater do not include fisheries (because the scenario modeling assumes the fisheries output changes for these countries), output changes for the other countries include all industries. *Data Source* Sorted out according to GTAP simulation results

13.7 Conclusions and Suggestions

13.7.1 Major Conclusions

Using the GTAP model and applying it to various scenarios regarding the potential effects of discharging nuclear wastewater from the Daiichi plant, we conclude the following:

1. From a macroeconomic point of view, the negative effects of nuclear wastewater discharge will reduce the GDP and welfare of most affected countries, and the loss will increase with the expansion of nuclear wastewater. While several countries' welfares is positive, all three simulations show that the impact of nuclear wastewater discharges would result in a global welfare loss, with a maximum welfare loss of $219.8 billion in Scenario 3. From a country perspective, in the case of trade impact, (S1), among countries in the region affected by nuclear wastewater discharge, Canada may be the only beneficiary country, while South Korea may be the most significant victim country. In the case of industry impact, (S2), China and Japan are the most affected countries. When the two influence effects are superimposed, (S3), China and South Korea may be the most affected countries. Among the countries outside the region affected by nuclear wastewater discharges, Norway is a country that is likely to benefit from all three scenarios.
2. From a trade point of view, the terms of trade vary from country to country in different scenarios. First, among the countries in the affected area, South Korea and Japan both suffer from deteriorating terms of trade in the simulation results of the three scenarios and thus suffer losses, which adversely affect their foreign trade conditions; The U.S. and Vietnam have improved terms of trade in all three simulations, thereby improving foreign trade. The terms of trade of the regions outside the affected area of nuclear wastewater are all positive in the three simulation results, indicating that the discharge of nuclear wastewater might have adverse effects on the countries within the affected area. However, it would improve the foreign trade situation of the unaffected regions. Secondly, in terms of trade balance, when the impact of nuclear wastewater reaches the maximum (the third stage), China's trade surplus and the United States' trade deficit would be further expanded.
3. From a sectoral output point of view, nuclear wastewater discharge would reduce output in most countries around the world. Among them, Vietnam has the largest reduction in the proportion of total output, followed by Indonesia and China. The largest increase in the proportion of output is in South Korea. Affected by combined negative effects brought by nuclear wastewater discharge, (S3), the reduction in the export of aquatic products and the decline in fishery output in the relevant countries or regions would cause changes in the output of different sectors in each country. While the output of national fisheries in areas affected by nuclear wastewater discharges decreases, fisheries in locations outside the affected area increase in output. Meanwhile, the resources of each country would be transferred to other sectors with relatively comparative advantages, but most

countries or regions would suffer losses from such passive resource redistribution, resulting in a decline in total output.

13.7.2 Policy Suggestions

According to the analysis results of this paper, Japan's nuclear wastewater discharge into the sea would bring certain economic losses to several countries in the world. Not only Japan but also China, South Korea, and Vietnam would be greatly affected. Shaun Burnie, a nuclear power expert at the international environmental protection organization Greenpeace Japan Office, believes that discharging nuclear wastewater into the sea is not the "only option" for its treatment. There were five options for treating nuclear wastewater that have been proposed by the ALPS (Multi-nuclide removal equipment) Subcommittee: geosphere injection, discharge into the sea, vapor release, hydrogen release, and underground burial. At the same time, factors such as costs and feasibility of each project were considered. Among them, the cost of discharging into the ocean would cost about 3.4 billion yen, while the Vapor release program would cost tens of billions of yen, and the Underground burial program would cost hundreds of billions of yen (Subcommittee on Handling of the ALPS Treated Water, 2020). Finally, Japan chose the method of discharge into the sea, which is the most convenient, the lowest cost, and the lowest risk. Based on the analysis of this paper, although this method is the most convenient, it might not only threaten the health and safety of human beings and the marine ecological environment but also bring certain losses to the economies of neighboring countries.

Therefore, according to the study results, the authors suggest that, first of all, countries should take certain preventive measures in advance according to the possible impact of nuclear wastewater discharge to minimize the damage, for example, strengthening the monitoring and supervision means for the safety of aquatic products. Second, other safer approaches can be taken to address the Fukushima nuclear wastewater problem, such as Greenpeace's proposal that long-term storage is feasible (Shaun, 2020). In addition, Japan can better solve the problem of nuclear wastewater by learning from the successful experience of other countries (Liu & Liu, 2019) and by using social forces, technology, and civil organizations to expand and enrich the means of nuclear wastewater governance (Jin, 2021). Third, countries should strengthen international cooperation and adopt a negotiated approach to prevent Japan from discharging nuclear waste water into the ocean in order to safeguard the rights and interests of each country. In addition, a technical working group should be formed internationally to carry out the Fukushima nuclear wastewater disposal plan, follow-up implementation, and international assessment and supervision (Wu, 2022).

13.7.3 Limitations

First of all, this paper only conducts a preliminary simulation and estimation of the economic impact of nuclear wastewater discharge on relevant countries. As each country's economic development, industrial conditions, resources, etc. are different, the impact on specific industries and imports and exports vary. In future analysis, more detailed analysis can be made for the specific situation of a certain country according to research needs. Second, due to the limitation of the model, this paper assumes that the market is completely cleared. However, as a matter of fact, because of the differences among different sectors, the labor and capital, and other resources cannot be transferred freely. Therefore, this paper may underestimate the economic loss. Third, because Japan has not yet discharged nuclear wastewater into the sea, the data of the direct economic losses induced by the discharge are not available, so various scenarios can only be assumed at present. Fourth, global warming has become an important obstacle to the sustainable development of the human economy and society. In recent years, the control of pollutants and greenhouse gas emissions has also attracted great attention. (Mi et al., 2017; 2018). Since nuclear wastewater discharge might also have a certain impact on the carbon emissions of related countries, that would also be the direction for future research in this article.

Acknowledgements Yanmin Liu and Jihong Chen also made great contributions to this manuscript. We express our heartfelt thanks to them. This research also was supported by: National Natural Science Foundation of China (72474127) and National Social Science Fund Major Bidding Project of China (18ZDA052). Part of this chapter was published:Guo J, Liu Y, Wu X, Chen J. Assessment of the impact of fukushima nuclear wastewater discharge on the global economy based on GTAP. Ocean & Coastal Management 2022;228:106296. https://doi.org/https://doi.org/10.1016/j.ococoaman.2022.106296. The publication has been conducted with copyright permission.

Appendix A

See Tables 13.7, 13.8, 13.9and 13.10

Table 13.7 Industry classification in the GTAP model

New sector	Old sector description
Cereals	Processed rice, wheat, Cereal grains.
Fruits vegetable nuts	Vegetables, fruit, nuts, etc.
Fishery	Fishery.
Milk products	Raw milk, Dairy products.
Meat products	Bovine meat products, Meat products.
Animal products	Animal products.

(continued)

Table 13.7 (continued)

New sector	Old sector description
Oils and sugars	Oil seeds, Vegetable oils and fats, Sugar
Food products	Food products.
Other agricultural products	Other crops, Cattle and sheep and Other animals, Wool, silk-worm cocoons, Forestry, Beverages and tobacco products, etc.
Textile and Apparel	Textiles, Wearing apparel.
Light Industry	Leather and Wood products, Paper products, publishing.
Mechanical equipment	Machinery and equipment.
Mining industry	Coal, Oil, Gas, Minerals, Petroleum, coal products and Mineral products.
Chemical industry	Chemical products, Basic pharmaceutical products, Rubber and plastic products.
Transportation Equipment	Motor vehicles and parts, Transport equipment.
Electronic product	Computer, electronic and optic, Electrical equipment.
Steel and metal	Ferrous metals, Metals, Metal products.
Public service and construction industry	Electricity, Gas manufacture, distribution, Water, Construction.
Manufactures	Manufactures.
Other service industries	Trade, Accommodation, Food and service, Transport, Communication, Financial and Other services, Public Administration and defe, Education, etc.

Data Source The GTAP model

Table 13.8 The top 20 countries in GDP in 2019 (unit: Billion dollars)

	GDP		GDP
USA	21433.226	Russia	1699.877
China	1427.994	Korea Rep	1646.739
Japan	5081.770	Australia	1396.567
Germany	3861.124	Spain	1393.491
India	2868.929	Mexico	1268.871
UK	2829.108	Indonesia	1119.191
France	2715.518	Netherlands	907.051
Italy	2003.576	Saudi Arabia	792.967
Brazil	1839.758	Turkey	761.425
Canada	1736.426	Switzerland	703.082

Data Source The World Bank[12]

[12] https://databank.worldbank.org/indicator/NY.GDP.MKTP.CD/1ff4a498/Popular-Indicators.

Table 13.9 World capture and aquaculture production of fish, crustaceans, molluscs, etc., by principal producers in 2019 (unit: t)

Capture production		Aquaculture production	
China	13,995,443	China	48,246,255
Indonesia	7,479,325	India	779,500
India	5,458,700	Indonesia	595,000
Russian Fed	4,974,162	Viet Nam	4,442,257
Peru	4,814,950	Bangladesh	2,488,600
USA	4,800,499	Egypt	1,641,949
Viet Nam	3,429,029	Norway	1,452,926
Japan	3,163,760	Chile	1,384,704
Norway	2,309,319	Myanmar	1,082,141
Philippines	2,053,681	Thailand	964,266
Chile	1,971,756	Philippines	858,277
Myanmar	1,951,120	Ecuador	695,790
Bangladesh	1,895,619	Brazil	599,550
Mexico	1,573,362	Japan	598,229
Thailand	1,542,465	Korea Rep	583,586
Malaysia	1,461,015	Iran	505,000
Morocco	1,458,594	USA	490,042
Korea Rep	1,358,252	Turkey	371,494
Iceland	905,429	Spain	308,025
Spain	879,802	Cambodia	305,408

Data Source FAO STATISTICS "Fishery and aquaculture statistics"[13]

Note For statistical purposes, the data for China do not include Hong Kong Special Administrative Region (Hong Kong SAR), Macao SAR and Taiwan Province of China

Appendix B

1. Simulation and results of loss ratio reduction

   See Tables 13.11, 13.12, 13.13, 13.14, and Figs. 13.6 and 13.7.

Appendix C

1. Simulation and results of loss ratio enlargement

   Tables 13.15, 13.16, 13.17, 13.18, and Figs. 13.8 and 13.9.

[13] https://doi.org/10.4060/cb7874t.

Table 13.10 International trade in fishery commodities by principal importers and exporters (unit: USD 1000)

Imports		Exports	
USA	23,316,681	China	20,074,881
China	17,935,553	Norway	11,994,519
Japan	15,128,022	Viet Nam	8,611,497
Spain	8,073,094	India	6,846,460
France	6,644,141	Chile	6,557,176
Italy	6,574,309	Thailand	5,838,038
Germany	5,826,688	Netherlands	5,677,082
Korea Rep	5,573,400	Canada	5,672,680
Sweden	5,266,272	USA	5,638,761
UK	4,537,998	Ecuador	5,518,082
Netherlands	4,498,132	Russian Fed	5,469,949
Denmark	3,903,028	Denmark	4,837,597
Thailand	3,675,119	Spain	4,694,657
China, H.Kong	3,450,032	Indonesia	4,498,061
Canada	3,171,982	Sweden	4,489,743
Poland	2,604,940	Peru	3,517,658
Portugal	2,427,666	UK	2,989,355
Russian	2,209,692	Germany	2,700,151
Belgium	2,116,260	Poland	2,506,710
Viet Nam	1,822,321	Iceland	2,360,426

Data Source Same as above

Table 13.11 Simulation scenarios of the impact of nuclear wastewater discharge on the economy

S1	Stage 1	Japan's exports of aquatic products will plunge by 60% Due to nuclear waste water discharge; Exports of meat, milk, and processed food will plunge by 30%
	stage 2	On the basis of stage 1, Exports of aquatic products from China, Korea Rep, Russia, and Canada will plunge by 30%
	stage 3	On the basis of stage 2, Exports of aquatic products from the United States, Mexico, Indonesia, and Viet Nam will plunge by 30%
S2	stage 1	The fishery output value of Japan's marine areas will drop by 40% because of nuclear wastewater discharge
	stage 2	On the basis of stage 1, the fishery output value of China, Korea Rep, Russia, and Canada's marine areas will drop by 20%
	stage 3	On the basis of stage 1, the fishery output value of the United States, Mexico, Indonesia and Viet Nam's marine areas will drop by 20%
S3	stage 1	In the 1 stage, the comprehensive impact of Scenario 1 and Scenario 2
	stage 2	In the 2 stage, the comprehensive impact of Scenario 1 and Scenario 2
	stage 3	In the 3 stage, the comprehensive impact of Scenario 1 and Scenario 2

Table 13.12 Changes in GDP of global countries and regions (Unit: %)

	Scenario 1			Scenario 2			Scenario 3		
	Stage 1	Stage 2	Stage 3	Stage 1	Stage 2	Stage 3	Stage 1	Stage 2	Stage 3
China	−0.0044	−0.006	−0.0093	−0.0006	−0.8351	−0.8538	0.0046	−0.8873	−0.9324
Japan	−0.0016	−0.0085	−0.0143	−0.1553	−0.1889	−0.1954	−0.1748	−0.2741	−0.0585
Korea Rep	−0.0052	−0.1403	−0.2295	0.0002	−0.251	−0.257	0.0082	2.060	−1.9304
Russia	−0.0003	−0.0009	−0.0008	0.0008	−0.1257	−0.1321	0.0009	−0.2353	−0.2546
USA	−0.0008	−0.0073	−0.0159	0.0001	0.0012	−0.0098	0.0011	−0.0043	−0.005
Canada	−0.001	−0.0035	0.0666	0.0006	−0.0227	−0.0327	0.0015	0.0156	−0.0288
Mexico	−0.0002	−0.0001	0.0005	0.0002	−0.0019	−0.0529	0.0002	−0.0051	−0.0434
Indonesia	−0.0006	−0.0102	−0.0103	0.0011	−0.0011	−0.7063	0.0009	0.0340	−0.6976
Viet Nam	−0.0486	−0.0547	−0.0728	−0.0053	0.0142	−1.2703	0.0697	0.1184	−1.1497
India	−0.0005	−0.0011	−0.0013	−0.0005	−0.0049	−0.0043	−0.0001	−0.0014	−0.0016
Peru	−0.0002	−0.0002	−0.0002	0.000	−0.001	−0.0011	−0.0002	−0.0015	−0.0009
Norway	0.0003	0.0038	0.0063	0.0038	0.024	0.0281	0.0027	0.0114	0.0378
Rest of World	−0.0007	−0.0015	−0.002	0.0001	0.0016	0.0014	0.0005	0.0057	0.0073

Data Source Sorted out according to GTAP simulation results

Table 13.13 Changes in the welfare of global countries and regions (unit: million dollars)

	Scenario 1			Scenario 2			Scenario 3		
	Stage 1	Stage 2	Stage 3	Stage 1	Stage 2	Stage 3	Stage 1	Stage 2	Stage 3
China	−438.7	−649.5	−967.8	−212.6	−94470.2	−96828.0	334.3	−100188.8	−104942.2
Japan	−395.1	−813.2	−1131.2	−7557.9	−10698.8	−11306.6	−8355.2	−13125.2	−6459.8
Korea Rep	−81.2	−2076.6	−3388.4	−61.7	−4872.9	−5036.2	57.4	28308.7	−29490.2
Russia	−6.0	−40.0	−32.7	57.0	−2355.9	−2518.2	74.0	−4150.8	−4373.1
USA	−105.8	−981.8	−2129.4	51.7	1653.3	−374.8	235.1	871.4	1058.8
Canada	16.4	−466.7	320.2	73.2	206.2	46.7	39.4	−212.0	−320.5
Mexico	7.0	42.9	−0.3	35.3	177.5	−378.5	32.1	82.3	−257.8
Indonesia	6.4	−41.7	−109.1	80.1	147.7	−5547.6	83.0	388.2	−5451.5
Viet Nam	−71.2	−75.3	−136.8	−52.4	139.7	−2158.0	61.3	115.6	−1769.9
India	2.9	10.1	11.1	26.3	321.8	391.4	31.1	312.6	557.3
Peru	3.2	8.6	11.2	10.5	68.1	77.3	10.9	43.4	88.3
Norway	30.9	205.1	332.5	161.7	946.2	1126.0	106.2	319.3	1438.6
Rest of World	−145.6	−218.3	−215.6	267.4	1053.6	757.8	434.1	1800.9	3052.7
Total	−1176.7	−5096.4	−7436.4	−7121.6	−107683.8	−121748.8	−6856.3	−85434.3	−146869.1

Data Source Sorted out according to GTAP simulation results

Table 13.14 Changes in terms of trade of global countries and regions in the third stage (unit: %)

	Scenario 1	Scenario 2	Scenario 3
	Terms of trade	Terms of trade	Terms of trade
China	0.0021	−0.0808	−0.0502
Japan	−0.0512	−0.2424	−0.3964
Korea Rep	−0.0262	−0.2248	−0.3802
Russia	−0.0044	0.0032	0.0937
USA	0.0294	0.0720	0.0915
Canada	−0.1754	0.1348	0.0487
Mexico	−0.0017	0.0785	0.0786
Indonesia	−0.0089	0.3718	0.3872
Viet Nam	0.0061	0.1956	0.2699
India	0.0075	0.1094	0.1355
Peru	0.0256	0.1864	0.2131
Norway	0.1700	0.5735	0.7300
Rest of World	0.0039	0.0076	0.0115

Data Source Sorted out according to GTAP simulation results

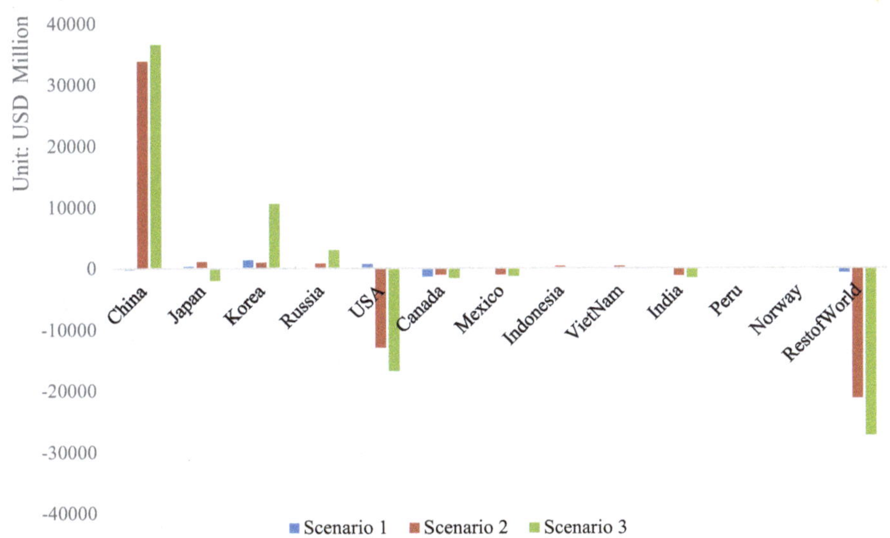

Fig. 13.6 Changes in the trade balance of global countries and regions in the third stage. *Data Source* Sorted out according to GTAP simulation results

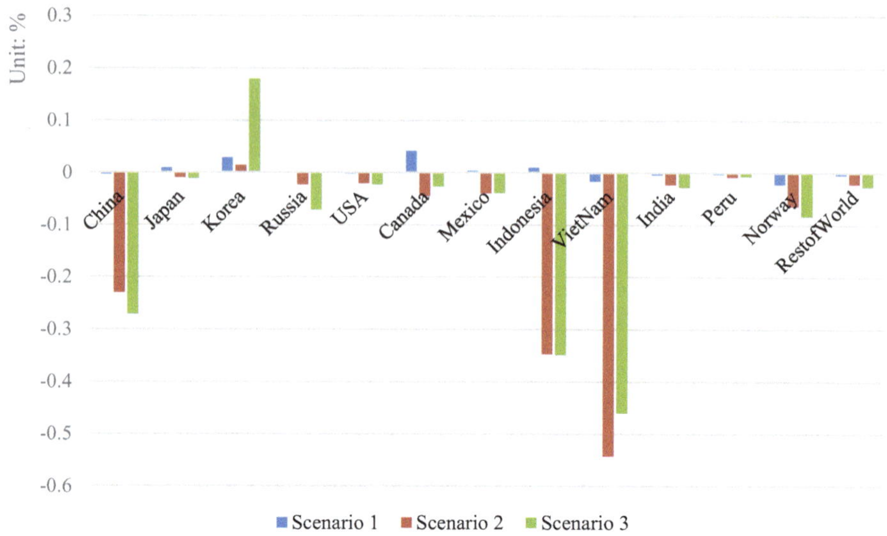

Fig. 13.7 The change ratios of the total output value of global countries and regions. *Data Source* Sorted out according to GTAP simulation results

Table 13.15 Simulation scenarios of the impact of nuclear wastewater discharge on the economy

S1	Stage 1	Japan's exports of aquatic products will plunge by 100% Due to nuclear waste water discharge; Exports of Meat, milk, processed food will plunge by 60%
	Stage 2	On the basis of stage 1, Exports of aquatic products from China, Korea Rep, Russia and Canada will plunge by 60%
	Stage 3	On the basis of stage 2, Exports of aquatic products from the United States, Mexico, Indonesia, and Viet Nam will plunge by 60%
S2	Stage 1	The fishery output value of Japan's marine areas will drop by 80% because of nuclear wastewater discharge
	Stage 2	On the basis of stage 1, the fishery output value of China, Korea Rep, Russia and Canada's marine areas will drop by 50%
	Stage 3	On the basis of stage 1, the fishery output value of the United States, Mexico, Indonesia and Viet Nam's marine areas will drop by 50%
S3	Stage 1	In the 1 stage, the comprehensive impact of Scenario 1 and Scenario 2
	Stage 2	In the 2 stage, the comprehensive impact of Scenario 1 and Scenario 2
	Stage 3	In the 3 stage, the comprehensive impact of Scenario 1 and Scenario 2

Table 13.16 Changes in GDP of global countries and regions (Unit: %)

	Scenario 1			Scenario 2			Scenario 3		
	Stage 1	Stage 2	Stage 3	stage 1	Stage 2	Stage 3	Stage 1	Stage 2	Stage 3
China	−0.0087	−0.0111	−0.0176	−0.0012	−2.0770	−2.1230	0.0103	−2.2233	−2.3444
Japan	−0.0043	−0.0172	−0.0294	−0.3105	−0.3939	−0.4098	−0.3527	−0.5724	0.1047
Korea Rep	−0.0084	−0.2724	−0.4465	0.0005	−0.6175	−0.6321	0.0178	5.7156	−5.1828
Russia	−0.0010	−0.0018	−0.0016	0.0015	−0.3077	−0.3237	0.0019	−0.5981	−0.6471
USA	−0.0016	−0.0142	−0.0314	0.0003	0.0030	−0.0241	0.0023	−0.0065	−0.0093
Canada	−0.0019	−0.0069	0.1334	0.0012	−0.0548	−0.0794	0.0032	0.0246	−0.0635
Mexico	−0.0008	−0.0001	0.0010	0.0004	−0.0049	−0.1304	0.0005	−0.0138	−0.1125
Indonesia	−0.0011	−0.0203	−0.0205	0.0021	−0.0032	−1.7480	0.0019	0.0902	−1.7532
Viet Nam	−0.5883	−0.0964	−0.1312	−0.0106	0.0378	−3.1447	0.1542	0.2591	−2.9726
India	−0.0012	−0.0022	−0.0024	−0.0009	−0.0119	−0.0104	−0.0001	−0.0029	−0.0039
Peru	−0.0004	−0.0004	−0.0004	−0.0001	−0.0023	−0.0027	−0.0004	−0.0038	−0.0021
Norway	0.0016	0.0073	0.0122	0.0075	0.0577	0.0677	0.0052	0.0214	0.0891
Rest of World	−0.0016	−0.0029	−0.0039	0.0001	0.0041	0.0036	0.0010	0.0143	0.0182

Data Source Sorted out according to GTAP simulation results

Table 13.17 Changes in welfare of global countries and regions (unit: million dollars)

	Scenario 1			Scenario 2			Scenario 3		
	Stage 1	Stage 2	Stage 3		Stage 1	Stage 2	Stage 3		Stage 1
China	−821.4	−1205.2	−1827.0	−425.3	−234891.6	−240702.2	770.7	−250945.6	−263490.4
Japan	−1020.3	−1600.4	−2261.4	−15115.8	−22896.4	−24396.7	−16817.8	−27004.0	−4609.7
Korea Rep	−152.9	−4033.1	−6592.3	−123.4	−11980.9	−12382.9	137.4	78998.7	−78802.3
Russia	−32.1	−78.6	−64.0	113.9	−5793.2	−6194.0	150.9	−10570.0	−11102.5
USA	−209.7	−1921.1	−4200.4	103.4	4090.7	−885.6	500.1	2515.5	3150.1
Canada	36.7	−929.8	643.7	146.4	502.5	111.5	73.6	−344.9	−431.8
Mexico	15.8	83.8	−1.2	70.6	424.6	−944.1	63.5	128.1	−679.0
Indonesia	19.8	−84.4	−218.9	160.2	327.7	−13764.5	166.2	916.6	−13721.1
Viet Nam	−941.1	−130.6	−251.5	−104.7	373.0	−5319.1	145.1	211.5	−4462.1
India	7.1	20.1	21.9	52.5	786.0	957.8	62.8	740.2	1362.0
Peru	7.8	16.8	22.1	21.0	164.1	186.8	21.6	90.8	202.0
Norway	110.9	396.8	650.1	323.3	2263.5	2706.9	203.7	474.0	3353.3
Rest of World	−320.2	−416.8	−415.3	534.7	2492.1	1758.9	901.6	4047.6	7177.4
Total	−3299.5	−9882.4	−14494.2	−14243.2	−264138.0	−298867.2	−13620.7	−200741.7	−362054.1

Data Source Sorted out according to GTAP simulation results

Table 13.18 Changes in terms of trade of global countries and regions in the third stage (unit: %)

	Scenario 1	Scenario 2	Scenario 3
China	0.0040	−0.1976	−0.1093
Japan	−0.0980	−0.5792	−0.9914
Korea Rep	−0.0510	−0.5519	−0.9612
Russia	−0.0086	0.0036	0.2414
USA	0.0582	0.1778	0.2254
Canada	−0.3510	0.3272	0.1598
Mexico	−0.0036	0.1907	0.2012
Indonesia	−0.0181	0.9040	0.9639
Viet Nam	0.0086	0.4928	0.7495
India	0.0144	0.2677	0.3330
Peru	0.0504	0.4514	0.4916
Norway	0.3323	1.3791	1.7026
Rest of World	0.0076	0.0179	0.0256

Data Source Sorted out according to GTAP simulation results

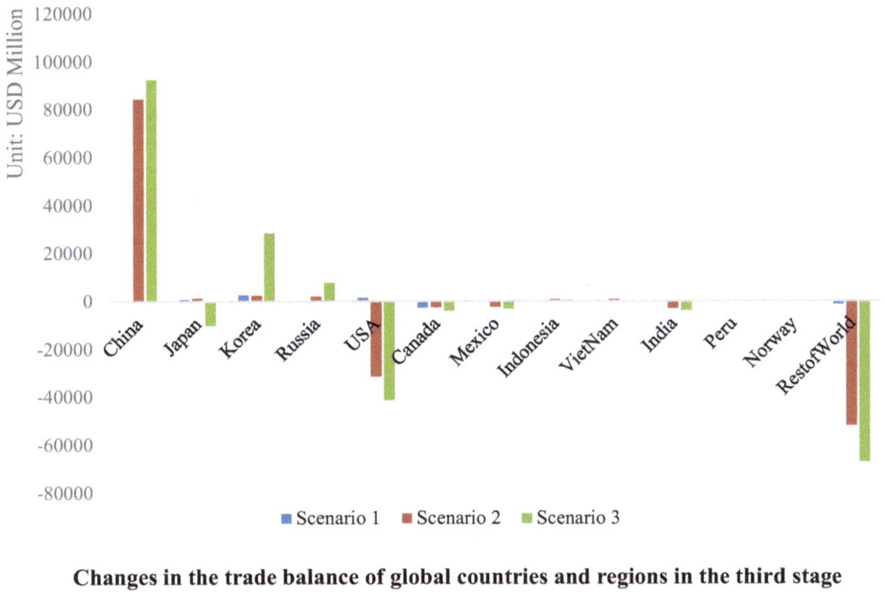

Fig. 13.8 Changes in the trade balance of global countries and regions in the third stage. *Data Source* Sorted out according to GTAP simulation results

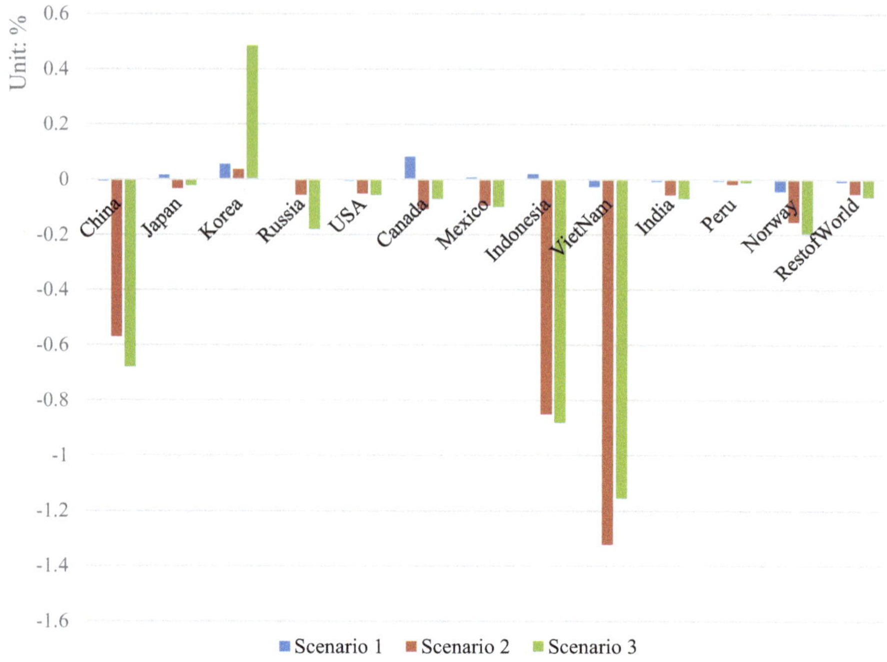

Fig. 13.9 The change ratios of the total output value of global countries and regions. *Data Source* Sorted out according to GTAP simulation results

References

Aguiar, A., Corong, E., & van der Mensbrugghe D. (2019a). *The GTAP Recursive Dynamic (GTAP-RD) model: Version 1.0.*Center for Global Trade Analysis, Purdue University. https://www.gtap.agecon.purdue.edu/resources/download/9871.pdf.

Aguiar, A., Corong, E., & van der Mensbrugghe, D. (2019b). *The GTAP-HS model technical specification and user manual*. Report Submitted to the U.S. Department of Commerce.

Akgul, Z., Villoria, N. B., & Hertel, T. W. (2016). GTAP-HET: Introducing firm heterogeneity into the GTAP model. *Journal of Global Economic Analysis, 1*(1), 111–180.

Bachev, H. (2021). *Agri-food impacts of Fukushima nuclear accident—Lessons learned 10 years after disaster* (MPRA Paper. No. 108041).

Boisvert, R. N., Kay, D., & Turvey, C. G. (2012). Macroeconomic costs to large scale disruptions of food production: The case of foot- and-mouth disease in the United States—ScienceDirect. *Economic Modelling, 29*(5), 1921–1930.

Buesseler, K. O. (2020). Opening the floodgates at Fukushima. *Science, 369*(6504), 621–622.

Burniaux, J.-M., & Truong, T. P. (2002). *GTAP-E: An energy-environmental version of the GTAP model* (GTAP Technical Paper No. 16:1–61).

Chen, S. M., & Bian, H. L. (2015). A country comparison of the economic effects of China-EU free trade area—based on the simulation evaluation of GTAP. *Chinese Journal of European Studies, 33*(2): 113-129+7–8.

Dai, J. H., & Yang, N. (2021). International public opinion analysis on Japan's nuclear wastewater discharge. *Contemporary World, 6*, 74–79.

Fisher, N. S., Beaugelin-Seiller, K., Hinton, T. G., et al. (2013). Evaluation of radiation doses and associated risk from the Fukushima nuclear accident to marine biota and human consumers of seafood. *Proceedings of the National Academy of Sciences of the United States of America., 110*(26), 10670–10675.

Grant, J. H., Hertel, T. W., & Rutherford, T. F. (2009). Dairy Tariff-quota liberalization: Contrasting bilateral and most favored nation reform options. *American Journal of Agricultural Economics, 91*(3), 673–684.

Hai, G. J. (2021). New impacts and countermeasures of nuclear waste water discharge from Japan on China's aquatic products trade. *EDP Sciences, 275*, 02033.

Hayashi, F., Sanpei, M., Ohira, T., et al. (2019). Changes in the mental health status of adolescents following the Fukushima Daiichi nuclear accident and related factors: Fukushima health management survey. *Journal of Affective Disorders, 260*, 432–439.

He, Y. M., & Zhou, F. M. (2021). Analysis and countermeasures of China's Aquatic products trade competitiveness. *Modern Business Trade Industry, 42*(2), 39–41.

Hertel, T. W. (1997). *Global Trade Analysis: Modeling and applications.* GTAP Books.

Hertel, T. W., Itakura, K., & McDougall, R. (2000). *GTAP model version 5.0.* Global Trade Analysis Project (GTAP), Department of Agricultural Economics, Purdue University, West Lafayette, IN, Center for Global Trade Analysis. https://www.gtap.agecon.purdue.edu/resources/res_display.asp?RecordID=575.

Hillberry, R., Hummels, D. (2013). Chapter 18: Trade elasticity parameters for a computable general equilibrium model. *Handbook of Computable General Equilibrium Modeling, 1*, 1213–1269.

Huff, K. M. & Hertel, T. W. (2001). *Decomposing welfare changes in GTAP* (GTAP Technical Paper No. 5). Global Trade Analysis Project (GTAP), Department of Agricultural Economics, Purdue University, West Lafayette, IN.

Ianchovichina, E., McDougall, R., & Hertel, T. W. (1999, June 20–22). *A disequilibrium model of international capital mobility*. The Second Annual Conference on Global Economics Analysis, Denmark.

Iwagami, S., Onda, Y., Tsujimura, M., et al. (2017). Vertical distribution and temporal dynamics of dissolved (137) Cs concentrations in soil water after the Fukushima Dai-ichi Nuclear Power Plant accident. *Environmental Pollution, 230*, 1090–1098.

Jin, Y. (2021). The Issue of Fukushima Nuclear wastewater discharge to the sea and its countermeasures. *Contemporary World, 6*, 68–73.

Ko, J., & Ito, S. (2017). Simulation for Japan-Korea FTA and Its economic impacts on agriculture: A CGE approach. *Journal- Faculty of Agriculture Kyushu University, 62*(1), 283–294.

Li, X. X., Cai, H. L., Cai, S. F., et al. (2020). Research on the future development prospects and potential impact of RCEP—Based on the GTAP model. *Macroeconomics, 7*, 165–175.

Liu, J. G., Wang, X. Y., Tan, Z. J., et al. (2021). A tripartite evolutionary game analysis of Japan's nuclear wastewater discharge. *Ocean & Coastal Management, 214*, Article 105896.

Liu, Y., Guo, X. Q., Li, S. W., et al. (2021). Discharge of treated Fukushima nuclear accident contaminated water: Macroscopic and microscopic simulations. *National Science Review, 9*(1), 1–3.

Liu, Y. F., & Liu, J. W. (2019). Japan planned to discharge Fukushima nuclear wastewater into the Pacific. *Ecological Economy, 35*(11), 1–4.

Lotze, H. K., Guest, H., O'Leary, J., et al. (2018). Public perceptions of marine threats and protection from around the world. *Ocean & Coastal Management, 152*, 14–22.

Lu, Y. L., Yuan, J. J., Du, D., et al. (2021). Monitoring long-term ecological impacts from release of Fukushima Radiation water into ocean. *Geography and Sustainability, 2*(2), 95–98.

Maliszewska, M., Mattoo, A., & Dominique, V. (2020). *The potential impact of COVID-19 on GDP and trade: A preliminary assessment* (World Bank Policy Research Working Paper, 9211) (pp. 1–24).

Matsuo, Y., Nakai, K., Tatsuta, N., et al. (2021). Using the larvae of caddisfly as a biomonitor to assess the spatial distribution and effective half-life of radiocesium in riverine environments in Fukushima. *Japan. Physics Open, 6*, Article 100060.

McDougall, R. (2003). *A new regional household demand system for GTAP (revision 1)* (GTAP Technical Paper No. 20). Global Trade Analysis Project (GTAP), Purdue University, West Lafayette, IN. https://www.gtap.agecon.purdue.edu/resources/download/1593.pdf.

Men, W. (2021). Discharge of contaminated water from the Fukushima Daiichi Nuclear Power Plant Accident into the Northwest Pacific: What is known and what needs to be known. *Marine Pollution Bulletin, 173*(Part A), 112984.

Merz, S., Steinhauser, G., & Hamada, N. (2013). Anthropogenic radionuclides in Japanese food: Environmental and legal implications. *Environmental Science & Technology, 47*(3), 1248–1256.

Mi, Z., Meng, J., Green, F., et al. (2018). China's "exported carbon" peak: patterns, drivers, and implications. *Geophysical Research Letters, 45*, 4309–4318.

Mi, Z., Meng, J., Guan, D., et al. (2017). Chinese CO2 emission flows have reversed since the global financial crisis. *Nature Communications, 8*, 1712.

Moriyama, N., Nakayama, C., Orui, M., et al. (2020). Associated factors related to participation in general health checkup and survey of the effect of low-dose radiation exposure on health of residents of Fukushima Prefecture after the Fukushima Daiichi nuclear power plant accident. *Preventive Medicine Reports, 20*, Article 101214.

Narayanan, B. G., Hertel, T. W., & Horridge, J. M. (2010). Disaggregated data and trade policy analysis: The value of linking partial and general equilibrium models. *Economic Modelling, 27*(3), 755–766.

Niisoe, T. (2021). An iterative application of the Green's function approach to estimate the time variation in 137Cs release to the atmosphere from the Fukushima Daiichi Nuclear Power Station. *Atmospheric Environment, 254*, Article 118380.

Normile, D. (2021). *Japan Plans to Release Fukushima's Wastewater into the Ocean*. https://www.science.org/content/article/japan-plans-release-fukushima-S-contaminated-water-ocean.

Periáñez, R., Qiao, F. L., Zhao, C., et al. (2021). Opening Fukushima floodgates: Modelling 137-Cs impact in marine biota. *Marine Pollution Bulletin, 170*, Article 112645.

Pu, X. J., Jiang, Q. Y., & Fan, B. (2022). Chinese public opinion on Japan's nuclear wastewater discharge: A case study of Weibo comments based on a thematic model. *Ocean & Coastal Management., 225*, Article 106188.

Qian, S., Song, Y., Zhang, M., Hu, H., & Zhang, W. (2025). Evaluating the economic and social benefits of global energy interconnection based on GTAP model. *Energy, 335*, Article 138260.

Shaun B. (2020). *Stemming the tide 2020 the reality of the Fukushima Radioactive water crisis*. Greenpeace Germany. https://www.greenpeace.org/static/planet4-japan-stateless/2020/10/5e303093-greenpeace_stemmingthetide2020_fukushima_radioactive_water_crisis_en_final.pdf.

Stenberg, L. C., & Siriwardana, M. (2015). Measuring the economic impacts of trade liberalisation on forest products trade in the Asia-Pacific region using the GTAP model. *International Forestry Review, 17*(4), 498–509.

Suarez, J. M. D., Cicin-Sain, B., Wowk, K., et al. (2014). Ensuring survival: Oceans, climate and security. *Ocean & Coastal Management, 90*, 27–37.

Subcommittee on Handling of the ALPS Treated Water. (2020). *Report of the subcommittee on handling of the ALPS treated water* (pp. 1–57). https://www.meti.go.jp/english/earthquake/nuclear/decommissioning/pdf/20200210_alps.pdf.

Tang, W. L., & Liu, Y. Y. (2020). Global Macroeconomic Analysis Under the Impact of the Epidemic—Based on GTAP Model. *Economic Forum, 9*, 86–93.

Tateda, Y., Tsumune, D., Tsubono, T., et al. (2016). Status of 137Cs contamination in marine biota along the Pacific coast of eastern Japan derived from a dynamic biological model two years simulation following the Fukushima accident. *Journal of Environmental Radioactivity, 151*(Part 2), 495–501.

Wada, T., Fujita, T., Nemoto, Y., et al. (2016). Effects of the nuclear disaster on marine products in Fukushima: An update after five years. *Journal of Environmental Radioactivity, 164*, 312–324.

Walmsley, T. L., Betina, V. D., & Robert, A. M. (2006). A base case scenario for dynamic GTAP model. *GTAP Resource, 417*, 1–14.

Wang, Y., Guo, J., Wu, X., & Wu, Y. (2025). The synergistic impact of the EU's provision of low-carbon technical assistance under the carbon border adjustment mechanism: A GTAP model-based study. *Structural Change and Economic Dynamics, 75*, 703–16.

Wang, Y. N. (2021). Six issues that cannot be ignored in Fukushima nuclear wastewater. *China Economic Weekly, 8*, 96–100.

Wright, A., Stacey, N., & Holland, P. (2006). The cooperative framework for ocean and coastal management in the Pacific Islands: Effectiveness, constraints and future direction. *Ocean & Coastal Management, 49*(9/10), 739–763.

Wu, Z. (2022). Study on the Legal Issues of Fukushima Nuclear Wastewater discharge into the sea. *Proceedings of the 2022 International Conference on Creative Industry and Knowledge Economy, 651*, 91–95.

Xiang, L. J. (2021). Japan decided to discharge nuclear wastewater into the sea. *Ecological Economy, 37*(6), 1–4.

Xu, X., Wang, Y., Yu, L., Gao, G., Liu, X., Jiang, M., et al. (2025). Impact assessment of reactive carbon tariff and proactive carbon tax on China's energy and power industries. *Environmental Impact Assessment Review, 114*, Article 107965.

Yang, B. S., Yin, K. D., Li, X. M., et al. (2021). Graph model under grey and unknown preferences for resolving conflicts on discharging Fukushima nuclear wastewater into the ocean. *Journal of Cleaner Production, 332*, Article 130019.

Yang, F., Boysen-Urban, K., Brockmeier, M., et al. (2017). Impact of increasing agricultural domestic support on China's food prices considering incomplete international agricultural price transmission. *China Agricultural Economic Review, 9*(4), 535–557.

Zahra, Z. G., Nematollah, M. S., & Bahaeddin, N. (2019). Economic evaluation of the effects of exerting green tax on the dispersion of bioenvironmental pollutants based on multi-regional general equilibrium model (GTAP-E). *Energy Sources* (Part A), 1–12.

Zer, O. O., Ozden, A., & Kan, M. (2020). Impact of global climate change on agricultural production: balkan countries in 2050. *Journal of Environmental Protection and Ecology, 21*(1), 151–158.

Zhang, X. X., Qu, T. K., & Wang Y. C. (2022). Optimal strategies for stakeholders of Fukushima nuclear waste water discharge in Japan. *Marine Policy, 135*, 104881.

Zhao, C., Wang, G., Zhang, M., et al. (2021). Transport and dispersion of tritium from the radioactive water of the Fukushima Daiichi nuclear plant. *Marine Pollution Bulletin, 169*, Article 112515.

Zhou, L. L., & Zhang, L. Y. (2020a). Study of Reconstruction of Chinese Global Value Chains Participation on COVID-19. *Review of Industrial Economics, 6*, 5–15.

Zhou, L. L., & Zhang, L. Y. (2020b). Would Trade Liberalization Improve Residents' Food Consumption Structure— Based on GTAP Model. *Journal of International Trade, 5*, 28–41.

Zhu, Q. R., Sun, M. S., & Yang, W. D. (2020). Assessment of impact of COVID-19 epidemic on China's economy: An empirical study based on GTAP model. *Statistics & Decision, 36*(21), 91–96.

Chapter 14
The Impact of Japanese Nuclear Wastewater Discharge into the Sea on the Global Economy: Input–Output Model Approach

14.1 Introduction

With the development of international trade, the intrinsic linkage of the world economy is getting closer and closer. Significant events occurring in a particular country or region will not only affect the local socio-economy but will also generate a chain reaction in the global economic system, ultimately affecting the entire world economy. On April 13, 2021, the Japanese government announced that they would dump approximately 1.25 million tons of nuclear wastewater from the Fukushima Daiichi nuclear power plant into the Pacific Ocean after two years, which has raised concerns among governments and people. On the one hand, the nuclear wastewater discharged into the Pacific Ocean will damage the marine ecosystem, which will affect the global ecosystem. On the other hand, radionuclides (e.g., Cs-137, etc.) in nuclear wastewater will be deposited in marine organisms such as fish and algae, which may eventually be transmitted to humans through the food chain (Niizeki et al., 2020; Morishita et al., 2019; Wang et al., 2018; Yu et al., 2019; Tsuboi et al., 2015; Ishimaru et al., 2019). Its effects have been studied by scholars, such as Zhao et al. (2021), who simulated the dispersion of nuclear wastewater discharged from the Tohoku region of Japan in the Pacific Ocean and international waters; Xiang (2021) and Jin (2021) studied the severe implications of the event. Dai and Yang (2021) analyzed the international public opinion of the event, etc. However, few scholars have explored the economic impact of Japan's nuclear wastewater discharge on countries worldwide and the associated effects of the event. This paper conducts a study on this subject, which can complement the research in this area.

The Fukushima nuclear accident in Japan in 2011 hit the country's agriculture and fishing industry hard. With consumers rejecting local seafood due to radioactive contamination (Aruga & Wakamatsu, 2018; Tsutom & Hiroki, 2018), the reputation of the origin was damaged, and uncontaminated products were negatively evaluated (Matsumoto & Harvey, 2020). It is difficult for consumers to have information on ocean currents' movement and consider the impact of Japanese nuclear wastewater

X. Wu, *Environmental Disaster in China*,
https://doi.org/10.1007/978-981-92-0910-1_14

on other countries. Only the direct effects of atomic wastewater on Japanese fisheries and the indirect effects due to the transmission of global production networks are considered here. Environmental pollution such as nuclear wastewater can affect direct consumption in agriculture, fisheries, or forestry (Suris-Regueiro & Santiago, 2018), and the direct effects of nuclear wastewater on Japanese fisheries include two aspects: first, consumers refuse to buy seafood from contaminated waters due to food safety concerns (Kudo & Nagaya, 2017; Tajima et al., 2016), second, producers predict that consumers will buy less and future sales of products will decrease, and producers will actively reduce their output to avoid losses. In the disaster impact analysis methods, the Leontief input–output model is more widely used (Okuyama & Santos, 2014). This model describes the input–output relationships among sectors of an economic system in equilibrium, and it provides a rational framework for assessing the impact of disruptive events on interdependent economic systems (Santos, 2010). The Inoperability Input–Output Model (IIM), obtained by extending the Leontief input–output model, can be used to describe the chain reaction of disruptive events on interdependent systems. For example, Montibeler and Oliveira (2018) used IIM to estimate the impact of labor shortages on national productivity due to the dengue epidemic in Brazil. Lenzen et al. (2019) quantified the economic losses in Australia due to disruptions in the supply of intermediate inputs of firms caused by tropical cyclones. Compared with the CGE model and others, the IIM has a relatively simple structure and more evident logic and is more practical and widely used. Therefore, the inoperability input–output model (IIM) is adopted to measure the impact of Japanese nuclear wastewater discharge on Japanese fisheries and the global industrial economy from the perspective of demand and output, respectively.

The remaining parts are organized as follows: the first part is a literature review; the second part presents the theoretical model; the third part is the data sources; the fourth part is the simulation scenarios. The fifth part is the results calculated and organized according to the simulation scenarios; the sixth part is a summary.

14.2 Literature Review

14.2.1 Literature Review on Economic Losses from Nuclear Events

To assess the economic impact of nuclear events, scholars at home and abroad have done more exploration. For example, Higashi (2020) studied the effect of the Japan Earthquake in 2011 on the efficiency of supply and demand matching in the regional labor market. Matsumoto and Harvey (2020) used the Dixit-Stiglitz demand framework to quantify reputation loss and measured the shrinking market share of local agricultural products after the Fukushima Daiichi Nuclear Disaster. Tanaka and Managi (2016) estimated the economic losses caused by radioactive contamination from the Fukushima nuclear power plant accident using the Hedonic model.

Dadakas and Tatsi (2021) used the gravity model to study the impact of international market restrictions on Japanese agricultural trade and consumer food safety concerns about Japanese agricultural products on global agricultural trade flows after the Fukushima nuclear accident. It is argued that the adverse effects of Japanese agricultural exports and imports will eventually be transmitted to the global agricultural trade market. Hendricks et al. (2020) believed that the nuclear accident caused global supply chain disruptions and negatively affected the supply chains of companies related to the nuclear industry. Through a comparative study of the Capital Asset Pricing Model (CAPM), Fama–French three-factor, and five-factor models, Alonso-Conde and Rojo-Suárez (2020) found significant nuclear disasters worldwide were associated with expected returns on numerous assets. Park and Park (2020) found that nuclear tests caused exchange rate depreciation in Korea in the short run.

14.2.2 Literature Review on IIM Applications

The inoperability input–output model (IIM) is an economic model that describes the interrelated relationships within the system sectors. The inoperability variable measures the degree to which a system fails to achieve the desired output. IIM is widely used to study the impact of disturbing events on infrastructure systems, economic systems, transportation systems, etc. For example, Xu et al. (2021) and Svegrup et al. (2019) used IIM to analyze the impact of external disturbance events on various system sectors at different stages. Samimi et al. (2020) assessed the impact of floods on six types of infrastructure and showed a ranking of the infrastructure affected. Zhang et al., (2018a, 2018b) proposed an optimal resource allocation model to facilitate infrastructure recovery based on IIM simulation results. Xian and Jeong (2018) demonstrated that dependence within the system influences system recovery time. Since economic sectors are interconnected, IIM can also be used to measure the overall economic effects of factor shortages. Poudineh and Jamasb (2017) explored the direct and indirect impact and social costs of power supply disruptions to the economic system. Pagsuyoin and Santos (2021) evaluated the effect of water supply disruptions on the regional economy. Sugrue et al. (2021) showed an interdependence between steel producers and the water transportation system of raw materials and proved chain disruptions in the water system can result in production losses for steel companies. Wan et al. (2024) assessed dynamic energy supply resilience using China's oil supply disruption incidents as case studies. Miura et al. (2025) employed an IIM to quantify the indirect economic losses incurred by the suspension of New York City's underground rail services. Sarkar et al. (2025) created a IIM to assess how disruptions in the Electronics and Photonics Sector affect India's economy in light of the Red Sea Crisis. Zhang et al. (2025) examined the impact of US-China trade demand shocks on industrial chain resilience.In addition, IIM has been used to estimate the degree of degradation of fixed assets for predictive maintenance. For example, Tamssaouet et al., (2019, 2020) used IIM to assess the degree of system aging and remaining useful life to make system failure predictions. Zhang et al.,

(2018a, 2018b) used IIM to predict the damage to buildings in a community after an earthquake. Lee et al. (2018) simulated the path of earthquake affecting system functions in the power sector. Sellevg (2021) used IIM to analyze the interdependence among Norway's industrial sectors, identify the key industrial sectors that affect national security, and enhance the level of national security by strengthening the resilience of these sectors.

14.3 Model Description

14.3.1 Leontief Input–output Model

Leontief (1936) proposed the input–output model to describe the interdependence among sectors in an economy, and many scholars have since extended and expanded the model. For example, Haimes and Jiang (2001) proposed the inoperability input–output model (IIM) to study the chain reaction of output failure caused by natural or man-made events on interconnected economic systems. On this basis, Santos and Haimes (2004) proposed the Demand-Reduction Inoperability Input–Output Model (or Demand-Reduction IIM). Leung et al. (2007) proposed the Output-Demand Mixed Inoperability Input–Output Model (or Output-Demand Mixed IIM). Based on the Demand-Reduction IIM and the Output-Demand Mixed IIM, since the demand side and output side have advantages and disadvantages, this paper conducts simulation and calculation from the demand side and output side, respectively, then compares the results.

The Leontief equilibrium equation assumes that there are n sectors in an economy, and the input–output model integrates the input–output data such as production and consumption of n sectors, and it reflects the linkages among the economic sectors through a linear equation:

$$x = Ax + c \Longleftrightarrow x_i = \sum_{j=1}^{n} a_{ij}x_j + c_i \Leftrightarrow x_i = \sum_{j=1}^{n} x_{ij} + c_i \tag{14.1}$$

where x is a vector of $n \times 1$, and x_i represents the total output of department i; the technical coefficient matrix A is the matrix of $n \times n$. Its element, $a_{ij} = \frac{x_{ij}}{x_j}$, represents the proportion of the product quantity invested by department i to department j in the total output of department i, and x_{ij} represents the product quantity invested by department i to department j; c is the vector of $n \times 1$, and c_i represents the final demand for the products of department i.

14.3.2 Demand-Based Inoperability Input–Output Model

Haimes and Jiang (2001) proposed the Inoperability Input–Output Model (IIM), which measures the effect of a perturbation on the overall economic system by using the inoperability of the initial system. Inoperability is the degree to which a system fails to achieve its intended productivity and ranges from 0 to 1. When inoperability is 0, the system is fully operational, and there is no change in system productivity. When inoperability is 1, the system is completely damaged and unproductive. Santos and Haimes (2004) proposed the Demand-Reduction Inoperability Input–Output Model (Demand-Reduction IIM) to describe the situation in which a shock event causes a change in consumer psychology, resulting in a decrease in consumer demand for products and services, and ultimately a reduction in output. The Demand-Reduction Inoperability Input–Output Model is:

$$q = A^*q + c^* \Longleftrightarrow q = (I - A^*)^{-1}c^* \tag{14.2}$$

where q is the inoperability variable, which represents the proportion of output decline due to reduced demand, expressed as a percentage of system productivity before the event. I is the identity matrix, A^* is the interdependence matrix, and c^* represents the proportion of final demand reduction caused by event shocks.

The derivation process of this model is as follows:

$$\hat{x} = A\hat{x} + \hat{c} \tag{14.3}$$

Formula (14.3) represents the Leontief input–output model before the event. $\hat{x}$ represents the total sectoral output before the event, A is the technology coefficient matrix and $\hat{c}$ represents the final demand for sectoral products before the event.

$$\tilde{x} = A\tilde{x} + \tilde{c} \tag{14.4}$$

Formula (14.4) represents the Leontief input–output model after the event. $\tilde{x}$ represents the total output of the department after the event, and $\tilde{c}$ represents the final demand for sectoral products after the event.

δc represents the change in final demand before and after the event, that is $\delta c = \hat{c} - \tilde{c}$, δx represents the change in total output before and after the event, that is $\delta x = \hat{x} - \tilde{x}$. Substituting Eq. (14.3) from Eq. (14.4) results in the following equation:

$$\delta x = A\delta x + \delta c \tag{14.5}$$

Based on the Leontief equilibrium equation, the Demand-Reduction IIM can be deduced as follows:

$$(diag(\hat{x}))^{-1}\delta x = [(diag(\hat{x}))^{-1}A(diag(\hat{x}))][(diag(\hat{x}))^{-1}\delta x] + (diag(\hat{x}))^{-1}\delta c \tag{14.6}$$

According to Eq. (14.2) and (14.6), it can be obtained:

$$c^* = [diag(\hat{x})]^{-1}\delta c \tag{14.7}$$

$$A^* = [diag(\hat{x})]^{-1}A[diag(\hat{x})] \tag{14.8}$$

$$q = [diag(\hat{x})]^{-1}\delta x \tag{14.9}$$

14.3.3 Supply-Based Inoperability Input–output Model

Event shocks not only affect final demand but also affect output directly. For example, when a producer predicts that demand for a product will decrease, he will reduce production accordingly to reduce loss. Another example is that if plants and equipment break down and cannot produce, the output will fall. To analyze such scenarios, Leung et al. (2007) proposed the Output-Demand Mixed Inoperability Input–Output Model based on the Demand-Reduction IIM. The model divides the studied variables into exogenous variables (output) and endogenous variables (demand or supply). The exogenous variables include q_k and $c_i^*(i = 1,2,\ldots,k-1,k+1,\ldots,n.)$, whose values are given; the endogenous variables include $q_i(i = 1,2,\ldots,k-1,k+1,\ldots,n.)$ and c_k^*, whose values are derived from the model.

Assuming that the output of sector k is affected, the proportion of output reduction (i.e. inoperability) is represented by the exogenous variable q_k. The equilibrium equation $q = A^*q + c^*$ for the Demand-Reduction IIM can be expressed as follows:

$$\begin{cases} (1-a_{11}^*)q_1 - a_{12}^*q_2 \cdots - a_{1(k-1)}^*q_{k-1} - a_{1(k+1)}^*q_{k+1} \cdots - a_{1n}^*q_n - 0c_k^* = c_1^* + a_{1k}^*q_k \\ -a_{21}^*q_1 + (1-a_{22}^*)q_2 \cdots - a_{2(k-1)}^*q_{k-1} - a_{2(k+1)}^*q_{k+1} \cdots - a_{2n}^*q_n - 0c_k^* = c_2^* + a_{2k}^*q_k \\ \vdots \\ -a_{k1}^*q_1 - a_{k2}^*q_2 \cdots - a_{k(k-1)}^*q_{k-1} - a_{k(k+1)}^*q_{k+1} \cdots - a_{kn}^*q_n - c_k^* = -(1-a_{kk}^*)q_k \\ \vdots \\ -a_{n1}^*q_1 - a_{n2}^*q_2 \cdots - a_{n(k-1)}^*q_{k-1} - a_{n(k+1)}^*q_{k+1} \cdots + (1-a_{nn}^*)q_n - 0c_k^* = c_n^* + a_{nk}^*q_k \end{cases} \# \tag{14.10}$$

Putting the k th equation in the last line, it can be obtained that:

$$
\begin{cases}
(1-a_{11}^*)q_1 - a_{12}^*q_2 \cdots - a_{1(k-1)}^*q_{k-1} - a_{1(k+1)}^*q_{k+1} \cdots - a_{1n}^*q_n - 0c_k^* = c_1^* + a_{1k}^*q_k \\
-a_{21}^*q_1 + (1-a_{22}^*)q_2 \cdots - a_{2(k-1)}^*q_{k-1} - a_{2(k+1)}^*q_{k+1} \cdots - a_{2n}^*q_n - 0c_k^* = c_2^* + a_{2k}^*q_k \\
\vdots \\
-a_{(k-1)1}^*q_1 - a_{(k-1)2}^*q_2 \cdots + (1-a_{(k-1)(k-1)}^*)q_{k-1} - a_{(k-1)(k+1)}^*q_{k+1} \cdots - a_{(k-1)n}^*q_n - 0c_k^* = c_{k-1}^* + a_{(k-1)k}^*q_k \\
-a_{(k+1)1}^*q_1 - a_{(k+1)2}^*q_2 \cdots - a_{(k+1)(k-1)}^*q_{k-1} + (1-a_{(k+1)(k+1)}^*)q_{k+1} \cdots - a_{(k+1)n}^*q_n - 0c_k^* = c_{k+1}^* + a_{(k+1)k}^*q_k \\
\vdots \\
-a_{n1}^*q_1 - a_{n2}^*q_2 \cdots - a_{n(k-1)}^*q_{k-1} - a_{n(k+1)}^*q_{k+1} \cdots + (1-a_{nn}^*)q_n - 0c_k^* = c_n^* + a_{nk}^*q_k \\
-a_{k1}^*q_1 - a_{k2}^*q_2 \cdots - a_{k(k-1)}^*q_{k-1} - a_{k(k+1)}^*q_{k+1} \cdots - a_{kn}^*q_n - c_k^* = -(1-a_{kk}^*)q_k
\end{cases} \#
\tag{14.11}
$$

The above formula can also be expressed as:

$$
\begin{pmatrix}
1-a_{11}^* & \cdots & -a_{1(k-1)}^* & -a_{1(k+1)}^* & \cdots & -a_{1n}^* & 0 \\
\vdots & \ddots & \vdots & \vdots & \ddots & \vdots & \vdots \\
-a_{(k-1)1}^* & \cdots & 1-a_{(k-1)(k-1)}^* & -a_{(k-1)(k+1)}^* & \cdots & -a_{(k-1)n}^* & 0 \\
-a_{(k+1)1}^* & \cdots & -a_{(k+1)(k-1)}^* & 1-a_{(k+1)(k+1)}^* & \cdots & -a_{(k+1)n}^* & 0 \\
\vdots & \ddots & \vdots & \vdots & \ddots & \vdots & \vdots \\
-a_{n1}^* & \cdots & -a_{n(k-1)}^* & -a_{n(k+1)}^* & \cdots & 1-a_{nn}^* & 0 \\
-a_{k1}^* & \cdots & -a_{k(k-1)}^* & -a_{k(k+1)}^* & \cdots & -a_{kn}^* & -1
\end{pmatrix}
\begin{pmatrix} q_1 \\ \vdots \\ q_{k-1} \\ q_{k+1} \\ \vdots \\ q_n \\ c_k^* \end{pmatrix}
=
\begin{pmatrix}
1 \cdots 0 & 0 \cdots 0 & a_{1k}^* \\
\vdots \ddots \vdots & \vdots \ddots \vdots & \vdots \\
0 \cdots 1 & 0 \cdots 0 & a_{(k-1)k}^* \\
0 \cdots 0 & 1 \cdots 0 & a_{(k+1)k}^* \\
\vdots \ddots \vdots & \vdots \ddots \vdots & \vdots \\
0 \cdots 0 & 0 \cdots 1 & a_{nk}^* \\
0 \cdots 0 & 0 \cdots 0 & -(1-a_{kk}^*)
\end{pmatrix}
\begin{pmatrix} c_1^* \\ \vdots \\ c_{k-1}^* \\ c_{k+1}^* \\ \vdots \\ c_n^* \\ q_k \end{pmatrix}
\tag{14.12}
$$

The equation above represents the situation when the exogenous output of a sector (k sector) changes. If there are changes in the output of r sectors in an economic system containing n sectors, the general model can be expressed as:

$$
\begin{pmatrix} (I-A^*)_q & 0 \\ (I-A^*)_{\bar{q}} & -I \end{pmatrix}
\begin{pmatrix} q \\ \bar{c}^* \end{pmatrix}
=
\begin{pmatrix} I & -(I-A^*)_{q'} \\ 0 & -(I-A^*)_{\bar{q}'} \end{pmatrix}
\begin{pmatrix} c^* \\ \bar{q} \end{pmatrix}
\iff Bb = Dd
\tag{14.13}
$$

In the above equation, inoperability is divided into q and $\overline{q}$. q is the inoperability vector of $(n-r)$ sectors, whose value is derived from the model. $\overline{q}$ is the inoperability vector of r sectors, which is an exogenous variable. The final demand is divided into c^* and $\overline{c}^*$. c^* is the final demand perturbation vector of $(n-r)$ sectors and it is an exogenous variable. $\overline{c}^*$ is the final demand perturbation vector of r sectors, whose value is derived from the model. The matrix $(I-A^*) = \{\alpha_{ij}\}$ in the inoperability input–output model is divided into four parts, namely, the $(n-r) \times (n-r)$ matrix $(I - A^*)_q$, the subscript $i-j$ of its elements $\{\alpha_{ij}\}$ are the serial number of $(n-r)$ sectors; the $r \times (n-r)$ matrix $(I - A^*)_{\bar{q}}$, the subscript i of its elements $\{\alpha_{ij}\}$ is the serial number of r sectors, the subscript j is the serial number of $(n-r)$ sectors;the $(n-r) \times r$ matrix $-(I - A^*)_{q'}$, the subscript i of its elements $\{\alpha_{ij}\}$ is the serial number of $(n-r)$ sectors, and the subscript j represents the serial number of r sectors; the $r \times r$ matrix $-(I - A^*)_{\bar{q}'}$, the subscript $i-j$ of its elements $\{\alpha_{ij}\}$ are the serial number of r sectors.

The vector $d = \begin{pmatrix} c^* \\ \bar{q} \end{pmatrix}$ is an exogenous variable, and vector $b = \begin{pmatrix} q \\ \bar{c}^* \end{pmatrix}$ can be solved according to equation $b = B^{-1}Dd$, that is, changes in output in (n - r) sectors and in final demand for products in r sectors can be found as a result of changes in output in r sectors and in final demand for products in (n - r) sectors.

14.4 Data Sources

The input–output data were obtained from the World Input–Output Database (WIOD) of World Input–Output Table 2014, published in 2016, covering data for 43 countries and regions (including 28 EU countries and 15 other major countries and regions in the world) and 56 sectors. Data related to Japanese fisheries were obtained from the official websites of the Statistics Bureau of Japan and the Ministry of Agriculture, Forestry and Fisheries of Japan.

14.5 Scenario Simulation

14.5.1 Impact of Reduced Demand

According to the simulation analysis of the extent of nuclear wastewater dispersion in the Pacific Ocean at different times (Zhao et al., 2021), the first contaminated area is the northeastern coastal region of Japan. It mainly involves Hokkaido, the Tohoku region (including Aomori, Iwate, Akita, Miyagi, Yamagata, and Fukushima prefectures), and the Kanto region (including Tokyo, Kanagawa, Chiba, Saitama, Ibaraki, Tochigi, and Gunma prefectures). Japan's fisheries include marine fisheries (including marine aquaculture) and inland fisheries (including inland aquaculture). According to data released by the official website of the Statistics Bureau of Japan, the output value of marine fisheries accounted for 93% of the total output value of fisheries in 2018. According to data released by the official website of the Ministry of Agriculture, Forestry and Fisheries of Japan, the value of fisheries in these regions accounted for about 37% of the total value of Japan's national output in 2019. Scenario 1: assuming that consumers no longer directly consume seafood from these regions, the final demand for Japanese fisheries decreases by 34.4%, i.e., in formula (14.2) $c^{*T} = (34.4\%, 0, \ldots, 0)$.

After two years, nuclear wastewater will contaminate the entire eastern coast of Japan. According to the Japanese classification of fishery sea areas, the affected areas include Hokkaido Pacific North, Pacific North, Pacific Central, Pacific South, and East China Sea areas. According to the output value of each sea area published on the Ministry of Agriculture, Forestry and Fisheries of Japan, the proportion of fishery output in these sea areas to the total output value is about 72%. Scenario 2: assuming that consumers no longer consume seafood directly from the eastern sea areas of Japan, the final demand for Japanese fisheries decreases by 66.2%, that is, in formula (14.2) $c^{*T} = (66.2\%, 0, \ldots, 0)$.

14.5.2 *The Impact of Reduced Supply*

The fishery producers realize that the nuclear wastewater contamination will cause consumers to reduce their purchases (scenario 1) and take the initiative to reduce their output to avoid losses. Therefore, scenario 3 assumes that the Japanese fishery output decreases by 34.4%, and the final demand of other countries and industries remains unchanged, that is, in formula (14.13) $d = (0, \ldots, 0{,}34.4\%)$.

Similarly, scenario 4 assumes that Japan's fishery output decreases by 66.2%, and the final demand from other countries and industries remained unchanged, that is, in formula (14.13) $d = (0, \ldots, 0{,}66.2\%)$.

14.6 Results Analysis

14.6.1 *Impact of Reduced Demand*

14.6.1.1 Scenario 1: Final Demand for Japanese Fisheries Decreases by 34.4%

Based on scenario 1, the calculated results for the top 20 countries (regions) in total output loss are shown in Table 14.1. Due to the interconnectedness of industries in various countries around the world, nuclear wastewater contamination caused a direct loss of 34.4% to Japanese fisheries, resulting in a total global economic output loss of US$11.533 billion. Japan lost US$9.495 billion in total output, accounting for 82.33% of the total global loss. Other 19 countries (regions) with large losses in total output are China (US$247.86 million), the United States of America (US$207.45 million), Australia (US$104.40 million), the Republic of Korea (US$92.98 million), Russian Federation (US$72.94 million), Indonesia (US$49.40 million), United Kingdom (US$43.85 million), Taiwan, China (US$36.21 million),

Germany (US$35.44 million), Canada (US$33.09 million), India (US$24.12 million), France (US$21.51 million), Brazil (US$20.27 million), Italy (US$20.18 million), Netherlands (US$11.11 million), Mexico (US$10.82), Spain (US$9.94 million), Norway (US$9.88 million), Switzerland (US$9.37 million). The total losses of the top 20 countries and regions are US$10.556 billion, accounting for about 91.53% of the total global losses.

The total output loss ratio is calculated from the total output lost in each country and the total output when it was not damaged, which is the total national output functional loss rate Q_1(Q_1 = nationaloutputloss/totalnationaloutput). The total global output function loss rate is about 0.000781%, and Japan has the highest total output function loss rate, about 0.11%.

Table 14.1 Top 20 most affected countries (regions)

	Country (region)	Total output loss (US$ million)	Country (region)	Functional loss rate Q_1
1	Japan	9494.85	Japan	0.109530%
2	China	247.86	Slovak Republic	0.007164%
3	United States of America	207.45	Portugal	0.003614%
4	Australia	104.40	Luxembourg	0.002966%
5	Korea, Republic of	92.98	Korea, Republic of	0.002882%
6	Russian Federation	72.94	Taiwan, China	0.002731%
7	Indonesia	49.40	Indonesia	0.002157%
8	United Kingdom	43.85	Italy	0.001284%
9	Taiwan, China	36.21	Estonia	0.001184%
10	Germany	35.44	Canada	0.001018%
11	Canada	33.09	Russian Federation	0.000830%
12	India	24.12	Sweden	0.000807%
13	France	21.51	Cyprus	0.000782%
14	Brazil	20.27	Australia	0.000670%
15	Italy	20.18	Finland	0.000670%
16	Netherlands	11.11	United States of America	0.000665%
17	Mexico	10.82	Slovenia	0.000646%
18	Spain	9.94	Germany	0.000643%
19	Norway	9.88	Hungary	0.000609%
20	Switzerland	9.37	Lithuania	0.000607%
	ROW	977.38		
	Total	11,533.04		

Countries with large total output losses and high functional loss rates are considered countries with high sensitivity to nuclear wastewater. The sensitivity ranking of 44 countries is obtained by ranking the results in two dimensions: total output loss and functional loss rate. The top 10 countries (regions) with high sensitivity to nuclear wastewater are Japan, the Republic of Korea, Indonesia, Taiwan (China), Russia, Australia, the United States of America, Canada, Italy, and Germany.

Based on scenario 1, the top 20 affected industries are shown in Table 14.2. The 34.4% reduction in final demand for fishing and aquaculture in Japan results in a decrease of 1.34% in global output, which is about US$5.72 billion, accounting for 49.58% of the total global loss. Other 19 industries with significant output losses are: Manufacture of coke and refined petroleum products (US$707.25 million), Mining and quarrying (US$578.93 million), Manufacture of food products, beverages and tobacco products (US$560.51 million), Wholesale trade, except of motor vehicles and motorcycles (US$383.16 million), Manufacture of other transport equipment (US$368.31 million), Financial service activities, except insurance and pension funding(US$225.49 million), Manufacture of chemicals and chemical products (US$204.42 million), Manufacture of textiles, wearing apparel and leather products (US$203.98 million), Electricity, gas, steam and air conditioning supply (US$173.95 million), Accommodation and food service activities (US$173.40 million), Manufacture of basic metals (US$166.89 million), Land transport and transport via pipelines (US$156.17 million), Manufacture of rubber and plastic products (US$150.09 million), Crop and animal production, hunting and related service activities (US$143.92 million), Other professional, scientific and technical activities; veterinary activities (US$115.60 million), Retail trade, except of motor vehicles and motorcycles (US$114.47 million), Manufacture of computer, electronic and optical products (US$90.01 million), Wholesale and retail trade and repair of motor vehicles and motorcycles (US$88.26 million), Manufacture of fabricated metal products, except machinery and equipment (US$83.77 million). The total loss of the top 20 industries amounts to US$10.41 billion, accounting for about 90.23% of the total global loss.

This paper can calculate the functional loss rate of industry output Q_2(Q_2 = totalindustrialoutputloss/totalindustrialoutput) based on the output loss of each industry and the output when it was not damaged. The highest functional loss rate is found in the fishing and aquaculture industry, with 1.34%.

Industries with large total output losses and high functional loss rates are considered industries with high sensitivity to nuclear wastewater. Ranking the results in terms of total output loss and functional loss rate, the top 10 industries with high sensitivity to nuclear wastewater are as follows: Fishing and aquaculture; Manufacture of coke and refined petroleum products; Manufacture of other transport equipment; Mining and quarrying; Manufacture of food products, beverages and tobacco products; Manufacture of textiles, wearing apparel and leather products; Wholesale trade, except of motor vehicles and motorcycles; Other professional, scientific and technical activities; veterinary activities; Manufacture of rubber and plastic products; Financial service activities, except insurance and pension funding.

Table 14.2 Top 20 most affected industries

No	Industry	Total output loss (US$ million)	Industry	Functional loss rate Q_2
1	Fishing and aquaculture	5718.04	Fishing and aquaculture	1.339026%
2	Manufacture of coke and refined petroleum products	707.25	Manufacture of other transport equipment	0.025204%
3	Mining and quarrying	578.93	Manufacture of coke and refined petroleum products	0.018366%
4	Manufacture of food products, beverages and tobacco products	560.51	Activities of households as employers; undifferentiated goods- and services-producing activities of households for own use	0.014256%
5	Wholesale trade, except of motor vehicles and motorcycles	383.16	Other professional, scientific and technical activities; veterinary activities	0.010161%
6	Manufacture of other transport equipment	368.31	Mining and quarrying	0.009709%
7	Financial service activities, except insurance and pension funding	225.49	Water transport	0.008636%
8	Manufacture of chemicals and chemical products	204.42	Manufacture of rubber and plastic products	0.008489%
9	Manufacture of textiles, wearing apparel and leather products	203.98	Manufacture of food products, beverages and tobacco products	0.008158%
10	Electricity, gas, steam and air conditioning supply	173.95	Manufacture of textiles, wearing apparel and leather products	0.007334%
11	Accommodation and food service activities	173.40	Advertising and market research	0.007261%
12	Manufacture of basic metals	166.89	Wholesale and retail trade and repair of motor vehicles and motorcycles	0.006329%

(continued)

Table 14.2 (continued)

No	Industry	Total output loss (US$ million)	Industry	Functional loss rate Q_2
13	Land transport and transport via pipelines	156.17	Manufacture of paper and paper products	0.005463%
14	Manufacture of rubber and plastic products	150.09	Printing and reproduction of recorded media	0.005140%
15	Crop and animal production, hunting and related service activities	143.92	Wholesale trade, except of motor vehicles and motorcycles	0.004900%
16	Other professional, scientific and technical activities; veterinary activities	115.60	Financial service activities, except insurance and pension funding	0.004847%
17	Retail trade, except of motor vehicles and motorcycles	114.47	Manufacture of chemicals and chemical products	0.004794%
18	Manufacture of computer, electronic and optical products	90.01	Accommodation and food service activities	0.004545%
19	Wholesale and retail trade and repair of motor vehicles and motorcycles	88.26	Manufacture of furniture; other manufacturing	0.004392%
20	Manufacture of fabricated metal products, except machinery and equipment	83.77	Land transport and transport via pipelines	0.003788%
	Other Industries	1126.40		
	Total	11,533.04		

14.6.1.2 Scenario 2: Final Demand for Japanese Fisheries Decreases by 66.2%

Based on scenario 2, the results for the top 20 affected countries (regions) are shown in Table 14.3. Nuclear wastewater contamination causes 66.2% of the direct loss of Japanese fisheries and indirectly results in a total global economic output loss of US$22.194 billion. Japan's total output loss is US$18.272 billion, accounting for about 82.33% of the total global loss. 19 other countries (regions) with large losses in total output are calculated: China (US$476.98 million), the United States of America (US$399.22 million), Australia (US$200.91 million), the Republic of Korea (US$178.92 million), Russian Federation (US$140.36 million), Indonesia (US$95.07 million), the United Kingdom (US$84.39 million), Taiwan (China) (US$69.67 million), Germany (US$68.20 million), Canada (US$63.69 million), India (US$46.41 million), France (US$41.40 million), Brazil (US$39.01 million), Italy (US$38.83 million), Netherlands (US$21.39 million), Mexico (US$20.82 million), Spain (US$19.12million), Norway (US$19.02 million), Switzerland (US$18.02 million). The total loss of the top 20 countries is US$20.313 billion, accounting for about 91.53% of the total global loss.

The global total output functional loss rate is 0.000781%, and Japan has the highest total output functional loss rate, which is about 0.21%.

The top 10 countries (regions) with high sensitivity to nuclear wastewater are Japan, the Republic of Korea, Indonesia, Taiwan (China), Russia, Australia, the United States of America, Canada, Italy, and Germany.

The result of the top 20 affected industries is shown in Table 14.4. The 66.2% reduction in final demand in the fishing and aquaculture industry in Japan results in a 2.58% reduction in the total global output of this industry, which is about US$11 billion, accounting for approximately 49.58% of the total global loss. Other 19 industries with significant output losses are: Manufacture of coke and refined petroleum products (US$1.36 billion), Mining and quarrying (US$1.11 billion), Manufacture of food products, beverages and tobacco products (US$1.08 billion), Wholesale trade, except of motor vehicles and motorcycles (US$737.36 million), Manufacture of other transport equipment (US$708.79 million), Financial service activities, except insurance and pension funding (US$433.95 million), Manufacture of chemicals and chemical products (US$393.39 million), Manufacture of textiles, wearing apparel and leather products (US$392.55 million), Electricity, gas, steam and air conditioning supply (US$334.75 million), Accommodation and food service activities (US$333.69 million), Manufacture of basic metals (US$321.17 million), Land transport and transport via pipelines (US$300.54 million), Manufacture of rubber and plastic products (US$288.84 million), Crop and animal production, hunting and related service activities (US$276.96 million), Other professional, scientific and technical activities; veterinary activities (US$222.46 million), Retail trade, except of motor vehicles and motorcycles (US$220.29 million), Manufacture of computer, electronic and optical products (US$173.21 million), Wholesale and retail trade and repair of motor vehicles and motorcycles (US$169.85 million), Manufacture of fabricated metal products, except machinery and equipment (US$161.21 million). The

Table 14.3 Top 20 most affected countries (regions)

	Country (region)	Total output loss (US$ million)	Country (region)	Functional loss rate Q_3
1	Japan	18,272.06	Japan	0.210781%
2	China	476.98	Slovak Republic	0.013786%
3	United States of America	399.22	Portugal	0.006955%
4	Australia	200.91	Luxembourg	0.005708%
5	Korea, Republic of	178.92	Korea, Republic of	0.005545%
6	Russian Federation	140.36	Taiwan, China	0.005257%
7	Indonesia	95.07	Indonesia	0.004151%
8	United Kingdom	84.39	Italy	0.002470%
9	Taiwan, China	69.67	Estonia	0.002278%
10	Germany	68.20	Canada	0.001958%
11	Canada	63.69	Russian Federation	0.001597%
12	India	46.41	Sweden	0.001553%
13	France	41.40	Cyprus	0.001505%
14	Brazil	39.01	Australia	0.001289%
15	Italy	38.83	Finland	0.001289%
16	Netherlands	21.39	United States of America	0.001280%
17	Mexico	20.82	Slovenia	0.001243%
18	Spain	19.12	Germany	0.001237%
19	Norway	19.02	Hungary	0.001173%
20	Switzerland	18.02	Lithuania	0.001169%
	ROW	1880.89		
	Total	22,194.39		

total loss of the top 20 industries amounts to US$20.03 billion, accounting for about 90.23% of the total global loss.

The functional loss rate Q_4 of industry output is calculated based on the amount of output lost and the output when undamaged in each industry. Among them, fishing and aquaculture have the highest functional loss rate of 2.58%.

The top 10 industries with high sensitivity to nuclear wastewater: Fishing and aquaculture; Manufacture of coke and refined petroleum products; Manufacture of other transport equipment; Mining and quarrying; Manufacture of food products, beverages and tobacco products; Manufacture of textiles, wearing apparel and leather products; Wholesale trade, except of motor vehicles and motorcycles; Other professional, scientific and technical activities; veterinary activities; Manufacture of

Table 14.4 Top 20 most affected industries

No	Industry	Total output loss (US$ million)	Industry	Functional loss rate Q_4
1	Fishing and aquaculture	11,003.91	Fishing and aquaculture	2.576847%
2	Manufacture of coke and refined petroleum products	1361.04	Manufacture of other transport equipment	0.048503%
3	Mining and quarrying	1114.11	Manufacture of coke and refined petroleum products	0.035344%
4	Manufacture of food products, beverages and tobacco products	1078.66	Activities of households as employers; undifferentiated goods- and services-producing activities of households for own use	0.027434%
5	Wholesale trade, except of motor vehicles and motorcycles	737.36	Other professional, scientific and technical activities; veterinary activities	0.019553%
6	Manufacture of other transport equipment	708.79	Mining and quarrying	0.018685%
7	Financial service activities, except insurance and pension funding	433.95	Water transport	0.016618%
8	Manufacture of chemicals and chemical products	393.39	Manufacture of rubber and plastic products	0.016336%
9	Manufacture of textiles, wearing apparel and leather products	392.55	Manufacture of food products, beverages and tobacco products	0.015699%
10	Electricity, gas, steam and air conditioning supply	334.75	Manufacture of textiles, wearing apparel and leather products	0.014114%
11	Accommodation and food service activities	333.69	Advertising and market research	0.013973%
12	Manufacture of basic metals	321.17	Wholesale and retail trade and repair of motor vehicles and motorcycles	0.012180%
13	Land transport and transport via pipelines	300.54	Manufacture of paper and paper products	0.010514%

(continued)

Table 14.4 (continued)

No	Industry	Total output loss (US$ million)	Industry	Functional loss rate Q_4
14	Manufacture of rubber and plastic products	288.84	Printing and reproduction of recorded media	0.009892%
15	Crop and animal production, hunting and related service activities	276.96	Wholesale trade, except of motor vehicles and motorcycles	0.009429%
16	Other professional, scientific and technical activities; veterinary activities	222.46	Financial service activities, except insurance and pension funding	0.009328%
17	Retail trade, except of motor vehicles and motorcycles	220.29	Manufacture of chemicals and chemical products	0.009226%
18	Manufacture of computer, electronic and optical products	173.21	Accommodation and food service activities	0.008747%
19	Wholesale and retail trade and repair of motor vehicles and motorcycles	169.85	Manufacture of furniture; other manufacturing	0.008453%
20	Manufacture of fabricated metal products, except machinery and equipment	161.21	Land transport and transport via pipelines	0.007290%
	Other Industries	2167.67		
	Total	22,194.39		

rubber and plastic products; Financial service activities, except insurance and pension funding.

14.6.1.3 Comparison Between Scenario 1 and Scenario 2

Scenarios 1 and 2 use the same model, and there is only a numerical difference between the two scenarios (34.4% and 66.2%). Comparing the simulation results of the two scenarios, it is found that (1) the same proportional relationship exists for some of the numerical results, the ratio is equal to 66.2% to 34.4%, that is, some results of scenario 2 are approximately 1.92 times that of scenario 1, such as the amount of total output loss and the rate of total output loss for the country and industry. (2) Some numerical results are the same, such as the ratio of total output loss of each country to total global loss. (3) Ranking results are the same, such as the ranking of countries and industries affected by nuclear wastewater and sensitivity rankings.

14.6.2 The Impact of Reduced Supply

14.6.2.1 Scenario 3: Japanese Fishery Output Decreases by 34.4%

Based on the assumptions of Scenario 3 $d = (0, \ldots, 0{,}66.2\%)$, vector $b = (q_1, \ldots, q_{k-1}, q_{k+1}, \ldots, q_n, c_k^*)$ can be solved, including the functional loss rate q of output for each industry in each country (except Japan's fishing and aquaculture industry) and the proportion c^* of change in final demand for Japan's fishing and aquaculture industry. The top 20 countries (regions) and industries in terms of total output loss, the change in global fishing and aquaculture output (except Japan), and the change in Japan's fishing and aquaculture final demand are shown in Tables 14.5, 14.6, 14.7, and 14.8.

In Table 14.5, 34.4% of reduction in output of fishing and aquaculture in Japan causes a total loss of US$5.567 billion of the total production of other countries and industries. The output of other industries excluding fishing and aquaculture in Japan decreases by US$3.619 billion, accounting for about 65% of the total global loss. Other 19 countries (regions) with large losses of other total output are China (US$236.91 million), the United States of America (US$198.29 million), Australia(US$99.79 million), the Republic of Korea (US$88.87 million), Russian Federation (US$69.72 million), Indonesia (US$47.22 million), the United Kingdom (US$41.92 million), Taiwan, China (US$34.61 million), Germany (US$33.88 million), Canada (US$31.63 million), India (US$23.05 million), France (US$20.56 million), Brazil (US$19.38 million), Italy (US$19.29 million), Netherlands (US$10.62 million), Mexico (US$10.34 million), Spain (US$9.50 million), Norway (US$9.45 million), Switzerland (US$8.95 million).

Based on the total output loss and the total output when it was not damaged in each country, the functional loss rate of national output Q_5 is calculated.

The top 10 countries (regions) with high sensitivity to nuclear wastewater are Japan, Australia, the Republic of Korea, Indonesia, Russian Federation, Taiwan, China, the United States of America, United Kingdom, Canada.

Table 14.5 Top 20 most affected countries (regions)

No	Country (region)	Total output loss (US$ million)	Country (region)	Functional loss rate Q_5
1	Japan	3618.71	Japan	0.041821%
2	China	236.91	Australia	0.003664%
3	United States of America	198.29	Taiwan, China	0.002835%
4	Australia	99.79	Indonesia	0.002754%
5	Korea, Republic of	88.87	Korea, Republic of	0.002611%
6	Russian Federation	69.72	Russian Federation	0.002062%
7	Indonesia	47.22	Luxembourg	0.001227%
8	United Kingdom	41.92	Norway	0.001131%
9	Taiwan, China	34.61	Canada	0.000973%
10	Germany	33.88	United Kingdom	0.000793%
11	Canada	31.63	Denmark	0.000771%
12	India	23.05	Ireland	0.000747%
13	France	20.56	China	0.000746%
14	Brazil	19.38	United States of America	0.000640%
15	Italy	19.29	Switzerland	0.000640%
16	Netherlands	10.62	Netherlands	0.000636%
17	Mexico	10.34	Estonia	0.000617%
18	Spain	9.50	Finland	0.000615%
19	Norway	9.45	Lithuania	0.000582%
20	Switzerland	8.95	Belgium	0.000581%
	Other countries	934.22		
	Total	5566.89		

The top 20 industries in terms of total output loss and functional loss rate are shown in Table 14.6. The manufacture of coke and refined petroleum products industry has the highest loss of US$676.01 million. The functional loss rate of industrial output Q_6 is calculated based on the total output loss and total output without damage of each industry. The top 10 industries with high sensitivity to nuclear wastewater: Manufacture of coke and refined petroleum products; Mining and quarrying; Manufacture of food products, beverages and tobacco products; Wholesale trade, except of motor vehicles and motorcycles; Manufacture of other transport equipment; Financial service activities, except insurance and pension funding; Manufacture of chemicals and chemical products; Manufacture of textiles, wearing apparel and leather products; Manufacture of rubber and plastic products; Other professional, scientific and technical activities, veterinary activities.

Table 14.6 Top 20 most affected industries

No	Industry	Total output loss (US$ million)	Industry	Functional loss rate Q_6
1	Manufacture of coke and refined petroleum products	676.01	Manufacture of other transport equipment	0.024091%
2	Mining and quarrying	553.37	Manufacture of coke and refined petroleum products	0.017555%
3	Manufacture of food products, beverages and tobacco products	535.76	Activities of households as employers; undifferentiated goods- and services-producing activities of households for own use	0.013626%
4	Wholesale trade, except of motor vehicles and motorcycles	366.24	Other professional, scientific and technical activities; veterinary activities	0.009712%
5	Manufacture of other transport equipment	352.05	Mining and quarrying	0.009280%
6	Financial service activities, except insurance and pension funding	215.54	Water transport	0.008254%
7	Manufacture of chemicals and chemical products	195.39	Manufacture of rubber and plastic products	0.008114%
8	Manufacture of textiles, wearing apparel and leather products	194.97	Manufacture of food products, beverages and tobacco products	0.007797%
9	Electricity, gas, steam and air conditioning supply	166.27	Manufacture of textiles, wearing apparel and leather products	0.007010%
10	Accommodation and food service activities	165.74	Advertising and market research	0.006940%
11	Manufacture of basic metals	159.52	Wholesale and retail trade and repair of motor vehicles and motorcycles	0.006050%
12	Land transport and transport via pipelines	149.27	Manufacture of paper and paper products	0.005222%

(continued)

Table 14.6 (continued)

No	Industry	Total output loss (US$ million)	Industry	Functional loss rate Q_6
13	Manufacture of rubber and plastic products	143.47	Printing and reproduction of recorded media	0.004913%
14	Crop and animal production, hunting and related service activities	137.56	Wholesale trade, except of motor vehicles and motorcycles	0.004683%
15	Other professional, scientific and technical activities; veterinary activities	110.49	Financial service activities, except insurance and pension funding	0.004633%
16	Retail trade, except of motor vehicles and motorcycles	109.42	Manufacture of chemicals and chemical products	0.004583%
17	Manufacture of computer, electronic and optical products	86.03	Accommodation and food service activities	0.004345%
18	Wholesale and retail trade and repair of motor vehicles and motorcycles	84.36	Manufacture of furniture; other manufacturing	0.004198%
19	Manufacture of fabricated metal products, except machinery and equipment	80.07	Land transport and transport via pipelines	0.003621%
20	Manufacture of machinery and equipment n.e.c	64.88	Warehousing and support activities for transportation	0.003591%
	Other Industries	1020.48		
	Total	5566.89		

Table 14.7 Changes in global fishing and aquaculture output (excluding Japan)

No	Country (region)	Total output loss (US$ million)	Country (region)	Functional loss rate q_1
1	China	1.57	Taiwan, China	0.022461%
2	Taiwan, China	0.8	Korea, Republic of	0.011772%
3	Korea, Republic of	0.62	Canada	0.002807%
4	Indonesia	0.55	United States of America	0.002541%
5	United States of America	0.51	Mexico	0.002331%
6	Canada	0.17	Indonesia	0.002246%
7	Australia	0.06	Australia	0.001986%
8	Norway	0.06	Ireland	0.001638%
9	Spain	0.05	Spain	0.001430%
10	Mexico	0.05	China	0.000952%
	Other countries	4.26		
	Total	8.70		

Table 14.8 Changes in final demand of fishing and aquaculture in Japan

No	Country	Reduction in final demand (US$ million)	Country	Proportion of reduction in final demand c_1^*
1	Japan	5215.83	Japan	32.88%

As shown in Table 14.7, the reduction of 34.4% in output of fishing and aquaculture in Japan results in a reduction of US$8.7 million in the total output of the fishing and aquaculture industry in other countries. China has the largest loss in this industry with US$1.57 million. The 5 countries (regions) with the highest sensitivity to nuclear wastewater in fishing and aquaculture are Taiwan, the Republic of Korea, the United States of America, Canada, and Indonesia.

As shown in Table 14.8, a 34.4% decrease in Japanese fishing and aquaculture output results in a decrease of US$5.216 billion in final demand of this industry in Japan, accounting for about 32.88% of Japan's total output of fishing and aquaculture industry and about 0.06% of Japan's total output of the whole industry.

14.6.2.2 Scenario 4: Japan's Fishery Output Declines by 66.2% and Final Demand from Other Countries and Industries Remains Unchanged

Simulation results based on Scenario 4 are shown in the following tables. In Table 14.9, the 66.2% decrease in the output of the Japanese fishing and aquaculture industry causes a decrease of US$10.713 billion in the total output of other countries and industries. Among them, the total output of industries (except fishing and aquaculture) in Japan decreases by US$6.964 billion, accounting for about 65% of the total global loss.

The top 20 industries in terms of total output loss and functional loss rate are shown in Table 14.10. The manufacture of coke and refined petroleum products industry loses the most, which is US$1.3 billion. And The top 10 industries of high sensitivity to nuclear wastewater: Manufacture of coke and refined petroleum products; Mining and quarrying; Manufacture of food products, beverages and tobacco products; Wholesale trade, except of motor vehicles and motorcycles; Manufacture of other transport equipment; Financial service activities, except insurance and pension funding; Manufacture of chemicals and chemical products; Manufacture of textiles,

Table 14.9 Top 20 most affected countries (regions)

No	Country (region)	Total output loss (US$ million)	Country (region)	Functional loss rate Q_7
1	Japan	6963.92	Japan	0.080481%
2	China	455.92	Australia	0.007050%
3	United States of America	381.59	Taiwan, China	0.005456%
4	Australia	192.03	Indonesia	0.005300%
5	Korea, Republic of	171.02	Korea, Republic of	0.005024%
6	Russian Federation	134.16	Russian Federation	0.003968%
7	Indonesia	90.87	Luxembourg	0.002361%
8	United Kingdom	80.66	Norway	0.002177%
9	Taiwan, China	66.6	Canada	0.001872%
10	Germany	65.19	United Kingdom	0.001527%
11	Canada	60.88	Denmark	0.001485%
12	India	44.36	Ireland	0.001438%
13	France	39.57	China	0.001436%
14	Brazil	37.29	United States of America	0.001232%
15	Italy	37.12	Switzerland	0.001232%
16	Netherlands	20.44	Netherlands	0.001223%
17	Mexico	19.9	Estonia	0.001188%
18	Spain	18.28	Finland	0.001183%
19	Norway	18.18	Lithuania	0.001121%
20	Switzerland	17.22	Belgium	0.001117%
	Other countries	1797.83		
	Total	10,713.03		

wearing apparel and leather products; Manufacture of rubber and plastic products; Other professional, scientific and technical activities, veterinary activities.

As shown in Table 14.11, the total loss of fishing and aquaculture in countries except Japan is US$16.75 million. China loses the most, at US$3.02 million. The 5 countries (regions) with the highest sensitivity to nuclear wastewater in fishing and aquaculture are Taiwan, the Republic of Korea, the United States of America, Canada, Indonesia.

In Table 14.12, the 66.2% of reduction in output of Japan's fishing and aquaculture industry results in a reduction in final demand for this industry in Japan of US$10.037 billion, accounting for about 63.28% of the total output of Japan's fishing and aquaculture and about 0.12% of the total output of the entire industry in Japan.

14.6.2.3 Comparison Between Scenario 3 and Scenario 4

The comparison results between scenarios 3 and 4 are similar to those of scenarios 1 and 2, i.e. (1) some of the numerical results have the same proportional relationship (about 1.92), such as the amount of total output loss and total output loss rate of countries and industries; (2) some of the numerical results are the same, such as the ratio of total output loss of each country to total global loss; (3) the ranking results are the same, such as the ranking of countries and industries affected by the nuclear wastewater and the ranking of sensitivity.

14.7 Conclusions

14.7.1 Main Conclusions

Based on the Leontief input–output model and the ideas of Haimes and Jiang (2001), Santos and Haimes (2004), and Leung et al. (2007), the inoperability input–output model (IIM) is applied to simulate the impact of Japanese nuclear wastewater discharge into the sea on the global economy from the perspectives of reduced demand and reduced output of Japanese fisheries. The results show that: ① The global indirect economic loss interval is [5.567, 22.194] billion USD due to damage to Japanese fishing and aquaculture. ② The top 10 countries (regions) and industries with high sensitivity to nuclear wastewater are identified.

14.7.2 Limitations and Prospects

First, this paper simulates the nuclear wastewater discharge event in Japan in 2023 by using the International Input–Output Tables 2014 published in 2016. There is a

Table 14.10 Top 20 most affected industries

No	Industry	Total output loss (US$ million)	Industry	Functional loss rate Q_8
1	Manufacture of coke and refined petroleum products	1300.94	Manufacture of other transport equipment	0.046361%
2	Mining and quarrying	1064.91	Manufacture of coke and refined petroleum products	0.033783%
3	Manufacture of food products, beverages and tobacco products	1031.02	Activities of households as employers; undifferentiated goods- and services-producing activities of households for own use	0.026222%
4	Wholesale trade, except of motor vehicles and motorcycles	704.8	Other professional, scientific and technical activities; veterinary activities	0.018690%
5	Manufacture of other transport equipment	677.49	Mining and quarrying	0.017859%
6	Financial service activities, except insurance and pension funding	414.78	Water transport	0.015885%
7	Manufacture of chemicals and chemical products	376.02	Manufacture of rubber and plastic products	0.015615%
8	Manufacture of textiles, wearing apparel and leather products	375.21	Manufacture of food products, beverages and tobacco products	0.015005%
9	Electricity, gas, steam and air conditioning supply	319.97	Manufacture of textiles, wearing apparel and leather products	0.013490%
10	Accommodation and food service activities	318.96	Advertising and market research	0.013356%
11	Manufacture of basic metals	306.98	Wholesale and retail trade and repair of motor vehicles and motorcycles	0.011642%
12	Land transport and transport via pipelines	287.26	Manufacture of paper and paper products	0.010050%

(continued)

Table 14.10 (continued)

No	Industry	Total output loss (US$ million)	Industry	Functional loss rate Q_8
13	Manufacture of rubber and plastic products	276.09	Printing and reproduction of recorded media	0.009455%
14	Crop and animal production, hunting and related service activities	264.73	Wholesale trade, except of motor vehicles and motorcycles	0.009013%
15	Other professional, scientific and technical activities; veterinary activities	212.63	Financial service activities, except insurance and pension funding	0.008916%
16	Retail trade, except of motor vehicles and motorcycles	210.56	Manufacture of chemicals and chemical products	0.008819%
17	Manufacture of computer, electronic and optical products	165.56	Accommodation and food service activities	0.008361%
18	Wholesale and retail trade and repair of motor vehicles and motorcycles	162.35	Manufacture of furniture; other manufacturing	0.008079%
19	Manufacture of fabricated metal products, except machinery and equipment	154.09	Land transport and transport via pipelines	0.006968%
20	Manufacture of machinery and equipment n.e.c	124.86	Warehousing and support activities for transportation	0.006910%
	Other Industries	1963.83		
	Total	10,713.03		

Table 14.11 Changes in global fishing and aquaculture output (excluding Japan)

No	Country (region)	Total output loss (US$ million)	Country (region)	Functional loss rate q_2
1	China	3.02	Taiwan, China	0.043225%
2	Taiwan, China	1.53	Korea, Republic of	0.022655%
3	Korea, Republic of	1.19	Canada	0.005403%
4	Indonesia	1.05	United States of America	0.004890%
5	United States of America	0.98	Mexico	0.004485%
6	Canada	0.33	Indonesia	0.004323%
7	Australia	0.12	Australia	0.003822%
8	Norway	0.11	Ireland	0.003151%
9	Spain	0.1	Spain	0.002751%
10	Mexico	0.09	China	0.001832%
	Other countries	8.21		
	Total	16.75		

Table 14.12 Changes in final demand of fishing and aquaculture in Japan

No	Country	Reduction in final demand (US$ million)	Country	Proportion of reduction in final demand c_2^*
1	Japan	10,037.45	Japan	63.28%

time lag in the data, and the data can be corrected through methods such as the RAS method to obtain more accurate simulation results.

Second, the development of the world input–output table used in this paper was funded by the Netherlands Science Foundation (NWO) and the European Commission Services. The data covers 28 EU member states and 15 other countries (regions). However, Japanese nuclear wastewater mainly affects countries and regions in Oceania, East Asia, North America and the Pacific coast of South America. Therefore, the results are somewhat underestimated. Further analysis can be developed in the future based on regional input–output tables.

Finally, the ocean currents dilute the nuclear wastewater and expand its impact. The discharge of nuclear wastewater from Japan may directly affect the output and demand of fisheries in other countries and regions along the Pacific Rim. Further studies can simulate the direct impact of nuclear wastewater on multiple industries in multiple countries and the indirect impact on the global economic system by changing the scenario assumptions.

Acknowledgements Yi Zhang and Xuehao Feng also made great contributions to this manuscript. We express our heartfelt thanks to them. This research also was supported by: National Natural Science Foundation of China (72474127) and National Social Science Fund Major Bidding Project of China (18ZDA052). Part of this chapter was published:Wu X, Zhang Y, Feng X. The impact of Japanese nuclear wastewater discharge into the sea on the global economy: Input-output model approach. Marine Pollution Bulletin 2023;192:115067. https://doi.org/https://doi.org/10.1016/j.marpolbul.2023.115067. The publication has been conducted with copyright permission.

References

Alonso-Conde, A. B., & Rojo-Suárez, J. (2020). Nuclear hazard and asset prices: Implications of nuclear disasters in the cross-sectional behavior of stock returns. *Sustainability, 12*(22), 9721.

Aruga, K., & Wakamatsu, H. (2018). Consumer perceptions toward seafood produced near the fukushima nuclear plant. *Marine Resource Economics, 33*(4), 373–386.

Dadakas, D., & Tatsi, S. (2021). Global agricultural trade impact of the 2011 triple disaster in Japan: A gravity approach*. *Australian Journal of Agricultural and Resource Economics.* https://doi.org/10.1111/1467-8489.12448

Dai, J. H., & Yang, N. (2021). Analysis of international public opinion on the discharge of nuclear wastewater in Japan. *Contemporary World, 06*, 74–79.

Haimes, Y. Y., & Jiang, P. (2001). Leontief-based model of risk in complex interconnected infrastructures. *Journal of Infrastructure Systems, 7*(1), 1–12.

Hendricks, K. B., Jacobs, B. W., & Singhal, V. R. (2020). Stock market reaction to supply chain disruptions from the 2011 great east Japan earthquake. *Manufacturing & Service Operations Management, 22*(4), 683–699.

Higashi, Y. (2020). Effects of region-specific shocks on labor market tightness and matching efficiency: Evidence from the 2011 Tohoku earthquake in Japan. *The Annals of Regional Science, 65*(4), 1–27.

Ishimaru, T., Tateda, Y., Tsumune, D., Aoyama, M., Hamajima, Y., Kasamatsu, N., Yamada,M., Yoshimura,T., Mizuno,T., & Kanda, J. (2019). Mechanisms of radiocesium depuration in Sebastes cheni derived by simulation analysis of measured 137Cs concentrations off southern Fukushima 2014–2016. *Journal of Environmental Radioactivity, 203*, 200–209.

Jin, Y. (2021). Problems of discharging nuclear wastewater from Fukushima. *Japan and Its Response. Contemporary World, 06*, 68–73.

Kudo, D., & Nagaya, K. (2017). Effects of matching and mismatching messages on purchase avoidance behavior following major disasters. *Psychology & Marketing, 34*(3), 335–346.

Lee, S., Hwang, S., Park, M., & Lee, H. (2018). Damage propagation from component level to system level in the electricity sector. *Journal of Infrastructure Systems, 24*(3), 04018016. https://doi.org/10.1061/(asce)is.1943-555x.0000437

Lenzen, M., Malik, A., Kenway, S., Daniels, P., & Geschke, A. (2019). Economic damage and spillovers from a tropical cyclone. *Natural Hazards and Earth System Sciences, 19*(1), 137–151.

Leontief, W. (1936). Quantitative input and output relations in the economic systems of the United States. *The Review of Economics and Statistics, 18*(3), 105–125.

Leung, M., Haimes, Y. Y., & Santos, J. R. (2007). Supply- and output-side extensions to the inoperability input-output model for interdependent infrastructures. *Journal of Infrastructure Systems, 13*(4), 299–310.

Matsumoto, S., & Harvey, D. (2020). Economic loss due to reputation damage: A new model and its application to Fukushima peaches. *Journal of Agricultural Economics, 71*(2), 581–600.

Miura, Y., Blackshaw, C. Y., Zhang, M. S., Mandli, K. T., & Deodatis, G. (2025). Coastal storm-induced flooding risk of the New York City subway amid climate change. *Transportation Research Part d: Transport and Environment, 149*, Article 104974.

Montibeler, E. E., & Oliveira, D. (2018). Dengue endemic and its impact on the gross national product of Brazilian's economy. *Acta Tropica, 178*, 318–326.

Morishita, D., Wada, T., Noda, T., Tomiya, A., Enomoto, M., Sato, T., Suzuki, S., & Kawata, G. (2019). Spatial and seasonal variations of radiocesium concentrations in an algae-grazing annual fish, ayu Plecoglossus altivelis collected from Fukushima Prefecture in 2014. *Fisheries Science, 85*(3), 561–569.

Niizeki, K., Wada, T., Nanba, K., Sasaki, K., Teramoto, W., Izumi, S., Nomura, H., & Inatomi, N. (2020). Estimating biological half-lives of 137Cs in a cyprinid fish Tribolodon hakonensis by a one-compartment model considering growth dilution effect. *Fisheries Science, 86*(5), 861–871.

Okuyama, Y., & Santos, J. R. (2014). Disaster impact and input-output analysis. *Economic Systems Research: Journal of the International Input-Output Association, 26*(1), 1–12.

Pagsuyoin, S. A., & Santos, J. R. (2021). Modeling regional impacts and resilience to water service disruptions in urban economies. *Environment and Planning b: Urban Analytics and City Science, 48*(5), 1058–1074.

Park, C., & Park, S. (2020). Rare disaster risk and exchange rates: An empirical investigation of south Korean exchange rates under tension between the two Koreas. *Finance Research Letters, 36*, Article 101314.

Poudineh, R., & Jamasb, T. (2017). Electricity supply interruptions: Sectoral interdependencies and the cost of energy not served for the Scottish economy. *The Energy Journal, 38*(1), 51–76.

Samimi, S., Ebrahimnejad, S., & Mojtahedi, M. (2020). Analysis of the susceptibility of interdependent infrastructures using fuzzy input–output inoperability model: The case of flood hazards in Tehran. *Natural Hazards: Journal of the International Society for the Prevention and Mitigation of Natural Hazards, 100*(1), 69–88.

Santos, J. R. (2010). Inoperability input-output modeling of disruptions to interdependent economic systems. *Systems Engineering, 9*(1), 20–34.

Santos, J. R., & Haimes, Y. Y. (2004). Modeling the demand reduction input-output (I-O) inoperability due to terrorism of interconnected infrastructures. *Risk Analysis, 24*(6), 1437–1451.

Sarkar, B. D., Gupta, L., & Jagtap, S. (2025). Resilient or fragile? Modelling economic disruptions in India's electronics sector due to the Red Sea crisis. *Journal of Transport Geography, 123*, Article 104121.

Sellevg, S. R. (2021). Changes in inoperability for interdependent industry sectors in Norway from 2012 to 2017. *International Journal of Critical Infrastructure Protection, 32*, Article 100405.

Sugrue, D., Martin, A., & Adriaens, P. (2021). Applied financial metrics to measure interdependencies in a waterway infrastructure system. *Journal of Infrastructure Systems, 27*(1), 05020010.

Suris-Regueiro, J. C., & Santiago, J. L. (2018). A methodological approach to quantifying socioeconomic impacts linked to supply shocks. *Environmental Impact Assessment Review, 69*, 104–110.

Svegrup, L., Johansson, J., & Hassel, H. (2019). Integration of critical infrastructure and societal consequence models: Impact on Swedish power system mitigation decisions. *Risk Analysis, 39*(9), 1970–1996.

Tajima, K., Yamamoto, M., & Ichinose, D. (2016). How do agricultural markets respond to radiation risk? Evidence from the 2011 disaster in Japan. *Regional Science and Urban Economics, 60*, 20–30.

Tamssaouet, F., Nguyen, K., & Medjaher, K. (2019). System-level prognostics under mission profile effects using inoperability input-output model. *IEEE Transactions on Systems, Man, and Cybernetics: Systems, 51*(8), 4659–4669.

Tamssaouet, F., Nguyen, K., Medjaher, K., & Orchard, M. E. (2020). Online joint estimation and prediction for system-level prognostics under component interactions and mission profile effects. *ISA Transactions, 113*, 52–63.

Tanaka, K., & Managi, S. (2016). Impact of a disaster on land price: Evidence from Fukushima nuclear power plant accident. *Singapore Economic Review, 61*(1), 1640003.

Tsuboi, J. I., Abe, S. I., Fujimoto, K., Kaeriyama, H., Ambe, D., Matsuda, K., Enomoto,M., Tomiya,A., Morita,T., & Ono, T. (2015). Exposure of a herbivorous fish to 134Cs and 137Cs from the riverbed following the Fukushima disaster. *Journal of Environmental Radioactivity, 141*, 32–37.

Tsutom, M., & Hiroki, W. (2018). Who refuses safe but stigmatized marine products due to concern about radioactive contamination? *Fisheries Science, 84*, 1119–1133.

Wan, K. D., Liu, B. Y., Fan, Y., & Ikonnikova, S. A. (2024). Modelling and assessing dynamic energy supply resilience to disruption events: An oil supply disruption case in China. *Energy Economics, 140*, Article 108013.

Wang, C. Y., Cerrato, R. M., & Fisher, N. S. (2018). Temporal changes in 137Cs concentrations in fish, sediments, and seawater off Fukushima Japan. *Environmental Science and Technology, 52*(22), 13119–13126.

Xian, H., & Jeong, C. E. (2018). Modeling the damage and recovery of interdependent critical infrastructure systems from natural hazards. *Reliability Engineering and System Safety, 177*, 162–175.

Xiang, L. J. (2021). Japan has decided to discharge nuclear wastewater into the sea. *Ecological Economy, 37*(06), 1–4.

Xu, W., Cong, J., Xiang, L., & Proverbs, D. (2021). Vulnerability analysis of interdependent infrastructure systems based on inoperability input–output models. *ASCE-ASME Journal of Risk and Uncertainty in Engineering Systems, Part a: Civil Engineering, 7*(3), 05021002.

Yu, S., Myouse, H., Akiyama, S., Matsumoto, A., & Arakawa, H. (2019). Temporal variation of radionuclide contamination of marine plants on the Fukushima coast after the east Japan nuclear disaster. *Environmental Science and Technology, 53*(16), 9370–9377.

Zhang, C., Kong, J. J., & Simonovic, S. P. (2018a). Restoration resource allocation model for enhancing resilience of interdependent infrastructure systems. *Safety Science, 108*, 169–177.

Zhang, W. L., Lin, P. H., & Wang, N. Y. (2018b). Probabilistic prediction of post-disaster functionality loss of community building portfolios considering utility disruptions. *Journal of Structural Engineering, 144*(4), 04018015.

Zhang, S. F., Xu, S. B., Huang, D. H., Zhang, Z. H., Zheng, Y. K., Cheng, C., et al. (2025). Impact of US–China trade demand shocks on industrial Chain resilience: A knowledge and risk management perspective. *Journal of Innovation & Knowledge, 10*, Article 100830.

Zhao, C., Wang, G., Zhang, M., Wang, G., & Qiao, F. (2021). Transport and dispersion of tritium from the radioactive water of the Fukushima Daiichi nuclear plant. *Marine Pollution Bulletin, 169*, 119–129.

Part III
Management Strategies for Environmental Disasters in China

Chapter 15
Study on the Allocation of SO_2 Emission Rights in the Yangtze River Delta City Agglomeration Region of China Based on Efficiency and Feasibility

15.1 Introduction

With the further deepening of climate warming and industrialization, human beings are faced with worsening air pollution and other problems (Elkhalfi et al., 2025; Luo & Wang, 2026; Orisaremi et al., 2025). The Donora smog event in the United States and the 12 major smog events in London are related to the excessive emission of SO_2 (Chen et al., 2020; Nemery et al., 2001; Shepherd & Mullins, 2019). Until now, some countries are still facing severe air pollution problems. Such as India, almost the entire population of India lives above the red line of air pollution set by the World Health Organization (Agrawal et al., 2021; Gulia et al., 2022; Janarthanan et al., 2021). Atmospheric pollutant causes more than 3 million premature deaths worldwide each year, which may double by 2050 (Lelieveld et al., 2015). The correlation between mortality from respiratory diseases and SO_2 concentration reached 0.95 when the SO_2 concentration in the air reached 10 $\mu g/m^3$ (Abdul-Wahab et al., 2018; Orellano et al., 2021).

The Yangtze River Delta, as a region with strong economic and industrial strength in China, is also densely populated. In recent years, as the integration process of the Yangtze River Delta advances deeper, further improvements for air quality and air pollutant emission reduction are receiving more and more attention from the government and society (Miao et al., 2022). From 2000 to 2020, the total emission of SO_2 in China was about 375,268,400 tons, of which the emission of the Yangtze River Delta was about 45,466,200 tons, accounting for 12.12% of the total national emission (National Bureau of Statistics, 2000–2021). SO_2 emissions in the Yangtze River Delta region have shown a decreasing trend year by year under the background of China's active air pollutant emission reduction policies and emission rights research. According to China's National Bureau of Statistics (2010–2020), SO_2 emissions in the Yangtze River Delta region were reduced by 88.22% from 2010 to 2020. But in fact, the allocation of SO_2 emission rights among cities in the Yangtze River Delta

X. Wu, *Environmental Disaster in China*,
https://doi.org/10.1007/978-981-92-0910-1_15

is unbalanced, and the potential for source emission reduction has not been fully explored (Hirano et al., 2016; Huang, 2018). At present, the allocation of emission targets in existing policies and policy implementation have been mainly based on the production-based emission accounting principle (Wang et al., 2022). 49.65% of cities in China have great pressure on emission reduction due to the unbalanced allocation of pollutant emission rights (Cheng et al., 2020). The main reason for this imbalance is that the population, pollutant transfer, and industrial structure are not fully considered by policymakers (He et al., 2016). Moreover, the allocation of emission rights on regional level can promote environmental collaborative governance and save emission reduction costs, compared with the allocation of emission rights on the national, provincial or industry levels (Chang et al., 2022; Liang et al., 2018). In addition, sulfate aerosol formed after SO_2 is oxidized in the air is one of the main components of $PM_{2.5}$, and a large amount of SO_2 emission will lead to a significant increase in the concentration of $PM_{2.5}$ (Zhao et al., 2022). At the same time, SO_2 produced by fossil fuel combustion and its associated pollutants (PM_{10}, $PM_{2.5}$, CO, NO_2, etc.) have similar spatial and temporal distribution characteristics and a positive correlation in concentration (Mi et al., 2019). Therefore, the research on the allocation of SO_2 emission rights oriented to emission reduction also has reference significance for the control of SO_2 co-pollutants. Therefore, this paper attempts to provide the empirical evidence for the allocation of SO_2 emission rights in the Yangtze River Delta city agglomeration. This will promote the integrated development of regional economy and environment.

The rest of this paper is conducted as follows: Sect. 15.2 reviews the literature about emission rights allocation. Some methods of DEA, shadow price, information entropy, and RDC are commented, and data sources of 27 cities in the Yangtze River Delta city agglomeration are introduced in Sect. 15.3. The results of SO_2 emission rights allocation in Yangtze River Delta city agglomeration are interpreted and discussed in Sect. 15.4. The conclusions and suggestions are summarized in Sect. 15.5.

15.2 Literature Review

During emission rights allocation, decision-makers allocate emission rights to different emission units based on specific rules made by the overall emission reduction target (Huang & Xu, 2020). A reasonable SO_2 emission allocation scheme can promote the growth of regional GDP and reduce the input cost at the same time (Zhao et al., 2021). For example, the pollutants emission are lower in countries that followed benchmarking allowance allocation (Peng et al., 2021). Power generation costs can be further reduced by optimizing pollutant emission reduction schemes (Li et al., 2014). In addition, some scholars have found that a reasonable SO_2 emission allocation scheme can effectively reduce the mortality of respiratory and cardiovascular diseases in the population (Xie et al., 2016).

Regarding the allocation principle of emission rights, efficiency first, fairness first, and efficiency-fairness first are most considered in the literature. For example, a DEA model based on the principle of maximum average efficiency is established to allocate the carbon emission quotas of various provinces in China (Guo et al., 2021; Pan & Pan, 2018). The thermal power industry has caused a large amount of coal consumption and pollutant emissions. In order to achieve emission reduction, the principle of efficiency allocation and the principle of sustainability are adopted (Gadre & Anandarajah, 2019; Sun et al., 2018; Tan et al., 2014). Some scholars proposed that fairness is the basis of carbon quota allocation on the world level, which should be considered first in the allocation of emission rights worldwide (Ding et al., 2010; Ju et al., 2021). Obviously, considering both efficiency and fairness can improve the feasibility of allocation schemes. For instance, EU member states and Chinese provinces have adopted a dual-objective allocation model that takes into account efficiency and fairness (Chiu et al., 2015; Guo et al., 2021; Zhao et al., 2021).

In the application of allocation methods, most studies use a single research method. For example, a popular method, DEA is usually applied to allocate emission rights in industries or provincial-level regions (Cui & Li, 2021; Mahdiloo et al., 2018; Xie et al., 2022). The maximum deviation method is used to solve the imbalance of emission rights allocation among cities. For example, this method is applied to allocate city emission rights in the Pearl River Delta region based on the principle of sustainability (Liang et al., 2018). The shadow price allocation method is applied to the carbon emission rights scheme of Chinese industries, which provides a reference for promoting regional cooperation on emission reduction (Yu et al., 2019). In addition, the marginal cost of CO_2 emission based on shadow prices is an important basis for measuring OECD global carbon reduction (Cui et al., 2022; Shen et al., 2021). The information entropy allocation method is also widely used. For example, the allocation scheme of carbon emission rights in China's power industry is divided into two stages: Firstly, the information entropy method is used to deal with the initial quota allocation, and then the ZGS-DEA method is used to reallocate the results of the first stage to achieve dual objective allocation (Cui et al., 2021). Secondly, the information entropy method integrates multiple indicators such as history, demand, contribution, security, and fairness into comprehensive allocation indicators. This method above provides a reference for China to achieve carbon neutrality (Tian et al., 2022).

Fairness is one of the key indicators to evaluate the rationality of the allocation scheme. Some scholars proposed that fairness is a constraint condition to establish the dual-objective DEA model (Guo et al., 2021; Kremantzis et al., 2022; Xie et al., 2022). However, increasing the number of constraints may reduce the feasible region of the solution and DEA model's output-input efficiency (Hu et al., 2019). Therefore, fairness evaluation indicators are introduced in this paper to solve this problem. The main indicators to measure fairness include the Gini coefficient and the RDC. The Gini coefficient is widely used to measure the size of the income gap. In recent years, the Gini coefficient is also used to measure the fairness of emission rights allocation (He et al., 2022; Sueyoshi et al., 2021). The RDC is usually used to measure the degree of deprivation and convey individual satisfaction with the allocation results to decision makers. Thus, using the RDC appropriately is key for decision makers to adjust their allocation schemes. In recent years, the RDC has been widely used in fairness evaluation (Lin et al., 2022; Yang et al., 2019), because it can better reflect the inequality of intra-group allocation than the Gini coefficient (Hey & Lambert, 1980b). Thereby the RDC is applied in this paper to evaluate the fairness of emission rights allocation within city agglomerations.

According to the analysis above, the existing literature on the emission rights allocation has the following limitations:

1. The allocation that simply pursues the maximization of overall efficiency usually sacrifices individual interests (Deng & Du, 2020). From a fair point of view, each emission unit should have the same emission rights. But if so, the overall efficiency may not be guaranteed, making it more difficult to complete the pollutant emission reduction scheme. The feasibility of allocation schemes is improved based on fairness and efficiency, but it failed to consider the GDP cost of emission reduction units. This may lead to higher economic costs.
2. Most studies are usually based on the allocation of provincial emission rights but ignore regional city economy and the environment's balanced and sustainable development.
3. As a developing country, China is faced with the problem of unbalanced industrial structure, economic structure, and population structure throughout the eastern and western regions. Per capita quota is one of China's main principles of provincial emission rights allocation (Liu et al., 2018; Zhou et al., 2021). However, compared with the eastern region, the western region has more heavy industry distribution and more pollutant emissions, while they have a smaller population and worse economic development. Therefore, the principle of per capita quota at the province level may lack fairness.

For decision-makers, the allocation of emission rights needs to ensure maximum efficiency while reducing regional economic losses under the established emission reduction targets (Li et al., 2022b). However, the pollutant emission units care more about fairness (Yang et al., 2021). Furthermore, the DEA allocation model can only work out the optimal allocation efficiency, while the shadow price allocation method could only lead to the minimum regional economic loss. But how to ensure optimal efficiency while minimizing economic losses? An information entropy allocation method integrating multiple allocation targets is proposed, and the RDC is applied to evaluate the fairness of the scheme. Together, it will take into account the principle of maximum efficiency, minimum economic loss and fairness.

Based on the analysis above, Firstly, this paper selects 27 cities in the Yangtze River Delta city agglomeration as the research objects. Secondly, the regional cities'

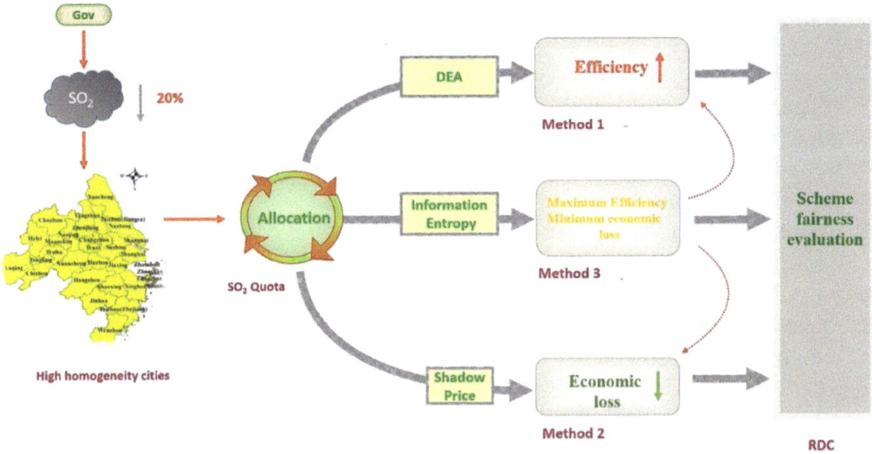

Fig. 15.1 Allocation method of SO_2 emission right in Yangtze River Delta city agglomeration

initial SO_2 emission quotas are figured out based on DEA and shadow price allocation. Thirdly, an information entropy allocation model is proposed to get the new SO_2 allocation scheme under the principle of maximum efficiency and minimum economic loss. Then, the RDC under the principle of fairness is applied to evaluate and verify the allocation results. Finally, the relevant policy suggestions are put forward.

The main contributions of this paper are as follows:

1. A three-objective information entropy allocation method based on DEA and shadow price is proposed. This method takes into account the principle of maximum efficiency and minimum economic loss, and allocates the SO_2 emission rights among cities in the Yangtze River Delta.
2. The allocation scheme proposed in this paper can better reflect the characteristics of air pollutants, which usually spill over into adjacent space, and the novel method embedded information entropy and relative deprivation coefficient can allocate SO_2 emission rights among units (Fig. 15.1).

15.3 Research Methods

First, the total amount of allocation is determined. The total planned future emissions of each city should be lower than the total historical emissions. Second, the allocation principle is determined. Different scholars have different understandings of allocation principle, which leads to different allocation schemes. Finally, a more reasonable and feasible allocation scheme is proposed, and its fairness is evaluated.

15.3.1 Estimation of Total Pollutant Emissions in the Yangtze River Delta City Agglomeration Based on Historical Emissions and Emission Reduction Tasks

During the 13th Five-Year Plan[1] period, the Chinese government issued a policy on energy conservation and emission reduction. The policy showed the actual emissions of various pollutants in each province in 2015 and the emission reduction ratio in 2020 (State Council of the PRC, 2017). After the announcement, the provinces set emission reduction targets applicable to all cities in the province during the 13th Five-year Plan period. Thus, the calculation method of the total emission quota of the Yangtze River Delta city agglomeration during the 13th Five-Year Plan period is proposed:

$$Y = \sum\nolimits_{i=1}^{n} (1 - \partial_i) D_i \tag{15.1}$$

where Y represents the total amount of pollutants to be allocated; $\partial_1 \ldots \partial_n$ represents the proportion of emission reduction for the n th city; D_i represents the historical pollutant emission of the i-th city.

15.3.2 DEA Allocation Model Based on Efficiency

Policymakers usually want to maximize the overall benefit while realizing the total emission target, that is, invest the least cost (including inputs such as human, resource consumption and fixed assets) to obtain the maximum desired output (GDP) and the least non-desired output (pollutants). Ahman et al. (2007) and Charnes et al. (1978) argued that non-desired output, as a negative output, should be as little as possible and they considered non-desired output as an input consumption The non-desired output is taken as one of the input elements. Quotas are generally regarded as positive indicators and each DMU wants to have as many quotas as possible, the formula is as follows (Berg et al., 1992; Sun et al., 2014):

[1] The 13th five-year plan generally refers to a plan on developing the national economy issued by the central government from 2015 to 2020.

$$
\begin{cases}
maxE = \frac{\sum_{j=1}^{n}(\sum_{r=1}^{k} a_r y_{rj} + \sum_{p=1}^{q} b_p z_{pj})}{\sum_{j=1}^{n}\sum_{i=1}^{m} c_i x_{ij}} \\
s.t. \\
\frac{\sum_{j=1}^{n}(\sum_{r=1}^{k} a_r y_{rj} + \sum_{p=1}^{q} b_p z_{pj})}{\sum_{j=1}^{n}\sum_{i=1}^{m} c_i x_{ij}} \le 1 \\
\frac{\sum_{r=1}^{k} a_r y_{rj} + \sum_{p=1}^{q} b_p z_{pj}}{\sum_{i=1}^{m} c_i x_{ij}} \le 1 \\
\sum_{j=1}^{n} y_{rj} = Y_r \\
y_{rj} \ge 0; j = 1 \dots n \\
b_p \ge 0; p = 1 \dots q \\
c_i \ge 0; i = 1 \dots m \\
a_r \ge 0; r = 1 \dots k
\end{cases}
\tag{15.2}
$$

where a_r is the weight of the final quota for the r-th non-desired output; b_p is the weight of pth desired output; c_i is the i-th input weight; x_{ij} is the i-th input of the j-th decision making unit (DMU); y_{rj} is the quota obtained for the r-th non-desired output of the j-th DMU; z_{pj} is the p-th desired output of the j-th DMU; Y_r represents the total quota for the r-th non-desired output.

The model is solved in three steps.

Step 1: Transform the above model into a linear model.

Let $a_r y_{rj} = m_{rj} \ge 0$, then $\sum_{j=1}^{n} a_r y_{rj} = \sum_{j=1}^{n} m_{rj} = a_r Y_r$, The model is transformed into:

$$
\begin{cases}
maxE = \frac{\sum_{j=1}^{n}(\sum_{r=1}^{k} m_{rj} + \sum_{p=1}^{q} b_p z_{pj})}{\sum_{j=1}^{n}\sum_{i=1}^{m} c_i x_{ij}} \\
s.t. \\
\frac{\sum_{j=1}^{n}(\sum_{r=1}^{k} m_{rj} + \sum_{p=1}^{q} b_p z_{pj})}{\sum_{j=1}^{n}\sum_{i=1}^{m} c_i x_{ij}} \le 1 \\
\frac{\sum_{r=1}^{k} m_{rj} + \sum_{p=1}^{q} b_p z_{pj}}{\sum_{i=1}^{m} c_i x_{ij}} \le 1 \\
\sum_{j=1}^{n} m_{rj} = a_r Y_r \\
m_{rj} \ge 0; j = 1 \dots n \\
b_p \ge 0; p = 1 \dots q \\
c_i \ge 0; i = 1 \dots m \\
a_r \ge 0; r = 1 \dots k
\end{cases}
\tag{15.3}
$$

Step 2: To facilitate calculation and discussion, the fractional planning of model (15.3) is transformed by the Charnes-Cooper transformation to an integer planning model (Charnes et al., 1978):

Let $\sum_{j=1}^{n}\sum_{i=1}^{m} c_i x_{ij} = \frac{1}{H}$, $Ha_r = A_r$, $Hb_p = B_p$, $Hc_i = C_i$, $Hm_j = M_j$,

Then the model (15.3) can be transformed into the model (15.4):

$$\begin{cases} maxE = \sum_{j=1}^{n}(\sum_{r=1}^{k} M_{rj} + \sum_{p=1}^{q} B_p z_{pj}) \\ s.t. \\ \sum_{j=1}^{n}\sum_{i=1}^{m} C_i x_{ij} = 1 \\ \sum_{j=1}^{n}(\sum_{r=1}^{k} M_{rj} + \sum_{p=1}^{q} B_p z_{pj}) \le 1 \\ \sum_{r=1}^{k} M_{rj} + \sum_{p=1}^{q} B_p z_{pj} - \sum_{i=1}^{m} C_i x_{ij} \le 0 \\ \sum_{j=1}^{n} M_{rj} = A_r Y_r \\ M_{rj} \ge 0; j = 1 \ldots n \\ B_p \ge 0; p = 1 \ldots q \\ C_i \ge 0; i = 1 \ldots m \\ A_r \ge 0; r = 1 \ldots k \end{cases} \quad (15.4)$$

Step 3: The optimal solution is calculated according to model (15.4) as: (A_r, B_p, C_i, M_{rj}), then the maximum allocation obtained for the j-th DMU is:

$$y_j = \frac{m_{rj}}{a_r} = \frac{M_{rj}}{A_r} \quad (15.5)$$

15.3.3 Shadow Price Allocation Model Based on Feasibility

When implementing allocation programs, policymakers must consider allocating limited quotas to economically productive cities to ensure both individual efficiency and the feasibility of allocation schemes. Non-desired and desired outputs usually follow the principle of weak disposability, i.e., non-desired outputs are usually reduced at the expense of desired outputs. The shadow price represents the scarcity of a resource at a certain optimum, which is interpreted as the marginal abatement cost of non-desired output in this paper. That is, the amount of desired output sacrificed for each unit of reduction in undesired output. In recent years, some scholars have applied shadow prices to the allocation of pollutant emission rights. The shadow price of each DMU is calculated through the input–output table, and the shadow price is used as the basis for quota allocation (Deng & Du, 2020; Wang et al., 2018). The higher the shadow price, the scarcer the emission quotas in the region, and the higher the expected output cost paid by pollutant emitters to reduce emissions. Therefore, priority should be given to allocate more quotas. This principle can further improve the feasibility of the allocation scheme and reduce the resistance of pollutant emission subjects to the allocation scheme. This paper establishes a shadow price model based on the directional distance function to calculate the shadow price of each DMU, and finally allocates pollutant emission quota in the same proportion according to the shadow price.

Suppose there are M input elements, and they are denoted as: $x = (x_1, \ldots, x_M) \in R_M^+$;$N$ desired output, and they are denoted as: $y = (y_1, \ldots, y_N) \in R_N^+$; J non-desired output, and they are denoted as: $z = (z_1, \ldots, z_J) \in R_J^+$. The change from desired output to non-desired output is represented by the directional vector $g = (g_y, -g_z)$, and $g \in R_N^+ \cdot R_J^+$. The directional vector can be set according to the study, and in this paper we set the directional vector as $g = (1, -1)$. Then the output-based directional distance function is:

$$\vec{D}_O\left(x, y, z;\ g_y, -g_z\right) = max\left\{\beta : \left(y + \beta.g_y, z - \beta.g_z\right) \in P(x)\right\} \tag{15.6}$$

where β is the value of the directional distance function and it indicates the gap between individual pollutant emission units relative to the frontier environmental technical level, that is, the maximum share of the expected output that can be increased with constant inputs. The larger the β, the lower the environmental technical efficiency of the pollutant emission unit, and vice versa. When $\beta = 0$, the pollutant emission unit is at the advanced environmental technical production level. $P(x)$ represents the environmental technical production feasibility set.

Based on the weak disposability principle of desired output and non-desired output, combining desired and non-desired outputs according to Lagrange theorem and envelope theorem, the formula for calculating shadow price is obtained:

$$p_j = -p_y \cdot \left[\frac{\frac{\partial \vec{D}_O\left(x, y, z;\ g_y, -g_z\right)}{\partial z}}{\frac{\partial \vec{D}_O\left(x, y, z;\ g_y, -g_z\right)}{\partial y}}\right] \tag{15.7}$$

where p_y is the price vector of desired output; p_j is the shadow price of non-desired output, i.e., the amount of GDP reduction lead by one unit reduction in pollutants.

Finally, pollutant emission quotas are allocated in equal proportions based on shadow prices:

$$y_j = \frac{p_j}{\sum_{j=1}^{n} p_j} Y \tag{15.8}$$

where y_j is the quota received by the j-th pollutant emission unit; Y is the total amount of pollutants to be allocated.

15.3.4 An Information Entropy Weighted Allocation Method Considering Efficiency and Feasibility

The methods for calculating index weights are usually divided into the analytic hierarchy process and the information entropy method (Cui et al., 2021; Wu et al., 2018). Compared with the analytic hierarchy process, the information entropy method can better avoid the subjectivity when assigning weights as well as can accurately and objectively determine the contribution of each indicator in the system to the total evaluation. The smaller the information entropy, the more informative and useful the indicator is in the pollutant allocation system, and the larger the weighting coefficient. Therefore, this method is adopted, and a comprehensive allocation scheme with both efficiency and feasibility is obtained by calculating the information entropy weight coefficients of allocation schemes in Sects. 15.3.2 and 15.3.3.

The steps to construct a pollutant allocation model using the information entropy weighting method are as follows:

Step 1: Establish the original matrix Q consisting of pollutant emission units (j rows) and allocation schemes (t columns), as shown in Table 15.1.

Step 2: The max–min method is used to standardize the values of each allocation element in the matrix to obtain the standardized matrix P.

$$X_{mn} = \frac{x_{mn} - min\{x_{mn}\}}{max\{x_{mn}\} - min\{x_{mn}\}} \tag{15.9}$$

Step 3: The standardized matrix P is weighted and assigned.

$$E(X)_j = -\frac{\sum_{i=1}^{j} p_{mn} ln p_{mn}}{ln(t)} \tag{15.10}$$

where

$$p_{mn} = \frac{x_{mn}}{\sum_{i=1}^{j} x_{mn}} \tag{15.11}$$

if $p_{mn} = 0, \lim_{p_{mn}\to 0} p_{mn} ln p_{mn} = 0.$

$$\omega_n = \frac{1 - E(X)_n}{\sum_{i=1}^{t} [1 - E(X)_n]} \tag{15.12}$$

The information entropy $E(X)_1, E(X)_2, \ldots, E(X)_j$ of each allocation element is calculated according to Eq. (15.10), and finally the information entropy value is substituted into Eq. (15.12) to obtain the information entropy weight coefficient ω_n of the n-th allocation element.

Step 4: Calculate the pollutant emission right allocation for the m-th DMU based on the information entropy weight coefficient ω_n.

Table 15.1 Original matrix

DMU to be allocated	Allocation factor 1	Allocation factor 2	…	Allocation factor t tTT1t
DMU_1	X_{11}	X_{12}	…	X_{1t}
DMU_2	X_{21}	X_{22}	…	X_{2t}
…	…	…		
DMU_j	X_{j1}	X_{j2}	…	X_{jt}

$$Y = \begin{bmatrix} y_1 = \omega_1 X_{11} + \omega_2 X_{12} + \cdots \omega_t X_{1t} \\ y_2 = \omega_1 X_{21} + \omega_2 X_{22} + \cdots \omega_t X_{2t} \\ \cdots \\ y_j = \omega_1 X_{j1} + \omega_2 X_{j2} + \ldots \omega_t X_{jt} \end{bmatrix} \tag{15.13}$$

where $y_1, y_2, \ldots, y_j$ are the quotas of the information entropy allocation.

15.3.5 Fairness Evaluation of Allocation Scheme Based on Per Capita RDC

Hey and Lambert (1980a) proposed the theory of relative deprivation to measure the feelings of deprivation generated by dissatisfaction and other emotions caused by the occurrence of unfairness in social clusters. The magnitude of relative deprivation is usually characterized by the deprivation coefficient, and the larger the deprivation coefficient, the stronger the deprivation feeling of an individual. This theory is applied to calculate the RDC of pollutant emission units to evaluate the fairness of different allocation schemes in this paper.

The RDC is calculated as follows.

Step 1: Assume that A and B are two comparison clusters, and the sample vectors of the two clusters are $a = (a_1, a_2, \ldots, a_m)$ and $b = (b_1, b_2, \ldots, b_n)$, a_s and b_t denote individual s in cluster *A* and individual t in cluster *B*. Compared individual s with individual t, the relative deprivation is:

$$d_{AB}(a_s, b_t) = \begin{cases} a_s - b_t, \ if \ a_s > b_t \\ 0 \qquad\qquad else \end{cases} \tag{15.14}$$

Step 2: Assume that the number of individuals in cluster *A* with emission right allocations greater than that of b_t is m_+, the mean of individuals greater than b_t is $\overline{a}_+$. The ratio of the total number of samples in cluster *A* greater than that of b_t to the total number of samples in cluster *A* is ϕ. Then the RDC in cluster *B* for cluster *A* is expressed as:

$$\begin{aligned} d(a, b_t) &= \frac{\sum_{i=1}^{m} d_{AB}(a_s, b_t)}{m} \\ &= \frac{m_+\overline{a}_+ \ - m_+ b_t}{m} \\ &= \phi(\overline{a}_+ - b_t) \end{aligned} \tag{15.15}$$

The above is the calculation method of the RDC among clusters. In this paper, the RDCs between cities within the Yangtze River Delta region need to be calculated, i.e., each city needs to be compared with other cities. Taking cluster of *A* as an example, formula (15.15) is transformed into $d(a_s, a_t)$, where $a_s \neq a_t$. Then the RDCs of each

Table 15.2 Warning interval for the RDC

RDC	Corresponding interval
$D < 0.2$	No warning
$0.2 \leq D < 0.3$	Mild warning
$0.3 \leq D < 0.4$	Moderate warning
$0.4 \leq D < 0.5$	Severe warning
$D > 0.5$	Serious warning

city are:

$$d(a_s, a_t) = \frac{\sum_{s=1}^{m} d_{AB}(a_s, a_t)}{m\overline{a}} = \frac{\phi(\overline{a}_+ - a_t)}{\overline{a}} \quad (15.16)$$

Gini coefficient is generally recognized as a measure of allocation fairness at home and abroad. Ren et al. showed that the Gini coefficient has many limitations in its application when decomposed by clusters, while the RDC can calculate both the relative deprivation of high-income clusters to low-income clusters and the relative deprivation of low-income clusters to high-income clusters (Ren & Shang, 2011). Yitzhaki (1979) studied the relationship between the RDC and the Gini coefficient and concluded that the Gini coefficient can be explained by the RDC. FU (2011) verified the relationship between the RDC and the Gini coefficient from the perspective of the Chinese condition. The results showed that the RDC can be used to characterize the Gini coefficient in China and proposed an early warning interval for the RDC (Table 15.2). The RDC's threshold in China is 0.5, which is applied to this paper, i.e., the feasibility of the allocation scheme decreases when the RDC of each city is greater than 0.5.

15.3.6 Regional Emission Reduction Pressure

To deeply understand which provincial regions need to increase emission reduction, we define the emission reduction pressure (ERP) based on historical emission data and emission right reallocation data of provincial regions:

$$ERP_i = \frac{U_l}{Y} - \frac{V_l}{\sum_{i=1}^{n} D_i} \quad (15.17)$$

where U_l is the l-th provincial region's historical pollutant emission; V_l is the l-th provincial region's allocation quotas. The positive and negative values of ERP indicate the pressure of emission reduction in each provincial region. When the ERP value is greater than zero, it means that the provincial region should increase emission reduction in the future, and vice versa.

15.3.7 Indicator Selection and Data Sources

In the Yangtze River Delta city agglomeration, most cities are located in the Yangtze River Delta plain area with relatively low altitudes (Wang et al., 2020). For example, among the cities in the Zhejiang Provincial region, only parts of Wenzhou and Jinhua are located in mountainous and hilly areas, and the rest of the cities are distributed in the northern Zhejiang plain and coastal plain. Cities in the Jiangsu Provincial region are all located in the Yangtze River Delta Plain. In the Anhui Provincial region, only parts of Anqing, Chizhou and Xuancheng are located in hilly areas, and the rest of the cities are distributed in the Yangtze River Delta Plain. In terms of climate and environment, the Yangtze River Delta city agglomeration belongs to the subtropical monsoon climate zone. The southeasterly wind prevails in summer and the northwesterly wind prevails in winter, and its dominant wind direction is relatively stable (Yu et al., 2022). In addition, the differences in precipitation, temperature and humidity among the Yangtze River Delta city agglomerations are also small. On the whole, the meteorological characteristics of most cities are relatively similar, and the homogeneity between cities is relatively high. Therefore, this paper studies the allocation of SO_2 emission rights under the condition that the cities in the Yangtze River Delta city agglomeration are independent of each other. A total of 27 cities in the Yangtze River Delta city agglomeration are selected as research objects in this paper and divided into 3 provincial regions and Shanghai, as shown in Fig. 15.2.

In DEA-based pollutant allocation models, input variables are usually indicators such as energy consumption, fixed asset inputs, and labor inputs, and output variables are usually GDP and pollutant production (Momeni et al., 2019; Xie et al., 2019). The input variables are energy consumption, city labor input, and city fixed asset input, and the output variables are city GDP and city SO_2 emissions. Energy consumption refers to the annual consumption of raw coal by city enterprises. Investment in fixed assets refers to the amount of fixed assets invested in the city in a whole year, and labor input is the total number of individuals employed, city units employed and private enterprises employed in each city throughout the year. Among the output variables, annual city GDP is the desired output, the larger the better, and annual city SO_2 emissions are considered as non-desired output, the smaller the better.

The three input variables and two output variables in this paper are all from the annual statistical yearbooks of cities in the Yangtze River Delta. In 2017, the Chinese government made clear the emission reduction targets of various provinces and cities in the next five years from 2015 levels (State Council of the PRC, 2017). Therefore, the data of 2015 are selected as the research object, and the emission rights for the next five years are allocated according to the emission reduction target announced by the Chinese government. Appendix A shows the raw data of raw coal consumption, fixed asset input, labor input, GDP and SO_2 emissions after standardization.

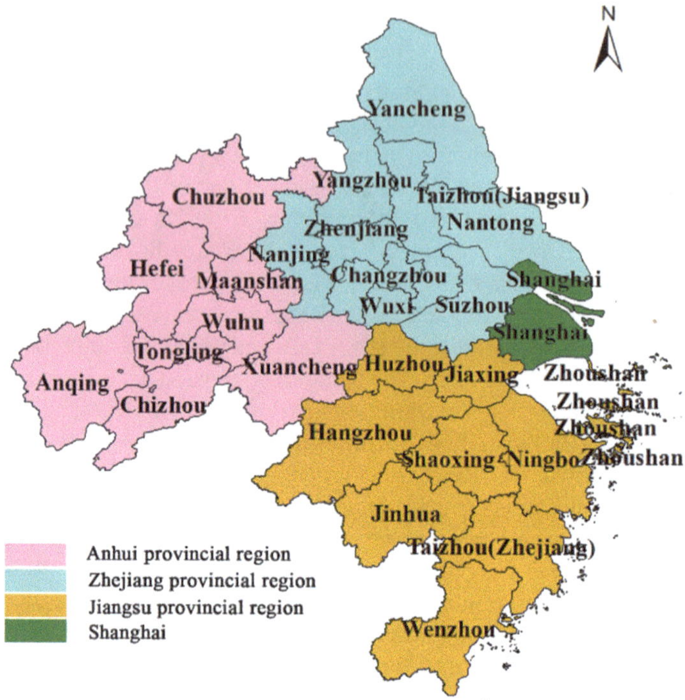

Fig. 15.2 27 cities in the Yangtze River Delta city agglomeration

15.4 Results and Discussions

15.4.1 Comparison of Allocation Results Based on Different Allocation Schemes

In 2015, the actual total SO_2 emission of the Yangtze River Delta city agglomeration was 1.4966 million tons. According to the emission reduction required by the central government, the SO_2 emission reduction ratio of the Yangtze River Delta city agglomeration by 2020 will be 20%. That is, after five years of emission reduction, the Yangtze River Delta's emission quotas in 2020 will be 1.2057 million tons. The methods of DEA, shadow price and information entropy proposed in this paper are applied to allocate the SO_2 emission quotas during the 13th Five-Year Plan period based on the input–output data of 2015. The allocation results are shown in Fig. 15.3.

DEA allocation method is based on efficiency maximization, that is, the higher the output-input ratio of the decision-making unit, the more emission rights will be allocated. Compared with shadow price allocation, the DEA allocation results are less volatile. Ningbo and Suzhou were allocated the highest amount, reaching 64.8 thousand tons, and Chizhou was the city with the lowest amount, only 24.8 thousand

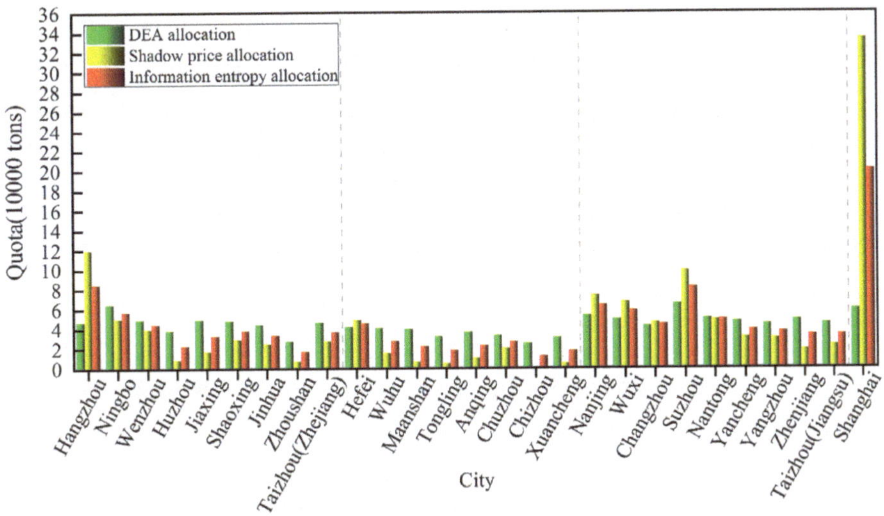

Fig. 15.3 Allocation results based on three allocation schemes

tons. In the Yangtze River Delta, the development speed of Anhui Province is slower than Shanghai, Zhejiang and Jiangsu. Its average city quota is only 34,900 tons, which is 75.7%, 70.5% and 59.1% of the average quota of Jiangsu, Zhejiang and Shanghai respectively.

The shadow price allocation method takes economic feasibility as the basic principle. Considering that the allocation of shadow price is aimed at protecting GDP, the more impact of emission reduction unit of pollutants on city GDP, the more quota will be allocated. It implies that cities with smaller losses to GDP from pollutant abatement should be allocated fewer quotas. To some extent, cities with smaller shadow prices may have to pay for cities with larger shadow prices to ensure the lowest total loss of regional economy. From the individual perspective, cities with higher shadow prices always expect more quotas, or they may be unsatisfied with the allocation scheme. The allocation results show that Shanghai has the highest allocation with 333,100 tons, indicating that Shanghai's GDP has a high dependence on SO_2 emissions. Overall, the average values of allocations in Anhui, Zhejiang and Jiangsu provinces are 13,600 tons, 35,800 tons and 48,400 tons, which means that the economic development of Jiangsu province is more dependent on SO_2 emissions.

The allocation scheme of information entropy considers multiple indicators. The importance of indicators and weight coefficients are determined by comparing the information provided by each allocation scheme, and finally the optimal scheme is formed by integrating multiple schemes. The information entropy allocation scheme takes into account the efficiency and makes sure that the impact of SO_2 emission reduction on the overall economy is minimized. The results show that Shanghai has the highest allocation of 206,600 tons, and Chizhou has the lowest allocation of

12,000 tons. These two cities belong to the two most and least economically developed cities in the Yangtze River Delta city agglomeration, with a GDP of 2,512,345 million and 544,474 million respectively in 2015, and this allocation result is more in line with expectations from the perspective of city economic development. Hangzhou and Suzhou have relatively similar city size and industrial structure, and their allocations are less different. On the whole, the average value of allocation of Anhui, Zhejiang and Jiangsu is 24.0 thousand tons, 40.8 thousand tons and 40.9 thousand tons, and the average allocation quotas of cities in Anhui Province is only 54.1% of the average allocation value of the Yangtze River Delta city agglomeration.

15.4.2 Fairness Evaluation of Allocation Scheme Based on Per Capita RDC

Table 15.3 shows the RDCs of the cities within the Yangtze River Delta city agglomeration under the three allocation schemes. (Pan et al., 2014) pointed out that cumulative emissions per capita can best reflect the principle of fairness and justice Therefore, the fairness and rationality of the allocation scheme are objectively measured by the per capita RDC. In order to present more visually the differences in quotas and corresponding per capita RDCs of each city under the three allocation schemes, a comparative analysis is made using GIS (Fig. 15.4). When the RDC is greater than 0.5, it shows serious a warning, indicating that the individual's resistance to the allocation scheme will cause the allocation scheme cannot be implemented. Based on the DEA, the RDC in Shanghai is 0.77, and the per capita RDCs in the rest of the cities are below 0.5, which indicates that considering the efficiency principle, only Shanghai is dissatisfied with the fairness of the allocation scheme. Based on the shadow price allocation method, six cities have relative deprivation factors exceeding 0.5, making this allocation scheme impossible to implement. The shadow price allocation is oriented to protect the economy, resulting in cities with large city population bases and less economic dependence on SO_2 emissions are deprived. The information entropy allocation method as a comprehensive allocation scheme, the RDC of all cities is less than 0.5, which means that all 27 cities in the Yangtze River Delta city agglomeration are acceptable to the fairness of the allocation scheme.

To examine the emission reduction pressure of each provincial region under the information entropy allocation scheme, we compared the proportion changes between historical SO_2 emissions in 2015 and the quotas of information entropy allocation scheme (Fig. 15.5). These changes reflect the pressure of SO_2 emission reduction in each province in the future (Table 15.4). The results show that the SO_2 emission of Jiangsu province in 2015 is 616,300 tons, accounting for 41.18% in the Yangtze River Delta city agglomeration, while it only accounts for 36.72% under the information entropy allocation scheme. Therefore, the information entropy allocation scheme will promote Jiangsu to increase emission reduction, Anhui and Zhejiang

Table 15.3 The RDC of three allocation schemes

Serial number	City	Population size (10 thousand people)	SO_2 per capita emission quotas (10^{-4} tons)			Per capita RDC		
			DEA allocation	Shadow price allocation	Information entropy allocation	DEA allocation	Shadow price allocation (Active set valid set)	Information entropy allocation
1	Hangzhou	723.55	65.33	165.30	116.98	0.39	0.00	0.03
2	Ningbo	586.57	110.42	85.26	97.42	0.16	0.10	0.05
3	Wenzhou	811.21	60.83	48.82	54.62	0.43	0.32	0.34
4	Huzhou	263.71	146.25	32.22	87.33	0.08	0.52	0.08
5	Jiaxing	349.48	140.53	47.71	92.58	0.09	0.33	0.06
6	Shaoxing	443.11	108.58	65.28	86.21	0.17	0.20	0.09
7	Jinhua	478.09	92.06	50.62	70.65	0.24	0.31	0.19
8	Zhoushan	97.36	280.00	73.63	173.37	0.00	0.15	0.00
9	Taizhou (Zhejiang)	597.49	77.82	46.02	61.39	0.31	0.35	0.27
10	Hefei	717.72	58.18	67.71	63.11	0.45	0.18	0.25
11	Wuhu	384.79	106.05	40.24	72.04	0.18	0.42	0.17
12	Maanshan	228.50	173.52	26.90	97.76	0.04	0.59	0.05
13	Tongling	170.43	189.54	25.63	104.85	0.03	0.61	0.04
14	Anqing	525.48	69.28	19.00	43.30	0.37	0.70	0.47
15	Chuzhou	449.06	73.86	45.26	59.08	0.34	0.36	0.29
16	Chizhou	161.61	153.60	0.00	74.24	0.06	1.00	0.16
17	Xuancheng	279.95	110.49	16.18	61.76	0.16	0.75	0.27
18	Nanjing	823.59	64.86	89.01	77.34	0.40	0.09	0.14

(continued)

Table 15.3 (continued)

Serial number	City	Population size (10 thousand people)	SO_2 per capita emission quotas (10^{-4} tons)			Per capita RDC		
			DEA allocation	Shadow price allocation	Information entropy allocation	DEA allocation	Shadow price allocation (Active set valid set)	Information entropy allocation
19	Wuxi	651.10	75.71	102.61	89.61	0.33	0.06	0.07
20	Changzhou	470.14	91.17	97.62	94.51	0.24	0.07	0.06
21	Suzhou	1061.60	61.03	92.50	77.29	0.43	0.08	0.14
22	Nantong	730.00	68.89	66.36	67.58	0.37	0.19	0.21
23	Yancheng	722.85	65.59	42.67	53.75	0.39	0.39	0.35
24	Yangzhou	448.36	98.80	67.23	82.49	0.21	0.18	0.11
25	Zhenjiang	317.65	154.30	58.49	104.80	0.06	0.25	0.04
26	Taizhou (Jiangsu)	464.16	96.43	49.74	72.31	0.22	0.32	0.17
27	Shanghai	2415.27	24.45	137.91	83.07	0.77	0.02	0.10

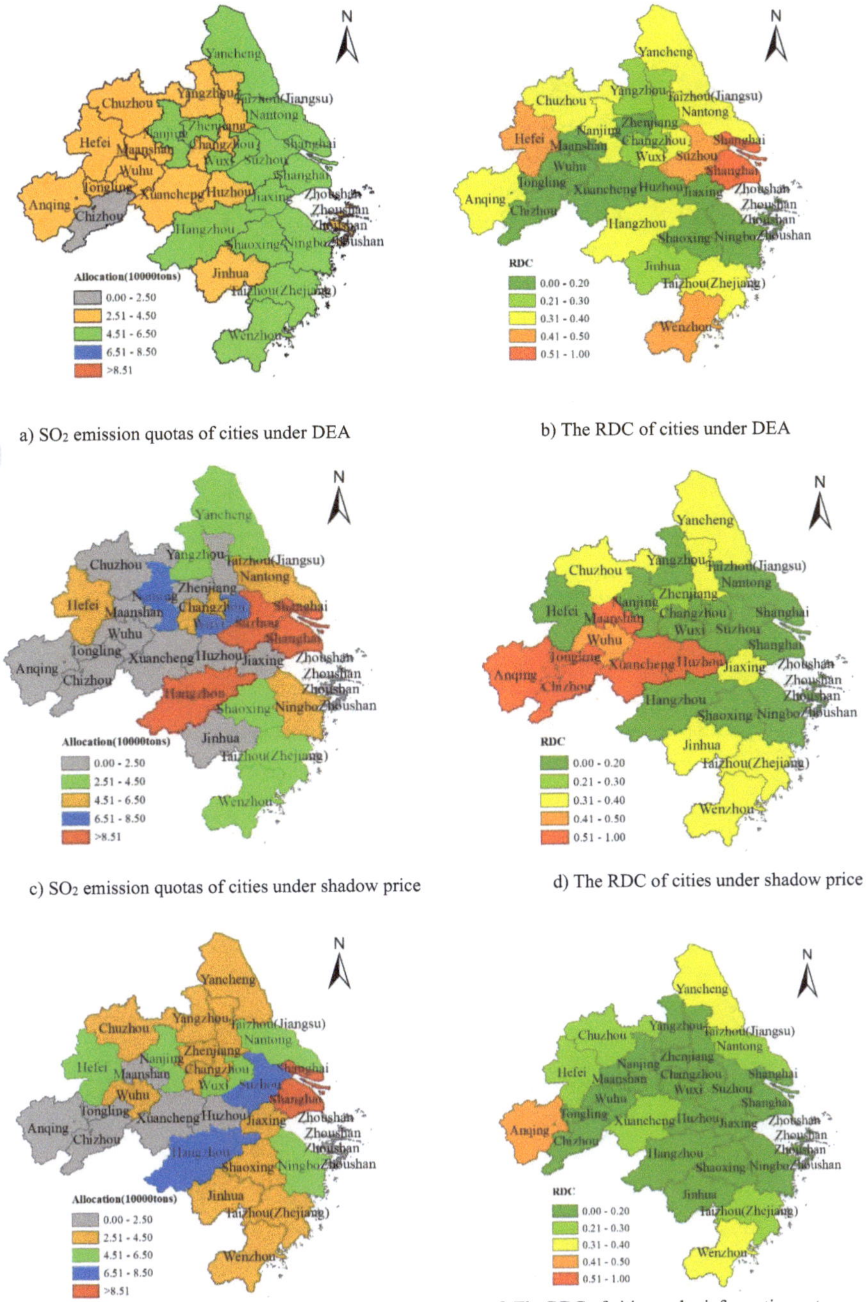

a) SO_2 emission quotas of cities under DEA

b) The RDC of cities under DEA

c) SO_2 emission quotas of cities under shadow price

d) The RDC of cities under shadow price

e) SO_2 emission quotas of cities under information entropy

f) The RDC of cities under information entropy

Fig. 15.4 SO_2 emission quotas and RDC of cities under different allocation methods

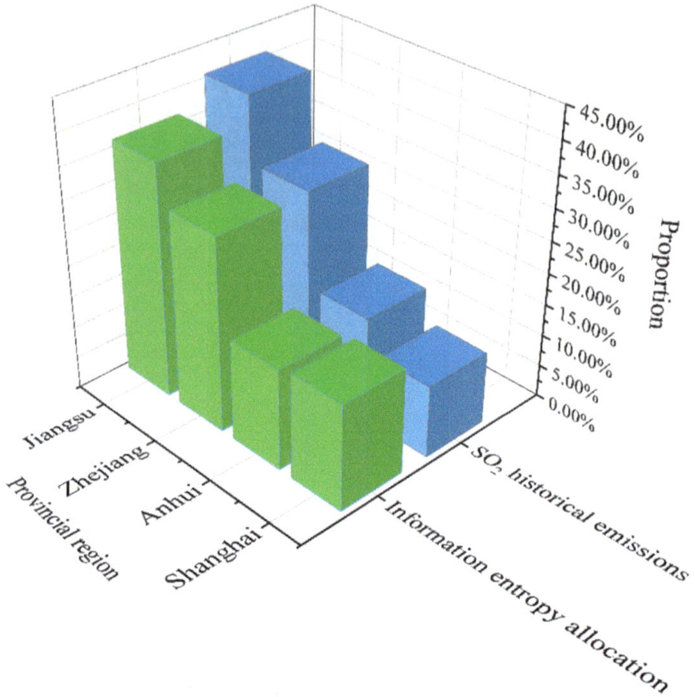

Fig. 15.5 Proportion of each provincial region in historical emissions and information entropy allocation

Table 15.4 SO_2 emission reduction pressure in various provincial regions

Provincial region	Emission reduction pressure (%)
Jiangsu	4.46
Zhejiang	0.47
Anhui	0.36
Shanghai	−5.29

provincial region's emission reduction pressure approaches zero, and Shanghai will have less pressure on emission reduction (from 11.43% to 16.72%).

15.4.3 Correlation Analysis

Which indicators will have a greater impact on the information entropy allocation quotas? To solve this problem, six indicators are selected including coal consumption, fixed asset input, labor input, GDP, population size and SO_2 emission, and the

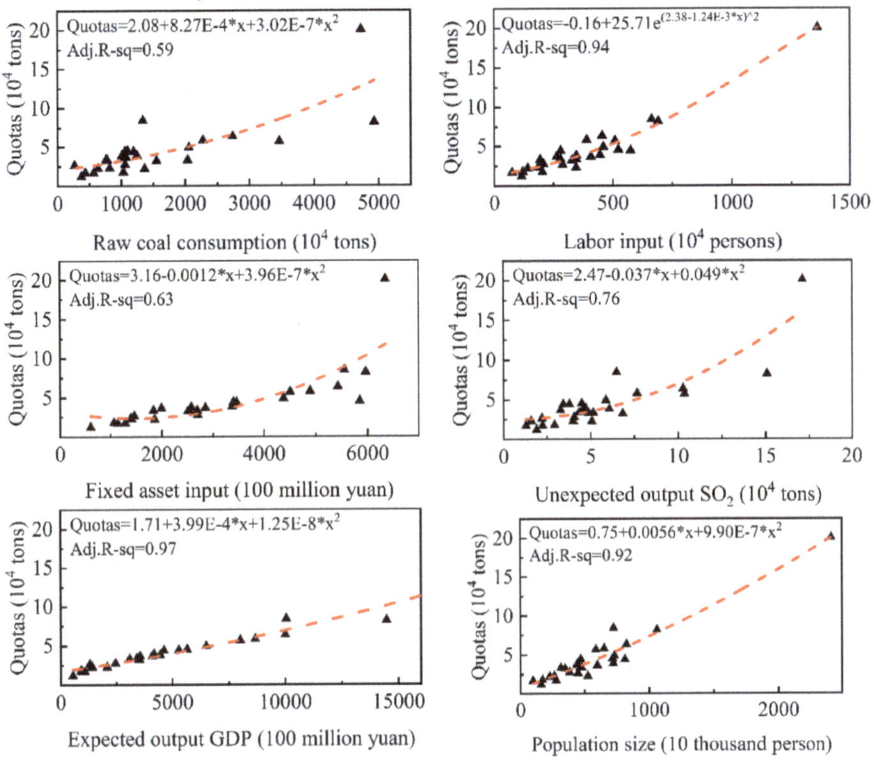

Fig. 15.6 The correlation between various indicators and information entropy allocation quotas

correlation between these indicators and allocated quotas are analyzed by calculating the Pearson correlation coefficient (Fig. 15.6).

The correlation analysis results show that GDP, labor input and population size are strongly correlated with allocation quotas, and the correlation coefficient between GDP and quota is 0.97. There is a medium correlation between fixed asset input, SO_2 emission and allocation quotas, and the correlation coefficients are 0.76 and 0.63 respectively. However, surprisingly, there is a weak correlation between coal consumption and allocation quotas, with a correlation coefficient of 0.59. This means that in the allocation scheme of this paper, more coal consumption does not mean more SO_2 emission quotas. Therefore, increasing coal consumption to obtain more allocation quotas is not encouraged in the allocation scheme of this paper.

This research's allocation results are similar to the carbon emission rights allocation in the Pearl River Delta city agglomeration based on the maximum deviation method (Liang et al., 2018), but there are some differences. Both allocation results show that the Pearson coefficients of GDP and population between carbon quotas are 0.98 and 0.94, respectively. In this paper, these two indicators are 0.97 and 0.92, respectively. The differences are as follows: Liang et al. (2018) only focused on the same province, and they selected GDP, population and historical emissions as the research indicators. However, this paper's research objects involve three provinces and one city in the Yangtze River Delta, and select seven research indicators as research indicators. Therefore, this paper's results are more transferable, applicable and robust.

15.5 Conclusions and Suggestions

15.5.1 Conclusions

A comprehensive allocation model combining maximum efficiency (DEA allocation method) and economic feasibility (shadow price allocation method) is established based on information entropy method. Three allocation schemes are evaluated through the per capita RDC. Finally, through the index correlation analysis, the following conclusions are drawn:

1. The DEA model fairness evaluation results show that the RDC of Shanghai is 0.77, and that of other cities is less than 0.5. The evaluation results of the shadow price allocation model show that the RDC of six cities is greater than 0.5. The fairness evaluation results of the information entropy allocation model show that the RDC of all cities is less than 0.5, which is the best reference scheme for the allocation of SO_2 emission rights in the Yangtze River Delta city agglomeration.
2. The results of the new allocation scheme show that Jiangsu provincial region's SO_2 emission reduction pressure is 4.46%, while Shanghai's emission reduction pressure is −5.29%. Anhui and Zhejiang provincial region's emission reduction pressure approaches zero.
3. GDP, labor input and city population size are strongly correlated with SO_2 emission quota. More emission quotas are obtained by increasing GDP, labor input and city population size, but the SO_2 quota for each city will not be increased as coal consumption increases.
4. Cities with high homogeneity have similar industrial scale and economic structure. Taking such regions as the research object can further promote the fairness and rationality of allocation scheme. The study of city agglomeration with high homogeneity has more realistic guiding significance.

This paper establishes a three-objective allocation scheme from the perspective of decision-makers and pollutant emission units, which further enriches the SO_2 emission right allocation method at the regional level, improves cities' satisfaction with the allocation results, and reduces the antagonism. The research results provide a new solution and analytical framework for rationally formulating SO_2 emission reduction targets in the Yangtze River Delta region, which can stimulate the emission reduction potential of regional city agglomerations and promote the sustainable development of the regional environment.

Moreover, this paper's emission allocation research method can also be applied to emission reduction research in other regions. The "Paris Agreement" in 2016, signed by a group of countries, sets different emission reduction targets for developed, developing and less developed countries. This is very similar to the research idea of this paper. "China's 13th five-year plan" proposes a scheme to restructure 19 city agglomerations, which are becoming an important carrier for China's regional integration development (Li et al., 2022a). Therefore, this paper's research ideas and methods are popularized and applied in the Pearl River Delta, Beijing-Tianjin-Hebei and other city agglomerations in the future. Compared with the provincial emission rights allocation, the research results are of great significance for establishing a fair and effective initial emission rights allocation frame, achieving the national total pollutant control target, and promoting high-quality development in China and the world.

Although the allocation scheme of pollutant emission rights at the regional level provides a new solution for decision-makers, there are still some limitations in this study to be further improved in the future. First, coal consumption is selected as the energy input index in this paper, but other factors will also lead to increasing SO_2 emissions, such as traffic emissions, which are also worthy of consideration. Second, this paper assumes that all cities are independent of each other, and the transfer of pollutants between cities will be considered in future studies. Besides, from a health perspective, cities with higher population densities deserve less SO_2 emission quotas. However, China is a developing country, and population density usually has a positive correlation with GDP. In other words, residents' health and sustainable economic development usually have weak disposability (Chaabouni & Saidi, 2017; Wang & Li, 2021). The allocation of SO_2 emission rights proposed in this paper is mainly based on city agglomerations' economic and environmental sustainability. How to further explore the relationship between population health, resources, environment and sustainable economic development in the study of SO_2 emission rights allocation is the focus of our future research.

15.5.2 Suggestions

Based on this research and the dynamics of pollutant allocation, some emission reduction recommendations for governments and other development agencies are as follows:

1. Encourage joint environmental governance between regional cities. Compared with previous studies focusing on the national, provincial and industry levels, the allocation of emission rights in regional city agglomerations can strengthen the coordinated development among cities and promote regional economic integration. It suggests that the governments take effective actions to improve relevant policies and regulations. For example, cities jointly build a sustainable ecological chain to promote emission reductions.

2. Scientific and reasonable selection of pollutant emission rights allocation subjects. Cities in most provinces in China show a huge difference in economic scale and industrial structure between north and south or east and west. There is a huge difference in industrial structure and economic scale between cities in the north and south of Jiangsu province. For example, as Xuzhou is the only heating city in Jiangsu province, it is obviously unreasonable to allocate SO_2 emission rights together with other non-heating cities in Jiangsu province. Therefore, it suggests that in the process of pollutant emission rights allocation to select regional city agglomeration with higher homogeneity (e.g., Yangtze River Delta city agglomeration, Pearl River Delta agglomeration, or Northwest city agglomeration, Northeast city agglomeration, etc.) as the allocation objects.
3. Further optimize the city industrial structure. The shadow price allocation model proposed in this paper is oriented to protect GDP, but it finds that maximizing economic efficiency and fairness are difficult to be achieved at the same time. The pollutant emission rights allocation is only a backward method for cities to passively reduce emissions at the expense of part of their GDP. It leads to a sharp increase in the cost of environmental treatment invested by enterprises if the industrial structure is not transformed as soon as possible, resulting in lower economic benefits.
4. For cities with similar SO_2 emission quotas should implement different emission reduction measures according to their meteorological conditions and geographical characteristics. For example, cities in coastal areas should appropriately reduce their emissions in summer to prevent excessive SO_2 from being brought to inland cities by the southwest monsoon from the Pacific Ocean, thus reducing the spatial spillover of pollutants from coastal cities. Duvivier and Xiong (2013) found that in order to reduce the pressure of environmental assessment, local governments have deliberately relaxed environmental control in the border areas of provinces (cities), thereby strategically "polluting neighbors". Therefore, the government should improve the environmental governance of the border area and curb the aggravation of the border pollution problem by means of ecological compensation.

Acknowledgements Jiqiang Zhao, Hongmei Zhao, and Zhijie Wang also made great contributions to this manuscript. We express our heartfelt thanks to them. This research also was supported by: National Natural Science Foundation of China (72474127) and National Social Science Fund Major Bidding Project of China (18ZDA052). Part of this chapter was published: Zhao J, Wu X, Guo J, Zhao H, Wang Z. Study on the Allocation of SO_2 Emission Rights in the Yangtze River Delta City Agglomeration Region of China Based on Efficiency and Feasibility. Sustainable Cities and Society 2022;87:104237. https://doi.org/10.1016/j.scs.2022.104237. The publication has been conducted with copyright permission.

References

Abdul-Wahab, S., Fadlallah, S., & Al-Rashdi, M. (2018). Evaluation of the impact of ground-level concentrations of SO_2, NOx, CO, and PM_{10} emitted from a steel melting plant on Muscat, Oman. *Sustainable Cities and Society, 38*, 675–683. https://doi.org/10.1016/j.scs.2018.01.048

Agrawal, G., Mohan, D., & Rahman, H. (2021). Ambient air pollution in selected small cities in India: Observed trends and future challenges. *IATSS Research, 45*(1), 19–30. https://doi.org/10.1016/J.IATSSR.2021.03.004

Ahman, M., Burtraw, D., Kruger, J., & Zetterberg, L. (2007). A Ten-Year Rule to guide the allocation of EU emission allowances. *Energy Policy, 35*(3), 1718–1730. https://doi.org/10.1016/j.enpol.2006.04.007

Berg, S. A., Førsund, F. R., & Jansen, E. S. (1992). Malmquist indices of productivity growth during the deregulation of Norwegian banking, 1980–89. *The Scandinavian Journal of Economics*, S211–S228. https://doi.org/10.2307/3440261

Chaabouni, S., & Saidi, K. (2017). The dynamic links between carbon dioxide (CO_2) emissions, health spending and GDP growth: A case study for 51 countries. *Environmental Research, 158*, 137–144. https://doi.org/10.1016/j.envres.2017.05.041

Chang, Y., Hu, P., Huang, Y., & Duan, Z. (2022). Effectiveness and heterogeneity evaluation of regional collaborative governance on haze pollution control: Evidence from 284 prefecture-level cities in China. *Sustainable Cities and Society, 86*, 104120. https://doi.org/10.1016/j.scs.2022.104120

Charnes, A., Cooper, W. W., & Rhodes, E. (1978). Measuring the efficiency of decision making units. *European Journal of Operational Research, 2*(6), 429–444. https://doi.org/10.1016/0377-2217(79)90229-7

Chen, J., Augenbroe, G., Zeng, Z., & Song, X. (2020). Regional difference and related cooling electricity savings of air pollutant affected natural ventilation in commercial buildings across the US. *Building and Environment, 172*, 106700. https://doi.org/10.1016/j.buildenv.2020.106700

Cheng, S., Fan, W., Meng, F., Chen, J., Cai, B., Liu, G., Liang, S., Song, M., Zhou, Y., & Yang, Z. (2020). Toward low-carbon development: Assessing emissions-reduction pressure among Chinese cities. *Journal of Environmental Management, 271*, 111036. https://doi.org/10.1016/j.jenvman.2020.111036

Chiu, Y.-H., Lin, J.-C., Su, W.-N., & Liu, J.-K. (2015). An efficiency evaluation of the EU's allocation of carbon emission allowances. *Energy Sources, Part b: Economics, Planning, and Policy, 10*(2), 192–200. https://doi.org/10.1080/15567249.2010.527900

Cui, L., Dong, R., Mu, Y., Shen, Z., & Xu, J. (2022). How policy preferences affect the carbon shadow price in the OECD. *Applied Energy, 311*, 118686. https://doi.org/10.1016/j.apenergy.2022.118686

Cui, Q., & Li, X.-Y. (2021). Which airline should undertake a large emission reduction allocation proportion under the "carbon neutral growth from 2020" strategy? An empirical study with 27 global airlines. *Journal of Cleaner Production, 279*, 123745. https://doi.org/10.1016/j.jclepro.2020.123745

Cui, X., Zhao, T., & Wang, J. (2021). Allocation of carbon emission quotas in China's provincial power sector based on entropy method and ZSG-DEA. *Journal of Cleaner Production, 284*, 124683. https://doi.org/10.1016/j.jclepro.2020.124683

Deng, X., & Du, L. (2020). Estimating the environmental efficiency, productivity, and shadow price of carbon dioxide emissions for the Belt and Road Initiative countries. *Journal of Cleaner Production, 277*, 123808. https://doi.org/10.1016/j.jclepro.2020.123808

Ding, Z. L., Duan, X. N., Ge, Q. S., & Zhang, Z. Q. (2010). On the major proposals for carbon emission reduction and some related issues. *Science China (Earth Sciences)* (02), 159–172.

Duvivier, C., & Xiong, H. (2013). Transboundary pollution in China: A study of polluting firms' location choices in Hebei province. *Environment and Development Economics, 18*(4), 459–483. https://doi.org/10.1017/S1355770X13000168

Elkhalfi, O., Chaabita, R., Ghoujdam, M., Zehraoui, K., EL Alaoui, H., & Belhaj, I. (2025). The impact of climate change on food security in the Middle East and North Africa: Challenges and adaptation strategies. *Journal of Agriculture and Food Research, 21*, 101963. https://doi.org/10.1016/j.jafr.2025.101963

Gadre, R., & Anandarajah, G. (2019). Assessing the evolution of India's power sector to 2050 under different CO_2 emissions rights allocation schemes. *Energy for Sustainable Development, 50*, 126–138. https://doi.org/10.1016/j.esd.2019.04.001

Gulia, S., Shukla, N., Padhi, L., Bosu, P., Goyal, S. K., & Kumar, R. (2022). Evolution of air pollution management policies and related research in India. *Environmental Challenges, 6*, 100431. https://doi.org/10.1016/j.envc.2021.100431

Guo, J., Zhao, M., Wu, X., Shi, B., & Gonzalez, E. D. S. (2021). Study on the distribution of PM emission rights in various provinces of China based on a new efficiency and equity two-objective DEA model. *Ecological Economics, 183*, 106956. https://doi.org/10.1016/j.ecolecon.2021.106956

He, S., Yan, T., & Zhou, H. (2016). Decomposition and spatial clustering analysis of China's SO_2 emissions. *Procedia Computer Science, 91*, 1064–1072. https://doi.org/10.1016/j.procs.2016.07.153

He, W., Liu, D., & Wang, C. (2022). Are Chinese provincial carbon emissions allowances misallocated over 2000–2017? Evidence from an extended Gini-coefficient approach. *Sustainable Production and Consumption, 29*, 564–573. https://doi.org/10.1016/j.spc.2021.11.007

Hey, J. D., & Lambert, P. J. (1980a). Relative deprivation and the Gini coefficient: Comment. *Quarterly Journal of Economics, 95*(3), 567–573. https://doi.org/10.2307/1885094

Hey, J. D., & Lambert, P. J. (1980b). Relative deprivation and the Gini coefficient: Comment. *The Quarterly Journal of Economics, 95*(3), 567–573. https://doi.org/10.2307/1885094

Hirano, Y., Ihara, T., & Yoshida, Y. (2016). Estimating residential CO_2 emissions based on daily activities and consideration of methods to reduce emissions. *Building and Environment, 103*, 1–8. https://doi.org/10.1016/j.buildenv.2016.02.021

Hu, Z., Yan, S., Li, X., Yao, L., & Luo, Z. (2019). Evaluating the oil production and wastewater treatment efficiency by an extended two-stage network structure model with feedback variables. *Journal of Environmental Management, 251*, 109578. https://doi.org/10.1016/j.jenvman.2019.109578

Huang, J.-T. (2018). Sulfur dioxide (SO_2) emissions and government spending on environmental protection in China—Evidence from spatial econometric analysis. *Journal of Cleaner Production, 175*, 431–441. https://doi.org/10.1016/j.jclepro.2017.12.001

Huang, Q., & Xu, J. (2020). Bi-level multi-objective programming approach for carbon emission quota allocation towards co-combustion of coal and sewage sludge. *Energy, 211*, 118729. https://doi.org/10.1016/j.energy.2020.118729

Janarthanan, R., Partheeban, P., Somasundaram, K., & Navin Elamparithi, P. (2021). A deep learning approach for prediction of air quality index in a metropolitan city. *Sustainable Cities and Society, 67*, 102720. https://doi.org/10.1016/j.scs.2021.102720

Ju, B.-G., Kim, M., Kim, S., & Moreno-Ternero, J. D. (2021). Fair international protocols for the abatement of GHG emissions. *Energy Economics, 94*, 105091. https://doi.org/10.1016/j.eneco.2020.105091

Kremantzis, M. D., Beullens, P., & Klein, J. (2022). A fairer assessment of DMUs in a generalised two-stage DEA structure. *Expert Systems with Applications, 187*, 115921. https://doi.org/10.1016/j.eswa.2021.115921

Lelieveld, J., Evans, J. S., Fnais, M., Giannadaki, D., & Pozzer, A. (2015). The contribution of outdoor air pollution sources to premature mortality on a global scale. *Nature, 525*(7569), 367–371. https://doi.org/10.1038/nature15371

Li, H., Mu, H., Gui, S., & Li, M. (2014). Scenario analysis for optimal allocation of China's electricity production system. *Sustainable Cities and Society, 10*, 241–244. https://doi.org/10.1016/j.scs.2013.03.005

Li, L., Ma, S., Zheng, Y., & Xiao, X. (2022a). Integrated regional development: Comparison of urban agglomeration policies in China. *Land Use Policy, 114*, 105939. https://doi.org/10.1016/j.landusepol.2021.105939

Li, R., Li, L., & Wang, Q. (2022b). The impact of energy efficiency on carbon emissions: Evidence from the transportation sector in Chinese 30 provinces. *Sustainable Cities and Society*, 103880. https://doi.org/10.1016/j.scs.2022.103880

Liang, C., Luo, A., & Zhong, Z. (2018). Knowledge mapping of medication literacy study: A visualized analysis using CiteSpace. *SAGE Open Medicine, 6*, 2050312118800199.

Lin, X., Wu, C.-H., Dong, Y., Chen, G. Z. X., Wei, W., & Duan, J. (2022). Psychological contract breach and destructive voice: The mediating effect of relative deprivation and the moderating effect of leader emotional support. *Journal of Vocational Behavior, 135*, 103720. https://doi.org/10.1016/j.jvb.2022.103720

Liu, C., Hong, T., Li, H., & Wang, L. (2018). From club convergence of per capita industrial pollutant emissions to industrial transfer effects: An empirical study across 285 cities in China. *Energy Policy, 121*, 300–313. https://doi.org/10.1016/j.enpol.2018.06.039

Luo, R., & Wang, N. (2026). Carbon emission quota allocation for 280 Chinese cities: Integrating machine learning and DEA with regional heterogeneity. *Expert Systems with Applications, 296*, 129036. https://doi.org/10.1016/j.eswa.2025.129036

Mahdiloo, M., Ngwenyama, O., Scheepers, R., & Tamaddoni, A. (2018). Managing emissions allowances of electricity producers to maximize CO_2 abatement: DEA models for analyzing emissions and allocating emissions allowances. *International Journal of Production Economics, 205*, 244–255. https://doi.org/10.1016/j.ijpe.2018.09.014

Mi, K., Zhuang, R., Zhang, Z., Gao, J., & Pei, Q. (2019). Spatiotemporal characteristics of $PM_{2.5}$ and its associated gas pollutants, a case in China. *Sustainable Cities and Society, 45*, 287–295. https://doi.org/10.1016/j.scs.2018.11.004

Miao, L., Liu, C., Yang, X., Kwan, M.-P., & Zhang, K. (2022). Spatiotemporal heterogeneity analysis of air quality in the Yangtze River Delta, China. *Sustainable Cities and Society, 78*, 103603. https://doi.org/10.1016/j.scs.2021.103603

Momeni, E., Lotfi, F. H., Saen, R. F., & Najafi, E. (2019). Centralized DEA-based reallocation of emission permits under cap and trade regulation. *Journal of Cleaner Production, 234*, 306–314. https://doi.org/10.1016/j.jclepro.2019.06.194

National Bureau of Statistics. (2000–2021). *China Statistical Yearbook*. Retrieved from http://www.stats.gov.cn/tjsj/ndsj/

National Bureau of Statistics of China. (2010–2020). *China Statistical Yearbook*. Retrieved from http://www.stats.gov.cn/

Nemery, B., Hoet, P. H., & Nemmar, A. (2001). The Meuse Valley fog of 1930: An air pollution disaster. *Lancet, 357*(9257), 704–708. https://doi.org/10.1016/S0140-6736(00)04135-0

Orellano, P., Reynoso, J., & Quaranta, N. (2021). Short-term exposure to sulphur dioxide (SO_2) and all-cause and respiratory mortality: A systematic review and meta-analysis. *Environment International, 150*(4), 106434. https://doi.org/10.1016/j.envint.2021.106434

Orisaremi, K. K., Chan, F. T. S., Fu, X., & U, T. S.-C. (2025). Optimal allocation of production quotas using an inverse DEA integrated with lean production capacity and negative data. *Expert Systems with Applications, 263*, 125745. https://doi.org/10.1016/j.eswa.2024.125745

Pan, W., & Pan, W. L. (2018). Research on quota allocation of provincial CO_2 emission rights in China based on energy efficiency. *Soft Science, 32*(06), 45–48.

Pan, X., Teng, F., & Wang, G. (2014). Sharing emission space at an equitable basis: Allocation scheme based on the equal cumulative emission per capita principle. *Applied Energy, 113*, 1810–1818. https://doi.org/10.1016/j.apenergy.2013.07.021

Peng, H., Qi, S., & Cui, J. (2021). The environmental and economic effects of the carbon emissions trading scheme in China: The role of alternative allowance allocation. *Sustainable Production and Consumption, 28*, 105–115. https://doi.org/10.1016/j.spc.2021.03.031

Ren, G., & Shang, J. (2011). The research on Gini index's subgroup decomposition method based on relative deprivation theory. *Journal of Quantitative & Technical Economics, 28*(08), 103–114. https://doi.org/10.13653/j.cnki.jqte.2011.08.011

Shen, Z., Bai, K., Hong, T., & Balezentis, T. (2021). Evaluation of carbon shadow price within a non-parametric meta-frontier framework: The case of OECD, ASEAN and BRICS. *Applied Energy, 299*, 117275. https://doi.org/10.1016/j.apenergy.2021.117275

Shepherd, A., & Mullins, J. T. (2019). Arthritis diagnosis and early-life exposure to air pollution. *Environmental Pollution, 253*. https://doi.org/10.1016/j.envpol.2019.07.054

State Council of the PRC. (2017). *Comprehensive work plan for energy conservation and emission reduction in the 13th five year plan*. Retrieved from http://www.gov.cn/zhengce/content/2017-01/05/content_5156789.htm

Sueyoshi, T., Qu, J., Li, A., & Liu, X. (2021). A new approach for evaluating technology inequality and diffusion barriers: The concept of efficiency Gini coefficient and its application in Chinese provinces. *Energy, 235*, 121256. https://doi.org/10.1016/j.energy.2021.121256

Sun, H., Tariq, G., Chen, H., Zhu, J., Liu, Y., & Wu, C. (2018). Allocation of carbon emission quotas to Chinese power enterprises. *Energy Procedia, 152*, 115–124. https://doi.org/10.1016/j.egypro.2018.09.068

Sun, J. S., Song, J., Miao, Y. J., & Ding, J. J. (2014). Allocation of emission permits using DEA from the perspectives of overall and individual rationality. *Systems Engineering, 32*(12), 74–80.

Tan, Z. F., Lei, T., Li, H. H., Ju, L. W., & Chen, Z. H. (2014). The impact of initial allocation of carbon emission rights on power generation replacement analysis model. *Applied Mechanics & Materials, 496–500*, 2760–2763. https://doi.org/10.4028/www.scientific.net/AMM.496-500.2760

Tian, M., Hu, Y.-J., Wang, H., & Li, C. (2022). Regional allowance allocation in China based on equity and efficiency towards achieving the carbon neutrality target: A composite indicator approach. *Journal of Cleaner Production, 342*, 130914. https://doi.org/10.1016/j.jclepro.2022.130914

Wang, C., Zhan, W., Liu, Z., Li, J., Li, L., Fu, P., Huang, F., Lai, J., Chen, J., Hong, F., & Jiang, S. (2020). Satellite-based mapping of the Universal Thermal Climate Index over the Yangtze River Delta urban agglomeration. *Journal of Cleaner Production, 277*, 123830. https://doi.org/10.1016/j.jclepro.2020.123830

Wang, E., Su, B., Zhong, S., & Guo, Q. (2022). China's embodied SO_2 emissions and aggregate embodied SO_2 intensities in interprovincial and international trade. *Technological Forecasting and Social Change, 177*, 121546. https://doi.org/10.1016/j.techfore.2022.121546

Wang, Q., & Li, L. (2021). The effects of population aging, life expectancy, unemployment rate, population density, per capita GDP, urbanization on per capita carbon emissions. *Sustainable Production and Consumption, 28*, 760–774. https://doi.org/10.1016/j.spc.2021.06.029

Wang, Y., Cheng, Y., Yang, G. C., & Dong, Y. (2018). Provincial decomposition of China's carbon emission rights under the constraint of 2020 and 2030 carbon intensity targets. *China Environmental Science, 38*(8), 3180–3188. https://doi.org/10.19674/j.cnki.issn1000-6923.2018.0339

Wu, S., Fu, Y., Shen, H., & Liu, F. (2018). Using ranked weights and Shannon entropy to modify regional sustainable society index. *Sustainable Cities and Society, 41*, 443–448. https://doi.org/10.1016/j.scs.2018.05.052

Xie, Q., Hu, P., Jiang, A., & Li, Y. (2019). Carbon emissions allocation based on satisfaction perspective and data envelopment analysis. *Energy Policy, 132*, 254–264. https://doi.org/10.1016/j.enpol.2019.05.024

Xie, Q., Xu, Q., Rao, K., & Dai, Q. (2022). Water pollutant discharge permit allocation based on DEA and non-cooperative game theory. *Journal of Environmental Management, 302*, 113962. https://doi.org/10.1016/j.jenvman.2021.113962

Xie, Y., Zhao, L., Xue, J., Hu, Q., Xu, X., & Wang, H. (2016). A cooperative reduction model for regional air pollution control in China that considers adverse health effects and pollutant reduction costs. *Science of the Total Environment, 573*, 458–469. https://doi.org/10.1016/j.scitotenv.2016.08.077

FU, Y. (2011). *Fairness measure of sustainable development: Theories, models and empirical studies on relative deprivation.* China Development Press.

Yang, X. Y., Hu, A., & Schieman, S. (2019). Relative deprivation in context: How contextual status homogeneity shapes the relationship between disadvantaged social status and health. *Social Science Research, 81*, 157–169. https://doi.org/10.1016/j.ssresearch.2019.03.011

Yang, Y., Yang, H., & Cheng, Y. (2021). Why is it crucial to evaluate the fairness of natural capital consumption in urban agglomerations in terms of ecosystem services and economic contribution? *Sustainable Cities and Society, 65*, 102644. https://doi.org/10.1016/j.scs.2020.102644

Yitzhaki, S. (1979). Relative deprivation and the Gini coefficient. *The Quarterly Journal of Economics, 93*(2), 321–324. https://doi.org/10.2307/1883197

Yu, A., You, J., Rudkin, S., & Zhang, H. (2019). Industrial carbon abatement allocations and regional collaboration: Re-evaluating China through a modified data envelopment analysis. *Applied Energy, 233–234*, 232–243. https://doi.org/10.1016/j.apenergy.2018.10.028

Yu, Z., Wang, Q., Xu, Y., Lu, M., Lin, Z., & Gao, B. (2022). Dynamic impacts of changes in river structure and connectivity on water quality under urbanization in the Yangtze River Delta plain. *Ecological Indicators, 135*, 108582. https://doi.org/10.1016/j.ecolind.2022.108582

Zhao, J., Wu, X., Guo, J., & Gao, C. (2022). Allocation of SO_2 emission rights in city agglomerations considering cross-border transmission of pollutants: A new network DEA model. *Applied Energy, 325*, 119927. https://doi.org/10.1016/j.apenergy.2022.119927

Zhao, L., Yuan, L., Yang, Y., Xue, J., & Wang, C. (2021). A cooperative governance model for SO_2 emission rights futures that accounts for GDP and pollutant removal cost. *Sustainable Cities and Society, 66*, 102657. https://doi.org/10.1016/j.scs.2020.102657

Zhou, H., Ping, W., Wang, Y., Wang, Y., & Liu, K. (2021). China's initial allocation of interprovincial carbon emission rights considering historical carbon transfers: Program design and efficiency evaluation. *Ecological Indicators, 121*, 106918. https://doi.org/10.1016/j.ecolind.2020.106918

Chapter 16
Allocation of SO_2 Emission Rights in City Agglomerations Considering Cross-Border Transmission of Pollutants: A New Network DEA Model

16.1 Introduction

Fossil energy resources including coal produce large amounts of SO_2 while promoting the development of human society, which has seriously damaged human health and the sustainable development of the environment. On the one hand, long-term exposure to the high concentration of SO_2 increases the risk of human respiratory diseases. For example, some scholars have found that SO_2 can promote the incidence of lung cancer (Karimi & Samadi, 2024; Meo et al., 2024; Sepehri et al., 2024). On the other hand, SO_2 is the main factor of acid rain, which has damaged the ecosystem and buildings (Liu et al., 2022b; Rosso et al., 2016). China is one of the few energy consumers with coal as the main energy resource. According to the latest disclosure of the China Coal Industry Association, China's domestic coal consumption accounts for 56% of total energy consumption in 2021. In addition, the *BP Statistical Review of World Energy 2020* shows that China, India and the United States are the top three coal consuming countries in the world. In recent years, energy systems in different countries are transitioning to more sustainable and low-carbon forms (Sun et al., 2022). China's renewable energy power generation also has increased rapidly, but the proportion of thermal power generation in 2020 is still as high as 71.19% (Jia et al., 2022). The Yangtze River Delta region has dense industrial clusters and large demand for fossil energy, making it a typical area of coal consumption and SO_2 emission in China. As a result, the Yangtze River Delta region becomes one of the three major acid rain areas in China (Li et al., 2021a). Since 2018, China's Ministry of Ecological Environment has issued the *Action Plan for Comprehensive Treatment of Air Pollution in Autumn and Winter in the Yangtze River Delta region* for many years. Controlling the total coal consumption and SO_2 emission reduction has become the important aspect of environmental governance in the Yangtze River Delta region.

It has been proved that setting the overall goal of emission reduction and allocating the emission rights of pollutants to the emission units can effectively promote

X. Wu, *Environmental Disaster in China*,
https://doi.org/10.1007/978-981-92-0910-1_16

pollutant emission reduction (Li et al., 2021b). For example, at the National Environmental Protection Work Conference in 1989, the Chinese government first put forward the policy of gradually implementing total pollutant emission control and allocation of pollutant emission rights. In order to reduce the harm of acid rain, the Chinese government revised the "*Air Pollution Prevention and Control Law*" in 2000, delineated acid rain areas and SO_2 control areas, and imposed strict SO_2 emission rights restrictions on the above two areas according to the actual situation. Up to now, "total control and gradual allocation" is still one of the main measures taken by the Chinese government to reduce SO_2 emissions. In China's *12th Five-Year Plan* and the *13th Five-Year Plan*, the SO_2 emission reduction targets (10% for 2011–2015 and 15% for 2016–2020) are clearly specified, and the governments at all levels are required to allocate the SO_2 emission rights to the pollutant emission enterprises by step decomposition (Wang et al., 2022). For example, in the *13th Five-Year Plan*, the SO_2 emission reduction ratio of Hangzhou and Nanjing is 23% and 20% respectively. Based on the overall carbon emission reduction target of China in 2015, He et al. (2021) reallocated and evaluated the economic performance of 15 carbon emission rights allocation schemes. The results show that, compared with the actual emission situation in 2015, the optimal allocation scheme can increase GDP by about 6.78% and save energy by about 3.01%. Therefore, making a reasonable and effective allocation scheme of SO_2 emission rights is an effective means to reduce SO_2 emission reduction.

At present, the allocation of pollutant emission rights is one of the hot issues in the field of emission reduction, but the existing researches still have some limitations. First of all, scholars mainly focus on the provincial and inter-industry levels in research on the allocation of pollutant emission rights. For example, 30 provinces in China are selected as the allocation objects to study the allocation of $PM_{2.5}$ and CO_2 emission rights (Guo et al., 2021; Zhou et al., 2021). Local governments allocate SO_2 emission quotas to typical thermal power enterprises (Qu et al., 2021). However, when studying the emission rights allocation among provinces, most scholars fail to consider the balanced development of regional city economy and the coordinated management of the environment. Since city agglomerations have the characteristics of industrial agglomeration and high homogeneity, and collaborative governance is the key to green production efficiency, it is more conducive to economic development and coordinated environmental governance by allocating pollutant emission rights to regional city agglomerations (Fang et al., 2022a; Yu et al., 2022). Secondly, in the selection of allocation methods and indicators, most scholars usually establish multi-objective allocation models based on DEA, information entropy and fixed cost allocation (FCAM). The allocation indicators mainly consider manpower, property power, financial resources and GDP (Cui et al., 2021; Gan et al., 2022; He et al., 2021). However, these indicators are primarily concentrated in the formation stage of pollutants, and few scholars put forward the allocation scheme of pollutant emission rights considered the stage of environmental governance (such as the incidence of respiratory diseases, the number of days of air quality reaching the standard and other environmental and health factors). Finally, scholars usually assume that the emission units are independent of each other when making the allocation scheme of pollutant emission rights, without considering the pollutant transmission among regions and the air self-purification capacity (Cheng et al., 2022; Zhang et al., 2021a). However, without considering the environmental factors such as wind direction, the optimal allocation result cannot be obtained. For example, the concentration of pollutants in the downwind area will be significantly higher than that in the upwind area, which is unfair from the perspective of residents' health.

To solve these problems, this paper proposes a two-stage network DEA model based on pollutant transmission, and applies it to the allocation of SO_2 emission rights in the Yangtze River Delta city agglomeration in China. Firstly, ISAM (Integrated Source Apportionment Method) in WRF-CMAQ is used to simulate and calculate the amount and proportion of SO_2 transmission between cities in the Yangtze River Delta city agglomeration. Secondly, a two-stage network DEA allocation model is established (SO_2 formation stage and SO_2 treatment stage). Each stage is given a corresponding weight coefficient, and the final quota is weighted by the quotas of the two stages. In addition, to reduce the city's economic losses caused by emission reduction, a constraint coefficient based on city GDP protection is introduced into the allocation model. Finally, the optimal allocation scheme is selected based on PRDC. The research framework is shown in Fig. 16.1.

The possible contributions of this paper are as follows:

1. Combining the WRF-CMAQ model with the network DEA allocation model for the first time, the new model takes into account the internal links of DMUs and the interaction characteristics between DMUs.
2. Different from previous studies, the final allocation quotas of each DMU are determined by the formation stage and treatment stage of SO_2, and the decision-makers can adjust the weight coefficient of each stage according to their preference and actual situation to determine the optimal allocation scheme. Ecological

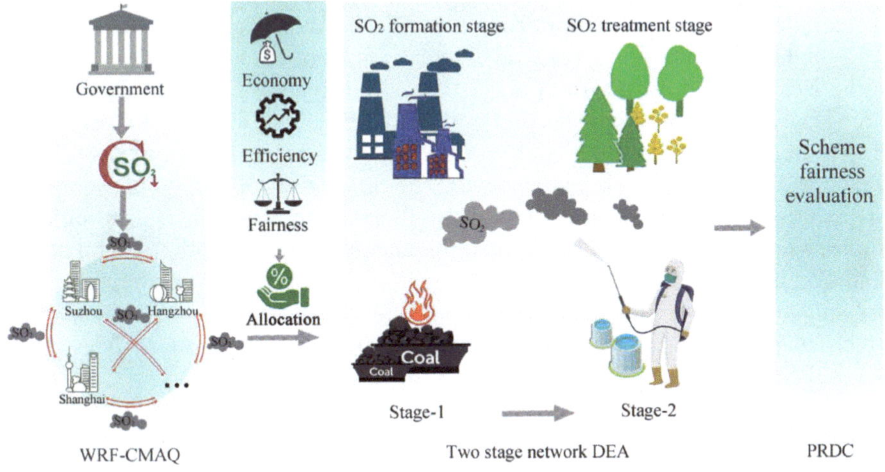

Fig. 16.1 Schematic diagram of the research framework: a new network DEA method

self-purification and health factors are introduced into the SO_2 treatment stage, prompting DMUs to attach importance to environmental treatment.
3. The allocation model established in this paper takes into account the three objectives of efficiency, fairness and economic protection, and provides a new idea for the allocation of pollutant emission rights and the coordinated treatment of environment in regional city agglomerations.

The rest of this paper is organized as follows: Sect. 16.2 is the literature review. The data source and model are discussed in Sect. 16.3. Section 16.4 lists the results and discussions, and Sect. 16.5 gives the conclusions and suggestions.

16.2 Literature Review

This section will review the related literature from three aspects: network DEA, pollutant emission rights allocation methods and allocation principles.

16.2.1 Network DEA

The traditional DEA model calculates the overall efficiency of DMU based on initial input and final output, without considering the input and output of intermediate links. To open the "black box" of the traditional DEA model and evaluate the efficiency of each link in DEA, Färe and Grosskopf (2000) proposed the network DEA model for the first time. Network DEA has been widely used because it avoids the low internal link efficiency caused by the pursuit of the maximum overall efficiency. For example, Lozano and Villa (2004) used the network DEA model to solve the problem of centralized resource allocation. Xiao et al. (2021) established a two-stage network DEA model to evaluate the ecological efficiency of 84 resource-based cities in China in the late financial crisis. An et al. (2019) introduced the Sharply value into the three-stage network DEA model, and evaluated the revenue allocation model based on resource sharing. Li et al. (2020) proposed a dynamic and network slack-based measure (DNSBM) model, and evaluated the differences of energy consumption, air pollution and environmental treatment efficiency in 30 provinces and cities in China, which helped to policies concerning energy allocation and emission reduction. Combining the WRF-CMAQ model with the network DEA model, this paper attempts to construct a network DEA model with a series–parallel hybrid structure, which can better reflect the interaction among DMUs. This new network DEA model is a useful supplement to the existing DEA models.

16.2.2 Allocation Method of Pollutant Emission Rights

The allocation methods of pollutant emission rights mainly include: optimization method, mixed method, game theory and index synthesis method (Zhou & Wang, 2016). The optimization method is mainly based on mathematical programming to calculate the optimal solution of the model. DEA allocation method is one of the most typical optimization methods (An et al., 2017; Pang et al., 2015; Zhou et al., 2014), which can not only solve linear programming problems, but also solve most nonlinear programming models through charnes-cooper transformation (Charnes et al., 1979). For example, Feng et al. (2015) proposed a two-step allocation scheme based on DEA to calculate the allocation of carbon emission rights among 21 Organization for Economic Co-operation and Development (OECD) member countries. Gan et al. (2022) studied the allocation of pollutant emission rights of provincial public buildings in China based on zero-sum game data envelopment analysis (ZSG-DEA). Shojaei and Mokhtar (2022) presented a DEA allocation method based on different efficient frontiers to deal with the allocation of carbon emission rights among the member countries of the Paris Climate Agreement.

The mixed allocation method is a comprehensive allocation method formed by combining various allocation methods. For example, Huang and Xu (2020) combined the comprehensive weighting method and the Karush–Kuhn–Tucker optimization method in the uncertain environment to explore the allocation of carbon emission rights in Zhejiang Province. Feng et al. (2018) established a double-layer allocation model based on cluster analysis and weighted voting method, and applied it to the research of carbon emission rights allocation in 30 provinces of China. The mixing of multiple allocation methods usually increases the complexity of the model. For instance, Yu et al. (2014) combined particle swarm optimization, fuzzy clustering and Sharply value method to study the allocation of provincial carbon emission rights in China.

The index synthesis method is to integrate two or more indexes to form a multi-objective allocation scheme. For example, He et al. (2022b) calculated China's provincial carbon emissions mismatch index based on the Gini coefficient, and put forward a new allocation scheme to avoid mismatch. Sun et al. (2013) integrated four indicators, including GDP, population, water resources capacity and water resources quantity based on information entropy, and investigated the allocation of chemical oxygen demand (COD) in seven river basins in China. Li et al. (2018) combined the indicators of population, GDP and historical emissions and put forward an allocation scheme of CO_2 emission rights in the Pearl River Delta region based on the maximum deviation method (MDM). It was found that the allocation result of the MDM method was better than that of the information entropy method and single indicator method. Tian et al. established a comprehensive index allocation method based on the equity of history, demand, contribution and guarantee, studied the allocation of carbon emission rights among provinces in China, and suggested that the Chinese government should adopt the allocation of carbon emission rights based on GDP preference (Tian et al., 2022).

The game theory takes into account the interests of allocation members by finding Nash equilibrium points among DMUs. For example, Ren et al. (2015) established a Stackelberg game allocation model to allocate carbon emission reduction tasks to suppliers and retailers respectively based on four scenarios. Liu et al. (2022a) proposed an allocation model based on the cooperative game, and applied it to the study of cost allocation of air pollutant control in the Yangtze River Delta. Shi et al. (2017) presented a cross-border air pollution cost analysis model based on game theory, which is applied to the cost allocation of pollutant emission reduction in Changsha, Zhuzhou and Xiangtan.

The index synthesis method can integrate various index standards (such as fairness and efficiency, etc.) to make the allocation plan more acceptable to each DMU. However, different evaluation criteria and weight coefficients usually lead to large differences in allocation results. Compared with the index synthesis method, the mixed method combines multiple allocation methods to establish a complex mixed allocation model to achieve the goal of considering multiple principles, but its allocation process lacks transparency. The advantage of the game theory method is to incorporate the negotiation and group consensus among DMUs into the allocation model, however, the disadvantage of this method is that it is too self-interested and may lead to overall inefficiency. Ensuring overall and individual efficiency is the main advantage of the optimization method. The allocation method that takes efficiency as the premise and takes into account various allocation principles makes up for the large differences in allocation results, the opaque allocation process and the low overall efficiency (An et al., 2019).

This paper takes the cross-border transmission of pollutants as the point of penetration, regards the control process of pollutants as two sub-processes of the formation stage and the treatment stage of pollutants, and studies the allocation of pollutant emission rights. In order to ensure that the overall efficiency of the pollutant control process and the efficiency of each sub-process are optimized at the same time, and to overcome the lack of transparency of the general allocation method, this paper establishes a two-stage network DEA model based on the optimization method to study the SO_2 emission rights of the Yangtze River Delta city agglomeration.

16.2.3 Principles of Allocation of Pollutant Emission Rights

The allocation principle of pollutant emission rights of property usually depends on the preference of decision-makers and the specific characteristics of DMUs. On the whole, the principles of efficiency, fairness and economic protection are widely used in the emission rights allocation model. For example, Cai and Ye (2019) established a ZSG-DEA allocation model based on the principle of per capita equity and economic protection and studied the allocation of CO_2 emission rights in 30 provinces of China. Some scholars believe that per capita cumulative emissions can best reflect fairness, and carbon emission rights should be allocated to countries based on this principle.

Developed countries should undertake more emission reduction tasks than developing countries (Pan et al., 2014). Based on the principle of GDP protection, Zhao et al. (2021) proposed an asymmetric Nash negotiation model to allocate the cost of SO_2 emission reduction in the Beijing-Tianjin-Hebei region. The results show that this model can promote the coordinated emission reduction of regional city agglomerations. Momeni et al. (2019) established a DEA allocation model based on the maximum overall efficiency, and studied the greenhouse gas licensing allocation in OECD countries. The results show that the model can reduce greenhouse gas emissions without reducing the expected output. Han et al. explored the optimal allocation scheme of China's energy cap policy in 2030 based on the principles of efficiency and fairness (Han et al., 2022). Based on the entropy method, Li et al. (2021b) established an allocation model considering the principles of fairness, efficiency and economic sustainability. They studied the allocation of CO_2 emission rights from regional to provincial level in China and formulated differentiated emission reduction schemes according to the emission reduction pressure of each province. In the research on the allocation of pollutant emission rights, it is easy to take care of one thing and lose another when considering the single principle. For example, simply considering the principle of optimal efficiency cannot ensure both the economic benefit and the fairness of the allocation scheme.

The DMUs are concerned with whether the allocation result is fair, while the central decision-maker is concerned with whether the allocation scheme achieves maximum efficiency. Therefore, fairness and efficiency jointly determine whether the allocation scheme can be implemented. In addition, the emission reduction of pollutants is usually based on the sacrifice of a certain economic cost. If the principle of economic protection is not considered, cities that are highly dependent on coal and other primary energy sources may suffer huge GDP losses due to emission reduction. Therefore, this paper takes efficiency, economy and fairness as the allocation principle of SO_2 emission rights, in order to promote the coordinated and sustainable development of economy and environment in the Yangtze River Delta city agglomeration.

16.3 Variables and Models

16.3.1 WRF-CMAQ/ISAM Mode System

WRF is a mesoscale meteorological model jointly developed by the National Center for Atmospheric Research (NCAR) of the United States and other departments. Its main function is to provide three-dimensional meteorological field data for air quality models (such as CMAQ, CAMx, etc.) that drive the transmission and diffusion of pollutants. CMAQ is a comprehensive model based on three-dimensional Eulerian atmospheric transmission and chemical process developed by the U.S. Environmental Protection Agency (USEPA). Its main function is to simulate the regional multi-scale chemical transport of pollutants. ISAM is a pollution source analysis tool for spatial allocation of non-point source emission lists based on geographic information data under CMAQ. Its main function is to track and calculate the contribution concentration values of different types of pollutants from different areas to the target area through pollutant tracking. CMAQ/ISAM is widely used because it can effectively reduce the uncertainty of source allocation of pollutants and obtain reliable source analysis results (Feng et al., 2022; Kitagawa et al., 2021; Li et al., 2019). The outflow area of pollutants is defined as the source city, and the inflow area is defined as the recipient city. The contribution rate of the source city to the recipient city is:

$$\eta_{jk} = \frac{R_{jk}}{\sum_{j=1}^{n} R_{jk}} \tag{16.1}$$

where η_{jk} represents the pollutant contribution rate of the j-th city to the k-th city, R_{jk} represents the pollutant concentration contribution value of the j-th city to the k-th city, $\sum_{j=1}^{n} R_{jk}$ represents the sum of the pollutant concentration contribution values of n cities to the k-th city (including the concentration contribution value of the k-th city to itself).

16.3.2 *Calculation of the Overall Pollutant Emission Reduction Target*

Decision-makers usually set emission reduction targets for the next stage according to historical emissions. Taking *China's 13th Five-Year Plan* as an example, the national emission reduction target for 2015–2020 was formulated based on the pollutant emissions in 2015, and the overall target was allocated to all provinces. Then, each province assigned the emission reduction task to all cities in it (The State Council, 2017). The total emission quotas of the regional city agglomeration in the next stage is:

$$E = \sum_{j=1}^{n} (1 - \partial_j) Q_j \tag{16.2}$$

E represents the total emission quotas of the city agglomeration in the next stage, ∂_j represents the proportion of pollutant emission reduction in the j-th city, Q_j indicates the historical pollutant discharge in the j-th city.

16.3.3 Two-Stage Network DEA Allocation Model Considering Economy and Efficiency

16.3.3.1 Model Establishment

In the existing studies, the quota of pollutant emission right is usually determined by the formation stage of pollutants, while the treatment stage of pollutants is rarely introduced into the research of emission right allocation (Cui et al., 2021; Guo et al., 2021; Zhou & Wang, 2016). However, the treatment effect of pollutants will directly affect the health of human respiratory system and social production efficiency (Cao et al., 2022). Therefore, this paper proposes a two-stage DEA allocation model, in which each DMU's pollutant emission right quota is jointly determined by the pollutant formation stage and the pollutant treatment stage. The two stages are given different weight coefficients, and the final objective function is the sum of the efficiency weights of the two stages. Considering the atmospheric transmission factors of pollutants, the pollutants formed in the first stage of each DMU are not only the input to its own second stage, but also the input to other DMUs in the second stage. In other words, pollutants transmit among all DMUs (Fig. 16.2). The initial quota of pollutant emission rights is essentially a kind of free resource. For DMUs, the quota should be as much as the expected output. Similarly, as the unexpected output, pollutants should be as little as possible. Therefore, it can be regarded as an input (You & Yan, 2011).

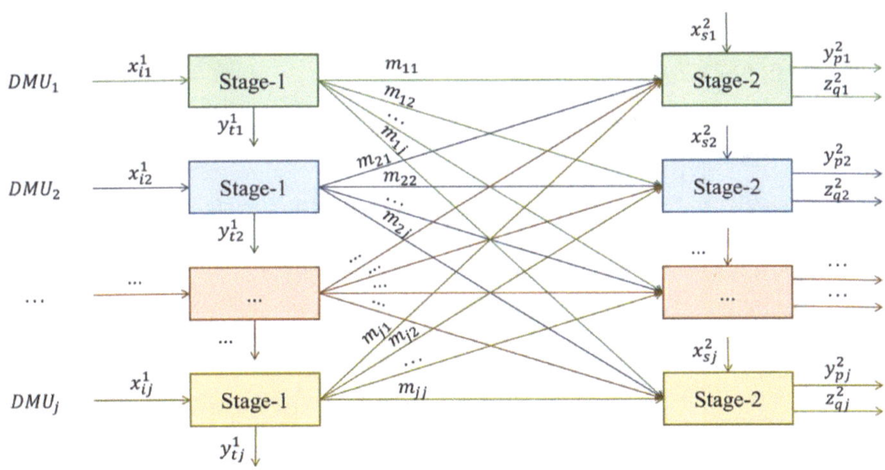

Fig. 16.2 The two-stage network DEA based on atmospheric pollutant transport

Based on the above analysis, firstly, this paper establishes a two-stage network DEA allocation model with maximum efficiency as the objective function, which as shown in model (16.3):

$$\begin{cases} max\phi = \alpha_1 \dfrac{\sum_{j=1}^{n}(\sum_{r=1}^{h} a_r^1 e_{rj}^1 + \sum_{t=1}^{g} b_t y_{tj}^1)}{\sum_{j=1}^{n}(\sum_{i=1}^{u} c_i x_{ij}^1 + \sum_{f=1}^{v}\sum_{k=1}^{n} d_f m_{jk,f})} + \alpha_2 \dfrac{\sum_{j=1}^{n}(\sum_{r=1}^{h} a_r^2 e_{rj}^2 + \sum_{p=1}^{w} \theta_p y_{pj}^2)}{\sum_{j=1}^{n}(\sum_{s=1}^{\tau} \rho_s x_{sj}^2 + \sum_{q=1}^{\varsigma} \lambda_q z_{qj}^2 + \sum_{f=1}^{v}\sum_{k=1}^{n} d_f m_{kj,f})} \\ s.t. \\ \dfrac{\sum_{j=1}^{n}(\sum_{r=1}^{h} a_r^1 e_{rj}^1 + \sum_{t=1}^{g} b_t y_{tj}^1)}{\sum_{j=1}^{n}(\sum_{i=1}^{u} c_i x_{ij}^1 + \sum_{f=1}^{v}\sum_{k=1}^{n} d_f m_{jk,f})} \le 1 \\ \dfrac{\sum_{j=1}^{n}(\sum_{r=1}^{h} a_r^2 e_{rj}^2 + \sum_{p=1}^{w} \theta_p y_{pj}^2)}{\sum_{j=1}^{n}(\sum_{s=1}^{\tau} \rho_s x_{sj}^2 + \sum_{q=1}^{\varsigma} \lambda_q z_{qj}^2 + \sum_{f=1}^{v}\sum_{k=1}^{n} d_f m_{kj,f})} \le 1 \\ \dfrac{\sum_{r=1}^{h} a_r^1 e_{rj}^1 + \sum_{t=1}^{g} b_t y_{tj}^1}{\sum_{i=1}^{u} c_i x_{ij}^1 + \sum_{f=1}^{v}\sum_{k=1}^{n} d_f m_{jk,f}} \le 1 \\ \dfrac{\sum_{r=1}^{h} a_r^2 e_{rj}^2 + \sum_{p=1}^{w} \theta_p y_{pj}^2}{\sum_{s=1}^{\tau} \rho_s x_{sj}^2 + \sum_{q=1}^{\varsigma} \lambda_q z_{qj}^2 + \sum_{f=1}^{v}\sum_{k=1}^{n} d_f m_{kj,f}} \le 1 \\ \alpha_1 + \alpha_2 = 1 \\ \sum_{j=1}^{n} e_{rj}^1 = E_r \\ \sum_{j=1}^{n} e_{rj}^2 = E_r \\ a_r^1, a_r^2 \ge 0; r = 1,2,\ldots,h \quad e_{rj}^1, e_{rj}^2 \ge 0; j = 1,2,\ldots,n \\ b_t \ge 0; t = 1,2,\ldots,g \quad \theta_p \ge 0; p = 1,2,\ldots,w \\ c_i \ge 0; i = 1,2,\ldots,u \quad \rho_s \ge 0; s = 1,2,\ldots,\tau \\ \lambda_q \ge 0; q = 1,2,\ldots,\varsigma \quad d_f \ge 0; f = 1,2,\ldots,v \\ \alpha_1, \alpha_2 \ge 0 \end{cases} \tag{16.3}$$

ϕ is the overall efficiency value. α_1 and α_2 represent the weight coefficients of the first stage and the second stage respectively, which are set according to the preference of decision-makers or the actual situation. a_r^1 is the weight coefficient of the quota for the r-th pollutant in the first stage. e_{rj}^1 is r-th pollutant quota of the j-th DMU in the first stage. b_t indicates the weight coefficient of the t-th expected output in the first stage. y_{tj}^1 is the t-th expected output value of the j-th DMU in the first stage. c_i is the i-th input weight coefficient in the first stage. x_{ij}^1 shows the i-th input value of the j-th DMU in the first stage. d_f is the weight coefficient of the f-th pollutant emission value in the first stage. $m_{jk,f}$ represents the transmission value of the f-th pollutant from the j-th city (stage 1) to the k-th city (stage 2). a_r^2 is the weight coefficient of the quota for the r-th pollutant in the second stage. e_{rj}^2 represents the r-th pollutant quota of the j-th DMU in the second stage. θ_p is the weight coefficient of the p-th expected output in the second stage. y_{pj}^2 is the p-th expected output value of the j-th DMU in the second stage. ρ_s is the s-th input weight coefficient in the second stage (excluding pollutant input). x_{sj}^2 indicates the s-th input value of the j-th DMU in the second stage. λ_q is the weight coefficient of the q-th undesirable output in the second stage. z_{qj}^2 is the q-th undesirable output value of the j-th DMU in the second stage. d_f represents the weight coefficient of the f-th pollutant input from the first stage to the second stage. m_{kjf} is the transmission value of the f-th pollutant from the k-th city (the first stage) to the j-th city (the second stage).

Then, the constraint conditions based on the principle of GDP protection are introduced into the model (16.3). The main methods of establishing multi-objective DEA model include the multi-stage method, introducing constraint conditions method and multi-objective function method and so on (Estellita Lins et al., 2004; Li & Reeves, 1999). For example, when introducing population indicators into the DEA allocation model, Guo et al. (2021) considered that the smaller the difference between quota proportion and population proportion, the better. However, during linear optimization, an extreme situation may occur that the quota ratio is approximately equal to the population ratio. Therefore, this paper limits the difference between the quota proportion and GDP proportion of the j-th DMU in each stage to a certain constraint range, and the constraint coefficient can be adjusted according to the preference of decision makers or the actual situation:

$$\left|\frac{e_{rj}^1}{E_r} - \frac{\mu_j}{\sum_{j=1}^{n} \mu_j}\right| \leq \gamma \tag{16.4}$$

$$\left|\frac{e_{rj}^2}{E_r} - \frac{\mu_j}{\sum_{j=1}^{n} \mu_j}\right| \leq \gamma \tag{16.5}$$

where μ_j represents the GDP of the j-th DMU and $\gamma \in [0,1]$ is the constraint coefficient of GDP protection.

Finally, we substitute formulas (16.4) and (16.5) into formula (16.3), and obtain a two-stage network DEA model (16.6) that takes into account efficiency and economic principles:

$$
\begin{cases}
max\phi = \alpha_1 \frac{\sum_{j=1}^{n}(\sum_{r=1}^{h} a_r^1 e_{rj}^1 + \sum_{t=1}^{g} b_t y_{tj}^1)}{\sum_{j=1}^{n}(\sum_{i=1}^{u} c_i x_{ij}^1 + \sum_{f=1}^{v}\sum_{k=1}^{n} d_f m_{jk,f})} + \alpha_2 \frac{\sum_{j=1}^{n}(\sum_{r=1}^{h} a_r^2 e_{rj}^2 + \sum_{p=1}^{w} \theta_p y_{pj}^2)}{\sum_{j=1}^{n}(\sum_{s=1}^{\tau} \rho_s x_{sj}^2 + \sum_{q=1}^{\varsigma} \lambda_q z_{qj}^2 + \sum_{f=1}^{v}\sum_{k=1}^{n} d_f m_{kj,f})} \\
s.t. \\
\frac{\sum_{j=1}^{n}(\sum_{r=1}^{h} a_r^1 e_{rj}^1 + \sum_{t=1}^{g} b_t y_{tj}^1)}{\sum_{j=1}^{n}(\sum_{i=1}^{u} c_i x_{ij}^1 + \sum_{f=1}^{v}\sum_{k=1}^{n} d_f m_{jk,f})} \le 1 \\
\frac{\sum_{j=1}^{n}(\sum_{r=1}^{h} a_r^2 e_{rj}^2 + \sum_{p=1}^{w} \theta_p y_{pj}^2)}{\sum_{j=1}^{n}(\sum_{s=1}^{\tau} \rho_s x_{sj}^2 + \sum_{q=1}^{\varsigma} \lambda_q z_{qj}^2 + \sum_{f=1}^{v}\sum_{k=1}^{n} d_f m_{kj,f})} \le 1 \\
\frac{\sum_{r=1}^{h} a_r^1 e_{rj}^1 + \sum_{t=1}^{g} b_t y_{tj}^1}{\sum_{i=1}^{u} c_i x_{ij}^1 + \sum_{f=1}^{v}\sum_{k=1}^{n} d_f m_{jk,f}} \le 1 \\
\frac{\sum_{r=1}^{h} a_r^2 e_{rj}^2 + \sum_{p=1}^{w} \theta_p y_{pj}^2}{\sum_{s=1}^{\tau} \rho_s x_{sj}^2 + \sum_{q=1}^{\varsigma} \lambda_q z_{qj}^2 + \sum_{f=1}^{v}\sum_{k=1}^{n} d_f m_{kj,f}} \le 1 \\
\alpha_1 + \alpha_2 = 1 \\
\sum_{j=1}^{n} e_{rj}^1 = E_r \\
\sum_{j=1}^{n} e_{rj}^2 = E_r \\
|\frac{e_{rj}^1}{E_r} - \frac{\mu_j}{\sum_{j=1}^{n}\mu_j}| \le \gamma \\
|\frac{e_{rj}^2}{E_r} - \frac{\mu_j}{\sum_{j=1}^{n}\mu_j}| \le \gamma \\
a_r^1, a_r^2 \ge 0; r = 1,2,\ldots,h \quad e_{rj}^1, e_{rj}^2 \ge 0; j = 1,2,\ldots,n \\
b_t \ge 0; t = 1,2,\ldots,g \quad \theta_p \ge 0; p = 1,2,\ldots,w \\
c_i \ge 0; i = 1,2,\ldots,u \quad \rho_s \ge 0; s = 1,2,\ldots,\tau \\
\lambda_q \ge 0; q = 1,2,\ldots,\varsigma \quad d_f \ge 0; f = 1,2,\ldots,v \\
0 \le \gamma \le 1 \quad \alpha_1, \alpha_2 \ge 0
\end{cases}
\tag{16.6}
$$

The final quota of the r-th unexpected output of the j-th DMU is:

$$
e_{rj} = \alpha_1 e_{rj}^1 + \alpha_2 e_{rj}^2 \tag{16.7}
$$

16.3.3.2 Model Solution

Obviously, the above model has nonlinear characteristics. The nonlinear model can be solved through heuristic algorithm, but the algorithm is used to approximately estimate the result, which is difficult to obtain the global optimum (Glover & Sueyoshi, 2009). Therefore, this paper adopts the charnes-cooper transformation to transform the nonlinear model into a linear model in three steps (Charnes et al., 1979).

Step 1: Let $a_r^1 e_{rj}^1 = l_{rj}^1 \geq 0, a_r^2 e_{rj}^2 = l_{rj}^2 \geq 0$, Then we can have $\sum_{j=1}^{n} a_r^1 e_{rj}^1 = \sum_{j=1}^{n} l_{rj}^1 = a_r^1 E_r$, $\sum_{j=1}^{n} a_r^2 e_{rj}^2 = \sum_{j=1}^{n} l_{rj}^2 = a_r^2 E_r$, the model (16.6) is transformed into model (16.8):

$$\begin{cases} max\phi = \alpha_1 \frac{\sum_{j=1}^{n}(\sum_{r=1}^{h} l_{rj}^1 + \sum_{t=1}^{g} b_t y_{tj}^1)}{\sum_{j=1}^{n}(\sum_{i=1}^{u} c_i x_{ij}^1 + \sum_{f=1}^{v}\sum_{k=1}^{n} d_f m_{jk,f})} + \alpha_2 \frac{\sum_{j=1}^{n}(\sum_{r=1}^{h} l_{rj}^2 + \sum_{p=1}^{w} \theta_p y_{pj}^2)}{\sum_{j=1}^{n}(\sum_{s=1}^{\tau} \rho_s x_{sj}^2 + \sum_{q=1}^{\varsigma} \lambda_q z_{qj}^2 + \sum_{f=1}^{v}\sum_{k=1}^{n} d_f m_{kj,f})} \\ s.t. \\ \frac{\sum_{j=1}^{n}(\sum_{r=1}^{h} l_{rj}^1 + \sum_{t=1}^{g} b_t y_{tj}^1)}{\sum_{j=1}^{n}(\sum_{i=1}^{u} c_i x_{ij}^1 + \sum_{f=1}^{v}\sum_{k=1}^{n} d_f m_{jk,f})} \leq 1 \\ \frac{\sum_{j=1}^{n}(\sum_{r=1}^{h} l_{rj}^2 + \sum_{p=1}^{w} \theta_p y_{pj}^2)}{\sum_{j=1}^{n}(\sum_{s=1}^{\tau} \rho_s x_{sj}^2 + \sum_{q=1}^{\varsigma} \lambda_q z_{qj}^2 + \sum_{f=1}^{v}\sum_{k=1}^{n} d_f m_{kj,f})} \leq 1 \\ \frac{\sum_{r=1}^{h} l_{rj}^1 + \sum_{t=1}^{g} b_t y_{tj}^1}{\sum_{i=1}^{u} c_i x_{ij}^1 + \sum_{f=1}^{v}\sum_{k=1}^{n} d_f m_{jk,f}} \leq 1 \\ \frac{\sum_{r=1}^{h} l_{rj}^2 + \sum_{p=1}^{w} \theta_p y_{pj}^2}{\sum_{s=1}^{\tau} \rho_s x_{sj}^2 + \sum_{q=1}^{\varsigma} \lambda_q z_{qj}^2 + \sum_{f=1}^{v}\sum_{k=1}^{n} d_f m_{kj,f}} \leq 1 \\ \alpha_1 + \alpha_2 = 1 \\ \sum_{j=1}^{n} l_{rj}^1 = a_r^1 E_r \\ \sum_{j=1}^{n} l_{rj}^2 = a_r^2 E_r \\ |\frac{l_{rj}^1}{a_r^1 E_r} - \frac{\mu_j}{\sum_{j=1}^{n} \mu_j}| \leq \gamma \\ |\frac{l_{rj}^2}{a_r^2 E_r} - \frac{\mu_j}{\sum_{j=1}^{n} \mu_j}| \leq \gamma \\ a_r^1, a_r^2 \geq 0; r = 1,2,\ldots,h \quad l_{rj}^1, l_{rj}^2 \geq 0; j = 1,2,...,n \\ b_t \geq 0; t = 1,2,\ldots,g \quad \theta_p \geq 0; p = 1,2,\ldots,w \\ c_i \geq 0; i = 1,2,\ldots,u \quad \rho_s \geq 0; s = 1,2,\ldots,\tau \\ \lambda_q \geq 0; q = 1,2,\ldots,\varsigma \quad d_f \geq 0; f = 1,2,\ldots,v \\ 0 \leq \gamma \leq 1 \quad \alpha_1, \alpha_2 \geq 0 \end{cases} \tag{16.8}$$

Step 2: Convert fractional planning into algebraic expression planning.
Let $\sum_{j=1}^{n}(\sum_{i=1}^{u} c_i x_{ij}^1 + \sum_{f=1}^{v}\sum_{k=1}^{n} d_f m_{jk,f}) = \frac{1}{H^1}(H^1 > 0)$,

$$\sum_{j=1}^{n}\left(\sum_{s=1}^{\tau}\rho_s x_{sj}^2 + \sum_{q=1}^{\varsigma}\lambda_q z_{qj}^2 + \sum_{f=1}^{v}\sum_{k=1}^{n} d_f m_{kj,f}\right) = \frac{1}{H^2}(H^2 > 0), H^1 a_r^1 = A_r^1,$$

$H^2a_r^2 = A_r^2$, $H^1l_r^1 = L_r^1$, $H^2l_r^2 = L_r^2$, $H^1b_t = B_t$, $H^2\theta_p = F_p$, $H^1c_i = C_i$, $H^1d_f = D_f^1$, $H^2d_f = D_f^2$, $H^2\rho_s = P_s$, $H^2\lambda_q = I_q$, the model (16.8) is transformed into model (16.9):

$$\begin{cases} max\phi = \alpha_1\sum_{j=1}^{n}(\sum_{r=1}^{h}L_{rj}^1 + \sum_{t=1}^{g}B_t y_{tj}^1) + \alpha_2\sum_{j=1}^{n}(\sum_{r=1}^{h}L_{rj}^2 + \sum_{p=1}^{w}F_p y_{pj}^2) \\ s.t. \\ \sum_{j=1}^{n}(\sum_{i=1}^{u}C_i x_{ij}^1 + \sum_{f=1}^{v}\sum_{k=1}^{n}D_f^1 m_{jk,f}) = 1 \\ \sum_{j=1}^{n}(\sum_{s=1}^{\tau}P_s x_{sj}^2 + \sum_{q=1}^{\varsigma}I_q z_{qj}^2 + \sum_{f=1}^{v}\sum_{k=1}^{n}D_f^2 m_{kj,f}) = 1 \\ \sum_{j=1}^{n}(\sum_{r=1}^{h}L_{rj}^1 + \sum_{t=1}^{g}B_t y_{tj}^1) \le 1 \\ \sum_{j=1}^{n}(\sum_{r=1}^{h}L_{rj}^2 + \sum_{p=1}^{w}F_p y_{pj}^2) \le 1 \\ \sum_{r=1}^{h}L_{rj}^1 + \sum_{t=1}^{g}B_t y_{tj}^1 - \sum_{i=1}^{u}C_i x_{ij}^1 - \sum_{f=1}^{v}\sum_{k=1}^{n}D_f^1 m_{jk,f} \le 0 \\ \sum_{r=1}^{h}L_{rj}^2 + \sum_{p=1}^{w}F_p y_{pj}^2 - \sum_{s=1}^{\tau}P_s x_{sj}^2 - \sum_{q=1}^{\varsigma}I_q z_{qj}^2 - \sum_{f=1}^{v}\sum_{k=1}^{n}D_f^2 m_{kj,f} \le 0 \\ \sum_{j=1}^{n}L_{rj}^1 = A_r^1 E_r \\ \sum_{j=1}^{n}L_{rj}^2 = A_r^2 E_r \\ |\frac{L_{rj}^1}{E_r} - \frac{A_r^1\mu_j}{\sum_{j=1}^{n}\mu_j}| \le \gamma A_r^1 \\ |\frac{L_{rj}^2}{E_r} - \frac{A_r^2\mu_j}{\sum_{j=1}^{n}\mu_j}| \le \gamma A_r^2 \\ \alpha_1 + \alpha_2 = 1 \\ A_r^1, A_r^2 \ge 0; r = 1,2,\ldots,h \quad L_{rj}^1, L_{rj}^2 \ge 0; j = 1,2,\ldots,n \\ B_t \ge 0; t = 1,2,\ldots,g \quad F_p \ge 0; p = 1,2,\ldots,w \\ C_i \ge 0; i = 1,2,\ldots,u \quad P_s \ge 0; s = 1,2,\ldots,\tau \\ I_q \ge 0; q = 1,2,\ldots,\varsigma \quad D_f^1, D_f^2 \ge 0; f = 1,2,\ldots,v \\ 0 \le \gamma \le 1 \quad \alpha_1, \alpha_2 \ge 0 \end{cases} \tag{16.9}$$

Step 3: Calculate a group of optimal solutions based on the model (16.9):

$$(A_r^1, A_r^2, L_{rj}^1, L_{rj}^2, B_t, F_p, C_i, P_s, I_q, D_f^1, D_f^2)$$

The final allocation of the j-th DMU is:

$$e_{rj} = \alpha_1 e_{rj}^1 + \alpha_2 e_{rj}^2 = \alpha_1\frac{l_{rj}^1}{a_r^1} + \alpha_2\frac{l_{rj}^2}{a_r^2} = \alpha_1\frac{L_{rj}^1}{A_r^1} + \alpha_2\frac{L_{rj}^2}{A_r^2} \tag{16.10}$$

16.3.4 Fairness Evaluation of the Allocation Scheme Based on PRDC

The fairness of an allocation scheme is the key to whether the scheme can be accepted by DMUs. Currently, the main methods of introducing the principle of fairness into the allocation scheme include adding constraints and fairness evaluation. (Antosiewicz et al., 2022; Gan et al., 2022; Zhou et al., 2021). Compared with adding constraints, fairness evaluation can more intuitively quantify the fairness of allocation schemes. Based on the quantitative results of scheme fairness, decision-makers can obtain the optimal scheme through evaluation-adjustment-reevaluation. At present, the RDC and the Gini coefficient are two widely used fairness evaluation indicators (He et al., 2022a; Lin et al., 2022; Yang et al., 2019). The Gini coefficient is usually used to measure income inequality between groups. Compared with the Gini coefficient, the RDC has more advantages in evaluating the income inequality of members within a group. Yitzhaki (1994) believed that the Gini coefficient could be explained by the relative deprivation coefficient. Fu Yun (2011) studied the application of the Gini coefficient and the RDC in China's income allocation, and proposed that the threshold of relative deprivation coefficient suitable for China's national conditions is 0.5. If $0.5 < \text{RDC} \leq 1$, it means that the allocation scheme cannot be accepted by DMUs. When $0 \leq \text{RDC} \leq 0.5$, the allocation scheme is fair. In addition, some scholars proposed that the per capita quota is the best indicator to reflect fairness because of the disparity in population size among provinces (Cai & Ye, 2019; Pan et al., 2014). Therefore, this paper introduces the PRDC to evaluate the fairness of the allocation scheme.

Let X and Y be two sample groups, and the corresponding sample vectors are $\boldsymbol{x}' = \left(x_1', x_2', \ldots, x_\Gamma'\right)$ and $\boldsymbol{y}' = \left(y_1', y_2', \ldots, y_\Lambda'\right)$. x_ω' and y_ϖ' represent individuals in X and Y, respectively. The relative sense of deprivation among groups can be expressed as:

$$U_{XY}\left(x_\omega', y_\varpi'\right) = \begin{cases} x_\omega' - y_\varpi', \textit{if} \quad x_\omega' > y_\varpi' \\ 0 \qquad \textit{else} \end{cases} \tag{16.11}$$

The relative deprivation coefficient is calculated as follows:

$$\begin{aligned} U\left(x', y_\varpi'\right) &= \frac{\sum_{i=1}^{\Gamma} U_{XY}\left(x_\omega', y_\varpi'\right)}{\Gamma} \\ &= \frac{\Gamma_+ \overline{x_+'} - \Gamma_+ y_\varpi'}{\Gamma} \end{aligned} \tag{16.12}$$

Let $\frac{\Gamma_+}{\Gamma} = \beta_+$, and then:

$$U\left(x', y_\varpi'\right) = \beta_+\left(\overline{x_+'} - y_\varpi'\right) \tag{16.13}$$

where Γ_+ is the number of individuals in group X whose DMU allocation is greater than y_ϖ', $\overline{x_+'}$ indicates the average of individuals in group X whose DMU allocation is greater than y'_{ϖ}, and β_{+} means the proportion of DMUs with a quota greater than y'_{ϖ} in group X.

When the equation is applied to the regional city agglomeration, let x''_{ω} and x''_{ϖ} ($\omega \neq \varpi$) respectively represent the per capita quotas of the ω-th and ϖ-th DMUs in the city agglomeration; $\overline{x''_{+}}$ indicates the per capita quota of DMUs that is larger than x''_{ϖ} in the city agglomeration; β^{+} is the ratio of the number of DMUs whose per capita quota is larger than x''_{ϖ} to the total number of DMUs; $\overline{x''}$ is the average per capita quota of all DMUs. Then, the PRDC of the ϖ-th DMU in the city agglomeration is:

$$U_{PRDC}\left(x''_{\omega}, x''_{\varpi}\right) = \beta^{+}\frac{\overline{x''_{+}} - x''_{\varpi}}{\overline{x''}} \tag{16.14}$$

16.3.5 Index Selection and Data Sources

16.3.5.1 Research Object

What needs to be done first is to determine the research object. Allocation of emission rights among regional city agglomerations with higher homogeneity is closer to reality and much fair than the provincial allocation (Li et al., 2021b). The Yangtze River Delta, located at 32°34′ to 27°83′ north latitude and 115°46′ to 123°25′ east longitude, is the most economically developed area along the eastern coast of China. It has many industrial clusters, and is also a typical coal consumption and SO_2 emission area in China. Therefore, this paper selects 27 cities in the Yangtze River Delta as research objects, including Shanghai (SH), Zhejiang provincial region (Hangzhou (HZ), Ningbo (NB), Wenzhou (WZ), Huzhou (HuZ), Jiaxing (JX), Shaoxing (SX), Jinhua (JH), Zhoushan (ZS), Taizhou (TZ (ZJ)), Anhui provincial region (Hefei (HF), Wuhu (WH), Maanshan (MAS), Tongling (TL), Anqing (AQ), Chuzhou (CZ), Chizhou (ChZ) and Xuancheng (XC)) and Jiangsu provincial region (Nanjing (NJ), Wuxi (WX), Changzhou (CZ (JS)), Suzhou (SZ), Nantong (NT), Yancheng (YC), Yangzhou (YZ), Zhenjiang (ZJ) and Taizhou (TZ)) (Fig. 16.3).

Then, the emission reduction target is calculated. In 2017, the Chinese government announced the pollutant emission reduction target of the 13th Five-Year Plan period based on the emissions in 2015. Therefore, this paper allocates pollutant emission rights based on the SO_2 emission data of the Yangtze River Delta city agglomeration in 2015. The calculation result of the formula (16.2) shows that the emission of SO_2 in the Yangtze River Delta city agglomeration in 2015 was about 1.5 million tons, and the emission reduction target of the 13th Five-Year Plan was 20%. It means the emission of SO_2 should be controlled within 1.2 million tons by the end of 2020.

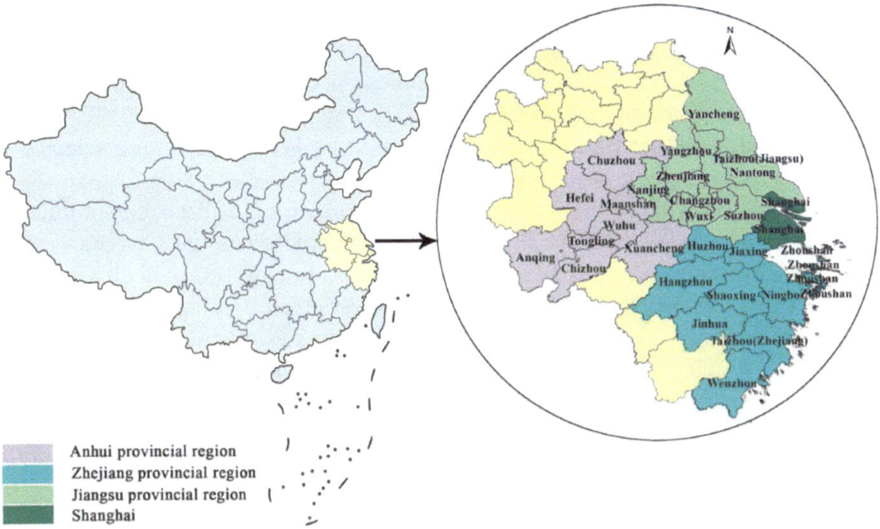

Fig. 16.3 Research object

16.3.5.2 WRF-CMAQ/ISAM Mode Setting and Model Verification

In this paper, WRF v4.0 and CMAQ v5.3.2 models are selected to simulate the mass concentration of pollutants in the Yangtze River Delta city agglomeration, and the ISAM tool is used to analyze the source of SO_2. CMAQ adopts CB06 as the gas phase chemical scheme and AERO6 as the aerosol scheme. The anthropogenic basic emission list is the high-resolution MEIC list published by Tsinghua University, and the simulation time is from January 1 to December 31, 2015, and the simulation area is 27 cities in the Yangtze River Delta Group. In this paper, the data from 471 monitoring stations in the Yangtze River Delta city agglomeration are selected to verify the simulation performance. The meteorological data comes from China's National Meteorological Center (NMC), which provides the hourly average monitoring data of the Yangtze River Delta city agglomeration. Statistical parameters used for model verification mainly include Pearson coefficient (R), normalized mean bias (NMB), mean fractional bias (MFB), mean fractional error (MFE) and normalized mean error (NME). According to the research of Boylan and Russell (2006), if both $|MFB| \leq 60\%$ and $|MFE| \leq 75\%$, the performance of the model meets the simulation requirements.

16.3.5.3 Two-Stage Network DEA Variable Setting

Table 16.1 summarizes the two-stage input–output variable settings in the related literature. For the selection of input–output indicators in the first stage, most scholars choose manpower, property power and financial resources as the input indicators, and select GDP and pollutants as the output indicator (Cai & Ye, 2019; Feng et al., 2015; Zhou & Wang, 2016). Like most scholars, we set coal consumption, city labor and investment in fixed assets as the input indicators of this paper, and the expected output and unexpected output are GDP and SO_2 emissions, respectively (for unified standards and units, the SO_2 emissions simulated by WRF-CMAQ are selected in this paper).

The allocation in the second stage is based on the treatment effect of city pollutants. At present, most studies focus on the evaluation of environmental governance efficiency. For example, Wu et al. (2021) took various pollutants, rainfall, city greening rate and waste gas treatment investment as input variables, and regarded the days of air quality reaching the standard and $PM_{2.5}$ emissions as output variables to evaluate China's provincial environmental governance efficiency. Wang and Feng (2020) used the two-stage DEA model to evaluate the waste gas treatment efficiency of China's industrial sector. In the second stage, they took waste gas emissions, capital investment and waste gas treatment facilities investment as input indicators, and the purification rate of waste gas as the output indicator. Li et al. (2020) established a two-stage network DEA to evaluate China's provincial pollutant emission reduction efficiency. In the second stage, the treatment cost of various pollutants was selected as the input index, and the annual reduction of various pollutants was selected as the output index. As we all know, the significance of environmental governance is, in part, to reduce the harm of pollutants to human health. For example, pollutants can obviously promote the onset of respiratory diseases. Scholars have found that SO_2 can inhibit the apoptosis of cancer cells and promote the incidence of respiratory diseases in children after entering the body. SO_2 and $PM_{2.5}$ particles formed by its oxidation can both increase the incidence of lung cancer (Meo et al., 2024; Nascimento et al., 2020; Zhang et al., 2021b). Therefore, health factors should also be considered in the allocation model. City environmental self-purification capacity is usually regarded as an input index, such as city green area, forest coverage and precipitation. The index of city environmental self-purification capacity should be artificially controllable. For example, compared with the uncontrollable precipitation, the government can take measures to increase city greening area and forest coverage. Furthermore, compared with city greening, forests can absorb more SO_2. Based on the above analysis, this paper selects the first stage's SO_2 emissions (including contributions from other cities), financial investment in environmental protection, the number of employees in environmental protection, and per capita forest area as the second stage's input indicators. Non-expected output and expected output indicators are the lung cancer incidence and the number of days with air quality above Grade 2. The input–output variables of the two stages in this paper as shown in Table 16.2. The descriptive statistics for the above variables of the two-stage network DEA allocation model are listed in Table 16.3.

The values of α_1 and α_2 are usually set according to the preference of decision-makers. For example, when evaluating oil production efficiency and wastewater

Table 16.1 The two-stage input–output variable settings in the related literature

Literatures	Domain	Input	Desirable output	Undesirable output
Wu et al. (2016b)	Emission rights allocation	Capital	Value added	CO_2
		Labor		
		Energy		
Momeni et al. (2019)	Emission rights allocation	Population	GDP	CO_2
		Labor		
		Capital		
		Energy		
Guo et al. (2021)	Emission rights allocation	Capital	GDP	$PM_{2.5}$
		Labor		
		Energy		
Zhou et al. (2014)	Environmental efficiency evaluation	Capital	GDP	CO_2
		Labor		
		Energy		
Wu et al. (2021)	Environmental efficiency evaluation	Various pollutants	Days above AQI level 2	$PM_{2.5}$
		Investment of waste gas treatment		
		Greening ratio		
Li et al. (2020)	Environmental efficiency evaluation	Assets	GDP	SO_2
		Labor		CO_2
		Energy		NO_2
		Treatment facilities		
Chu et al. (2020)	Bank fixed cost allocation	Labor	Interest income	Bad debts
		Fixed assets	Non-interest income	
		Operations costs		
Wu et al. (2016a)	Environmental efficiency evaluation	Coal expenses	GDP	Wastewater
		Capital	Comprehensive values	Solid wastes
		Labor		Unclean residuals
		Treatment investments		
		Solid (water) wastes		
Wang and Feng (2020)	Environmental efficiency evaluation	Capital	Industrial gross output	CO_2
		Labor		Waste gas
		Energy	Waste removal rate	Solid waste

(continued)

Table 16.1 (continued)

Literatures	Domain	Input	Desirable output	Undesirable output
		Waste material		
		Waste removal operating expenses		

Table 16.2 The input–output variables of the two-stage in this paper

Different stages	Variable category	Variables name	Data source	Unit
Stage-1	Input variables	Coal consumption	China City Statistical Yearbook 2016[a]	10,000 tons
		Investment in fixed assets	China City Statistical Yearbook 2016	100 million yuan
		City labor	China City Statistical Yearbook 2016	10,000 persons
	Output variables	GDP	China City Statistical Yearbook 2016	100 million yuan
		SO_2	WRF-CMAQ simulation	ug/m^3
Stage-2	Input variables	SO_2	WRF-CMAQ simulation	ug/m^3
		Financial investment in environmental protection	China City Statistical Yearbook 2016	10,000 yuan
		Number of employees in the environmental protection	China City Statistical Yearbook 2016	10,000 persons
		Forest area per capita	Local ecological departments	m^2
	Output variables	Lung cancer incidence[b]	Provincial Cancer Registry Annual Reports 2018 and Centers for Disease Control and Prevention	1/100000
		Number of days with air quality above Grade 2	www.aqistudy.cn	days

[a]The 2016 edition of the China City Statistical Yearbook released the 2015 data
[b](1) The annual report statistics are based on county-level data. In this paper, the average of county-level data is taken as the city lung cancer incidence. (2) The 2018 edition of the annual report publishes 2015 data. (3) Anqing, Yangzhou, and Nanjing didn't publish lung cancer incidence rates in 2015, and the values for these three cities were replaced by the mean values of adjacent counties

Table 16.3 The descriptive statistics of variables

Different stages	Variable	Unit	Mean	Max	Min	SD
Stage-1	Raw coal consumption	10,000 tons	1518	4931	266	1178
	Labor	10,000 persons	378	1362	75	251
	Investment in fixed assets	100 million yuan	3077	6353	601	1685
	GDP	100 million yuan	5198	25,123	545	5117
Stage-2	Output SO_2	ug/m^3	105,229	273,813	7798	68,001
	Financial investment in environmental protection	10,000 yuan	202,956	1,043,500	33,419	218,835
	Forest area per capita	m^3	763	3084	28	699
	Number of employees in the environmental protection	person	12,933	90,623	2539	16,413
	Input SO_2	ug/m^3	105,229	227,505	25,652	62,723
	Lung cancer incidence	1/100000	61	98	36	16
	Number of days with air quality above Grade 2	days	270	345	218	31

treatment efficiency, Hu et al. (2019) considered that oil production and wastewater treatment are equally important. Therefore, they set $\alpha_1 = \alpha_2 = 0.5$. Wu et al. (2016a) adopted the two-stage network DEA to evaluate China's provincial environmental efficiency, and the range of α_1 was 0.1–0.9, with an increment of 0.1. Similarly, this paper adopts their method and generates 9 allocation schemes. Then, the fairness of the schemes is evaluated by the PRDC, and the optimal allocation scheme is selected according to the evaluation results. The Pearson's coefficient (R) is one of the parameters that characterize the correlation between variables, and it is generally considered that when $|R| \geq 0.7$, strong correlation is achieved (Akoglu, 2018). In order to give consideration to the principle of economic protection, the constraint coefficient of economic protection is set as $\gamma=1 - R$ $(R > 0)$, that is $\gamma \in (0,0.3)$.

Table 16.4 The reliability parameter verification of the CMAQ model

Parameter	Minimum (%)	Maximum (%)
NMB	−15.56	28.57
MFB	−26.99	46.77
MFE	18.93	52.87
NME	13.07	33.62

16.4 Results and Discussion

16.4.1 Verification of WRF-CMAQ Simulation Results

To verify the reliability of WRF-CMAQ simulation results, firstly, we verify the deviation and error between simulation values and monitoring values (Table 16.4). The results show that $|MFB| \leq 60\%$ and $|MFE| \leq 75\%$, proving that the simulation results are good.

Then, the correlation between the monitored values and the simulated values is verified. As CMAQ simulation takes a long time, this paper converts hourly data into daily data and then verifies the simulation results (Fig. 16.4a). Finally, we verify the correlation between SO_2 simulation value and the yearbook value (Fig. 16.4b). SO_2 simulation value is the sum of local emissions and the contribution of other cities to the region. The results demonstrate that the R values obtained from the above two verifications are greater than 0.7, indicating that the simulation results are good.

16.4.2 WRF-CMAQ Simulation Results

WRF-CMAQ simulation results reveal that SO_2 transmission exists among 27 cities in the Yangtze River Delta city agglomeration. Most of the 27 cities are located in the middle and lower reaches of the Yangtze River plain, so the SO_2 transmission is sensitive to wind direction. In order to intuitively show the wind direction of each city and the SO_2 transmission law under the influence of wind direction, we have drawn the wind direction rose diagram and SO_2 transmission diagram (Fig. 16.5). As shown in Fig. 16.5, the dominant wind direction of the Yangtze River Delta city agglomeration is northwest wind, southeast wind and northeast wind. Combined with the SO_2 transmission diagram, it can be seen that most source cities have a great influence on neighboring cities in the southeast and northwest. For example, influenced by the dominant wind direction, Nanjing has a great influence on Chuzhou in the northwest and Changzhou in the southeast. To further quantify the SO_2 transmission among 27 cities, Table 16.5 lists the mutual contribution rates of SO_2 transmission among cities. The results show that the contribution rate of Nanjing to Chuzhou is 9.14%, and that of Nanjing to Changzhou is 9.41%. Northeast wind has a significant impact on cities in the eastern coastal areas, such as Shanghai, Nantong and Zhoushan.

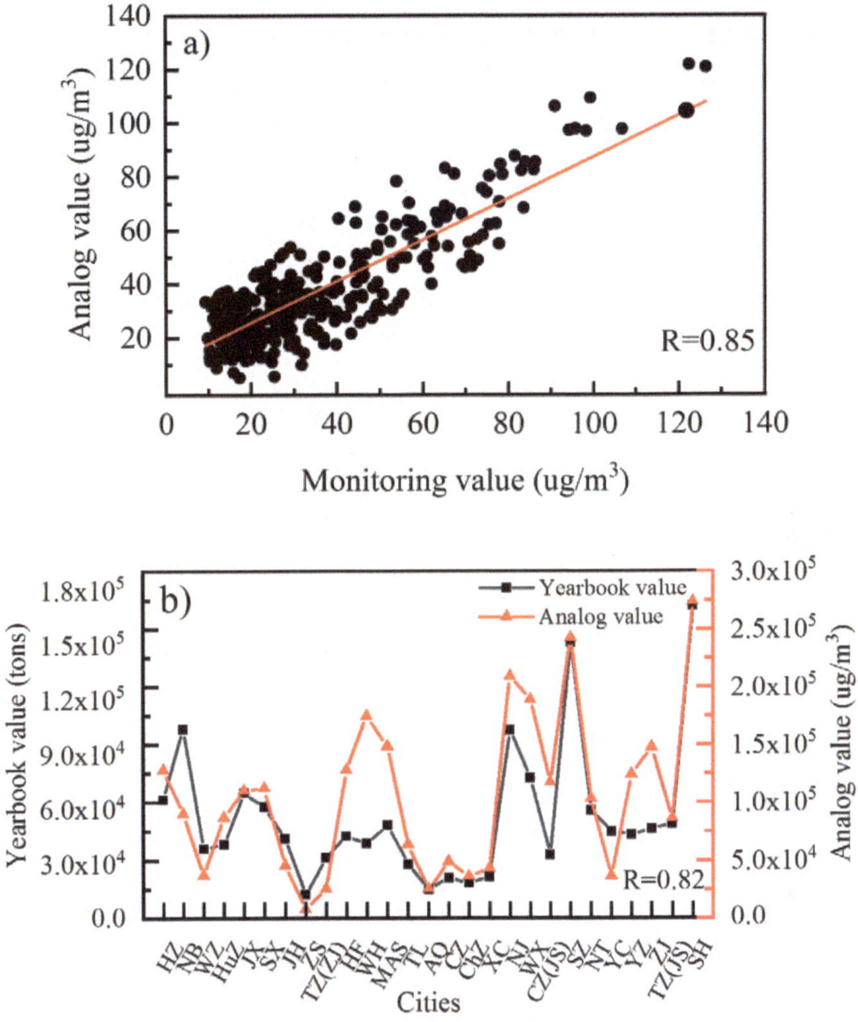

Fig. 16.4 The correlation verification of simulation results

Shanghai is affected by the northeast wind, northwest wind and southwest wind at the same time. Its SO_2 contribution rate to local areas is 62.93%, that to Jiaxing in the southwest is 11.37%, that to Suzhou in the northwest is 6.61% and that to Zhoushan in the southeast is 17.92%.

From the simulation results of this paper, we can see that SO_2 in the recipient city also comes from other areas besides the Yangtze River Delta city agglomeration. In addition, the geographical location of a city has a significant impact on its own SO_2 source. For example, since Zhoushan is surrounded by the sea on three sides, the

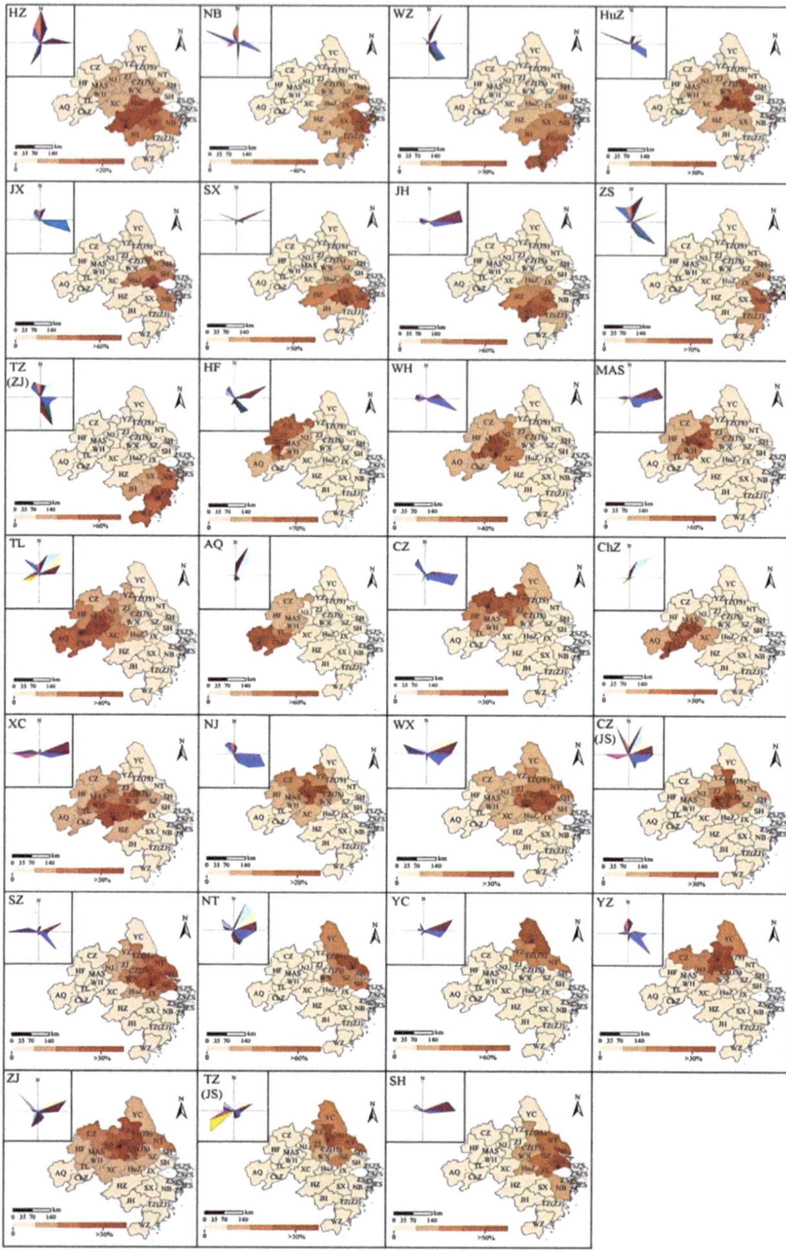

Fig. 16.5 Wind direction rose and SO_2 transmission diagram of each source city in 2015. *Note* Indicates source city

Table 16.5 The contribution rate of SO_2 transmission in the Yangtze River Delta city agglomeration

%	AQ	ChZ	CZ	HF	MAS	TL	WH	XC	CZ (JS)	NJ	NT	SZ	TZ (JS)
AQ	30.03	4.21	0.32	3.01	0.98	6.46	2.77	0.90	0.52	1.07	0.35	1.34	0.32
ChZ	12.80	27.26	0.41	2.75	0.95	7.67	2.84	1.55	0.59	1.26	0.52	1.95	0.40
CZ	0.28	0.15	21.25	16.86	2.97	0.23	1.37	0.42	1.18	9.14	0.76	1.91	1.02
HF	0.50	0.23	1.70	64.87	1.85	0.41	1.81	0.38	0.42	2.05	0.27	0.82	0.30
MAS	0.13	0.39	0.41	1.65	44.17	0.88	34.34	1.40	0.97	3.76	0.30	1.01	0.33
TL	1.76	8.36	0.42	7.71	2.22	34.25	18.30	2.24	0.71	1.61	0.42	1.56	0.35
WH	0.26	10.65	0.30	1.60	5.62	19.30	46.05	3.90	0.77	1.93	0.35	1.31	0.31
XC	0.29	3.02	0.48	1.41	2.63	1.79	11.37	30.25	2.12	4.19	0.96	4.15	0.73
CZ (JS)	0.05	0.10	0.35	0.30	2.17	0.15	1.11	3.20	26.88	9.41	1.48	6.02	2.38
NJ	0.14	0.21	9.38	1.68	18.49	0.35	5.80	2.02	2.24	33.77	0.71	2.09	1.00
NT	0.02	0.02	0.10	0.12	0.13	0.03	0.11	0.07	0.97	0.49	40.60	14.98	8.99
SZ	0.02	0.02	0.07	0.10	0.12	0.03	0.13	0.25	1.30	0.34	3.16	43.07	3.23
TZ (JS)	0.03	0.05	0.41	0.22	0.34	0.06	0.27	0.15	13.44	1.76	3.76	5.79	21.67
WX	0.03	0.05	0.16	0.18	0.37	0.07	0.38	1.28	18.94	1.39	2.06	13.28	1.79
YC	0.07	0.05	0.62	0.45	0.33	0.07	0.26	0.10	0.99	1.65	6.70	3.84	8.48
YZ	0.07	0.07	6.22	0.50	0.68	0.10	0.43	0.14	1.95	14.63	2.00	2.64	5.64
ZJ	0.04	0.07	0.83	0.27	1.38	0.10	0.68	0.37	4.03	40.34	0.98	2.28	2.33
SH	0.01	0.01	0.03	0.05	0.04	0.01	0.04	0.03	0.14	0.10	2.21	14.81	0.18
HZ	0.08	0.11	0.18	0.40	0.37	0.14	0.50	1.25	0.47	0.69	0.52	2.53	0.29
HuZ	0.07	0.10	0.20	0.36	0.50	0.15	0.63	6.34	1.32	1.22	0.87	5.99	0.51
JX	0.01	0.02	0.05	0.09	0.07	0.02	0.09	0.11	0.28	0.22	0.43	7.13	0.19
JH	0.04	0.05	0.12	0.22	0.18	0.06	0.21	0.23	0.32	0.42	0.38	1.57	0.20

(continued)

Table 16.5 (continued)

%	AQ	ChZ	CZ	HF	MAS	TL	WH	XC	CZ (JS)	NJ	NT	SZ	TZ (JS)
NB	0.01	0.01	0.06	0.09	0.08	0.02	0.08	0.06	0.22	0.23	0.37	1.79	0.18
SX	0.02	0.02	0.07	0.11	0.09	0.03	0.10	0.12	0.24	0.29	0.30	1.48	0.12
TZ (ZJ)	0.02	0.02	0.10	0.13	0.11	0.03	0.10	0.08	0.31	0.35	0.41	1.65	0.23
WZ	0.02	0.02	0.10	0.13	0.12	0.03	0.11	0.08	0.20	0.28	0.27	1.09	0.16
ZS	0.02	0.02	0.09	0.15	0.12	0.02	0.11	0.06	0.22	0.29	1.02	2.72	0.27

%	WX	YC	YZ	ZJ	SH	HZ	HuZ	JX	JH	NB	SX	TZ (ZJ)	WZ	ZS
AQ	0.86	0.16	0.44	0.54	1.05	0.97	0.45	0.35	0.32	0.25	0.49	0.05	0.07	0.01
ChZ	1.17	0.21	0.49	0.59	1.69	2.24	0.65	0.58	0.89	0.54	1.30	0.13	0.19	0.02
CZ	1.44	0.51	3.34	2.50	1.53	0.79	0.53	0.52	0.11	0.32	0.41	0.03	0.03	0.01
HF	0.58	0.15	0.60	0.68	0.72	0.48	0.29	0.25	0.08	0.15	0.22	0.02	0.02	0.01
MAS	0.95	0.11	0.49	0.98	0.82	0.72	0.58	0.38	0.08	0.22	0.32	0.02	0.03	0.01
TL	1.12	0.15	0.48	0.69	1.18	1.25	0.71	0.47	0.26	0.30	0.58	0.05	0.07	0.01
WH	1.04	0.11	0.40	0.69	1.03	1.14	0.73	0.45	0.18	0.29	0.50	0.04	0.05	0.01
XC	3.35	0.28	0.77	1.30	3.65	5.26	3.53	1.72	0.81	1.15	2.25	0.16	0.23	0.04
CZ (JS)	11.27	0.33	2.42	19.17	3.01	1.68	1.99	1.30	0.15	0.65	0.89	0.06	0.06	0.02
NJ	2.05	0.33	3.82	4.56	1.55	1.00	0.84	0.64	0.11	0.36	0.50	0.04	0.04	0.01
NT	4.57	1.17	1.04	1.41	17.26	0.38	0.18	0.43	0.07	0.43	0.31	0.05	0.04	0.03
SZ	28.39	0.18	0.30	0.51	6.61	1.46	5.15	3.02	0.12	0.71	0.86	0.05	0.05	0.03
TZ (JS)	6.84	2.49	14.63	17.49	2.55	0.55	0.42	0.50	0.07	0.33	0.36	0.04	0.03	0.01
WX	40.05	0.22	0.79	2.45	4.06	1.74	3.10	1.75	0.14	0.73	0.98	0.07	0.05	0.03
YC	2.18	35.47	5.12	1.82	3.72	0.46	0.26	0.41	0.07	0.36	0.32	0.05	0.04	0.02
YZ	2.00	1.75	38.89	8.56	1.73	0.55	0.36	0.42	0.07	0.30	0.34	0.03	0.03	0.01

(continued)

Table 16.5 (continued)

%	WX	YC	YZ	ZJ	SH	HZ	HuZ	JX	JH	NB	SX	TZ (ZJ)	WZ	ZS
ZJ	2.55	0.33	12.17	25.03	1.27	0.59	0.51	0.45	0.06	0.26	0.34	0.03	0.02	0.01
SH	0.70	0.09	0.09	0.10	62.93	0.46	0.16	3.25	0.07	1.65	0.51	0.08	0.06	0.16
HZ	1.08	0.14	0.29	0.41	3.97	53.47	3.05	3.09	3.78	2.11	10.32	0.26	0.37	0.06
HuZ	3.75	0.18	0.42	0.74	5.89	17.41	33.85	6.61	0.66	1.87	3.57	0.15	0.16	0.06
JX	0.93	0.07	0.17	0.23	11.37	14.37	14.17	36.83	0.30	3.48	4.74	0.10	0.08	0.09
JH	0.63	0.09	0.24	0.31	3.14	8.82	0.73	0.95	57.49	1.94	6.78	0.92	0.92	0.07
NB	0.53	0.09	0.19	0.22	6.24	7.72	0.43	4.12	0.44	48.44	16.86	2.28	0.43	0.97
SX	0.48	0.07	0.18	0.23	3.40	19.16	0.95	1.71	8.02	2.98	56.73	0.45	0.29	0.07
TZ (ZJ)	0.63	0.12	0.26	0.31	4.06	1.38	0.39	0.84	2.26	10.57	2.65	41.48	16.67	0.60
WZ	0.43	0.09	0.17	0.20	2.29	0.86	0.25	0.44	0.81	3.12	0.98	5.91	62.77	0.18
ZS	0.66	0.22	0.21	0.21	17.92	1.04	0.27	1.78	0.18	25.20	1.42	0.92	0.39	10.33

SO_2 contribution rate of the Yangtze River Delta city agglomeration to Zhoushan is only 65.84%. Influenced by the northwest wind, the contribution rate of Ningbo to Zhoushan reaches 25.2%. Zhenjiang is located in the middle of the Yangtze River Delta city agglomeration, and the contribution rate of areas outside the Yangtze River Delta city agglomeration is about 2.7%. Similarly, Anqing is the westernmost city in the Yangtze River Delta, and the contribution rate of the city agglomeration is only 58.3%.

16.4.3 Allocation Results of the Two-Stage Network DEA

The calculation results of two-stage network DEA allocation model show that the objective function value is 1, that is, the overall efficiency value and the two-stage efficiency value are both optimal. The two-stage allocation results (Fig. 16.6) show that there is a big gap between the SO_2 emission rights quotas of different cities in the two stages. Among them, Shanghai's quotas reach the maximum in both stages (134,300 tons and 91,900 tons, respectively), and the difference between the two stages is also large. Chizhou has the lowest allocation in the first stage, only 12,400 tons, while the allocation in the second stage is 74,400 tons, second only to Shanghai. The main reason is that in the SO_2 treatment stage, the financial investment in environmental protection, the number of environmental protection employees and the incidence of lung cancer are relatively low, and the number of days with air quality above Grade 2 is the largest. Therefore, Chizhou performs well in the pollutant treatment stage. This further shows that when studying the allocation of pollutant emission rights, the impact of pollutant treatment on the allocation results cannot be ignored. The quotas of Suzhou and Nanjing in the first stage are 111,300 tons and 70,800 tons, while the quotas in the second stage only account for 61.6% and 57.4% of the first stage respectively. That is, Suzhou and Nanjing perform poorly in the SO_2 treatment stage. The main reason is that compared with other cities, Nanjing has fewer days with air quality above grade two, and Suzhou has a higher incidence of lung cancer. In summary, most cities have a significant performance gap between the first stage and the second stage, which further proves the rationality of introducing the SO_2 treatment stage into the allocation model in this paper. In terms of provincial regions, the order of average quotas of provincial regional cities in the two stages is: Shanghai > Jiangsu > Zhejiang > Anhui (the first stage), Shanghai > Zhejiang > Jiangsu > Anhui (the second stage). It can be seen that Zhejiang has better performance in the SO_2 treatment stage than Jiangsu, and the performance of Anhui in both stages is poor. Overall, the standard deviations of the quotas in the two stages are 2.76 and 1.64 respectively. The main reason is that in the second stage, SO_2 transmission reduces the quota gap between cities, resulting in a small standard deviation. It further proves the necessity of co-governance of pollutant in city agglomeration.

Table 16.6 shows nine allocation results of SO_2 emission rights in Yangtze River Delta city agglomeration under different values of α_1. The results confirm that the

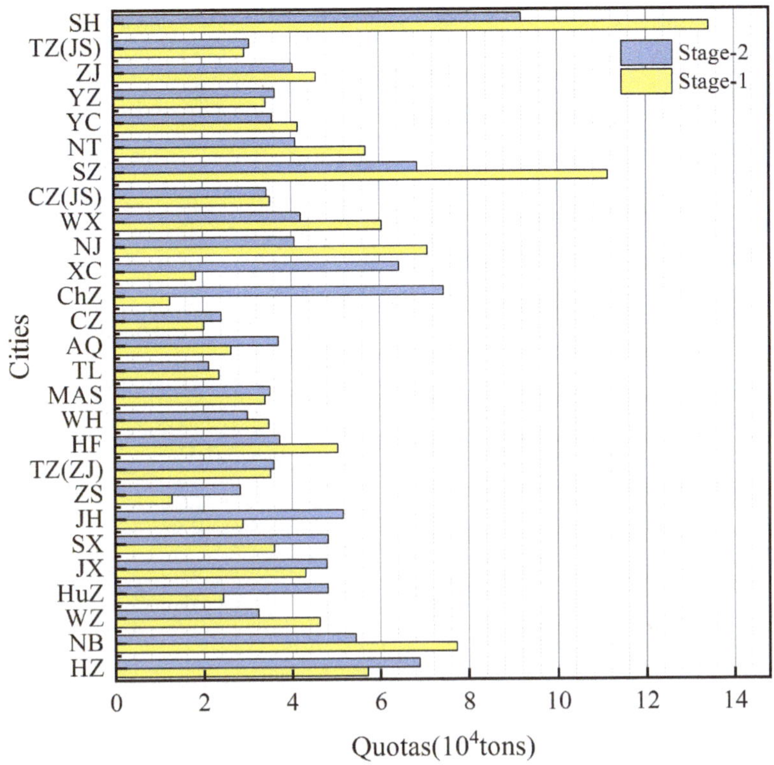

Fig. 16.6 Allocation of SO_2 emission right in two stages

quotas of 13 cities increase with the increase of α_1, and the quotas of 14 cities decrease with the increase of α_1. Different cities have different performances in the two stages. With the increase of α_1, the quotas of cities with better performance in the SO_2 treatment stage will decrease. For example, when $\alpha_1 = 0.1$, Chizhou's quotas are 68,200 tons. When $\alpha_1 = 0.9$, Chizhou's quotas are only 27% of that when $\alpha_1 = 0.1$. Similarly, Suzhou's advantage in the first stage was greater than that in the second stage. When α_1 increases from 0.1 to 0.9, the quotas increase by 46.78%. From the provincial perspective, 78% of cities in the Zhejiang provincial region perform better in the second stage than in the first stage, 63% of cities in the Anhui provincial region perform better in the second stage, only 22% of cities in Jiangsu provincial region perform better in the second stage, and Shanghai performs better in the first stage than in the second stage. Assuming that the two stages are of equal importance, Jiangsu and Shanghai need to strengthen SO_2 treatment, while Zhejiang and Anhui should focus on the first stage.

Table 16.6 Allocation of SO_2 emission rights under different weight coefficients

Serial number	DMU	α_1								
		0.1	0.2	0.3	0.4	0.5	0.6	0.7	0.8	0.9
1	HZ	6.77	6.65	6.53	6.41	6.30	6.18	6.06	5.94	5.83
2	NB	5.66	5.89	6.12	6.34	6.57	6.80	7.03	7.26	7.49
3	WZ	3.37	3.51	3.65	3.79	3.93	4.07	4.21	4.34	4.48
4	HuZ	4.56	4.33	4.09	3.86	3.62	3.38	3.15	2.91	2.68
5	JX	4.73	4.68	4.63	4.59	4.54	4.49	4.44	4.39	4.34
6	SX	4.69	4.57	4.45	4.33	4.21	4.08	3.96	3.84	3.72
7	JH	4.94	4.71	4.48	4.25	4.02	3.80	3.57	3.34	3.11
8	ZS	2.67	2.51	2.36	2.20	2.05	1.89	1.74	1.58	1.42
9	TZ (ZJ)	3.58	3.57	3.56	3.56	3.55	3.54	3.53	3.52	3.51
10	HF	3.85	3.98	4.11	4.25	4.38	4.51	4.64	4.77	4.91
11	WH	3.04	3.09	3.14	3.18	3.23	3.28	3.33	3.37	3.42
12	MAS	3.49	3.48	3.47	3.46	3.44	3.43	3.42	3.41	3.40
13	TL	2.14	2.16	2.19	2.21	2.23	2.26	2.28	2.30	2.33
14	AQ	3.59	3.48	3.38	3.27	3.17	3.06	2.95	2.85	2.74
15	CZ	2.37	2.33	2.29	2.25	2.21	2.17	2.13	2.09	2.05
16	ChZ	6.82	6.20	5.58	4.96	4.34	3.72	3.10	2.48	1.86
17	XC	5.97	5.51	5.05	4.59	4.13	3.67	3.21	2.75	2.29
18	NJ	4.37	4.67	4.97	5.27	5.57	5.88	6.18	6.48	6.78
19	WX	4.40	4.58	4.77	4.95	5.13	5.32	5.50	5.69	5.87
20	CZ (JS)	3.45	3.45	3.46	3.47	3.47	3.48	3.49	3.49	3.50
21	SZ	7.29	7.71	8.14	8.57	9.00	9.42	9.85	10.28	10.70
22	NT	4.26	4.42	4.58	4.74	4.90	5.06	5.21	5.37	5.53
23	YC	3.63	3.69	3.74	3.80	3.86	3.92	3.98	4.03	4.09
24	YZ	3.61	3.59	3.57	3.55	3.53	3.51	3.49	3.47	3.45
25	ZJ	4.10	4.15	4.20	4.25	4.31	4.36	4.41	4.46	4.52
26	TZ (JS)	3.05	3.04	3.03	3.02	3.01	3.00	2.98	2.97	2.96
27	SH	9.62	10.04	10.47	10.89	11.31	11.74	12.16	12.58	13.01

16.4.4 Fairness Evaluation of Allocation Scheme

Figure 16.7 gives the PRDC evaluation results of SO_2 emission rights allocation in Yangtze River Delta city agglomeration under different values of α_1. The results show that with the increase of α_1, the number of cities with PRDC ≥ 0.5 is gradually decreasing. When $\alpha_1 = 0.1$, the PRDC values of six cities are greater than 0.5. When $\alpha_1 = 0.7$, 0.8 or 0.9, the PRDC values of all cities are less than 0.5, which are the feasible schemes. On the whole, with a small population, Chizhou performs well in the SO_2 treatment stage. The PRDC values of the nine allocation schemes are close to 0. However, due to the poor performance of Chuzhou in both stages, PRDC ≥ 0.5 for $\alpha_1 = 0.1 \sim 0.6$. Similarly, although Shanghai displays the best performance in both stages, the PRDC values are less than 0.5 only when $\alpha_1 = 0.7, 0.8$ or 0.9 due to its large population. From the provincial level, the PRDC of Shanghai is higher than those of other regions, followed by the Jiangsu provincial region, and the PRDC of Zhejiang province is lower than those of other regions. To sum up, the PRDC evaluation results show that the three allocation schemes reach the fairness standard when $\alpha_1 = 0.7, 0.8$ or 0.9.

Figure 16.8 shows three feasible allocation schemes that simultaneously meet the principle of maximum efficiency, economic protection and fairness. Overall, most cities have relatively small changes in their quotas across the three allocation schemes. With the increase of α_1, the quotas of Chizhou and Xuancheng decrease by 40.03% and 28.66%, while the quotas of other cities change by less than 20%. Shanghai, Suzhou and Ningbo have relatively large allocations, all of which are greater than 70,000 tons, while Chizhou, Tongling, Chuzhou and Zhoushan have relatively small allocations, all of which are less than 25,000 tons. Nanjing and Hangzhou, the capitals of Jiangsu and Zhejiang, respectively, are similar in terms of city scale, economy, industry structures, and their quotas are also close to each other.

According to the average allocation results, the Yangtze River Delta city agglomeration can be roughly divided into three categories. The quotas of the first class cities are much higher than the average quota: Shanghai, Suzhou, Ningbo, Nanjing, Hangzhou and Wuxi. The quotas of the second class cities are near the average level: Nantong, Hefei, Zhenjiang, Jiaxing, Wenzhou, Yancheng, Shaoxing, Taizhou, Changzhou, Yangzhou and Maanshan. The quotas of the third class cities are far

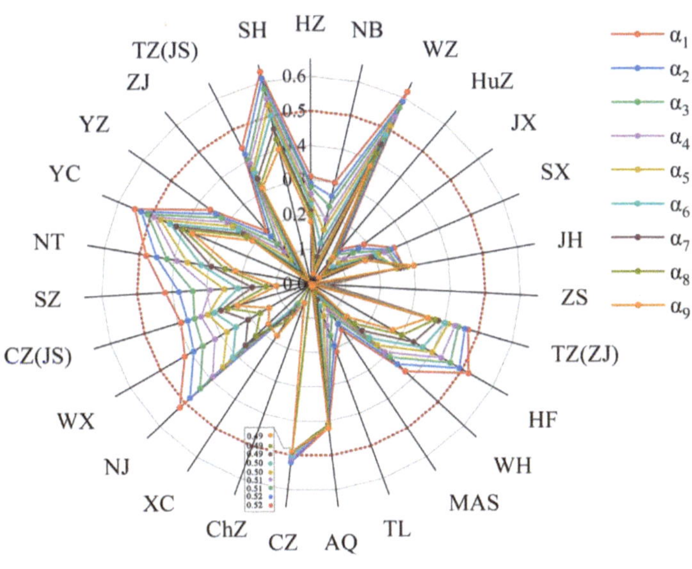

Fig. 16.7 The PRDC values under 9 allocation schemes

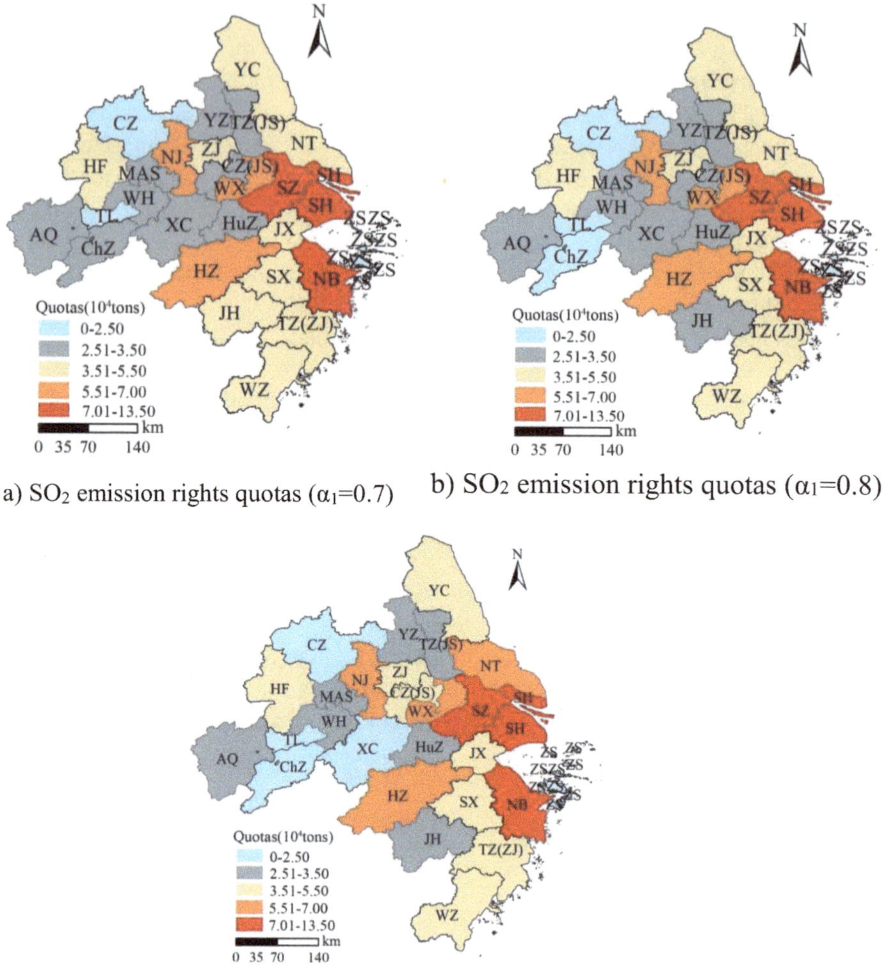

a) SO_2 emission rights quotas (α_1=0.7)

b) SO_2 emission rights quotas (α_1=0.8)

c) SO_2 emission rights quotas (α_1=0.9)

Fig. 16.8 Three optimal allocation schemes

less than the average level: Wuhu, Jinhua, Taizhou, Huzhou, Anqing, Xuancheng, Chizhou, Tongling, Chuzhou and Zhoushan. From the perspective of provincial regions, Anhui provincial region has a higher proportion of cities which belong to the third class, and Jiangsu provincial region has a higher proportion of cities which belong to the first class.

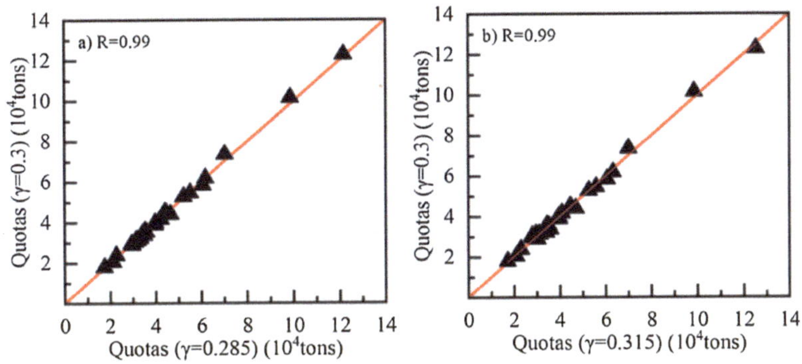

Fig. 16.9 Sensitivity test of two-stage network DEA allocation model

16.4.5 Model Sensitivity Analysis

The higher the sensitivity of the model, the worse the robustness of the model. On the contrary, the model is robust. The model with better robustness has better universality and transferability. In order to verify the sensitivity of allocation results to the changes of model parameters, this paper studies the changes of allocation results by adjusting the value of γ ($\pm 5\%$). Firstly, we recalculate the allocation results by taking $\gamma = 0.285$ and $\gamma = 0.315$ respectively. Secondly, take the allocation result when $\alpha_1 = 0.7$ as an example to verify the sensitivity (Fig. 16.9). The result indicates that when the value of γ fluctuates within $\pm 5\%$, the average fluctuation of each DMU's quotas is within 0.1%. Furthermore, we conducted a paired-sample t-test between the allocation results obtained when $\gamma = 0.285$ and $\gamma = 0.315$ and the original allocation results, respectively. The results showed that the p-values of the two groups of samples were both greater than 0.05, that is, there was no significant difference between the samples. Therefore, the model in this paper has good robustness and transferability.

16.5 Conclusions and Suggestions

16.5.1 Conclusions

In this paper, a two-stage network DEA allocation model based on pollutant transmission is proposed, and the allocation of SO_2 emission rights in China's Yangtze River Delta city agglomeration is taken as an empirical case. Different from previous studies, this paper combines WRF-CMAQ with DEA allocation model for the first time, and introduces environmental factors and health factors. The scheme finally takes into account the three principles of efficiency, economic protection and fairness. The main conclusions are as follows:

1. For the research on the allocation of SO_2 emission rights in city agglomerations, we cannot ignore such factors as regional SO_2 mutual transmission, environmental self-purification and health factors. WRF-CMAQ simulation results reveal the law of SO_2 transmission in the Yangtze River Delta city agglomeration, and provide a basis for the coordinated treatment of SO_2 in the Yangtze River Delta.
2. Compared with the single-stage DEA allocation model, the two-stage network allocation model introduces the pollutant treatment stage, which promotes the treatment of city pollutants while achieving SO_2 emission reduction.
3. When α_1 is 0.7, 0.8 or 0.9, the per capita relative deprivation coefficient of the three allocation schemes is less than 0.5. This conclusion can provide a reference for decision-makers who can determine the optimal allocation scheme according to their preference or actual situation.
4. The results of sensitivity analysis showed that when the value of γ changes within $\pm 5\%$, the average fluctuation of allocation results is within 0.1%. With good robustness and transferability, this model can be applied to pollutant allocation in different regional city agglomerations, such as Pearl River Delta city agglomeration, Beijing-Tianjin-Hebei city agglomeration, etc.

16.5.2 Suggestions

Based on the above research and findings, it is suggested that the Chinese government should consider the principles of efficiency, fairness and economic protection in addition to historical emission factors when allocating pollutant emission rights in regional city agglomerations, so as to be better committed to the environment-economy-energy sustainable development. Considering that the cross-border transmission effect of pollutants has not been considered in the actual city agglomeration pollution treatment, it is recommended that decision-makers could establish an effective cross-regional and cross-departmental environmental co-governance mechanism based on regional city agglomerations to jointly deal with the increasingly serious environmental pollution problems. Since city pollutants have the characteristics of cross-regional transmission, the impact of pollutant emission source cities on neighboring cities is more significant. The significance of pollutant treatment is, in part, to reduce the impact of the environment on residents' health. Therefore, the health of every resident in a city agglomeration is affected not only by the pollutants in the city where they live, but also by other cities.

Pollutant treatment has a great impact on the allocation results of pollutant emission rights. In the research on the allocation of pollutant emission rights, more attention should be paid to the pollutant treatment stage. For example, Cities such as Chizhou and Xuancheng that perform well in the pollutant treatment stage should be allocated more quotas to further encourage other cities to pay more attention to pollutant treatment.

In addition, policymakers can also optimize and adjust the energy structure, such as building an energy optimization model, and conducting a comparative analysis of energy transition with developed countries (Fang et al., 2020, 2022b), so as to fundamentally reduce the consumption of primary energy and improve the environment.

16.5.3 Limitations and Future Research

There are still some limitations in this paper, which can be solved in future research. First, although coal is the major source of SO_2 emission, it is undeniable that transportation and catering industries also contribute to SO_2 emission. If the above factors are introduced into the allocation model, it will be closer to the actual situation. Then, although SO_2 has been proved to be one of the culprits of lung cancer, the impact of pollutants on human health is a long-term and complicated process. Therefore, when selecting health indicators for the allocation model, we should consider more environmental and cyclical factors in the future.

Acknowledgements Jiqiang Zhao and Chao Gao also made great contributions to this manuscript. We express our heartfelt thanks to them. This research also was supported by: National Natural Science Foundation of China (72474127) and National Social Science Fund Major Bidding Project of China (18ZDA052). Part of this chapter was published: Zhao J, Wu X, Guo J, Gao C. Allocation of SO_2 emission rights in city agglomerations considering cross-border transmission of pollutants: A new network DEA model. Applied Energy 2022;325:119927. https://doi.org/10.1016/j.apenergy.2022.119927. The publication has been conducted with copyright permission.

Appendix A: Input-Output Variables in the Stage-1

Serial number	Cities	Raw coal consumption (10,000 tons)	Labor (10,000 persons)	Investment in fixed assets (100 million yuan)	GDP (100 million yuan)	Output SO_2 (ug/m^3)	Proportion of emission reduction in 2020 (%)
1	HZ	1339.76	663.03	5556.32	10,050.21	127,972.76	23
2	NB	3466.08	509.5	4506.58	8003.61	90,138.37	17
3	WZ	1192.48	573.98	3456.39	4618.08	36,890.28	15
4	HuZ	818.18	184.48	1402.64	2084.26	87,094.17	23
5	JX	1555.31	328.91	2513.82	3517.81	110,381.35	21
6	SX	1008.59	346.8	2582.84	4465.97	112,384.80	22

(continued)

(continued)

Serial number	Cities	Raw coal consumption (10,000 tons)	Labor (10,000 persons)	Investment in fixed assets (100 million yuan)	GDP (100 million yuan)	Output SO_2 (ug/m^3)	Proportion of emission reduction in 2020 (%)
7	JH	762.80	346.01	1836.16	3402.34	45,682.79	21
8	ZS	566.90	74.5	1134.76	1092.85	7797.93	3
9	TZ (ZJ)	1076.66	407.32	1996.03	3553.85	25,473.87	13
10	HF	1097.18	523.8	5851.9	5660.27	127,916.64	23.1
11	WH	1061.85	207.1	2709.19	2457.32	174,681.56	17.9
12	MAS	1365.82	139.98	1859.84	1365.3	148,769.52	23.1
13	TL	1033.90	119.04	1062.93	911.6	63,681.59	17.6
14	AQ	635.82	345.99	1385.58	1417.43	24,841.39	16
15	CZ	265.51	288.7	1457.6	1305.7	49,028.07	13.2
16	ChZ	369.99	114.85	600.54	544.74	36,183.86	5
17	XC	433.68	202.9	1283.36	971.46	43,143.89	8.5
18	NJ	2741.14	455	5425.98	10,015.73	209,159.07	20
19	WX	2278.42	390	4888.55	8681.37	189,132.79	22
20	CZ (JS)	1061.46	281	3398.97	5281.31	117,435.21	20
21	SZ	4931.33	691.4	5965.44	14,468.68	242,078.93	20
22	NT	2062.15	460	4376.03	6498.29	103,410.22	20
23	YC	1244.73	445.7	3372.89	4181.71	36,433.57	18
24	YZ	1060.40	264.5	2856.82	4099.91	124,424.70	20
25	ZJ	2039.36	193.1	2541.07	3088.47	147,642.30	20
26	TZ (JS)	782.89	281.3	2693.75	3494.93	85,596.92	22
27	SH	4728.13	1361.51	6352.7	25,123.45	273,812.81	20

Appendix B: Input-Output Variables in the Stage-2

Serial number	Cities	Financial investment in environmental protection (10,000 yuan)	Forest area per capita (m^3)	Number of employees in the environmental protection (person)	Iutput SO_2 (ug/m^3)	Lung cancer incidence (1/100000)	Number of days with air quality above Grade 2 (days)
1	HZ	445,749	1495.95	29,531	49,964.40	67.80	242
2	NB	288,768	802.59	16,024	85,163.35	94.00	293
3	WZ	91,228	894.15	7144	34,572.00	43.20	312
4	HuZ	128,212	1123.35	5918	76,951.73	73.28	218
5	JX	155,136	485.40	10,326	178,888.83	87.68	235
6	SX	93,249	1009.49	10,411	103,900.13	80.23	270
7	JH	138,467	1398.16	15,361	43,836.85	69.20	268
8	ZS	33,419	750.66	5400	34,919.46	50.81	332
9	TZ (ZJ)	101,208	949.78	9487	36,770.09	51.17	319
10	HF	198,309	456.07	9885	112,378.49	55.05	248
11	WH	99,212	313.21	4747	125,569.00	53.30	282
12	MAS	57,305	287.06	3102	209,328.21	58.79	274
13	TL	77,312	171.87	2539	70,381.66	50.73	284
14	AQ	56,316	1150.14	5400	31,523.71	50.81	307
15	CZ	64,104	499.63	5641	61,549.59	36.46	263
16	ChZ	69,213	3083.95	4002	25,652.03	48.69	345
17	XC	50,386	2559.80	3280	39,412.15	47.53	292
18	NJ	315,700	236.42	20,086	129,737.04	54.90	235
19	WX	344,448	190.74	11,028	151,547.32	59.26	234
20	CZ (JS)	146,771	345.94	11,776	132,910.38	53.72	258
21	SZ	714,250	241.05	21,830	217,104.58	76.41	249
22	NT	155,328	339.45	10,940	146,257.85	80.65	247
23	YC	193,905	611.33	10,639	49,450.37	57.03	264
24	YZ	159,822	567.43	8160	114,870.78	47.95	248
25	ZJ	163,900	323.98	8825	227,505.24	48.74	263
26	TZ (JS)	94,600	297.98	7093	144,883.74	49.57	261
27	SH	1,043,500	28.20	90,623	206,160.38	98.07	258

References

Akoglu, H. (2018). User's guide to correlation coefficients. *Turkish Journal of Emergency Medicine, 18*, 91–93. https://doi.org/10.1016/J.TJEM.2018.08.001

An, Q., Wen, Y., Ding, T., & Li, Y. (2019). Resource sharing and payoff allocation in a three-stage system: Integrating network DEA with the Shapley value method. *Omega, 85*, 16–25. https://doi.org/10.1016/J.OMEGA.2018.05.008

An, Q., Wen, Y., Xiong, B., Yang, M., & Chen, X. (2017). Allocation of carbon dioxide emission permits with the minimum cost for Chinese provinces in big data environment. *Journal of Cleaner Production, 142*, 886–893. https://doi.org/10.1016/J.JCLEPRO.2016.07.072

Antosiewicz, M., Fuentes, J. R., Lewandowski, P., & Witajewski-Baltvilks, J. (2022). Distributional effects of emission pricing in a carbon-intensive economy: The case of Poland. *Energy Policy, 160*, 112678. https://doi.org/10.1016/J.ENPOL.2021.112678

Boylan, J. W., & Russell, A. G. (2006). PM and light extinction model performance metrics, goals, and criteria for three-dimensional air quality models. *Atmospheric Environment, 40*, 4946–4959. https://doi.org/10.1016/J.ATMOSENV.2005.09.087

Cai, W., & Ye, P. (2019). A more scientific allocation scheme of carbon dioxide emissions allowances: The case from China. *Journal of Cleaner Production, 215*, 903–912. https://doi.org/10.1016/J.JCLEPRO.2019.01.043

Cao, Y., Wang, Q., & Zhou, D. (2022). Does air pollution inhibit manufacturing productivity in Yangtze River Delta, China? Moderating effects of temperature. *Journal of Environmental Management, 306*, 114492. https://doi.org/10.1016/J.JENVMAN.2022.114492

Charnes, A., Cooper, W. W., & Rhodes, E. (1979). Measuring the efficiency of decision-making units. *European Journal of Operational Research, 3*, 339. https://doi.org/10.1016/0377-2217(79)90229-7

Cheng, Y., Gu, B., Tan, X., Yan, H., & Sheng, Y. (2022). Allocation of provincial carbon emission allowances under China's 2030 carbon peak target: A dynamic multi-criteria decision analysis method. *Science of the Total Environment, 837*, 155798. https://doi.org/10.1016/J.SCITOTENV.2022.155798

Chu, J., Wu, J., Chu, C., & Zhang, T. (2020). DEA-based fixed cost allocation in two-stage systems: Leader-follower and satisfaction degree bargaining game approaches. *Omega, 94*, 102054. https://doi.org/10.1016/J.OMEGA.2019.03.012

Cui, X., Zhao, T., & Wang, J. (2021). Allocation of carbon emission quotas in China's provincial power sector based on entropy method and ZSG-DEA. *Journal of Cleaner Production, 284*, 124683. https://doi.org/10.1016/J.JCLEPRO.2020.124683

Estellita Lins, M. P., Angulo-Meza, L., & Moreira Da Silva, A. C. (2004). A multi-objective approach to determine alternative targets in data envelopment analysis. *Journal of the Operational Research Society, 55*, 1090–1101. https://doi.org/10.1057/palgrave.jors.2601788

Fang, G., Gao, Z., Tian, L., & Fu, M. (2022a). What drives urban carbon emission efficiency? – Spatial analysis based on nighttime light data. *Applied Energy, 312*, 118772. https://doi.org/10.1016/j.apenergy.2022.118772

Fang, G., Tian, L., & Yang, Z. (2020). The construction of a comprehensive multidimensional energy index. *Energy Economics, 90*, 104875. https://doi.org/10.1016/j.eneco.2020.104875

Fang, G., Wang, L., Gao, Z., Chen, J., & Tian, L. (2022b). How to advance China's carbon emission peak?—A comparative analysis of energy transition in China and the USA. *Environmental Science and Pollution Research, 29*, 71487–71501. https://doi.org/10.1007/s11356-022-20958-9

Färe, R., & Grosskopf, S. (2000). Network DEA. *Socio-Economic Planning Sciences, 34*, 35–49. https://doi.org/10.1016/S0038-0121(99)00012-9

Feng, C., Chu, F., Ding, J., Bi, G., & Liang, L. (2015). Carbon Emissions Abatement (CEA) allocation and compensation schemes based on DEA. *Omega, 53*, 78–89. https://doi.org/10.1016/J.OMEGA.2014.12.005

Feng, S., Jiang, F., Wang, H., Shen, Y., Zheng, Y., Zhang, L., et al. (2022). Anthropogenic emissions estimated using surface observations and their impacts on $PM_{2.5}$ source apportionment over the

Yangtze River Delta, China. *Science of the Total Environment, 828*, 154522. https://doi.org/10.1016/J.SCITOTENV.2022.154522

Feng, Z., Tang, W., Niu, Z., & Wu, Q. (2018). Bi-level allocation of carbon emission permits based on clustering analysis and weighted voting: A case study in China. *Applied Energy, 228*, 1122–1135. https://doi.org/10.1016/J.APENERGY.2018.07.005

Gan, L., Ren, H., Cai, W., Wu, K., Liu, Y., & Liu, Y. (2022). Allocation of carbon emission quotas for China's provincial public buildings based on principles of equity and efficiency. *Building and Environment, 216*, 108994. https://doi.org/10.1016/J.BUILDENV.2022.108994

Glover, F., & Sueyoshi, T. (2009). Contributions of Professor William W. Cooper in operations research and management science. *European Journal of Operational Research, 197*, 1–16. https://doi.org/10.1016/J.EJOR.2008.08.011

Guo, J., Zhao, M., Wu, X., Shi, B., & Santibanez Gonzalez, E. D. R. (2021). Study on the distribution of PM emission rights in various provinces of China based on a new efficiency and equity two-objective DEA model. *Ecological Economics, 183*, 106956. https://doi.org/10.1016/J.ECOLECON.2021.106956

Han, R., Li, J., & Guo, Z. (2022). Optimal quota in China's energy capping policy in 2030 with renewable targets and sectoral heterogeneity. *Energy, 239*, 121971. https://doi.org/10.1016/j.energy.2021.121971

He, W., Liu, D., & Wang, C. (2022a). Are Chinese provincial carbon emissions allowances misallocated over 2000–2017? Evidence from an extended Gini-coefficient approach. *Sustainable Production and Consumption, 29*, 564–573. https://doi.org/10.1016/J.SPC.2021.11.007

He, W., Yang, Y., & Gu, W. (2022b). A comparative analysis of China's provincial carbon emission allowances allocation schemes by 2030: A resource misallocation perspective. *Journal of Cleaner Production, 361*, 132192. https://doi.org/10.1016/J.JCLEPRO.2022.132192

He, W., Zhang, B., Li, Y., & Chen, H. (2021). A performance analysis framework for carbon emission quota allocation schemes in China: Perspectives from economics and energy conservation. *Journal of Environmental Management, 296*, 113165. https://doi.org/10.1016/J.JENVMAN.2021.113165

Hu, Z., Yan, S., Li, X., Yao, L., & Luo, Z. (2019). Evaluating the oil production and wastewater treatment efficiency by an extended two-stage network structure model with feedback variables. *Journal of Environmental Management, 251*, 109578. https://doi.org/10.1016/J.JENVMAN.2019.109578

Huang, Q., & Xu, J. (2020). Bi-level multi-objective programming approach for carbon emission quota allocation towards co-combustion of coal and sewage sludge. *Energy, 211*, 118729. https://doi.org/10.1016/J.ENERGY.2020.118729

Jia, Z., Wen, S., & Sun, Z. (2022). Current relationship between coal consumption and the economic development and China's future carbon mitigation policies. *Energy Policy, 162*, 112812. https://doi.org/10.1016/J.ENPOL.2022.112812

Karimi, B., & Samadi, S. (2024). Long-term exposure to air pollution on cardio-respiratory, and lung cancer mortality: A systematic review and meta-analysis. *Journal of Environmental Health Science and Engineering, 22*(1), 75–95. https://doi.org/10.1007/s40201-024-00900-6

Kitagawa, Y. K. L., Pedruzzi, R., Galvão, E. S., de Araújo, I. B., Alburquerque, T. T. de A., Kumar, P., et al. (2021). Source apportionment modelling of $PM_{2.5}$ using CMAQ-ISAM over a tropical coastal-urban area. *Atmospheric Pollution Research, 12*, 101250. https://doi.org/10.1016/J.APR.2021.101250

Li, C., Liu, X., Meng, M., Zhai, L., Zhang, B., Jia, Z., et al. (2021a). The use of Biolog Eco microplates to compare the effects of sulfuric and nitric acid rain on the metabolic functions of soil microbial communities in a subtropical plantation within the Yangtze River Delta region. *CATENA, 198*, 105039. https://doi.org/10.1016/J.CATENA.2020.105039

Li, L., Li, Y., Ye, F., & Zhang, L. (2018). Carbon dioxide emissions quotas allocation in the Pearl River Delta region: Evidence from the maximum deviation method. *Journal of Cleaner Production, 177*, 207–217. https://doi.org/10.1016/J.JCLEPRO.2017.12.205

Li, R., Mei, X., Wei, L., Han, X., Zhang, M., & Jing, Y. (2019). Study on the contribution of transport to $PM_{2.5}$ in typical regions of China using the regional air quality model RAMS-CMAQ. *Atmospheric Environment, 214*, 116856. https://doi.org/10.1016/J.ATMOSENV.2019.116856

Li, X. B., & Reeves, G. R. (1999). A multiple criteria approach to data envelopment analysis. *European Journal of Operational Research, 115*, 507–517. https://doi.org/10.1016/S0377-2217(98)00130-1

Li, Y., Chiu, Y. h., Wang, L., Zhou, Y., & Lin, T. Y. (2020). Dynamic and network slack-based measure analysis of China's regional energy and air pollution reduction efficiencies. *Journal of Cleaner Production, 251*, 119546. https://doi.org/10.1016/J.JCLEPRO.2019.119546

Li, Z., Zhao, T., Wang, J., & Cui, X. (2021b). Two-step allocation of CO_2 emission quotas in China based on multi-principles: Going regional to provincial. *Journal of Cleaner Production, 305*, 127173. https://doi.org/10.1016/J.JCLEPRO.2021.127173

Lin, X., Wu, C.-H., Dong, Y., Chen, G. Z. X., Wei, W., & Duan, J. (2022). Psychological contract breach and destructive voice: The mediating effect of relative deprivation and the moderating effect of leader emotional support. *Journal of Vocational Behavior, 135*, 103720. https://doi.org/10.1016/J.JVB.2022.103720

Liu, X., Wang, W., Wu, W., Zhang, L., & Wang, L. (2022a). Using cooperative game model of air pollution governance to study the cost sharing in Yangtze River Delta region. *Journal of Environmental Management, 301*, 113896. https://doi.org/10.1016/J.JENVMAN.2021.113896

Liu, Z., Shi, Z., Wei, H., & Zhang, J. (2022b). Acid rain reduces soil CO_2 emission and promotes soil organic carbon accumulation in association with decreasing the biomass and biological activity of ecosystems: A meta-analysis. *CATENA, 208*, 105714. https://doi.org/10.1016/J.CATENA.2021.105714

Lozano, S., & Villa, G. (2004). Centralized resource allocation using data envelopment analysis. *Journal of Productivity Analysis, 22*, 143–161. https://doi.org/10.1023/B:PROD.0000034748.22820.33

Meo, S. A., Salih, M. A., Alkhalifah, J. M., Alsomali, A. H., & Almushawah, A. A. (2024). Environmental pollutants particulate matter ($PM_{2.5}$, PM_{10}), Carbon Monoxide (CO), Nitrogen dioxide (NO_2), Sulfur dioxide (SO_2), and Ozone (O_3) impact on lung functions. *Journal of King Saud University—Science, 36*(7), 103280. https://doi.org/10.1016/j.jksus.2024.103280

Momeni, E., Lotfi, F. H., Saen, R. F., & Najafi, E. (2019). Centralized DEA-based reallocation of emission permits under cap and trade regulation. *Journal of Cleaner Production, 234*, 306–314. https://doi.org/10.1016/J.JCLEPRO.2019.06.194

Nascimento, A. P., Santos, J. M., Mill, J. G., Toledo de Almeida Albuquerque, T., Reis Júnior, N. C., Reisen, V. A., et al. (2020). Association between the incidence of acute respiratory diseases in children and ambient concentrations of SO_2, PM_{10} and chemical elements in fine particles. *Environmental Research, 188*, 109619. https://doi.org/10.1016/J.ENVRES.2020.109619

Pan, X., Teng, F., & Wang, G. (2014). Sharing emission space at an equitable basis: Allocation scheme based on the equal cumulative emission per capita principle. *Applied Energy, 113*, 1810–1818. https://doi.org/10.1016/J.APENERGY.2013.07.021

Pang, R. Z., Deng, Z. Q., & Chiu, Y. H. (2015). Pareto improvement through a reallocation of carbon emission quotas. *Renewable and Sustainable Energy Reviews, 50*, 419–430. https://doi.org/10.1016/J.RSER.2015.05.022

Qu, Y., Yuan, Y., Guo, L., & Li, Y. (2021). Initial allocation of SO_2 emission rights based on the combination weighting method: Evidence from China's thermal power plants. *Environmental Science and Pollution Research, 29*, 14743–14753.

Ren, J., Bian, Y., Xu, X., & He, P. (2015). Allocation of product-related carbon emission abatement target in a make-to-order supply chain. *Computers & Industrial Engineering, 80*, 181–194. https://doi.org/10.1016/J.CIE.2014.12.007

Rosso, F., Jin, W., Pisello, A. L., Ferrero, M., & Ghandehari, M. (2016). Translucent marbles for building envelope applications: Weathering effects on surface lightness and finishing when exposed to simulated acid rain. *Construction and Building Materials, 108*, 146–153. https://doi.org/10.1016/J.CONBUILDMAT.2016.01.041

Sepehri, B., Darbani, R., Mesgari-Abbasi, M., Kheirouri, S., Shanehbandi, D., Khordadmehr, M., & Alizadeh, M. (2024). The effects of short-time air pollution, SO_2, and ozone on biochemical, histo-pathological, oxidative stress, and carcinogenesis related genes expressions in the liver of the rats. *Human & Experimental Toxicology, 43*, 09603271241263569. https://doi.org/10.1177/09603271241263569

Shi, G. M., Wang, J. N., Fu, F., & Xue, W. B. (2017). A study on transboundary air pollution based on a game theory model: Cases of SO_2 emission reductions in the cities of Changsha, Zhuzhou and Xiangtan in China. *Atmospheric Pollution Research, 8*, 244–252. https://doi.org/10.1016/J.APR.2016.09.003

Shojaei, T., & Mokhtar, A. (2022). Carbon mitigation by quota allocation. *Journal of Environmental Management, 304*, 114097. https://doi.org/10.1016/J.JENVMAN.2021.114097

Sun, J., Wen, W., Wang, M., & Zhou, P. (2022). Optimizing the provincial target allocation scheme of renewable portfolio standards in China. *Energy, 250*, 123699. https://doi.org/10.1016/j.energy.2022.123699

Sun, T., Zhang, H., & Wang, Y. (2013). The application of information entropy in basin level water waste permits allocation in China. *Resources, Conservation and Recycling, 70*, 50–54. https://doi.org/10.1016/J.RESCONREC.2012.10.001

The State Council. (2017). The Central People's Government of the People's Republic of China. http://www.gov.cn/zhengce/content/2017-01/05/content_5156789.htm

Tian, M., Hu, Y.-J., Wang, H., & Li, C. (2022). Regional allowance allocation in China based on equity and efficiency towards achieving the carbon neutrality target: A composite indicator approach. *Journal of Cleaner Production, 342*, 130914. https://doi.org/10.1016/j.jclepro.2022.130914

Wang, E., Su, B., Zhong, S., & Guo, Q. (2022). China's embodied SO_2 emissions and aggregate embodied SO_2 intensities in interprovincial and international trade. *Technological Forecasting and Social Change, 177*, 121546. https://doi.org/10.1016/J.TECHFORE.2022.121546

Wang, M., & Feng, C. (2020). Regional total-factor productivity and environmental governance efficiency of China's industrial sectors: A two-stage network-based super DEA approach. *Journal of Cleaner Production, 273*, 123110. https://doi.org/10.1016/J.JCLEPRO.2020.123110

Wu, J., Yin, P., Sun, J., Chu, J., & Liang, L. (2016a). Evaluating the environmental efficiency of a two-stage system with undesired outputs by a DEA approach: An interest preference perspective. *European Journal of Operational Research, 254*, 1047–1062. https://doi.org/10.1016/J.EJOR.2016.04.034

Wu, J., Zhu, Q., & Liang, L. (2016b). CO_2 emissions and energy intensity reduction allocation over provincial industrial sectors in China. *Applied Energy, 166*, 282–291. https://doi.org/10.1016/J.APENERGY.2016.01.008

Wu, X., Ji, Z., Gong, Y., Chen, Y., & Toloo, M. (2021). Haze emission efficiency assessment and governance for sustainable development based on an improved network data envelopment analysis method. *Journal of Cleaner Production, 317*, 128424. https://doi.org/10.1016/J.JCLEPRO.2021.128424

Xiao, H., Wang, D., Qi, Y., Shao, S., Zhou, Y., & Shan, Y. (2021). The governance-production nexus of eco-efficiency in Chinese resource-based cities: A two-stage network DEA approach. *Energy Economics, 101*, 105408. https://doi.org/10.1016/J.ENECO.2021.105408

Yang, X. Y., Hu, A., & Schieman, S. (2019). Relative deprivation in context: How contextual status homogeneity shapes the relationship between disadvantaged social status and health. *Social Science Research, 81*, 157–169. https://doi.org/10.1016/J.SSRESEARCH.2019.03.011

Yitzhaki, S. (1994). Economic distance and overlapping of distributions. *Journal of Econometrics, 61*, 147–159. https://doi.org/10.1016/0304-4076(94)90081-7

You, S., & Yan, H. (2011). A new approach in modelling undesirable output in DEA model. *Journal of the Operational Research Society, 62*, 2146–2156. https://doi.org/10.1057/jors.2011.1

Yu, Q., Li, M., Li, Q., Wang, Y., & Chen, W. (2022). Economic agglomeration and emissions reduction: Does high agglomeration in China's urban clusters lead to higher carbon intensity? *Urban Climate, 43*, 101174. https://doi.org/10.1016/J.UCLIM.2022.101174

Yu, S., Wei, Y. M., & Wang, K. (2014). Provincial allocation of carbon emission reduction targets in China: An approach based on improved fuzzy cluster and Shapley value decomposition. *Energy Policy, 66*, 630–644. https://doi.org/10.1016/J.ENPOL.2013.11.025

Yun Fu. (2011). *Fairness measure of sustainable development: Theories, models and empirical studies on relative deprivation*. China Development Press.

Zhang, J., Jin, W., Yang, G. l., Li, H., Ke, Y., & Philbin, S. P. (2021a). Optimizing regional allocation of CO_2 emissions considering output under overall efficiency. *Socio-Economic Planning Sciences, 77*, 101012. https://doi.org/10.1016/J.SEPS.2021.101012

Zhang, Y., Kong, X., Li, M., Yin, Y., & Lin, W. (2021b). The development of a biotin-guided and mitochondria-targeting fluorescent probe for detecting SO_2 precisely in cancer cells. *Talanta, 225*, 121992. https://doi.org/10.1016/J.TALANTA.2020.121992

Zhao, L., Yuan, L., Yang, Y., Xue, J., & Wang, C. (2021). A cooperative governance model for SO_2 emission rights futures that accounts for GDP and pollutant removal cost. *Sustainable Cities and Society, 66*, 102657. https://doi.org/10.1016/J.SCS.2020.102657

Zhou, H., Ping, W., Wang, Y., Wang, Y., & Liu, K. (2021). China's initial allocation of interprovincial carbon emission rights considering historical carbon transfers: Program design and efficiency evaluation. *Ecological Indicators, 121*, 106918. https://doi.org/10.1016/J.ECOLIND.2020.106918

Zhou, P., Sun, Z. R., & Zhou, D. Q. (2014). Optimal path for controlling CO_2 emissions in China: A perspective of efficiency analysis. *Energy Economics, 45*, 99–110. https://doi.org/10.1016/J.ENECO.2014.06.019

Zhou, P., & Wang, M. (2016). Carbon dioxide emissions allocation: A review. *Ecological Economics, 125*, 47–59. https://doi.org/10.1016/J.ECOLECON.2016.03.001

Chapter 17
Efficiency Assessment and Emission Governance for Haze Based on an Improved Network Data Envelopment Analysis Method

17.1 Introduction

In recent years, haze in developing countries has caused widespread social concerns and policy and decision makers hope to select and implement strategies towards sustainable development (Santibanez & Huisingh, 2015; Santibanez Gonzalez et al., 2019). In China, haze has been listed among top ten important events of weather and climate (China Meteorological Administration, 2011). Once formed, the haze is hard to disappear and can lead to particularly serious harm to the urban environment and intensively negative impact on society (Cao et al., 2016). $PM_{2.5}$ concentration is an essential variable of haze weather. According to the World Health Organization (WHO), $PM_{2.5}$ is defined as fine particulate, which can enter the human's lungs (Englert, 2004). Additionally, studies in epidemiology have demonstrated its harm to health, and the average ensemble years of life lost have a complex correlation with the total population and the $PM_{2.5}$ concentration (Cao et al., 2016). Therefore, how to effectively reduce haze has become a crucial problem.

Different from the single pollutants such as CO_2, SO_2, O_3, and NO_X, haze is the product of a multi-stage and complicated physical–chemical reaction (Wang, 2017). The formation of haze and its treatment include two operational sub-processes: in the first sub-process, related pollutants are produced in production activities for economic benefits, and in the second sub-process, comprehensive utilization of output values and pollutants discharged after treatment are obtained through environmental management. Notice that the efficiencies of two sub-processes cannot be assessed separately in haze formation or haze control. In general, decision-makers may not have the exact information for environmental performance, ones' preference over the two sub-processes. The preference refers to the decision makers' attitude to the scheme due to the restriction of objective conditions and the ambiguity of the decision maker's cognition of things (Bui et al., 2012). The preference has an impact on the performance of the two sub-processes and the operational system. If two operational sub-processes are separately measured or without considering the

X. Wu, *Environmental Disaster in China*,
https://doi.org/10.1007/978-981-92-0910-1_17

preference information, misleading results often may be derived (Guo et al., 2017). Thereby our research question is: how to build a model with preference information by combining the meteorological condition and atmospheric environment to evaluate the haze governance efficiency considering haze formation and control, and to identify the main factors that affect the efficiency differences?

To answer this question, an improved second-order cone programming (SOCP) approach is developed to solve this challenging nonlinear model. In this study, the system of haze formation and haze control is divided into two processes: production sub-process (stage 1) and governance sub-process (stage 2). The performance of the operational process for governing $PM_{2.5}$ emission can be evaluated by network DEA. Our study particularly proposes an additive aggregation network DEA framework where the weight is viewed as a degree of attention to a certain stage to evaluate the performance of haze formation and management. DEA (Cooper et al., 2007) has been used to evaluate the performances of entities engaged in a large range of activities in different contexts. Our model captures the two-stage network relationship of haze formation, haze control, and the efficiency of different Decision Making Units (DMUs) based on multiple input and output data. The empirical study about 31 provinces in China shows that the proposed DEA model can provide information on rankings and efficiency changes in haze formation and haze control processes among various regions.

This study makes the following contributions: (1) This study proposes an additive aggregation network DEA framework, where the weight is viewed as a degree of attention to a particular stage to evaluate the performance of formation and control of haze, to understand haze governance by differing from the conventional physical–chemical measures to tackle haze; (2) We improved a second order cone programming (SOCP) approach and applied it to two-stage network DEA models for haze governance efficiencies evaluation; (3) Based on a sample set from 31 provinces in China, we empirically test how different combinations of inputs and outputs affect the efficiency of haze governance. This information obtained from this analysis is valueable for decision makers adjusting their preferences on the information of the haze formation and control to achieve the goal of haze governance.

The rest of this paper is organized as follows. Section 17.2 presents the literature review. Section 17.3 discusses the variables and model. Section 17.4 is empirical analysis. The last section provides concluding remarks.

17.2 Literature Review

(1) Research on the application of DEA in environmental performance evaluation

In recent years, Data envelopment analysis (DEA) has been used in environmental efficiency evaluation (e.g., Song et al., 2018; Sueyoshi & Yuan, 2015; Sueyoshi et al., 2016; Zhou et al., 2008). It mainly considers two aspects: (1) Research on treatment methods of undesirable outputs (e.g., pollutants). For example, Fare et al. (1989)

proposed the first DEA model to deal with environmental pollutants (undesirable outputs). Since then, a large number of studies on environmental efficiency evaluation have been conducted. (2) Research on the application of the DEA environmental efficiency evaluation model. For example, Liu and Wu (2015) took 30 regional transportation departments in China as the research object, took CO_2 as an undesirable output, and used the slack-based DEA model to evaluate the CO_2 emission efficiency. Moutinho et al. (2017) evaluated the ecological efficiency of cities in Germany and France. Song et al.(2018) took the power generation industry in China as the research object and used the SBE-DDF model to evaluate the environmental efficiency. Song et al. (2019) constructed the Meta-Frontier Non-radial Angle Efficiency (MNAE) model and used the improved model to evaluate the environmental efficiency of regions in China. Zhang et al. (2024) employed a cross-efficiency network DEA approach to evaluate the low-carbon efficiency of China's rail-water intermodal transport. Chen et al. (2025) employed a Slack-Based Measurement (SBM) DEA model, treating CO_2 as an undesirable output, to evaluate the operational and environmental efficiency of 26 major Chinese ports.

From the existing studies, the research on the emission efficiencies of air pollutants, especially $PM_{2.5}$, are rare. Moreover, the generation of $PM_{2.5}$ involves complex physical and chemical processes, which are also influenced by meteorological and environmental factors (Pratsini, 1999; Zhang et al., 2014). Therefore, SO_2, NO_X, smoke (powder) and dust are also the intermediate inputs in the $PM_{2.5}$ generation process. How to incorporate meteorological and environmental factors into the assessment of $PM_{2.5}$ emission efficiency is worth in-depth study. Considering that the components and generation of atmospheric pollutant $PM_{2.5}$ are very complex (Zhang et al., 2014), this paper intends to build a two-stage DEA model which included meteorological and environmental factors to evaluate the emission efficiency of $PM_{2.5}$.

(2) Additive aggregation network DEA and its application

Additive efficiency aggregation introduced by Chen et al. (2009), is a method representing overall efficiency of a DMU as additive weighted average efficiencies of individual stages or components that make up the DMU. Then Cook et al. (2010) extend the additive efficiency aggregation to examine more general network structures. However, additive two-stage DEA has its own computational issues as demonstrated by Despotis et al. (2016) and Ang and Chen (2016). In order to isolate the impact of weights on the overall efficiency, Guo et al. (2017) recommend to use fixed weights to define overall efficiency.

Additive network DEA cannot be usually converted into the linear model. Lim and Zhu (2013) and Ang and Chen (2016) develop two parametric methods to solve the nonlinear models. However, the solutions of such methods are still approximated optimal efficiency scores. Recently, Chen and Zhu (2017) develop a second order cone programming (SOCP) approach to solve such nonlinear models. By using the convex optimization method, optimal solutions can be derived and the computation workload can be reduced. However, the weights indicating the preferences over the two stages in additive models are viewed as endogenous variables in SOCP transformation in Chen and Zhu (2017). Guo et al. (2017) shows that the weights, determined by the preferences of decision makers, are exogenous. Yu et al. (2025) compared optimal solutions obtained through different branching strategies within the literature. In this paper, we directly transform the additive network DEA model into a single SOCP problem rather than a sum of two SOCP sub-problems in Chen and Zhu (2017).

17.3 Variables and Models

17.3.1 Variables and Measures

A key in using the DEA model to study the emission efficiency of haze is how to select input and output variables. Based on the previous relevant researches (Sueyoshi & Yuan, 2015), according to the composition and source of $PM_{2.5}$, this study selects examined variables. The variables and data sources for each stage are described below.

17.3.1.1 Input and Output Variables in the First Sub-process

(1) Input variables in the first sub-process

In haze production and governance efficiency evaluation, capital and labor inputs are indispensable. The input variables of traditional production functions include capital and labor, and the output variable is GDP (Cobb & Douglas, 1928). Different from previous literature, this study considers environmental and energy variables relevant to $PM_{2.5}$ production to the input side of the production function. It extends the production function expression as follows:

$$Y_i = A_i K_i^{\alpha} L_i^{\beta} C_i^{\gamma} \tag{17.1}$$

In the equation, Y_i represents the output in area *i*; A_i represents the technology in area *i*; K_i represents the capital input in area *i*; L_i represents the labor input in area *i*; C_i represents the environmental and energy input relevant to $PM_{2.5}$ production in area *i*, and α, β, and γ are output elasticities of environmental and energy input variables relevant to capital, labor and $PM_{2.5}$ production.

According to the sources and constituents of $PM_{2.5}$, two variables (coal consumption and car ownership) are selected as environmental and energy input variables.

Coal Consumption. $PM_{2.5}$ samples of different seasons between 2009 and 2010 in Beijing are analyzed, and coal is found to account for 18% among all the sources that influence $PM_{2.5}$ in Beijing in different seasons. According to emission form, the contribution of coal to atmospheric environment $PM_{2.5}$ can be divided into two types: the directly discharged primary particulate matter is emitted into the atmosphere in the gaseous form such as SO_2, NO_X and VOC, and it forms the second particulate matter (Li, 2009; Zhang et al., 2014). Therefore, coal consumption variables should be added as the input variables in the first sub-process.

Car Ownership. Motor vehicle pollution has become an important source of air pollution in China, also an important reason that causes haze (Zhang et al., 2014). Researches show that vehicle exhaust emission is the main contributors to $PM_{2.5}$ in urban areas (Kinney et al., 2000). Therefore, the sum number of civil and private car ownership is adopted as an input variable.

(2) Output variables in the first sub-process

Carbon emission. Coal consumption and car usage will produce a large amount of carbon. Coal-related carbon dioxide emission has become the main source of carbon dioxide emission in China (Zhang, 2011). Researches of the International Energy Agency (IEA) also show that in 2006, the contribution of the transport sector to the world's carbon dioxide emission is 23%. The carbon emission data are estimated according to IPCC (2006) balancing method. It is shown in Appendix.

Soot the sources of $PM_{2.5}$ are dust, building dust, soot, smelting dust, sulfate and automobile dust, among which soot contributes 30.34% to $PM_{2.5}$. The soot of coal burning contains a large amount of $PM_{2.5}$ fine particles, which will significantly influence the environment and health (Gieré et al., 2006). Therefore, soot is chosen as an output variable in the first sub-process.

Sulfides and nitrogen oxides. The secondary sulfate has an important impact on the formation of haze, and the decrease of sulfur emission has a preferable correlation with the reduction of haze frequency and the decrease of $PM_{2.5}$ concentration (Zhang et al., 2014). The complex reaction is made by SO_2 and NO_x emission from coal and other pollutants in the air, and in this process, gaseous pollutants convert into solid pollutants (Gieré et al., 2006). It is the main cause of $PM_{2.5}$ increase. Therefore, SO_2 and NOx emission are taken as output variables in the first sub-process and as input variables in the second sub-process.

17.3.1.2 Input and Output Variables in the Second Sub-Process

(1) Input variables in the second sub-process

Investment of waste gas treatment. A non-radial DEA model is adopted to evaluate the exhaust emission efficiency of each province, and it is found that industrial departments of many provinces should reduce pollution through technology investment (Wang et al., 2015). The research adopted provincial panel data from 2001 to 2012 and panel data models to analyze the fundamental driving forces of $PM_{2.5}$ emission at the regional level in China and the results showed that energy efficiency improvement has more significant potentials to mitigate $PM_{2.5}$ emission in the central and western regions than that in the eastern regions due to its low level of technology (Xue et al., 2014). The research examined the relationship between environmental protection investment and emission of industrial waste gas and found the investment of waste gas treatment had decreased air pollution emission (Dong et al., 2013). So the investment of waste gas treatment is taken as an input variable in the second sub-process.

Meteorological factors. It is found that meteorological factors significantly influence atmospheric particulates (Wang et al., 2015). Relative humidity, rainfall, and temperature are key factors affecting $PM_{2.5}$ concentration (Jones et al., 2010). Meteorological factors, particularly precipitation in different seasons, greatly influence the concentration of particulates (Li et al., 2009). So the amount of precipitation is taken as an input variable in the second sub-process.

Greening ratio. Many scholars have studied the regulation and reduction function of vegetation to particulate matter (Liu et al., 2017). Vegetation is important to reduce particular matter like $PM_{2.5}$ in the air (Jin et al., 2014). The leaf surface has a certain wettability and roughness, which is favorable for the deposition of $PM_{2.5}$ and other particles (Beckett et al., 2000). Yang et al. (2008) estimated the reduction function of roof greening on air pollutants in Chicago. Thereby we use the greening ratio of built-up area as an input.

(2) Output variables in the second sub-process

Haze evaluation index. The main components of haze are PM. The study has adopted annual average $PM_{2.5}$ from 2001 to 2010 based on population weighting that has been developed by Battelle Memorial Institute and Center for International Earth Science Information Network. Two factors can then be considered to have a weighted approach: the first is the land area of each province, and the second is the initial atmospheric environmental capacity of $PM_{2.5}$ in each province under the $PM_{2.5}$ standard constraint.

Atmospheric environment capacity. The maximum allowable discharge of $PM_{2.5}$ at a time in 31 provinces is considered as the data source of the research (Xue et al., 2014). Then we consider factors like land areas of each province and atmospheric environment capacity and handle $PM_{2.5}$ concentration. So a new $PM_{2.5}$ variable can be considered as the output variable of this research. The detailed calculation process is shown in Table 17.1 (in Appendix).

Atmospheric environmental assessment variables. Atmospheric environmental assessment refers to the factors that measure the quality of the atmospheric environment (Yang & Wang, 2017). This study selects the number of days when the air quality reaches or is better than the second level as the output in the second sub-process to analyze the influencing factors of air quality.

Above all, the study selects input variables such as capital, labor, car ownership, coal consumption, GDP, SO_2 emission, NO_X emission, soot emission, carbon emission, precipitation, investment of waste gas treatment, greening ratio, $PM_{2.5}$ emission and the number of days when the air quality reaches or is better than the second level. The definition and calculation method of each variable are shown in Table 17.1(in Appendix).

17.3.2 Additive Aggregation Network DEA

Figure 17.1 presents a two-stage network process for haze control. There are a set of n DMUs and that for each DUM$_j$ ($j = 1,2,\ldots,n$). In stage 1, m_1 inputs $x_{i_1j}^1$ ($i_1 = 1, \ldots, I_1$) produce R_1 desirable outputs $y_{r_1j}^1$ ($r_1 = 1, \ldots, R_1$) and D undesirable outputs z_{dj} ($d = 1, \ldots, D$). In addition, the desirable outputs $y_{r_1j}^1$ leave the system but the undesirable outputs z_{dj} will be further disposed in stage 2 with extra m_2 inputs $x_{i_2j}^2$ ($i_2 = 1, \ldots, I_2$) to obtain R_2 undesirable outputs $y_{r_2j}^2$ ($r_2 = 1, \ldots, R_2$), and R_3 desired outputs $y_{r_3j}^3$ ($r_3 = 1, \ldots, R_3$). These D outputs z_{dj} are referred to as intermediate measures or intermediate productions. Different from conventional network DEA structures, the intermediate measures z_{dj} and outputs $y_{r_2j}^2$ are undesirable outputs in Fig. 17.1.

Because of the different economic values of desirable and undesirable outputs, two different processing methods need to be adopted. Seiford and Zhu (2002) present that the undesirable outputs can be transformed into desirable outputs and then treated as ordinary desirable outputs in conventional DEA. In network DEA, Fukuyama and Weber (2010) use directional distance function to increase the desirable outputs and decrease undesirable outputs simultaneously. To decrease undesirable outputs, Wu et al., (2016a, 2016b) add a negative sign to intermediate measures (undesirable outputs) in Stage 1. However, the intermediate measures are considered as ordinary inputs of Stage 2 instead of undesirable inputs in Wu et al., (2016a, 2016b). Note that as in Kao and Hwang (2008) and Liang et al. (2010), the weights for the intermediate measures are assumed to be equal for the two stages. That is, the technique of dealing with undesirable outputs in Wu et al., (2016a, 2016b) cannot follow such a regularity of network DEA.

In DEA, as presented in Chen et al. (2012), higher levels of outputs usually indicate better performance. We here treat the intermediate measures in the two stages by taking the reciprocals of the emission levels of Carbon, Soot, Sulfides, and nitrogen oxides. We can thereby obtain that equal weights for the intermediate

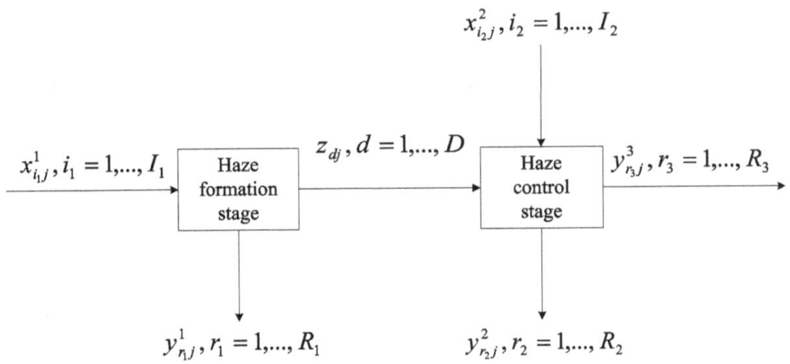

Fig. 17.1 Two-stage network structure for haze governance

measures for the two stage. Similarly, we use the reciprocal of $PM_{2.5}$ in our analysis to fit the network DEA use.

Typically, using the CCR model of Charnes et al. (1978), we can define the efficiency of Stage 1 and Stage 2 respectively as:

$$\begin{aligned} &\max \frac{\sum_{d=1}^{D} \eta_d z_{do} + \sum_{r_1=1}^{s_1} u_{r_1} y_{r_1 o}^1}{\sum_{i_1=1}^{m_1} v_{i_1} x_{i_1 o}^1} \\ &s.t. \frac{\sum_{d=1}^{D} \eta_d z_{dj} + \sum_{r_1=1}^{s_1} u_{r_1} y_{r_1 j}^1}{\sum_{i_1=1}^{m_1} v_{i_1} x_{i_1 j}^1} \le 1, \ \forall j \\ &\eta_d, v_{i_1}, u_{r_1} \ge \varepsilon \, \forall d, i_1, r_1 \end{aligned} \tag{17.2}$$

and

$$\begin{aligned} &\max \frac{\sum_{r_2=1}^{s_2} u_{r_2} y_{r_2 o}^2 + \sum_{r_3=1}^{s_3} u_{r_3} y_{r_3 o}^3}{\sum_{d=1}^{D} \eta_d z_{do} + \sum_{i_2=1}^{m_2} v_{i_2} x_{i_2 o}^2} \\ &s.t. \frac{\sum_{r_2=1}^{s_2} u_{r_2} y_{r_2 j}^2 + \sum_{r_3=1}^{s_3} u_{r_3} y_{r_3 j}^3}{\sum_{d=1}^{D} \eta_d z_{dj} + \sum_{i_2=1}^{m_2} v_{i_2} x_{i_2 j}^2} \le 1, \forall j \\ &u_{r_2}, u_{r_3}, \eta_d, v_{i_2} \ge \varepsilon \, \forall r_2, r_3, d, i_2 \end{aligned} \tag{17.3}$$

where $\eta_d, v_{i_1}, u_{r_1}, u_{r_2}, u_{r_3}, v_{i_2}$ are positive weights, by incorporating the small non-Archimedean ε in models (2) and (3). By Charnes-Cooper transformation (Charnes & Cooper, 2010), the two models above can be transformed to linear models.

Under the technique of additive efficiency aggregation, the overall efficiency is a weighted average of two stage efficiencies. The weight is regarded as a preference over two stages. In general, as presented in Guo et al. (2017), one may not have the exact information on the weight. Following the technique of Guo et al. (2017), we take weight as a parameter and change it from 0 to 1. By doing so, we can obtain different information of overall efficiency and two stage efficiency scores under different preferences of decision makers.

The general additive aggregation two-stage network DEA model for DMU$_o$ is proposed as follows:

$$\max \alpha \frac{\sum_{d=1}^{D} \eta_d z_{do} + \sum_{r_1=1}^{s_1} u_{r_1} y^1_{r_1 o}}{\sum_{i_1=1}^{m_1} v_{i_1} x^1_{i_1 o}} + (1-\alpha) \frac{\sum_{r_2=1}^{s_2} u_{r_2} y^2_{r_2 o} + \sum_{r_3=1}^{s_3} u_{r_3} y^3_{r_3 o}}{\sum_{d=1}^{D} \eta_d z_{do} + \sum_{i_2=1}^{m_2} v_{i_2} x^2_{i_2 o}} \tag{17.4}$$

$$s.t. \frac{\sum_{d=1}^{D} \eta_d z_{dj} + \sum_{r_1=1}^{s_1} u_{r_1} y^1_{r_1 j}}{\sum_{i_1=1}^{m_1} v_{i_1} x^1_{i_1 j}} \leq 1, \forall j \tag{4a}$$

$$\frac{\sum_{r_2=1}^{s_2} u_{r_2} y^2_{r_2 j} + \sum_{r_3=1}^{s_3} u_{r_3} y^3_{r_3 j}}{\sum_{d=1}^{D} \eta_d z_{dj} + \sum_{i_2=1}^{m_2} v_{i_2} x^2_{i_2 j}} \leq 1, \forall j \tag{4b}$$

$$\eta_d, v_{i_1}, u_{r_1}, u_{r_2}, u_{r_3}, v_{i_2} \geq \varepsilon \, \forall i_1, r_1, r_2, r_3, d, i_2$$

In model (17.4), there are two special cases: $\alpha = 1$ and $\alpha = 0$. In such cases, the decision makers focus on the information of one stage, and the other one is not considered. When $\alpha = 1$, model (17.4) is reduced to model (17.2), since the constraints of (17.4b) are redundant to model (17.2). Similarly, model (17.4) becomes model (17.3) when $\alpha = 0$.

Note that model (17.4) is non-linear and cannot be transformed to a linear program. To enhance the computational capability of such network models, Lim and Zhu (2013) (see also Guo et al. (2017)) and Ang and Chen (2016) respectively propose two heuristic methods to obtain approximate optimal solutions. Zhang and Chen (2018) prove that the two parametric linear approaches are equivalent in basic network structure. Although we can obtain the efficiency sores by using two approaches, a large amount of calculations needs to be undertaken and the global optimal solutions cannot be guaranteed.

Recently, Chen and Zhu (2017) develop a second-order cone programming (SOCP) approach to solve nonlinear network DEA models. In Chen and Zhu (2017), additive two-stage network DEA model is a cone structure in disguise and the nonlinear problem can be transformed into a SOCP problem. Compared with the above parametric linear methods, the transformed SOCP problem can be directly solved in convex optimization, without the need for solving a series of linear programs.

Note that the weight α in model (17.4) is viewed as an endogenous variable in SOCP transformation in Chen and Zhu (2017). However, the weight α is exogenous in model (17.4). As shown in the above, the weight can be determined by the preference of decision makers. The transformed SOCP problem of Chen and Zhu (2017) is a sum of two SOCP sub-problems. We here improve the method of Chen and Zhu (2017) to solve model (17.4), and show model (17.4) can be solved by a single SOCP sub-problem.

Theorem. Model (17.4) can be converted to a linear model with a single second-order cone constraint.

Proof. We first apply Charnes-Cooper transformation (Charnes & Cooper, 2010) to model (17.4). Specifically, let $t = \frac{1}{\sum_{i_1=1}^{m_1} v_{i_1} x_{i_1 o}^1}$ and set $\tilde{v}_{i_1} = tv_{i_1}, \tilde{v}_{i_2} = tv_{i_2}, \tilde{\eta}_d = t\eta_d$, $\tilde{u}_{r_1} = tu_{r_1}, \tilde{u}_{r_2} = tu_{r_2}, \tilde{u}_{r_3} = tu_{r_3}$.. Then model (17.4) is converted into the following model.

$$\max \alpha(\sum_{d=1}^{D} \tilde{\eta}_d z_{do} + \sum_{r_1=1}^{s_1} \tilde{u}_{r_1} y_{r_1 o}^1) + (1-\alpha)\frac{\sum_{r_2=1}^{s_2} \tilde{u}_{r_2} y_{r_2 o}^2 + \sum_{r_3=1}^{s_3} \tilde{u}_{r_3} y_{r_3 o}^3}{\sum_{d=1}^{D} \tilde{\eta}_d z_{do} + \sum_{i_2=1}^{m_2} \tilde{v}_{i_2} x_{i_2 o}^2}$$

$$s.t. \sum_{i_1=1}^{m_1} \tilde{v}_{i_1} x_{i_1 o}^1 = 1$$

$$\sum_{d=1}^{D} \tilde{\eta}_d z_{dj} + \sum_{r_1=1}^{s_1} \tilde{u}_{r_1} y_{r_1 j}^1 - \sum_{i_1=1}^{m_1} \tilde{v}_{i_1} x_{i_1 j}^1 \le 0, \forall j$$

$$\sum_{r_2=1}^{s_2} \tilde{u}_{r_2} y_{r_2 o}^2 + \sum_{r_3=1}^{s_3} \tilde{u}_{r_3} y_{r_3 o}^3 - \sum_{d=1}^{D} \tilde{\eta}_d z_{do} - \sum_{i_2=1}^{m_2} \tilde{v}_{i_2} x_{i_2 o}^2 \le 0, \ \forall j$$

$$\tilde{\eta}_d, \tilde{v}_{i_1}, \tilde{u}_{r_1}, \ \tilde{u}_{r_2}, \tilde{u}_{r_3}, \tilde{v}_{i_2} \ge \varepsilon \ \ \forall i_1, r_1, r_2, r_3, d, i_2 \tag{17.5}$$

Model (17.5) is still nonlinear. We transform this model to a SOCP problem as follows. After introducing a new variable ϕ, we convert model (17.5) into model (17.6) with linear and hyperbolic constraints.

$$\max \alpha(\sum_{d=1}^{D} \tilde{\eta}_d z_{do} + \sum_{r_1=1}^{s_1} \tilde{u}_{r_1} y_{r_1 o}^1) + (1-\alpha)\phi \tag{17.6}$$

$$s.t. \phi(\sum_{d=1}^{D} \tilde{\eta}_d z_{do} + \sum_{i_2=1}^{m_2} \tilde{v}_{i_2} x_{i_2 o}^2) \le \sum_{r_2=1}^{s_2} \tilde{u}_{r_2} y_{r_2 o}^2 + \sum_{r_3=1}^{s_3} \tilde{u}_{r_3} y_{r_3 o}^3 \tag{17.6a}$$

$$\sum_{i_1=1}^{m_1} \tilde{v}_{i_1} x_{i_1 o}^1 = 1 \tag{17.6b}$$

$$\sum_{d=1}^{D} \tilde{\eta}_d z_{dj} + \sum_{r_1=1}^{s_1} \tilde{u}_{r_1} y_{r_1 j}^1 - \sum_{i_1=1}^{m_1} \tilde{v}_{i_1} x_{i_1 j}^1 \le 0, \forall j \tag{17.6c}$$

$$\sum_{r_2=1}^{s_2} \tilde{u}_{r_2} y_{r_2 o}^2 + \sum_{r_3=1}^{s_3} \tilde{u}_{r_3} y_{r_3 o}^3 - \sum_{d=1}^{D} \tilde{\eta}_d z_{do} - \sum_{i_2=1}^{m_2} \tilde{v}_{i_2} x_{i_2 o}^2 \le 0, \ \forall j \tag{17.6d}$$

$$\tilde{\eta}_d, \tilde{v}_{i_1}, \tilde{u}_{r_1}, \tilde{u}_{r_2}, \tilde{u}_{r_3}, \tilde{v}_{i_2} \ge \varepsilon \, \forall i_1, r_1, r_2, r_3, d, i_2$$

Since $\tilde{\eta}_d, \tilde{u}_{r_2}, \tilde{u}_{r_3}, \tilde{v}_{i_2} > 0$ and $\phi > 0$, the nonlinear constraint (6a) is equivalent to the following positive semidefinite matrix

$$\begin{pmatrix} 1 & \sum_{d=1}^{D} \tilde{\eta}_d z_{do} + \sum_{i_2=1}^{m_2} \tilde{v}_{i_2} x_{i_2 o}^2 \\ \phi & \sum_{r_2=1}^{s_2} \tilde{u}_{r_2} y_{r_2 o}^2 + \sum_{r_3=1}^{s_3} \tilde{u}_{r_3} y_{r_3 o}^3 \end{pmatrix} \succ= 0, \tag{17.7}$$

Since square matrixes (7) and the following (8) have the same characteristic polynomial and eigenvalues, the two matrices are equivalent (see also Chen & Zhu, 2017).

$$\begin{pmatrix} 1 & \frac{1}{2}(\sum_{d=1}^{D} \tilde{\eta}_d z_{do} + \sum_{i_2=1}^{m_2} \tilde{v}_{i_2} x_{i_2 o}^2 + \phi) \\ \frac{1}{2}(\sum_{d=1}^{D} \tilde{\eta}_d z_{do} + \sum_{i_2=1}^{m_2} \tilde{v}_{i_2} x_{i_2 o}^2 + \phi) & \sum_{r_2=1}^{s_2} \tilde{u}_{r_2} y_{r_2 o}^2 + \sum_{r_3=1}^{s_3} \tilde{u}_{r_3} y_{r_3 o}^3 \end{pmatrix} \succ= 0, \tag{17.8}$$

From semidefinite matrix (8), we have the following inequality

$$\alpha\gamma - \beta^2 \ge 0 \Leftrightarrow (\frac{\alpha+\gamma}{2})^2 \ge (\frac{\alpha-\gamma}{2})^2 + \beta^2 \tag{17.9}$$

where $\alpha = 1$, $\beta = \frac{1}{2}(\sum_{d=1}^{D} \tilde{\eta}_d z_{do} + \sum_{i_2=1}^{m_2} \tilde{v}_{i_2} x_{i_2 o}^2 + \phi)$ and $\gamma = \sum_{r_2=1}^{s_2} \tilde{u}_{r_2} y_{r_2 o}^2 + \sum_{r_3=1}^{s_3} \tilde{u}_{r_3} y_{r_3 o}^3$.

Note that (9) is equivalent to the following second-order cone structure under l_2 norm

$$\left\| \begin{matrix} \sum_{d=1}^{D} \tilde{\eta}_d z_{do} + \sum_{i_2=1}^{m_2} \tilde{v}_{i_2} x_{i_2 o}^2 + \phi \\ 1 - \sum_{r_2=1}^{s_2} \tilde{u}_{r_2} y_{r_2 o}^2 - \sum_{r_3=1}^{s_3} \tilde{u}_{r_3} y_{r_3 o}^3 \end{matrix} \right\|_2 \le 1 + \sum_{r_2=1}^{s_2} \tilde{u}_{r_2} y_{r_2 o}^2 + \sum_{r_3=1}^{s_3} \tilde{u}_{r_3} y_{r_3 o}^3 \tag{17.10}$$

Model (17.4) can be equivalently converted to model (17.11) in SOCP.

$$\max \alpha(\sum_{d=1}^{D} \tilde{\eta}_d z_{do} + \sum_{r_1=1}^{s_1} \tilde{u}_{r_1} y_{r_1 o}^1) + (1-\alpha)\phi \tag{17.11}$$

$$s.t. \left\| \begin{matrix} \sum_{d=1}^{D} \tilde{\eta}_d z_{do} + \sum_{i_2=1}^{m_2} \tilde{v}_{i_2} x_{i_2 o}^2 + \phi \\ 1 - \sum_{r_2=1}^{s_2} \tilde{u}_{r_2} y_{r_2 o}^2 - \sum_{r_3=1}^{s_3} \tilde{u}_{r_3} y_{r_3 o}^3 \end{matrix} \right\|_2 \leq 1 + \sum_{r_2=1}^{s_2} \tilde{u}_{r_2} y_{r_2 o}^2 + \sum_{r_3=1}^{s_3} \tilde{u}_{r_3} y_{r_3 o}^3 \quad (17.11a)$$

$$\sum_{i_1=1}^{m_1} \tilde{v}_{i_1} x_{i_1 o}^1 = 1 \quad (17.11b)$$

$$\sum_{d=1}^{D} \tilde{\eta}_d z_{dj} + \sum_{r_1=1}^{s_1} \tilde{u}_{r_1} y_{r_1 j}^1 - \sum_{i_1=1}^{m_1} \tilde{v}_{i_1} x_{i_1 j}^1 \leq 0 \forall j \quad (17.11c)$$

$$\sum_{r_2=1}^{s_2} \tilde{u}_{r_2} y_{r_2 o}^2 + \sum_{r_3=1}^{s_3} \tilde{u}_{r_3} y_{r_3 o}^3 - \sum_{d=1}^{D} \tilde{\eta}_d z_{do} - \sum_{i_2=1}^{m_2} \tilde{v}_{i_2} x_{i_2 o}^2 \leq 0 \, \forall j \quad (17.11d)$$

$$\tilde{\eta}_d, \tilde{v}_{i_1}, \tilde{u}_{r_1}, \tilde{u}_{r_2}, \tilde{u}_{r_3}, \tilde{v}_{i_2} \geq \varepsilon \, \forall i_1, r_1, r_2, r_3, d, i_2 \quad (17.11e)$$

Note that the constraints (17.11b)–(17.11e) are linear in model (17.11). Our model (17.11) is different from SOCP model (24) of Chen and Zhu (2017). The weight α is an endogenous variable and its optimal value is determined by model itself in model (24) of Chen and Zhu (2017). In addition, there are two constraints with cone structure in Chen and Zhu (2017). As shown in the above, however, the weight α is exogenous in model (17.11). In addition, our model is a linear model with a single second-order cone constraint.

17.4 Empirical Analysis

We set weight $\alpha = \frac{i}{10}$, i = 1, 2,..., 9 (Guo et al., 2017), and calculate the efficiency scores of the system and two divisions (haze formation and control) with the available data. The efficiency values of 31 provinces and cities in China are obtained. See the results in Table 17.3 (in Appendix). According to the calculation results, the evaluation results of additive aggregation network DEA can be obtained, as shown in Figs. 17.2–17.4.

17.4.1 Overall Efficiency Comparison

As we can see from Fig. 17.2 and Table 17.3 (in Appendix), there exists significant difference in the overall efficiency among different provinces, which can be categorized into two groups. The first group contains two decision units including Inner Mongolia and Tibet. Their overall efficiency values are the highest, with the

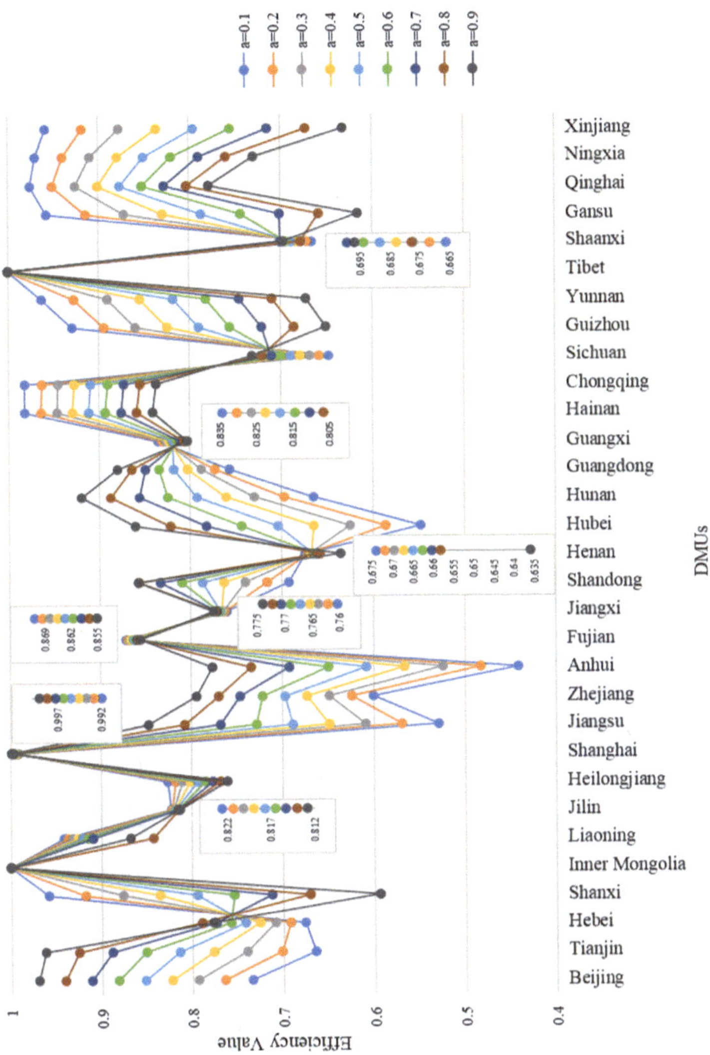

Fig. 17.2 Overall efficiency value distribution

efficiency value reaching 1 and the efficiency value not changing with the weight α. The second group includes the remaining provinces. The overall efficiency value is less than 1 and changes with the weight α.

We examine the relationship between alpha (weight) and the overall efficiency. Figure 17.2 and Table 17.3 (in Appendix) show, for most DMUs, overall efficiency can decrease or increase relatively sharp when alpha increases by 0.1, and can be divided into three cases. The first case shows that the overall efficiency e monotonically increases with alpha. Taking Beijing as an example, the overall efficiency e for DUM Beijing monotonically increases when alpha increases by 0.1. The second case shows that the overall efficiency e monotonically decreases with the increase of alpha. Taking Shanxi as an example, the overall efficiency e for DUM Shanxi monotonically decreases when alpha increases by 0.1. The last case shows that the change in the overall efficiency e for some DMUs may not be "gradual" in relation to the changing weight α. DMUs, like Hebei, Liaoning and Shaanxi are the case above. Considering DMU Heibei in Table 17.3 (in Appendix) when we vary alpha by 0.1, the overall efficiency increases first with the increase of weight α, and decreases when alpha vary from 0.8 to 0.9. This indicates that the overall efficiency variation is mainly affected by the change in stage efficiencies.

17.4.2 Efficiency Comparison in Each Stage

Viewing from the relative efficiency value e_1 in the Fig. 17.3, the efficiency in the first sub-process does not change with the weight α, and nine curves coincide. There exist significant gaps in the efficiencies of haze formation among different provinces, which can be categorized into three groups. The first group contains five units including Beijing, Tianjin, Inner Mongolia, Shanghai and Tibet. Their efficiency values in first stage are the highest, with the efficiency values reaching 1 and they are situated at the efficiency frontier, where the ratio of output to input in DMU is the highest. The second group includes Hunan and the other 12 provinces. The efficiency in haze formation of 13 provinces is higher than the nation's average level 0.809 and they are situated very close to the efficiencies frontier. 13 provinces including Guangxi are listed in the third group. The emission efficiencies in these provinces are lower than the nation's average level and they are situated very far from the frontier, among which the input efficiency in Shanxi is the lowest, with a value 0.573.

In the second sub-process, the efficiency value e_2 in the first category including Inner Mongolia and the other 8 provinces is 1, and the value of efficiency does not change with the weight α. The second category includes Beijing and the other 15 provinces. The efficiency values in haze control stage of these 16 provinces do not affect by the weight $\alpha_i = \frac{i}{10}, i = 1, 2, ..., 9$, the 16 curves coincide in the figure. The third category contains Hebei and the other 5 provinces. Their efficiency in the second sub-process is constant when the weight $\alpha_i = \frac{i}{10}, i = 1, 2, ..., 7$, and the efficiency value changes when the weight $\alpha = 0.8$ or $\alpha = 0.9$.

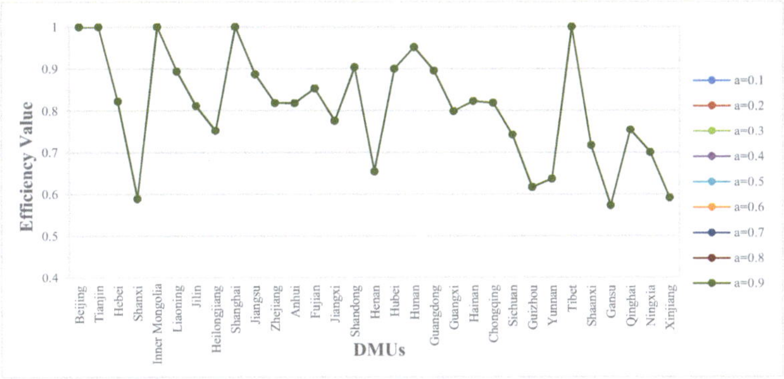

Fig. 17.3 Distribution of Efficiency Value e_1 in the First Sub-process

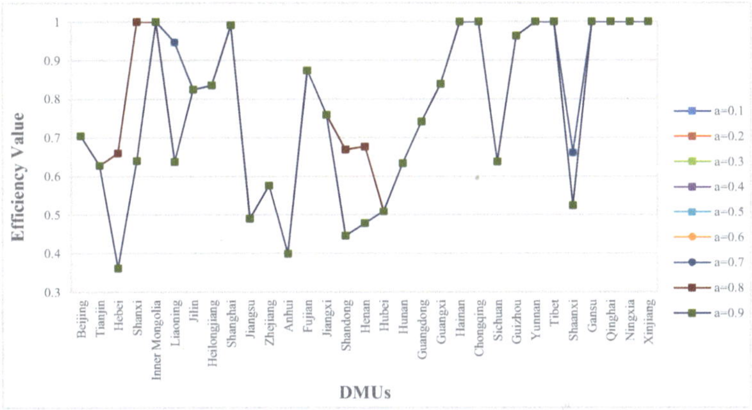

Fig. 17.4 Distribution of Efficiency Value e_2 in the Second Sub-process

There exists relatively obvious difference in the efficiencies between the first sub-process and the second sub-process. Except for Inner Mongolia and Tibet, where the efficiency values in both stages are "1", there exists a relatively large gap between other provinces, where the efficiency of haze formation is higher than that of haze control. In these provinces such as Beijing, the efficiency of haze formation in the first sub-process is higher than that in the second sub-process, which means the ratio of output (GDP, carbon emission, SO_2 emission, NO_X emission and smoke and dust emission) in the first sub-process to input (capital, labor, car ownership, coal consumption) is relatively high.

We can find that most of the provinces with relatively low efficiency in haze formation are inland provinces with relatively slow economic growth and poor weather condition, including Gansu. The provinces with high haze management efficiency can be categorized into two groups. The first group contains provinces in the central and eastern parts of China with a backward economic development. These areas have a small amount of haze formation, thus there is less difficulty in haze governance. Shanghai is in the second group. These regions also suffer from high rainfall and frequent occurrence of windy days since they are located in coastal regions. Therefore, benefited from good weather conditions, there is little difficulty in haze governance and management. Based on the analysis above, the main factors determining efficiency in the two stages are economic development and ecological conditions (including weather, climate, and environment).

Note that weight α can be determined by the preference of decision makers to achieve higher efficiency in this study. As we find that weight α may not affect the calculation of stage efficiency values and the variation in the overall efficiency resulting from using different weights can be associated with constant stage efficiencies. If the decision makers pay full attention to the first sub-process, that is, the weight α is defined as 1, the stage efficiency e_2 is nonexistent. On the contrary, if the decision makers pay full attention to the second sub-process where the weight α is defined as 0, the stage efficiency e_1 is nonexistent. Decision makers can choose the stage efficiencies to maximize the overall efficiency score. That is, the decision makers can choose the "appropriate" pair of weights that maximizes the overall efficiency score.

17.4.3 Effect of Decision Maker's Preferences on Efficiency

We present the results of two typical samples of DMUs, Hebei and Liaoning. The weight α plays a decisive role of the overall efficiency of Hebei and Liaoning provinces. Figure 17.5 shows the efficiency variation when alpha changes from 0.1 to 0.9 in Hebei. When alpha changes from 0.1 to 0.8, overall efficiency shows relatively large change (e increases from 0.68 to 0.79). This is because the variation in overall efficiency is a result of variation in stage efficiencies (e_2 is constant 0.66 and e_1 slightly changes from 0.82261 to 0.82266). When alpha changes from 0.8 to 0.9, overall efficiency shows a relatively large change (e decreases from 0.79 to 0.78). This is because of a sharp decrease in second-stage efficiency (e_2 decreases from 0.66 to 0.36). Note that when alpha changes from 0.1 to 0.8, overall efficiency shows a slight change (e decreases from 0.94 to 0.84) in Fig. 17.6. This is because the variation in overall efficiency is a result of variation in stage efficiencies (e_1 almost has no change and e_2 changes from 0.95 to 0.64). When alpha changes from 0.8 to 0.9, overall efficiency shows a relatively large change (e increases from 0.84 to 0.87). Note that the overall efficiency e is mainly influenced by the second efficiency of e_2. The exogenous variable, weight α, can be determined by the preference of decision makers to achieve higher efficiency. Decision makers consider the information of the haze formation and control stage to achieve the goal.

The Hebei provincial government should increase the preference towards the haze formation efficiency in the first stage. The decision makers can promulgate the environmental governance policy to control the emission of sulfur dioxide, and limit the urban motor vehicles, which try to control the haze emission by cutting

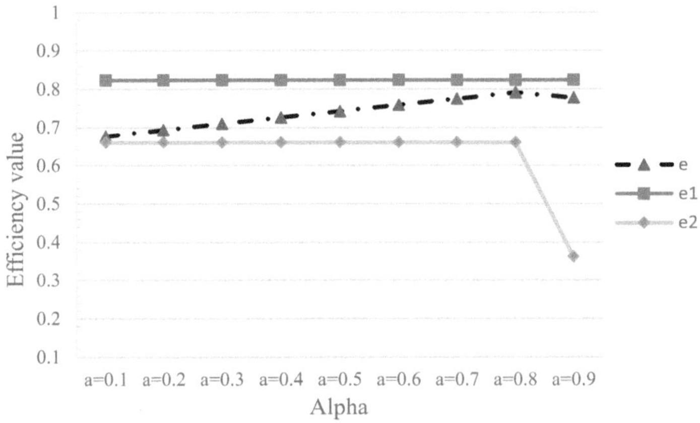

Fig. 17.5 The efficiency values in Hebei province

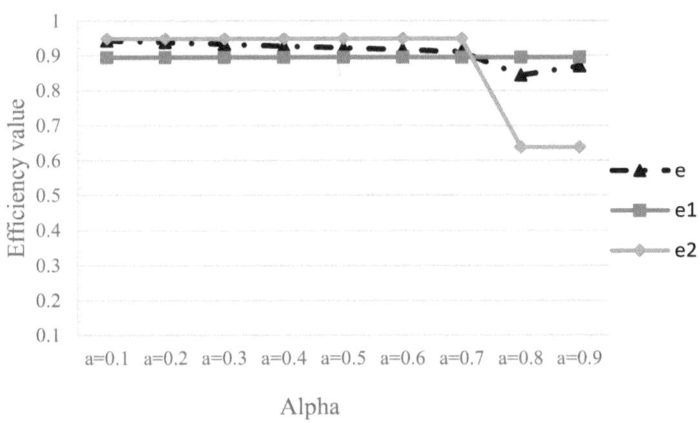

Fig. 17.6 The efficiency values in Liaoning province

the source indicators which cause the haze. Liaoning provincial government can increase the preference towards the haze governance efficiency in the second stage. The decision makers can increase the proportion of investment in waste gas pollution to control air pollution. The government takes the lead in the construction of a public green space environment in Liaoning, clarifies the responsibility for making clear the responsibilities for afforestation of enterprises and individuals in the provinces, and increases green areas to increase greening ratio. These measures can improve haze governance efficiency.

17.5 Concluding Remarks

In this paper, for the first time, a network DEA model for evaluating the performance of haze emission is built. At the same time, the additive aggregation network data envelopment analysis with undesirable intermediate measures and undesirable outputs is proposed.

The empirical results of this paper show that the emission efficiencies of different emission units (provinces) fluctuate in two stages of haze emission. Some provinces have higher efficiency in the formation stage of haze, but a lower efficiency in the control stage. The results tell us that in the process of haze governance, government departments should take different measures at different stages with specific targets, to better reduce haze emissions. In addition, this research shows that different weights play essential roles in the overall efficiency and stage efficiency (Chen et al., 2012; Guo et al., 2017), we can solve the problem of haze pollution by changing the decision maker preferences, and this may be a direction for future application of our proposed model. To those provinces with the overall efficiency for DUMs monotonically increasing when alpha increases by 0.1, the decision makers can pay higher attention in the haze formation efficiency in the first stage and take measures to improve first stage efficiency. To the provinces with the overall efficiency for DUMs monotonically decreasing when alpha increases by 0.1, the decision makers should pay more attention on the haze governance in the second stage and reduce the air pollution by increasing the proportion of investment, and by increasing the greening ratio to improve the second efficiency.

Admittedly, there exists space for further study. First of all, owing to the existence of trans-boundary transmission of haze, how to define cooperation and competition between provinces with trans-boundary transmission remains uncertain. Second, due to the difference in the development among provinces, we cannot assess environmental government without considering the economic and social development. Therefore, researches can take both efficiency and equality into consideration, to evaluate the emission efficiency of different units and to determine the emission reduction in outputs.

Acknowledgements Zhiyong Ji, Yeming Gong, Yufeng Chen, and Mehdi Toloo also made great contributions to this manuscript. We express our heartfelt thanks to them. This research also was supported by: National Natural Science Foundation of China (72474127) and National Social Science Fund Major Bidding Project of China (18ZDA052). Part of this chapter was published:Wu X, Ji Z, Gong Y, Chen Y, Toloo M. Haze emission efficiency assessment and governance for sustainable development based on an improved network data envelopment analysis method. Journal of Cleaner Production 2021c;317:128424. https://doi.org/https://doi.org/10.1016/j.jclepro.2021.128424. The publication has been conducted with copyright permission.

Appendix 1

See (Tables 17.1, 17.2 and 17.3).

Table 17.1 Description of input and output variables*

Stage	Variable category	Variable name	Accounting content	Data source	Unit
The first sub-process	Input variables	Capital	Fixed assets investment in the whole society	*China Statistical Yearbook 2015*	100 million yuan
		Labor	Employed population	*China Statistical Yearbook 2015*	10,000 persons
		Car ownership	Car ownership of civil and private cars	*China's auto market almanac 2015*	10,000 units
		Coal consumption		*China Energy Statistical Yearbook 2015*	10,000 tons
	Output variables	GDP	Gross Domestic Product	*State Statistical Bureau 2015*	100 million yuan
		Carbon emission		To account according to the methods in Appendix 1	
		SO_2 emission		*China Statistical Yearbook on Environment 2015*	Tons
		NO_X emission		*China Statistical Yearbook on Environment 2015*	Tons
		Soot emission		*China Statistical Yearbook on Environment 2015*	Tons

(continued)

Table 17.1 (continued)

Stage	Variable category	Variable name	Accounting content	Data source	Unit
The second sub-process	Input variables	Carbon emission		To account according to the methods in Appendix 1	
		SO_2 emission		*China Statistical Yearbook on Environment 2015*	Tons
		NO_X emission		*China Statistical Yearbook on Environment 2015*	Tons
		Soot emission		*China Statistical Yearbook on Environment 2015*	Tons
		Precipitation	Annual precipitation	*State Statistical Bureau 2015*	Millimeter
		Investment of waste gas treatment	Investment completed in waste gas treatment projects	*China Energy Statistical Yearbook 2015*	100 million yuan
		Greening ratio	Green coverage rate of built-up area	*China Statistical Yearbook on Environment 2015*	%
	Output variables	$PM_{2.5}$ emission	Population weighting $PM_{2.5}$ concentration in each province *the land area in each province/ the primary atmospheric environmental capacity of $PM_{2.5}$ bound by the national $PM_{2.5}$ standard	*The data came from China Statistical Yearbook on Environment 2015and* Wang et al. (2015); The formula referred to Wu et al., (2016a, 2016b)	
		Number of days when the air quality reaches or is better than the second level		*China Statistical Yearbook 2015*	Days

Note Only environmental capacity data of each province in 2010 are available. To ensure the feasibility and completeness of the data, in addition to the data of 2010 for $PM_{2.5}$ emission variables, we use data of 2014 for the other variables. Data in 2014 are shown in Table 17.2

Table 17.2 The original data

DMU	Capital (100 million Yuan)	Labor force (10,000 people)	Car ownership (10,000))	Coal consumption (10,00 tons)	GDP (100 million Yuan)	SO_2 emission (t.)	NO_X emission (t.)	Soot emission (t.)	Carbon emission(t.)
Beijing	6924.23	755.86	966.62	1736.54	21,330.83	78,906.03	150,955.14	57,372.45	10,374,334,362
Tianjin	10,518.19	295.51	509.29	5027.28	15,726.93	209,200	282,300	139,511.47	20,034,636,270
Hebei	26,671.92	656.18	1764.98	29,635.54	29,421.15	1,189,902.56	1,512,468.92	1,797,683.42	91,228,396,696
Shanxi	12,354.53	452.09	791.38	37,587.43	12,761.49	1,208,225.28	1,069,859.77	1,506,778.02	83,330,264,650
Inner Mongolia	17,591.83	301.45	642.63	36,465.97	17,770.19	1,312,436.37	1,258,281.03	1,021,510.38	80,768,506,870
Liaoning	24,730.8	665.17	937.02	18,002.27	28,626.58	994,597.4	901,964.07	1,120,701.44	72,061,262,032
Jining	11,339.62	334.42	526.75	10,379.34	13,803.14	372,256	549,246.84	475,132.99	27,547,194,701
Heilongjiang	9828.99	450.88	587.81	13,595.53	15,039.38	472,248.4	730,584.86	793,548.57	37,138,071,681
Shanghai	6016.43	648.88	438.33	4895.78	23,567.7	188,149.16	332,790.47	141,650.1	25,351,388,731
Jiangsu	41,938.62	1602.4	2022.93	26,912.61	65,088.32	904,740.62	1,232,552.49	763,678.35	80,879,250,176
Zhejiang	24,262.77	1102.68	1882	13,824.37	40,173.03	574,012.17	687,850.05	379,666.45	43,787,271,206
Anhui	21,875.58	521.74	757.86	15,786.98	20,848.75	492,965.67	807,304.79	652,781.96	40,050,837,023
Fujian	18,177.86	654.64	716.13	8198.3	24,055.76	355,957.2	411,661.5	367,902.63	28,503,278,119
Jiangxi	15,079.26	465.26	521.85	7477.31	15,714.63	534,414.92	540,114.93	462,330.57	21,394,264,516
Shandong	42,495.55	1266.34	2541.87	39,561.73	59,426.59	1,590,236.96	1,593,310.72	1,208,102.46	1.27943E + 11
Henan	30,782.17	1108.89	1745.05	24,249.88	34,938.24	1,198,182.07	1,422,013.46	882,102.59	63,369,774,643

(continued)

Table 17.2 (continued)

DMU	Capital (100 million Yuan)	Labor force (10,000 people)	Car ownership (10,000))	Coal consumption (10,00 tons)	GDP (100 million Yuan)	SO_2 emission (t.)	NO_X emission (t.)	Soot emission (t.)	Carbon emission(t.)
Hubei	22,915.3	706.8	771.87	11,887.83	27,379.22	583,759.24	580,221.62	504,006.07	36,321,023,923
Hunan	21,242.92	597.9	819.19	10,899.51	27,037.32	623,688.6	552,773.06	496,165.87	30,836,483,005
Guangdong	26,293.93	1973.28	2481.67	17,013.71	67,809.85	730,146.58	1,122,111.74	449,548.89	60,752,546,791
Guangxi	13,843.22	401.46	583.4	6796.51	15,672.89	466,588.74	442,398.76	402,934.62	23,532,265,913
Hainan	3112.23	101.52	136.76	1018.3	3500.72	32,563.91	95,001.9	23,171.24	5,992,939,638
Chongqing	12,285.42	414.47	427.62	6095.78	14,262.6	526,944.28	355,017.97	226,130.56	15,335,979,629
Sichuan	23,318.57	808.75	1243.02	11,045.39	28,536.66	796,401.53	585,438.61	428,629.91	36,554,045,895
Guizhou	9025.75	304.75	452.19	13,117.6	9266.39	925,787.1	491,070.7	377,856.14	29,053,741,520
Yunnan	11,498.53	419.57	803.52	8674.67	12,814.59	636,683.23	498,879.94	366,818.71	23,555,447,704
Tibet	1069.23	32.54	52.48	0.001	920.83	4249.87	48,343.52	13,889.88	124.5933333
Shaanxi	17,191.92	516.52	716.54	18,375.34	17,689.94	780,954.38	705,755.99	709,137.44	48,931,332,033
Gansu	7884.13	264.74	327.27	6715.87	6836.82	575,648.72	418,399.8	345,810.85	21,557,583,060
Qinghai	2861.23	63.19	122.94	1816.51	2303.32	154,276	134,518.07	239,866.81	5,362,116,677
Ningxia	3173.79	73.25	168.66	8857	2752.1	377,055.54	404,032.35	239,170.83	20,594,142,088
Xinjiang	9447.74	316.65	477.81	16,0880.3	9273.46	852,981.33	862,792.27	813,915.92	46,225,260,022

(continued)

Table 17.2 (continued)

DMU	Precipitation (ml)	Investment of waste gas treatment (100 million Yuan)	Greening ratio (%)	$PM_{2.5}$ emission	Number of days when the air quality reaches or is better than the second level (days)
Beijing	461.5	62,281	49.1	86	168
Tianjin	441.4	151,108	34.9	83	175
Hebei	294.8	779,803	41.9	124	97
Shanxi	428.7	230,646	40.1	72	197
Inner Mongolia	394.8	708,403	39.8	46	240
Liaoning	362.9	326,098	40.1	74	190
Jining	446	153,382	35.8	68	239
Heilongjiang	415.8	160,834	36	72	241
Shanghai	1295.3	98,400	38.4	52	278
Jiangsu	1091.1	383,361	42.6	74	188
Zhejiang	1359.9	414,013	40.8	65	216
Anhui	1180.2	149,590	41.2	83	151
Fujian	1628	183,548	42.8	34	340
Jiangxi	1431.8	101,458	44.6	52	294
Shandong	521.4	1,281,351	42.8	87	107
Henan	551.6	464,979	38.3	88	135
Hubei	1208.6	232,512	37.9	82	177
Hunan	1386.8	133,227	38.6	74	224
Guangdong	2234	271,987	41.4	49	282
Guangxi	1234.7	106,049	39.3	49	292
Hainan	1861.3	55,649	41.3	23	346
Chongqing	1452.1	34,278	40.6	65	246

(continued)

Table 17.2 (continued)

DMU	Precipitation (ml)	Investment of waste gas treatment (100 million Yuan)	Greening ratio (%)	$PM_{2.5}$ emission	Number of days when the air quality reaches or is better than the second level (days)
Sichuan	975	164,883	37.5	77	216
Guizhou	1562	169,146	34	48	301
Yunnan	1078.3	134,079	38.1	35	350
Tibet	637.8	1453	43.8	25	321
Shaanxi	660.2	255,980	40.5	77	172
Gansu	355.6	137,981	30.8	61	247
Qinghai	446.5	48,768	31.6	63	261
Ningxia	169.2	219,777	38	53	255
Xinjiang	300.9	278,911	36.8	61	202

Table 17.3 Results

	DMUs	Alpha = 0.1			Alpha = 0.2			Alpha = 0.3			Alpha = 0.4			Alpha = 0.5		
		e	e1	e2	e	e1	e2	e	e1	e2	e	e1	e2	e	e1	e2
1	Beijing	0.73441	1.00000	0.70490	0.76392	1.00000	0.70490	0.79343	1.00000	0.70490	0.82294	1.00000	0.70490	0.85245	1.00000	0.70490
2	Tianjin	0.66504	1.00000	0.62782	0.70226	1.00000	0.62782	0.73947	1.00000	0.62782	0.77669	1.00000	0.62782	0.81391	1.00000	0.62782
3	Hebei	0.67618	0.82261	0.65991	0.69245	0.82262	0.65991	0.70873	0.82262	0.65991	0.72500	0.82264	0.65991	0.74127	0.82264	0.65991
4	Shanxi	0.95889	0.58892	1.00000	0.91779	0.58893	1.00000	0.87668	0.58894	1.00000	0.83558	0.58894	1.00000	0.79447	0.58895	1.00000
5	Inner Mongolia	1.00000	1.00000	1.00000	1.00000	1.00000	1.00000	1.00000	1.00000	1.00000	1.00000	1.00000	1.00000	1.00000	1.00000	1.00000
6	Liaoning	0.94213	0.89383	0.94750	0.93677	0.89383	0.94750	0.93140	0.89383	0.94750	0.92603	0.89383	0.94750	0.92067	0.89384	0.94750
7	Jilin	0.82268	0.81170	0.82390	0.82146	0.81170	0.82390	0.82024	0.81170	0.82390	0.81902	0.81170	0.82390	0.81780	0.81170	0.82390
8	Heilongjiang	0.82665	0.75175	0.83497	0.81833	0.75175	0.83497	0.81001	0.75175	0.83497	0.80168	0.75175	0.83497	0.79336	0.75175	0.83497
9	Shanghai	0.99204	1.00000	0.99115	0.99292	1.00000	0.99115	0.99381	1.00000	0.99115	0.99469	1.00000	0.99115	0.99558	1.00000	0.99115
10	Jiangsu	0.52950	0.88705	0.48978	0.56923	0.88705	0.48978	0.60896	0.88705	0.48978	0.64868	0.88705	0.48978	0.68841	0.88705	0.48978
11	Zhejiang	0.60053	0.81859	0.57630	0.62476	0.81859	0.57630	0.64899	0.81859	0.57630	0.67322	0.81859	0.57630	0.69745	0.81859	0.57630
12	Anhui	0.44086	0.81790	0.39897	0.48275	0.81790	0.39897	0.52465	0.81790	0.39897	0.56654	0.81790	0.39897	0.60843	0.81790	0.39897
13	Fujian	0.87192	0.85329	0.87399	0.86985	0.85329	0.87399	0.86778	0.85329	0.87399	0.86571	0.85329	0.87399	0.86364	0.85329	0.87399
14	Jiangxi	0.76062	0.77578	0.75893	0.76230	0.77578	0.75893	0.76399	0.77578	0.75893	0.76567	0.77578	0.75893	0.76736	0.77578	0.75893
15	Shandong	0.69254	0.90379	0.66907	0.71602	0.90382	0.66907	0.73950	0.90383	0.66907	0.76297	0.90384	0.66906	0.78645	0.90384	0.66907
16	Henan	0.67438	0.65369	0.67668	0.67208	0.65369	0.67668	0.66979	0.65371	0.67668	0.66749	0.65371	0.67668	0.66520	0.65371	0.67668
17	Hubei	0.54754	0.89994	0.50838	0.58669	0.89994	0.50838	0.62585	0.89994	0.50838	0.66501	0.89994	0.50838	0.70416	0.89994	0.50838
18	Hunan	0.66528	0.95149	0.63348	0.69708	0.95149	0.63348	0.72888	0.95149	0.63348	0.76068	0.95149	0.63348	0.79249	0.95149	0.63348
19	Guangdong	0.75689	0.89544	0.74149	0.77228	0.89544	0.74149	0.78768	0.89544	0.74149	0.80307	0.89544	0.74149	0.81847	0.89544	0.74149
20	Guangxi	0.83466	0.79865	0.83866	0.83066	0.79865	0.83866	0.82666	0.79865	0.83866	0.82266	0.79865	0.83866	0.81866	0.79865	0.83866

(continued)

Table 17.3 (continued)

	DMUs	Alpha = 0.1			Alpha = 0.2			Alpha = 0.3			Alpha = 0.4			Alpha = 0.5		
		e	e1	e2	e	e1	e2	e	e1	e2	e	e1	e2	e	e1	e2
21	Hainan	0.98228	0.82276	1.00000	0.96455	0.82276	1.00000	0.94683	0.82276	1.00000	0.92910	0.82276	1.00000	0.91138	0.82276	1.00000
22	Chongqing	0.98186	0.81862	1.00000	0.96372	0.81862	1.00000	0.94559	0.81862	1.00000	0.92745	0.81862	1.00000	0.90931	0.81862	1.00000
23	Sichuan	0.64791	0.74206	0.63744	0.65837	0.74206	0.63744	0.66883	0.74206	0.63744	0.67929	0.74206	0.63744	0.68975	0.74206	0.63744
24	Guizhou	0.92897	0.61639	0.96371	0.89424	0.61639	0.96371	0.85951	0.61639	0.96371	0.82478	0.61639	0.96371	0.79005	0.61639	0.96371
25	Yunnan	0.96364	0.63640	1.00000	0.92728	0.63640	1.00000	0.89092	0.63640	1.00000	0.85456	0.63640	1.00000	0.81820	0.63640	1.00000
26	Tibet	1.00000	1.00000	1.00000	1.00000	1.00000	1.00000	1.00000	1.00000	1.00000	1.00000	1.00000	1.00000	1.00000	1.00000	1.00000
27	Shaanxi	0.66640	0.71639	0.66084	0.67196	0.71641	0.66084	0.67751	0.71641	0.66084	0.68307	0.71642	0.66084	0.68863	0.71642	0.66084
28	Gansu	0.95729	0.57286	1.00000	0.91457	0.57286	1.00000	0.87186	0.57286	1.00000	0.82914	0.57286	1.00000	0.78643	0.57286	1.00000
29	Qinghai	0.97537	0.75367	1.00000	0.95073	0.75367	1.00000	0.92610	0.75367	1.00000	0.90147	0.75367	1.00000	0.87684	0.75367	1.00000
30	Ningxia	0.96995	0.69955	1.00000	0.93991	0.69955	1.00000	0.90986	0.69955	1.00000	0.87982	0.69955	1.00000	0.84977	0.69955	1.00000
31	Xinjiang	0.95910	0.59096	1.00000	0.91819	0.59096	1.00000	0.87728	0.59094	1.00000	0.83638	0.59095	1.00000	0.79548	0.59096	1.00000

	DMUs	Alpha = 0.6			Alpha = 0.7			Alpha = 0.8			Alpha = 0.9		
		e	e1	e2	e	e1	e2	e	e1	e2	e	e1	e2
1	Beijing	0.88196	1.00000	0.70490	0.91147	1.00000	0.70490	0.94098	1.00000	0.70490	0.97049	1.00000	0.70490
2	Tianjin	0.85113	1.00000	0.62782	0.88835	1.00000	0.62782	0.92556	1.00000	0.62782	0.96278	1.00000	0.62782
3	Hebei	0.75756	0.82265	0.65991	0.77383	0.82265	0.65991	0.79010	0.82265	0.65991	0.77654	0.82266	0.36152
4	Shanxi	0.75336	0.58894	1.00000	0.71226	0.58895	1.00000	0.67116	0.58894	1.00000	0.59408	0.58895	0.64025
5	Inner Mongolia	1.00000	1.00000	1.00000	1.00000	1.00000	1.00000	1.00000	1.00000	1.00000	1.00000	1.00000	1.00000
6	Liaoning	0.91530	0.89383	0.94750	0.90993	0.89383	0.94750	0.84258	0.89385	0.63754	0.86822	0.89385	0.63754

(continued)

Table 17.3 (continued)

	DMUs	Alpha = 0.6			Alpha = 0.7			Alpha = 0.8			Alpha = 0.9		
		e	e1	e2	e	e1	e2	e	e1	e2	e	e1	e2
7	Jilin	0.81658	0.81170	0.82390	0.81536	0.81170	0.82390	0.81414	0.81170	0.82390	0.81292	0.81170	0.82390
8	Heilongjiang	0.78504	0.75175	0.83497	0.77672	0.75175	0.83497	0.76840	0.75175	0.83497	0.76008	0.75175	0.83497
9	Shanghai	0.99646	1.00000	0.99115	0.99735	1.00000	0.99115	0.99823	1.00000	0.99115	0.99912	1.00000	0.99115
10	Jiangsu	0.72814	0.88705	0.48978	0.76787	0.88705	0.48978	0.80759	0.88705	0.48978	0.84732	0.88705	0.48978
11	Zhejiang	0.72168	0.81859	0.57630	0.74591	0.81859	0.57630	0.77013	0.81859	0.57630	0.79436	0.81859	0.57630
12	Anhui	0.65033	0.81790	0.39897	0.69222	0.81790	0.39897	0.73412	0.81790	0.39897	0.77601	0.81790	0.39897
13	Fujian	0.86157	0.85329	0.87399	0.85950	0.85329	0.87399	0.85743	0.85329	0.87399	0.85536	0.85329	0.87399
14	Jiangxi	0.76904	0.77578	0.75893	0.77073	0.77578	0.75893	0.77241	0.77578	0.75893	0.77410	0.77578	0.75893
15	Shandong	0.80993	0.90383	0.66907	0.83340	0.90383	0.66907	0.85688	0.90383	0.66907	0.85803	0.90385	0.44566
16	Henan	0.66290	0.65371	0.67668	0.66061	0.65372	0.67668	0.65831	0.65371	0.67668	0.63615	0.65372	0.47796
17	Hubei	0.74332	0.89994	0.50838	0.78247	0.89994	0.50838	0.82163	0.89994	0.50838	0.86079	0.89994	0.50838
18	Hunan	0.82429	0.95149	0.63348	0.85609	0.95149	0.63348	0.88789	0.95149	0.63348	0.91969	0.95149	0.63348
19	Guangdong	0.83386	0.89544	0.74149	0.84926	0.89544	0.74149	0.86465	0.89544	0.74149	0.88005	0.89544	0.74149
20	Guangxi	0.81466	0.79865	0.83866	0.81065	0.79865	0.83866	0.80665	0.79865	0.83866	0.80265	0.79865	0.83866
21	Hainan	0.89366	0.82276	1.00000	0.87593	0.82276	1.00000	0.85821	0.82276	1.00000	0.84048	0.82276	1.00000
22	Chongqing	0.89117	0.81862	1.00000	0.87303	0.81862	1.00000	0.85490	0.81862	1.00000	0.83676	0.81862	1.00000
23	Sichuan	0.70021	0.74206	0.63744	0.71068	0.74206	0.63744	0.72114	0.74206	0.63744	0.73160	0.74206	0.63744
24	Guizhou	0.75532	0.61639	0.96371	0.72058	0.61639	0.96371	0.68585	0.61639	0.96371	0.65112	0.61639	0.96371
25	Yunnan	0.78184	0.63640	1.00000	0.74548	0.63640	1.00000	0.70912	0.63640	1.00000	0.67276	0.63640	1.00000
26	Tibet	1.00000	1.00000	1.00000	1.00000	1.00000	1.00000	1.00000	1.00000	1.00000	1.00000	1.00000	1.00000

(continued)

Table 17.3 (continued)

	DMUs	Alpha = 0.6			Alpha = 0.7			Alpha = 0.8			Alpha = 0.9		
		e	e1	e2	e	e1	e2	e	e1	e2	e	e1	e2
27	Shaanxi	0.69419	0.71642	0.66084	0.69975	0.71642	0.66084	0.67805	0.71643	0.52452	0.69724	0.71643	0.52452
28	Gansu	0.74372	0.57286	1.00000	0.70100	0.57286	1.00000	0.65829	0.57286	1.00000	0.61558	0.57286	1.00000
29	Qinghai	0.85220	0.75367	1.00000	0.82757	0.75367	1.00000	0.80294	0.75367	1.00000	0.77830	0.75367	1.00000
30	Ningxia	0.81973	0.69955	1.00000	0.78968	0.69955	1.00000	0.75964	0.69955	1.00000	0.72959	0.69955	1.00000
31	Xinjiang	0.75457	0.59096	1.00000	0.71367	0.59096	1.00000	0.67276	0.59096	1.00000	0.63186	0.59096	1.00000

*In Table 17.3, e denotes the overall efficiency of the system, e_1 denotes the efficiency of the first sub-process, and e_2 denotes the efficiency of the second sub-process. The weight α is viewed as an endogenous variable in SOCP transformation, and the preference of decision makers can determine the weight. Here we set weights $\alpha = \frac{i}{10}$, $i = 1, 2, \ldots, 9$.

Appendix 2: Estimation of Carbon Emission

IPCC (2006) provides sectoral approaches and reference methods for estimating fossil energy carbon emission based on the method of valuation. This study uses the method to estimate the industrial energy carbon emission, and the formula is as follows:

$$\sum_{i=1}^{n} (CO_2)_i = \sum_{i=1}^{n} E_i \times NCV_i \times CEF_i \times COF_i \times \frac{44}{12}$$

where E is the physical quantity of energy consumption, i the energy type, NCV the average low calorific value of the energy (IPCC is called the net calorific value), CEF the carbon emission factor's per unit calorific value equivalent, COF the carbon oxidation factor (only a small fraction of the carbon in fossil fuels will not be oxidized during the combustion process, while 99% to 100% of the carbon is oxidized, so the default value is 1), and 44 and 12 are the molecular weights of CO_2 and C respectively. This article only estimates carbon emission from the industrial level. The primary energy carbon emission is accounted for by the principle of terminal energy consumption. According to the expression above, when estimating industrial energy CO_2 emission, we need all kinds of industrial energy consumption. We need to set the average low calorific value, carbon emission factor, carbon oxidation factor, and other parameters of various types of energy. The relevant parameters are shown in Table 17.4. The average low calorific value of each fossil fuel is taken from *the China Energy Statistical Yearbook*, where the net calorific value of briquette, other petroleum products, and other coking products are taken from IPCC (2006).

Table 17.4 Parameter setting in the estimation of industrial energy of carbon emission

Fuel type	Average low calorific value	Carbon oxidation rate	Carbon emission factor
	KJ/Kg	%	t/TJ
Raw coal	20,908	100	25.8
Coke	28,435	100	29.2
Crude oil	41,816	100	20
Gasoline	43,070	100	18.9
Kerosene	43,070	100	19.5
Diesel	42,625	100	20.2
Fuel oil	41,816	100	21.1
Natural gas	38,931	100	15.3

Source General principles for calculating the total energy consumption (GB/T2589–2008), Table 1.2 and Table 1.3 in Volume II, Chapter 1 of the 2006 IPCC Guidelines for National Greenhouse Gas Inventories (IPCC, 2006)

According to the expression above and various types of fuel consumption of each province in China in 2014, each province's carbon emission can be calculated in 2014 (Table 17.4).

References

Ang, S., & Chen, C. M. (2016). Pitfalls of decomposition weights in the additive multi-stage DEA model. *Omega, 58*, 139–153. https://doi.org/10.1016/j.omega.2015.05.008

Beckett, K. P., Freer-Smith, P., & Taylor, G. (2000). Effective tree species for local air quality management. *Journal of Arboriculture, 26*, 12–19.

Bui, L. T., Abbass, H. A., Barlow, M., & Bender, A. (2012). Robustness against the decision-maker's attitude to risk in problems with conflicting objectives. *IEEE Transactions on Evolutionary Computation, 16*, 1–19. https://doi.org/10.1109/TEVC.2010.2051443

Cao, C., Lee, X., Liu, S., Schultz, N., Xiao, W., Zhang, M., & Zhao, L. (2016). Urban heat Islands in china enhanced by haze pollution. *Nature Communications, 7*, 12509. https://doi.org/10.1038/ncomms12509

Charnes, A., & Cooper, W. W. (2010). Programming with linear fractional functionals. *Naval Research Logistics, 9*, 181–186. https://doi.org/10.1002/nav.3800090303

Charnes, A., Cooper, W. W., & Rhodes, E. (1978). Measuring the efficiency of decision making units. *European Journal of Operational Research, 2*, 429–444. https://doi.org/10.1016/0377-2217(78)90138-8

Chen, C., Zhu, J., Yu, J. Y., & Noori, H. (2012). A new methodology for evaluating sustainable product design performance with two-stage network data envelopment analysis. *European Journal of Operational Research, 221*, 348–359. https://doi.org/10.1016/j.ejor.2012.03.043

Chen, J. R., Choi, J. W., & Seo, Y. J. (2025). Environmental efficiency assessment of coastal ports in China: Implications for sustainable port management. *Marine Pollution Bulletin., 211*, Article 117436. https://doi.org/10.1016/j.marpolbul.2024.117436

Chen, K., & Zhu, J. (2017). Second order cone programming approach to two-stage network data envelopment analysis. *European Journal of Operational Research, 262*, 231–238. https://doi.org/10.1016/j.ejor.2017.03.074

Chen, Y., Cook, W. D., Li, N., & Zhu, J. (2009). Additive efficiency decomposition in two-stage DEA. *European Journal of Operational Research, 196*, 1170–1176. https://doi.org/10.1016/j.ejor.2008.05.011

China Meteorological Administration. (2011). *The top 10 weather and climate events at home and abroad in 2011 have been announced.* http://www.cma.gov.cn/2011xwzx/2011xqxxw/2011xqxyw/201112/t20111230_157762.html.

Cobb, C. W., & Douglas, P. H. (1928). A theory of production. *American Economic Review, 18*, 139–165.

Cook, W. D., Zhu, J., Bi, G., & Yang, F. (2010). Network DEA: Additive efficiency decomposition. *European Journal of Operational Research, 207*, 1122–1129. https://doi.org/10.1016/j.ejor.2010.05.006

Cooper, W. W., Seiford, L. M., Tone, K., & Zhu, J. (2007). Some models and measures for evaluating performances with DEA: Past accomplishments and future prospects. *The Journal of Productivity Analysis, 28*, 151–163. https://doi.org/10.1007/s11123-007-0056-4

Despotis, D. K., Koronakos, G., & Sotiros, D. (2016). Composition versus decomposition in two-stage network DEA: A reverse approach. *The Journal of Productivity Analysis, 45*, 71–87. https://doi.org/10.1007/s11123-014-0415-x

Dong, X. Lin., & Yang, M. Y. (2013). Relationship between industrial waste gas emission and control investment based on pollution control investment index. *Journal of Environmental & Earth Sciences, 35*, 113–118. https://doi.org/10.3969/j.issn.1672-6561.2013.03.011

Englert, N. (2004). Fine particles and human heath—A review of epidemiological studies. *Toxicology Letters, 149*(1–3), 235–242. https://doi.org/10.1016/j.toxlet.2003.12.035

Fare, R., Grosskopf, S., Lovell, C. A. K., & Pasurka, C. (1989). Multilateral productivity comparisons when some outputs are undesirable: A nonparametric approach. *The Review of Economics and Statistics, 71*, 90–98. https://doi.org/10.2307/1928055

Fukuyama, H., & Weber, W. L. (2010). A slacks-based inefficiency measure for a two-stage system with bad outputs. *Omega, 38*, 398–409. https://doi.org/10.1016/j.omega.2009.10.006

Giere, R., Blackford, M., & Smith, K. (2006). TEM study of PM2.5 emitted from coal and tire combustion in Athermal power station. *Environmental Science and Technology, 40*, 6235–6240.

Gomes, E. G., & Lins, M. P. E. (2008). Modelling undesirable outputs with zero sum gains data envelopment analysis models. *The Journal of the Operational Research Society, 59*, 616–623. https://doi.org/10.1057/palgrave.jors.2602384

Guo, C., Shureshjani, R. A., Foroughi, A. A., & Zhu, J. (2017). Decomposition weights andoverall efficiency in two-stage additive network DEA. *European Journal of Operational Research, 257*, 896–906. https://doi.org/10.1016/j.ejor.2016.08.002

Harding, A. (2006). Integrated natural resource management: Linking productivity, the environment and development. *The Journal of the Operational Research Society, 56*, 750–751. https://doi.org/10.1057/palgrave.jors.2601918

IPCC. (2006). 2006 IPCC guidelines for national greenhouse gas inventories. In Eggleston, H. S, Buendia, L., Miwa, K., Ngara, T., Tanabe, K. (Eds.), *IGES, Hayama* (prepared by the National Greenhouse Gas Inventories Programme).

Jin, S. J., Guo, J. K., Wheeler, S., Kan, L. Y., & Che, S. Q. (2014). Evaluation of Impacts of trees on PM2.5 dispersion in urban streets. *Atmospheric Environment, 99*, 277–287. https://doi.org/10.1016/j.atmosenv.2014.10.002

Jones, A. M., Harrison, R. M., & Baker, J. (2010). The wind speed dependence of the concentrations of airborne particulate matter and NOx. *Atmospheric Environment, 44*, 1682–1690. https://doi.org/10.1016/j.atmosenv.2010.01.007

Jun, L. I., Bao, S. C., De, L. X., Ping, D. S., Jing, G., Yan, W., & An, X. X. (2009). Non-parameter statistical analysis of impacts of meteorological conditions on PM concentration in Beijing. *Research of Environmental Sciences, 22*, 663–669.

Kao, C., & Hwang, S. N. (2008). Efficiency decomposition in two-stage data envelopment analysis: An application to non-life insurance companies in Taiwan. *European Journal of Operational Research, 185*, 418–429. https://doi.org/10.1016/j.ejor.2006.11.041

Kinney, P. L., Aggarwal, M., Northridge, M. E., Janssen, N. A. H., & Shepard, P. (2000). Airborne concentrations of PM 2.5 and diesel exhaust particles on Harlem sidewalks: A community-based pilot study. *Environmental Health Perspectives, 108*, 213–218. https://doi.org/10.2307/3454436

Liang, L., Cook, W. D., & Zhu, J. (2010). DEA models for two-stage processes: Game approach and efficiency decomposition. *Naval Research Logistics, 55*, 643–653. https://doi.org/10.1002/nav.20308

Lim, S., & Zhu, J. (2013). Integrated data envelopment analysis: Global vs. local optimum. *European Journal of Operational Research, 229*, 276–278. https://doi.org/10.1016/j.ejor.2013.02.023

Liu, H., Fang, C., Zhang, X., Wang, Z., Bao, C., & Li, F. (2017). The effect of natural and anthropogenic factors on haze pollution in Chinese cities: A spatial econometrics approach. *Journal of Cleaner Production, 165*, 323–333. https://doi.org/10.1016/j.jclepro.2017.07.127

Liu, X., & Wu, J. (2015). Energy and Environmental Efficiency analysis of china's regional transportation sectors: A slack-based DEA approach. *Energy Systems, 8*, 747–759. https://doi.org/10.1007/s12667-015-0164-6

Moutinho, V., Madaleno, M., Robaina, M., & Villar, J. (2017). Advanced scoring method of eco-efficiency in European cities. *Environmental Science and Pollution Research, 25*, 1637–1654. https://doi.org/10.1007/s11356-017-0540-y

Pratsini, E. (1999). Frontier analysis in air pollution. *The Journal of the Operational Research Society, 50*, 526–530. https://doi.org/10.1057/palgrave.jors.2600718

Santibanez Gonzalez, E. D., Koh, L., Leung, J. (2019). Towards a circular economy production system: Trends and challenges for operations managemdent. International Journal of Production Research, 7209–7218. https://doi.org/10.1080/00207543.2019.1656844

Santibanez-Gonzalez, E. D. R., & Huisingh, D. (2015). Announcement about an exciting opportunity for Operations Researchers! *Journal of Cleaner Production, 86*. https://doi.org/10.1016/j.jclepro.2014.08.060

Seiford, L. M., & Zhu, J. (2002). A response to comments on modeling undesirable factors in efficiency evaluation. *European Journal of Operational Research, 142*, 16–20. https://doi.org/10.1016/j.ejor.2003.09.018

Song, M. L., Peng, J., Wang, J., & Zhao, J. (2018). Environmental efficiency and economic growth of china: A ray slack-based model analysis. *European Journal of Operational Research, 269*, 51–63. https://doi.org/10.1016/j.ejor.2017.03.073

Song, M. L., & Wang, J. (2018). Environmental efficiency evaluation of thermal power generation in china based on a slack-based endogenous directional distance function model. *Energy, 161*, 325–336. https://doi.org/10.1016/j.energy.2018.07.158

Song, M. L., Wang, S. H., Lei, L., & Zhou, L., (2019). Environmental efficiency and policy change in china: a new meta-frontier non-radial angle efficiency evaluation approach. *Process Safety and Environmental Protection, 121*, 281–289. https://doi.org/10.1016/j.psep.2018.10.023

Sueyoshi, T., & Yuan, Y. (2015). China's regional sustainability and diversified resource allocation: Dea environmental assessment on economic development and air pollution. *Energy Economics, 49*, 239–256. https://doi.org/10.1016/j.eneco.2015.01.024

Sueyoshi, T., Yuan, Y., & Goto, M. (2016). A literature study for DEA applied to energy and environment. *Energy Economics, 62*, 104–124. https://doi.org/10.1016/j.eneco.2016.11.006

Wang, P., Cao, J. J., Tie, X. X., Wang, G. H., Li, G. H., Hu, T. F., Wu, Y. T., Xu, Y. S., Xu, G. D., Zhao, Y. Z. (2015). Impact of Meteorological Parameters and Gaseous Pollutants on PM2.5 and PM10 Mass Concentrations during 2010 in Xi'an, china. *Aerosol and Air Quality Research, 15*, 11. https://doi.org/10.4209/aaqr.2015.05.0380.

Wu, J., Zhu, Q., Ji, X., Chu, J., & Liang, L. (2016a). Two-stage network processes with shared resources and resources recovered from undesirable outputs. *European Journal of Operational Research, 251*, 182–197. https://doi.org/10.1016/j.ejor.2015.10.049

Wu, X. H., Tan, L., & Guo, J. (2016b). A study of allocative efficiency of PM2.5, emission rights based on a zero sum gains data envelopment model. *Journal of Cleaner Production, 113*, 1024–1031. https://doi.org/10.1016/j.jclepro.2015.11.025

Xue, W. B., Fei, F. U., Wang, J. N., Ke-Bi, H. E., Yu, L., & Yang, J. T. (2014). Modeling study on atmospheric environmental capacity of major pollutants constrained by PM2.5 compliance of Chinese cities. *China Environmental Science, 34*, 2490–2496.

Yang, J., Yu, Q., & Gong, P. (2008). Quantifying air pollution removal by green roofs in Chicago. *Atmospheric Environment, 42*, 7266–7273. https://doi.org/10.1016/j.atmosenv.2008.07.003

Yang, Z., & Wang, J. (2017). A new air quality monitoring and early warning system: Air quality assessment and air pollutant concentration prediction. *Environmental Research, 158*, 105–117. https://doi.org/10.1016/j.envres.2017.06.002

Yu, A., Zhou, F., & Zheng, Z. (2025). Performance analytics for the value chain process of international airports: A dynamic network measure with the shared and unsplittable links. *Annals of Operations Research, 351*, 3–33. https://doi.org/10.1007/s10479-024-05988-5

Zhang, C., Liu, H., Bressers, H. T. A., & Buchanan, K. S. (2011). Productivity growth and environmental regulations—Accounting for undesirable outputs: Analysis of china's thirty provincial regions using the malmquist–luenberger index. *Ecological Economics, 70*, 2369–2379. https://doi.org/10.1016/j.ecolecon.2011.07.019

Zhang, L. Y., & Chen, Y. (2018). Equivalent solutions to additive two-stage network data envelopment analysis. *European Journal of Operational Research, 264*, 1189–1191. https://doi.org/10.1016/j.ejor.2017.07.042

Zhang, N., & Choi, Y. (2014). A note on the evolution of directional distance function and its development in energy and environmental studies 1997–2013. *Renewable & Sustainable Energy Reviews, 33*, 50–59. https://doi.org/10.1016/j.rser.2014.01.064

Zhang, R., Li, Q., & Zhang, R. N. (2014). Meteorological conditions for the persistent severe fog and haze event over eastern china in January 2013. *Science China Earth Sciences, 51*, 26–35. https://doi.org/10.1007/s11430-013-4774-3

Zhang, W., Wu, X., & Chen, J. (2024). Low-carbon efficiency analysis of rail-water multimodal transport based on cross efficiency network DEA approach. *Energy, 305*, Article 132348. https://doi.org/10.1016/j.energy.2024.132348

Zhou, P., Ang, B. W., & Poh, K. L. (2008). A survey of data envelopment analysis in energy and environmental studies. *European Journal of Operational Research, 189*, 1–18. https://doi.org/10.1016/j.ejor.2007.04.042

Chapter 18
Study on the Distribution of PM Emission Rights in Various Provinces of China Based on a New Efficiency and Equity Two-Objective DEA Model

18.1 Introduction

With the continuous development of global industrialization, many countries in the world have experienced air pollution problems. Historically, the Mas valley smog in the 1930s, the London smog in 1952 and the photochemical smog in Los Angeles in 1943 have all brought serious impacts on the local economy and environment (Lucier, 2003; Nemery et al., 2001; SCAQMD, 2003). In recent years, air pollution has become particularly serious in some developing countries. In mid-November 2016, the average concentration of fine particulate matter (*PM*) in the air in Tehran, Iran exceeded 150 μg per cubic meter, resulting in more than 400 deaths (Xinhua News Agency, 2016). According to the report of Global Burden of Disease (GBD), it is estimated that about 2 million people in India died prematurely due to air pollution in 2016 (GBD, 2015, Mortality & Causes of Death Collaborators, 2016).

In the past 20 years, air pollution in China, with *PM* as a typical representative, has been increasing continuously. 2013 was the worst year for haze pollution in 52 years, according to China's state of the environment bulletin released in 2013. In the winter of 2016, a large area of haze appeared in the central and eastern regions. According to remote sensing monitoring data, the haze area once reached 1.88 million square kilometers (Xinhua News Agency, 2016). At the beginning of 2018, the Beijing-Tianjin-Hebei region and surrounding areas experienced heavy regional air pollution for nearly 10 days, and more than 60 cities implemented regional emergency linkage measures (Xinhua News Agency, 2018). During the 'two sessions' in 2019, moderate and heavy pollution weather continues to occur in Beijing and other places (The Environmental Protection Agency, 2019), and the $PM_{2.5}$ concentration in some areas even exceeds 300 μg/m^3. Air pollution interferes with daily traffic, reduces inbound tourism and endangers health, which has aroused great concern from the whole society (Chen et al., 2025; Ji et al., 2025; Ni et al., 2019; Wang et al., 2018a, 2018b). How to effectively control air pollution has become an important issue for the government, academia and the public.

X. Wu, *Environmental Disaster in China*,
https://doi.org/10.1007/978-981-92-0910-1_18

From the experience at home and abroad, there are two major types of air pollutants governance, one is the source control (Rosa et al., 2019). For example, on the basis of analyzing the composition of air pollutants, sources of pollution are found, and then methods such as shutdown, production limit and technical transformation are adopted to reduce the emission of pollutants from the source. The difficulty lies in how to analyze the components and trace the pollution sources. The second category is socio-economic means. Such as the implementation of emission permit system, levy taxes, carry out emission rights trading. In the second approach, when the total amount of pollution is set, the key lies in how to allocate the emission to each emitter.

According to a research report released in January of 2014, less than 1% of the 500 largest cities in China had achieved the air quality standards recommended by the World Health Organization. In addition, China accounts for the seven out of the ten most polluted cities in the world (Zhang & Crooks, 2012). To address such important public goods as air pollutants, there are two types of regulations: one is tax regulation (Baumol & Oates, 1988), and the other is emission right trading (Nordhaus, 2005). Emissions right trading regulations work by first setting an environmental goal at a national, or sometimes regional/local level, imposing a limit on the overall amount of pollution that sources are allowed to emit into the environment. This environmental goal (limit) is a critical part of an emissions trading program. For instance, in the international arena, carbon emissions right regulations intend to control the total emission amount of haze typical components by allocating the initial emission rights at the local level (country, region, province), and trade surplus emission rights on the market. The key question in this type of regulation is what will the initial haze emission right (limit) be for each province respectively? Emission rights of economic entities can be settled in the form of emission permits, and surplus emission rights can also be traded. In this context, the method of stipulating emission rights in the form of permits is known as the initial allocation of emission rights (Burton & Sanjour, 1969, 1970).

Then the allocation of air pollutant emission rights is a key component in emission trading programs. The environmental governance practices of China do not consider or define how the allocation of air pollutant emission rights should be established for each of the provinces. The Article 21 of the Air Pollution Prevention and Control Law of the People's Republic of China (revised in 2018) stipulates that the specific methods for determining the total volume control objectives and the corresponding control indicators will be formulated by the ecological environment department of the State Council in conjunction with the relevant departments.

However, as far as we know, the question of how to determine the emission rights for each province (municipality) and what the goals will be for them remains unclear. Based on the above considerations, this paper contributes to solving the emission rights problem by proposing a bi-objective DEA model to assign initial pollutant emission rights to various regions of the country. Undoubtedly, this research contributes in an innovative way to the implementation and improvement of the pollutant emission rights program in China.

To assign initial pollutant emission rights to various regions of the country under a total emissions target., this paper proposes a *DEA* model from two perspectives of fairness and efficiency. Then, taking *PM* as an example, the emission rights of each province are allocated on the basis of evaluating the emission performance of each province. To verify the rationality of the methods proposed in this work, the results obtained are compared with the distribution results based on the entropy of the information. The empirical conclusions can provide new ideas and basis for the practical work of reducing *PM* emissions in China.

18.2 Research Review

18.2.1 PM Emission Trading Research

Air pollutants include primary pollutants (such as sulfur dioxide, nitrogen dioxide, carbon monoxide, particulate matter, etc.) and secondary pollutants (such as sulfate aerosols, nitrate aerosols, ozone, photochemical oxidants, and active intermediates, etc.). Among them, *PM* is a typical representative of air pollution, including both primary and secondary pollutants (Cativiela-Campos et al., 2025; Fu et al., 2016; Yousefi et al., 2025).

Scholars believe that an important prerequisite for launching *PM* emission trading is to determine the emission of *PM* (Wang et al., 2014). The chemical constituents of *PM* are very complex. From the perspective of the formation process, it can be generally divided into primary aerosol and secondary aerosol, which causes difficulties in the total accounting of *PM* (Fu et al., 2016). However, in view of the total amount of *PM* generated in one time, the Ministry of Environmental Protection issued the Technical Guide for Compiling the Primary Source Emission Inventory of Atmospheric Fine Particulate Matter (pilot) (hereinafter referred to as the Guide) in 2014, which introduced the accounting technology and method of primary $PM_{2.5}$ emission sources and emissions. In addition, some typical industries have also issued $PM_{2.5}$ emissions accounting standards. For example, the Third IMO GHG Study of the International Maritime Organization (2014) gives the total amount of *PM* emitted by ships using different fuels. It can be seen that the total amount of *PM* formed at one time can be accounted for. Then, in the case of a given amount, emission rights can be allocated to each emission unit. It is worth mentioning that this paper uses *PM* concentration, land area and other data to put forward a formula for calculating the total emission of $PM_{2.5}$ (including primary and secondary $PM_{2.5}$), which provides a data basis for the verification of initial emission rights.

Although there is no *PM* emission trading market in China, some scholars have discussed the possibility of *PM* emission trading. For example, Yue and Ma (2016) put forward the idea of controlling haze through emission right trading. Wu et al. (2016) discussed the framework of $PM_{2.5}$ emission rights trading. Ye (2016) studied the pricing of $PM_{2.5}$ emission trading products. Wu et al., 2016 used the Zero Sum Gains (ZSG) *DEA* model to study the allocation of $PM_{2.5}$ emission rights of provinces and cities in China. Some scholars have done excellent researches on the design of sewage policies;they believe that effective environmental regulations should make the marginal damage of pollution equal to the marginal cost of abatement in space (Muller & Mendelsohn, 2009); under different resource conditions, different market-oriented emission policies should be designed (Fowlie & Muller, 2018).

18.2.2 Research on Allocation of Emission Rights of Air Pollutants Based on Efficiency and Equity

At present, research on the allocation of air pollutant emission rights can be divided into efficiency priority and equity orientation (Wang & Zhou, 2018). Most efficiency-first models consider the optimal economic objective, and many scholars use *DEA* model to conduct research (Zhou & Wang, 2016; Zhu et al., 2025). The whole system can be divided into four categories: the first category is the centralized resource allocation model proposed by Lai (2007). Bian (2010) using *DEA* model to allocate pollutant emission amount, sets pollutant emission quota as the decision variable and obtains the quota of each decision-making unit. The second type is $ZSG-DEA$ model. For example, Wu et al. (2016) adopts the $ZSG-DEA$ model to allocate PM emission rights of each province under the condition that the total resource is controlled and the input and output of each decision-making unit interact with each other. The third type is the parametric *DEA* resource allocation model, which introduces the efficiency frontier constructed by parametric methods. For example, when Wang and Li (2012) studied the efficiency of air pollution control in China, they used the super-efficiency *DEA* model to analyze the relative efficiency of industrial sulfur dioxide, industrial smoke and dust as input and output variables. Zheng et al. (2007) established a regional reduction indicator allocation model based on *DEA* model. The fourth type is the fixed cost allocation model based on *DEA*. For example, Wang and Li (2013), based on DEA method, take efficiency as the target, allocate the total amount of carbon dioxide emission to each province to improve the energy saving and emission reduction efficiency of each emission unit.

There is also a wealth of equity-oriented research. For example, Ma et al. (2006) combined with the linear programming model, divided the regions into different management zones and functional areas, constructed the inter-regional air pollutant transfer matrix, and distributed the SO_2 in Jinan, reflecting the fairness of capacity allocation of each functional area. Proceeding from the principle of fairness, Li et al. (2010) adopted the Logarithmic Mean Divisia Index model to systematically decompose the SO_2 emission intensity in eastern, central and western regions of China according to the changes of sulfur dioxide influencing factors in different periods in different regions, and developed a fair and flexible emission reduction plan. Lin and Fu (2017) established an axiomatic system of "fairness" based on the environmental conflict analysis theory, and designed a fair distribution method of total pollutant discharge to meet the axiomatic system. Wang et al. (2010) introduced "equal weight function" and "equal amount of pollution discharge" in the allocation of pollution distribution rights to ensure the reality and fairness of the distribution results. Mao and Li (1999) hold that in the process of pollutant reduction and distribution, attention should be paid to both the distribution results and the fairness of the distribution process. Janssen and Rotmans (1995) believe that population size is one of the important factors in the allocation of emission rights. Kverndokk (1995) also supports the allocation of pollutant emission rights based on population size, and believes that this allocation method is operational. In fact, the principle of per capita emissions is also the main form of distribution proposed by most developing countries in pursuit of equity in distribution schemes, which is worth learning from this paper.

In recent years, DEA were used to evaluate the efficiency of PM emission increasingly. Li et al. (2010) calculated the environmental governance efficiency (EGE) of China's iron and steel enterprises by a slack-based global data envelopment analysis (DEA) model and a global Malmquist-Luenberger index model. Wu et al. (2019) evaluated the haze emission efficiencies based on a new coopetition data envelopment analysis model. With bi-objectives of operations and marketing, Yin et al. (2020) proposed a two stage network structure model and analyzed the performances of 68 Taiwanese hotels in 2011. Considering the bi-objectives of source allocation and environmental performance evaluation, Jiang et al. (2020) constructed a new DEA model. However, efficiency and equity, as important dimensions to allocate the resources, seldom been introduced in previous researches.

The choice between equity and efficiency is one of the most difficult choices in social policy design and resource allocation (Okun, 1975). Usually, in setting optimal pollution strategies tensions between efficiency and equity arise (Levy et al., 2007). In 1998, the Environmental Protection Agency (EPA) established a relationship between fairness and environmental justice. It defines environmental justice as the "fair treatment of people of all races, income, and culture with respect to the development, implementation, and enforcement of environmental laws, regulations, and policies. Fair treatment means that no population, due to policy or economic disempowerment, is forced to bear a disproportionate burden of the negative human health or environmental impacts of pollution or other environmental consequences resulting from industrial, municipal, and commercial operations or the execution of federal, state, local and tribal programs and policies.'' (EPA, 1998). Although there are a variety of interpretations for equity, in this paper equity refers to the equality in the distribution of benefits among people or entities at risk (Levy et al., 2007). In line with this, the magnitude and distribution of benefits depend on the geographic distribution of emission controls, i.e., the emission rights allocated to the provinces. In this paper, we use the term fairness as a proxy for equity.

In the problem setting discussed in this paper, environmental equity considerations should be incorporated as an essential ingredient in the discussion on how to allocate emissions rights. In this paper, environmental equity refers to the proportional distribution of pollution and environmental benefits among communities so that no group is more affected than another by environmental hazards. According to Solomon and Lee (2000) "unrestricted emission trading does not decrease and may exacerbate environmental inequities". Recognizing the impact of PM emissions on human health(Lelieveld et al., 2015), the issue of how to measure environmental equity in the context of the allocation of allowances has not been adequately discussed in the published literature on decision-making approaches. In a power plant air pollution control setting, Levy et al. (2007) propose to quantify the risk of inequality within a context of health benefits and in this way provide decision-makers with a quantitatively based tool "to construct an optimal efficiency–equality frontier". They propose several other indicators to incorporate the dimension of environmental equity in decision-making problems. In the context of determining water ecological carrying capacity, Chen et al. (2020) propose the use of GINI for measuring the regional fairness. However, the proportion of PM emissions related to the population is a simple and easy to collect data to start the analysis and discussion of the impacts of considering the balance of the equity-efficiency dimensions in the allocation of emission rights. On the other hand, few studies are using efficiency-based DEA methods to assess the initial emission rights of air pollutants (such as Wu et al., 2016, 2019).

In summary, to the best of our knowledge, there are not published literature addressing the balance of fairness-efficiency in the decision-making problem of allocation emission rights. This article innovatively constructs the efficiency and fairness-oriented DEA model, and conducts empirical research, which is a useful supplement to the existing research.

18.3 Allocation Method of PM Emission Rights

18.3.1 DEA Model with Consideration of Efficiency and Equity

First of all, the economic output efficiency should be guaranteed first. The objective function is to maximize the overall average efficiency value, that is, the larger the ratio of output to input, the better, or at least the ratio should not be reduced. Secondly, the principle of per capita equity should be reflected as much as possible, that is, the proportion of *PM* distribution in the total *PM* emission of the whole country should be as close as possible to the proportion of the population of the province in the total population of the whole country.[1] Finally, ensure the optimal allocation of *PM* emissions nationwide.

[1] Considering that PM mainly affects human health(Lelieveld et al., 2015), when designing fairness indicators, the proportion of PM emissions in each province is equal to the proportion of population as an indicator of fairness. Of course, other indicators can also be considered, such as capital per capita, land, and atmospheric environmental capacity. However, for readers and policy makers, these indicators are not as direct, practical and easy to calculate as the population ratio indicators, so this indicator is chosen.

In order to achieve the primary goal of optimal efficiency, *DEA* model is built according to step 1:

$$\begin{cases} E = Max\frac{1}{n}\sum_{q=1}^{n} eq \\ s.t. \\ eq = \dfrac{\sum_{r=1}^{s} \alpha ryrq}{fq + \sum_{i=1}^{m} \beta ixiq} \\ 0 \le eq \le 1,\ q = 1, 2, \ldots, n \\ \alpha\, r \ge \varepsilon,\ \beta i \ge \varepsilon,\ i = 1, 2, \ldots, m,\ \ r = 1, 2, \ldots, s \\ \sum_{q=1}^{n} fq = F \end{cases} \tag{18.1}$$

Step 1 assumes that there are n decisive units DMU; DMU_q $(q = 1, 2, \ldots, n)$ is the q_{th} decision unit. And m and s respectively represent the number of types of input and output indicators. *yrq* is the r_{th} $(r = 1, 2, ..., s)$ output of DMU_q, which is expressed by the GDP of each province, and αr is the r_{th} weight corresponding to the output index. Similarly, x_{iq} is the i_{th} $(i = 1, 2, 3)$ input index of DMU_q x_{1q}, x_{2q}, x_{3q} represent the capital input, labor input and energy input of each province respectively, and βi is the $i-th$ corresponding weight of the input factor index. *eq* is the efficiency value of a single DMU_q.

fq represents $PM_{2.5}$ emission allocated to DMU_q; *F* represents the total amount of $PM_{2.5}$ emissions available for distribution; ε is an infinitesimal constant, and it is usually calculated at $\varepsilon=10^{-6}$ (non-archimedes infinitesimal decimal). In the *DEA* model, considering the non-archimedes infinitesimal, it is required that the weight coefficient of each input index and output index must be greater than 0, so as to ensure the effectiveness of *DEA*.

In step 1, the optimal efficiency value *eq* of each province in China can be obtained, and then the maximum value E^* of the overall average efficiency in the objective function can be obtained by weighted average.

In order to ensure the overall efficiency is not reduced, the results *E* of step 1 are incorporated into the new plan in subsequent steps, that is, $\frac{1}{n}\sum_{q=1}^{n} e_q \ge E^*$ is one of the necessary constraints.

Secondly, find the value range of *fq*. Take *Minfq* and *Maxfq* as objective function respectively, step 2 of *DEA* model is constructed:

$$\begin{cases} M = Minfq/N = Maxfq \\ s.t. \\ eq = \dfrac{\sum\limits_{r=1}^{s} \alpha ryrq}{fq + \sum\limits_{i=1}^{m} \beta ixiq} \\ 0 \leq eq \leq 1, q = 1, 2, \ldots, n \\ \sum\limits_{q=1}^{n} fq = F \\ fq \geq 0, q = 1, 2, \ldots, n \\ \alpha r \geq \varepsilon, \beta i \geq \varepsilon, i = 1, 2, \ldots, m, r = 1, 2, \ldots, s \\ \dfrac{1}{n} \sum\limits_{q=1}^{n} eq \geq E^{*} \end{cases} \tag{18.2}$$

Step 2 considers actually two linear programming models, the maximum and minimum values of *fq*, can be obtained respectively. Let the minimum value of *fq* is M, the maximum value is N, then for all DMU_q, values(M and N) of *fq* can be obtained respectively; If for a DMU_q, $M = N$. Only one value ($M = N$) of *fq* can be obtained. The allocation of the $PM_{2.5}$ emission right of the DMU_q is uniquely determined.

Finally, the allocation *fq* of the Decision Making Unit is calculated. The principle of equity is introduced so that the ratio of $PM_{2.5}$ distribution (*fq*) to the total amount of $PM_{2.5}$ emissions (F) in each province is as close as possible to the proportion of the population of the provinces to the total population of the country, as expressed in a function as follows:

$$\begin{cases} Min \sum\limits_{q=1}^{n} \left| \dfrac{fq}{F} - \varphi q \right| \\ \varphi q = \dfrac{pq}{P} \\ P = \sum\limits_{q=1}^{n} pq \end{cases} \tag{18.3}$$

In the function, *p* is the total population of the country, *pq* is the population of the $q - th$ province. The function $Min \sum\limits_{q=1}^{n} \frac{fq}{F} - \varphi q$ indicates that the proportion of $PM_{2.5}$ emission of each province in the total national emission is close to the proportion of population of that province in the total national population, that is, the fair target. Get to step 3:

$$\begin{cases} Z = Min\sum_{q=1}^{n}\left|\frac{fq}{F}-\varphi q\right| \\ s.t. \\ N \geq fq \geq M\,, q = 1, 2, \ldots, n \\ eq = \frac{\sum_{r=1}^{s}\alpha ryrq}{fq + \sum_{i=1}^{m}\beta ixiq} \\ 0 \leq \text{eq} \leq 1, q = 1, 2, \ldots, n \\ \sum_{q=1}^{n} fq = F \\ fq \geq 0, q = 1, 2, \ldots, n \\ \alpha r \geq \varepsilon, \beta i \geq \varepsilon,\ i = 1, 2, \ldots, m,\ r = 1, 2, \ldots, s \\ \frac{1}{n}\sum_{q=1}^{n} \text{eq} \geq \text{E*} \end{cases} \tag{18.4}$$

In step 3, φq and F are known quantity, and Z^* and fq can be calculated. Z^* represents the optimal value of objective function in step 3. By solving step 3, some decisions making units may obtain multiple fq values, and satisfy fq by step 2:

$$M \leq fq \leq N \tag{18.5}$$

So the only solution of fq can be determined.

The allocation process in steps 1–3 is as follows: In step 1, the optimal value $E*$ of the overall average efficiency is obtained by taking the maximization of the overall efficiency as the objective function in the case that the total $PM_{2.5}$ emission in China is determined, but the $PM_{2.5}$ emission value of each decision-making unit is not determined. Steps 2 and 3, in ensuring the overall average efficiency does not reduce, with equity (the proportion of $PM_{2.5}$ emission of each province in the total national emission is close to the proportion of population of that province in the total national population) as the target function to distribute the specific emissions of the provinces. Emissions of individual decision-making units once established; the efficiency of the unit value can be obtained. Finally, the corresponding efficiency values $e_q*(q = 1, 2, \ldots, 30)$ of 30 provinces can be obtained in turn. After all, these values obtained through the weighted average efficiency, and the overall average, should be greater than or equal to the overall efficiency of the average value of model 1, thus improving both efficiency and fairness.

Combining all fq values obtained in step 2 and step 3, that is, the $PM_{2.5}$ emission amounts of 30 provinces with both efficiency and equity goals. The above three steps constitute *DEA* model with both efficiency and equity goals in mind.

18.3.2 Information Entropy Method

The information entropy method was first proposed by Shannon, which can be used to assign the weight of decision-making indicators, and it can better reflect the differentiation level of indicators (see for example Song et al., 2017; Zeng & Wang, 2017). Also the weight obtained by the information entropy methods is more objective than the weight obtained by the analysis hierarchy process (Yang & Ren, 2019). Therefore, this paper chooses it as the method of comparative study.

By comparing the amount of information provided by index data, the information entropy method determines the relative importance of indexes, and then gives weight to each index objectively. Finally, the specific emission amount of $PM_{2.5}$ is allocated according to the weight. The specific steps are as follows:

First, establish the original data matrix A composed of n row allocation units and m column indicators, as shown in Table 18.1:

Second, in order to eliminate unit differences of indicators, the original data in matrix A should be standardized to obtain the standardized matrix B of indicators. In this paper, the maximum-minimum method is adopted to standardize the indicator Xij:

$$Xij = \frac{xij - Min\{xij\}}{Max\{xij\} - Min\{xij\}}; i = 1, 2, ..., n; j = 1, 2, ..., m. \tag{18.6}$$

Third, the index in the standardized matrix B is weighted. Firstly, the entropy value $H(Xj)$ of each indicator is calculated according to formula (18.7), and then the calculated entropy value $H(Xj)$ is substituted into formula (18.8) to get the information entropy weight of each indicator:

$$H(Xj) = -k \sum_{i=1}^{n} Pijln(Pij) \tag{18.7}$$

$$\text{in which } k = -\frac{1}{ln(n)}, Pij = \frac{X_{ij}}{\sum_{i=1}^{n} X_{ij}}(0 \le H(Xj) \le 1), W_j = \frac{1 - H(X_j)}{\sum_{j-1}^{m} [1 - H(X_j)]} \tag{18.8}$$

Table 18.1 Original Matrix A

Index	Factor indicator 1	Factor indicator 2	…	Factor indicator 1
DMU_1	X_{11}	$X12$	…	X_{1m}
DMU_2	X_{21}	X_{22}	…	X_{2m}
…	…	…	…	…
DMU_n	X_{n1}	X_{n2}	…	X_{nm}

$H(X_j)$ Is the information entropy of the index j, and n represents the number of units to be allocated, W_j is the information entropy weight of the index j, $\sum_{j=1}^{n} W_j=1$.

Fourth, the W_j is calculated according to the factor indexes j, the total amount is distributed among each unit according to formula (18.9):

$$S_j = \left(\sum_{j=1}^{m} X_j \times W_j\right)^{\left(\frac{2}{3}\right)} \tag{18.9}$$

So S_i is the $PM_{2.5}$ emission shared by the i_{th} province.

18.4 Selection of Indicators and Data

18.4.1 Index Selection

The selection of input and output indicators should have a clear theoretical basis and economic meaning. In the generation of *PM* and the evaluation of governance efficiency, capital and labor input are essential. The input side of the traditional production function includes capital and labor, while the output side is GDP. Later, scholars added energy, environment, climate and other factors affecting GDP output on the input side (Wang et al., 2018). This paper also uses this idea for reference and adds environmental and energy indicators related to *PM* output to the input side of the production function. As $PM_{2.5}$ is a typical representative of *PM* (Wang et al., 2018),$PM_{2.5}$ index is added on the output side to expand the production function expression as follows:

$$Y_k = A_k K_k^{\alpha} L_k^{\beta} C_k^{\gamma} \tag{18.10}$$

In the function, Y_k is the output of region k, A_k is the generalized Technology of region k, K_k represents the capital input of region k, L_k represents the labor input of region k, C_k represents the environmental and energy input related to $PM_{2.5}$ output of region k, α, β and γ represents the output elasticity of environmental and energy input indexes related to capital, labor and $PM_{2.5}$ output respectively. Timmermans et al. (2017) found that atmospheric pollutants mainly come from the consumption of various energy sources. According to the composition and source of $PM_{2.5}$ and the availability of data, coal consumption, oil and natural gas are selected as environmental and energy investment indicators.

The input and output indicators used below are described as follows.

(1) Input indicators: in addition to labor and capital, three energy consumption of coal, oil and natural gas are also selected as input indicators.

(2) Expected output index: the GDP of each province is selected as the expected output index to reflect the economic output of each province.
(3) Non-expected output indicators: the expected output is positively correlated with the efficiency value, while the non-expected output and input are negatively correlated with the efficiency value. In other words, the fewer input elements, the more expected output, and the less non-expected output, the closer the overall efficiency value is to the optimal. Based on the principle of directional consistency, this paper puts the emission of $PM_{2.5}$, which is not expected to produce, in the denominator as an input factor (Wang et al., 2018). On the one hand, considering the simplicity of calculation, the input factors in the denominator change in the same direction as the non-expected output; On the other hand, $PM_{2.5}$ emission right is regarded as a rare resource, which is regarded as an input indicator independent of labor, capital and energy.

18.4.2 Data Sources

This paper collected data on capital stock, labor input, energy consumption, regional gross domestic product (GDP) and $PM_{2.5}$ concentration in 30 provinces in 2016. The data of capital input, labor input and provincial GDP are all from China Statistical Yearbook 2017, the official website of China Statistics Bureau and the official website of provincial statistics bureau. Energy consumption selects three representative fossil energy consumption of coal, oil and natural gas, to represent the structural characteristics of energy consumption in each province.

The capital is fixed asset investment, which can be expressed by the capital stock at the end of the previous year; Labor input is the sum of "urban unit employment number", "private enterprise employment number" and "individual employment number". Energy consumption is expressed in terms of the consumption of three fossil fuels of coal, oil and natural gas. Among them, coal is the sum of coal and coke, and oil is the sum of crude oil, gasoline, kerosene, diesel oil and fuel oil.

$PM_{2.5}$ is usually presented in the form of concentration. Based on the consideration of atmospheric environment capacity, this paper constructs the total index of $PM_{2.5}$. Atmospheric environmental capacity refers to the maximum amount of certain air pollutants allowed to be discharged in the region under the constraint of a certain environmental target (Wu et al., 2016). Referring to the research of Wu et al (2016), $PM_{2.5}$ emissions that are not expected to be produced are converted through atmospheric environment capacity (Wu et al., 2016) to obtain the total $PM_{2.5}$ data of a certain region.

The above input–output indicators and data are shown in Table 18.2.

Table 18.2 Input and output indicators

Indicator classification	Indicator name	Indicator description
Input factor	Capital	Capital stock at the end of 2016
	Labor force	Individual employment, employment in urban units and employment in private enterprises
	Coal	Coal, coke
	Oil	Crude oil, gasoline, kerosene, diesel, fuel oil
	Gas	Natural gas consumption
Expected output	GDP	Gross output value of each province
Undesired output	$PM_{2.5}$ row Volume	$\frac{\text{population -weighted } PM_{2.5} \text{ concentration in Chinese provinces} \times \text{Land area of each province}}{\text{atmospheric environmental capacity of } PM_{2.5} \text{ under the national } PM_{2.5} \text{ compliance}}$

18.5 Distribution Process and Results of $PM_{2.5}$ Emission Rights Between Provinces in China

18.5.1 Data and Normalization

18.5.1.1 Original Data

The raw and standardized data of GDP, capital input, energy consumption (coal, oil and natural gas), labor force and $PM_{2.5}$ emission of each province in 2016 are shown in Table 18.3.

18.5.1.2 Weight Assignment

The information entropy method needs to assign a value to the weight. According to formula (18.6)–(18.9), the information entropy of each indicator is calculated to determine the relative importance of the indicator. On the basis of standardized processing of the data in Table 18.3, the entropy weight of each indicator (Wj) is calculated, and the results are shown in Table 18.4.

When the weight is assigned by information entropy, the larger the information entropy value $H(Xj)$ of the indicator is, the smaller the entropy weight value Wj is, indicating that the information provided by the indicator is less, and the smaller the difference between provinces of the indicator is, and the less important it is.

The weight calculated according to the above Table 18.4 is as follows: $PM_{2.5}concentration > oil > capital > gas > coal > labor > GDP$, $PM_{2.5}$ concentration to a certain extent, reflects the $PM_{2.5}$ emissions intensity of pollutant discharging unit, the weight of $PM_{2.5}$ is the largest showing that the information provided by the index is more and provincial difference are obvious. The information entropy weight value of energy (coal, oil and gas) and capital index is smaller,

Table 18.3 Raw and standardized input–output indicator data for each province in 2016

Serial number	Province	Coal		Oil		Natural gas		Labor force	
		Raw Values (10,000 tons)	Standardized Values	Raw Values (10,000 tons)	Standardized Values	Raw Values (One hundred million cubic meters)	Standardized Values	Raw Values (10,000 people)	Standardized Values
1	Beijing	847.80	0.00	2063.00	0.10	162.30	0.89	1838.30	0.31
2	Tianjin	5117.50	0.01	2205.80	0.11	74.50	0.38	495.70	0.06
3	Hebei	36,185.10	0.81	3183.50	0.17	70.50	0.36	1621.30	0.27
4	Shanxi	37,819.10	0.84	792.00	0.03	69.40	0.36	984.00	0.15
5	Inner Mongolia	38,310.70	0.86	1237.60	0.05	45.10	0.21	818.50	0.12
6	Liaoning	19,936.80	0.43	9196.00	0.53	50.60	0.25	1387.80	0.23
7	Jilin	9894.40	0.21	1632.20	0.08	21.50	0.08	937.30	0.15
8	Heilongjiang	14,218.60	0.31	3029.90	0.16	38.00	0.17	787.40	0.12
9	Shanghai	5222.60	0.01	4841.60	0.27	79.00	0.41	1822.00	0.31
10	Jiangsu	31,888.30	0.71	6167.80	0.35	172.70	0.95	4611.40	0.82
11	Zhejiang	14,278.00	0.31	4854.90	0.27	87.80	0.46	3626.70	0.64
12	Anhui	16,893.30	0.37	1709.00	0.08	39.20	0.18	1573.20	0.26
13	Fujian	7435.20	0.15	3316.10	0.18	48.60	0.24	1818.10	0.31
14	Jiangxi	8457.40	0.17	1584.60	0.08	20.00	0.07	1370.10	0.23
15	Shandong	44,657.60	1.00	16,938.20	1.00	98.60	0.52	3588.10	0.63
16	Henan	26,272.40	0.58	2358.30	0.12	92.80	0.49	2319.80	0.40
17	Hubei	12,781.20	0.27	3079.40	0.17	41.50	0.19	2284.50	0.39
18	Hunan	12,404.00	0.26	2279.80	0.12	28.30	0.12	1303.40	0.21
19	Guangdong	16,917.80	0.37	8972.00	0.52	167.80	0.92	5595.40	1.00

(continued)

Table 18.3 (continued)

Serial number	Province	Coal		Oil		Natural gas		Labor force	
		Raw Values (10,000 tons)	Standardized Values	Raw Values (10,000 tons)	Standardized Values	Raw Values (One hundred million cubic meters)	Standardized Values	Raw Values (10,000 people)	Standardized Values
20	Guangxi	7638.70	0.16	2330.90	0.12	12.90	0.03	1104.80	0.18
21	Hainan	1015.30	0.00	1455.90	0.07	41.30	0.19	279.00	0.03
22	Chongqing	6058.70	0.12	828.20	0.03	89.30	0.47	1472.30	0.24
23	Sichuan	10,505.90	0.22	3108.00	0.17	181.60	0.10	2193.40	0.38
24	Guizhou	13,896.20	0.30	873.60	0.03	17.10	0.05	894.70	0.14
25	Yunnan	8373.30	0.17	1045.70	0.04	7.70	0.00	1167.60	0.19
26	Shaanxi	20,511.20	0.45	2533.20	0.13	98.20	0.52	1029.70	0.16
27	Gansu	6926.20	0.14	1885.00	0.10	26.40	0.11	676.40	0.10
28	Qinghai	2190.90	0.03	334.50	0.00	46.30	0.22	144.80	0.00
29	Ningxia	9115.70	0.19	811.50	0.03	22.40	0.09	223.40	0.01
30	Xinjiang	19,789.40	0.43	3409.10	0.19	132.40	0.72	655.10	0.10
	Maximum values	44,657.60	1.00	16,938.20	1.00	181.60	0.95	5595.40	1.00
	Minimum values	847.80	0.00	334.50	0.00	7.70	0.00	144.80	0.00
	Arithmetic averages	15,518.64	0.33	3268.58	0.18	69.46	0.33	1620.81	0.27

(continued)

Table 18.3 (continued)

Serial number	Province	Capital		GDP		$PM_{2.5}$ concentration	
		Raw Values (100 million yuan)	Standardized Values	Raw Values (100 million yuan)	Standardized Values	Raw Values (micrograms per cubic meter)	Standardized Values
1	Beijing	7943.90	0.09	25,669.10	0.3-	72.00	0.97
2	Tianjin	12,779.40	0.19	17,885.40	0.20	70.50	0.95
3	Hebei	31,750.00	0.57	32,070.50	0.38	70.40	0.94
4	Shanxi	14,198.00	0.21	13,050.40	0.13	60.10	0.76
5	Inner Mongolia	15,080.00	0.23	18,128.10	0.20	36.40	0.32
6	Liaoning	6692.30	0.06	22,246.90	0.25	46.20	0.50
7	Jilin	13,923.20	0.21	14,776.80	0.16	42.70	0.44
8	Heilongjiang	10,648.40	0.14	15,386.10	0.16	35.30	0.30
9	Shanghai	6755.90	0.07	28,178.70	0.33	45.50	0.49
10	Jiangsu	49,663.20	0.93	77,388.30	0.96	57.70	0.71
11	Zhejiang	30,276.10	0.54	47,251.40	0.57	41.00	0.41
12	Anhui	27,033.40	0.47	24,407.60	0.28	53.10	0.63
13	Fujian	23,237.40	0.40	28,810.60	0.34	27.10	0.16
14	Jiangxi	19,694.20	0.33	18,499.00	0.20	44.00	0.46
15	Shandong	53,322.90	1.00	68,024.50	0.84	59.90	0.75
16	Henan	40,415.10	0.74	40,471.80	0.48	73.50	1.00
17	Hubei	30,011.70	0.53	32,665.40	0.38	66.40	0.87
18	Hunan	28,353.30	0.50	31,551.40	0.37	48.20	0.54
19	Guangdong	33,303.60	0.69	80,854.90	1.00	31.70	0.24
20	Guangxi	18,236.80	0.30	18,317.60	0.20	37.00	0.34
21	Hainan	3890.50	0.01	4053.20	0.01	18.60	0.00
22	Chongqing	16,048.10	0.25	17,740.60	0.19	53.00	0.63

(continued)

Table 18.3 (continued)

Serial number	Province	Capital		GDP		$PM_{2.5}$ concentration	
		Raw Values (100 million yuan)	Standardized Values	Raw Values (100 million yuan)	Standardized Values	Raw Values (micrograms per cubic meter)	Standardized Values
23	Sichuan	28,812.00	0.51	32,934.50	0.39	48.00	0.54
24	Guizhou	13,204.00	0.19	11,776.70	0.12	34.00	0.28
25	Yunnan	16,119.40	0.25	14,788.40	0.16	26.20	0.14
26	Shaanxi	20,825.30	0.35	19,399.60	0.22	59.30	0.74
27	Gansu	9664.00	0.12	7200.40	0.06	39.30	0.38
28	Qinghai	3528.10	0.00	2572.50	0.00	39.30	0.38
29	Ningxia	3794.30	0.01	3168.60	0.01	45.40	0.49
30	Xinjiang	10,287.50	0.14	9649.70	0.09	61.80	0.79
	Maximum values	53,322.90	1.00	80,854.90	1.00	73.50	1.00
	Minimum values	3528.10	0.00	2572.50	0.00	18.60	0.00
	Arithmetic averages	19,983.07	0.33	25,963.96	0.30	48.12	0.54

Table 18.4 Weights of input–output indicators

Weights	GDP	$PM_{2.5}$ concentration	Coal	Oil	Natural gas	Labor force	Capital
$H(Xj)$	0.62	0.54	0.59	0.55	0.59	0.60	0.58
1-$H(Xj)$	0.39	0.47	0.41	0.45	0.41	0.40	0.42
Wj	0.13	0.16	0.14	0.15	0.14	0.14	0.14

and the index weight of gross output value is the smallest. Therefore, it can be judged that more attention should be paid to the difference of $PM_{2.5}$ emission in different provinces when allocating $PM_{2.5}$ emission rights by using the information entropy method.

18.5.2 Allocation of PM Emission Rights Between Provinces in China

According to the *DEA* model (see formula (18.1)–(18.5)) and the information entropy method (see formula (18.6)–(18.9)) with both efficiency and equity, two sets of inter-provincial distribution schemes of $PM_{2.5}$ emission amount are obtained by substituting the above index data for calculation. The results are shown in 18.5.

Table 18.5 $PM_{2.5}$ allocation results based on *DEA* model and information entropy method

Province	Initial emission of $PM_{2.5}X_i$	*DEA* allocated emissions X_i	Emissions from the information entropy method	Efficiency Hi at initial discharge	Efficiency after distribution of *DEA*
	①	②	③	④	⑤
Beijing	0.17	0.17	0.17	1.00	1.00
Tianjing	0.16	0.16	0.15	1.00	0.99
Hebei	0.22	0.17	0.26	0.82	0.91
Shanxi	0.09	0.04	0.11	0.69	0.89
Inner Mongolia	0.23	0.09	0.33	0.99	1.00
Liaoning	0.07	0.05	0.06	0.64	0.73
Jilin	0.10	0.08	0.09	0.79	0.87
Heilongjiang	0.15	0.15	0.16	0.80	0.80
Shanghai	0.00	0.00	0.00	0.99	0.88
Jiangsu	0.08	0.11	0.07	0.98	0.91
Zhejiang	0.08	0.11	0.08	0.86	0.83
Anhui	0.14	0.16	0.15	0.77	0.74
Fujian	0.03	0.04	0.03	0.83	0.81
Jiangxi	0.13	0.15	0.14	0.73	0.73
Shandong	0.09	0.08	0.09	0.70	0.74
Henan	0.20	0.19	0.21	0.84	0.88

(continued)

(continued)

Province	Initial emission of $PM_{2.5}X_i$	*DEA* allocated emissions X_i	Emissions from the information entropy method	Efficiency Hi at initial discharge	Efficiency after distribution of *DEA*
Hubei	0.20	0.19	0.20	0.79	0.80
Hunan	0.18	0.28	0.17	1.00	1.00
Guangdong	0.03	0.10	0.03	1.00	0.89
Guangxi	0.05	0.05	0.05	0.77	0.76
Hainan	0.00	0.00	0.00	0.42	0.66
Chongqing	0.10	0.11	0.10	0.74	0.76
Sichuan	0.25	0.36	0.26	0.74	0.73
Sichuan	0.07	0.04	0.09	0.72	0.99
Yunnan	0.04	0.03	0.04	0.74	0.84
Shaanxi	0.19	0.18	0.18	0.68	0.81
Gansu	0.26	0.20	0.34	0.44	0.81
Qinghai	2.04	2.00	2.05	0.40	0.76
Ningxia	0.18	0.09	0.30	0.44	0.95
Xinjiang	2.10	1.19	2.48	0.41	0.79
Arithmetic averages of efficiency value E				0.76	0.93
Standard deviation of efficiency values					
Coefficient of variation of efficiency value					

Province	Efficiency H_i after information entropy method	Initial $PM_{2.5}$ concentration Zi	$PM_{2.5}$ concentration Zpi after *DEA* distribution	$PM_{2.5}$ concentration Zqi after information entropy method
		⑦	⑧	⑨
Beijing	1.00	72.01	72.09	72.00
Tianjing	0.99	70.48	73.13	69.44
Hebei	0.89	70.41	42.55	50.06
Shanxi	0.88	60.12	26.43	34.52
Inner Mongolia	0.99	36.44	14.4	20.44
Liaoning	0.73	46.23	36.61	36.01
Jilin	0.90	42.70	34.32	30.77
Heilongjiang	0.80	35.31	35.31	37.92
Shanghai	0.88	45.47	76.87	79.01
Jiangsu	0.92	57.67	72.67	68.2
Zhejiang	0.83	41.02	55.34	50.45
Anhui	0.73	53.12	50.01	52.09
Fujian	0.81	27.14	37.64	37.98
Jiangxi	0.737	44.03	47.86	49.50

(continued)

(continued)

Province	Efficiency H_i after information entropy method	Initial $PM_{2.5}$ concentration Zi	$PM_{2.5}$ concentration Zpi after *DEA* distribution	$PM_{2.5}$ concentration Zqi after information entropy method
Shandong	0.74	59.83	57.89	58.6
Henan	0.87	73.45	70.1	73.05
Hubei	0.85	66.37	64.72	63.09
Hunan	1.00	48.22	49.31	49.01
Guangdong	0.89	31.71	63.44	70.62
Guangxi	0.78	37.02	38.86	37.99
Hainan	0.65	18.56	18.09	19.91
Chongqing	0.76	53.01	53.07	53.09
Sichuan	0.73	48.01	51.15	53.37
Sichuan	0.98	34.02	17.74	22.92
Yunnan	0.84	26.23	20.17	20.18
Shaanxi	0.81	59.29	55.31	53.01
Gansu	0.80	39.27	19.52	25.62
Qinghai	0.76	39.31	37.71	37.80
Ningxia	0.94	45.41	14.89	25.62
Xinjiang	0.793	61.78	34.18	40.36
Arithmetic averages of efficiency value E	0.87		44.71	46.42
Standard deviation of efficiency values			17.28	18.85
Coefficient of variation of efficiency value			0.39	0.41

As can be seen from Table 18.5, from the initial emission efficiency value, in 2016, the average $PM_{2.5}$ emission efficiency value of each province in China was 0.7575, and the difference between the initial emission efficiency values of each province was obvious. It can be roughly divided into three categories. The first category is Beijing, Guangdong, Tianjin and Hunan, whose existing efficiency is above the average level. In the second category, the emission efficiency of Hebei, Inner Mongolia, Henan, Hubei, Guangxi, Jilin, Heilongjiang, Shanghai, Jiangsu, Zhejiang, Anhui and Fujian are all around the overall average efficiency. The third category is Shanxi, Liaoning, Jiangxi, Shandong, Hainan, Chongqing, Sichuan, Guizhou, Yunnan, Shaanxi, Gansu, Qinghai, Ningxia, Xinjiang, etc. The efficiency values of these 14 provinces are all lower than the average efficiency.

On the basis of *DEA* model and information entropy weight method, there are obvious differences in inter-provincial allocation of emission rights, but the former is fairer, more efficient and more feasible.

(1) The distribution scheme of *DEA* model is more efficient. It can be seen from the mean value E of the efficiency value in the penultimate row of Table 18.5:

  1) 1)$E_{DEA} > E_{\text{information } entropy} > E_{original}$, On the one hand, no matter *DEA* model based on efficiency and fairness or information entropy weight assignment method, $PM_{2.5}$ emissions after distribution have improved the overall average efficiency of the whole country, that is, the overall performance level after distribution is higher than that before distribution.
  2) 2)$E_{DEA} > E_{\text{information } entropy}$, indicating that for the information entropy method, *DEA* model have better effect on improving the overall output efficiency. That is to say, in order to achieve the same efficiency value, the $PM_{2.5}$ emission reduction required by *DEA* model is lower than the information entropy method, and the emission reduction task is light, which means the emission reduction cost is lower.

(2) The distribution scheme of *DEA* model is more feasible. The variation coefficient can be used as an effective indicator to measure the difference of $PM_{2.5}$ distribution results between provinces. The values of variation coefficients of the two groups can be obtained by comparing the standard deviation of $PM_{2.5}$ data (⑧–⑨ line) allocated by each group with its mean value. According to the calculation, in the distribution scheme based on *DEA* model, the variation coefficient of emission quota distribution results of each province is 0.39, and in the distribution scheme based on information entropy method, the variation coefficient is 0.41, which has a slightly larger fluctuation degree. This indicates that *DEA* model alleviates the problem of large gap of information entropy method in resource allocation to some extent, and each allocated unit is more likely to accept the scheme with relatively small gap.

(3) The distribution scheme of *DEA* model is fairer. The *Gini* coefficient method is used to further compare the fairness of the *DEA* model, which gives consideration to efficiency and fairness, and the information entropy method in $PM_{2.5}$ distribution. Here, "fair" is defined as "fair distribution per capita". The population parameter was used as the evaluation index to calculate the $PM_{2.5}$ distribution of the unit population of each province (the ratio of $PM_{2.5}$ distribution of each province to the population of the province), and then the cumulative percentage of population of each province and $PM_{2.5}$ distribution was calculated by using formula (18.11)–(18.12) in the order from small to large. Then, according to formula (18.12), the environmental *Gini* coefficient value of $PM_{2.5}$ distribution in each province is obtained.

$$X(i) = X(i-1) + M(i) / \sum_{\mathrm{i}=1}^{n} M(i) \tag{18.11}$$

$$Y(i) = Y(i-1) + Pi / \sum_{\mathrm{i}=1}^{n} Pi \tag{18.12}$$

$$G = 1 - \sum_{i=1}^{n} (X(i) - X(i-1))(Y(i) - Y(i-1)) \tag{18.13}$$

In the function, i represents the $i - th$ province; $X(i)$ represents the cumulative percentage of evaluation indicators (i.e., population parameters); $Y(i)$ represents the cumulative percentage of $PM_{2.5}$ distribution in each province; n represents the number of provinces to be evaluated; $M(i)$ represents the evaluation index value (i.e. population parameter); $P_{(i)}$ represents the $PM_{2.5}$ emissions allocated by each province; G represents the environmental *Gini* coefficient with population as evaluation index. So when $i = 1, X_{(i-1)} = 0 Y_{(i-1)} = 0$.

According to the calculation method of *Gini* coefficient, the distribution results of provinces based on *DEA* model and information entropy method are substituted into formula (18.11) and formula (18.12), and the two groups of $X_{(i)1}, X_{(i)2}$ (i.e. the cumulative percentage of the total population of each province) and $Y_{(i)1}, Y_{(i)2}$ (i.e. the cumulative percentage of $PM_{2.5}$ distribution) are calculated, as shown in Table 18.6.

As shown in Table 18.6, the environmental *Gini* coefficients corresponding to the *DEA* model and the information entropy weighting method are 0.47 and 0.51 respectively with the population as the index. Therefore, the environmental *Gini* coefficient obtained by *DEA* model is smaller and fairer when population parameter is used as an evaluation index.

18.6 Summary and Conclusions

The problem of allocating air pollutant emission rights that simultaneously incorporates equity and efficiency dimensions is a new and challenging problem. First, the distribution of air pollutant emission rights is related to the characteristics of air pollutants. The generation, combination, and transmission of air pollutants have a great relationship with the resources and environment of the places where air pollutants are generated. Therefore, scholars mostly start from the production and source of air pollutants to study local governance. Few scholars have carried out research on the allocation of emission rights based on the evaluation of the emission performance of each emission unit from a macro perspective. The second is related to national conditions. Literatures have mostly studied market-driven emissions trading issues, but there are few studies on the allocation of command-controlled air pollutant emissions rights. In addition, to the best of our knowledge, there are not published literature addressing the balance of fairness-efficiency in the decision-making problem of allocation emission rights. This article innovatively constructs the efficiency and fairness-oriented DEA model, and conducts empirical research to validate the proposed method. It evaluates the emission performance of air pollutants from a macro perspective, distributes emission rights, and takes China, the world's most populous country, as an example to carry out empirical research and validate the proposed approach.

Table 18.6 Population and $PM_{2.5}$ parameters calculated by *DEA* model and information entropy method

Province	Population parameter *j*		Cumulative percentage of population		$PM_{2.5}$ Distribution		$PM_{2.5}$ Cumulative percentage		$PM_{2.5}$/population	
	DEA	Information entropy	$X(i)1$	$X(i)2$	*DEA*	Information entropy	$Y(i)1$	$Y(i)2$	*DEA*	Information entropy
Zhejiang	5590	5590	0.041	0.675	19.52	50.45	0.015	0.476	0.004	0.009
Henan	9532	9532	0.110	0.446	35.31	73.05	0.041	0.286	0.004	0.008
Yunnan	4771	4771	0.145	0.035	18.09	20.18	0.054	0.015	0.004	0.004
Shaanxi	3813	3813	0.172	0.900	14.89	53.01	0.066	0.715	0.004	0.014
Shanxi	3682	3682	0.199	0.702	14.40	34.52	0.076	0.501	0.004	0.009
Guangdong	10,999	10,999	0.279	0.187	47.86	70.62	0.112	0.107	0.004	0.006
Hebei	7470	7470	0.333	0.327	42.55	50.06	0.144	0.198	0.006	0.007
Anhui	6196	6196	0.378	0.576	38.86	52.09	0.173	0.391	0.006	0.008
Hubei	5885	5885	0.421	0.819	37.64	63.09	0.201	0.619	0.006	0.011
Sichuan	8262	8262	0.481	0.273	55.31	53.37	0.242	0.162	0.007	0.007
Heilongjiang	3799	3799	0.509	0.776	26.43	37.92	0.262	0.574	0.007	0.010
Shandong	9947	9947	0.581	0.107	72.67	58.60	0.316	0.057	0.007	0.006
Jiangxi	4592	4592	0.614	0.852	36.61	49.50	0.343	0.655	0.008	0.011
Hunan	6822	6822	0.664	0.377	55.34	49.01	0.384	0.233	0.008	0.007
Jiangsu	7999	7999	0.722	0.634	70.10	68.20	0.437	0.440	0.009	0.009
Beijing	2173	2173	0.738	0.979	20.17	72.00	0.452	0.905	0.009	0.033
Chongqing	3048	3048	0.760	0.939	34.18	53.09	0.477	0.782	0.011	0.017
Tianjin	1562	1562	0.771	0.996	17.74	69.44	0.490	0.973	0.011	0.045

(continued)

Table 18.6 (continued)

Province	Population parameter *j*		Cumulative percentage of population		$PM_{2.5}$ Distribution		$PM_{2.5}$ Cumulative percentage		$PM_{2.5}$/population	
	DEA	Information entropy	$X(i)1$	$X(i)2$	*DEA*	Information entropy	$Y(i)1$	$Y(i)2$	*DEA*	Information entropy
Fujian	3874	3874	0.799	0.730	50.01	37.98	0.528	0.528	0.013	0.010
Liaoning	4378	4378	0.831	0.531	57.89	36.01	0.571	0.354	0.013	0.008
Guangxi	4838	4838	0.866	0.481	64.72	37.99	0.619	0.313	0.013	0.008
Guizhou	3555	3555	0.892	0.213	53.07	22.92	0.659	0.124	0.015	0.006
Xinjiang	2398	2398	0.909	0.917	37.71	40.36	0.687	0.744	0.016	0.017
Jilin	2733	2733	0.929	0.872	51.15	30.77	0.725	0.677	0.019	0.011
Shanghai	2420	2420	0.947	0.964	49.31	79.01	0.762	0.853	0.020	0.033
Inner Mongolia	2520	2520	0.965	0.499	63.44	20.44	0.809	0.328	0.025	0.008
Gansu	2610	2610	0.984	0.749	76.87	25.62	0.866	0.547	0.030	0.010
Hainan	917	917	0.991	0.946	34.32	19.91	0.892	0.796	0.037	0.022
Ningxia	675	675	0.996	0.984	73.13	25.62	0.946	0.923	0.108	0.038
Qinghai	593	593	1.000	1.000	72.09	37.80	1.000	1.000	0.122	0.064

In particular a two-objective *DEA* model is constructed to allocate the $PM_{2.5}$ emission rights of 30 provinces in China in 2016, and then an environmental *Gini* coefficient based on population evaluation index is constructed to compare the two schemes.

Policy and decision-makers addressing emissions trading programs face the problem of setting an environmental target at the national, or sometimes regional/local levels. As part of deployment of those programs, it is necessary to impose a limit on the total amount of pollution that sources located in different parts of a country can emit into the environment. Our proposal can help policy and decision makers to define and quantitatively assign these limits. In addition, the proposed approach can help decision and policy-makers to lead with the growing public concern for environmental justice and equity, providing a tool to analyze the impact of balancing efficiency and equity in the discussion about how to assign emission rights.

The main conclusions are as follows:

(1) The two-objective *DEA* model constructed in this paper takes population, environmental bearing capacity and other factors into consideration, which is more reasonable and equitable than the allocation scheme that only pursues maximum efficiency; Meanwhile, it enriches the method research of *DEA* model and can provide analytical tools for the performance evaluation of resource and environmental policies.
(2) The distribution scheme based on *DEA* method is fairer, with higher efficiency and less volatility. Compared with the actual emission at the beginning, $PM_{2.5}$ emission should be reduced in the central and western regions and economically backward regions, such as Yunnan, Ningxia, Inner Mongolia, Liaoning, Jilin, Anhui, Gansu, Henan, Hubei, Hainan, Guizhou, Hebei, Shaanxi, Shandong, Qinghai, Shanxi and Xinjiang. Eastern and economically developed regions, such as Beijing, Tianjin, Shanghai, Jiangsu, Zhejiang, Fujian, Jiangxi, Guangdong and Sichuan provinces, are entitled to share more $PM_{2.5}$ emissions than before.[2]

Further research can be carried out from the following aspects in the future.

(1) More evaluation indicators are included. Due to the availability and completeness of data, the gross regional product (GDP) is selected as the main expected

[2] From the perspective of comparison of previous relative researches, Sueyoshi and Yuan (2015) constructed a DEA model based on efficiency orientation, using PM2.5 and PM10 as undesired outputs, and evaluated the environmental performance of each province. It is believed that the Chinese government should allocate its economic resources to Northwestern cities such as Lanzhou, Xining, Yinchuan and Urumqi, and strengthen environmental protection in Beijing, Tianjin, Shanghai, and Chongqing. Wu et al. (2016) constructed an efficiency-oriented ZSG-DEA model, used PM2.5 as an undesired output, and evaluated the environmental performance of each province; Wu et al. (2019) constructed a cross-efficiency DEA model and evaluated the PM2.5 emission performance of various provinces in China. These papers are only efficiency-oriented and do not consider the principle of fairness. Because of the different DEA models constructed, their conclusions are difficult to directly compare with this article. This article does not only consider a more comprehensive perspective, but also obtains more reasonable conclusions.

output index. However, the development of different industries in different provinces is quite different, so the objectives and characteristics of regional and industrial development planning should also be quantified and included in the evaluation system (Grunewald et al., 2014).

(2) More detailed classification of emission bodies. In the future, big data resources and big data analysis technology can be used to make a more detailed division of emission subjects of industries, and conduct more in-depth efficiency assessment studies.

(3) Different from GHG, although PM has inter-regional transmission relationships, most of the components of PM will settle locally. In particular, each province in China has a large area (The average area of the province is 300,000 square kilometers; of course, if the evaluation unit is a city, the transmission between cities needs to be considered), and then PM is easier to settle in the province. Based on the above considerations, this article does not consider the problem of inter-provincial flow of pollution sources, but uses provinces as evaluation units to allocate emission rights.

(4) From the perspective of the government and the market, there are two main types of policies to promote haze emission reduction. One is government-led emission reduction based on administrative forces; the other is market-led emission reduction based on economic policy guidance and stimulation. The method proposed in this article is more suitable for the first scheme. That is to say, we must first evaluate the emission performance of each emission unit, consider the resources and conditions of each emission unit, and then propose a new emission target. But it should be pointed out that the method proposed in this article needs to collect more basic data, and if the data collected is complete and the indicators are reasonable it will have a greater impact on the results.

(5) China is a country where the central government centralizes power and local governments determine the direction and degree of local social and economic development. The formulation of air pollution control policies and the allocation of pollution rights are all led by the government. Taking into account China's national conditions and the desirability of popularizing and solving real problems based on the research findings, this paper proposes a command-and-control emission right allocating method. The design of decentralized policies based on economic interests is a guarantee for the sustainable development of emissions trading, and also an important stimulus for the forces that move the market and the private sector (enterprises and other pollutants). China has carried out some attempts in this line. For example, emission trading centers have been established in Zhejiang and Hunan Province. However, the transactions are not active, possibly because the government has too much power to allocate resources, resulting in insufficient private vitality. Therefore, in the future, we should gradually cultivate and guide the establishment of a market-oriented pollution emission trading center. Academic research in this area should be based on field investigations and rational design of models and methods; it is necessary to measure the costs and benefits of emissions, set a price range for emissions trading, and design reasonable trading mechanisms and methods.

(6) The method proposed in this article assumes that the central government or the higher-level government takes the lead and each pollutant subject unconditionally obeys the allocation of emissions rights. Since this method does not involve the issue of emission rights trading, it does not consider the cost of conducting emission rights trading among multiple entities. In future research, the allocation of emission rights can be combined with emission rights trading to construct an emission rights allocation and trading model that combines command control and market transactions.

Acknowledgements Mengke Zhao, Beibei Shi, and Ernesto D.R. Santibanez Gonzalez also made great contributions to this manuscript. We express our heartfelt thanks to them. This research also was supported by: National Natural Science Foundation of China (72474127) and National Social Science Fund Major Bidding Project of China (18ZDA052). Part of this chapter was published:Guo J, Zhao M, Wu X, Shi B, Santibanez Gonzalez E. D. R. Study on the distribution of PM emission rights in various provinces of China based on a new efficiency and equity two-objective DEA model. Ecological Economics 2021;183:106956. https://doi.org/10.1016/j.ecolecon.2021.106956. The publication has been conducted with copyright permission.

References

Baumol, W. J., & Oates, W. E. (1988). *The theory of environmental policy* (2nd ed.). Cambridge University Press.

Bian, Y. (2010). Pollutant emission quota allocation based on DEA. *O. R. Transactions., 14*(4), 75–82.

Burton, E., & Sanjour, W. (1969). Multiple source analysis of air pollution abatement strategies. *Federal Accountant, XVIII*, 48–69.

Burton, E. S., & Sanjour, W. A. (1970). Simulation approach to air pollution abatement program planning. *Socio-Economic Planning Sciences, 4*, 147–159.

Cativiela-Campos, B., Ruiz-Sobremazas, D., Rodulfo-Cárdenas, R., Barrasa, A., Sánchez-Santed, F., Colomina, M. T., Aschner, M., López-Granero, C. (2025). What are the consequences of PM air pollution exposure on elderly behavior? A systematic review. *Environmental Pollution*, 126279.

Chen, Y. Z., Lu, H. W., Li, J., Yang, Y. Y., Xia. J. (2020). Multi-criteria decision making and fairness evaluation of water ecological carrying capacity for inter-regional green development. Environmental Science and Pollution Research. Chu, J., Jiang, H., Wu, J., Liu, H., 2020. Better resource utilization: A new DEA bi-objective resource reallocation approach considering environmental efficiency improvement. *Computers & Industrial Engineering*, 144, 106504.

Chen, C. Y., Huang, K. Y., Chen, C. C., Chang, Y. H., Li, H. J., Wang, T. H., Yang, P. C., 2025. The role of PM2. 5 exposure in lung cancer: mechanisms, genetic factors, and clinical implications. EMBO Molecular Medicine, 17(1), 31–40.

Environmental Protection Agency, Office of Federal Activities. (1998). *Final guidance for incorporating environmental justice concerns in EPA's NEPA compliance analyses*. U.S. Government Printing Office, Washington, D.C.

Fu, Y., Apk, T., & Liao, H. (2016). Impacts of historical climate and land cover changes on fine particulate matter ($PM_{2.5}$) air quality in East Asia between 1980 and 2010. *Atmospheric Chemistry and Physics, 16*, 10369–10383.

Gao, Q., Hao, M. L., & Gong, L. P. (2014). Research on establishing $PM_{2.5}$ emissions trading framework. *Chinese Journal of Environmental Management, 4*, 13–17.

GBD 2015 mortality and causes of death collaborators. (2016). Global, regional, and national life expectancy, all-cause mortality, and cause-specific mortality for 249 causes of death, 1980–2015: A systematic analysis for the global burden of disease study 2015. *Lancet, 388*(10053), 1459–1544.

Genave, A., Blancard, S., & Sabine G. (2020). An assessment of energy vulnerability in small island developing states. *Ecological Economics, 171.*

Grunewald, N., Jakob, M., & Mouratiadou, I. (2014). Decomposing inequality in CO_2 emissions: The role of primary energy carriers and economic sectors. *Ecological Economics, 100*, 183–194.

Janssen, M., & Rotmans, J. (1995). Allocation of fossil CO_2 emissions rights quantifying cultural perspectives. *Ecological Economics, 13*, 65–79.

Ji, J. S., Dominici, F., Gouveia, N., Kelly, F. J., & Neira, M. (2025). Air pollution interventions for health. *Nature Medicine*, 1–13.

Kverndokk, S. (1995). Tradeable CO_2 emission permits: Initial distribution as a justice problem. *Environmental Values, 4*(2), 129–148.

Lai, Y. B. (2007). The optimal distribution of pollution rights in the presence of political distortions. *Environmental and Resource Economics, 36*(3), 367–388.

Lelieveld, J., Evans, J. S., Fnais, M., Giannadaki, D., & Pozzer, A. (2015). The contribution of outdoor air pollution sources to premature mortality on a global scale. *Nature, 525*, 367–371.

Levy, J. I., Wilson, A. M., & Zwack, L. M. (2007, May). Quantifying the efficiency and equity implications of power plant air pollution control strategies in the United States. *Environ Health Perspect, 115*(5), 743–750.

Li, H. L., Zhu X. H., Chen J. Y., & Jiang F.T. (2019). Environmental regulations, environmental governance efficiency and the green transformation of China's iron and steel enterprises. *Ecological Economics, 165*, 106397.

Li, L., Bi, J., Yang, J. T., & Gang, Y. (2010). Regional difference decomposition analysis of sulfur dioxide emission intensity in China. *Chinese Journal of Population, Resources and Environment, 20*(115), 34–38.

Lin, W., & Fu, G. W. (2017). A fair distribution model of total emissions based on axiom system. *Journal of Environmental Sciences, 3*, 35–37.

Lucier, G. (2003). When smoke ran like water: Tales of environmental deception and the battle against pollution. *BMJ Clinical Research, 85*(1), 114–115.

Ma, X. M., Wang, D. H., Yi, Z. B., Yin, Y. P., Ma, X. M., & Wang, D. H. (2006). Research on calculation and allocation method of total allowable emissions of urban air pollutants. *Acta Scientiarum Naturalium Universitatis Pekinensis, 42*(2), 271–279.

Mao, Z. P., & Li, H. E. (1999). Solution to the problem of rational allocation of pollutants reduction in total quantity control. *Journal of Water Resources and Water Engineering, 1*, 25–30.

Nemery, B., Hoet, P. H., & Nemmar, A. (2001). The Meuse valley fog of 1930: An air pollution disaster. *Lancet, 357*(9257), 704–708.

Ni, L. L., Wang, D., Singh, V. P., Wu, J. F., Wang, Y. K., Tao, Y. W., Liu, J. F., Zou, Y., & He, R. (2019). A hybrid model-based framework for estimating ecological risk. *Journal of Cleaner Production, 225*, 1230–1240.

Nordhaus, W. D. (2005). *Life after Kyoto: Alternative approaches to global warming*. National Bureau of Economic Research.

Okun, A. M. (1975). *Equality and efficiency, the big tradeoff*. The Brookings Institution.

Rosa, V., Carla, A., Payam, D., Marta, C., Ángel, O., Alejandra, G., Ana, R., Pablo, C., Xavier, B., Ana, Z., Elisa, C., Bénédicte, J., Jaume, R., & Gregory, W. (2019). Association of residential air pollution, noise, and greenspace with initial ischemic stroke severity. *Environmental Research, 179*, Article 108725.

SCAQMD. (2003). *The southland's war on smog: Fifty years of progress toward clean air*. http://www.aqmd.gov/news-mobile/archives/history/marchcov.html

Solomon, B. D., & Lee, R. (2000). Emissions trading systems and environmental justice. *Environ, 42*, 32–45.

Song, M., Zhu, Q., Peng, J., & Santibanez Gonzalez, E. D. R. (2017). Improving the evaluation of cross efficiencies: A method based on Shannon entropy weight. *Computers and Industrial Engineering, 112*, 99–106. https://doi.org/10.1016/j.cie.2017.07.023

The Environmental Protection Agency. (2019). *Beijing $PM_{2.5}$ and air quality index query*. http://www.pm25x.com/c-ity/beijing.html.

The National People's Congress of the People's Republic of China. (2018). *Air pollution prevention and control law of the people's Republic of China* (revised in 2018). http://www.npc.gov.cn/npc/sjxflfg/201906/daae57a178344d39985dcfc563cd4b9b.shtml

Timmermans, R., Kranenburg, R., Manders, A., Hendriks, C., Segersa, A., Dammers, E., Zhang, Q., Wang, L., Liu, Z., Zeng, L., Denier, H., & Schaap, M. (2017). Source apportionment of $PM_{2.5}$ across China using LOTOS-EUROS. *Atmospheric Environment*, 164.

Wang, D., Wang, Y. K., Singh, V. P., Zhu, J. Y., Jiang, L. L., Zeng, D. B., Liu, D. F., Zeng, X. K., Wu, J. C., Wang, L. C., & Zeng, C. F. (2018a). Ecological and health risk assessment of PAHs, OCPs, and PCBs in Taihu Lake basin. *Ecological Indicators, 92*, 171–180.

Wang, K., & Li, M. J. (2013). DEA modeling and application of carbon emission quota allocation. *Journal of Beijing Institute of Technology (Social Science Edition), 4*(2), 7–13.

Wang, H., & Zhou, P. (2018). Assessing global CO_2 emission inequality from consumption perspective: An index decomposition analysis. *Ecological Economics, 154*, 257–271.

Wang, J. N., Zhang, B., Wu, Y. Y., & Guo, M. (2014). Paid use and transaction of China's emission rights: Practice and prospects. *Environmental Protection, 42*(14), 22–25.

Wang, L., Li, W., Sun, Y., Tao, M. H., Xin, J. Y., Tao, S., Li, X. R., Zhang, N., Ying, K., & Wang, Y. S. (2018). $PM_{2.5}$ characteristics and regional transport contribution in five cities in southern north China plain, during 2013–2015. *Atmosphere, 9*(4), 157.

Wang, Q. G., Li, Z. Y., Chen, Z. P., & Cheng, W. (2010). Initial allocation method of emission rights in total control area. *China Environmental Science, 20*(1), 68–72.

Wang, Q., & Li, M. Q. (2012). Evaluation of air pollution control efficiency in China based on DEA method. *China Environmental Science, 5*(5), 942–946.

Wu, X. H., Tan, L., Guo, J., & Zhu, W. (2016). A Study of Allocative Efficiency of Air Pollutant Emission Rights Based on a Zero Sum Gains Data Envelopment Model: Taking PM2.5 as an Example. *J. Cleaner Prod., 113*, 1024–1031.

Wu, X. H., Chen, Y. F., & Guo, J. (2019). Study of haze emission efficiency based on new coopetition data envelopment analysis. *Expert Systems*. https://doi.org/10.1111/exsy.12466

Xinhua News Agency. (2016). *The smog has been aggravated in consecutive days*. The time for the suspension of classes in Tehran School. http://mini.eastday.com/a/161117080211853.html.

Xinhua News Agency. (2016). *This round of smog relief on the 22nd, the area of ash is 1.88 million square kilometers*. http://www.xinhuanet.com//politics/2016-12/21/c_1120163267.html

Xinhua News Agency. (2018). *The longest pollution time in recent years! The smog in Hebei and other places will continue until 22nd*. http://www.xinhuanet.com/2018-01/19/c_129795094.html

Yang, L., & Ren, J. (2019). Low-carbon comprehensive evaluation of coal-fired power generation enterprises based on analytic hierarchy process. *Research of Environmental Sciences, 1*, 153–162.

Ye, F. (2016). *Pricing of PM2.5 emission trading products* (Doctoral dissertation), Zhejiang University.

Yin, P. Z., Chu, J. F., Wu, J., Ding, J. J., Yang, M., & Wang, Y. H. (2020). A DEA-based two-stage network approach for hotel performance analysis: An internal cooperation perspective. *Omega, 93*, Article 102035.

Yousefi, R., Wang, F., Shaheen, A., Ge, Q., Kaveh-Firouz, A., Kaskaoutis, D. G., & Arshad, A. (2025). Long-term perspectives on land-use changes and air pollution policies in Iran: A comparative analysis of regional and global patterns in atmospheric PM2. 5. *Journal of Environmental Management, 375*, 124346.

Yue, L. P., & Ma, R. G. (2016). Innovation of smog treatment based on emission rights accounting. *The Journal of Humanities, 4*, 38–44.

Zhang, Q. F., & Crooks, R. (2012). *Toward an environmentally sustainable future: Country environmental analysis of the people's Republic of China*, 11. China Financial & Economic Publishing House.

Zheng, P. N., Chen, H. B., Chen, X. G., & Zhang, Y. B. (2007). Research on regional reduction index allocation based on DEA model. *Chinese Journal of Environmental Engineering, 11*, 135–141.

Zhou, P., & Wang, M. (2016). Carbon dioxide emissions allocation: A review. *Ecological Economics, 125*, 47–59.

Zhu, M., Du, Y., Sha, M., Jiang, Y., Wu, Z., & Han, B. (2025). Multi-criteria carbon emission allowance allocation for liner shipping companies in China: An integrated approach based on composite indicators and zero-sum gains DEA. *Ocean & Coastal Management, 270*, Article 107909.

Zeng, W. H., & Wang, H. H. (2017). Information environment pollution index based on information entropy and its application. *China Environmental Science, 37*(2), 768–776.

Chapter 19
Efficiency Evaluation and PM Emission Reallocation of China Ports Based on Improved DEA Models

19.1 Introduction

As the great artery of the national economy, the waterway is the most important part of transportation around the world, but it can negatively impact the environment. According to *Third IMO GHG Study 2014-Final Report* from the International Maritime Organization (IMO),[1] international shipping, domestic shipping, and fishing produce about 1.45 million tons of PM as calculated by the top-down method, which is 7% higher in 2011 than in 2008. Especially in China, where waterway transportation contributes to more frequent foggy and hazy weather at those ports. As a result, these unusual weather patterns attract society's attention. In 2017, the report from *China Ports Yearbook (CPY)* states that seven Chinese ports are the top ten in the world. This is exciting economic news but also alarming because it may exacerbate environmental changes in those ports. Although economic and social development is still the main task of national work at this present stage in China, the government has started tackling environmental problems and has taken actions to protect the environment such as releasing tighter regulation to limit the air pollutant emission near the harbor. For example, on Nov 30th 2018, the Chinese government enacted emission control regulations to protect the blue sky, facilitate green shipping, port development, and energy savings.

However, port pollution is not only an important environmental problem in China but also in other countries or cities. Because of that, most of them set up their strategies (which are not limited to PM) to reduce air pollution. For instance, Ulsan Metropolitan City defines its policy as a low carbon strategy to greenhouse gas (GHG) and to act as a global stronghold for the green industry (Mat et al., 2016). In some Emission Control Areas (ECAs), like the U.S. coast, the Baltic Sea and the English Channel, the government enacts a rule that vessels must burn low sulfur fuel when they enter these areas (Davarzani et al., 2016). Similarly, Yangtze River

[1] Abbreviations used in this paper are listed in Appendix A1 (Table 19.3).

X. Wu, *Environmental Disaster in China*,
https://doi.org/10.1007/978-981-92-0910-1_19

Delta cities in China have issued environmental policies to jointly limit air pollutants such as *Obligatory Targets of Environmental Protection* from Anhui Province (Yang et al., 2019). The IMO's Marine Environmental Protection Committee agrees that the sulfur content in vessel fuel is required to be abated to 0.50% (5000 ppm) to protect the environment beginning first January 2020, known as Regulation 14 (Cullinane & Bergqvist, 2014). Because of these strict rules, ports have to continue to increase investment to meet the government's requirements, negatively impacting profits.

The above facts raise some questions about how much governments should limit air pollutants and what ports can do to follow a more stringent set of rules. To solve these problems, scholars carried out some researches. For example, Chang and Park (2016) measured the output losses of Korean ports when the government imposed different levels of air pollution regulation. They suggested that this was a reason to issue a more temperate emission target by the local government at the beginning because the foregone cargo traffic increased more quickly than the emission reduction. Gobbi et al. (2019) tried to compare the impact of port emissions such as NO_2, SO_2, and PM_{10} in Civitavecchia. The study illustrated that both ultrafine particles and black carbon had a great impact in high concentrations at ports with non-environmental regulation (NER). One way for the port to comply with regulatory requirements was to cut down the emissions of auxiliary engines at berth. Yu et al. (2019) developed a multi-objective model to provide strategic planning to present an efficient way to use shore-side electricity. The other was to allow the substitution of air pollutants between different locations to maximize the expected output (Chen, 2013; He et al., 2018). These studies, however, did not make a comprehensive analysis of output loss with respect to different ports' strategies.

Thus, to answer this question, three improved DEA models are established to measure port efficiency under the circumstances of environmental regulation (ER), non-environmental regulation (NER), and reallocation of PM emission permits through inter-ports cooperation. A comprehensive production model is extended from the directional distance function of Färe et al. (2007a, 2007b) to incorporate both ER and NER and to allow all the inputs and outputs to be simultaneously and disproportionately changed. Then, a comprehensive emission model is generalized from the Chang and Park (2016) industry emission target model to allow non-proportional changes in inputs. Next, a comprehensive reallocation model is improved from the Zhou et al. (2014) spatial allocation model to depict the output distance at different PM emission targets with the port's cooperative strategy. A preference co-efficient is set for ports in order to choose favorable decisions for all three improved models. Further, the data collected from the 11 Chinese main ports from 2012 to 2016 are divided into four areas for calculation. Finally, the empirical result is summarized to put forward correlative suggestions to port strategies and government policies.

The remainder of the paper is structured as follows: Section 19.2 (Literature review) is a review of studies about the environmental pollution of the transportation sector based on DEA models. Section 19.3 (Model) firstly introduces the existing

DEA models (Färe model, Chang and Park model, and Zhou model). Then, three improved DEA models (the comprehensive production model, the comprehensive emission model, and the comprehensive reallocation model) are constructed to assess the environmental efficiency of ports under the circumstances of ER, NER, and reallocation of PM emission permits. In Sect. 19.4 (Case study), the improved DEA models are applied to calculate the environmental efficiency of 11 Chinese ports between 2012 and 2016. Section 19.5 (Conclusion) summarizes the paper and puts forward relevant policy suggestions and makes a recommendation for future research.

19.2 Literature Review

The literature has been proliferated to investigate the environmental pollution in transportation over the past decades. However, DEA becomes one of the most widely used non-parametric methods to evaluate public programs since an important study by Charnes et al. (1978).

Many scholars have evaluated the environmental efficiency of transportation systems (TS) from multiple perspectives. For example, He et al. (2018) centered on carbon dioxide and calculated the environmental efficiency of several major sectors including transportation part at the Chinese provincial level with non-separable non-expected output. Chang et al. (2013) investigated the environmental efficiencies of transportation systems (TS) of Chinese provinces. From the result, they found that the government could lower much carbon emissions at the provincial level because of the inefficiency of the Chinese transportation industry. Chang et al. (2014) measured the economic and environmental efficiency of a large number of global airlines. Sun et al. (2017) calculated the environmental efficiency of Chinese 17 ports, and their results suggested that policies should be implemented by following their respective circumstances to overcome their shortcomings. Mavi et al. (2019) measured transport efficiency with the Malmquist productivity index based on an improved DEA model. Egilmez and Park (2014) calculated the pollution emissions such as carbon emissions of manufacturing sectors, and compared their environmental performance and economic performance by an improved hierarchical methodology. Li et al. (2016) assessed the regional TS efficiency in China to dig out the reason of inefficiency. Pina and Torres (2001) analyzed the transport efficiency under the urban TS of public sectors and private sectors in the European Union. Baran and Górecka (2019) evaluated transport efficiency of road and rail freight in different counties in the European Union and analyzed the correlation between transport efficiency, GDP and CO_2 emissions. However, most of these studies evaluated the environmental efficiency based on only one year of data at the city or provincial level and rarely focused on the port. Chen et al. (2025) evaluated the operational and environmental efficiency of 26 Chinese coastal ports through the SBM-DEA model, specifying cargo throughput as desirable output and CO2 emissions as an undesirable output. Kong et al. (2025) used a novel framework combining cross-hierarchical data envelopment analysis and cross-efficiency to assess the sustainability of 18 ports in China during the period 2017 to 2020.

Some scholars carried out studies on energy efficiency and sustainability efficiency. Most of them concentrate on assessing the energy efficiency of various TS in cities. For example, Song et al. (2014) obtained the tendency of energy consumption of transportation and petroleum in Shanghai from 2000 to 2010 to dig out the influence factors on them. Hu et al. (2011) estimated the energy efficiency of lots of cities including Taipei, Yilan, and Taoyuan in Taiwan between 1998 and 2007 based on an improved four-stage DEA approach, and their result showed that the local governments must commit to changing inefficient cars to green energy cars with the central government help. Tian et al. (2020) assessed the sustainability efficiency of TS in Shaanxi province in China. Liu et al. (2016) calculated both energy and environment efficiency of TS including road and railway at the Chinese provincial level. The study showed that the energy structure and cleaner energy should be continuously adjusted and used in China. Cui and Li (2014) examined the energy efficiency of TS by proposing a novel three-stage DEA model. The study found that the energy efficiency of transportation is dependent on two points (transport structure and management measures). Guo et al. (2017) established an index system, easy to use to select both input and output, and provided an improved SBM model to assess the efficiency evaluation of the potential decrease of energy and pollution emission in China. Ramanathan (2000) obtained energy efficiency under different TS in India. Considering bad output, Li et al. (2016) evaluated transportation efficiency and sought the factors of transportation inefficiency in China. To design green-energy planning of transportation schemes for an air-conditioning manufacturer in China, Ji et al. (2016) constructed an improved model for suppliers, manufacturers, and retailers.

Recently, performance assessment is another hot topic discussed by scholars. Xiong et al. (2019) studied performance improvement of the transportation sector based on an improved distance friction minimization model Wu et al., 2016a, 2016b) constructed a novel parallel DEA model to assess the environmental efficiency of TS of all Chinese provinces along with energy consumption, and noticed that there were big dissimilarities between passenger and freight transportation subsystems. Particularly, they found that local governments should pay more attention to improving freight transportation. Lee et al. (2014) selected 11 main ports in the world to evaluate their environmental performance with oxide, sulfur oxide and carbon dioxide as undesirable output. Barros and Peypoch (2009) employed both operational and financial variables to compute the operational performance of the Association of European Airlines between 2000 to 2005. Zhou et al. (2013) designed an improved DEA model to obtain the CO_2 emission performance at the Chinese provincial level under the transport sector. Bostian et al. (2018) measured the environmental performance with application in a paper company based on a novel directional distance model obtaining maximum expected outputs and minimum bad outputs. Lan and Lin (2005) constructed an improved DEA model to include environmental effects and then applied in performance evaluation of railways based on a newly four-stage DEA model to contribute to some policy suggestions. To monitor manufacturers' emissions from freight transport, Holden et al. (2016) constructed a newly context-dependent non-expected output DEA approach.

Considering a measurement of the output loss between ER and NER, (Färe et al., 2007a, 2007b) established a newly DEA model with the application in electric power plants to calculate the technical efficiency. Färe et al., 2007a, 2007b also calculated the output loss between ER and NER with a newly directional environmental distance approach. Comparing with the observed GDP and simulated GDP, Färe et al. (2012) assessed the production technology of different OECD countries with different emissions targets. Chang and Park (2018) investigated whether ECAs' regulations influenced the efficiencies of ports operation in these areas. The results proved that ECA regulations reflected the views of the stakeholders that could hurt port efficiency. Haralambides and Gujar (2012) assessed the port service production for both expected outputs and non-expected outputs based on a novel eco-DEA approach. To measure the pollution abatement cost change indexes, Cui et al. (2018) constructed a newly DEA approach named Dynamic Environmental approach to split the indexes into small parts.

Inspired by the previous work, some scholars have studied the emission reallocation questions. To cope with emission permits reallocation, Lozano et al. (2009) proposed a DEA approach that had three phases and examined it in the paper industry. Wu et al., 2016a, 2016b) developed a novel DEA model for environmental performance improvement permitting resource reallocation between different decision-making units (DMUs) and allowing emission target change. Zhou et al. (2014) set up a number of centralized DEA approaches to optimize CO_2 emissions at various conditions. The result showed that the Chinese government might use a modest emission reduction policy to achieve GDP growth and CO_2 emissions reduction. Chu et al. (2024) introduced a dynamic bargaining game approach for carbon emission reduction allocation, applying it to analyze the distribution among 27 EU member states.

While few papers study both the output loss and PM emission reallocation in the port sector by using the DEA model, this paper attempts to measure them simultaneously to achieve the desired behavior for ports and provide some helpful suggestions to the government.

We will generalize the Färe et al., 2007a, 2007b) model to set environmental control parameters to incorporate both ER and NER, permit non-equal proportional change of input index, expected output and non-expected output index in existing models (Chang & Park, 2016; Färe et al., 2007a, 2007b), establish a comprehensive reallocation model to measure expected output distance at different pollution emission targets based on Zhou et al. (2014) spatial allocation model, and set preference co-efficient in all improved models for ports to choose favorable decisions. Last, the models are applied to 11 Chinese main ports for measurement.

19.3 Model

19.3.1 Environmental Technology

Considering a basic production process (port operation), producing expected outputs (standard container throughput) it inevitably delivers non-expected outputs such as suspended PM emissions or other emissions. The basic model called environmental technology can measure the expected output respectively under ER and NER.

Let $x = (x_1, \ldots, x_N) \in \Re_+^N$ denote the input vector, $y = (y_1, \ldots, y_S) \in \Re_+^S$ the vector of expected outputs, and $p = (p_1, \ldots, p_M) \in \Re_+^M$ the vector of non-expected outputs. Then the environmental technology introduced by Färe et al. (2005) can be given as:

$$T = \{(x, y, p) : x \text{ can produce } (y, p)\} \tag{19.1}$$

The environmental technology can be determined as output sets $P(x)$ or input sets $L(y, p)$:

$$(x, y, p) \in T \iff (y, p) \in P(x) \iff x \in L(y, p) \tag{19.2}$$

This environmental technology assumes two axioms which are null-jointness and weak disposability, as described in Eq. (19.3) and Eq. (19.4), respectively.

$$(y, p) \in P(x) \text{ and } p = 0 \text{ imply } y = 0 \tag{19.3}$$

$$(y, p) \in P(x) \text{ and } 0 \leq \theta \leq 1 \text{ imply } (\theta y, \theta p) \in P(x) \tag{19.4}$$

Then, the environmental technology can be formulated via the DEA model as an environmental production function. Assume there are $k = 1, \ldots, K$ DMUs with inputs vector $x_k = (x_{1k}, \ldots, x_{Nk})$, expected output vector y_k, and non-expected outputs vector $p_k = (p_{1k}, \ldots, p_{Mk})$ for DMU$_k$. z_k ($k = 1, \ldots, K$) denote the intensity variables. The environmental production model for port k' with single output under ER introduced by Färe et al., 2007a, 2007b) is formulated as:

$$F\left(x^{k'}, p^{k'}\right) = \max \sum_{k=1}^{K} z_k y_k$$
$$st \begin{cases} \sum_{k=1}^{K} z_k x_{nk} \leqslant x_{nk'} & n = 1, \ldots, N \quad \text{(i)} \\ \sum_{k=1}^{K} z_k p_{mk} = p_{mk'} & m = 1, \ldots, M \quad \text{(ii)} \\ z_k \geqslant 0 & k = 1, \ldots, K \quad \text{(iii)} \end{cases} \tag{19.5}$$

Furthermore, in order to measure the port's performance, Färe et al., 2007a, 2007b defined the directional output distance function (DODF) as:

$$\vec{D}(x, y, p; g_y, g_p) = Sup\{\beta : (y + \beta g_y, p - \beta g_p) \epsilon P(x)\} \tag{19.6}$$

where the vector $g = (g_y, g_p)$ indicates the direction that the port would like to adjust the expected outputs and non-expected outputs, and β describes the maximum increase of non-expected outputs. Assume expected outputs vector for DMU_k is $y_k = (y_{1k}, \ldots, y_{Sk})$. Then the DODF can be evaluated for the port k' with multiple outputs under ER by the following DEA model:

$$\vec{D}\left(x^{k'}, y^{k'}, p^{k'}; g_y, g_p\right) = \max \beta^{k'}$$

$$s.t. \begin{cases} \sum_{k=1}^{K} z_k y_{sk} \geqslant y_{sk'} + \beta^{k'} g_{y_s} & s = 1, \ldots, S \quad \text{(i)} \\ \sum_{k=1}^{K} z_k p_{mk} = p_{mk'} - \beta^{k'} g_{p_m} & m = 1, \ldots, M \quad \text{(ii)} \\ \sum_{k=1}^{K} z_k x_{nk} \leqslant x_{nk'} & n = 1, \ldots, N \quad \text{(iii)} \\ z_k \geqslant 0 & k = 1, \ldots, K \quad \text{(iv)} \end{cases} \tag{19.7}$$

Eq. (19.7) represents that the port could simultaneously and proportionally expand multiple expected outputs and reduce non-expected outputs. Referring to Färe et al., 2007a, 2007b, the environmental production function and DODF under ER can be easily changed to NER to replace the strict equality of non-expected outputs constraints Eq. (19.5)–ii and Eq. (19.7)–ii with Eq. (19.8) and Eq. (19.9).

$$\sum_{k=1}^{K} z_k p_{mk} \leqslant p_{mk'} \tag{19.8}$$

$$\sum_{k=1}^{K} z_k p_{mk} \leqslant p_{mk'} - \beta^{k'} g_{p_m} \tag{19.9}$$

Let $\vec{DU}$ be a DODF under the NER. The output loss between ER and NER can be calculated by

$$\vec{DPAC} = \vec{DU}\left(x^{k'}, y^{k'}, p^{k'}; g_y, g_p\right) - \vec{D}\left(x^{k'}, y^{k'}, p^{k'}; g_y, g_p\right) \tag{19.10}$$

19.3.2 Previous Industry Emission Target Model

To better imitate the realistic world, Chang and Park (2016) extended the Färe et al., 2007a, 2007b model by adding the maximum PM emission target limits on the total emissions of industries. This model does not require proportionally change in expected outputs and non-expected outputs. It is defined as:

$$\rho = \max \sum_{k=1}^{K} \beta_k y_k$$

$$s.t. \begin{cases} \sum_{h=1}^{K} z_{hk} y_h \geqslant y_k + \beta_k g_{y_k} & k = 1, \ldots, K & \text{(i)} \\ \sum_{h=1}^{K} z_{hk} p_h = p_k - \tau_k g_{p_k} & k = 1, \ldots, K & \text{(ii)} \\ \sum_{h=1}^{K} z_{hk} x_{nh} \leqslant x_{nk} & n = 1, \ldots, N; k = 1, \ldots, K & \text{(iii)} \\ \sum_{k=1}^{K} \sum_{h=1}^{K} z_{hk} p_h \leqslant P & & \text{(iv)} \\ z_{hk} \geqslant 0, \beta_k, \tau_k \geqslant 0 & k = 1, \ldots, K & \text{(v)} \end{cases} \tag{19.11}$$

where P denotes the PM emission target for the industry. It can be controlled by the government to evaluate the maximum output of port k at the PM emission target P. The output loss can be calculated by

$$\rho_{loss} = \rho\left(x, y, b; \sum_{k=1}^{K} p_k\right) - \rho(x, y, b; P) \tag{19.12}$$

$$\text{for } \overline{P_{LB}} \leq P \leq \sum_{k=1}^{K} p_k \tag{19.13}$$

where $\overline{P_{LB}}$ is the lower bound of P, and $\sum_{k=1}^{K} p_k$ denotes the total emission in the industry.

19.3.3 Previous Spatial Allocation Model

Furthermore, to ensure not exceeding the upper limit of the PM emission target, port could adopt a cooperative strategy to permit the substitution of PM emission between different DMUs to achieve the maximum of the aggregate expected outputs. Assume σ as a control coefficient of PM emission (between zero to one) imposed by the government. Then the spatial reallocation model (SRM) is expressed by Zhou et al. (2014) as:

$$\text{SRM} = \max \sum_{k=1}^{K} \hat{y}_k$$

$$s.t. \begin{cases} \sum_{h=1}^{K} z_{hk} x_{nh} \leqslant x_{nh} & n = 1, \ldots, N; k = 1 \ldots, K & \text{(i)} \\ \sum_{h=1}^{K} z_{hk} y_h \geqslant \hat{y}_k & k = 1, \ldots, K & \text{(ii)} \\ \sum_{h=1}^{K} z_{hk} p_h = \hat{p}_k & k = 1, \ldots, K & \text{(iii)} \\ \sum_{h=1}^{K} \hat{p}_h = \sigma \cdot \sum_{h=1}^{K} p_h & & \text{(iv)} \\ z_{hk} \geqslant 0, \hat{y}_k \geqslant 0, \hat{p}_k \geqslant 0 & h, k = 1, \ldots, K & \text{(v)} \end{cases} \tag{19.14}$$

where $\hat{p}_k$ and $\hat{y}_k$ are decision variables, z_{hk}, $\hat{y}_k$ and $\hat{p}_h$ denote respectively intensity variables, the expected outputs after PM emission reallocation between cooperative

DMUs, and the optimal amount of PM emission permit to each DMUs. Eq. (19.14)–(iv) could be controlled by the government to re-draft the number of total emissions to measure how much production should be reduced.

19.3.4 The Comprehensive Production Model

Eq. (19.7), Eq. (19.11) and Eq. (19.14) are all useful models to measure the output loss under ER and NER and to deal with emission reallocation question, but there are still some problems. First, Eq. (19.7) and Eq. (19.11) do not allow non-equal proportional change of input index, expected output and non-expected output index. However, the automated port may allow inputting less manpower or berths to get more standard container throughput with less PM emission. Second, three equations do not consider ports to adjust the direction vector in their favor. Third, Eq. (19.14) does not provide a distance function like DODF to measure the output loss directly. Last, Eq. (19.11) and Eq. (19.14) can only measure single expected output and non-expected output at the same time. Thus, to generalize these functions, suppose that there are $k = 1, \ldots, K$ DMUs and then we improve Eq. (19.6) to describe the automated port operation as follows:

$$\vec{D}\left(x, y, p; g_x, g_y, g_p\right) = Sup\left\{\beta : \left(y + \beta g_y, p - \tau g_p\right) \epsilon P(x - \alpha g_x)\right\} \tag{19.15}$$

where β, τ and α indicate respectively the maximum feasible expansion of expected outputs $y_k = (y_{1k}, \ldots, y_{Sk})$, contraction of non-expected outputs $p_k = (p_{1k}, \ldots, p_{Mk})$, and inputs $x_k = (x_{1k}, \ldots, x_{Nk})$. The direction vector can be expressed as

$$g = \left(g_{x_k} = \psi_{nk}, g_{y_k} = \mu_{sk}, g_{p_k} = \zeta_{mk}\right). \tag{19.16}$$

We also assume

$$\sum\nolimits_{n}^{N} \psi_{nk} + \sum\nolimits_{s}^{S} \mu_{sk} + \sum\nolimits_{m}^{M} \zeta_{mk} = 1, (\psi_{nk} \geq 0, \mu_{sk} \geq 0, \zeta_{mk} \geq 0) \tag{19.17}$$

where g denotes the direction vector, ψ, μ and ζ are preference co-efficient controlled by port.[2] When Eq. (19.15) equals zero, it means that the product plan is situated on the frontier of the output sets. Then the comprehensive production model (CPM), incorporating ER and NER, can be formulated as:

$$\overrightarrow{\text{CMP}}(\delta) = max \sum\nolimits_{k=1}^{K} \beta_k$$

[2] Färe et al. (2013) assume the direction vector ($\psi + \mu + \zeta = 1$) endogenously. In this paper, we suppose the direction vector could be decided by ports in their best interests.

$$
s.t.\begin{cases}
\sum_{h=1}^{K} z_{hk} y_{sh} \geqslant y_{sk} + \beta_k g_{y_k} & s = 1, \ldots, S; k = 1, \ldots, K \quad (i)\\
(1-\delta)\cdot \sum_{h=1}^{k} z_{hk} p_{mh} = (1-\delta)\cdot (p_{mk} - \tau_k g_{p_k}) & m = 1, \ldots, M; k = 1, \ldots, K \quad (ii)\\
\delta \cdot \sum_{k=1}^{K} z_{hk} p_{mh} \geqslant \delta \cdot (p_{mk} - \tau_k g_{p_k}) & m = 1, \ldots, M; k = 1, \ldots, K \quad (ii)\\
\sum_{h=1}^{K} z_{hk} x_{nh} \leqslant x_{nk} - \alpha_k g_{x_k} & n = 1, \ldots, N; k = 1, \ldots, K \quad (iv)\\
z_{hk} \geqslant 0,\ \beta_k,\ \tau_k,\ \alpha_k \geqslant 0,\ \delta = 0, 1 & h; k = 1, \ldots, K \quad (v)
\end{cases}
\tag{19.18}
$$

where δ denotes the environmental control parameter. If $\delta = 1$, the model represents NER. If $\delta = 0$, it expresses ER. The direction vector follows $g = (g_{x_k} = \psi_{nk}, g_{y_k} = \mu_{sk}, g_{p_k} = \zeta_{mk} | \sum_{n}^{N} \psi_{nk} + \sum_{s}^{S} \mu_{sk} + \sum_{m}^{M} \zeta_{mk} = 1, \psi_{nk} \geq 0, \mu_{sk} \geq 0, \zeta_{mk} \geq 0)$ is decided by the port. The comprehensive model allows all the inputs and outputs to be contracted or improved disproportionately by using different coefficients (β, τ, α) and incorporates both ER and NER. Then the output loss between ER and NER can be measured by the following function:

$$
\vec{\Delta y} = \vec{CPM}(1) - \vec{CPM}(0) \tag{19.19}
$$

Figure 19.1 provides graphic explanations for Eq. (19.18). Assume there is an observed port (Point A) inside output set $P(x - \alpha g_x)$. It can choose any north-western directions to increase its technical efficiency to the production frontier. In Fig. 19.1, Point A increases the technical efficiency along the vertical direction to Point B, which locates at the frontier of output set $P(x - \alpha g_x)$. Then, the distance AB measures the needed increasable expected output from technical inefficiency to technical efficiency under ER. Considering the observed port (Point A) being projected to point B', the distance AB' measures the needed increasable expected output under NER. Therefore, the distance BB' is calculated by Eq. (19.19) to evaluate the output loss between ER and NER. Figure 19.1 illustrates that environmental production technology produces fewer outputs than non-environmental production technology.

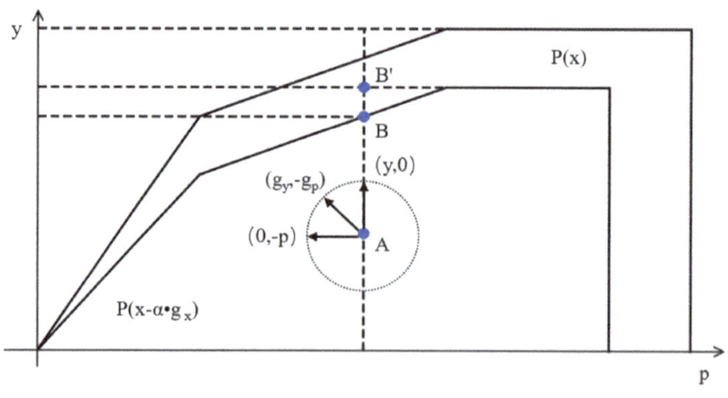

Fig. 19.1 The comprehensive production model and DODF

19.3.5 The Comprehensive Emission Model

Then, we improve the industry emission target model by forming a comprehensive emission model (CEM).

$$\overrightarrow{\text{CEM}}(\omega) = max \sum_{k=1}^{K} \beta_k$$

$$s.t \begin{cases} \sum_{h=1}^{K} z_{hk} y_{sh} \geqslant y_{sk} + \beta_k g_{y_k} & s = 1, \ldots, S; k = 1, \ldots, K \quad \text{(i)} \\ \sum_{h=1}^{K} z_{hk} p_{mh} = p_{mk} - \tau_k g_{p_k} & m = 1, \ldots, M; k = 1, \ldots, K \quad \text{(ii)} \\ \sum_{h=1}^{K} z_{hk} x_{nh} \leqslant x_{nk} - \alpha_k g_{x_k} & n = 1, \ldots, N; k = 1, \ldots, K \quad \text{(iii)} \\ \sum_{h=1}^{K} \sum_{h=1}^{K} p_{mh} \leqslant \omega \cdot \sum_{k=1}^{K} p_{mk} & m = 1, \ldots, M \quad \text{(iv)} \\ z_{hk} \geqslant 0, \beta_k, \tau_k, \alpha_k \geqslant 0 & h, k = 1, \ldots, K \quad \text{(v)} \end{cases} \tag{19.20}$$

where ω (between zero to one) denotes the coefficient of actual PM emission $(\sum_{k=1}^{K} p_{mk})$, which is controlled by the government. The direction vector follows $g = \left(g_{x_k} = \psi_{nk}, g_{y_k} = \mu_{sk}, g_{p_k} = \zeta_{mk} \mid \sum_{n}^{N} \psi_{nk} + \sum_{s}^{S} \mu_{sk} + \sum_{m}^{M} \zeta_{mk} = 1, \psi_{nk} \geq 0, \mu_{sk} \geq 0, \zeta_{mk} \geq 0\right)$,which is decided by the port. To achieve a reasonable environmental policy, coefficient ω could be used to help the government to check the change of increasable expected outputs with adjusting PM emission targets. Non-equal proportional change of input index, expected output and non-expected output index are allowed in a comprehensive emission model. This model can change multiple pollution targets at the same time.

19.3.6 The Comprehensive Reallocation Model

Next, a comprehensive reallocation model (CRM), improved from Zhou et al. (2014), is established to measure the maximum increasable expected output in the context of PM emission reallocation of ports under different PM emission targets.

$$\overrightarrow{\text{CRM}}(\omega) = \max \sum_{k=1}^{K} (\beta_{1k} - \beta_{2k})$$

$$s.t. \begin{cases} \sum_{h=1}^{K} z_{hk} y_{sh} \geqslant y_{sk} + (\beta_{1k} - \beta_{2k}) g_{y_k} & s = 1, \ldots, S; k = 1, \ldots, K \quad \text{(i)} \\ \sum_{h=1}^{K} z_{hk} p_{mh} = p_{mk} - (\tau_{1k} - \tau_{2k}) g_{p_k} & m = 1, \ldots, M; k = 1, \ldots, K \quad \text{(ii)} \\ \sum_{h=1}^{K} z_{hk} x_{nh} \leqslant x_{nk} - (\alpha_{1k} - \alpha_{2k}) g_{x_k} & n = 1, \ldots, N;\ k = 1, \ldots, K \quad \text{(iii)} \\ \sum_{h=1}^{K} g_{p_h} (\tau_{1k} - \tau_{2k}) = (1 - \omega) \cdot \sum_{h=1}^{K} p_{mh} & m = 1, \ldots, M \quad \text{(iv)} \\ z_{hk}\ ,\ \beta_{1k}, \beta_{2k}, \tau_{1k}, \tau_{2k}, \alpha_{1k}, \alpha_{2k} \geqslant 0 & k = 1, \ldots, K \quad \text{(v)} \end{cases} \tag{19.21}$$

where $(\beta_{1k} - \beta_{2k}), (\tau_{1k} - \tau_{2k})$ and $(\alpha_{1k} - \alpha_{2k})$[3] denote the change in expected outputs, non-expected outputs, and inputs after PM emission reallocation. ω (between zero to one) stands for the coefficient controlled by the government of actual PM emission $(\sum_{h=1}^{K} p_{mh})$. The direction vector follows $g = \left(g_{x_k} = \psi_{nk}, g_{y_k} = \mu_{sk}, g_{p_k} = \zeta_{mk} \mid \sum_n^N \psi_{nk} + \sum_s^S \mu_{sk} + \sum_m^M \zeta_{mk} = 1, \psi_{nk} \geq 0, \mu_{sk} \geq 0, \zeta_{mk} \geq 0\right)$, which is decided by the port.

Figure 19.2 provides a simple graphic illustration of Eq. (19.21). Suppose there are three different ports (Points *A*, *B*, *C*) lying inside the production frontier. To achieve a maximum sum of standard container throughput, the three points increase the technical efficiency firstly to point to *R*, *F* and *T*. Then an alliance is formed between the three ports to permit reallocation of PM emission. Points *R*, *F* and *T* move to the allocation efficiency Point *F*. The distance *CF*, *BF* and *AF* represent potential increasable expected output after PM emission reallocation along the direction vector. So, the total maximum expansion of the expected output of the alliance is calculated by Model (21). If the government requires a stricter PM emission target for the port industry, the alliance has to meet the requirement by finding a new optimal solution that minimizes the total distance.[4]

Then, the potential output loss between cooperative strategy and uncooperative strategy can be measured by

$$\overrightarrow{\Delta y'} = \overrightarrow{CRM}(\omega) - \overrightarrow{CEM}(\omega) \tag{19.22}$$

where the left side is the comprehensive reallocation model and the right side is the comprehensive emission model. It can be used to evaluate the output loss associated with cooperative strategy and uncooperative strategy at different PM emission targets. Eq. (19.22) takes a value equal to or greater than zero at the condition of the same ports' preferences. If the result is equal to zero, it means that there is no room for improvement with a cooperative strategy. If the value is greater than zero, it is beneficial for ports to ally to seek a maximum sum of expected outputs.

In the next section, the data of 11 Chinese main ports are applied to these models to evaluate the efficiencies and assess the effects of different strategies.

[3] In the reallocation process, Eq. (19.21) allows DMUs to decrease or increase the expected outputs, non-expected outputs or inputs to reach the maximum sum of expected outputs of the alliance that means the coefficient may be negative or positive. Eq. (19.21) can be easily changed to the northwestern model by defining $(\beta_{1k} - \beta_{2k}) \geq 0$, $(\tau_{1k} - \tau_{2k}) \geq 0$, and $(\alpha_{1k} - \alpha_{2k}) \geq 0$.

[4] With issuing the stricter emission regulation, the value of the Eq. (19.21) will change from positive to negative.

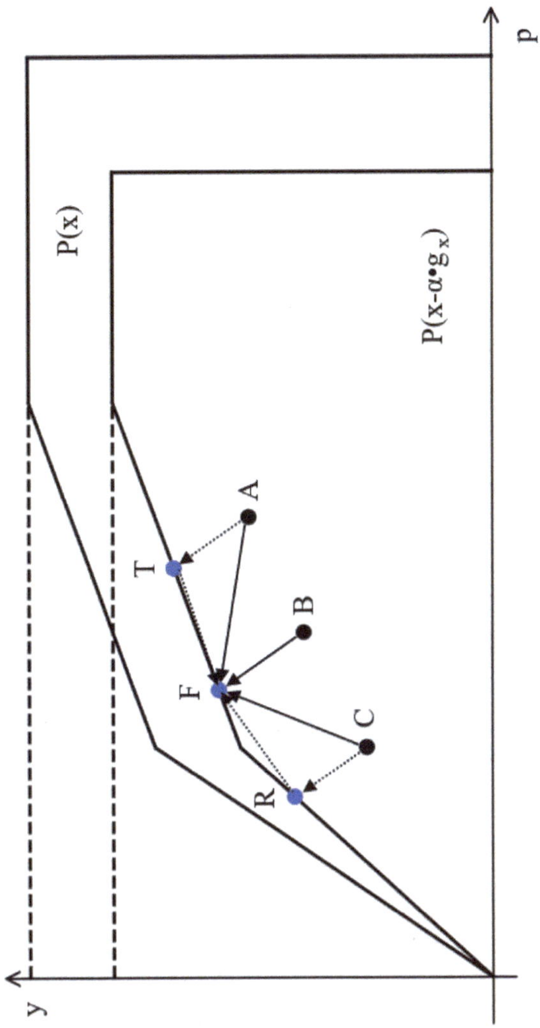

Fig. 19.2 The comprehensive reallocation model

19.4 Case Study

19.4.1 Description of Case Ports

There are about 20[5] major ports with container-handling capacity surpassing two-million twenty-feet equivalent unit (Teu) in China, which are located in the Bohai Bay, Yangtze River Delta, and Pearl River Delta. Eleven[6] of them are selected to evaluate the efficiency based on the improved DEA models because they are classified as above designated size ports and account for about 70%–85%[7] of standard container throughput among all Chinese ports from 2012 to 2016. Figure 19.3 shows the geographical location of these 11 ports.

Considering possible cooperative strategies between different ports to maximize the sum of standard container throughput, the 11 ports are divided into 4 areas for simulation. The 4 areas are shown in detail in Table 19.1. The grouping is mainly based on geographical proximity and cooperation which has already been formed between ports in reality. For example, Tianjin port and Dalian port have signed a framework agreement on strategic in 2013.

19.4.2 Data Collection

Data for the 11 Chinese ports from 2012 to 2016 are collected to measure single expected output (standard container throughput) and single non-expected output (PM emission[8]). The inputs consist of labor, terminal length, berth quantity, and total assets. Table 19.2 lists the summary statistics of the 11 Chinese ports data.

The data of PM emission is estimated in this paper. The amount of PM emission in a port is split into two parts (vessel operation and terminal operation) to consider all the factors and calculated by

$$PM_{total} = PM_{vessel} + PM_{terminal} \tag{19.23}$$

For the vessel part, to reflect the influence of vessel operation and total transport distance on PM emission in a port, we will not calculate it based on a short distance (after vessel entering port[9]); Instead, we use water freight turnover volume to measure it. Then, it is formulated by:

---

[5] The number is collected from CPY 2012–2016.

[6] Eleven ports are listed companies in the Shanghai Stock Exchange that the data is easily obtainable.

[7] The figure is calculated from CPY 2012–2016.

[8] In this study, we mainly focus on PM emission and disregard other pollution such as NO_X and CO_2.

[9] Chang and Park (2016) evaluated CO_2 emission after vessel entering port from a specific location (harbor limit) to a quay wall.

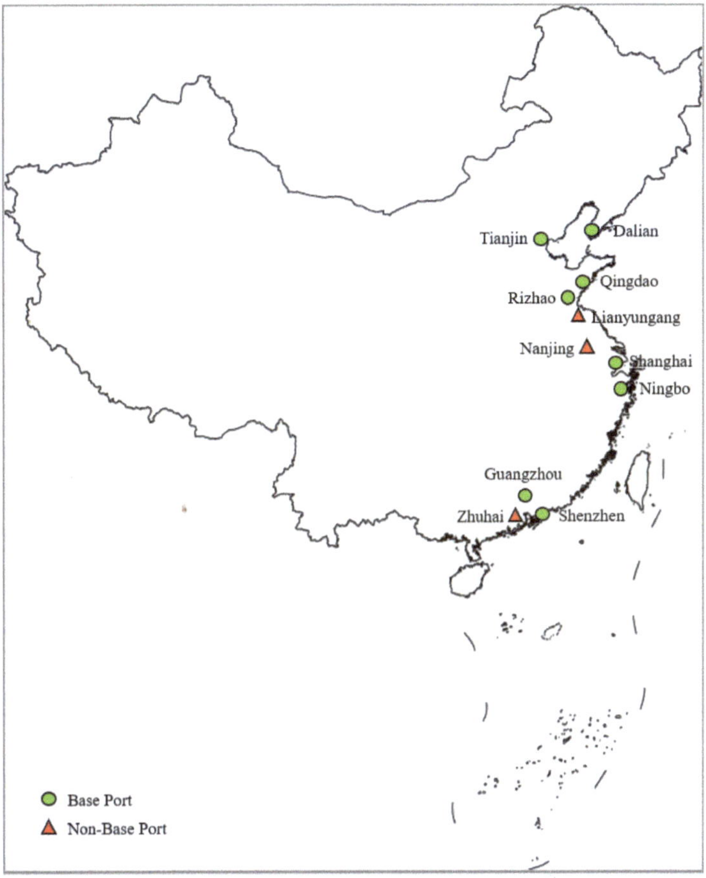

Fig. 19.3 Geographical locations of 11 Chinese ports

Table 19.1 Eleven Chinese ports in four areas

Area	Ports
1	Shanghai, Ningbo, Nanjing
2	Qingdao, Rizhao, Lianyungang
3	Tianjin, Dalian
4	Zhuhai, Guangzhou, Shenzhen

$$PM_{vessel} = WFT * EF_i * H \div K_i \tag{19.24}$$

where WFT denotes water freight turnover volume, EF_i the emission factor of specific fuel i, H the transportation consumption of standard coal per kiloton · n mile by ocean carrier, and K_i the standard coal coefficient of type i. To approximately evaluate vessel activities in transportation, EF_i is chosen of large-sized ships based on PM emission

Table 19.2 Summary statistics (11 Chinese ports, 2012–2016)

	Variable	Unit	Mean	Minimum	Maximum	Standard deviation
Input	Labor	Workers	6147.1	529.0	20748.0	5389.0
	Terminal length	Meter	36844.7	11236.0	86089.0	21851.9
	Berth quantity	Pcs	258.9	50.0	624.0	203.4
	Total assets	RMB (in hundred millions)	261.5	10.5	1167.9	264.6
Expected output	Standard container throughput	10,000 Teu	1317.7	81.3	3713.0	1004.8
Non-expected output	PM emission	Ton	6700.8	339.7	28771.3	7663.4

of 2.7% Sulphur content HFO fuel with slow speed diesel from *Third IMO GHG Study 2014-Final Report.*

Freight throughput is selected to evaluate the PM emission for the terminal part considering that loading and discharging is the main activity of the port. Then the $PM_{terminal}$ can be calculated in a similar way by

$$PM_{terminal} = FT * EF_i' * H' \div K_i' \tag{19.25}$$

where FT represents the freight throughput, and H' stands for the port consumption of standard coal per million tons. Energy consumption of $PM_{terminal}$ is converted to general coal consumption (K_i') supposing that the primary machines (cranes) are driven by electricity with thermal power. Since there is no standard PM emission factor of coal, it is collected from Shen et al. (2015) under laboratory conditions.

All data sources for both inputs and outputs are listed in Appendix A2 (Table 19.4).

19.4.3 Results and Discussions

In this section, the environmental efficiency of port is first calculated under ER and NER with different kinds of preferences decided by the ports. Then, the efficiency is reassessed at different pollution regulation targets in the reallocation model. Finally, we can compare the output loss between PM reallocation and non-reallocation at

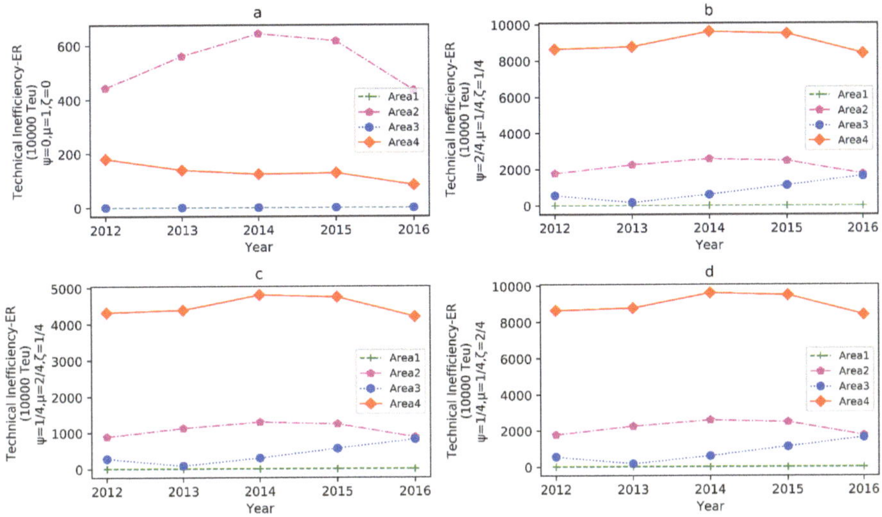

Fig. 19.4 Technical inefficiency under ER (2012–2016)

the same PM emission target. With these results, the efficiency of different areas is shown to provide some helpful suggestions to ports and the government.

Figure 19.4 shows the results for Eq. (19.18) when $\delta = 0$. The maximum potential increasable standard container throughput is calculated by the comprehensive production model of four kinds of preferences under ER for four areas from 2012 to 2016. The preference co-efficient ψ equally divides into the inputs index to make calculation easier.[10] As can be seen clearly, Fig. 19.4 (a) shows that Area 1 and Area 3 have zero standard container throughput to increase between 2012 to 2016 if they make the decisions as ($\psi = 0$, $\mu = 1$, $\zeta = 0$), which means they have higher technical efficiency. It is a slow downward trend from 2012 to 2016 for Area 4 that reflects it becomes more and more efficient. The most inefficient area is Area 2. If areas allow change of input index, expected output and non-expected output index to develop into technical efficiency, Fig. 19.4 (b, c, d) depicts that all areas have more potential increasable expected output to grow. The potential increase mainly follows the direction vector of expected output. For example, the growth of expected output with the same weight ($\mu = \frac{1}{4}$) in Fig. 19.4 (b, d) is two times than that in Fig. 19.4 (c) ($\mu = \frac{2}{4}$). However, it is noted that all areas follow the same trend regardless of the preferences chosen by the ports. In this case, only Area 1 has higher technical efficiency all the time and Area 4 becomes the most inefficient area.

[10] This approach is used in this study.

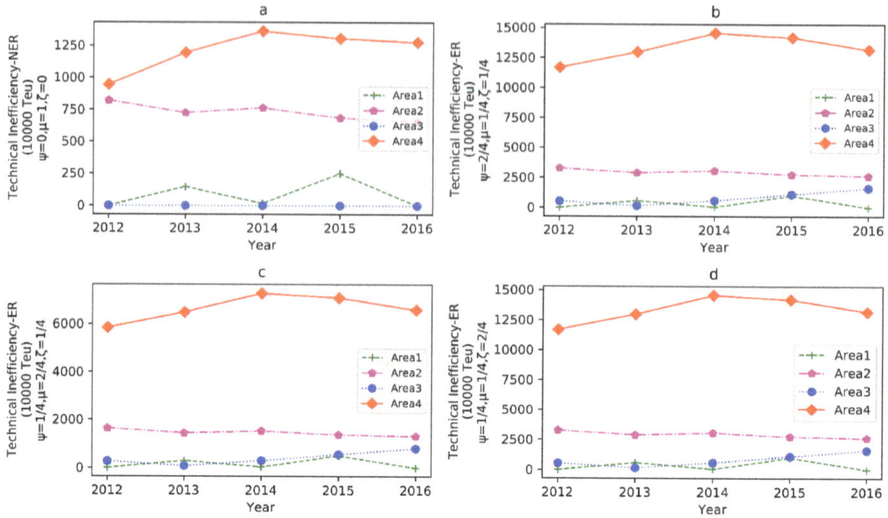

Fig. 19.5 Technical inefficiency under NER (2012–2016)

Figure 19.5 describes the results for Eq. (19.18) when $\delta = 1$. The maximum potential increasable standard container throughput is measured by the comprehensive production model of four kinds of preferences under NER for the four areas from 2012 to 2016. Figure 19.5 shows that no matter what preferences areas are chosen, in order to obtain technical efficiency, Area 4 has the most potential expected output to increase, which means it has the most technical inefficiency, while Area 2 comes in second. Area 1 still performs well under NER and its technical efficiency remains unchanged in most years. It is obtained similar conclusions from Fig. 19.5: Few growths of areas are expected in the vertical direction ($\psi = 0, \mu = 1, \zeta = 0$). Except for vertical direction, all areas still follow the same trend regardless of preferences chosen by the ports.

Figure 19.6 reveals the output loss between ER and NER of the four kinds of preferences (Eq. (19.19)). We can clearly see that it has a dramatic increase in an output loss of Area 4 before 2014 and then stays high until 2016, implying that Area 4 bears a huge loss under ER than others in any directions. One potential reason is that the port resources have not been fully developed in Area 4 resulting in technical inefficiency. In particular, Area 3 has zero output loss from 2012 to 2016, which represents that it keeps the same technical inefficiency under ER and NER in any directions. Importantly, Fig. 19.6 shows that all areas have the same output loss trend no matter what preference areas chosen to respond to ER.

Figure 19.7 shows the results of Eq. (19.20) about the potential increasable standard container throughput for each area at different PM emission targets (ω from 1 to 0.3) of the four kinds of preferences. Figure 19.7 (a, e, i, m) represent that areas

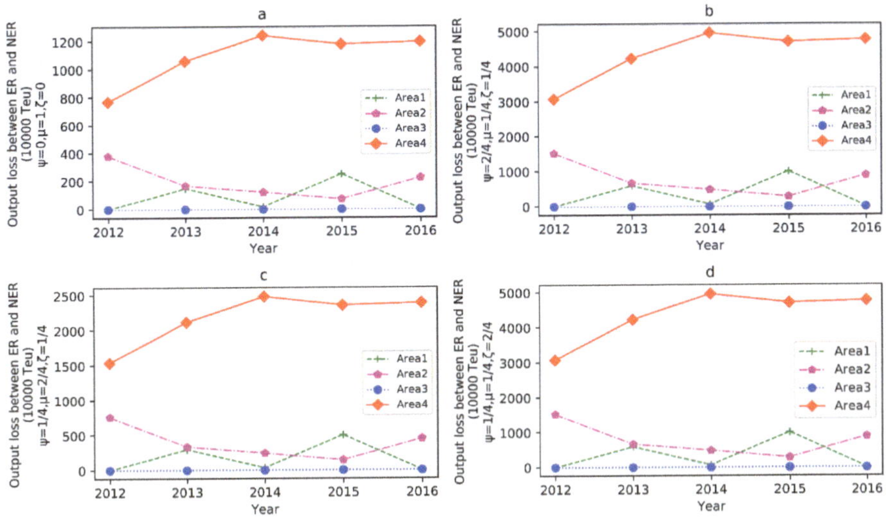

Fig. 19.6 Output loss between ER and NER (2012–2016)

choose the vertical direction to respond to the government's ER. The rest of the pictures in Fig. 19.7 depicts that areas choose other directions to promote expected output under different PM emission targets. It can be obtained the following results from Fig. 19.7. First, even if stringent emission standard is in effect, almost areas still have potential increasable expected output as long as they are willing not to choose a vertical direction. Second, if the government raises the standards to promote emission reductions, all areas must abate activities to meet the pollution regulation. Third, Area 1 has zero increasable expected output from 2012 to 2016 with an initial emission target ($\omega = 1$) in any direction, and hence it has the highest technical efficiency again. Last, all areas have the same trend to decrease the expected output with the stricter ER except for the vertical direction.

Figure 19.8 shows the results of Eq. (19.21) about the potential increasable standard container throughput after PM reallocation of the four kinds of preferences at different PM emission targets (ω from 1 to 0.3), compared with the actual expected output. Figure 19.8 (a, e, i, m) represent that areas choose the vertical direction in the PM reallocation model. They have the same shape as Fig. 19.7 (a, e, i, m) due to no change of PM emissions to each area. In other directions, it can be easily observed that all areas have huge potential increasable expected output after PM reallocation in most years. With raising the emission limit by the government, areas have the same trend to reduce the expected output in their preferred directions and value is from positive to negative. Comparing the two kinds of preference $\left(\psi = \frac{2}{4}, \mu = \frac{1}{4}, \zeta = \frac{1}{4}\right)$ and $\left(\psi = \frac{1}{4}, \mu = \frac{2}{4}, \zeta = \frac{1}{4}\right)$, all areas can efficiently reduce the amount of potential increasable expected output in a reallocation of PM emission with more weight given

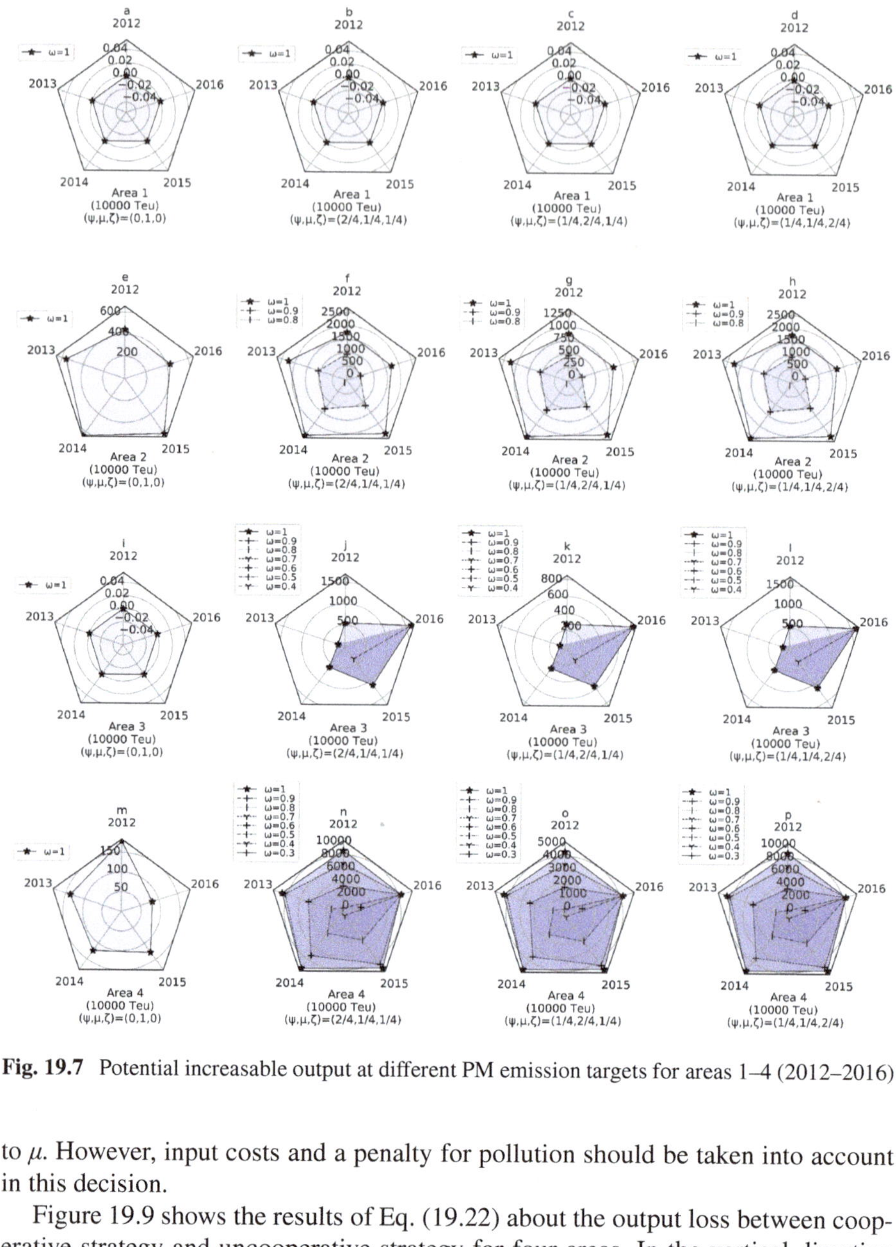

Fig. 19.7 Potential increasable output at different PM emission targets for areas 1–4 (2012–2016)

to μ. However, input costs and a penalty for pollution should be taken into account in this decision.

Figure 19.9 shows the results of Eq. (19.22) about the output loss between cooperative strategy and uncooperative strategy for four areas. In the vertical direction (Fig. 19.9 (a, e, i, m)), all areas have zero output loss because the value is the same between cooperative strategy and non-cooperative strategy. In other directions, output loss is foreseeable to increase in Area 1, Area 3 and Area 4 in most years. A possible

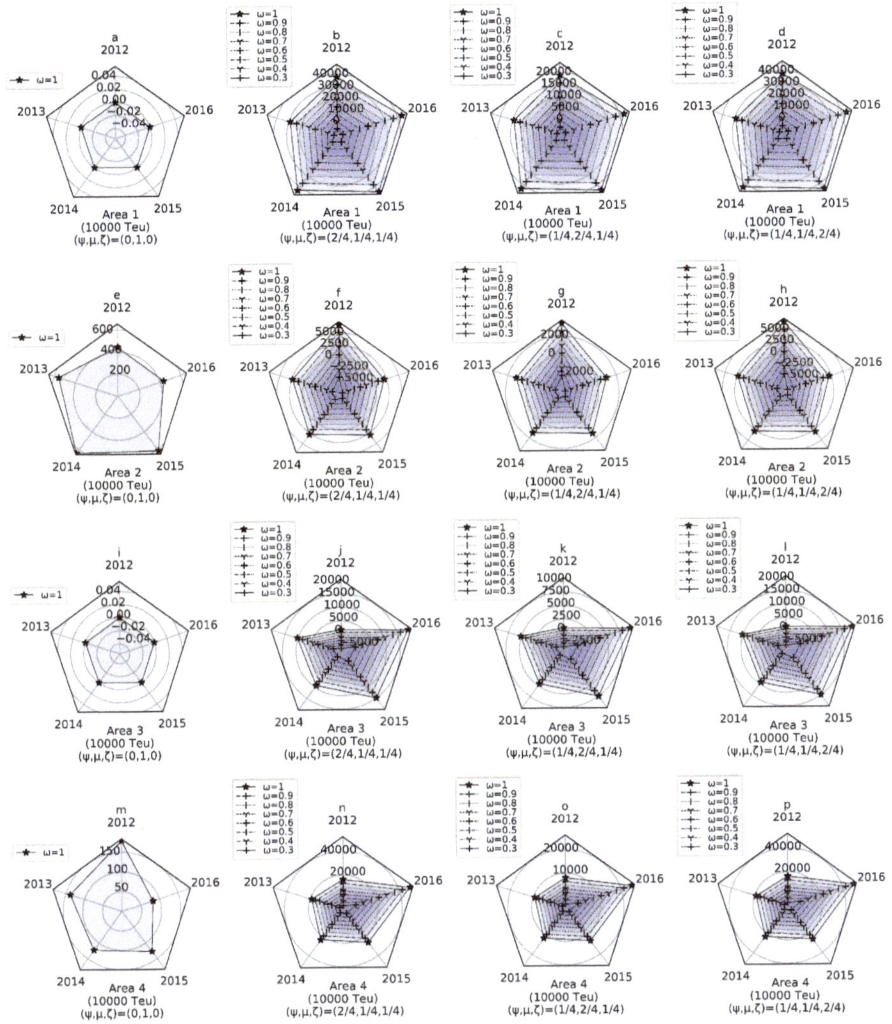

Fig. 19.8 Potential increasable output after PM reallocation at different PM emission targets for areas 1–4 (2012–2016)

explanation is that some members may become more and more inefficient and hence result in the false improvement with cooperation. For Area 2, the cooperation strategy is only valid in 2012. Therefore, efficient ports in Area 2 may consider dropping out of the alliance to seek new partners or the alliance kicks out the relatively inefficient

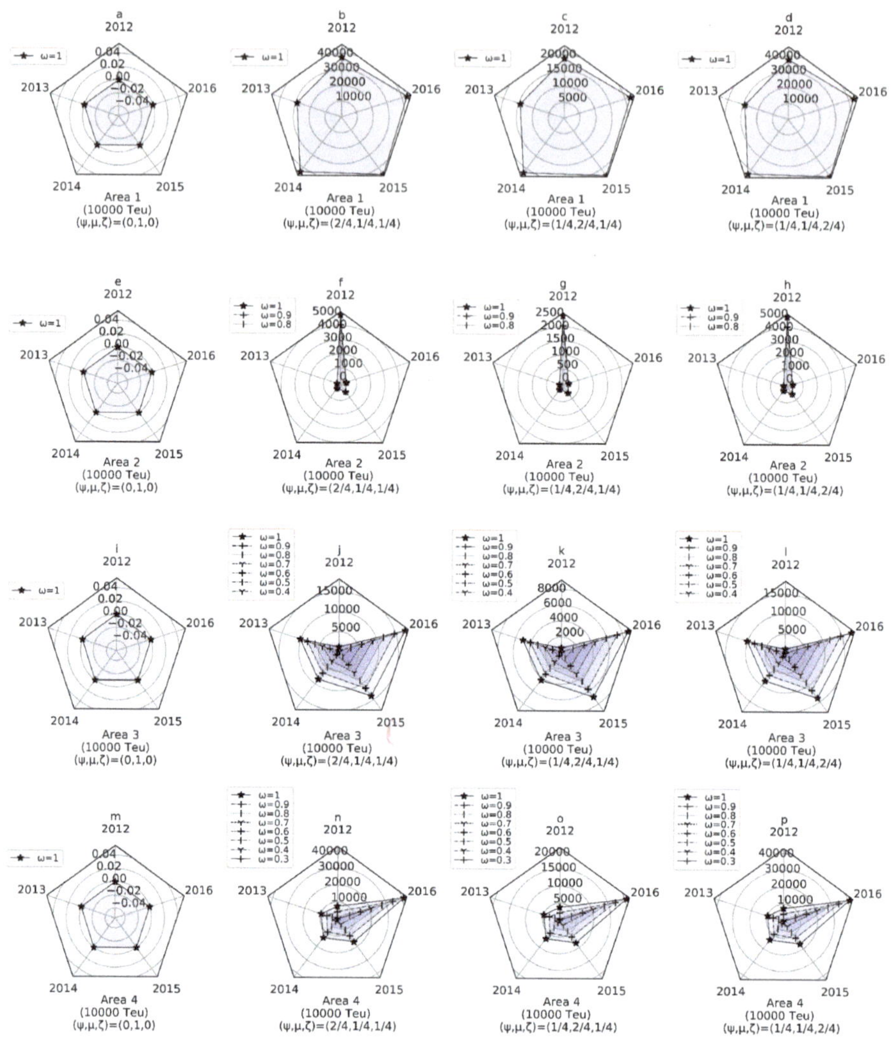

Fig. 19.9 The output loss between cooperation and non-cooperation strategies for areas 1–4 (2012–2016)

members to optimize the sum of the standard container throughput. Figure 19.9 also shows that cooperative strategy will lose ground to its uncooperative strategy (the value of output loss will gradually reduce to zero) when reducing the PM emission required by the government.

19.5 Conclusion

In this study, to contribute some suggestions to the ports and government, a comprehensive production model, a comprehensive emission model, and a comprehensive reallocation model are developed to evaluate the environmental efficiency of 11 Chinese main ports. First, setting environmental control parameters in models to shift them between ER and NER easily. Second, non-equal proportional change of input index, expected output and non-expected output index is allowed in the improved models. Third, the preference co-efficient is set for ports to choose favorable decisions. Last, many earlier studies investigate and measure other emissions like NO_x or CO_2 but we mainly focus on PM emission (calculated by water freight turnover volume and freight throughput), a top concern in China nowadays.

All models are applied to the 11 Chinese ports, and these ports are divided into four areas by using the port data from 2012 to 2016. Based on the findings, it is suggested: First, eastern ports in China (Area 1) generally have higher technical efficiency due to efficient production with relatively few input resources. To increase efficiency, the ocean carrier and port should perform their best to reduce transportation and port consumption of specific energy. For example, using cleaner vessels and improving port management is a way to approach the target.

Second, the port alliance can effectively increase the sum of expected output under ER. However, there usually exists a competition between the ports of the closest cities. The cities always make policies to attract more vessels to develop local GDP and reduce competitors' cargo volume. Although the competition mechanism will bring benefits to the port, its drawback is obvious. The port may abate some activities or increase more costs in pollution governance to keep current expected output with a tighter environmental policy issued by the government. If administrators can set up an effective mechanism to trigger spontaneous cooperation for ports, compared with uncooperative strategy, the alliance will possibly increase the total expected output under current pollution regulation or reduce the loss once the government makes a stricter emission policy. Although some members in the alliance will see reductions of expected output after cooperation, the whole alliance can be more resilient to deal with the foreseeable stricter environmental regulation.

Third, the ports cannot accept an overly restrictive pollution emission target no matter whether they ally or not. This is because, if the allowable discharge amount of the pollution emission is below a certain value, a port can only produce limited or even zero expected output. Thus, the local government or central government needs to balance the profit and damage caused by environmental pollution in the process of economic production to make a reasonable emission strategy. In this strategy, the current situation in China such as port facilities, vessel conditions, and international standards should be taken into account. In the opposite side, to some extent, inefficiency ports and old vessels are forced to be transformed by the strict standard.

Fourth, the ports have multiple decisions to deal with the ER. However, it will cause different results when ports assign different weights to input index, expected output index and non-expected output index. If the penalty for pollution emission is not high enough, ports usually assign more weights to the expected output index to ignore the penalty without decreasing the expected output. Therefore, the government needs to evaluate the validity of the environment policy to restrain ports' preferences.

Last but not least, the government may consider issuing some economic incentive policies to encourage ocean carriers and port enterprises to stimulate the use of high technologies in a greener way, thus reducing pollution emissions. Water transportation is the main mode of carrying goods for those ports and has a significant impact on the environment. Thus, it is necessary for the government to issue flexible policies to manage both vessels and ports. For example, cleaner energy, efficient waste gas treatment system, or other new technologies could be applied to those vessels and ports through government incentives. In other words, economic incentive policies, to a certain extent, can prevent environmental deterioration or even improve it.

There are still some limitations to this paper. First, it does not provide a method to determine the weight for preference co-efficient of ports exogenously. Further research is needed to assess multi-influenced factors such as the penalty for pollution emission and input costs. Second, the top-down method for calculating PM emissions is not very accurate. The bottom-up approach is recommended to achieve precise values. However, the bottom-up method is not addressed in this paper because our study covers 11 Chinese ports and most data like vessels' or cranes' power may not be obtained. Third, more valuable inputs may be included in the models for simulation if they can be readily available in the future. Finally, it is interesting to have a more in-depth discussion on the reallocation of emission permits by applying game theory to reveal the dilemma of ports' cooperative and uncooperative strategy.

Acknowledgements Zhijie Wang, Guo Wei, and Thomas A. Dooling also made great contributions to this manuscript. We express our heartfelt thanks to them. This research also was supported by: National Natural Science Foundation of China (72474127) and National Social Science Fund Major Bidding Project of China (18ZDA052). Part of this chapter was published:Wang Z, Wu X, Guo J, Wei G, Dooling T. A. Efficiency evaluation and PM emission reallocation of China ports based on improved DEA models. Transportation Research Part D: Transport and Environment 2020;82:102317. https://doi.org/10.1016/j.trd.2020.102317. The publication has been conducted with copyright permission.

Appendix

A1 Abbreviation

Table 19.3 Abbreviations in this paper

Full name	Abbreviation
China Ports Yearbook	CPY
Data envelopment analysis	DEA
Decision-making units	DMUs
Directional output distance function	DODF
Emission Control Areas	ECAs
Environmental regulation	ER
International Maritime Organization	IMO
Non-environmental regulation	NER
Particulate Matter	PM
Transportation systems	TS
Twenty-feet equivalent unit	Teu

A2 Data sources

Table 19.4 Data sources for inputs and outputs

Sources	Year	Data
Annual Report	2012–2016	Labor
		Total assets
The National Bureau of Statistics of the People's Republic of China	2012–2016	Terminal length
		Berth quantity
China Marine Statistical Yearbook China Ports Yearbook	2012–2016	Standard container throughput
Statistical Communique of The People's Republic of China On the National Economic and Social Development	2012–2016	Water freight turnover volume
The National Bureau of Statistics of the People's Republic of China		
China Ports Yearbook	2012–2016	Freight throughput

(continued)

Table 19.4 (continued)

Sources	Year	Data
Third IMO GHG Study 2014-Final Report	2014	EF_i (Emission factor of fuel oil i)[11]
Shen et al. (2015)	2015	EF_i'(Emission factor of general coal i)[12]
Statistics on the Development of Transportation Industry	2012–2016	H (transportation consumption of standard coal)
		H' (port consumption of standard coal)
China Energy Statistical Yearbook	2012–2016	K_i (The standard coal coefficient of fuel oil i)
		K_i' (The standard coal coefficient of general coal i)[13]

References

Baran, J., & Górecka, A. K. (2019). Economic and environmental aspects of inland transport in EU countries. *Econ. Res. Istraz., 32*, 1037–1059. https://doi.org/10.1080/1331677X.2019.1578680

Barros, C. P., & Peypoch, N. (2009). An evaluation of European airlines' operational performance. *International Journal of Production Economics, 122*, 525–533. https://doi.org/10.1016/j.ijpe.2009.04.016

Bostian, M., Färe, R., Grosskopf, S., Lundgren, T., & Weber, W. L. (2018). Time substitution for environmental performance: The case of Swedish manufacturing. *Empirical Economics, 54*, 129–152. https://doi.org/10.1007/s00181-016-1180-7

Chang, Y. T., & Park, H. (2016). Measuring foregone output under industry emission reduction target in the transportation sector. *Transportation Research Part D Transport and Environment, 49*, 138–153. https://doi.org/10.1016/j.trd.2016.08.017

Chang, Y.T., Park, H. (Kevin), Lee, S., Kim, E., 2018. Have Emission Control Areas (ECAs) harmed port efficiency in Europe? *Transportation Research Part D Transport and Environment, 58*, 39–53. https://doi.org/10.1016/j.trd.2017.10.018.

Chang, Y. T., Park, H., Jeong, J., & Lee, J. (2014). Evaluating economic and environmental efficiency of global airlines: A SBM-DEA approach. *Transportation Research Part D Transport and Environment, 27*, 46–50. https://doi.org/10.1016/j.trd.2013.12.013

Chang, Y. T., Zhang, N., Danao, D., & Zhang, N. (2013). Environmental efficiency analysis of transportation system in China: A non-radial DEA approach. *Energy Policy, 58*, 277–283. https://doi.org/10.1016/j.enpol.2013.03.011

[11] EF_i is chosen of large-sized ships based on PM emission of 2.7% Sulphur content HFO fuel with slow speed diesel from Third IMO GHG Study 2014-Final Report.

[12] Since there is no standard PM emission factor of coal, it is collected from SHEN et al. (2015) under laboratory conditions.

[13] Energy consumption of $PM_{terminal}$ is converted to general coal consumption (K_i') supposing that the primary machines (cranes) are driven by electricity with thermal power.

Charnes, A., Cooper, W. W., & Rhodes, E. (1978). Measuring the efficiency of decision making units. *European Journal of Operational Research, 2*, 429–444. https://doi.org/10.1016/0377-2217(78)90138-8

Chen, C. M. (2013). A critique of non-parametric efficiency analysis in energy economics studies. *Energy Economics, 38*, 146–152. https://doi.org/10.1016/j.eneco.2013.03.009

Chen, J.-R., Choi, J.-W., & Seo, Y.-J. (2025). Environmental efficiency assessment of coastal ports in China: Implications for sustainable port management. *Marine Pollution Bulletin., 211*, 117436. https://doi.org/10.1016/j.marpolbul.2024.117436

Chu, J., Hou, T., Li, F., & Yuan, Z. (2024). Dynamic bargaining game DEA carbon emissions abatement allocation and the Nash equilibrium. *Energy Economics., 134*, 107622. https://doi.org/10.1016/j.eneco.2024.107622

Cui, Q., & Li, Y. (2014). The evaluation of transportation energy efficiency: An application of three-stage virtual frontier DEA. *Transportation Research Part D Transport and Environment, 29*, 1–11. https://doi.org/10.1016/j.trd.2014.03.007

Cullinane, K., & Bergqvist, R. (2014). Emission control areas and their impact on maritime transport. *Transportation Research Part D Transport and Environment, 28*, 1–5. https://doi.org/10.1016/j.trd.2013.12.004

Davarzani, H., Fahimnia, B., Bell, M., & Sarkis, J. (2016). Greening ports and maritime logistics: A review. *Transportation Research Part D Transport and Environment, 48*, 473–487. https://doi.org/10.1016/j.trd.2015.07.007

Egilmez, G., & Park, Y. S. (2014). Transportation related carbon, energy and water footprint analysis of U.S. manufacturing: An eco-efficiency assessment. *Transportation Research Part D Transport and Environment, 32*, 143–159. https://doi.org/10.1016/j.trd.2014.07.001

Färe, R., Grosskopf, S., Margaritis, D., & Weber, W. L. (2012). Technological change and timing reductions in greenhouse gas emissions. *Journal of Productivity Analysis, 37*, 205–216. https://doi.org/10.1007/s11123-011-0232-4

Färe, R., Grosskopf, S., Noh, D.-W., & Weber, W. (2005). Characteristics of a polluting technology: Theory and practice. *Journal of Econometrics, 126*(2), 469–492. https://doi.org/10.1016/j.jeconom.2004.05.010

Färe, R., Grosskopf, S., & Pasurka, C. A. (2007a). Pollution abatement activities and traditional productivity. *Ecological Economics, 62*, 673–682. https://doi.org/10.1016/j.ecolecon.2006.08.014

Färe, R., Grosskopf, S., & Pasurka, C. A. (2007b). Environmental production functions and environmental directional distance functions. *Energy, 32*, 1055–1066. https://doi.org/10.1016/j.energy.2006.09.005

Färe, R., Grosskopf, S., & Whittaker, G. (2013). Directional output distance functions: endogenous directions based on exogenous normalization constraints. *Journal of Productivity Analysis, 40*, 267–269. https://doi.org/10.1007/s11123-012-0333-8

Gobbi, G. P., Di Liberto, L., & Barnaba, F. (2019). Impact of port emissions on EU-regulated and non-regulated air quality indicators: The case of Civitavecchia (Italy). *Science Total Environment, 134984*. https://doi.org/10.1016/j.scitotenv.2019.134984

Guo, X., Zhu, Q., Lv, L., Chu, J., & Wu, J. (2017). Efficiency evaluation of regional energy saving and emission reduction in China: A modified slacks-based measure approach. *Journal of Cleaner Production, 140*, 1313–1321. https://doi.org/10.1016/j.jclepro.2016.10.021

Haralambides, H., & Gujar, G. (2012). On balancing supply chain efficiency and environmental impacts: An eco-DEA model applied to the dry port sector of India. *Marit. Econ. Logist., 14*, 122–137. https://doi.org/10.1057/mel.2011.19

He, Q., Han, J., Guan, D., Mi, Z., Zhao, H., & Zhang, Q. (2018). The comprehensive environmental efficiency of socioeconomic sectors in China: An analysis based on a non-separable bad output SBM. *Journal of Cleaner Production, 176*, 1091–1110. https://doi.org/10.1016/j.jclepro.2017.11.220

Holden, R., Xu, B., Greening, P., Piecyk, M., & Dadhich, P. (2016). Towards a common measure of greenhouse gas related logistics activity using data envelopment analysis. *Transportation Research Part A Policy and Practice, 91*, 105–119. https://doi.org/10.1016/j.tra.2016.06.001

Hu, J. L., Lio, M. C., Yeh, F. Y., & Lin, C. H. (2011). Environment-adjusted regional energy efficiency in Taiwan. *Applied Energy, 88*, 2893–2899. https://doi.org/10.1016/j.apenergy.2011.01.068

Ji, X., Wu, J., & Zhu, Q. (2016). Eco-design of transportation in sustainable supply chain management: A DEA-like method. *Transportation Research Part D Transport and Environment, 48*, 451–459. https://doi.org/10.1016/j.trd.2015.08.007

Kiani Mavi, R., Fathi, A., Farzipoor Saen, R., & Kiani Mavi, N. (2019). Eco-innovation in transportation industry: A double frontier common weights analysis with ideal point method for Malmquist productivity index. *Resources, Conservation and Recycling, 147*, 39–48. https://doi.org/10.1016/j.resconrec.2019.04.017

Kong, Y., Tian, X., Sun, J., & Zhou, H. (2025). Charting sustainable vistas: Analysis of internal and external sustainability performance of Chinese ports. *Transport Policy, 161*, 31–43. https://doi.org/10.1016/j.tranpol.2024.11.016

Lan, L. W., & Lin, E. T. J. (2005). Measuring railway performance with adjustment of environmental effects, data noise and slacks. *Transport metrica, 1*, 161–189. https://doi.org/10.1080/18128600508685645

Lee, T., Yeo, G. T., & Thai, V. V. (2014). Environmental efficiency analysis of port cities: Slacks-based measure data envelopment analysis approach. *Transport Policy, 33*, 82–88. https://doi.org/10.1016/j.tranpol.2014.02.009

Li, T., Yang, W., Zhang, H., & Cao, X. (2016). Evaluating the impact of transport investment on the efficiency of regional integrated transport systems in China. *Transport Policy, 45*, 66–76. https://doi.org/10.1016/j.tranpol.2015.09.005

Liu, Z., Qin, C. X., & Zhang, Y. J. (2016). The energy-environment efficiency of road and railway sectors in China: Evidence from the provincial level. *Ecological Indicators, 69*, 559–570. https://doi.org/10.1016/j.ecolind.2016.05.016

Lozano, S., Villa, G., & Brännlund, R. (2009). Centralised reallocation of emission permits using DEA. *European Journal of Operational Research, 193*, 752–760. https://doi.org/10.1016/j.ejor.2007.07.029

Mat, N., Cerceau, J., Shi, L., Park, H. S., Junqua, G., & Lopez-Ferber, M. (2016). Socio-ecological transitions toward low-carbon port cities: Trends, changes and adaptation processes in Asia and Europe. *Journal of Cleaner Production, 114*, 362–375. https://doi.org/10.1016/j.jclepro.2015.04.058

Pina, V., & Torres, L. (2001). Analysis of the efficiency of local government services delivery. An application to urban public transport. *Transportation Research Part A Policy and Practice, 35*, 929–944. https://doi.org/10.1016/S0965-8564(00)00033-1

Ramanathan, R. (2000). A holistic approach to compare energy efficiencies of different transport modes. *Energy Policy, 28*, 743–747. https://doi.org/10.1016/S0301-4215(00)00072-0

Shen, G., Tao, S., Wei, S., Chen, Y., Zhang, Y., Huang, Y., Zhu, D., Yuan, C., Wang, H., Wang, Y., Pei, L., Liao, Y., Duan, Y., Wang, B., Wang, R., Lv, Y., Li, W., & Zheng, X. (2015). *NIH Public Access, 47*, 2998–3005. https://doi.org/10.1021/es304599g.Field

Song, M., Wu, N., & Wu, K. (2014). Energy consumption and energy efficiency of the transportation sector in Shanghai. *Sustain., 6*, 702–717. https://doi.org/10.3390/su6020702

Sun, J., Yuan, Y., Yang, R., Ji, X., & Wu, J. (2017). Performance evaluation of Chinese port enterprises under significant environmental concerns: An extended DEA-based analysis. *Transport Policy, 60*, 75–86. https://doi.org/10.1016/j.tranpol.2017.09.001

Tian, N., Tang, S., Che, A., & Wu, P. (2020). Measuring regional transport sustainability using super-efficiency SBM-DEA with weighting preference. *Journal of Cleaner Production, 242*, 118474. https://doi.org/10.1016/j.jclepro.2019.118474

Wu, J., Chu, J., An, Q., Sun, J., & Yin, P. (2016). Resource reallocation and target setting for improving environmental performance of DMUs: An application to regional highway transportation systems in China. *Transportation Research Part D Transport and Environment, 61*, 204–216. https://doi.org/10.1016/j.trd.2016.09.002

Wu, J., Zhu, Q., Chu, J., Liu, H., & Liang, L. (2016). Measuring energy and environmental efficiency of transportation systems in China based on a parallel DEA approach. *Transportation Research Part D Transport and Environment, 48*, 460–472. https://doi.org/10.1016/j.trd.2015.08.001

Xiong, B., Chen, H., An, Q., & Wu, J. (2019). A multi-objective distance friction minimization model for performance assessment through data envelopment analysis. *European Journal of Operational Research, 279*, 132–142. https://doi.org/10.1016/j.ejor.2019.05.007

Yang, W., Yuan, G., & Han, J. (2019). Is China's air pollution control policy effective? Evidence from Yangtze River Delta cities. *Journal of Cleaner Production, 220*, 110–133. https://doi.org/10.1016/j.jclepro.2019.01.287

Yu, J., Voß, S., & Tang, G. (2019). Strategy development for retrofitting ships for implementing shore side electricity. *Transportation Research Part D Transport and Environment, 74*, 201–213. https://doi.org/10.1016/j.trd.2019.08.004

Zhou, G., Chung, W., & Zhang, X. (2013). A study of carbon dioxide emissions performance of China's transport sector. *Energy, 50*, 302–314. https://doi.org/10.1016/j.energy.2012.11.045

Zhou, P., Sun, Z. R., & Zhou, D. Q. (2014). Optimal path for controlling CO2 emissions in China: A perspective of efficiency analysis. *Energy Economics, 45*, 99–110. https://doi.org/10.1016/j.eneco.2014.06.019

Chapter 20
Impact of Energy Structure Adjustment and Environmental Regulation on Air Pollution in China: Simulation and Measurement Research by the Dynamic General Equilibrium Model

20.1 Introduction

Air pollution is a common problem faced by developing countries in the process of industrialization. With the increase of the level of industrialization, air pollution is a serious problem in China. Severe air pollution not only harms the physical and mental health of residents, destroys the accumulation of human capital (Chang et al., 2016), but also increases China's economic burden. This economic burden is manifested in both direct economic loss and the long-term developing barrier of cities. For example, according to a report by OECD at the International Transport Forum's 2014 summit, heavy air pollution costs China nearly 1.4 trillion a year, which are about three times of India. The Financial Times in 2013 also pointed out that many foreigners have to leave Beijing because of heavy air pollution, which leads to brain drain and reduces development potential of Beijing.

In order to improve air quality, the Chinese government has taken measures such as adjusting the energy structure and strengthening environmental control to control air pollution, with remarkable results. In 2019, the Ministry of Ecology and Environment of China issued *China Air Quality Improvement Report (2013–2018).* The report shows that the concentration of a number of air pollutants in China has dropped significantly, and that the national air quality has improved significantly; the average concentration of $PM_{2.5}$ in 74 cities fell by 42%, and the average concentration of $PM_{2.5}$ in the Beijing-Tianjin-Hebei, Yangtze River Delta and Pearl River Delta regions fell by 48%, 39% and 32% respectively from 2013. Air pollution control measures improved air quality in recent years (Song et al., 2018, 2019). But what kinds of governance measures are more effective in improving air quality? How to achieve continuous air pollution control? Exploring these issues is crucial to formulating air pollution control policies in the future.

To answer the above questions, this study uses a dynamic general equilibrium (DGE) model that includes time and space dimensions to explore the dynamic impact

X. Wu, *Environmental Disaster in China*,
https://doi.org/10.1007/978-981-92-0910-1_20

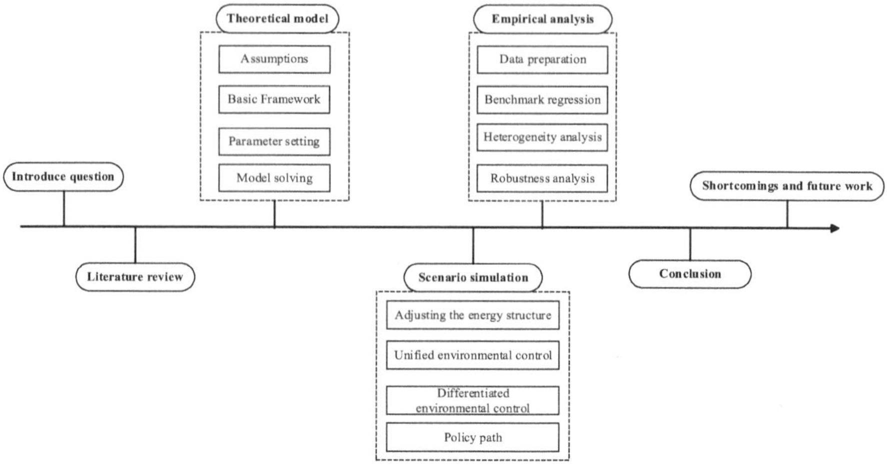

Fig. 20.1 Research framework

of adjusting energy structure and government environmental regulations on air pollution. For the empirical part, panel data from 286 cities at prefecture-level and above in China from 2010 to 2018 are used to discuss the heterogeneous impact of energy structure and government environmental regulation on urban air pollution of different scales and development periods. The research results show that both government environmental regulation and energy structure adjustment can help control air pollution; however, in different scenarios, different tools and methods are used, and the governance results have certain differences. The research ideas of this study are shown in Fig. 20.1.

The possible innovations in this study are as follows: (1) By theoretically combing the formation logic of air pollution and the influence mechanism of air pollution control measures on air pollution, and comparing the policy effects under different control measures and scenarios, this study provides some theoretical and empirical analysis results to solve the contradiction between air pollution control and economic growth, to retard the regional short-sightedness of government in air pollution control, and to adjust air pollution control measures. (2) The study found that the use of administrative orders to adjust the energy structure and reduce air pollution will not only reduce the economic output but also the governance effect will not be sustainable; there may be "retaliatory pollution" behavior afterward. (3) Considering the time variation of government environmental control intensity from the dynamic perspective, this study expands the perspective of analyzing the policy effect of air pollution control measures. It is found that the policy path of "fast first and slower" is the best choice. The rest of this study is: the second part is a literature review; the third part is a theoretical model; the fourth part covers scenario simulation; the fifth part deals with an empirical analysis, and finally the full text is summarized.

20.2 Literature Review

Regarding the reasons for the decline in air pollution in China, the existing literature roughly gives three explanations. First, the consumption of traditional energy sources such as coal is generally considered as direct source of air pollution. For example, Ma and Zhang (2014) used a spatial lag panel data model to study the relationship between air pollution and economic growth and energy structure. They found that for every 1% increase in the proportion of coal in the energy consumption structure, the $PM_{2.5}$ concentration increased by 0.064% on average. Hence, some scholars found that structure adjustment of energy consumption improved air quality recently (Yang et al., 2014; Modis, 2019). For example, Zhang et al. (2014) studied the impact of energy efficiency improvements in the iron and steel industry on air pollution. They found that energy efficiency improvements have reduced $PM_{2.5}$ emissions in the iron and steel industry by 3%. It should be stressed here that some scholars also pointed out that the improvement of energy efficiency will reduce the relative price of energy, reduce energy costs, stimulate energy consumption, and then increase air pollution. Therefore, improving energy efficiency requires supporting resource taxes and government regulatory constraints (Allan et al., 2014; Sancho, 2010). Other scholars believe that optimizing the industrial structure and energy structure and increasing the proportion of clean energy and renewable energy in energy consumption can reduce air pollution (Chuai et al., 2012; Joo et al., 2015; Sharifi et al., 2019). However, some scholars believe that China's coalbased energy structure makes economic growth have a rigid demand for energy consumption (Yang, 2011), which is difficult to change in the short term. The research of Song et al. (2020) provided evidence of the "rigid" energy structure in China, Song et al. (2020) studied the relationship between LCCP (China's lowcarbon city pilot) policy and urban ecological environment and found a positive relation of the two variables. But from the view of action mechanisms, the positive relation was only due to technological innovation but not the improvement of energy utilization efficiency in China. Using microdata from coal mines between 2013 and 2022, Wang et al. (2026) evaluated the air pollution reduction effects resulting from energy structure transition, indicating that an effective energy structure transition could achieve synergistic reductions in air pollutant.

Second, government environmental control has a positive impact on reducing air pollution. The government can effectively improve air quality by levying environmental taxes, increasing R&D subsidies and other economic measures, and strengthening environmental protection legislation and other legal measures to improve the environment (Chay & Greenstone, 2005; Correia et al., 2013). For example, because the air pollution from US manufacturing declined between 1990 to 2008 and the output of US manufacturing increased obviously, Shapiro and Walker (2018) constructed a model with heterogeneous firms and endogenous pollution abatement to explain the phenomenon. Their research showed that the environmental regulation of government plays an important role in the declining trend. Zhang et al. (2019) studied the effect of comprehensive government measures on firm's emission reduction behavior in the manufacturing sectors of China. They found that if the government environmental control and financial tightening policies were implemented together, the effect of enterprise's emission reduction would be reduced. It implied that the relaxation of financial constraints would improve the policy performance of government environmental regulation. To research the causal relationship between environmental regulation and air quality improvement, Wang et al. (2019) used DID (Difference in Difference) model by regarding the implementation of New Ambient Air Quality Standards as a policy shock. Their study results confirmed that environmental regulation would reduce air pollution in urban areas, and government should set more detailed environmental regulation measures to weaken the haze. Using *China's Key Cities for Air Pollution Control* as a quasi-natural experiment, Luo and Xiang (2024) evaluated the regulatory effects on air pollution control. The results indicate that while air quality improved in regulated areas, water pollution increased, reflecting a substitute effect. Fan et al. (2025) compared the air quality improvement role of three Chinese government environmental control policy using prefecture-level panel data from 2008–2021: *Environmental Protection Vertical Reform* (EPVR), *Environmental Protection Administrative Interviews* (EPAI), and *Central Environmental Protection Inspection* (CEPI). They found that all three policies can reduce air pollution, particularly the environmental regulations CEPI imposed by the central government.

However, scholars are quite different in the choice of government environmental control indicators. For instance, Keller and Levinson (2002) used pollution tax rates and pollution control costs as indicators to measure government environmental regulation. And then, some scholars pointed out that the pollution tax rate was affected by the level of local economic development, industrial structure and environmental protection technology. The changes in the tax rate could hardly reflect changes in the intensity of government environmental control. Therefore, some scholars used the environmental protection legislation of the local People's Congress of China as a proxy variable for government environmental regulation (Bao et al., 2013); some scholars also used the frequency and proportion of terms related to environmental governance in government work reports as substitute variables for government environmental regulation (Chen & Chen, 2018; Chen et al., 2016).

Third, the transfer of high-polluting companies to areas with lower production and environmental costs could also improve the air quality in the region (Autor & Hanson, 2013; Pierce & Schott, 2016). For example, Levinson and Taylor (2008) studied the correlation between foreign trade and environmental pollution from the perspective of international trade, and they found that in international trade, pollution-intensive companies indeed moved to regions with low labor costs and loose environmental controls. Leal et al. (2021) paid attention to the problem of FDI (foreign direct investment) and globalization, and used the autoregressive distributed lag model to solve this problem. They found that if no measures eliminated the transfer of companies from a country with strict environmental regulation to one with loose environmental regulation, FDI and globalization may damage the environment of areas with loose environmental regulation and would be inefficient in economic growth. Since globalization may trigger environmental pollution through the movement of dirty industries to developing countries, Bilgili et al. (2020) studied the relationship between environmental sustainability and globalization of Turkey by using Markov regime-switching models. They found that financial globalization, political globalization, trade globalization, human capital, and the capital stock would reduce the ecological footprint growth of Turkey. Other studies have shown that due to the spatial spillover characteristics of air pollution, there were obvious air pollution spillover effects between regions (Ma & Zhang, 2014) and even between countries (Hosseini et al., 2013). Within a certain range, air pollution overflow may aggravate the air pollution in the place where the polluting company was located before, and weaken the air pollution control effect in the place where the polluting company was located before.

From the previous literature, scholars have conducted extensive research on the reasons for the decline in air pollution, but the theoretical mechanism of air pollution control was still relatively scarce. It was worth mentioning that Andersen (2016) developed a general equilibrium model with a monopolistic competition market structure, and regarded pollution emissions as a byproduct of production in the model. Cherniwchan et al. (2017) provided a new method to decompose the changes of pollution emissions at the theoretical level. Though the works above were effective, their researches paid little attention to specific spatial variations of pollutants in theoretical aspects. Since air pollution had a cumulative effect in time and a spillover effect in space, the corresponding theoretical analysis framework needed to include both time and space dimensions. In view of this, this study introduces the concept of "iceberg cost" (Krugman, 1991), and then incorporates the spatial (regional) dimension in the theoretical analysis (An & Zhou, 2013; Ben Kheder & Zugravu, 2012). After that, to construct the theoretical model with both time and space dimensions, this study draws on the substantial work of Desmet and Hansberg (2014) and Ye and An (2017), and introduces air pollution and energy consumption variables into the dynamic general equilibrium (DGE) model that includes time and space dimensions, thus providing a new analysis method for environmental economics. Moreover, the DGE model constructed in this study is different from the research framework of Desmet and Hansberg (2014) and Ye and An (2017) in two points: (1) In the theoretical model constructed in this study, companies no longer bear the cost of leasing land, but they need to bear the cost of energy consumption to increase output; (2) The companies are divided into heavy pollution enterprises and light pollution enterprises, mainly considering the pollution emissions of the industrial sector.

20.3 Theoretical Model

20.3.1 Model Assumptions

Based on the theoretical framework of Desmet and Hansberg (2014) and Ye and An (2017), this study adds the following assumption. (1) To introduce energy consumption into the model, the enterprise's costs are changed. The costs were changed from "labor hire, land auction, and technological innovation" to "labor hire, energy consumption, and technological innovation", separately. (2) The energy here refers to traditional energy such as coal and oil. For simplicity of calculation, the energy supply of each region is unitized into 1. Assuming that in the original theoretical framework, energy is always allocated to companies that pay higher energy prices. (3) Enterprises need to consume a certain amount of energy for production. When one company has acquired energy from one region, others should look for energy supply from other regions to maintain production. Hence, there is also only one enterprise in each area, and the enterprises are divided into heavy pollution enterprises and light pollution enterprises.

20.3.2 Basic Model Framework

The spatial dimension of the model is represented by a linear interval. The linear interval [0,1] is equally divided into several intervals, each interval is represented by l, and a subinterval represents an area. The period t represents the time dimension of the model. Labor and enterprises are the main actors of the model. The total labor supply is $\bar{L}$. The labor force owns the ownership of the enterprises and determines energy distribution, and the labor force survival period is infinite. Pollution emissions are formed in the production process of enterprises, but the air pollution concentrations of each region are also determined by the selfcleaning capacity of regional air and emission reduction capacity of enterprises. Figure 20.2 reports the basic framework of the model.

The following briefly introduces the model structure and equilibrium state of the t period. The other periods are the same as the t period.

As far as labor is concerned, the utility of labor is determined by the products of the heavy and light-polluting enterprises. All the products consumed are produced in the current period; the preference relationship of the labor to the products of the heavy and light-polluting enterprises meets the CES preference. Specifically, the labor utility

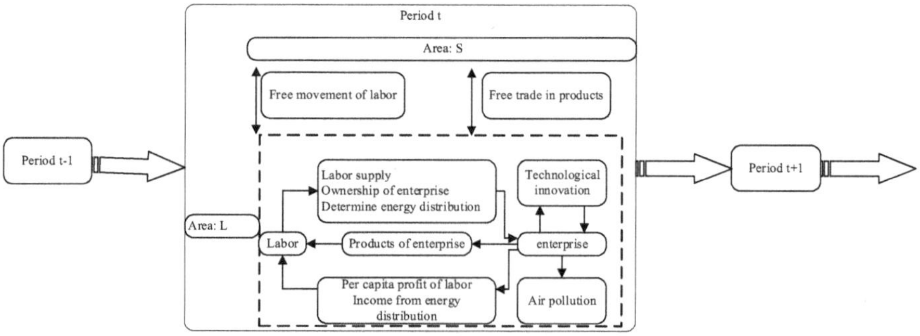

Fig. 20.2 Basic framework of the model

of area l is expressed as $U = \left\{f_h[d_h(l,t)]^{\alpha} + f_m[d_m(l,t)]^{\alpha}\right\}^{\frac{1}{\alpha}}$. In the formula, $d_i(l,t)$ represents the consumption of labor for the products of type i enterprise, $i = h, m$, where h means heavy pollution enterprise, and m means light pollution enterprise. f_h and f_m respectively represent the preference of the labor force for the products of heavy and light pollution enterprises, $f_h > 0, f_m > 0$. α is a substitute parameter, $\alpha \leq 1$ and $\alpha \neq 0$. The utility maximization of labor force in area l can be expressed as:

$$\max_{\{d_i(l,t)\}} U \tag{20.1}$$

$$\text{s.t. } p_h(l,t)d_h(l,t) + p_m(l,t)d_m(l,t) \leq w(l,t) + \sigma(t)/\bar{L} + \pi(t)/\bar{L} \tag{20.2}$$

Here, (2) is the budget constraint of labor, and the intertemporal savings and borrowing behavior of labor is not considered. $p_i(l,t)$ represents the products price of type i enterprise, $i = h - m$. $w(l,t)$ is the labor wage, with $\pi(t)/\bar{L}$ and $\sigma(t)/\bar{L}$ representing the per capita profit of labor and the income obtained by labor from energy distribution. The equilibrium utility of labor is solved by Lagrangian multiplier method.

As far as the enterprise is concerned, the enterprise needs to determine the number of labor force employed, technological innovation and energy consumption in the current period. In the production process of a company, air pollution emissions are unexpected output of the company. Air pollution has a negative impact on the environment, economy and residents' health. In general, the output of enterprise in area l is $Y_i(l,t) = (1+\theta)L_i(l,t)^{\epsilon_i}[I_i^{+}(l,t)]^{1-\epsilon_i}$, where, $L_i(l,t)$ represents the labor force employed by the enterprise, $I_i^{+}(l,t)$ means the enterprise's technological innovation level, ϵ_i is the contribution of labor to output, $0 < \epsilon_i < 1$, and θ refers to the contribution of energy consumption to the company's output. What needs to be explained is that enterprise technology has the characteristics of technology diffusion between regions, and that enterprises face a market structure of perfect competition. The profit maximization of an enterprise can be expressed as:

$$\max\left\{p_i(l,t)Y_i(l,t) - w(l,t)L_i(l,t) - \psi_i - \sigma_i(l,t)\right\} \tag{20.3}$$

where, $w(l,t)L_i(l,t)$ represents labor employment cost, $\sigma_i(l,t)$ is energy consumption cost, and ψ_i is innovation cost. The Lagrangian multiplier method can be used to obtain the equilibrium solution of $L_i(l,t)$ and ψ_i when the enterprise profit is maximized.

Since technology has the characteristics of diffusion, inter-regional products have free trade, enterprises will face a perfectly competitive market structure. Meanwhile, enterprises are the receivers of product prices, and the profit of the firm equilibrium state $\pi(t) = 0$. Therefore, substituting the equilibrium solutions $L_i^{*}(l,t)$ and ψ_i^{*} into Eq. (20.3), the energy consumption cost paid by the enterprise is obtained:

$$\sigma_i(l, t) = p_i(l, t)Y_i^*(l, t) - w(l, t)L_i^*(l, t) - \psi_i^* \tag{20.4}$$

It should be noted that since each region only has 1 unit of energy, the energy cost borne by the enterprise is equal to the energy price. Therefore, the equilibrium energy price is $\sigma^*(l, t) = \max\sigma_h(l, t), \sigma_m(l, t)$. The result of energy distribution in area l among enterprises is expressed as:

$$\theta(l, t) = \begin{cases} 1, \ \sigma_h(l, t) \geq \sigma_m(l, t) \\ 0, \ \sigma_h(l, t) < \sigma_m(l, t) \end{cases} \tag{20.5}$$

Because there is only one company in each region, the energy distribution results also indicate the regional distribution of companies. $\theta(l, t) = 1$ or 0 mean that the heavy-polluting enterprises or light-polluting enterprises located in area l, separately.

The expression of air pollution concentration of area l is:

$$\text{poll}(l,t) = A_1\text{poll}(l, t-1) + \frac{A_2Y_h^*(l,t)}{\left(I_h^+(l,t)\right)^{\eta_h}} + \frac{A_3Y_m^*(l,t)}{\left(I_m^+(l,t)\right)^{\eta_m}} \tag{20.6}$$

where, $A_1\text{poll}(l, t-1)$ represents the pollutant residue in t−1 period, and A_1 is the natural air purification capacity; $\frac{A_2Y_h^*(l,t)}{(I_h^+(l,t))^{\eta_h}}$ is the pollution discharge of heavily polluting enterprises in the t period; $\frac{A_3Y_m^*(l,t)}{(I_m^+(l,t))^{\eta_m}}$ is the pollution discharge of light-polluting enterprises. η_h and η_m respectively represent the inhibitory effects of technological innovation of heavy-polluting enterprises and light-polluting enterprises on pollution emissions; the higher the technical level, the lower the pollution emissions, $\eta_h, \eta_m \in (0, 1)$. A_2 and A_3 separately represent the impact of the heavy and light polluting enterprises on local environmental quality during period t, $A_2, A_3 > 0$.

In addition, in areas with severe air pollution, the morbidity rate of children and the elderly is higher. Adult labor is forced to spend more time to take care of the elderly and children (Aragon et al., 2017). For the increased hospitalization costs and time costs caused by air pollution, the real wage of labor in heavily polluted areas will be decreased. Therefore, heavily polluting companies need to pay higher wages to retain the workers. One of our investigations found that, for air pollution, the Yangzi Petrochemical Company in Nanjing, China had to pay its workers higher wages than nearby light pollution enterprises. It is specifically manifested as the difference in wages between heavy-polluting companies and light-polluting companies, namely $(1-A_4)w_h(l, t) = w_m(l, t) = w(l, t)$. A_4 represents the negative impact of air pollution on labor, $0 < A_4 < 1$.

The inter-regional links are manifested in the free flow of labor and free trade in products. Moreover, the free flow of labor and free trade in products also make the model reach an equilibrium state. The free flow of labor between regions achieves labor market equilibrium, and the free trade of products between regions achieves product market equilibrium. In equilibrium, the equilibrium condition of the labor market is expressed as $\int_0^1 L^*(l,t)dl = \overline{L}$, where,$L^*(l,t)$ represents the equilibrium labor market demand of area l, and the expression is $L^*(l,t) = \Theta(l,t)L_h^*(l,t) + (1-\Theta(l,t))L_m^*(l,t)$.

The product market equilibrium conditions of each region are the same. Taking region l as an example, the product market equilibrium conditions in region l are:

$$\left[Y(l,t) - \frac{\psi_i{}^*}{p_i(l,t)}\right] + \left[Q_i(l-\Delta l,t) - \kappa|Q_i(l-\Delta l,t)|\right] = d_i(l,t)L(l,t) \qquad (20.7)$$

In the formula, the right side is the product market demand, and the left side is the product market supply.$Q_i(l-\Delta l,t)$ represents the excess supply in the adjacent area,$\kappa|Q_i(l-\Delta l,t)|$ represents the iceberg cost (Krugman, 1991) in the process of transporting products from the adjacent area to area l, such as transportation cost, reasonable loss etc.$\left[Y(l,t) - \frac{\psi_i{}^*}{p_i(l,t)}\right]$ represents the product supply of area l itself, in which innovation costs are deducted, the total output $Y(l,t)$ is equal to $\theta(l,t)Y_h(l,t) + (1-\Theta(l,t))Y_m(l,t)$.

In addition, although there is free trade between regions, the prices of products from different production locations in a certain region are different. Specifically, assuming that the product price of a light-polluting company is valuable, $p_m(l,t) = 1$; the inter-regional transaction price of a heavy-polluting company's product meets the iceberg cost (Krugman, 1991) principle, and the expression is $p_f(s,t) = e^{\kappa|l-s|}p_f(l,t)$.$\kappa$ is a parameter for measuring the cost of the iceberg, $\kappa > 0$.$|l-s|$ represents the distance between area l and area s, and the price of products of heavily polluting enterprises in area s increases as the distance between the two areas increases.

20.3.3 Parameter Setting and Model Solving

This study has carried out the following parameter calibration: (1) According to Chen and Chen (2016), the contribution of China's industrial coal to output is between 0.229 and 0.313, and the value of θ is set to the average value of this research: 0.271. (2) Referring to Ye and An (2017), the annual average value of China's environmental pollution control investment in GDP accounted for about 1.32% (2001–2017), and science and education expenditures accounted for about 4.5% of GDP in 2017; the ratio of the two is adopted as the positive externality of light-polluting enterprise technology, with η_m set to 0.3. The positive externality of technology of heavily polluting enterprises should be less than the positive externality of technology of polluting enterprises (including heavy and light pollution enterprises) by 0.2, with the value of η_h set to 0.15. (3) According to Ye and An (2017), the coefficient of impact of polluting enterprise unit output on air quality is 0.0052. Taking into account the difference between heavy polluting enterprises and light polluting enterprises,

Table 20.1 Parameter setting

Parameter	Calibration value	Implication
θ	0.271	Contribution of energy consumption to economic output
f_h	0.6	Contribution of heavy-polluting enterprise products to labor utility
f_m	1.4	Contribution of light-polluting enterprise products to labor utility
η_h	0.15	Positive externality of heavy-polluting enterprise technology to environment quality
η_m	0.3	Positive externality of light-polluting enterprise technology to environment quality
A_2	0.0058	Negative externality of heavy-polluting enterprise to environment quality
A_3	0.0046	Negative externality of light-polluting enterprise to environment quality
A_4	0.08	Negative impact of air pollution on the health of labor force

Sources Collated according to the literatures by author

the value of A_2 is set to 0.0058, and A_3 to 0.0046. (4) Yang and Xu (2020) believe that a 1% increase in air pollution will lead to a 0.0842% increase in labor wage distortion, with A_4 set to 0.08 in this study. The parameter settings in this study are shown in Table 20.1. Other parameters refer to Desmet and Rossi (2014) and Ye and An (2017) settings.

In the process of solving the model, the linear interval [0,1] is equally divided into 500 sub-regions. [1,200] is the south, [201,300] is the middle, and [301,500] is the north; the labor supply is 100, and the number of simulation periods is 110. Matlab software is used to solve the model.

20.4 Scenario Simulation

20.4.1 Scenario 1 no Air Pollution Control Measures

The first is about the simulation of no air pollution control measures. Observing Fig. 20.3, it can find that the air pollution concentration gradually rises over time; the air pollution in the northern region is the most serious, and the air pollution in the southern region is lower than that in the northern region, which is in line with the actual situation in China. This shows that when controlling air pollution, the government should first strengthen air pollution control in the northern region. The above results are also valid in the set2 scenario of adjusting the negative externality parameters of enterprise production to the environment. It is indicated that the simulation results have certain robustness. Figure 20.4 reports the simulated air quality, equilibrium utility (reflecting the economic welfare status), and average changes in labor

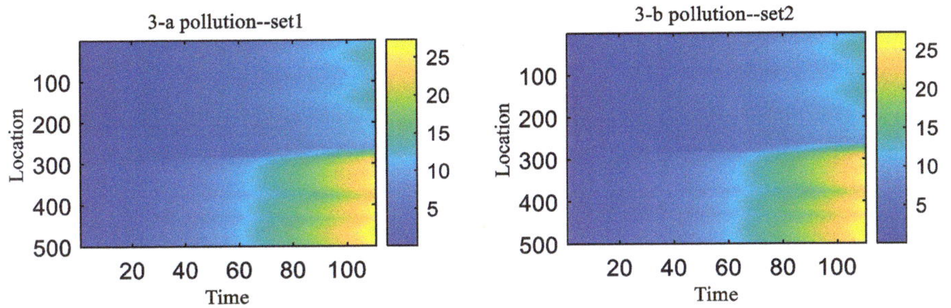

Fig. 20.3 Regional dynamic change of air pollution concentration. $F_2 = 0.0058$, $F_3 = 0.0046 F_2 = 0.006$, $F_3 = 0.0044$ (*Source* drawn by the author based on the Matlab simulation results). *Notice* (1) Time in the figure is periods and increases from left to right. Location in the figure is areas and respectively reflects north, central and south from bottom to top; (2) The color from blue to yellow indicates that the concentration of air pollution is increasing and air quality is declining; (3) parameters of set1 scenario are and parameters of set2 scenario are

wages under the set1 and set2 scenarios. As shown in Fig. 20.4a, the concentration of air pollution has gradually increased over time, and the overall environmental quality of the economy has deteriorated. As shown in Fig. 20.4b, as air pollution increases, the equilibrium utility gradually increases. The reason is that technological innovation and technological diffusion have increased economic output, accumulated national wealth, and improved equilibrium utility. It also shows that the increase in equilibrium utility is positively correlated with the increase in air pollution. As shown in Fig. 20.4c, with technological innovation and economic growth, the income of enterprises increases, and the labor wage paid by enterprises gradually rises over time; heavy pollution companies pay higher wages to compensate for the negative effects of air pollution on labor.

In general, during the simulation process, the overall level of economic welfare gradually improved; but the price of economic development was the increase in air pollution, the decline in air quality, and the damage to the health of residents. How can we curb the growth of air pollution? Next, this study discusses the dynamic impact of adjusting energy structure and government environmental regulation on air pollution.

20.4.2 Scenario Simulation 2 Energy Structure Adjustment

It is supposed that the government issues an administrative order to reduce consumption of traditional energy and increase use of clean energy. The purpose of the policy is to adjust the energy structure and control air pollution. Here, this study lowers the contribution value θ of energy consumption to output in the output function of the enterprise (θ is 0.271, 0.2, 0, -0.2 in turn) to reflect the gradual increase in the intensity of restricting energy consumption. To reflect the effect of developing clean

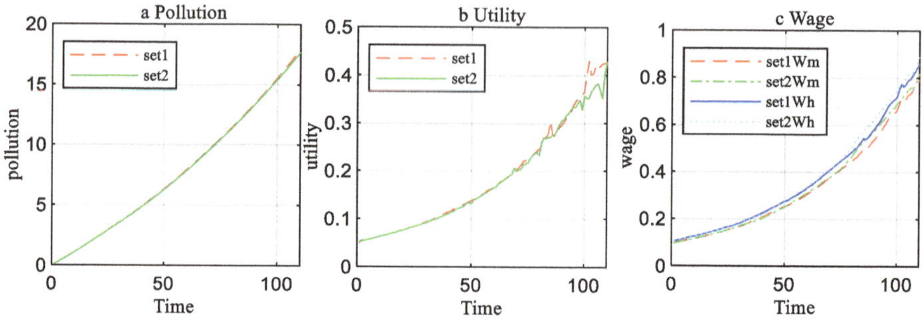

Fig. 20.4 The change of air quality, labor utility and labor wage over time. (*Source* drawn by the author based on the Matlab simulation results). *Notice* (1) set1 and set2 in the figure are same as Fig. 20.3; (2) 4a reports the change of regional mean air pollution concentration, the higher the value of pollution in the figure, the more serious the air pollution; (3) 4b reports the change of labor utility, the higher the value of utility in the figure, the higher the equilibrium utility; (4) 4c reports the change of labor wage, the higher the value of wage in the figure, the higher the labor income. Wm is the labor wage paid by light-polluting enterprise and Wh is the labor wage paid by heavy-polluting enterprise

energy, the base values of η_h, η_m (increase by 1%, 5%, 10% in turn) are adjusted. Then, this study analyzes the corresponding air pollution, equilibrium utility and economic output status, see Fig. 20.5.

As shown in Fig. 20.5a, as energy consumption restrictions are strengthened, the increase in air pollution has been suppressed. Comparing the reduction rate of air pollution with different values of θ, it is found that the greater the restriction on energy consumption, the more obvious the restraining effect on air pollution. It is indicated that adjusting the energy structure does help to improve economic air quality. As shown in Fig. 20.5b, the adjustment of the energy structure has no significant effect on the equilibrium utility. The reason may be that the labor force tends to pay more attention to factors such as wage levels and air quality, and does not respond significantly to the adjustment of the energy structure. As shown in Fig. 20.5c, the adjustment of the energy structure has significantly reduced the equilibrium output of the economy. As the intensity of adjustment increases, the greater the decrease in the equilibrium output. The possible reason is that it is difficult for companies to develop alternative energy sources and promote technological innovation in a short period of time. Therefore, as restrictions on energy consumption increase, the economic output of enterprises declines and the economic growth rate decreases. The key to restoring economic output is to encourage companies to carry out "green" technological innovation and develop alternative energy sources.

It should be noted that related studies have shown that through administrative orders to shut down polluting enterprises, and control of air pollution has become an "image project" for Chinese governments at all levels during the politically sensitive period. It has poor persistence and is prone to "retaliatory pollution" behavior afterwards (Shi et al., 2016). Supposing that in the 110 simulation periods, θ takes -0.2 (η_h, η_m takes 0.165, 0.33) in the first 60 periods and 0.271 (η_h, η_m takes 0.15, 0.3)

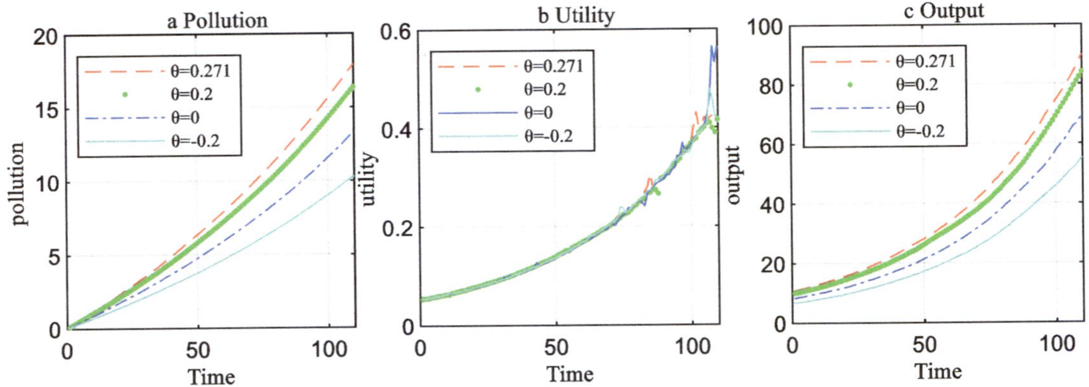

Fig. 20.5 The change of air quality, labor utility and output under energy structure adjustment Scenario. $\theta\theta = 0.3\theta$ (*Source* drawn by the author based on the Matlab simulation results). *Notice* (1) in the figure reflects the contribution of energy consumption to economic output, means no restrictions on corporate energy consumption. A lower value means that the firm uses less coal energy, and a lower energy consumption contribution to the economic output; (2) 5c reports the change of the regional mean economic output, the higher the value of output in the figure, the higher the regional mean economic output

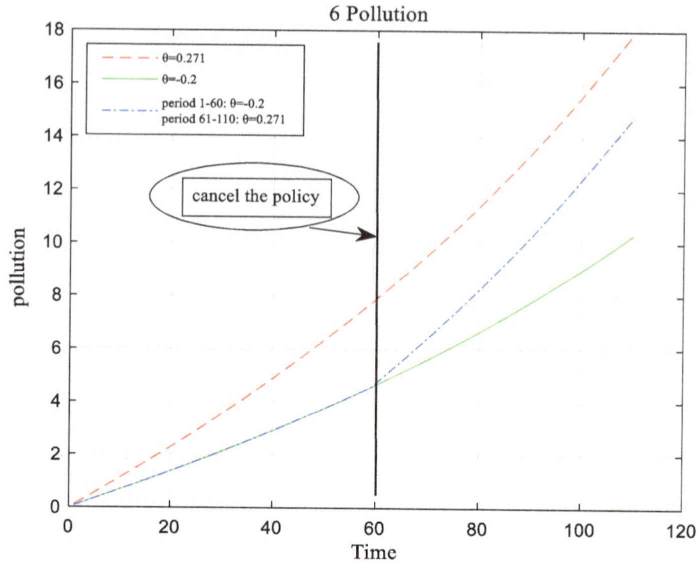

Fig. 20.6 Measuring the post-event "retaliatory pollution" behavior $\theta = -0.2, \eta_{\mathrm{h}} = 0.165, \eta_{\mathrm{m}} = 0.33 \theta = 0.271, \eta_{\mathrm{h}} = 0.15, \eta_{\mathrm{m}} = 0.3$ (*Source* drawn by the author based on the Matlab simulation results). *Notice* "cancel the policy" in the figure means that the government stops the policy of restricting the energy consumption of regions and developing clean energy. In period 1 to 60, means the government restricts energy consumption of regions with the maximum intensity, and encourages the development of clean energy with the maximum intensity. And in the period 61 to 110 reflects the government has no restriction on energy consumption of regions and has no encouragement on developing clean energy

in the next 50 periods to test whether this post-event "retaliatory pollution" behavior exists. Looking at Fig. 20.6, when the government abolished the policy of restricting energy consumption, companies did have "retaliatory pollution" behavior, and the level of air pollution increased rapidly. This shows that adjusting the energy structure by using administrative order for reducing air pollution is not sustainable. Adjusting the energy structure should pay attention both to improving energy efficiency and increasing the proportion of alternative energy and clean energy consumption. So how to maintain stable economic growth while controlling pollution, and realize continuous air pollution control? Next, this study will explore the impact of government environmental regulations on air pollution.

20.4.3 Scenario 3 Government Environmental Regulation

It is supposed that the government implements environmental control on pollution discharge. Here, environmental control mainly refers to economic or legal means such as collecting environmental taxes and strengthening environmental legislation. In formula (20.6), introduce environmental control $(1-(x)^{\upsilon})$, where x represents the government environmental control intensity, $0 < x < 0.5$. A value of x less than 0.5 indicates that the government is carrying out environmental control with low intensity. The parameter ν represents the returning elasticity of pollution investment, $0 < \nu < 1$. The formula (20.6) is changed to $\mathrm{poll}(l,t) = A_1\mathrm{poll}(l,t-1)+(\frac{A_2 Y_h^*(l,t)}{(I_h^+(l,t))^{\eta_h}}+\frac{A_3 Y_m^*(l,t)}{(I_m^+(l,t))^{\eta_m}})(1-(x)^{\upsilon})$. This formula reflects the air pollutants remaining in the area after natural purification and government control.

Two situations are considered here: One is the government's unified environmental control over each region; the second is that the government controls the northern regions where heavily polluting companies are concentrated, leaving other regions free. Then, the dynamic change of air pollution, equilibrium utility, and economic output in the two situations are analyzed. Figure 20.7 shows the situation of unified environmental control. As shown in Fig. 20.7a, as the intensity of environmental control increases, air quality gradually improves, and the upward trend of air pollution slows down. Comparing the decline of air pollution when x takes different values, it is found that the inhibitory effect of environmental control on air pollution is decreasing. Unlike the situation of adjusting the energy structure, as shown in Fig. 20.7b and c, while reducing air pollution, government environmental control did not result in a significant decrease in equilibrium utility and economic output. This is different from the situation in which the adjustment of the energy structure leads to a decline in economic output. It is indicated that government environmental regulation is a "win–win" measure to promote economic growth and environmental improvement.

Although unified environmental control can promote "green" growth, is this environmental improvement caused by the transfer of polluting companies? Due to the spatial spillover effect of air pollution, if the unified environmental control only results in the transfer of polluting companies to "polluted paradise", then the air pollution in the places where polluting companies are relocated may still reduce the environmental quality of the places where they are transferred. Moreover, the initial

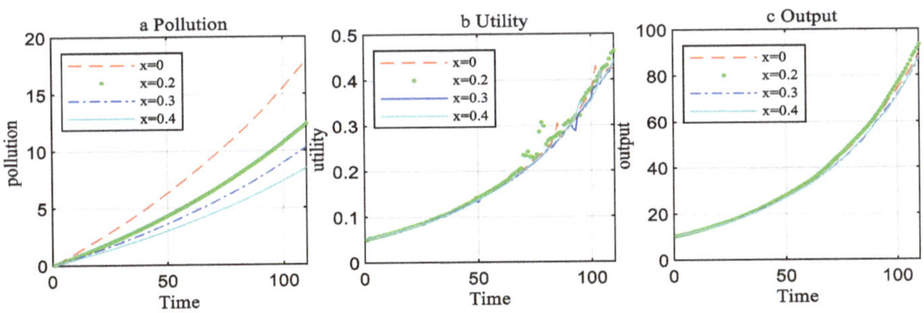

Fig. 20.7 The change of air pollution, labor utility and output under unified environmental control Scenario xxx = 0 (*Source* drawn by the author based on the Matlab simulation results). *Notice* in the figure means environmental control intensity, the higher the value, the higher the environmental control intensity, means no environmental control

goal of government environmental control is not to transfer air pollution between regions within the economy, but to improve the overall environmental quality. Therefore, the transfer of polluting companies may lead to unsustainable air pollution control effects.

There is a question to discuss here. It is supposed that the government implements environmental control in the heavily polluted northern region, and leaves it to other regions. Figure 20.8 reports the changes in the regional distribution of air pollution under differentiated government environmental controls. As shown in Figs. 20.8a, b and c, from a numerical point of view, as the intensity of environmental control in the northern region increased from 0.2 to 0.4, the overall environmental quality has improved; however, from the perspective of regional distribution, the air pollution gap between the north and south has gradually narrowed. When the intensity of environmental control in the northern region is 0.4, the degree of air pollution in some southern regions has exceeded that in the north. The reason is that due to the higher intensity of pollution control in the northern region, some heavily polluting companies in the northern region will move to the south where environmental control is weaker, resulting in the deterioration of the environment in the southern region, and the southern region has become a "polluted paradise".

As shown in Figs. 20.8d, e and f, from a numerical point of view, unified environmental control has improved the overall air quality; from the perspective of regional distribution, unified environmental control has not significantly changed the distribution pattern of air pollution in the north. There has been no obvious shift of polluting companies. This shows that the unified government environmental control does not encourage the transfer of polluting companies, but has improved the company's "green" production technology through environmental protection legislation and increased R&D investment and strengthened the company's environmental awareness. It has effectively suppressed air pollution and has good sustainability. In general, the coordinated government environmental control between regions can continuously and effectively improve air quality, and has a less negative impact on economic growth.

20.4.4 Scenario 4 Policy Path

The previous simulation analysis has shown that government environmental regulation could achieve a "win–win" of economy and environment. However, the government environmental regulation discussed earlier is static during the simulation period. As the number of periods increases, the intensity of environmental control is fixed at a certain value, which is contrary to reality. With economic development, the government's emphasis on environmental issues is gradually increasing (Chen & Chen, 2018). In the early stages of economic development, the government tends to pay more attention to economic growth and ignore environmental protection. Therefore, for this part, it is assumed that the government's environmental control intensity is gradually increased according to three paths (fixed proportional growth, first fast and

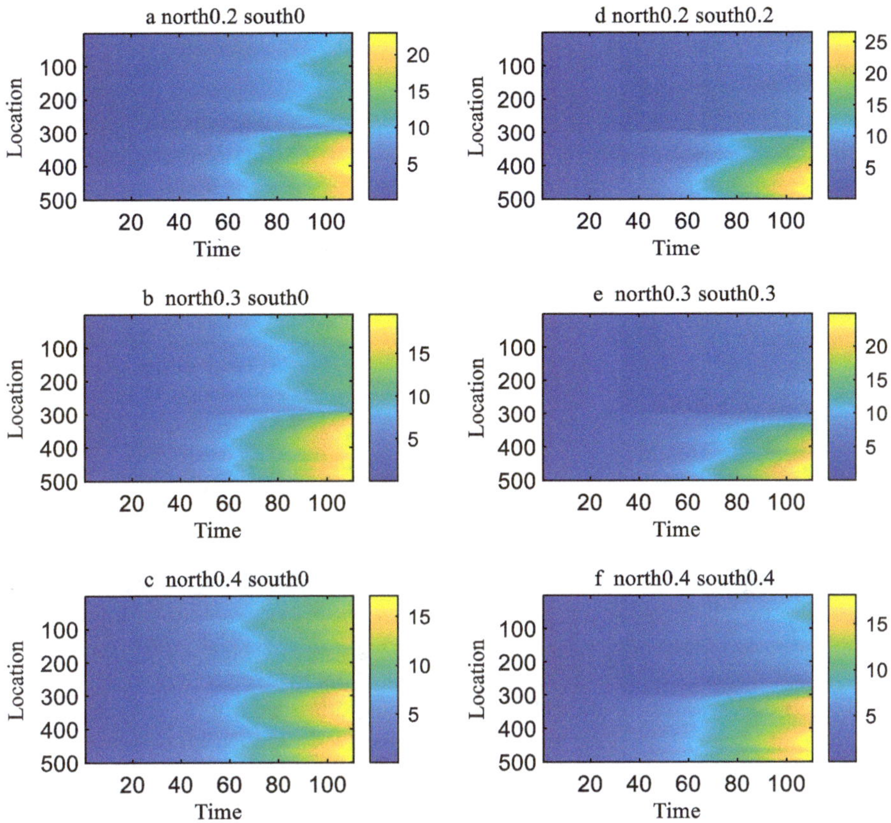

Fig. 20.8 The change of air quality under unified and differentiated government environmental controls Scenario (*Source* drawn by the author based on the Matlab simulation results). *Notice* (1) In the Fig. 20.9a, north 0.2 means that the environmental control intensity in the heavily polluted northern region is 0.2, and south 0 means no environmental control in the southern region; (2) the same value of north and south reflects the unified environmental control over regions, the different value of north and south reflects the differentiated environmental control over regions

then slow, first slow and then fast). Then, this study analyzes the corresponding air pollution and equilibrium utility status.

Figure 20.9 reports the simulation results, and Fig. 20.9a is a schematic diagram of the three policy paths. As shown in Fig. 20.9b, the policy path of fast first and slower is the optimal policy path choice, which has the best suppression effect on air pollution. It is better than static environmental control and adjustment of energy structure. Figure 20.9b also showed that policy path of "fast first and slow latter" could make air pollution concentration reached the inflection point faster than policy path of "fixed ratio" and "slow first and fast latter". Comparing with reality and observing the results of policy path of "fast first and slow latter" in Fig. 20.9b, this study found China located at the right of the inflection point and air pollution level was declining. The possible reasons are that, first of all, the early increase in the intensity

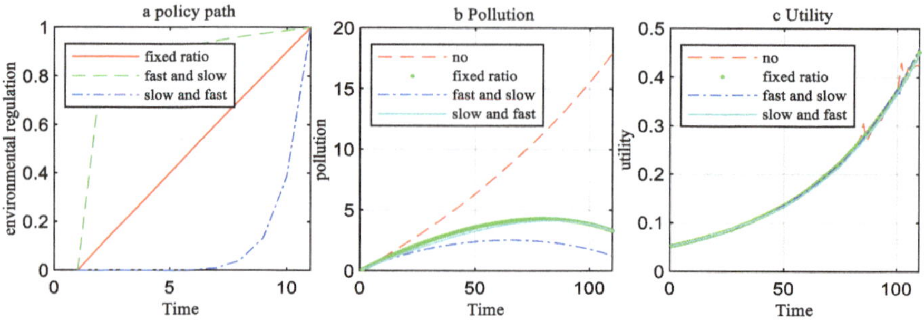

Fig. 20.9 The change of air quality and labor utility under different policy paths (*Source* drawn by the author based on the Matlab simulation results). *Notice* (1) Fig. 20.9a reports three policy paths that environmental control intensity respectively increases by fixed ratio, fast and slow, slow and fast speed; (2) In the Fig. 20.9b and c, "no" means no environmental control, "fixed ratio" means environmental control intensity increase by fixed ratio speed, "fast and slow" means environmental control intensity increase by first fast and then slow speed, "slow and fast" means environmental control intensity increase by first slow and then fast speed

of environmental control will signal to companies that the government pays attention to environmental issues. After that, companies expect that over time, the intensity of government environmental control will be higher. Therefore, some companies will quickly adjust production decisions, promote "green" technological innovation, strengthen pollution end treatment, and reduce pollution emissions. Another part of the companies may withdraw from the local market to reduce air pollution in the region. Eventually, with the decline in corporate pollution emissions, the overall air quality in the region has improved, and the rate of increase in environmental control intensity has gradually decreased. Therefore, the policy path of fast first and slower should be the optimal policy choice in the three policy paths. The conclusion in Fig. 20.9c is consistent with that in Fig. 20.7b. Environmental regulation has not a significant negative impact on the equilibrium utility in the process of reducing air pollution.

Based on the above analysis, this article proposes the following hypotheses: Adopting appropriate methods to adjust the energy structure and reasonable government environmental control can continuously and effectively alleviate air pollution, improve environmental quality, and promote green economic growth.

20.5 Empirical Analysis

20.5.1 Model Setting

In order to test the rationality of the simulation results, the empirical part constructed a regression model with the average urban air pollution concentration as the explained variable, the government environmental regulation and energy structure as the main explanatory variables:

$$\ln Poll_{it} = \alpha + \beta_1 \ln Gov_{it} + \beta_2 \ln Str_{it} + X_{it}{\prime}\phi + c_i + y_t + \epsilon_{it} \quad (20.8)$$

where, the subscript i represents the city, and t represents the year; $Poll_{it}$ represents the average concentration of air pollution in city i in year t; Gov_{it} represents the government environmental regulation faced by city i in year t; the larger the value of Gov_{it}, the stronger the government environmental regulation. It is predicted that β_1 is negative, which means that increasing environmental control will reduce air pollution.Str_{it} represents the energy structure of city i in year t. It is predicted that β_2 is positive, indicating that the deterioration of the urban energy structure will aggravate air pollution.$X_{it}{\prime}$ represents the matrix of control variables, including population, urban GDP, road freight volume and annual precipitation. ϕ represents the coefficient vector of the control variables.c_i represents the fixed effect of the city, y_t represents the fixed effect of the year, and ϵ_{it} is the error term.

20.5.2 Data Source and Description

This study selects the panel data of 286 cities at prefecture-level and above in China from 2010 to 2018 as sample data. The index mainly used to explain the average concentration of variable air pollution is $PM_{2.5}$. Among them, the data for 2010–2016 comes from SEDAC,[1] and the data of 2017–2018 comes from *the China City Statistical Yearbook*; control variables include urban population, urban GDP, urban road freight volume, and the annual precipitation of cities. The first three data mainly come from *the China City Statistical Yearbook*. The data of annual precipitation of cities comes from NOAA.[2]

Regarding government environmental control indicators, drawing on the research of Chen and Chen (2018), this study multiplied the number or proportion of words related to pollution control in the provincial government work report with the proportion of the output value of the secondary industry of the prefecture-level city. Then, the factor above would be an alternative variable for the government environmental control indicators of prefecture-level cities. The process is as follows: (1) the 2010–2018 provincial government work reports are manually collected; statistics and calculation are made of the number of words related to pollution control in the work reports of each province, the number of words and the proportion of words in the total words. The vocabulary related to pollution control includes environmental protection, environmental protection, pollution, energy consumption, emission reduction, pollution discharge, ecology, green, low carbon, air, and environment. (2) The number of

[1] Sources: NASA Socioeconomic Data and Applications Center. Address: https://doi.org/10.7927/H4ZK5DQS.

[2] Sources: NOAA (National Oceanic and Atmospheric Administration). Address: https://www.noaa.gov/.

pollution-related words, the number of words and the proportion of words in the total number of words in the provincial government work report are multiplied by the proportion of the output value of the secondary industry in the prefecture-level city. (3) Eventually, this study obtained three proxy variables of government environmental control. The three variables are environmental control variable 1 (number of words), environmental control variable 2 (number of words in vocabulary), and environmental control variable 3 (proportion of words in the number of words), as shown in Table 20.2.

The urban energy structure is expressed by the ratio of urban coal energy consumption to urban GDP. However, China's *City Statistical Yearbook* did not publish data on coal energy consumption in various cities. The National Bureau of Statistics

Table 20.2 Statistical description of the main variables in 2010–2018

Variables	Units	Mean	Minimum	Maximum	Standard deviation
$PM_{2.5}$	ug/m3	37.7295	4.6764	116	16.2174
Environmental control variable 1 (ECV-1)	number of words	36.5374	5.1488	82.9361	13.023
Urban energy structure (UES)	10 thousand tons/10 million yuan	0.0972	0.0011	0.9901	0.0972
Population	10 thousand people	447.7129	19.5	3397	318.0665
GDP	Hundred million	1942.905	29.3909	26,914.56	2745.111
Per capita GDP (P-GDP)	10 thousand yuan	4.5878	0.5304	539.9914	4.5878
Road freight volume (RFV)	10 million tons	11.581	0	554.203	17.8809
Annual precipitation of cities (APC)	Millimeter	1032.93	59.42954	2722.844	541.7612
Environmental control variable 2 (ECV-2)	Number of words in vocabulary	75.3906	10.2976	170.937	26.9704
Environmental control variable 3 (ECV-3)	Proportion of words in the number of words	0.4331	0.0708	1.0915	0.1399

Notice (1) Urban energy structure (UES) is the ratio of urban coal consumption to the urban GDP. The ratio only included coal consumption, but not included other energy consumption. The ratio reflects the contribution of coal consumption to the urban economy and reflects the urban energy structure in some extent. The higher ratio means that the greater contribution of coal consumption to the urban economy and the higher proportion of coal consumption in urban energy consumption. (2) GDP and per capita GDP (P-GDP) are deflated using the price deflator only published data on coal energy consumption at the provincial level. According to the research of Ma and Zhang (2014), the coal consumption of some high coal consumption industries in China's secondary industry accounts for more than 90% of the national coal consumption. This shows that the secondary industry is China's main coal energy consumption industry. Therefore, in order to construct urban coal energy consumption indicators, this study uses the ratio of the output of the secondary industry in the city to the total output of the province's secondary industry as the weight to allocate the provincial coal energy consumption to each city. The process is as follows: (1) Calculating the output of the secondary industry in each city and the proportion of the output of the secondary industry in each city in the total output of the secondary industry in the corresponding province; (2) Using this proportion as the weight to calculate the amount of the total coal energy consumption of the corresponding province that should be shared by each city; (3) Dividing the coal energy consumption shared by each city by the city's GDP output value, and using this ratio as a proxy variable for the city's energy structure. This index indicated that the greater the value of urban energy structure, the worse the urban energy structure.

The statistical description of the main variables is shown in Table 20.2. By comparing air pollution data from the north and south regions, it is found that $PM_{2.5}$ concentrations in the north were on average 9.62 higher than those in the south. This is consistent with the results of theoretical model simulation.

20.5.3 Benchmark Regression Results

Table 20.3 reports the benchmark regression results. In the case of controlling the fixed effects of cities and years, the estimated coefficient in column (1) shows that the increase in the intensity of environmental control significantly reduces urban air pollution, and for every 1% increase in the intensity of environmental control, urban air pollution decreases by an average of 0.0766%. The estimated coefficient in column (2) indicates that the higher pollution is usually related to the worse energy structure. For every 1% increase in the contribution of coal consumption to output in the energy structure, urban air pollution will increase by 0.254% on average. This is basically in line with expectations. In the regression of column (3), environmental regulation and energy structure variables are added, and the results are similar to the estimates in columns (1) and (2). Taking into account the estimation bias that may be caused by the omitted variables, the control variables are added to the regression of column (4). It is found that the restraining effect of environmental regulation on air pollution and the negative impact of the worse energy structure on environmental quality still exist. Considering the possible lag effect of policy implementation, the regression of column (5) lags the environmental control variables by one period. The results show that environmental control still significantly reduces urban air pollution. For every 1% increase in the intensity of environmental control, urban air pollution decreases by an average of 0.0814%, showing that the lag effect of environmental control has a more obvious improvement effect on urban air quality. In order to reflect the interaction of government environmental control and urban energy structure, the regression of column (6) added the cross item of government environmental control and urban energy structure. It implied that if the government could promote the energy structure adjustment measure and environmental control measures simultaneously, the effect of environmental control on urban air quality will be further strengthened. Due to the added lag term in column (5), the column (7) used the first difference item of Lag. ECV-1 as the IV and adopted the 2SLS method to estimate. And the regression result is similar to the bench analysis.

20.5.4 Heterogeneity Analysis

For the different city sizes and development periods, there may be some differences in air pollution control. Therefore, in accordance with the "Notice on Adjusting the Standards for City Size Classification" issued by the State Council in 2014, the sample is divided into small and medium cities, large cities, mega cities and different development periods, taking into account the heterogeneity of environmental regulation and energy structure to air pollution influences.

Table 20.4 reports the regression results of the heterogeneity analysis. Observing the regression results of the first three columns, it is found that environmental control has significantly reduced air pollution in cities of different sizes. Environmental control has the most obvious inhibitory effect on air pollution in big cities. For every 1% increase in environmental control, air pollution in big cities will drop by 0.104% on average. However, due to the more complex industrial structure, the control of air pollution in megacities is more difficult, and the effect of environmental control on air pollution is slightly less. The regression results of energy structure variables show that the worse energy structure has a certain negative impact on the air quality of cities of different sizes. Observing the environmental control regression results in columns 4–5, it is found that the effect of environmental control on air pollution in 2016–2018 is better than that in 2010–2015, which is consistent with China's recent strengthening of environmental protection and the emphasis on the construction of ecological civilization. The energy structure regression results in columns 4–5 showed that the negative impact of worse energy structure on air quality in 2016–2018 is less than that in 2010–2015. It indicates that energy structure has a slight improvement of China in recent years.

In short, Table 20.4 reports the regression results of the heterogeneity analysis. Observing the regression results of the first three columns, it is found that environmental control has significantly reduced air pollution in cities with different sizes. Strengthening corporate technological innovation, actively researching and developing new and renewable energy sources are of great significance to improve China's energy structure adjustment and control air pollution of China.

Table 20.3 Benchmark regression results

Variables	$PM_{2.5}$						
	(1)	(2)	(3)	(4)	(5)	(6)	(7)
ECV-1	–0.0766***		–0.0781***	–0.0782***		–0.210***	
	(0.0166)		(0.0160)	(0.0158)		(0.0635)	
UES		0.254***	0.255***	0.244***	0.231***	0.407***	0.200***
		(0.0338)	(0.0340)	(0.0361)	(0.0382)	(0.0826)	(0.0270)
Lag. ECV-1					–0.0814***		–0.139***
					(0.0175)		(0.0431)
ECV-1 × UES						–0.0470**	
						(0.0206)	
Control variables	No	No	No	Yes	Yes	Yes	Yes
FE of the city	Yes	Yes	Yes	Yes	Yes	Yes	Yes
FE of the year	Yes	Yes	Yes	Yes	Yes	Yes	Yes
Constant	3.496***	3.754***	3.999***	3.914***	3.858***	4.372***	4.003***
	(0.0706)	(0.0856)	(0.103)	(0.128)	(0.121)	(0.253)	(0.162)
Observations	2,527	2,527	2,527	2,527	2,246	2,527	1,966
Adjusted R^2	0.848	0.855	0.856	0.857	0.845	0.858	0.834
Craggdonald Wald F statistic							594.269

Notice (1) robust standard error in parentheses; (2) ***, **, * respectively express significant at the significance level of 1%, 5% and 10%

Table 20.4 Heterogeneity analysis

Variables	(1)	(2)	(3)	(4)	(5)
	small and medium cities	large cities	mega cities	2010–2015 year	2016–2018 year
ECV-1	–0.0566	–0.104***	–0.0439**	–0.0281**	–0.224***
	(0.109)	(0.0194)	(0.0213)	(0.0113)	(0.0418)
UES	0.135	0.323***	0.0462	0.169***	0.134***
	(0.0848)	(0.0385)	(0.0477)	(0.0484)	(0.0518)
Control variables	Yes	Yes	Yes	Yes	Yes
FE of the city	Yes	Yes	Yes	Yes	Yes
FE of the year	Yes	Yes	Yes	Yes	Yes
Constant	4.735***	4.604***	4.058***	3.652***	4.023***
	(1.156)	(0.175)	(0.168)	(0.136)	(0.214)
Observations	93	1,538	896	1,686	841
Adjusted R^2	0.878	0.858	0.849	0.955	0.813

Notice (1) the sample is divided into different city sizes using "Notice on Adjusting the Standards for City Size Classification" issued by the State Council in 2014; (2) robust standard error in parentheses; (3) ***, **, * respectively express significant at the significance level of 1%, 5% and 10%

20.5.5 Robustness Analysis

In order to further verify the reliability of the conclusion, this study has done the following robustness test, with the corresponding results reported in Table 20.5. Considering that the mean estimate is greatly affected by special values and abnormal data, the regression of column (1) excludes the data of the four municipalities in China, and the regression of column (2) excludes the top 5% and bottom 5% of the $PM_{2.5}$ value in the sample data. It was found that environmental control still significantly suppressed air pollution, and the negative impact of the worse energy structure on air quality is still valid, which is basically consistent with the conclusion of the benchmark regression. Taking into account the possible bias in the selection of control variables, column (3) shows the estimation when the GDP output value of the control variable was replaced by per capita GDP; it is found that the regression result is consistent with the benchmark regression results. The regressions in columns (4) and (5) replace the environmental control variables with environmental control variable 2 (total words of vocabulary) and environmental control variable 3 (proportion of words), respectively; it is found that the negative relationship between environmental regulation and air pollution is still valid, and the estimated coefficient is basically stable. Considering that the controlled variable may also have a reverse causal relationship with the explained variable, leading to estimation bias, the estimation in column (6) lags all the controlled variables by one period; it is found that the regression results are basically unchanged. Since some control variables are increasing simultaneously, like population, GDP, and freight volume, in order to reduce possible multicollinearity, the estimation in column (7) used the first order difference term of the three variables above, and it was found that the results did not change significantly.

20.6 Conclusion

This study conducts research from both theoretical and empirical aspects. In the theoretical part, energy consumption and government environmental control are added to the DGE model to explore the dynamic impact of adjusting energy structure and government environmental control on air pollution. The research findings are as follows: the northern part of China is the worst-hit area, and the air quality in the south is slightly better; while adjusting the energy structure to suppress air pollution, it will reduce economic output and slow down economic growth; reasonable government environmental control will promote economic "green" growth, improving environmental quality and economic welfare; differentiated environmental control will make the southern part of the country a "polluting paradise", unified environmental control will not lead to the transfer of polluting companies, and it is sustainable effective air pollution control; the path of environmental control, which is faster first and slower later, has a better effect on air pollution control. In the empirical part, using panel data of 286 cities at prefecture-level and above in China from 2010 to 2018, this study explored the impact of government environmental regulation and energy structure on air pollution. The research findings are as follows: Government environmental control and improvement of energy structure can help reduce urban air pollution; environmental control can help reduce air pollution in cities of different sizes, and over time, government environmental control has gradually improved its effectiveness in controlling air pollution; the deterioration of the energy structure has a negative impact on the environmental quality of cities of different sizes and periods.

The possible policy implications of the article are: (1) Reasonably increasing the intensity of government environmental control is a "win–win" strategy to control air pollution and promote sustainable economic growth. From the perspective of policy path, increasing the intensity of environmental control should follow the policy path of fast first and slow later, deliver correct policy signals to enterprises, encourage enterprises to "green" technological innovation, and improve their environmental responsibility awareness. (2) According to the 2019 BP World Energy Statistical Yearbook and the data from the World Bank and the National Bureau of Statistics in China, the proportion of China's coal in primary energy consumption increased by 0.86% in 2018. Compared with developed countries in the same period, China's coal consumption accounted for about 4.5 times that of the United States, 2.5 times that of Japan, and 2.8 times that of Germany. If the effects of population were striped out, it was still found that China's per capita coal consumption rose by 0.48% in 2018. In contrast, per capita coal consumption of the United States declined by 4.91% and

Table 20.5 Robustness analysis

Variables	(1)	(2)	(3)	(4)	(5)	(6)	(7)
	Exclude four municipalities	Exclude abnormal data	Change control variables	Change explanatory variables	Change explanatory variables	Lag control variables	Adjusted control variables
ECV-1	–0.0902***	–0.0638***	–0.0930***			–0.139***	–0.146***
	(0.0152)	(0.0130)	(0.0151)			(0.0210)	(0.0216)
UES	0.261***	0.198***	0.251***	0.239***	0.235***	0.219***	0.247***
	(0.0380)	(0.0300)	(0.0358)	(0.0365)	(0.0362)	(0.0359)	(0.0386)
ECV-2				–0.0837***			
				(0.0146)			
ECV-3					–0.0423**		
					(0.0172)		
Control variables	Yes	Yes	Yes	Yes	Yes	Yes	Yes
FE of the city	Yes	Yes	Yes	Yes	Yes	Yes	Yes
FE of the year	Yes	Yes	Yes	Yes	Yes	Yes	Yes
Constant	4.078***	3.846***	4.097***	4.086***	3.714***	4.487***	4.118***
	(0.119)	(0.105)	(0.114)	(0.123)	(0.107)	(0.146)	(0.116)
Observations	2,491	2,273	2,527	2,527	2,527	2,246	2,246
Adjusted R^2	0.861	0.831	0.860	0.860	0.858	0.854	0.850

Notice (1) robust standard error in parentheses; (2) ***, **, * respectively express significant at the significance level of 1%, 5% and 10%

Japan declined by 1.64% in 2018. China is still facing severe challenges in adjusting the energy structure; in the process of adjusting the energy structure and improving air quality, it should focus on the research and development of alternative and renewable energy sources to change energy structure. (3) Pollution control strategies should be designed that first north and then south, first focus heavy industries and then light, and regional linkage. From a regional perspective, pollution in the northern region is more serious, and the air pollution control in the north should be the first priority. From an industry perspective, heavy-polluting enterprises emit more pollution, and pollution control of heavy-polluting enterprises, such as thermal power, steel, coal, and petrochemical industries, should be strengthened. In the process of governance, regional linkages should be strengthened to avoid excessive differences in environmental control between regions, which will trigger polluting companies to move to a "polluting paradise" and make economies unable to obtain full air pollution control benefits.

Possible shortcomings of this study include: (1) The energy consumption in the model mainly refers to traditional energy consumption, without adding other energy sources. (2) Energy endowment distribution's differences between sub-regions were not considered in the model of this study. Compared with the characteristics of the regional heterogeneity of energy endowment distribution in reality, this hypothesis may underestimate the extent of haze reduction in the regions with abundant energy endowment, and overestimate the extent of haze reduction in areas with poor energy endowments. (3) The labor endowment in the model remains unchanged during the simulation process, and does not reflect the impact of population agglomeration or population growth on air pollution. (4) Many assumptions of the theoretical model are not imperfect, such as the free flow of labor force in the region, the market in a state of complete competition and so on. (5) The parameter setting of the model mainly comes from the literature collation, without extensive investigation and field experiment. The parameter setting may have limitations. (6) Due to the availability of samples, the government environmental control variables include not only the control of air pollution but also the control of three wastes. These shortcomings are left for serious follow-up study.

Acknowledgements Huai Deng, Hua Li, and Yiming Guo also made great contributions to this manuscript. We express our heartfelt thanks to them. This research also was supported by: National Natural Science Foundation of China (72474127) and National Social Science Fund Major Bidding Project of China (18ZDA052). Part of this chapter was published:Wu X, Deng H, Li H, Guo Y. Impact of Energy Structure Adjustment and Environmental Regulation on Air Pollution in China: Simulation and Measurement Research by the Dynamic General Equilibrium Model. Technological Forecasting and Social Change 2021;172:121010. https://doi.org/https://doi.org/10.1016/j.techfore.2021.121010. The publication has been conducted with copyright permission.

References

Acemoglu, D., Aghion, P., Bursztyn, L., et al. (2012). The environment and directed technical change. *American Economic Review., 102*(1), 131–166.

Allan, G., Lecca, P., & McGregor, P., et al. (2014). The economic and environmental impact of a carbon tax for Scotland: A computable general equilibrium analysis. *Ecological Economics*, 40–50.

An, H. S., & Zhou, Y. X. (2013). Research on the inter-regional ecological compensation: An analysis based on new economic geography. *The Journal of World Economy., 36*(2), 117–136.

Andersen, D. C. (2016). Credit constraints, technology upgrading, and the environment. *Journal of the Association Environmental and Resource Economists., 3*(2), 283–319.

Aragon, F. M., Miranda, J. J., & Oliva, P. (2017). Particulate matter and labor supply: The role of caregiving and non-linearities. *Journal of Environmental Economics and Management*, 295–309.

Autor, D. H., Dorn, D., & Hanson, G. H. (2013). The China syndrome: Local labor market effects of import competition in the United States. *American Economic Review., 103*(6), 2121–2168.

Bao, Q., Shao, M., & Yang, D. L. (2013). Environmental regulation, provincial legislation and pollution emission in China. *Economic Research Journal., 48*(12), 42–54.

Ben Kheder, S., & Zugravu, N. (2012). Environmental regulation and French firms location abroad: An economic geography model in an international comparative study. *Ecological Economics*, 48–61.

Bilgili, F., Ulucak, R., Kocak, E., et al. (2020). Does globalization matter for environmental sustainability? Empirical investigation for Turkey by Markov regime switching models. *Environmental Science and Pollution Research., 27*(1), 1087–1100.

Chang, T., Zivin, J. G., Gross, T., et al. (2016). Particulate pollution and the productivity of pear packers. *American Economic Journal-Economic Policy., 8*(3), 141–169.

Chay, K. Y., & Greenstone, M. (2005). Does air quality matter? Evidence from the housing market. *Journal of Political Economy., 113*(2), 376–424.

Chen, S. Y., & Chen, D. K. (2016). Energy structure, Haze governance and sustainable growth. *Journal of Environmental Economics., 1*(1), 59–75.

Chen, S. Y., & Chen, D. K. (2018). Air pollution, government regulations and high-quality economic development. *Economic Research Journal., 53*(2), 20–34.

Chen, Z., Kahn, M. E., & Liu, Y., et al. (2016). The consequences of spatially differentiated water pollution regulation in China. *National Bureau of Economic Research*, 468–485.

Cherniwchan, J., Copeland, B. R., & Taylor, M. S. (2017). Trade and the environment: New methods, measurements, and results. *Annual Review of Economics, 9*, 59–85.

Chuai, X. W., Huang, X. J., Wang, & W. J., et al. (2012). Spatial econometric analysis of carbon emissions from energy consumption in China. *Journal of Geographical Sciences*, 630–642.

Correia, A. W., Pope, C. A., III., Douglas, D. W., et al. (2013). The effect of air pollution control on life expectancy in the United States: An analysis of 545 US counties for the period of 2000 to 2007. *Epidemiology, 24*(1), 23–31.

Desmet, K., & Rossi-Hansberg, E. (2014). Spatial development. *American Economic Review., 104*(4), 1211–1243.

Fan, L. Y., Yao, S., Zhou, C., & Richardsonbarlow, C. (2025). Environmental regulatory policies and local air quality: Evidence from a comparison of environmental policy effects in China. *Journal of Cleaner Production*, 518.

Hosseini, H. M., & Kaneko, S. (2013). Can environmental quality spread through institutions? *Energy Policy*, 312–321.

Joo, Y. J., Kim, C. S., & Yoo, S. H. (2015). Energy consumption, Co-2 emission, and economic growth: Evidence from Chile. *International Journal of Green Energy., 12*(5), 543–550.

Keller, W., & Levinson, A. (2002). Pollution abatement costs and foreign direct investment inflows to U. S. States. *Review of Economics and Statistics*, 566–576.

Krugman, P. (1991). Increasing returns and economic geography. *Journal of Political Economy., 99*(3), 483–499.

Leal, P. H., Caetano, R. V., & Marques, A. C. (2021). Is the relocation of polluting industries prompted by FDI flow and stock, globalization, corruption and regulation? *International Journal of Environmental Research and Public Health, 18*(4).

Levinson, A., & Taylor, S. M. (2008). Unmasking the pollution haven effect. *International Economic Review., 49*(1), 223–254.

Luo, W. J., & Xiang, X. Y. (2024). Unsynchronized legislation and unintended pollution: Estimating regulation-induced substitution in China. *Environmental & Resource Economics, 87*(3), 731–760.

Ma, L. M., & Zhang, X. (2014). The spatial effect of China's Haze pollution and the impact from economic change and energy structure. *China Industrial Economics*, 19–31.

Modis, T. (2019). Forecasting energy needs with logistics. *Technological Forecasting and Social Change*, 135–143.

Nordhaus, W. D. (2010). Economic aspects of global warming in a Postcopenhagen environment. *Proceedings of the National Academy of Sciences of the United States of America., 107*(26), 11721–11726.

Pierce, J. R., & Schott, P. K. (2016). The surprisingly swift decline of US manufacturing employment. *American Economic Review., 106*(7), 1632–1662.

Sancho, F. (2010). Double dividend effectiveness of energy tax policies and the elasticity of substitution: A CGE appraisal. *Energy Policy, 38*(6), 2927–2933.

Shapiro, J. S., & Walker, R. (2018). Why is pollution US manufacturing declining? The role of environmental regulation, productivity, and trade. *American Economic Review., 108*(12), 3814–3854.

Sharifi, M., Pool, J. K., & Jalilvand, M. R., et al. (2019). Forecasting of advertising effectiveness for renewable energy technologies: A neural network analysis. *Technological Forecasting and Social Change,* 154–161.

Shi, Q. L., Guo, F., & Chen, S. Y. (2016). "Political Blue Shy" in Fog and Haze governance-evidence from the local annual "Two Sessions" in China. *China Industrial Economics*, 40–56.

Song, M. L., Wang, S. H., Lei, L., & Zhou, L. (2019). Environmental efficiency and policy change in china: A new meta-frontier non-radial angle efficiency evaluation approach. *Process Safety and Environmental Protection., 121*, 281–289.

Song, M. L., Peng, J., Wang, J. J., & Zhao, J. J. (2018). Environmental efficiency and economic growth of China: A Ray slackbased model analysis. *European Journal of Operational Research*, 51–63.

Song, M. L., Zhao, X., & Shang, Y. P. (2020). The impact of lowcarbon city construction on ecological efficiency: Empirical evidence from quasi-natural experiments. *Resources Conservation and Recycling*, 157.

Stern, N. (2006). *Review on the economics of climate change.* NBER Working Paper.

Wang, K. L., Yin, H. C., & Chen, Y. W. (2019). The effect of environmental regulation on air quality: A study of new ambient air quality standards in China. *Journal of Cleaner Production., 215*, 268–279.

Wang, Y. R., Zhang, J., Song, X. T., Wang, A. D., & Li, J. J. (2026). Spatiotemporal air pollutant and carbon footprints of China's coal sector and their synergetic mitigation pathway towards 2035. *Resources Conservation and Recycling*, 224.

Yang, Z. H. (2011). Study on the dynamic relationship between economic growth, energy consumption and carbon dioxide emission. *The Journal of World Economy., 34*(6), 100–125.

Yang, Y., Cai, W. J., & Wang, C. (2014). Industrial CO2 intensity indigenous innovation and R&D spillovers in China's provinces. *Applied Energy., 131*(10), 117–127.

Yang, J., & Xu, L. (2020). How does China's air pollution influence its labor wage distortions? Theoretical and empirical analysis from the perspective of spatial spillover effects. *Science of the Total Environment*, 745.

Ye, J. Z., & An, H. S., 2017. Can environmental taxes effectively control air pollution? *China Industrial Economics*, 54–74.

Zhang, D. Y., Du, W. C., Zhuge, L. Q., et al. (2019). Do financial constraints curb firm's efforts to control pollution? Evidence from Chinese manufacturing firms. *Journal of Cleaner Production.*, *215*, 1052–1058.

Zhang, X. H., Wu, L. Q., & Zhang, R., et al. (2013). Evaluating the relationship among economic growth, energy consumption, air emissions and air environmental protection investment in China. *Renewable and Sustainable Energy Reviews*, 259–270.

Zhang, S. H., Worrell, E., & Crijns-Graus, W., et al. (2014). Cobenefits of energy efficiency improvement and air pollution abatement in the Chinese iron and steel industry. *Applied Energy,* 333–345.

Part IV
Policy Effectiveness on Environmental Disasters in China

Chapter 21
Does Promoting New Urbanization Help Mitigate Environmental Pollution: Evidence from China's Pilot Policies for New Urbanization

21.1 Introduction

In the four decades since China began its reform and opening up, its urbanization process has been unparalleled in speed and scope globally (Yu, 2021). By the end of 2022, the urban population rate had soared to an impressive 65.22%, marking a significant 47.3% increase since 1978. The rapid expansion of urban areas, nearly tripling in size, has been a crucial driving force behind China's economic growth and social development (Yan et al., 2023). Nevertheless, the conventional approach to urbanization has predominantly focused on numerical and scale-oriented metrics related to cities and towns, neglecting the essential dimension of high-quality urban development (Zhao & Wang, 2022a). This neglect poses a serious threat to the sustainable development of the economy, society, and environment (Zhang et al., 2020). Consequences of this neglect include the exacerbation of the heat island effect (Chen et al., 2016), environmental degradation (Xu et al., 2020), energy and resource shortages (Zhang et al., 2020), heightened reliance on fossils, increased air and noise pollution, accumulation of solid waste, and higher levels of domestic waste emissions (Xu et al., 2020). Emphasizing the gravity of the situation, the 2022 Environmental Performance Index by Yale and Columbia Universities ranks China at a concerning 160th out of 180 countries and region (Nie et al., 2023), with a score of merely 28.4 points. This ranking starkly contrasts with China's standing as a major economic and population power.

Chinese governments at all levels prioritize addressing the challenges posed by traditional urbanization and actively advocate for the advancement of new urbanization. The New Urbanization Plan (2014–2020), issued by the Central Committee of the Communist Party of China (CPC) Central Committee and the State Council in March 2014, stresses the importance of focusing on the urbanization of people, supported by comprehensive carrying capacity, utilizing city clusters as the primary form, and driven by innovation in institutions and mechanisms. This approach aims to facilitate economic transformation and upgrading, as well as harmonious social

X. Wu, *Environmental Disaster in China*,
https://doi.org/10.1007/978-981-92-0910-1_21

progress. Moreover, the *Key Tasks for the Construction of New Townships and Urban-Rural Integration and Development for the Year 2020*, released by the National Development and Reform Commission (NDRC) in 2020, underscores the significance of adhering to the new development concept and expediting the implementation of the new urbanization strategy that prioritizes human urbanization and quality improvement.

New urbanization diverges significantly from traditional urbanization, prioritizing people-centered, green, low-carbon, and sustainable development principles. Its primary goal is to shift from a land-centered approach to a people-centric urbanization model, transitioning from an economic growth-driven urbanization to an ecological urbanization that promotes harmonious coexistence between humans and nature (Chen et al., 2016). By enhancing the quality of urbanization, new urbanization aims to achieve coordinated development among large, medium, and small cities, and promote balanced economic, social, and environmental progress.

Despite nearly a decade since the implementation of China's new urbanization policy, there remains a lack of empirical studies exploring its impact on environmental pollution. To address this gap, this study utilizes panel data from 186 Chinese cities from 2012 to 2021 and employs a multi-period difference-in-differences model. The study aims to investigate the effects of the policy, its mechanisms of action, and the heterogeneity of its impact on environmental pollution within the context of new urbanization construction. The findings are intended to provide insights into the policy's effectiveness and offer guidance for evaluating its implementation outcomes. Key innovations of this study are as follows: (1) This study analyzes the relationship between the implementation of the National New Urbanization Comprehensive Pilot Policy and environmental pollution, expanding the research on the influencing factors of environmental pollution and the policy's implementation effects. (2) By treating the implementation of the National New Urbanization Comprehensive Pilot Policy as an exogenous shock, this study effectively mitigates potential endogeneity concerns, allowing for a more accurate assessment of the policy's impact on environmental pollution. (3) The study enriches existing research by broadening the scope of mechanism analysis and incorporating mediating factors such as energy structure optimization and green technological innovation. Additionally, it considers the heterogeneity of urban location and resource endowment, providing insights for policymakers on mitigating environmental pollution in the process of new urbanization development.

The subsequent sections are structured as follows: Sect. 2 reviews relevant studies, Sect. 3 delves into the institutional background and influence mechanism, Sect. 4 outlines the research design, Sect. 5 presents the empirical results, and the final Section concludes the research and provides policy implications.

21.2 Literature Review

21.2.1 Urbanization and New Urbanization

The term urbanization was initially introduced in Cerdá's seminal work, *An Introduction to Urbanization*, published in 1867 (Cerdá, 1867; Gao et al., 2022). Uttara et al. defined urbanization as a broad phenomenon encompassing the overall growth in population and industrialization within a settlement. This definition includes an increase in both the quantity and size of cities, accompanied by the migration of individuals from rural to urban areas (Uttara et al., 2012).

The distinction between new urbanization and traditional urbanization is primarily evident in three aspects: First, concerning objectives, new urbanization stresses the coordinated development of the economy, society, culture, and urban-rural integration (Shang et al., 2018), while traditional urbanization typically focuses on economic growth, often neglecting social and human factors. Second, regarding content, new urbanization adopts a people-centered approach, striving for equality, happiness, sustainability, and efficiency (Lin & Zhu, 2021), and encompasses various aspects like industry, population, space, society, and the environment (Zhao & Wang, 2022b). In contrast, traditional urbanization prioritizes urban population growth and urban area expansion. Third, in terms of driving forces, traditional urbanization is primarily propelled by a single industrial driver, whereas new urbanization is driven by factors such as urbanization, industrialization, and agricultural modernization.

21.2.2 Influencing Factors of Environmental Pollution

Regarding the sources of environmental pollution, some scholars emphasize the expansion of economic output scale, leading to increased energy consumption and the generation of industrial wastewater, exhaust emissions, soot, domestic waste, and other pollutants. This form of pollution can be mitigated by reducing the size of economic output (Rao et al., 2020). Meanwhile, others argue that environmental pollution stemming from an irrational industrial structure, such as an overly high proportion of secondary or manufacturing industries. Addressing this issue requires optimizing the production structure and eliminating low-efficient and energy-intensive and industries, thereby alleviating the pressure on environmental pollution (Zhang & Ran, 2023). The third perspective argues that outdated production technologies contribute to higher pollution levels due to their inefficiency and excessive resource consumption. Additionally, factors such as international trade, foreign direct investment (Chaudhry et al., 2022), environmental regulation (Shapiro & Walker, 2018), the use of digital technology (Gobbo et al., 2018; Li et al., 2020), the level of financial development (Wang et al., 2022), economic growth, and transportation, among others, can either exacerbate or mitigate environmental pollution.

21.2.3 Relationship Between Urbanization and Environmental Pollution

In recent decades, urbanization has emerged as a significant global social transformation, attracting considerable attention due to its profound environmental implications (Ahmad et al., 2021; Zhang et al., 2023). The relationship between urbanization and environmental pollution has been widely debated in the literature, with scholars presenting differing views on its nature and mechanisms. Some studies argue that urbanization intensifies pollution due to increased industrial activities and higher energy consumption. The demographic and industrial changes associated with urbanization contribute to increased energy and resource consumption (Hao et al., 2020). The combustion of fossil fuels like coal and oil releases pollutants such as sulfur dioxide, solid waste (Zhao et al., 2021), carbon dioxide, and PM2.5 (Wang et al., 2020), leading to environmental degradation (Xu et al., 2020), climate change, land erosion, and water quality deterioration. However, other studies present an opposing view, arguing that urbanization acts as a mitigating force against environmental pollution. The compact city theory suggests that concentrated urban forms, by minimizing commuting distances and fostering shared public transportation and facilities, can lower air pollution, energy consumption, and pollution emissions compared to sprawling low-density cities (Liu et al., 2020). Scholars like Seto et al. argue that environmental regulations, technological advancements, and shifts in economic structures driven by urbanization can contribute to pollution reduction (Seto et al., 2010). In China, urbanization has demonstrated positive environmental impacts by improving infrastructure utilization efficiency (Chen et al., 2019), reducing residential and industrial emissions, and enhancing energy efficiency (Shen et al., 2022). Beyond these two perspectives, some research adopts a nonlinear approach, introducing the concept of an inverted U-shaped relationship between urbanization and environmental pollution. Based on the Environmental Kuznets Curve hypothesis (Suki et al., 2020), this viewpoint suggests that as urbanization levels increase, environmental pollution experiences an initial rise followed by a subsequent decline (Dinda, 2004). The theory of ecological and environmental modernization further proposes that while environmental quality may deteriorate in the early and middle stages of urban development, technological innovation and industrial upgrading, spurred by urbanization, ultimately lead to environmental improvement (Huber, 2000). Empirical studies by Sun et al. and Xu et al. validate this U-shaped relationship (Sun et al., 2022), demonstrating that as urbanization progresses, industrial pollution emissions initially increase before decreasing (Xu et al., 2020). Additionally, Awan et al. also presented empirical evidence supporting this U-shaped relationship, highlighting the non-linear impact of urbanization on CO_2 emissions in Malaysia (Awan et al., 2022).

21.2.4 Evaluation of New Urbanization Promotion Policies

Existing studies have extensively examined the economic, social, and ecological impacts of the new urbanization pilot programs. (1) Economic Advantages: New urbanization initiatives stimulate urban economic growth and enhance city competitiveness by optimizing industrial structures, upgrading infrastructure, and elevating educational standards. (2) Social Benefits: These policies foster high-quality development in both urban and rural areas, enhancing living conditions, narrowing the income disparity between urban and rural populations, and promoting social equity (Ma & Shi, 2023). (3) Ecological Impact: New urbanization policies advocate for green and low-carbon urban transformations (Xu et al., 2024), enhancing ecological quality (Li et al., 2023). They emphasize resource efficiency, boost urban green energy efficiency, and support sustainable development (Shao & Wang, 2023).

Focusing on the impact of new urbanization on environmental pollution, the majority of scholars believe that new urbanization has a suppressive effect on pollution emissions (Wu et al., 2021). Cui and Cao examined the impact and mechanisms of new urbanization on carbon emission performance using panel data from 283 Chinese cities from 2003 to 2020, employing a difference-in-differences approach, suggest that the new urbanization pilot programs have led to a reduction in carbon emissions and an improvement in carbon emission efficiency (Cui & Cao, 2023). Zhao and Wang investigated the intrinsic relationship between new urbanization and air pollution using panel data from 30 Chinese provinces between 2005 and 2018. Their results indicate that new urbanization has significantly reduced local air pollution (Zhao & Wang, 2022b). Li et al. treated China's new urbanization policy as a quasi-natural experiment to evaluate its direct impact on carbon emission intensity, indirect mechanisms, and spatial spillover effects. Their findings suggest that the policy effectively reduced carbon emission intensity both locally and in neighboring cities (Li et al., 2023).

The existing literature predominantly focuses on the relationship between traditional urbanization and environmental pollution, with limited attention to the connection between new urbanization and environmental pollution, particularly its impact on industrial sulfur dioxide (SO_2) pollution. While some studies have empirically investigated how new urbanization mitigates environmental pollution using spatial econometric models, as evidenced by the research of Zhang and Cai, there remains a paucity of investigations utilizing the event study method to analyze this correlation (Zhang & Cai, 2022). Most of the current literature relies on national or provincial data, neglecting empirical analyses at the city level. By using data at the prefecture-level city scale, this study provides more precise alignment with the implementation units of the new urbanization policy compared to provincial-level data, offering stronger empirical support for differentiated policy formulation. In view of these factors, an empirical study is conducted based on data from cities in China.

21.3 Institutional Context and Impact Mechanisms

21.3.1 Institutional Background

To realize human-centered and ecologically sustainable urbanization, the 18th National Congress of the CPC introduced the concept of new urbanization for the first time. In 2014, China's NDRC issued the National New Urbanization Plan (2014–2020), which outlined the development principles of "ecological civilization, green and low-carbon", setting forth a comprehensive strategy for China's new urbanization. Subsequently, in February 2015, the NDRC identified 62 cities and towns in Jiangsu and Anhui provinces, along with Ningbo, as the initial pilot areas for this new urbanization model. A second batch of 15 pilot cities and towns was announced in November of the same year, followed by a third batch of 111 cities and towns, including Beijing's Shunyi District, in December 2016. All the three batches of pilot cities have since commenced their pilot construction phase. According to the "14th Five-Year Plan" Implementation Plan for New Urbanization, China is in a rapid urbanization stage, but there is still a need to enhance the quality of this process. The plan emphasizes the promotion of new urbanization, adaptation to emerging urban development trends, and the creation of livable, resilient, innovative, smart, green, and humanistic cities.

Addressing environmental pollution has been a pivotal aspect of new urbanization in all three national pilot programs and in China's "14th Five-Year Plan" for new urbanization. Based on the experiences from comprehensive national pilot projects on new urbanization published by the NDRC, the pilot areas have made significant strides in curbing the unchecked development of high energy consumption and high pollution projects, promoting the low-carbon transformation of cities, implementing clean energy heating, and fostering the development of green buildings.

21.3.2 Analysis of Theoretical Mechanisms

The advancement of new urbanization is expected to bring about improvements in the energy structure and promote the increased utilization of clean energy sources, thereby mitigating environmental pollution. The National New Urbanization Plan (2014–2020) underscores the core tenets of "ecological civilization, green, and low-carbon development". Pilot cities for new urbanization will elevate production costs for high-energy-consuming and polluting enterprises by imposing environmental taxes and fines, incentivizing the adoption of clean energy and advanced energy-saving practices to optimize energy structure. Furthermore, these pilot cities will facilitate the transition from coal-dominated energy consumption to cleaner alternatives, such as natural gas, renewable energy, and electrification, through the implementation of a clean energy quota system and intelligent energy consumption monitoring policies. These strategic measures are designed to optimize the city's energy structure, reduce reliance on fossil fuels and other environmentally harmful energy sources in the manufacturing sector, and promote the use of clean energy through green technological innovations, ultimately reducing pollutant emissions. Therefore, the first hypothesis is proposed:

H1: The new urbanization pilot policy can reduce environmental pollution through energy structure optimization.

New urbanization plays a crucial role in fostering urban green technological innovation (Liu & Zhou, 2025) and reducing environmental pollution by facilitating the agglomeration of talents and production factors. The National Plan for New-Type Urbanization (2014–2020) emphasizes that cities should pursue innovation-driven development by leveraging scientific, technological, educational, and human resources, in line with the latest trends in scientific and technological progress and industrial transformation. The implementation of new urbanization has resulted in the concentration of talents and production factors, alongside the enhancement of regional infrastructure. This concentration has led to intensified exchanges of talents, knowledge, and technology within agglomeration areas (Pan et al., 2024), fueling continuous innovation and promoting the development of clean production technologies. Additionally, heightened competition within these clusters compels enterprises to ramp up their R&D and innovation efforts. For enterprises characterized by high pollution and energy consumption, green technology innovation incentivizes the adoption of clean production technologies, accelerating their transformation and upgrading processes. New urbanization supports low-carbon technology research and development through R&D subsidies and accelerates the adoption of environmental technologies via the green patent fast-track policy. Moreover, high-tech enterprises can also capitalize on their financial advantages to seize early opportunities in future technological production, thereby enhancing production efficiency and reducing environmental pollution Based on these analyses, the second hypothesis is proposed (Fig. 21.1):

H2: The new urbanization pilot policy can reduce environmental pollution through green technological innovation.

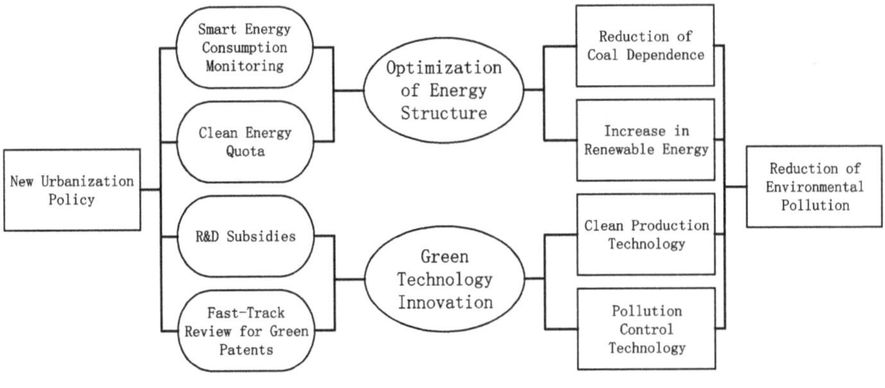

Fig. 21.1 The influence path of new urbanization on environmental pollution

21.4 Research Design

21.4.1 Model Design

The concept of new urbanization was initially introduced at the 18th National Congress of the CPC, followed by the implementation of various policies aimed at supporting its development. The establishment of comprehensive pilot areas for new urbanization has provided a valuable opportunity to evaluate the impact of these policies on environmental pollution. This assessment involves comparing environmental pollution levels before and after the implementation of pilot policies. Nevertheless, such a comparison may be affected by various factors beyond the pilot policy itself that also affect environmental pollution. The difference-in-differences (DID) method, distinct for its ability to eliminate the effects of these confounding factors, is particularly effectively in addressing the issues like omitted variables and endogeneity in the regression analysis. In addition, the multi-period DID model is particularly suited for capturing the long-term and cumulative effects of new urbanization policies. It accommodates variations in the timing of policy implementation, allowing for a dynamic tracking of policy effects over time. Moreover, its successful application in similar studies (Li, Zheng, et al., 2023). Therefore, the multi-period DID method is employed herein to assess the impact of new urbanization pilot policies on urban environmental pollution.

The pilot policies were rolled out in three batches, with the first batch enacted in December 2014, followed by the second and third batches in late 2015 and 2016 respectively. The policy implementation years were thus set as 2015, 2016, and 2017, accordingly. In this study, a city designated as a comprehensive pilot city for new urbanization in the year of policy enactment is assigned a value of 1 for that year and subsequent years; otherwise, it is assigned a value of 0. The pilot cities constitute the treatment group, while non-pilot cities form the control group, coded as 1 and 0, respectively. The pilot areas span multiple administrative levels, including provinces, prefecture-level cities, county-level cities, and townships. To ensure data consistency, focuses on prefecture-level cities as the primary unit of analysis, excluding regions where pilot areas are limited to counties or townships. Additionally, Beijing, Tianjin, Shanghai, and Chongqing are excluded due to their unique administrative structure and distinct economic development patterns. Furthermore, we exclude cities with severe data deficiencies.[1] The prefecture-level cities within pilot provinces are designated as the treatment group. After considering data availability and administrative division adjustments, a final sample of 186 prefecture-level cities is obtained, with 84 cities in the treatment group and 102 cities in the control group. The study period is set from 2012 to 2021 to encompass the Plan's implementation period from 2014

[1] Cities with severe data deficiencies include Chaohu, Sansha, Danzhou, Bijie, Tongren, Lhasa, Rikaze, Changdu, Linzhi, Shannan, Naqu, Haidong, Tulufan, Hami, Laiwu, Longnan, Xiangyang, Baishan, Zhongwei,etc.

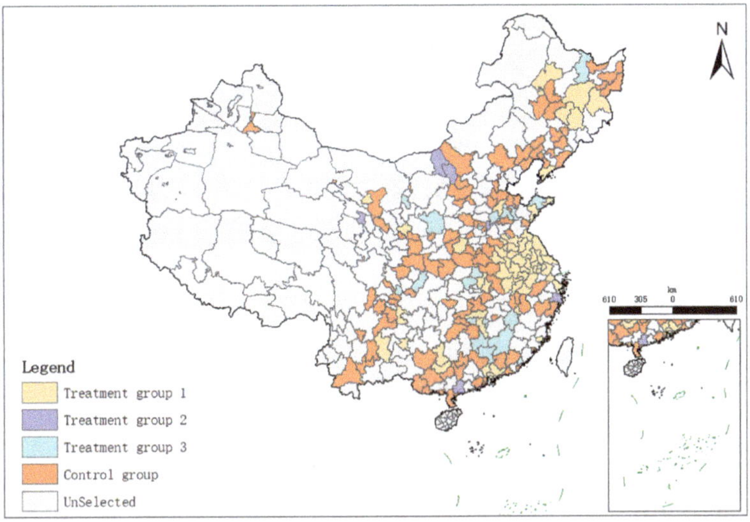

Fig. 21.2 Study area and locations

to 2020. The selection criteria for the treatment and control groups are detailed in Fig. 21.2.

Building upon the aforementioned analysis, a two-way fixed effects multi-period DID model is established for further analysis:

$$Y_{it} = \alpha_0 + \alpha_1 DID_{it} + \alpha_2 X_{it} + \lambda_t + \mu_i + \varepsilon_{it} \tag{21.1}$$

$$DIDit = Treatit \times Postit \tag{21.2}$$

In the above equation, the subscript *i* represents the city *i*, *t* represents the year *t*, and the dependent variable Y_{it} is the level of environmental pollution. DID_{it} represents the dummy variable for the new urbanization pilot policy, which is the interaction term b*etween* $Treat_{it}$ *and* $Post_{it}$. $Treat_{it}$ refers to the study sample. When $Treat_{it} = 0$, it signifies cities that have not implemented the new urbanization pilot policy, categorized as the control group; when $Treat_{it} = 1$, it designates cities that have implemented the new urbanization pilot policy, categorized as the treatment group. $Post_{it}$ is a time grouping variable, whose value is set to 1 if the city i is on the list of pilot cities in year t, and to 0 if it is not. When $Post_{it} = 0$, it indicates that the pilot policy has not been implemented; when $Post_{it} = 1$, it indicates that the pilot policy has been implemented. X_{it} is a series of control variables, mainly including economic development level, population density, degree of openness, industrialization level, financial development, and government behavior, population size, infrastructure construction. λ_t is the time fixed effect, μ_i is the city fixed effect, and ε_{it} is the random error term.

21.4.2 *Variable Setting and Description*

21.4.2.1 Explained Variables

The primary variable of interest is environmental pollution, which is quantified using SO_2 emissions. Higher levels of SO_2 emissions signify more severe environmental pollution. The specific reasons for selecting SO_2 as the dependent variable are as follows:

1. Scientific of indicators

SO_2 is a major contributor to air pollution and acid rain and often coexists with other pollutants such as nitrogen oxides and particulate matter. The emission of SO_2 not only results in significant economic losses but also poses serious threats to public health. Its concentration level reflects not only the presence of a single pollutant but also serves as an indirect indicator of other associated pollutants, thereby providing insights into the overall environmental pollution status of a given region. In existing studies, scholars often use SO_2 to measure the extent of environmental pollution (Yan et al., 2022).

2. Representativeness of indicators

The development of urbanization is often accompanied by industrialization, which requires substantial energy consumption. China's energy structure has historically been dominated by coal. This heavy reliance on fossil fuels, particularly coal, has led to significant SO_2 emissions. Moreover, the concentration of population and the increasing demand for transportation during the urbanization process further contribute to the rise in SO_2 emissions. To address this issue, China identified SO_2 as a benchmark pollutant for environmental management in its 11th Five-Year Plan. The 13th Five-Year Plan further reinforced this commitment by setting a binding target to reduce total SO_2 emissions by 15%.

The spatiotemporal variation of SO_2 emissions in the sample cities is illustrated in Fig. 21.3. In the figure, a color gradient is used to represent SO_2 pollution levels, with lighter colors indicating areas with lower pollution and darker colors indicating areas with more severe pollution.

21.4.2.2 Core Explanatory Variables

To assess the impact of new urbanization pilot policies on urban environmental pollution, the core explanatory variable is whether the sample city is designated as a pilot city for new urbanization. The binary variable DID is assigned a value of 1 for pilot cities, and 0 for non-pilot cities. This study includes 84 cities in the treatment group (pilot cities) and 102 in the control group.

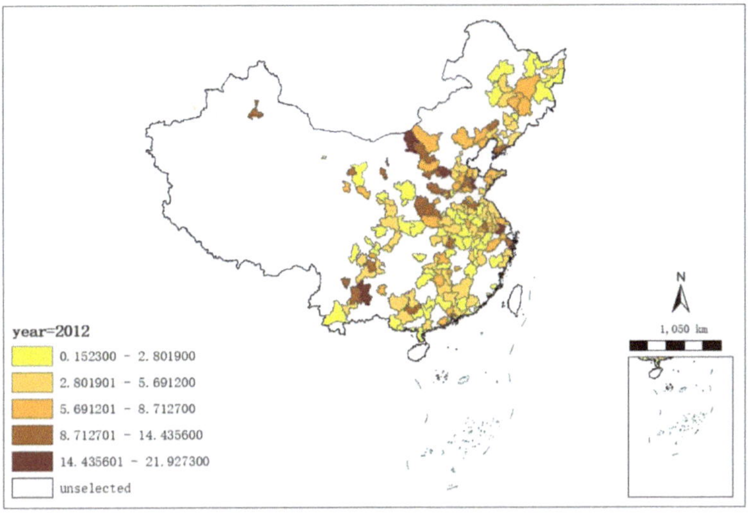

(a).

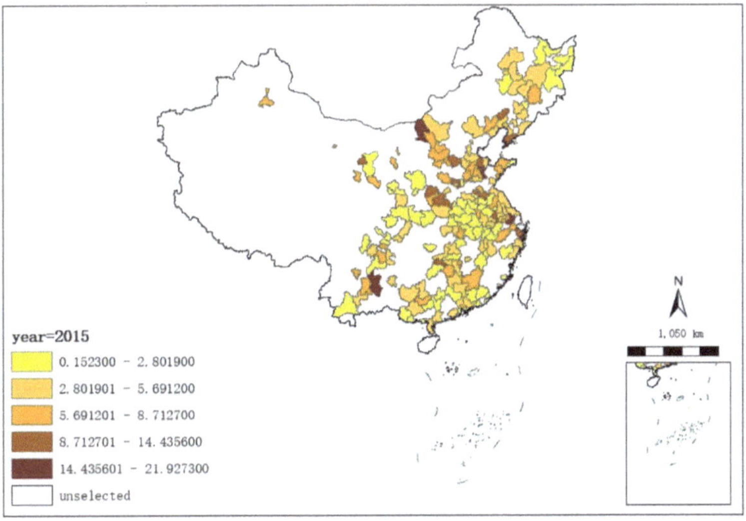

(b).

Fig. 21.3 **a** Spatial Distribution of SO_2 Emissions in Sample Cities for 2012 (in 10,000 tons). **b** Spatial Distribution of SO_2 Emissions in Sample Cities for 2015 (in 10,000 tons). **c** Spatial Distribution of SO_2 Emissions in Sample Cities for 2018 (in 10,000 tons). **d** Spatial Distribution of SO_2 Emissions in Sample Cities for 2021 (in 10,000 tons)

21.4.2.3 Control Variables

To account for external factors and mitigate estimation biases stemming from omitted variables, the study incorporates the following control variables based on existing literature on environmental pollution determinants:

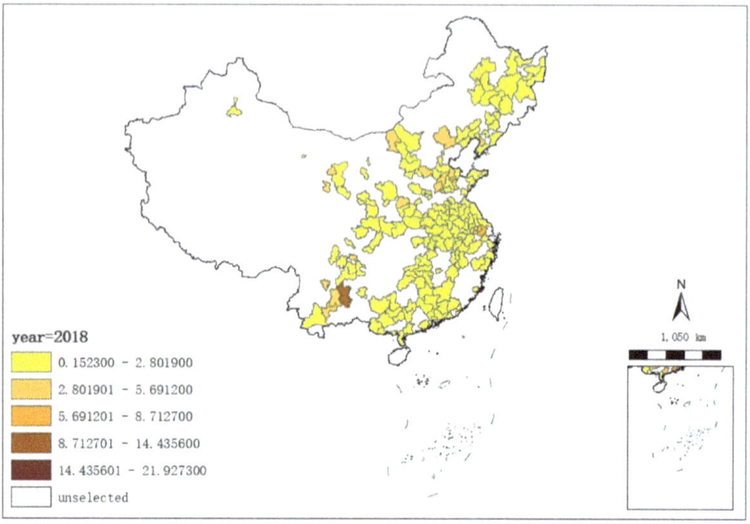

(c).

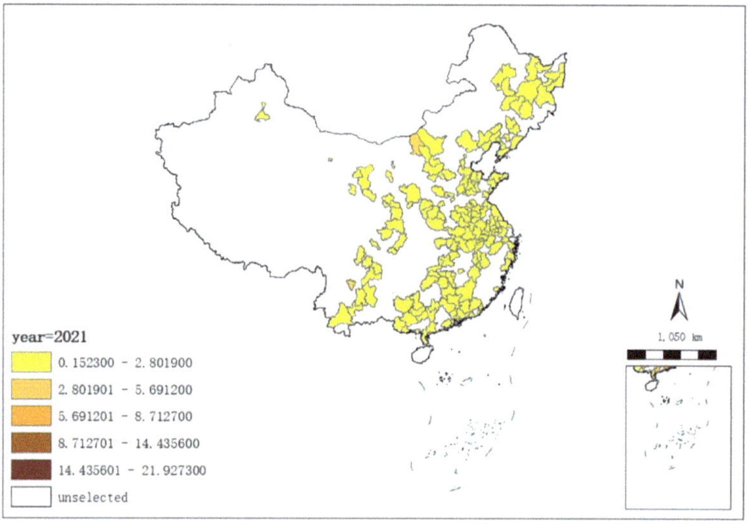

(d).

Fig. 21.3 (continued)

1. The level of economic development (*ln_rgdp*) is represented by the logarithm of real gross regional product per capita. The Environmental Kuznets Curve theory, proposed by Grossman and Krueger, posits a non-linear relationship between economic development and environmental pollution (Grossman & Krueger, 1995).

2. Population density (*ln_density*) is calculated as the logarithm of the city's population divided by its administrative area. It reflects the level of economic agglomeration to some extent, which can enhance productivity and mitigate environmental pollution. However, excessive agglomeration may lead to such issues as resource wastage and traffic congestion, exerting pressure on the environment.
3. Degree of Openness to Foreign Capital (*fdi*) is measured as the ratio of foreign capital utilization to the gross domestic product (GDP) in a given year. Foreign investment typically facilitates technological spillovers by introducing advanced production technologies and management systems to host countries. This will contribute to improvements in environmental quality through demonstration-learning effects. However, it may also lead to pollution issues as some foreign-funded enterprises introduce highly polluting industries to host countries.
4. Level of Industrialization (*industry*) is measured as the ratio of the secondary sector value added to GDP. In China, the secondary industry contributes 60–70% of SO_2 emissions, the impact of industrial development on regional environmental pollution is significant, as emissions of pollutants such as SO_2, smoke, and wastewater from industrial production are major sources of urban environmental pollution (Lou et al., 2021).
5. Financial Development (*finance*) is represented by the balance of RMB loans from regional financial institutions as a percentage of GDP. The variable *finance* is used to measure fiscal support within the new urbanization pilot program, primarily based on local government budget reports and the allocation of special funds. Financial Development can help redirect investment away from high-energy-consuming and high-polluting industries toward more environmentally friendly sectors, thereby promoting the sustainable development of the socio-economy (Sun & Chen, 2022).
6. Government Interventions (*gov*) is represented by fiscal expenditure as a share of GDP. Government interventions influence resource allocation, energy and resource utilization efficiency, and consequently, pollutant emissions.
7. Population size (*ln_pop*) is measured as the logarithm of the population. The expansion of urban scale is typically accompanied by population growth (Richardson, 1973), which in turn leads to increased energy consumption, industrial production, and transportation demand, resulting in higher pollution emissions.
8. (8) Infrastructure construction (*inf*) plays a crucial role in supporting the development of new urbanization, facilitating the cross-regional flow of production factors and improving economic efficiency. However, large-scale infrastructure construction, particularly industrial facilities powered by coal, may also become a significant source of pollution emissions. Under the competitive pressure of the Chinese government's "championship" policy, the vast majority of fixed asset investments have been directed towards infrastructure expansion. Therefore, the ratio of fixed asset investment to GDP is used as an indicator of the level of urban infrastructure investment (Chen & Chu, 2022).

21.4.2.4 Mechanism Variables

(1) Energy Structure Optimization (*ln_pec*)

Generally, energy can be divided into two categories: traditional fossil fuels, such as coal and oil, and clean energy, such as electricity. The former directly emits SO_2 during usage, while the latter does not. Moreover, the consumption of fossil fuels like coal is a primary contributor to SO_2 pollution. Since detailed data on coal consumption for individual cities is not available, and given the high correlation between electricity consumption and overall energy use, the logarithm of urban per capita electricity consumption is used to represent the city's energy structure (*ln_pec*). As coal power is the primary source of electricity in China, higher per capita electricity consumption is associated with increased pollution emissions (Guo & Liang, 2022; Li & Xu, 2018).

(2) Green Technological Innovation (*pgpt*)

The per capita number of green patent applications serves as a measure of green technological innovation (*pgpt*). Patent data is quantifiable, publicly available, and reflects the technological advancements and innovative capabilities of firms or research institutions in specific fields. Specifically, green patents encompass technologies and inventions related to environmental protection and sustainable development, making them an effective measure of the level of green innovation. It reflects the level of green technological innovation in a city, with a higher count indicating a more advanced level of innovation (Liu et al., 2024).

21.4.3 Data Source

The research samples comprise observations from 186 Chinese prefecture-level cities spanning the years 2012 to 2021. The purpose of the study is to evaluate the responses of these cities to new urbanization pilot programs. Green patent application data is obtained from the State Intellectual Property Office. Additional data sources include the *China Regional Statistical Yearbook*, *China Urban Statistical Yearbook*, and the EPS (Express Professional Superior) data platform. In cases of missing data, linear interpolation techniques[2] are employed in addition to annual bulletins of each city. Variables containing outliers are reduced[3] to ensure the robustness of the analysis

---

[2] The fundamental formula for linear interpolation is as follows:

Given two known data points (x_0,y_0) and (x_1,y_1), the interpolated value y at any x within the interval [x_0,x_1] is computed using the following equation:

$y = y0 + \frac{(y1-y0)}{(x1-x0)} \times (x - x0)$where:$x_0$,$x_1$ are adjacent known data points, satisfying $x_0 < x < x_1$;
y_0,y_1 are the corresponding known values of x_0 and x_1;
x is the point to be interpolated, and y is the estimated interpolated value.

[3] "outliers are reduced "refers to the adjustment of extreme values during data preprocessing through winsorization at the first and 99th percentiles, which constrains variable values within this range to mitigate the influence of extreme values and enhance the robustness of the estimation.

Table 21.1 Descriptive statistics for key variables

Variables	Description	Mean	SD	Min	Max
SO_2	SO_2 emissions	3.256	3.227	0.0777	15.60
Finance	Financial development	0.990	0.598	0.135	7.450
Fdi	Degree of openness to foreign investment	0.0177	0.0172	2.88e-05	0.199
ln_density	Population density	−3.319	0.776	−5.662	−1.288
Gov	Government actions	0.192	0.0843	0.0439	0.675
Industry	Level of industrialization	46.73	9.487	19.70	71.17
ln_rgdp	Economic development level	10.77	0.554	9.596	12.06
ln_pop	Population size	5.881	0.626	3.807	7.066
Inf	Infrastructure development	0.799	0.274	0.150	2.032

(Li et al., 2022). Descriptive statistics for each variable are presented in Table 21.1. The unit of SO_2 emissions is ten thousand tons, with a mean value of 3.256. The maximum value of 15.60 is significantly higher than the minimum value of 0.0777, indicating considerable variation in SO_2 emissions across different regions. From the ratio of the standard deviation to the mean, SO_2 emissions and FDI demonstrate the greatest regional heterogeneity, while economic development level, population size, and industrialization level are more consistently distributed across regions.

21.5 Empirical Findings

21.5.1 Benchmark Regression Analysis

In line with the constructed benchmark model, urban SO_2 emissions are incorporated into Model (1) as the explanatory variable. Table 21.2 reports the estimation outcomes of the DID model, with Columns (1), (2), (3), and (4) respectively presenting the impact of new urbanization construction on urban environmental pollution across various specifications. Column (1) presents the estimation results without controlling for control variables and fixed effects. Column (2) shows the estimation results after controlling for city and year fixed effects. Columns (3) and (4) report the estimation results with control variables included, based on the specifications in Columns (1) and (2), respectively. The regression results in Table 21.2 reveal that the estimated coefficients of new urbanization pilot policies are significantly negative at the 1% level across all columns. This implies that new urbanization construction contributes to a certain reduction in urban environmental pollution. Notably, the significance of the DID coefficients persists even after accounting for control variables and fixed effects. This robustness analysis demonstrates the reliability and stability of the findings. Specifically, in Column (4), after being selected as a national pilot for comprehensive new urbanization, the SO_2 concentration in the region decreased by an average of 52.86%. During the implementation of the national comprehensive new urbanization pilot policy, the pilot program incentivized local governments to change their urbanization development philosophy, abandoning the growth-driven expansion model and instead focusing more on the quality of urbanization.

Table 21.2 Baseline regression results

	(1)	(2)	(3)	(4)
Variables	SO_2	SO_2	SO_2	SO_2
Did	−1.2866***	−0.5069***	−1.2564***	−0.5286***
	(0.1666)	(0.1670)	(0.1929)	(0.1663)
ln_rgdp			−0.3623	2.3419***
			(0.2536)	(0.4571)
ln_density			−1.0295***	0.1529
			(0.1328)	(1.3050)
Fdi			13.1662***	−0.8232
			(4.1984)	(3.7507)
Industry			0.1019***	−0.0073
			(0.0104)	(0.0158)
Finance			0.7583***	0.2958
			(0.1688)	(0.2017)
Gov			−10.7602***	4.7201**
			(1.3116)	(2.0652)
ln_pop			1.3654***	−0.0604
			(0.1540)	(0.7730)
Inf			−1.7469***	1.2288***
			(0.2810)	(0.3405)
Year fixed	N	Y	N	Y
City fixed	N	Y	N	Y
Constant	3.5767***	3.3822***	−6.2653*	−22.7877**
	(0.1032)	(0.0570)	(3.4162)	(9.2181)
Observations	1431	1431	1431	1431
R-squared	0.0298	0.8025	0.2322	0.8130

Robust standard errors in parentheses.
*** $p < 0.01$, ** $p < 0.05$, * $p < 0.1$

21.5.2 Robustness Tests

21.5.2.1 Parallel Trend Test

The validity of the DID model hinges on the absence of time trend differences between the control and treatment groups before the policy intervention. This prerequisite necessitates the trends of the environmental pollution index for both groups run parallel leading up to the introduction of the new urbanization pilot policy. To demonstrate that the control and treatment groups satisfy the parallel trends assumption, this study employs the event study method proposed by Jacobson et al. to explore the dynamic effects (Jacobson et al., 1993). The model is constructed as follows:

$$Y_{it} = \delta_0 + \delta_1 \sum_{k=-3}^{3} DID_{t+k} + \delta_2 X_{it} + \lambda_t + \mu_i + \varepsilon_{it} \quad (21.3)$$

Model (3) is consistent with Model (1) and delves into assessing the policy's impacts over the three years preceding and succeeding the commencement of new urbanization pilot project. To mitigate multicollinearity concerns, data from a single period preceding the pilot are excluded. Figure 21.4 presents the results. The lack of significance in the regression coefficients prior to the pilot policy implies that there were no notable disparities in the changes between the treatment and control groups prior to the pilot project. This confirms that the parallel trend assumption is true prior to the implementation of the new urbanization pilot policy. Moreover, the results indicate that the new urbanization policy exerts a negligible effect on urban environmental pollution levels during its implementation year. The discernible decline in urban environmental pollution is only statically significant in the initial year following the pilot construction, suggesting a temporal lag in the effectiveness of the new urbanization pilot policy in curtailing urban environmental pollution. Over time, the pilot policy has had a positive impact. A possible reason for this is that with the implementation and promotion of the new urbanization policy, the pilot policy has gradually matured, contributing to a reduction in SO_2 emissions.

21.5.2.2 Placebo Test

To assess the robustness of the model, a re-estimation is performed by randomly selecting 123 cities from the research samples and designating them as a pseudo-experimental group. If the estimated coefficients on the placebo interaction terms are not significantly different from zero, it suggests that the model is not substantially biased by omitted variables. Conversely, significant deviations of the estimated coefficients from zero indicate a potential bias in the model's design. Figure 21.5 presents the kernel densities and corresponding p-value distributions of the estimated coefficients for 500 randomly generated treatment groups. The results reveal that the

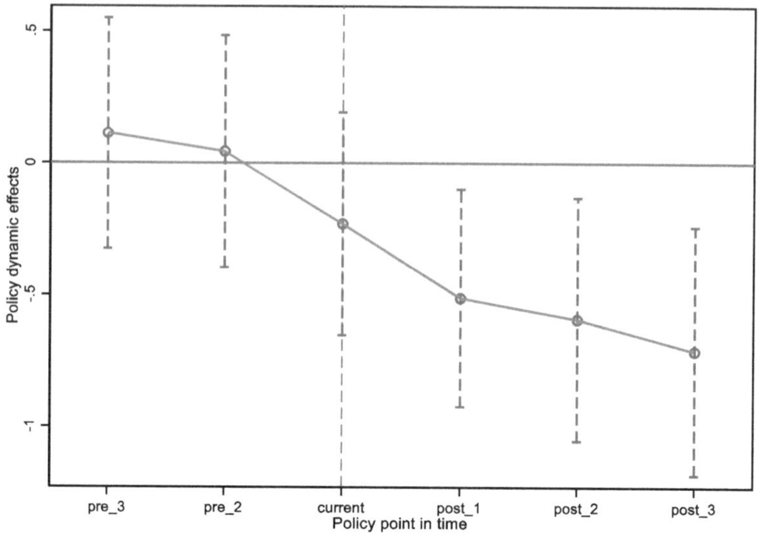

Fig. 21.4 Results of parallel trend test

majority of p-values exceed 0.1, and the mean of the regression coefficients approximates zero. This suggests that crucial variables were not omitted form the model, and the baseline regression results successfully withstand the placebo test.

21.5.2.3 Adding Control Variables

To ensure the robustness of the results, several additional variables are introduced. Scientific research investment (*rrd*) is quantified by the ratio of scientific expenditures to local fiscal expenditures. Internet penetration (*ln_internet*) is measured by the logarithm of the total number of Internet users. Human capital (*hc*) is quantified by the ratio of the number of students enrolled in general higher education institutions to the region's total population. The regression results, as presented in Table 21.3, demonstrate that the coefficient of the DID variable in column (1) is significantly negative at the 1% level. This finding indicates that the efficacy of the new urbanization pilot policy in reducing urban environmental pollution, aligning with the previous results.

21.5.2.4 Lag Period Test

Given the phased nature of pilot policy implementation, the effects of the policy itself may exhibit a lag. The impact of the previous year's implementation of the new urbanization pilot policy on urban environmental pollution level might become evident with a lag of one or two years. To address this potential lag effect, the

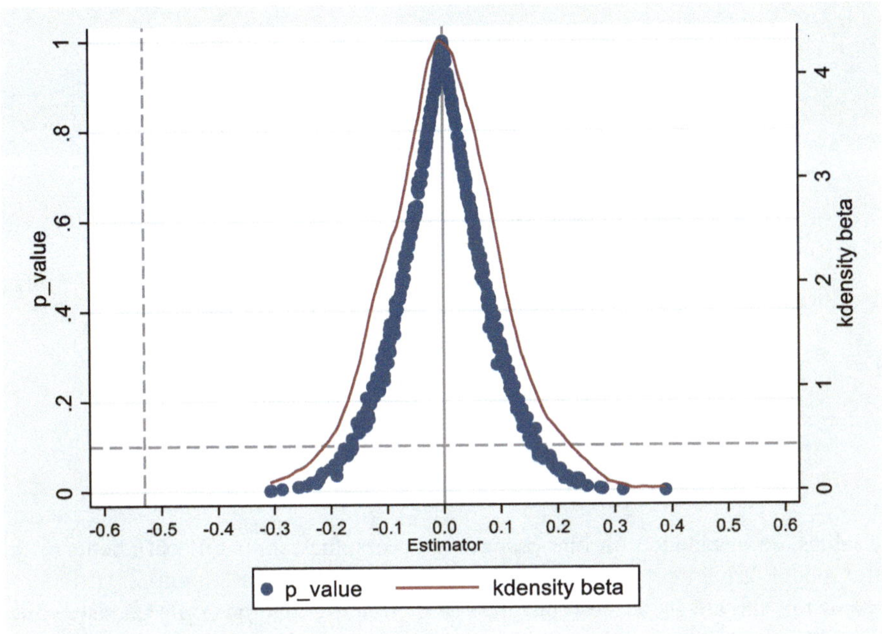

Fig. 21.5 Placebo test results

Table 21.3 Adding control variables

	(1)
Variables	SO_2
Did	−0.5381***
	(0.1660)
Controls	Y
Year fixed	Y
City fixed	Y
Constant	−14.8735
	(9.6730)
Observations	1431
R-squared	0.8152

Robust standard errors in parentheses.
*** $p < 0.01$, ** $p < 0.05$, * $p < 0.1$

environmental pollution variable SO_2 1–2 years post-policy implementation is introduced as an additional explanatory variable (designated as F1. SO_2 and F2. SO_2, respectively). This inclusion aims to investigate the impact of the pilot policy on environmental pollution. The specific regression results are detailed in columns (2) and (3) of Table 21.4. The findings reveal that irrespective of whether the explanatory

Table 21.4 Lag period test

	(1)	(2)	(3)
Variables	SO_2	F1. SO_2	F2. SO_2
Did	−0.5286***	−0.5070***	−0.3792**
	(0.1663)	(0.1696)	(0.1713)
Controls	Y	Y	Y
Year fixed	Y	Y	Y
City fixed	Y	Y	Y
Constant	−22.7877**	−31.9714***	−37.6314***
	(9.2181)	(10.2691)	(11.3678)
Observations	1431	1248	1067
R-squared	0.8130	0.8147	0.8100

Robust standard errors in parentheses.
*** $p < 0.01$, ** $p < 0.05$, * $p < 0.1$

variables are introduced for one year or two years, the estimated coefficients of the DID model remain significantly negative. This consistent result supports the core conclusion and attests to the robustness of the finding that the comprehensive pilot policy on new urbanization significantly mitigates urban environmental pollution.

21.5.2.5 Replacement of Explanatory Variables

To ensure the stability of the baseline regression results and reduce potential bias resulting from the qualitative selection of variables, a replacement of variable was undertaken. Replacing SO_2 emissions with the annual average PM2.5 concentration, industrial wastewater discharge, and NO_X emissions as the dependent variable. The regression results are presented in Table 21.5. The results reveal that the coefficient of DID in column (1) is significantly negative at the 5% level, while the coefficients in column (2) and (3) are significantly negative at the 10% level.

This suggests that substituting SO_2 emissions with urban PM2.5 concentrations industrial wastewater discharge, and NO_X emissions as the dependent variable to measure environmental pollution levels maintains the significant suppression of urban environmental pollution by the new urbanization pilot policy. These findings align closely with the previous results, confirming the consistency of the observed trends.

21.5.3 Impact Mechanism Test

The preceding analysis demonstrates that new urbanization construction effectively curtails urban environmental pollution. To delve into the underlying pathway and

Table 21.5 Replaced the interpreted variable

	(1)	(2)	(3)
Variables	PM2.5	Water	NO_X
Did	−0.8540**	−456.2903*	−7169.9580*
	(0.4002)	(243.5521)	(3735.5212)
Controls	Y	Y	Y
Year fixed	Y	Y	Y
City fixed	Y	Y	Y
Constant	82.9310***	28,030.9417*	−21,595.0506
	(26.3232)	(16,717.6651)	(156,261.5953)
Observations	1431	1352	826
R-squared	0.9432	0.9127	0.7879

Robust standard errors in parentheses.
*** $p < 0.01$, ** $p < 0.05$, * $p < 0.1$

inherent transmission mechanism, we explore the hypothesis that this impact is mediated through the optimization of energy structure and innovation in green technology. Building upon the methodology proposed by Wen Zhonglin (2004), we extend model (1) and introduce the following mediating effect models.

Model (4) is a regression equation of the pilot construction of new urbanization on the mediating variables. Model (5) is a regression equation of the pilot construction of new urbanization and the mediating variables jointly on the urban environmental pollution level. The mediation test involves the following steps: (1) Regression of Multiplicative Term with Mechanism Variable (Model 4): If the coefficient β_1 of the multiplicative term is significant, it indicates a substantial impact of new urbanization pilot policies on the mechanism variable.(2) Joint Regression of Multiplicative Term and Mechanism Variable with Explanatory Variables (Model 5): If the coefficient γ_2 of the mechanism variable is significant, and the coefficient γ_1 of the multiplicative term is insignificant or significantly reduced, it establishes that the new urbanization pilot policies reduce environmental pollution through the mediating variable M_{it} The model is set up as follows:

$$M_{it} = \beta_0 + \beta_1 DID_{it} + \beta_2 X_{it} + \lambda_t + \mu_i + \varepsilon_{it} \tag{21.4}$$

$$Y_{it} = \gamma_0 + \gamma_1 DID_{it} + \gamma_2 M_{it} + \gamma_3 X_{it} + \lambda_t + \mu_i + \varepsilon_{it} \tag{21.5}$$

Where M_{it} is the mediating variable (optimization of energy structure and innovation of green technology). The other variables are defined as in model (1). The product of β_1 and γ_2 signifies the mediating effect, indicating the impact of new urbanization on environmental pollution through the intermediary variable. This mediation analysis aims to provide deeper insights into the nuanced mechanisms through which new urbanization affects urban environmental pollution.

Table 21.6 Mechanism test—energy structure optimization

	(1)	(2)	(3)
Variables	SO_2	ln_pec	SO_2
Did	−0.5286***	−0.0613*	−0.5088***
	(0.1663)	(0.0326)	(0.1649)
ln_pec			0.3228***
			(0.1198)
Controls	Y	Y	Y
Year fixed	Y	Y	Y
City fixed	Y	Y	Y
Constant	−22.7877**	13.2241***	−27.0571***
	(9.2181)	(1.7140)	(9.3484)
Observations	1431	1431	1431
R-squared	0.8130	0.9100	0.8140

Robust standard errors in parentheses.
*** $p < 0.01$, ** $p < 0.05$, * $p < 0.1$

21.5.3.1 Energy Structure Optimization

The logarithm of urban per capita electricity consumption (*ln_pec*) is used as an indicator of energy structure optimization to assess its impact on environmental pollution. In China, urban electricity consumption is predominantly driven by coal power, and an increase in electricity consumption leads to higher consumption of fossil energy, resulting in environmental pollution. The results presented in columns (2) and (3) of Table 21.6 affirm that the new urbanization pilot policy significantly reduces per capita electricity consumption, thereby alleviating urban environmental pollution.

21.5.3.2 Green Technological Innovation

The number of urban per capita green invention patents (*pgpt*) is used as an indicator to measure the level of green technological innovation in cities. The results presented in columns (2) and (3) of Table 21.7 confirm that the implementation of new urbanization significantly enhances the level of urban green technological innovation, which, in turn, reduces environmental pollution.

Table 21.7 Mechanism test—green technology innovation

	(1)	(2)	(3)
Variables	SO_2	Pgpt	SO_2
Did	−0.5286***	0.4300***	−0.4280**
	(0.1663)	(0.0913)	(0.1682)
Pgpt			−0.2339***
			(0.0592)
Controls	Y	Y	Y
Year fixed	Y	Y	Y
City fixed	Y	Y	Y
Constant	−22.7877**	21.6592***	−17.7212*
	(9.2181)	(7.5111)	(9.2454)
Observations	1431	1431	1431
R-squared	0.8130	0.8916	0.8163

Robust standard errors in parentheses.
*** $p < 0.01$, ** $p < 0.05$, * $p < 0.1$

21.5.4 Further Discussion: Heterogeneity Analysis

21.5.4.1 Heterogeneity Analysis of City Location

The significant differences in the resource environment and economic foundations across the eastern, central, and western regions may engender disparities in the response, implementation, and effectiveness of the new urbanization pilot policy in the cities within these regions, consequently influencing the environmental pollution level. Accordingly, the sample cities are divided into eastern and central-western clusters to analyze the regional heterogeneity. The empirical results, presented in Table 21.8, reveal a significant negative estimated coefficient for the eastern region, signifying that new urbanization construction in the eastern region greatly reduces environmental pollution. In contrast, the estimated coefficients for the central and western regions are insignificant, suggesting that new urbanization construction does not exert a significant impact on curbing environmental pollution in these areas. This disparity can be ascribed to the advantage of policy support and favorable geographical positioning enjoyed by eastern cities, which are superior in economic prowess, innovation resources, and factor allocation, allowing for better support and enhanced policy implementation. Conversely, the central and western regions, characterized by slower economic progress, higher dependence on traditional resources, and suboptimal energy and resource utilization efficiencies, struggle to leverage the pollution-reducing effect of the pilot policy.

Table 21.8 Heterogeneity analysis of urban locations

	SO_2	SO_2
Variables	Eastern	Midwest
Did	−1.5727***	
	(0.2956)	
Did		0.0710
		(0.2024)
Controls	Y	Y
Year fixed	Y	Y
City fixed	Y	Y
Constant	23.5600	−43.2198***
	(33.7352)	(10.8419)
Observations	512	903
R-squared	0.8421	0.8050

Robust standard errors in parentheses.
*** $p < 0.01$, ** $p < 0.05$, * $p < 0.1$

21.5.4.2 Analysis of Heterogeneity of Urban Resource Endowment

Cities categorized as resource-based cities and non-resource-based cities exhibit contrasting resource endowments and levels of resource utilization, potentially giving rise to distinct challenges in the course of new urbanization construction and variations in the effectiveness of their pilot projects. To further investigate the impact of new urbanization construction on environmental pollution levels in cities with varied natural resource endowments, the 186 cities are divided into two groups based on the National Sustainable Development Plan for Resource-based Cities (2013–2020) issued by the State Council in 2013: 69 resource-based cities and 117 non-resource-based cities. The results, presented in Table 21.9, indicate that in non-resource-based cities, new urbanization construction significantly reduces the environmental pollution level. This inhibitory effect is, however, insignificant in resource-based cities. This discrepancy is attributable to the reliance of non-resource-based cities on their industrial strengths, energy structure, and technological innovation for economic development, as well as the high dependence of resource-based cities on local natural resources like coal, oil, and natural gas. The absence of incentives to enhance energy or resource use efficiency through industrial transformation, energy structure optimization, and technological innovation in resource-based cities results in lower resource utilization efficiency and exacerbated environmental pollution.

Table 21.9 Heterogeneity analysis of urban resource endowment

	SO_2	SO_2
Variables	Resource-based cities	Non-resource cities
Did_	−0.4722	
	(0.3366)	
Did		−0.3732**
		(0.1782)
Controls	Y	Y
Year fixed	Y	Y
City fixed	Y	Y
Constant	−47.9088***	16.9624
	(13.8753)	(11.8163)
Observations	526	874
R-squared	0.8229	0.8165

Robust standard errors in parentheses.
*** $p < 0.01$, ** $p < 0.05$, * $p < 0.1$

21.6 Conclusions and Policy Implications

Exploring the intrinsic relationship between China's new urbanization and environmental pollution holds paramount importance for mitigating urban environmental challenges and fostering sustainable development. By leveraging the three batches of national pilot policies on new urbanization as a quasi-natural experiment and drawing on panel data from 186 Chinese cities spanning 2012–2021, we employ the multi-period DID method to investigate the impact of China's new urbanization construction on environmental pollution. Additionally, we delve into its mechanism of action and regional heterogeneity.

21.6.1 Conclusions

1. New urbanization construction robustly and significantly alleviates urban environmental pollution, a finding supported by a series of robustness tests. Specifically, after the implementation of the national comprehensive new urbanization pilot policy, SO_2 emissions were reduced by 52.86%.
2. The impact mechanism test underscores that new urbanization construction effectively curbs urban environmental pollution by fostering energy structure optimization and green technological innovation.
3. Regional disparities revealed by the heterogeneity analysis demonstrate that the inhibitory impact of new urbanization construction on urban environmental pollution is more pronounced in the eastern region compared to the central and western

regions. Furthermore, urban resource endowment influences the impact, with non-resource-based cities experiencing a more pronounced mitigating effect than resource-based cities.

21.6.2 Revelations

1. The construction of new urbanization exerts a mitigating effect on environmental pollution in pilot cities. Building upon the experiences gained from new urbanization construction, it is imperative to further advance the implementation of the pilot policy on new urbanization. This will facilitate the formation of a pilot-propagation operational mechanism, which is essential for achieving high-quality urban development.
2. Optimizing the energy structure and fostering green technological innovation are key strategies for mitigating environmental pollution in the construction of new urban areas. Therefore, in formulating policies for new urban development, emphasis should be placed on optimizing energy structures, advocating green consumption practices, and reducing fossil fuel consumption by residents, enterprises, and other market stakeholders through tax incentives and subsidies. Additionally, a concerted effort should be directed towards bolstering the introduction and research and development of clean energy technologies to enhance the energy utilization efficiency. Embracing green technological innovation is imperative, which involves increasing investment in scientific research and development, fostering a conducive environment for technological innovation, and nurturing professionals in the field of green technology. Furthermore, enterprises should be incentivized to adopt green production practices in production, consumption, and end-of-pipe treatment, thereby augmenting resource utilization efficiency and reducing environmental pollution.
3. The pilot policy has not effectively stimulated the mitigating impact of new urbanization on environmental pollution in the central and western regions. Therefore, tailored new urbanization strategies should be devised in alignment with local conditions, accounting for the differences in geographic locations, population sizes, and advantageous industries in these regions. Differentiated development policies should be formulated to cater to local development requisites and environmental characteristics. In crafting new urbanization policies, there should be a focus on strengthening new urbanization construction in the central and western regions. This approach aims to promote the rational flow of resources, enhance the efficiency of resource allocation, and maximize its role in curbing environmental pollution.
4. The mitigating effect of the pilot policy on environmental pollution in resource-based cities is insignificant. Therefore, the implementation of new urbanization should be tailored based on the cities' resource endowments. Moreover, China should offer special financial support for the construction of new urbanization in areas exhibiting relatively high resource dependencies. This support

aims to prevent the influx of highly polluting enterprises and reduce the level of environmental pollution.

5. Urbanization is a significant global development trend, presenting both challenges and opportunities for different countries. As the world's largest developing country, China's new urbanization policies offer valuable insights for global urban development. China's new urbanization initiatives encompass various aspects, including people-centered urban planning, smart city development, green growth, and urban-rural integration. The implementation of these policies has yielded remarkable outcomes in improving residents' quality of life and promoting sustainable development. For instance, smart city construction and green development strategies can enhance urban management efficiency and environmental sustainability in other countries, while urban-rural coordination and housing policies can contribute to balanced resource allocation and improved housing conditions. However, when drawing lessons from China's experience, countries should consider their own social, economic, and environmental contexts to ensure the effectiveness and adaptability of these policies.

Acknowledgements Ling Li and Jiqiang Zhao also made great contributions to this manuscript. We express our heartfelt thanks to them. This research also was supported by: National Natural Science Foundation of China (72474127) and National Social Science Fund Major Bidding Project of China (18ZDA052). Part of this chapter was published: Li L, Guo J, Wu X, Zhao J. Does promoting new urbanization help mitigate environmental pollution: Evidence from China's pilot policies for new urbanization. Journal of Environmental Management 2025; 388: 125925. https://doi.org/10.1016/j.jenvman.2025.125925. The publication has been conducted with copyright permission.

References

Ahmad, M., Rehman, A., Shah, S. A. A., Solangi, Y. A., Chandio, A. A., & Jabeen, G. (2021). Stylized heterogeneous dynamic links among healthcare expenditures, land urbanization, and CO_2 emissions across economic development levels. *Science of The Total Environment, 753*, 142228.

Awan, A., Sadiq, M., Hassan, S. T., Khan, I., & Khan, N. H. (2022). Combined nonlinear effects of urbanization and economic growth on CO_2 emissions in Malaysia. An application of QARDL and KRLS. *Urban Climate, 46*, 101342.

Cerdá, I. (1867). *Teoría general de la urbanización, y aplicación de sus principios y doctrinas a la reforma y ensanche de Barcelona*. Imprenta Española.

Chaudhry, I. S., Yin, W. H., Ali, S. A., Faheem, M., Abbas, Q., Farooq, F., & Rahman, S. U. (2022). Moderating role of institutional quality in validation of pollution haven hypothesis in BRICS: A new evidence by using DCCE approach. *Environmental Science and Pollution Research, 29*(6), 9193–9202.

Chen, B., & Chu, L. (2022). Decoupling the double jeopardy of climate risk and fiscal risk: A perspective of infrastructure investment. *Climate Risk Management, 37*, 100448.

Chen, M., Liu, W., & Lu, D. (2016). Challenges and the way forward in China's new-type urbanization. *Land Use Policy, 55*, 334–339.

Chen, Y., Shen, L. Y., Zhang, Y., Li, H., & Ren, Y. T. (2019). Sustainability based perspective on the utilization efficiency of urban infrastructure - a China study. *Habital International, 93*, 102050.

Cui, H., & Cao, Y. (2023). How can market-oriented environmental regulation improve urban energy efficiency? Evidence from quasi-experiment in China's SO_2 trading emissions system. *Energy, 278*, 127660.

Dinda, S. (2004). Environmental Kuznets curve hypothesis: A survey. *Ecological Economics, 49*(4), 431–455.

Gao, Y. P., Guo, C. Y., & Wei, Y. (2022). Study on the dynamic evolution and synergistic relationship between urbanization and eco-efficiency through the selection of 76 agricultural counties and districts in Liaoning province as research units. *Frontiers In Ecology and Evolution, 10*, 975426.

Gobbo, J. A., Busso, C. M., Gobbo, S. C. O., & Carreão, H. (2018). Making the links among environmental protection, process safety, and industry 4.0. *Process Safety and Environmental Protection, 117*, 372–382.

Grossman, G. M., & Krueger, A. B. (1995). Economic growth and the environment. *The Quarterly Journal of Economics, 110*(2), 353–377.

Guan, Y., Xiao, Y., Rong, B., Lu, W., Zhang, N., & Qin, C. (2023). Assessing the synergy between CO2 emission and ambient PM2.5 pollution in Chinese cities: An integrated study based on economic impact and synergy index. *Environmental Impact Assessment Review., 99*(106989).

Guo, P., & Liang, D. (2022). Has the low-carbon pilot policy improved urban carbon emission efficiency?—A quasi-natural experiment based on low-carbon pilot cities. *Journal of Natural Resources, 37*(07), 1876–1892.

Hao, Y., Zheng, S. Q., Zhao, M. Y., Wu, H. T., Guo, Y. X., & Li, Y. W. (2020). Reexamining the relationships among urbanization, industrial structure, and environmental pollution in China-new evidence using the dynamic threshold panel model. *Energy Reports, 6*, 28–39.

Huber, J. (2000). Towards industrial ecology: Sustainable development as a concept of ecological modernization. *Journal of Environmental Policy & Planning, 2*(4), 269–285.

Jacobson, L. S., LaLonde, R. J., & Sullivan, D. G. (1993). Earnings losses of displaced workers. *The American Economic Review, 83*(4), 685–709.

Li, J., & Xu, B. (2018). "curse" or "blessing": How does resource abundance affect China's green economic growth? *Economic Research Journal, 53*(09), 151–167.

Li, Y., Dai, J., & Cui, L. (2020). The impact of digital technologies on economic and environmental performance in the context of industry 4.0: A moderated mediation model. *International Journal of Production Economics, 229*, 107777.

Li, S., Liu, Y., Elahi, E., Meng, X., & Deng, W. (2023). A new type of urbanization policy and transition of low-carbon society: A "local- neighborhood" perspective. *Land Use Policy, 131*, 106709.

Li, W., Li, D., & Yang, S. (2022). The impact of internet penetration on venture capital investments: Evidence from a quasi-natural experiment. *Journal of Corporate Finance, 76*, 102281.

Li, X., Zheng, Z., Shi, D., Han, X., & Zhao, M. (2023). New urbanization and carbon emissions intensity reduction: Mechanisms and spatial spillover effects. *Science of The Total Environment, 905*, 167172.

Lin, B., & Zhu, J. (2021). Impact of China's new-type urbanization on energy intensity: A city-level analysis. *Energy Economics, 99*, 105292.

Liu, X., Sun, T., & Feng, Q. (2020). Dynamic spatial spillover effect of urbanization on environmental pollution in China considering the inertia characteristics of environmental pollution. *Sustainable Cities and Society, 53*, 101903.

Liu, Y., Deng, W., Wen, H., & Li, S. (2024). Promoting green technology innovation through policy synergy: Evidence from the dual pilot policy of low-carbon city and innovative city. *Economic Analysis and Policy, 84*, 957–977.

Liu, Y., & Zhou, X. (2025). Does the new-type urbanization policy promote green technological innovation? Evidence from a quasi-natural experiment in China. *Applied Economics.*, 1–16.

Lou, L., Li, J., & Zhong, S. (2021). Sulfur dioxide (SO_2) emission reduction and its spatial spillover effect in high-tech industries: Based on panel data from 30 provinces in China. *Environmental Science and Pollution Research, 28*(24), 31340–31357.

Ma, Q., & Shi, F. (2023). New urbanization and high-quality urban and rural development: Based on the interactive coupling analysis of industrial green transformation. *Ecological Indicators, 156*, 111044.

Nie, C., Luo, W., Feng, Y., Chen, Z., & Health, P. (2023). The impact of economic growth target constraints on environmental pollution: Evidence from China. *International Journal of Environmental Research and Public Health, 20*(4), 2831.

Pan, Y., Teng, T., Wang, S., & Wang, T. (2024). Impact and mechanism of urbanization on urban green development in the Yangtze River Economic Belt. *Ecological Indicators, 158*, 111612.

Rao, C., Yan, B., Research, P. (2020). Study on the interactive influence between economic growth and environmental pollution. *Environmental Science Pollution Research, 27*, 39442–39465.

Richardson, H. W. (1973). Theory of the distribution of city sizes: Review and prospects. *Regional Studies, 7*(3), 239–251.

Seto, K. C., Sánchez-Rodríguez, R., Fragkias, M., & Resources. (2010). The new geography of contemporary urbanization and the environment. *Annual Review of Environment, 35*, 167–194.

Shang, J., Wang, Z., Li, L., Chen, Y., & Li, P. (2018). A study on the correlation between technology innovation and the new-type urbanization in Shaanxi province. *Technological Forecasting and Social Change, 135*, 266–273.

Shao, J., & Wang, L. (2023). Can new-type urbanization improve the green total factor energy efficiency? *Evidence from China. Energy, 262*, 125499.

Shapiro, J. S., & Walker, R. (2018). Why is pollution from US manufacturing declining? The roles of environmental regulation, productivity, and trade. *American Economic Review, 108*(12), 3814–3854.

Shen, Y., Zhang, F., Gong, M., & Huang, J. (2022). The urbanization–environmental pollution Nexus: An analysis based on a spatial perspective. *Emerging Markets Finance and Trade, 58*(8), 2355–2367.

Suki, N. M., Sharif, A., Afshan, S., & Suki, N. M. (2020). Revisiting the environmental Kuznets curve in Malaysia: The role of globalization in sustainable environment. *Journal of Cleaner Production, 264*, 121669.

Sun, H., & Chen, F. (2022). The impact of green finance on China's regional energy consumption structure based on system GMM. *Resources Policy, 76*, 102588.

Sun, Y., Li, H., Andlib, Z., & Genie, M. G. (2022). How do renewable energy and urbanization cause carbon emissions? Evidence from advanced panel estimation techniques. *Renewable Energy, 185*, 996–1005.

Uttara, S., Bhuvandas, N., & Aggarwal, V. (2012). Impacts of urbanisation on environment. *IJREAS, 2*.

Wang, S., Gao, S., Li, S., & Feng, K. (2020). Strategizing the relation between urbanization and air pollution: Empirical evidence from global countries. *Journal of Cleaner Production., 243*, 118615.

Wang, X., Wang, X., Ren, X., & Wen, F. (2022). Can digital financial inclusion affect CO_2 emissions of China at the prefecture level? Evidence from a spatial econometric approach. *Energy Economics, 109*, 105966.

Wen, Z., Chang, L., Hau, K. T., & Liu, H. (2004). Testing and application of the mediating effects. *Acta Psychologica Sinica, 36*(05), 614–620.

Wu, Y., Cui, Z., & Hu, C. (2021). Does the new urbanization influence air quality in China? *Frontiers in Environmental Science, 9*, 645010.

Xu, A., Song, M., Wu, Y., Luo, Y., Zhu, Y., & Qiu, K. (2024). Effects of new urbanization on China's carbon emissions: A quasi-natural experiment based on the improved PSM-DID model. *Technological Forecasting and Social Change, 200*, 123164.

Xu, F., Huang, Q., Yue, H., He, C., Wang, C., & Zhang, H. (2020). Reexamining the relationship between urbanization and pollutant emissions in China based on the STIRPAT model. *Journal of Environmental Management, 273*, 111134.

Yan, C., Li, H., & Li, Z. (2022). Environmental pollution and economic growth: Evidence of SO_2 emissions and GDP in China. *Frontiers in Public Health, 10*, 930780.

Yan, M., Zhao, J., Yan, S., & Zhu, M. (2023). Coupling coordination of new urbanization in Chinese urban agglomeration—Characteristics and driving factors. *Environmental Science and Pollution Research, 30*(55), 117082–117095.

Yu, B. (2021). Ecological effects of new-type urbanization in China. *Renewable and Sustainable Energy Reviews, 135*, 110239.

Zhang, D., Xu, J., Zhang, Y., Wang, J., He, S., & Zhou, X. (2020). Study on sustainable urbanization literature based on web of science, scopus, and China national knowledge infrastructure: A scientometric analysis in CiteSpace. *Journal of Cleaner Production, 264*, 121537.

Zhang, W., Xu, Y., Streets, D. G., & Wang, C. (2023). Can new-type urbanization realize low-carbon development? A spatiotemporal heterogeneous analysis in 288 cities and 18 urban agglomerations in China. *Journal of Cleaner Production, 420*, 138426.

Zhang, Y., & Cai, Q. (2022). Impact mechanism of new urbanization on environmental pollution: Empirical analysis based on spatial panel model. *Frontiers in Public Health, 10*, 928100.

Zhang, Y., & Ran, C. (2023). Effect of digital economy on air pollution in China? New evidence from the "National big Data Comprehensive Pilot Area" policy. *Economic Analysis and Policy, 79*, 986–1004.

Zhao, C., & Wang, B. (2022a). How does new-type urbanization affect air pollution? Empirical evidence based on spatial spillover effect and spatial Durbin model. *Environment International, 165*, 107304.

Zhao, C., & Wang, B. (2022b). How does new-type urbanization affect air pollution? Empirical evidence based on spatial spillover effect and spatial Durbin model. *Environment International, 165*, 107304.

Zhao, J., Zhao, Z., & Zhang, H. (2021). The impact of growth, energy and financial development on environmental pollution in China: New evidence from a spatial econometric analysis. *Energy Economics, 93*, 104506.

Chapter 22
Spatio-Temporal Impact of Major Events on Air Quality Based on Spatial Differences-in-Differences Model: Big Data Analysis from China

22.1 Introduction

According to the latest evaluation of WHO (2018), 90% of the world's population currently lives in areas where $PM_{2.5}$ levels exceed WHO limits (an average of 10 ug/m^3 per year). In developed countries and regions such as North America, Europe, the haze-related problems are well controlled, but haze still poses a serious threat in undeveloped regions such as in East Asia, South Asia and Africa (Rafaj et al., 2018).

In China, with the global warming and increasing calm weather days in recent decades, the frequent occurrence of haze has attracted wide attention of the government and all sectors of the society. The effective prevention and control of air pollution and the guarantee of air quality are closely related to the image of the state and government, as well as the health of the people (Matus et al., 2012). When faced with major political conferences or events, governments at all levels take various measures to control air pollution. For example, during the 2008 Beijing Olympic Games, the Ministry of Environmental Protection laid down *Measures for Air Quality Guarantee in Beijing for the 29th Olympic Games*, and the General Office of Hebei Provincial Government formulated *the Emergency Measures for Air Pollution Control under Extremely Adverse Meteorological Conditions* during the Olympic Games in the Hebei Province. These efforts were meant to ensure that the atmospheric environment quality during the Olympic Games met acceptable standards to ensure the normal progress of the competition. In addition, Beijing and its surrounding provinces implemented a series of temporary control measures for air pollution, including strengthening motor vehicle management, suspending part of a construction site operation, enhanced road cleaning, shutting down and limiting production of key pollution enterprises (Song et al., 2019, 2020). Other measure taken included the reduction in the discharge of organic exhaust and implementing emergency measures for pollution control under extreme adverse weather conditions. These measures achieved remarkable results. For example, from July 20th to September 20th in 2008, the total emission of air pollutants in Beijing dropped by

X. Wu, *Environmental Disaster in China*,
https://doi.org/10.1007/978-981-92-0910-1_22

34.98%, the best level in 14 years after 1995 (Chen et al., 2013). The momentary "political blue sky" also appeared during the APEC meetings, the 70th anniversary parade of anti-fascist victory, the Nanjing Youth Olympic Games and the Hangzhou G20 Summit. Whilst it seems a common practice to conduct temporary air quality control during major events (Shi et al., 2016; Wu et al., 2018, 2019a, 2019b), little is known about the real impact of these major events on the quality of air in these cities and the surrounding towns. This study therefore seeks to address the following questions how much impact do major events have on urban air quality? How long does the impact last? What are the spatial characteristics of the impact?

Very limited studies on the impact of major events on air quality have been carried using quantitative methods. In this regard, this paper adopts a quantitative approach by using data on the daily air quality index (AQI) and the concentration of individual pollutants constituting air quality indices of 140 cities from 2015 to 2017, to analyse the impact of major events on air quality in China. The data included details on major competitions, important conferences, and urban air quality from two dimensions of time and space. The data was analysed using a spatial differences-in-differences (DID).

The rest of this paper is arranged as follows. The second part deals with the literature review, followed by the data and empirical strategy, indicators and sources of data. The fourth section covers the empirical analysis, followed by the conclusion.

22.2 Literature Review

Some cities, both at home and abroad, endeavor to take temporary control measures to ensure air quality during major events, such as the Busan Asian Games in 2002, Delhi Federal Games in 2010, Beijing Olympic Games in 2008, Beijing APEC Conference in 2014, etc. Several authors have conducted thorough analyses and reported improvements in urban air quality during major events (Beig et al., 2013; He et al., 2016, 2025; Huang et al., 2015; Jia & Chen, 2019; Lee et al., 2005; Li et al., 2019a, 2019b; Liu et al., 2016; Wang, 2012; Wang & Liu, 2025; Wang & Xie, 2009; Zhao et al., 2016; Zuo & Yu, 2026). For example, Lee et al. (2005) investigated the significant concentration decrease of PM_{10}, CO, NO_2 and SO_2 in 13 air stations during the 24th Asian Games in Busan, Korea in 2002. Beig et al. (2013) found that Delhi adopted a series of measures, such as vehicle and traffic control, factory relocation and power plant emission reduction when the 2010 Commonwealth Games was held in India.

In addition, Wang and Xie (2009) evaluated the effect of environmental quality improvement by reducing traffic emissions during the 2008 Beijing Olympic Games. Based on the analysis of "Parade Blue" and "APEC Blue", Liu et al. (2016) found that the concentration of NO_2 decreased by 43% in the military parade, compared with the normal level, whilst a decrease of 21% in concentration of NO_2 was reported during the APEC conference. Huang et al. (2015) analysed the regional emission control effects and found the "APEC Blue" phenomenon in China. Zhao et al. (2016) made a comparison between the air pollutant concentration data in Beijing from August 1, 2015 to September 18, 2015 and the monitoring data of the same period in 2014. Based on the comparison of NO_2 concentration in Hangzhou during and before the G20 Summit, Zhao et al. (2017) made a conclusion that the concentration of NO_2 decreased significantly during the G20 Summit. Ngo et al. (2019) gave an interesting research about the effects of transboundary air pollution following major events in China on air quality in the USA.

Towards the end of 2019, COVID-19 as a major event, has had a huge impact on industrial production and human life. Many scholars have studied the impact of COVID-19 on air pollution emissions. For example, He et al. (2020) analysed the short-term impact of China's lockdown measures on urban air governance during COVID-19. Fan et al. (2020) studied the impact of China's prevention and control measures on air pollution during COVID-19. Bera et al. (2020) studied the impact of the COVID-19 blockade on air pollution in Kolkata (India). Other studies include Bogdan (2020), Dutheil et al. (2020), Li et al. (2020), etc. However, these studies only focused on the impact of a single event not multiple events on air pollution. Based on the above considerations, this paper explores a number of events, and their impact on air pollution to supplement previous studies.

Despite the findings from available literature on urban air quality, this paper holds that the existing studies (e.g. list some of the studies—Beig et al., 2013; Huang et al., 2015; Lee et al., 2005; Liu et al., 2016; Ngo et al., 2019; Wang & Xie, 2009; Wu et al., 2019c; Xu et al., 2019; Zhao et al., 2016) did not exclude the influence of seasonal trend, year difference, and regional effect on urban air quality. In addition, existing studies did not assess the effect of major events on urban air quality, and also failed to reveal continuous changes and associated spatial characteristics before and after the major events. In order to address these shortfalls, this study posits spatial differences-in-differences to provide a better estimation of the impact of major events on urban air quality. In this regard, the present study adopts the spatial DID models to investigate spatio-temporal impact of major events on urban air quality as new contribution to the current literature.

22.3 Data and Empirical Strategy

22.3.1 Variable and Data

22.3.1.1 The Air Quality Level

Air pollutants mainly include particulate matter, nitrogen oxides, carbon monoxide, sulfur dioxide, etc. (Li et al., 2019a, 2019b). In the present study, the daily average air quality index (AQI) and the concentration data of fine particulate matter ($PM_{2.5}$), inhalable particulate matter (PM_{10}), sulfur dioxide (SO_2), carbon monoxide (CO), nitrogen dioxide (NO_2) and ozone (O_3) are selected to measure the air pollution quality level.

22.3.1.2 Control Variable

In an attempt to exclude the influence of other factors (apart from major events) on air quality indicators, weather factors are added as control variables X_{it} in this paper. Weather variables mainly include the highest temperature (highest_t), lowest temperature (lowest_t), rainfall, density of snow and wind grade, etc., to control the impact of weather changes on haze level.

22.3.1.3 Sample Selection

The major events in this paper are defined as international or national sporting events and important political conferences in China. The major events considered were from January 2, 2015 to November 28, 2017 (as shown in Table 22.1).

The sample include 28 cities involved in these activities (9 host cities and 19 co-host cities). Based on the panel data of 140 cities over a span of 1094 days, 28 cities involved in major events (treatment group) were identified together with 112 control cities (control group) around them (as shown in Table 22.2 and Fig. 22.1).

22.3.1.4 Data Collection

In this paper, the daily AQI and the daily mean value of individual pollutant concentration are used as the object of analysis. AQI data was downloaded from the Environmental Protection Data Center of the People's Republic of China (http://datacenter.mee.gov.cn/websjzx/queryIndex.vm), and the individual pollutant concentration data was obtained from the national real-time publishing platform of urban air quality of China Environmental Monitoring Station (http://106.37.208.233:20035/). In order to ensure the slight fluctuation of the city fixed effects in the short term, the time span of the samples was not too long. Besides, due to the missing data, the final time span of samples is from January 2, 2015 to November 28, 2017.

In addition, meteorological data such as rainfall, temperature and wind grade are derived from the historical weather data of the cities provided by "2345 Weather Network". Specific indicators include the highest temperature (highest_t), lowest temperature (lowest_t), rainfall, density of snow and wind grade, etc. In terms of the wind grade, this paper adopts the mean treatment, such as 3.5 wind grade instead of 3–4 grade.

Table 22.1 Major events during January 2, 2015 and November 28, 2017

Major event	Start Date	End Date	Host Cities
The first youth movement of the people's Republic of China	October 18, 2015	October 27, 2015	Fuzhou
The Fifth Plenary Session of the 18th Communist Party of China (CPC) National Congress	October 26, 2015	October 29, 2015	Beijing
Fourth Meeting of China-Central and Eastern European Leaders	November 24, 2015	November 25, 2015	Suzhou
Fourteenth Meeting of the Council of Heads of Government of the Shanghai Cooperation Organization (SCO) Member States	December 14, 2015	December 15, 2015	Zhengzhou
Third Session of the Twelfth Chinese People's Political Consultative Conference (CPPCC) National Committee	March 3, 2015	March 13, 2015	Beijing
Third Session of the Twelfth National People's Congress	March 5, 2015	March 15, 2015	Beijing
Beijing International Association Of Athletics Federations (IAAF) World Track and Field Championship	August 22, 2015	August 30, 2015	Beijing
The 70th Anniversary Parade of the Victory of the War of Resistance Against Japan	September 3, 2015	September 3, 2015	Beijing
Boao Forum for Asia (BFA)	March 22, 2016	March 25, 2016	Haikou
Fourth Session of the Twelfth Chinese People's Political Consultative Conference (CPPCC) National Committee	March 3, 2016	March 14, 2016	Beijing
Fourth Session of the Twelfth National People's Congress	March 5, 2016	March 14, 2016	Beijing
Group of Twenty (G20) summit	September 4, 2016	September 5, 2016	Hangzhou
The Standing Committee of the Political Bureau of the Central Committee of the Communist Party of China (CPC)	January 10, 2017	January 10, 2017	Beijing
The Nineteenth National People's Congress of the Communist Party of China	October 18, 2017	October 24, 2017	Beijing
Five Sessions of the Twelfth Chinese People's Political Consultative Conference (CPPCC) National Committee	March 3, 2017	March 13, 2017	Beijing
Fifth Session of the Twelfth National People's Congress	March 5, 2017	March 15, 2017	Beijing
International Cooperation Summit Forum	May 14, 2017	May 15, 2017	Beijing
BRICS (Brazil, Russia, India, China, South Africa) Games	June 17, 2017	June 21, 2017	Guangzhou

(continued)

Table 22.1 (continued)

Major event	Start Date	End Date	Host Cities
Thirteenth National Games of the People's Republic of China	August 27, 2017	September 8, 2017	Tianjin
General Assembly of International Standards Organization	September 12, 2017	September 14, 2017	Beijing
BRICS (Brazil, Russia, India, China, South Africa) Summit	September 3, 2017	September 5, 2017	Xiamen
Thirteenth Hangzhou National Student Games	September 4, 2017	September 8, 2017	Hangzhou

Table 22.2 A list of cities involved in events and corresponding control cities

Cities involved in events (Treatment group)	Beijing, Changzhou, Foshan, Fuzhou, Guangzhou, Haikou, Hangzhou, Huzhou, Jinan, Jiaxing, Nanjing, Nantong, Ningbo, Quanhzou, Sanya, Xiamen, Shanghai, Suzhou, Taizhou, Tianjin, Wuzi, Wuhan, Xian, Zhenjiang, Zhengzhou, Zhongshan, Shenzhen, Shaoxing
Control cities (Control group)	Shijiazhuang, Zhangjiakou, Handan, Langfang, Baoding, Cangzhou, Xingtai, Chengde, Hengshui, Qinhuangdao, Tangshan, Xinxiang, Kaifeng, Jiaozuo, Hebi, Xuchang, Luoyang, Pingdingshan, Anyang, Puyang, Sanmenxia, Nanyang, Shangqiu, Xinyang, Zhoukou, Zhumadian, Luohe, Yancheng, Huaian, Suqian, Xuzhou, Lianyungang, Yangzhou, Nanping, Zhangzhou, Putian, Sanming, Longyan, Ningde, Wenzhou, Jinhua, Taizhou, Lishui, Zhoushan, Quzhou, Weinan, Tongchuan, Shangluo, Baoji, Ankang, Hanzhong, Yanan, Yulin, Taian, Liaocheng, Zibo, Dezhou, Binzhou, Qingdao, Weihai, Yantai, Dongying, Weifang, Zhizhao, Heze, Linyi, Zaozhuang, Jining, Xiaogan, Suizhou, Huangshi, Xianning, Jingzhou, Xiangyang, Yichang, Shiyan, Jingmen, Ezhou, Huanggang, Huizhou, Dongguan, Zhuhai, Jiangmen, Zhaoqing, Qingyuan, Shantou, Chaozhou, Jieyang, Shanwei, Zhanjiang, Maoming, Yangjiang, Shaoguan, Yunfu, Meizhou, Heyuann, Datong, Shuozhou, Taiyuan, Yangquan, Changzhi, Jincheng, Linfen, Suzhou, Bengbu, Chuzhou, Maanshan, Wuhu, Xuancheng, Huangshan, Qingyang, Changsha

22.3.2 *Empirical Strategy*

During the major event of a city, other cities without such events become the control group. In this regard, a double difference is shown between the occurrence and non-occurrence periods in the same city, as well as the host and non-host cities in the same period, which lays the foundation of the idea of spatial differences-in-differences.

Specifically speaking, the model is established as follows.

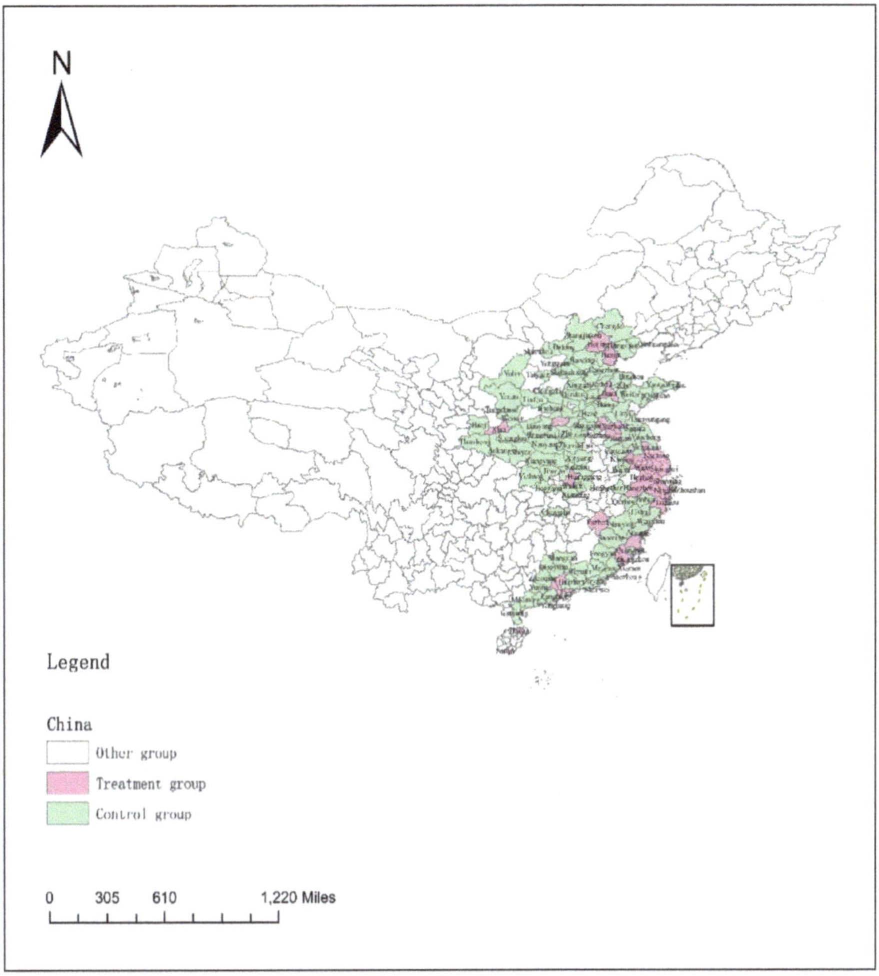

Fig. 22.1 Location of the cities in treatment group and control group

$$\begin{aligned} y_{it} = \beta_0 + \rho W y_{it} + \beta_1 event_i dt_{it1} + \beta_2 event_i dt_{it2} \\ + \beta_3 event_i dt_{it3} + \beta_4 dis\tan ce_{ij} dt_t + \lambda X_{it} + \mu_i + v_t + \varepsilon_{it} \end{aligned} \tag{22.1}$$

where subscript i represents the ith city, subscript t represents the date (year, month, day) of the data, ε_{it} represents the stochastic disturbance, and Y_{it} represents the air quality level of the ith city at the time t. W represents the spatial weight matrix (please see exlaination of W in Part of "3.2.1 Establishment of spatial weight matrix"). $event_i$ denotes the cities involved in the major event. If $event_i = 1$, the ith city is the one involved in the major event, otherwise $event_i = 0$. Owing to the different time of each major event. In order to investigate the duration of the impact of major events on urban air quality, the values of "1–5 days after events" and "6–10 days after events" were added to dt_{it}. Given the occurrence of major events in the ith city at the time t, $dt_{it1} = 1$, otherwise 0; at the time t after 1–5 days of the major event in the ith city, $dt_{it2} = 1$, otherwise 0; at the time t after 6–10 days of the major event in the ith city, $dt_{it3} = 1$, otherwise 0. Besides, β_1, β_2 and β_3 are used to analyze the duration of the impact of the major events on urban air quality.

Where $dis\tan ce_{ij}$ is a grouping variable, representing the distance of the jth city from where the major event takes place (in ith city) at the time t. dt_t denotes whether a major event occurs within the time t. Accordingly, the coefficient β_4, is the estimation of the impact of spatial scope of each major event on urban air quality and the difference of its impact on different cities.

A set of control variables X_{it} is added to exclude the influence of other factors on air quality indicators. μ_i refers to the regional dummy variable, which reflects the city- fixed effects that will not change in a short time. v_t represents a set of time-fixed effects used to control the effects of seasonal factors and human working hours on air quality. Seasonal factors mainly include dummy variables of the year, month in the year and the week in the year, and the effects of human working hours on air pollution include the dummy variables of the day in a week.

22.3.2.1 Establishment of Spatial Weight Matrix

Geographical location and distance are important factors that need to be considered. For the sake of comparison, adjacent weight matrix and inverse distance weight matrix are selected, and their definitions are as follows.

Set w_{ij} as the distance between region i and region j, and

$$w_{ij} = \begin{cases} 1, & Regionaisadjacenttoregionb \\ 0, & Regionaisnotadjacenttoregionb \end{cases} \tag{22.2}$$

Therefore, $W = \begin{bmatrix} w_{11} & \cdots & w_{1n} \\ \vdots & \vdots & \vdots \\ w_{n1} & \cdots & w_{nn} \end{bmatrix}$ is the adjacent weight matrix of n cities, and the main diagonal elements are all 0 (the distance between the same city is 0).

Define: $w_{ij} = \frac{1}{d_{ij}}$ d_{ij} is defined as geographical distance.

$W_d = \begin{bmatrix} w_{11} & \cdots & w_{1n} \\ \vdots & \vdots & \vdots \\ w_{n1} & \cdots & w_{nn} \end{bmatrix}$ is the inverse distance weight matrix, in which the main diagonal elements are also 0.

The rows of the spatial weight matrix need to be standardized. Each element in the matrix (denoted as $\tilde{w}_{ij}$) is divided by the sum of the elements in its row. Ensure that the sum of the elements in each row is 1, and the formula is:

$$w_{ij}{}' \equiv \frac{\tilde{w}_{ij}}{\sum_j \tilde{w}_{ij}} \tag{22.3}$$

22.3.2.2 The Selection of Control Group

The impact assessment in this paper is mainly based on cities where major events occurred during the period under review. In this sense, cities with major events are defined as treatment groups, and the ones around which no major event takes place are defined as control groups. A double difference is made between the two groups of cities.

22.3.2.3 Robustness Test

In order to further prove the robustness of the estimated results, the methods adopted are as follows.

(1) Transform window. Samples of 20, 30 and 40 days before and after major events are retained. The retained similar results account for the robustness of results.
(2) Using different controls. Different cities are selected as the control group to investigate the consistency of the findings.
(3) Placebo test. Under the condition of fictitious event time, the change of regressive results from significant to insignificant verifies the robustness of the previous conclusion.

22.4 Empirical Analysis

22.4.1 Exploratory Analysis

The daily air quality data of the 28 cities involved in the events are divided into two parts according to dt_{it}=1 or 0, that is to say, whether the major events take place in the ith city at the time t. Descriptive statistics are made for the main variables, and longitudinal comparison is made with the sample data of the 28 cities. The average and maximum concentrations of AQI and individual pollutant during major events were found to be lower than those during non-major events (Table 22.3). Among them, the average AQI of the 28 activities involved in the urban major events is 64.346, which is 10.3 lower than that without major events, decreasing by about 16.01%. Besides, the maximum AQI during major events is 331, much lower than that without major events. From the perspective of individual pollutant, the average concentration of fine particulate matter ($PM_{2.5}$), inhalable particulate matter (PM_{10}), sulfur dioxide (SO_2), carbon monoxide (CO), nitrogen dioxide (NO_2) and ozone (O_3) in the 28 cities were lower than those during non-major events. Therefore, based on preliminary judgement, the air quality during major events is better than that during non-major events.

The samples with major events are divided into two parts. One part includes the samples of cities involved in the major events, and the other involves the ones without major events for a horizontal comparison. The average AQI of the cities involved was 66.74, which is much lower than that of the cities without major events (Table 22.4). In terms of individual pollutant, the mean concentration values of fine particulate matter ($PM_{2.5}$), inhalable particulate matter (PM_{10}), sulfur dioxide (SO_2), carbon monoxide (CO), nitrogen dioxide (NO_2) and ozone (O_3) in the cities with major events are lower than those without major events. On this basis, it can be preliminarily demonstrated that the occurrence of major events can improve the urban air quality.

Table 22.5 presents the mean values of air quality indicators and control variables in the 28 cities during major events, 1–5 days before and after major events, as well as 6–10 days before and after major events. It can be seen that except ozone, AQI and the average values of five other single pollutants show a sudden increase within 1–5 days after the end of major events, and then decrease slowly within 6–10 days. This indicates that local governments may improve air pollution quality by adopting temporary measures in advance (e.g. 7 days). However, after major events, the air quality index rebounds significantly, even higher than before. In this regard, it can

Table 22.3 Non-event period and event period for 28 cities involved in the events

Variable	Units	Non-event period			Event period			Difference		
		Mean	Max	Min	Mean	Max	Min	Mean	Max	Min
AQI	Index number	74.65	500	11	64.346	331	15	10.30	169.00	−4.00
$PM_{2.5}$	μg/m^3	47.68	606	2.55	37.45	228	6.52	10.23	378.00	−3.97
PM_{10}	μg/m^3	78.96	838	6.67	61.86	365	15.73	17.10	473.00	−9.06
SO_2	μg/m^3	16.64	175	1.57	12.34	89	2.47	4.30	86.00	−0.90
NO_2	μg/m^3	38.52	161	3.36	33.87	87	7.45	4.65	74.00	−4.09
CO	μg/m^3	0.93	9	0	0.92	3	0	0.01	6.00	0.00
O_3	μg/m^3	112.45	363	2.25	106.85	270	31	5.60	93.00	−28.75
Highes_t	°C	22.96	41	−10.84	23.34	35	4.51			
Lowest_t	°C	15.47	31	−15.27	15.76	27	−4.84			
Rainfall	Dummy variable	0.39	1	0	0.33	1	0			
Density of snow	Dummy variable	0.01	1	0	0	0	0			
Wind grade	Ordinal number	3.1	11.5	1.7	2.99	6	2			

Table 22.4 The cities involved and cities not involved at event period

Variables	Cities involved			Cities not involved			Difference		
	Mean	Max	Min	Mean	Max	Min	Mean	Max	Min
AQI	66.74	423	16	81.12	500	10.01	−14.38	−77.00	5.99
$PM_{2.5}$	37.92	316	6.71	49.86	537	3.02	−11.94	−221.00	3.69
PM_{10}	64.83	398	16	88.15	675	5.05	−23.32	−277.00	10.95
SO_2	12.96	112	2.5	22.19	551	1.11	−9.23	−439.00	1.39
NO_2	34.18	110	7.12	34.72	144	2.13	−0.54	−34.00	4.99
CO	1.012	4	0	1.15	13	0	−0.14	−9.00	0.00
O_3	106.72	264	29.5	114.56	407	2.02	−7.84	−143.00	27.48
Highest_t	23.17	35	2.5	21.22	38	−8.6			
Lowest_t	15.36	27	−5.85	12.56	29	−22.5			
Rainfall	0.41	1	0	0.32	1	0			
Density of snow	0	0	0	0.02	1	0			
Wind grade	2.97	5.5	1.52	2.98	10	2.2			

be speculated that the temporary control has no long-term benefits for air quality. Therefore, the statistically significant difference needs to be further analyzed by establishing a model.

The whole sample of 140 cities are classified by month to calculate the monthly mean of seven air quality indicators. Figure 22.2 shows the trend of the average air quality in the cities from January to December. For ease of observation, the

Table 22.5 The mean values of main variables in 28 cities involved during different periods

Variables	6–10 days before	1–5 days before	Event period	1–5 days after	6–10 days after
AQI	76.15	76.39	74.61	93.84	82.11
$PM_{2.5}$	45.61	47.35	47.611	60.35	54.04
PM_{10}	74.85	74.15	78.88	91.57	89.72
SO_2	15.05	13.96	16.54	19.08	16.08
NO_2	35.41	36.31	38.44	48.01	43.53
CO	1.01	1.06	0.94	1.06	1.01
O_3	113.42	104.91	112.29	106.52	92.84
Highest_t	23.71	23.11	22.98	21.81	21.53
Lowest_t	16.31	15.92	15.48	13.82	14.31
Rainfall	0.311	0.37	0.41	0.29	0.39
Density of snow	0.021	0.011	0.01	0.01	0.01
Wind grade	3.01	1.01	3.11	2.75	2.92

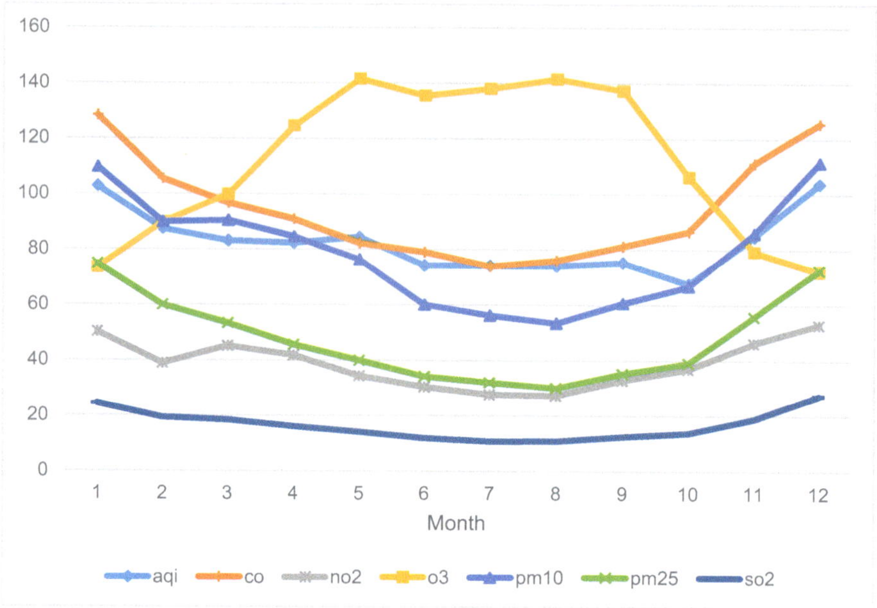

Fig. 22.2 The trend of the average air quality in cities from January to December. *Notes* Carbon monoxide concentration has been amplified 100 times to facilitate comparison with other pollutants

carbon monoxide concentration is magnified 100 times. As the Fig. 22.2 indicates, there are obvious seasonal characteristics in air quality. The AQI and most single pollutant concentration data in winter are higher than those in summer, while ozone concentration presents an opposite trend. In this sense, in order to eliminate the impact of other factors (excluding major events) on air quality, it is necessary to make some seasonal adjustments of air quality indicators, which accounts for the consideration of seasonal factors in the regression equation.

To illustrate the impact of major events on air quality, the event "the Fifth Plenary Session of the 18th Communist Party of China (CPC) National Congress" which took place in Beijing in October 26–29, 2015 was chosen as a case. In order to reflect the comprehensive effect of air pollution, the daily data of AQI is selected. To make a comparison before and after the event, the data of 10 days before and after the meeting were selected. To carry out the comparison between cities, Baoding in Hebei province, which is close to Beijing, was selected as the comparison sample (Fig. 22.3).

From the date comparison, AQI gradually decreases before the event and AQI is the lowest point in the event. After the event, AQI gradually rose. This shows that in order to hold the meeting, the government carried out air quality control, such as restricting the production of highly polluting factories, which gradually improved the air quality. But after the meeting, factories resumed production and air pollution rebounded in retaliation.

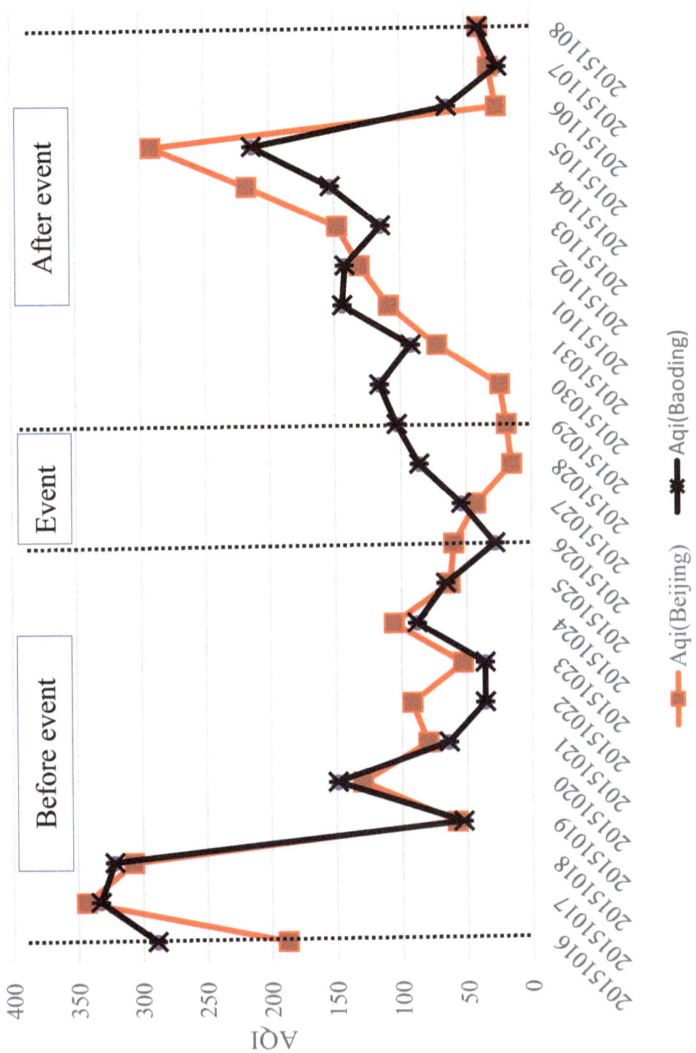

Fig. 22.3 Comparison of AQI series in Beijing and Baoding

In terms of urban comparison, the daily air pollution data of Beijing and Baoding have followed roughly the same trend. But in the days and days after the meeting, Beijing's air quality was significantly better than Baoding's. Interestingly, by November 2, 2015, the level of air pollution in Beijing had surpassed that in Baoding, and the trend of retaliatory rebound is obvious.

The above is a statistical descriptive analysis. Did the event have a significant impact on air quality? Whether there is a statistically significant difference between the cities where the event occurred and the surrounding city requires further quantitative analysis.

22.4.2 *Time Persistence Analysis of the Impacts of Major Events on Urban Air Quality*

22.4.2.1 Empirical Analysis Based on DID

In this paper, the 28 cities involved in hosting major events from January 2, 2015 to November 28, 2017, were selected as the treatment group, and the 112 prefecture-level cities participating in major events in the very province and adjacent provinces were taken as initial control cities. This constitutes a double difference between the occurrence and non-occurrence periods in the same city, as well as between the cities with and without major cities. On this basis, a spatial differences-in-differences model could be established directly. Regression based on Model 2 with AQI as the interpreted variable shows the remarkable effect of the model as a whole. An elaborate examination is made into the existence of inter-group heteroscedasticity, inter-group synchronous correlation and intra-group autocorrelation in the perturbation test of regression model. It is found that there are intergroup heteroscedasticity and intergroup synchronous correlation, but no intra-group autocorrelation. Therefore, the clustered standard error is used in the regression results in this paper. The regression coefficients of each variable are summarized in Table 22.6.

The observed coefficients during the major events are significantly negative, and the value of AQI is obviously lower than that of other periods at a significant level of 5% among the four models (Table 22.6). The model presents a better fitting effect with the addition of the city and temporal fixed effects. In terms of the dummy variables with city fixed effects, the significant coefficient of each region indicates that there are differences in air quality among the regions, and city-fixed effects should be added. However, the coefficients for most years, most months, most weeks of the year, and most days of the week are significant with respect to the dummy variables at each time (due to space limitations, the results are not reported), suggesting the existence of time-fixed effects, as well as more addition of dummy variables such as seasons and holidays. Therefore, the double fixed effect model is used for further analysis.

Table 22.6 Empirical results of AQI based on DID

Variables	Model 1	Model 2	Model 3	Model 4
Event period	−11.76***	−5.85**	−13.96***	−8.24**
	(2.78)	(2.71)	(2.64)	(3.24)
1–5 days after	4.12	7.05**	3.38***	5.51***
	(2.88)	(2.81)	(2.72)	(2.02)
6–10 days after	7.00**	4.85*	5.18***	1.84
	(2.89)	(2.83)	(2.74)	(5.01)
Wind grade	−4.78***	−4.95***	−0.94***	−0.81**
	(0.17)	(0.18)	(0.18)	(0.39)
Rainfall	−5.03***	−3.70***	−2.06***	0.61
	(0.32)	(0.32)	(0.31)	(0.44)
Density of snow	2.21**	5.71***	0.04	5.27**
	(1.17)	(1.16)	(1.1)	(2.07)
Highest_t	1.72***	2.26***	1.30***	2.06***
	(0.05)	(0.05)	(0.05)	(0.16)
Lowest_t	−3.92***	−4.11***	−3.24***	−1.50***
	(0.05)	(0.05)	(0.05)	(0.25)
Time fixed effects	No	Yes	No	Yes
City fixed effects	No	No	Yes	Yes
R-square	0.19	0.24	0.29	0.36
Number of observations	17,640	15,120	13,860	13,490

Notes(1) *, ** and *** are significant at 10%, 5% and 1% levels, respectively, with standard errors in parentheses, the same as in the other tables. (2) One of the reasons for the low value of R-Square is that PM is affected by many factors, such as energy consumption, industrial structure, economic conditions, etc. Since the focus of this paper is to analyze the impact of major events on PM, these factors are not included

According to the double fixed effect model, the AQI value drops by 8.24% during major events, equivalent to 10% of the average AQI during non-major events. In other words, during the period of major events, AQI decreases by about 10% on average compared with that when events do not occur. After 1–5 days or 6–10 days of the major events, AQI presents a sharp increase, even much higher than usual. For the weather variables, the significant coefficients display the rationality of the variable addition. Of all the weather factors, wind grade is significantly correlated with air quality, which is also true of rainfall and air quality, but the highest temperature presents a negative correlation with air quality. In fact, considering the higher temperature, less rainfall, and lower wind grade, pollutants, discharged into the atmosphere, are not easy to diffuse, which leads to the poor air quality. In this regard, it can be seen that the conclusions of the study are consistent with common sense.

Table 22.7 Cross-term coefficient of single atmospheric pollutant

	$PM_{2.5}$	PM_{10}	SO_2	NO_2	CO	O_3
Event period	−9.15**	−8.52**	1.56	−6.61***	−0.01	−15.95***
	(3.53)	(3.73)	(0.88)	(1.60)	(0.06)	(2.05)
1–5 days after	3.59**	3.86***	2.59***	1.00	0.01	−5.00*
	(1.51)	(1.30)	(0.51)	(1.05)	(0.03)	(2.62)
6–10 days after	1.57	6.50	1.57	1.34	−0.01	−20.31***
	(3.90)	(4.46)	(1.04)	(1.15)	(0.01)	(2.29)
Control variable	Yes	Yes	Yes	Yes	Yes	Yes
Time fixed effects	Yes	Yes	Yes	Yes	Yes	Yes
City fixed effects	Yes	Yes	Yes	Yes	Yes	Yes
R-square	0.40	0.44	0.46	0.54	0.47	0.45
Number of observations	17,640	17,640	17,640	17,640	17,640	17,640

In order to further explore the specific impact of the major events on individual air pollutants, the spatial differences-in-differences regression is carried out with each individual pollutant concentration as the explained variable in Regression Model 1. The regression results are shown in Table 22.7.

The long-term impact of major events on individual pollutants revealed very interesting results (Table 22.7). Despite a concentration decrease in PM_{10} and $PM_{2.5}$, there is a rebound after 1–5 days of the major event, even much higher than in the normal period. However, a slight difference is shown between the normal period and 6–10 days after the major event. Thus, temporary air management had no long-term effect on air improvement, but a retaliatory rebound appears after the end of the temporary measures, then returning to the normal state. Besides, an obvious rebound is shown in SO_2 after the major events. As for NO_2, the concentration decreases significantly in the process of major events and returns to the normal level after events. Another individual pollutant, CO, presents a slight decrease during the major events, but not obvious. Ozone is an exception, since the content is neither an indicator of government performance appraisal nor a public concern. Therefore, Ozone is less affected by the major events, but to a larger extent by seasonal factors, and the specific reasons needs to be investigated further (Li et al., 2019a, 2019b).

22.4.2.2 Further Investigation into Robust Test

Due to the short duration of the major events, the period without events had to be appropriately shortened in order to avoid other possible overlooked interference factors. On this basis, the robustness of the results are tested by changing the sample

Table 22.8 Empirical results in different sample windows

	Window	AQI	$PM_{2.5}$	PM_{10}	SO_2	NO_2	CO	O_3
Event period	20 days	−9.15**	−10.42**	−10.89**	0.46	−6.90***	−0.01	− 15.00***
	30 days	−9.92**	−11.08**	−11.32**	0.46	−7.36***	−0.04	−14.75**
	40 days	−10.40	−11.32**	−11.71**	0.55	−7.24***	−0.04	−14.67***
Before event	20 days	10.52***	7.56***	7.16**	3.37***	1.88*	0.04	−6.11**
	30 days	5.89***	3.71***	4.70***	2.87***	1.03	0.01	−5.00**
	40 days	5.81***	3.81**	3.31**	3.03***	1.17	0.02	−4.87**
After event	20 days	8.68	6.79	13.88**	2.15	2.05**	0.02	−18.00***
	30 days	6.41	5.50	3.45**	2.44	2.03	0.01	−16.04***
	40 days	5.30	4.73	11.57**	2.27	1.81*	0.02	−16.58***

Notes (1) In "Event period", "20 days, 30 days and 40 days" means that the event period were extended to 20 days, 30 days and 40 days, separately. (2) In "Before", "20 days, 30 days and 40 days" means that 20 days, 30 days and 40 days data prior to the event were collected, separately. (3) The coefficients for "Event period", "Before event", and "After event", come from the econometric analysis with data during "Event period", before "event", and after "event", and by using the formula (22.1)

window, and samples of 20, 30 and 40 days before and after the major events are retained respectively. After changing the different sample windows, the impact of the major events on air quality remains significant in the event-involved cities, and the air quality rebound after the major events. Moreover, this paper also assesses the robustness of the results with the help of a placebo that artificially sets the time for the major events. Specifically, one month ahead of schedule for the major events, it could be found that the cross-term coefficient was no longer significant, which indicates that the major events indeed contributed to the improvement in urban air quality. At this point, it is reasonable to believe that the above conclusions are robust and valid (Table 22.8).

22.4.3 Spatial Characteristics of the Impacts of Major Events on Urban Air Quality

22.4.3.1 Empirical Analysis Based on Spatial DID

With an aim to eliminate the bias of sample selection, the method of Spatial DID is adopted for further analysis. During the occurrence of each major event, the cities within 900 km from the host city are treated as the treatment group and the others as the control group. The cities beyond 900 km are searched to match the ones within 900 km. It should be noted that altogether there were 9 host cities owing to the unfixed location of the major events and the matching process is carried out separately, nine in total. Furthermore, the equilibrium test should be conducted each time, and then the spatial differences-in-differences method is applied in all successful matched cities. In this paper, the nearest neighbor matching in the caliper scope is used as the matching method, where the caliper is set to 0.01 with one-to-one playback matching. As the equilibrium test shows (Table 22.9), the standard deviation of almost all variables after matching decreases sharply, and the t-test results do not reject the original hypothesis that there is no systematic difference between the treatment group and the control group, which indicates a favorable result of matching.

After 9 times' matching, 120 cities were successfully matched and the regression results of the DID for the 120 cities, based on Model 3, are shown in Table 22.10. Combined with AQI and the regression results of individual pollutant concentrations, the major events significantly improved the air quality of cities within 800 km of the event-hosted cities. Moreover, as the distance increases, the improvement gets smaller and it is no longer significant beyond 800 km. The effect of the major events on the concentration of individual pollutants in cities becomes gradually apparent with the distance decreasing from major events, which is consistent with the results of DID and further verifies the robustness of the results to a large extent.

In addition, owing to the shorter period of the major events than that without their occurrences, the period without events should be appropriately shortened in order to avoid other possible interfering factors. After retaining the samples 30 days before

Table 22.9 DID applicability test

Variables	Matched or not	Mean		Bias (%)	Bias reduction (%)	P-value
		Treatment group	Control group			
Wind grade	Unmatched	2.9629	2.9326	8.70	−74.8	0.609
	Matched	2.9243	2.8713	15.20		0.250
Rainfall	Unmatched	0.0232	0.0056	197.8	87.3	0.000
	Matched	0.0163	0.0140	25.10		0.870
Density of snow	Unmatched	19.988	24.319	−198.8	94.6	0.000
	Matched	21.253	21.021	10.70		0.770
Highest_t	Unmatched	10.142	16.888	−226.9	95.1	0.000
	Matched	12.381	12.048	11.20		0.470
Lowest_t	Unmatched	0.2564	0.4616	−358.3	96.0	0.000
	Matched	0.3279	0.3197	14.20		0.260

Table 22.10 Spatial scope of event impact based on Spatial DID

	AQI	$PM_{2.5}$	PM_{10}	SO_2	NO_2	CO	O_3
< 100 km	−12.48***	−10.80***	−13.31**	0.44	−4.63***	−0.09	−16.14***
	(3.26)	(4.00)	(5.60)	(0.81)	(1.19)	(−0.09)	(2.68)
100–200 km	−10.11***	−9.90***	−9.90***	0.87	−2.79***	−0.06	−12.32***
	(2.82)	(3.27)	(4.13)	(0.94)	(0.86)	(0.06)	(1.74)
200–300 km	−10.66***	−9.78***	−13.16***	−2.65**	−3.45***	−0.07**	−9.04***
	(2.50)	(2.29)	(2.81)	(1.03)	(0.70)	(0.03)	(2.42)
300–400 km	−10.32***	−10.69***	−11.06***	−0.22	−0.95	−0.05*	−10.50***
	(1.91)	(1.86)	(2.18)	(1.07)	(0.76)	(0.04)	(2.04)
400–500 km	−7.39***	−7.57***	−6.25***	−1.25	−0.74	−0.06**	−8.20***
	(2.58)	(2.35)	(2.56)	(1.08)	(0.71)	(−0.04)	(1.86)
500–600 km	−4.90***	−4.93***	−2.48	1.08	−0.05	−0.03	0.07
	(1.57)	(1.23)	(1.61)	(1.37)	(0.50)	(0.03)	(1.59)
600–700 km	−6.68**	−7.12***	−5.60*	−2.09	0.04	−0.02	−2.80
	(2.70)	(1.90)	(3.22)	(2.19)	(0.68)	(0.03)	(1.89)
700–800 km	−3.29**	−5.99***	−3.16	−0.62	−0.57	−0.03	1.05
	(1.67)	(1.45)	(2.46)	(0.85)	(0.46)	(0.02)	(1.81)
800–900 km	−0.00	−1.35	1.59	−0.89	0.18	−0.01	−0.25
	(1.34)	(1.27)	(1.67)	(0.57)	(0.41)	(0.01)	(2.03)

and after the events and conducting the DID analysis of the 120 cities, conclusions can be drawn that the major events significantly improved the air quality of cities within 800 km of the host city, and the improvement in cities within 500 km is relatively remarkable. With regard to individual pollutants such as PM_{10} and $PM_{2.5}$, great improvement is presented in cities within 800 km of the event site, and the concentration of nitrogen oxides in cities within 300 km also improved significantly (Table 22.11).

22.4.3.2 Further Analysis of Robustness

In order to identify whether the impact of major events on urban air quality in different areas is sensitive to the sample window, samples of 20, 30 and 40 days before and after major events are retained respectively. The regression results were insensitive to the different sample windows (Table 22.12) and these indicate the robustness of the conclusions drawn.

Table 22.11 Empirical results in shorter sample window

	AQI	$PM_{2.5}$	PM_{10}	SO_2	NO_2	CO	O_3
< 100 km	−13.49***	−11.81***	−14.77**	0.35	−5.14***	−0.10	−16.40***
	(3.92)	(4.39)	(6.24)	(0.81)	(1.13)	(−0.09)	(2.64)
100–200 km	−10.64***	−10.23***	−10.55***	0.98	−3.41***	−0.06	−12.63***
	(3.39)	(3.46)	(4.40)	(0.90)	(0.86)	(0.06)	(1.90)
200–300 km	−10.37***	−9.62***	−12.32***	−2.36**	−3.80***	−0.08**	−8.08***
	(3.03)	(2.61)	(3.51)	(1.08)	(0.74)	(0.03)	(2.38)
300–400 km	−12.29***	−11.76***	−12.95***	0.29	−1.40	−0.05*	−10.50***
	(2.18)	(2.02)	(2.89)	(1.26)	(0.87)	(0.04)	(1.93)
400–500 km	−9.20***	−8.87***	−8.29***	−1.53	−0.98	−0.06**	−8.63***
	(3.08)	(2.61)	(2.89)	(1.13)	(0.73)	(−0.05)	(1.88)
500–600 km	−6.27***	−5.91***	−3.82**	0.98	−0.68	−0.03	0.26
	(1.62)	(1.27)	(1.52)	(0.75)	(0.50)	(0.03)	(1.59)
600–700 km	−7.88**	−8.31***	−7.26**	−1.91	0.48	−0.03	−3.38
	(2.64)	(1.89)	(3.21)	(2.03)	(0.67)	(0.03)	(2.01)
700–800 km	−7.88**	−7.05***	−4.42**	−0.54	−0.92	−0.02	1.41
	(2.67)	(1.37)	(2.31)	(0.88)	(0.46)	(0.02)	(1.81)
800–900 km	−0.56	−1.91	1.15	−0.78	0.16	−0.02	−1.23
	(1.51)	(1.39)	(1.78)	(0.57)	(0.44)	(0.01)	(2.15)

Table 22.12 Empirical results in different sample windows

	Window	AQI	$PM_{2.5}$	PM_{10}	SO_2	NO_2	CO	O_3
< 100 k	20 days	−12.06***	−10.16***	−13.11**	0.27	−4.42***	−0.04	−17.26***
	30 days	−13.49***	−11.81***	−14.77**	0.35	−5.14***	−0.10	−16.40***
	40 days	−13.59***	−11.63***	−14.90***	0.40	−4.90***	−0.10	−16.43***
100–200 km	20 days	−8.72***	−8.11***	−8.69**	1.75	−2.74***	−0.01	−13.16***
	30 days	−10.64***	−10.23***	−10.55***	0.98	−3.41***	−0.06	−12.63***
	40 days	−10.83***	−10.35***	−10.88***	0.96	−3.21***	−0.06	−12.58***
200–300 km	20 days	−7.84***	−7.10***	−9.43***	−0.96	−3.59***	−0.03	−8.50***
	30 days	−10.37***	−9.62***	−12.32***	−2.36**	−3.80***	−0.08**	−8.08***
	40 days	−10.65***	−9.73	−12.93***	−2.44	−3.57***	−0.08	−8.06***
300–400 km	20 days	−9.46***	−9.23***	−9.97***	2.72	−0.53	−0.00	−11.77***
	30 days	−12.29***	−11.76***	−12.95***	0.29	−1.4	−0.05*	−10.50***
	40 days	−12.10***	−11.53	−12.82***	0.05	−1.16	−0.06	−10.03***
400–500 km	20 days	−6.04***	−5.51***	−4.86**	−0.39	−0.46	−0.02	−10.35***
	30 days	−9.20***	−8.87***	−8.29***	−1.53	−0.98	−0.06**	−8.63***
	40 days	−8.95***	−8.75***	−8.31***	−1.71	−0.91	−0.08	−8.29***
500–600 km	20 days	−5.01***	−3.87***	−1.76	2.28	−0.70	−0.00	−1.88
	30 days	−6.27***	−5.91***	−3.82**	0.98	−0.68	−0.03	0.26
	40 days	−6.56***	−6.08***	−4.39***	0.66	−0.53	−0.03	0.08
600–700 km	20 days	−5.38**	−5.85***	−4.56	0.18	−0.34	0.00	−4.34**
	30 days	−7.88**	−8.31***	−7.26**	−1.91	0.48	−0.03	−3.38
	40 days	−8.16***	−8.28***	−7.46**	−2.26	−0.40	−0.03	−2.93
700–800 km	20 days	−2.81**	−4.93***	−2.11	−0.19	−0.99**	0.10	0.75

(continued)

Table 22.12 (continued)

	Window	AQI	$PM_{2.5}$	PM_{10}	SO_2	NO_2	CO	O_3
	30 days	−7.88**	−7.05***	−4.42**	−0.54	−0.92	−0.02	1.41
	40 days	−4.93***	−7.13	−4.89**	−0.69	−0.85*	−0.02	1.93
800–900 km	20 days	0.01	−1.05	2.34	−0.54	−0.41	−0.01	−1.12
	30 days	−0.56	−1.91	1.15	−0.78	0.16	−0.02	−1.23
	40 days	−0.98	−2.13	0.44	−0.89	−0.04	−0.01	−0.97

22.5 Conclusion

22.5.1 *Research Findings*

In this study, the daily air quality index (AQI) and the concentration of individual pollutants from January 2, 2015 to March 28, 2017 are empirically investigated. A summary of the research findings is as follows. The air quality of the cities involved in hosting the major events improved significantly during the occurrence, and it is warranted that the improvement is caused by the occurrence of the major events. From the perspective of individual pollutant concentrations, particulate matter exercises the most remarkable impact on the urban air quality, followed by nitrogen oxide, while less obvious impact is shown on the concentrations of carbon monoxide and ozone. After a period of the major events, the AQI value and the concentrations of individual pollutants present a rising trend with varying degrees, exceeding the normal level, which shows that the improvement of air quality brought by temporary control exhibits no sustained effect, even at the cost of retaliatory pollution. In this regard, the major events not only affected the air quality of the host city, but also affected the air quality of the surrounding areas. The spatial range of each major event, affecting the urban air quality, is within 800 km around the host city, and the farther away from the host city, the smaller the impact (Guo et al., 2020).

22.5.2 *Countermeasures and Suggestions*

China's environmental management system should be innovated systematically based on the experience of developed countries but set within the context of China's in air pollution joint prevention and control measures in recent years, such as during the Beijing Olympic Games, Shanghai World Expo, APEC, etc. On this basis, effective countermeasures should be taken as follows. First of all, a new system of atmospheric environmental management should be established with regional management as the main part and territorial management as the supplement, so as to avoid the "illusion" of governance effectiveness caused by local government "interference" (Yu et al., 2019). Secondly, great attention should be paid to the phenomenon of atmospheric transboundary transmission as a result of meteorological field factors, and a new mechanism of joint prevention and control of atmospheric pollution aimed at improving air quality in different regions should be comprehensively promoted (Xu et al., 2019). In addition, the total amount and proportion of the transboundary transport of atmospheric pollutants should be studied scientifically and reasonably for a further establishment of the regional ecological compensation management system of atmospheric environment. Fourthly, taking big data as the analysis resource, the early warning and emergency warning mechanism should be proposed (Yan et al., 2019). At last, taking China's ongoing current regional economic integrations (such as Beijing-Tianjin-Hebei integration, Yangtze river delta integration, Pearl river delta integration, etc.) as opportunities, regional unified standards for pollutant discharge, industrial access and law enforcement are proposed to be formulated to improve the urban air quality.

Acknowledgements Yingying Guo, Yinshan Tang, and Michael D. Dzandu also made great contributions to this manuscript. We express our heartfelt thanks to them. This research also was supported by: National Natural Science Foundation of China (72474127) and National Social Science Fund Major Bidding Project of China (18ZDA052). Part of this chapter was published: Guo J, Wu X, Guo Y, Tang Y, Dzandu M. D. Spatiotemporal impact of major events on air quality based on spatial differences-in-differences model: big data analysis from China. Natural Hazards 2021;107:2583–604. https://doi.org/10.1007/s11069-021-04517-y. The publication has been conducted with copyright permission.

References

Beig, G., Chate, D. M., Ghude, S. D., Mahajan, A. S., Srinivas, R., Ali, K., Ali, Sahu, S. K., Parkhi, N., Surendran, D., & Trimbake, H. R. (2013). Quantifying the effect of air quality control measures during the 2010 Commonwealth Games at Delhi, India. *Atmospheric Environment, 80*, 455–463. https://doi.org/10.1016/j.atmosenv.2013.08.012

Bera, B., Bhattacharjee, S., Shit, P. K., Sengupta, N., & Saha, S. (2020). Significant impacts of COVID-19 lockdown on urban air pollution in Kolkata (India) and amelioration of environmental health. *Environment Development and Sustainability*. https://doi.org/10.1007/s10668-020-00898-5

Bogdan, E. A. (2020). How a deadly pandemic cleared the air: Narratives and practices linking COVID-19 with air pollution and climate change. *Space and Culture, 23*(3), 293–300. https://doi.org/10.1177/1206331220938641

Chen, Y., Jin, G., Kumar, N., & Shi, G. (2013). The promise of Beijing: Evaluating the impact of the 2008 Olympic Games on air quality. *Journal of Environmental Management, 66*(3), 424–443. https://doi.org/10.1016/j.jeem.2013.06.005

Dutheil, F., Baker, J. S., & Navel, V. (2020). COVID-19 as a factor influencing air pollution. *Environmental Pollution, 263*, Article 114466. https://doi.org/10.1016/j.envpol.2020.114466

Fan, C., Li, Y., Guang, J., Li, Z., & Leeuw, G. D. (2020). The impact of the control measures during the covid-19 outbreak on air pollution in china. *Remote Sensing, 12*(10), 1613. https://doi.org/10.3390/rs12101613

Guo, J., Guo, Y. Y., & Wu, X. H. (2020). The effect of impact of major events on urban air quality based on PSM-DID. *Journal of Applied Statistics and Management, 12*, 1–15. https://doi.org/10.13860/j.cnki.sltj.20201219-011

He, G., Fan, M., & Zhou, M. (2016). The effect of air pollution on mortality in China: Evidence from the 2008 Beijing Olympic Games. *Journal of Environmental Economics and Management, 79*, 18–39. https://doi.org/10.1016/j.jeem.2016.04.004

He, G. J., Pan, Y. H., & Tanaka, T. (2020). The short-term impacts of COVID-19 lockdown on urban air pollution in China. *Nature Sustainability*. https://doi.org/10.1038/s41893-020-0581-y

He, L. Y., Hu, K., & Lin, X. (2025). The impact of collaborative governance on air Pollution: Evidence from high-frequency data in Pearl river delta, China. *Energy, 337*, Article 138711. https://doi.org/10.1016/j.energy.2025.138711

Huang, K., Zhang, X., & Lin, Y. (2015). The "APEC Blue" phenomenon: Regional emission control effects observed from space. *Atmospheric Research, 164*, 65–75. https://doi.org/10.1016/j.atmosres.2015.04.018

Jia, K., & Chen, S. (2019). Could campaign-style enforcement improve environmental performance? Evidence from China's central environmental protection inspection. *Journal of Environmental Management, 245*, 282–290. https://doi.org/10.1016/j.jenvman.2019.05.114

Lee, B. K., Jun, N. Y., & Lee, H. K. (2005). Analysis of impacts on urban air quality by restricting the operation of passenger vehicles during Asian Game events in Busan, Korea. *Atmospheric Environment, 39*(12), 2323–2338. https://doi.org/10.1016/j.atmosenv.2004.11.044

Li, B., Wang, F., Yin, H., & Li, X. (2019a). Mega events and urban air quality improvement: A temporary show? *Journal of Cleaner Production, 217*, 116–126. https://doi.org/10.1016/j.jclepro.2019.01.116

Li, K., Jacob, D. J., Liao, H., Shen, L., Zhang, Q., & Bates, K. H. (2019b). Anthropogenic drivers of 2013–2017 trends in summer surface ozone in China. *Proceedings of the National Academy of Sciences of the United States of America, 116*(2), 422–427. https://doi.org/10.1073/pnas.1812168116

Li, L., Li, Q., Huang, L., Wang, Q., Zhu, A. S., Xu, J., Liu, Z. Y., Li, H. L., Shi, L. S., Li, R., Azari, M., Wang, Y. J., Zhang, X. J., Liu, Z. Q., Zhu, Y. H., Zhang, K., Xue, S. H., Ooi, M. C. G., Zhang, D. P., & Chan, A. D. (2020). Air quality changes during the covid-19 lockdown over the Yangtze River delta region: An insight into the impact of human activity pattern changes on air pollution variation. *Science of The Total Environment, 732*, Article 139282. https://doi.org/10.1016/j.scitotenv.2020.138704

Liu, H., Liu, C., Xie, Z., Li, Y., Huang, X., Wang, S., Xu, J., & Xie, P. (2016). A paradox for air pollution controlling in China revealed by "APEC Blue" and "ParadeBlue." *Science Reports, 6*, 34408. https://doi.org/10.1038/srep34408

Matus, K., Nam, K. M., Selin, N. E., Lamsal, L. N., Reilly, J. M., & Paltsev, S. (2012). Health damages from air pollution in china. *Global Environmental Change, 22*(1), 55–66. https://doi.org/10.1016/j.gloenvcha.2011.08.006

Ngo, N. S., Zhong, N., & Bao, X. (2019). The effects of transboundary air pollution following major events in China on air quality in the U.S.: Evidence from Chinese New Year and sandstorms. *Journal of Environmental Management, 212*(15), 169–175. https://doi.org/10.1016/j.jenvman.2018.01.057

Rafaj, P., Kiesewetter, G., Gül, T., Schöpp, W., Cofala, J., Klimont, Z., Purohit, P., Heyes, C., Amann, M., Borken-Kleefeld, J., & Cozzi, L. (2018). Outlook for clean air in the context of sustainable development goals. *Global Environmental Change, 53*, 1–11. https://doi.org/10.1016/j.gloenvcha.2018.08.008

Shi, Q., Guo, F., & Chen, S. (2016). "Political Blue Sky" in fog and haze governance—Evidence from the local annual "Two Sessions" in China. *China Industrial Economics*, (5), 40–56. https://doi.org/10.19581/j.cnki.ciejournal.2016.05.003 (in Chinese).

Song, M., Zhu, S., Wang, J., & Wang, S. (2019). China's natural resources balance sheet from the perspective of government oversight: Based on the analysis of governance and accounting attributes. *Journal of Environmental Management, 248*, Article 109232.

Song, M., Zhu, S., Wang, J., & Zhao, J. (2020). Share green growth: Regional evaluation of green output performance in China. *International Journal of Production Economics, 219*, 152–163.

Wang, Q., & Liu, Y. (2025). Evaluating the impact of new energy demonstration city construction on urban air quality: Evidence from a DID analysis of Chinese cities. *Energy, 332*, Article 137108. https://doi.org/10.1016/j.energy.2025.137108

Wang, S. (2012). Efficiency of mitigation measures to reduce particulate air pollution—A case study during the Olympic Summer Games 2008 in Beijing, China. *Science of the Total Environment, 427*, 146–158. https://doi.org/10.1016/j.scitotenv.2012.04.004

Wang, T., & Xie, S. (2009). Assessment of traffic-related air pollution in the urban streets before and during the 2008 Beijing Olympic Games traffic control period. *Atmospheric Environment, 43*(35), 5682–5690. https://doi.org/10.1016/j.atmosenv.2009.07.034

Wu, X., Cao, Y., Xiao, Y., & Guo, J. (2018). Finding of urban rainstorm and waterlogging disasters based on microblogging data and the location-routing problem model of urban emergency logistics. *Annals of Operations Research*. https://doi.org/10.1007/s10479-018-2904-1

Wu, X., Chen, Y., Zhao, P., Guo, J., & Ma, Z. (2019a). Study of haze emission efficiency based on new co-opetition data envelopment analysis. *Expert Systems, 37*(4), Article e12466. https://doi.org/10.1111/exsy.12466

Wu, X., Wang, Z., Gao, G., Guo, J., & Xue, P. (2019b). Disaster probability, optimal government expenditure for disaster prevention and mitigation, and expected economic growth. *Science of the Total Environment, 709*, Article 135888. https://doi.org/10.1016/j.scitotenv.2019.135888

Wu, X., Xu, Z., Liu, H., Guo, J., & Zhou, L. (2019c). What are the impacts of tropical cyclones on employment? An analysis based on meta-regression. *Weather, Climate, and Society, 11*, 259–275. https://doi.org/10.1175/WCAS-D-18-0052.1

Xu, W., Sun, J., Liu, Y., Xiao, Y., Tian, Y., Zhao, B., & Zhang, X. (2019). Spatiotemporal variation and socioeconomic drivers of air pollution in China during 2005–2016. *Journal of Environmental Management, 245*(1), 66–75. https://doi.org/10.1016/j.jenvman.2019.05.041

Yan, L. X., Duarte, F., Wang, D., Zheng, S. Q., & Ratti, R. C. (2019). Exploring the effect of air pollution on social activity in China using geotagged social media check-in data. *Cities, 91*, 116–125. https://doi.org/10.1016/j.cities.2018.11.011

Yu, M., Zhu, Y., Lin, C., Wang, S., Xing, J., Jang, C., Huang, J., Huang, J., Jin, J., & Yu, L. (2019). Effects of air pollution control measures on air quality improvement in Guangzhou, China. *Journal of Environmental Management, 244*(15), 127–137. https://doi.org/10.1016/j.jenvman.2019.05.046

Zhao, H., Zheng, W., Xu, J., Wang, Z., Yuan, Y., Huang, J., & Chu, Z. (2016). Evaluation of the improvement of the air quality during the parade in Beijing. *Journal of Environmental Science, 36*(10), 2881–2889. https://doi.org/10.3969/j.issn.1000-6923.2016.10.001

Zhao, J., Luo, L., Zheng, Y., & Liu, H. (2017). Analysis on air quality characteristics and meteorological conditions in Hangzhou during the G20 summit. *Acta Scientiae Circumstantiae, 37*(10), 3885–3893. https://doi.org/10.13671/j.hjkxxb.2017.0195. (in Chinese).

Zuo, M., & Yu, S. (2026). Dual dividends of multilateral energy cooperation: Renewable investment surge and air quality improvements under the Belt and Road Energy Partnership. *Renewable Energy, 256*, Article 124164. https://doi.org/10.1016/j.renene.2025.124164

Chapter 23
Features and Comparative Research on Ecological Civilization Vocabularies in the Five-Year Plan of China: An Analysis Based on Semantic Phrases

23.1 Introduction

The "Outline of the Five-Year Plan for National Economic and Social Development" is a medium and long-term plan for economic and social development in China (Lin, 2021), which sets clear goals and directions for economic and social development (Group, 2020). The "Five-Year Plan" was started in 1953. So far, thirteen five-year plans have been implemented in China. It is driving China out of poverty, building a moderately prosperous society, and marching towards the dream of the great rejuvenation of the Chinese nation. The "Thirteenth Five-Year Plan" is a key stage for building a well-off society in an all-round way and an important node for achieving the first centenary goal. The "Fourteenth Five-Year Plan" will embark on a new journey of comprehensive modernization and building a grand blueprint for China's development in a new era. In the "Five-Year Plan", the ecological environment construction plays an important role, such as the research on the driving factors of energy consumption and intensity changes (Yan & Su, 2020), the energy system structure adjustment and renewable energy (Hong et al., 2013; Lo & Wang, 2013). In 2007, the 17th National Congress of the Communist Party of China (CPC) took the construction of ecological civilization as one of the goals of building a well-off society. The 13th Five-Year Plan promotes an innovative, coordinated, green, open, and shared development. During these five years, great achievements have been made in ecological environment, and the construction of ecological civilization has been integrated into all aspects of economy and society and the whole process of development. The 14th Five-Year Plan takes promoting high-quality development as the theme, and promotes the overall layout of economic, political, cultural, social and ecological civilization construction as a whole. It unswervingly implements the new development concept of innovation, coordination, green, openness and sharing. During the 13th Five-Year Plan and 14th Five-Year Plan, the construction of ecological civilization has been carried out in an all-around way and deepened continuously. Through a comparative study, we can understand the changing trends and laws of ecological

X. Wu, *Environmental Disaster in China*,
https://doi.org/10.1007/978-981-92-0910-1_23

civilization construction in different periods. Therefore, this paper chooses these two periods for research (Onan et al., 2016; Vega-Oliveros et al., 2019). Five-Year Plans are development plans with Chinese characteristics. Due to the development of massive texts and text mining tools, the analysis of the Five-Year Plans focuses on news commentary and expert opinions. Text mining tools can help to get the main idea of texts through keywords according to language features. Starting from the keyword frequency and the correlation between keywords, this paper makes an objective and quantitative analysis of the policy content. Through analysis of the text content, we can clarify the core and key points of the plans, and grasp the development theme and focus. This paper compares the focus change in different periods, explores the changing trends, and provides an objective and scientific basis for the implementation of policy formulation.

The following parts are arranged as follows. The first part is the literature review; the second part is the model introduction; the third part is the empirical structure; the last is the conclusions and prospect.

23.2 Literature Review

23.2.1 Overview of Quantitative Research on Ecological Civilization in Policy Texts

There are narrow and broad definitions of ecological civilization. In a narrow sense, ecological civilization involves ecosystem protection, environmental quality, and resource utilization (Lu et al., 2020; Wu et al., 2020). In a broad sense, ecological civilization covers the layout of land space, the ecological economy development, the ecological environment protection, the ecological culture construction (Cai et al., 2021; Huang & Wu, 2021), strategic emerging industries (Meng et al., 2021; Xu et al., 2025), economic development concept modes (Zhang et al., 2016). At present, the quantitative analysis of ecological civilization policy includes environmental pollution control, energy resource utilization and new energy development, and sustainable development policy effect evaluation (Thuestad, 2020). The green governance of the government has undergone logical transition in governance concepts, governance modes, governance supervision, and ecological civilization construction, showing that ecological protection and environmental governance policies keep pace with the times (Ran, 2020). Environmental governance needs the leadership of the government, the guidance of the news media, and the active participation of enterprises and the public in order to improve the overall environmental governance effect (Lam et al., 2019). Resources and energy industry policies involve resource utilization, new energy, energy structure transformation and international cooperation (Lu et al., 2019; Tong et al., 2021). Ecological economy not only promotes the upgrading and transformation of energy, but also improves the environment and help to realize targeted poverty alleviation (Zhang et al., 2018).

In the quantitative research on ecological civilization in policy texts, scholars at home and abroad often take the institutions, executors and the policy intensity as one of the weights or factors of quantitative analysis, and then conduct in-depth quantitative analysis combined with statistical data after coding conversion. There are few studies on the quantitative analysis of the policy text. The current research focuses ecological civilization in a narrow sense, or is limited to a certain field, such as resources (Du et al., 2021a, 2021b; Vikstrom et al., 2025), pollution (Lu et al., 2020), emission reduction (Baldwin et al., 2018; Jean-Philippe, 2019), etc. Combined with the research of scholars, taking the Guide for the Ecological Civilization Construction Standard System (2018–2020) as the framework, this paper focuses on the ecological civilization-related ecological environment, resources and energy, green economy and pollution control to conduct research.

23.2.2 Overview of Phrases and Topic Models

Key-phrase extraction can be divided into three categories: supervised, semi-supervised and unsupervised. Unsupervised automatic extraction technology includes statistical methods, machine learning algorithms, deep learning algorithms, and a combination of multiple algorithms. The statistical methods are mainly probabilistic models and Markov (Chang & Choi, 2006; Feng & Croft, 2001). Compared with the supervised manual classification method, it improves the efficiency of phrase generation and reduces labor costs. The development of artificial intelligence and the application of big data have accelerated the application of machine learning and deep learning algorithms, and the introduction of the attention mechanism has improved the accuracy of the contextual semantic features of phrases (Lei et al., 2021). Combined with the semantic network and the polycentric network, the keyword extraction method based on the hierarchical structure obtains keywords that can effectively reflect the internal feature and topic (Sung & Kim, 2019). It focuses on context-relevant keyword analysis, and continuously improves the accuracy of keyword extraction (Eirini & Grigorios, 2018), which promotes the development of text analysis. The existing research focuses on comprehensive methods, and continuously improves the accuracy of keywords and the relevance between contexts through the combination of various algorithms (Eirini & Grigorios, 2018). For example, the phrase extraction method based on part-of-speech rules, combined with statistical features and graph models, obviously improves the accuracy and recall rate of keywords (Haihua et al., 2022). In the unsupervised keyword algorithm, the title and abstract words in research papers are referred to as reference word vectors. The quality of keywords embedded in the content analysis is better than that of global representation (Papagiannopoulou & Tsoumakas, 2018).

Topic model is one of the most advanced techniques in text mining, and it was evolved from the earliest news topic tracking (Allan, 2002). The LDA (Latent Dirichlet Allocation) proposed by (Blei et al., 2003) is a classic model for text analysis. Based on the unsupervised "bag of words", the topic is explained by using the most frequent word in each topic to analyze the simple relationship between the document and the topic. Based on this, the author expands the topic model, hierarchical topic model (Xu et al., 2018), and mixed topic model (Michal et al., 2010). With the popularity of applications, there are limitations in the probabilistic reasoning topic models. Scholars attempt to build a neural topic model of generative confrontation network (Wang et al., 2019), a word-pair topic model suitable for social media (He et al., 2017), and a neural network topic model that combines word vectors and entity vectors (Zhao et al., 2021), automated pipelines further synergize machine learning with conventional topic modeling (Tahir et al., 2025), and expert-guided hierarchical topic modeling architectures (Calvo-Bartolomé & Arenas-García, 2025), etc. During the formulation and implementation of environmental policies, through text analysis of stakeholders (government, environmental protection organizations, news media), the results demonstrate that the government and environmental protection organizations are more concerned about pollutant emission and control, while the media are concerned about the impact of air pollution on health (Lam et al., 2019). Using the LDA theme model, the effectiveness and quantity intensity index of air pollution prevention and control policy are constructed. The results show that economic incentives and supportive policies have limited effect on emission reduction, and it is necessary to improve the legal effect of policies (Du et al., 2021a, 2021b). To understand the impact of COVID-19 space prevention policy on passengers' experience, the space policy review texts of 64 major hubs in the United States find that passengers have good experience of staff and shop services, but poor experience of space services. This result provides reference for improving the health policy of transportation hubs in the future (Park et al., 2022).

In summary, the key-phrase extraction statistical methods lack semantic relevance, ignore the contextual information, so the phrases obtained are not representative and accurate. The phrase extraction of machine and deep learning are mostly applied into search engines, translation, robot services, etc., which are professional and high in development costs. At the same time, it highly depends on the quality and scale of the training set, and the training set generation is marked manually, with high cost and low portability. There are limitations of lacking language information and topic quality and data sparsity in "bag-of-words" hypothesis of the LDA topic model. The fusion of word embedding and neural network improve semantic accuracy, but the overall model efficiency is reduced. Therefore, the topic quality is low, and there are problems such as small information granularity, low topic recognition, and fuzzy meaning of topic words. Meanwhile, due to the language features of Chinese, the above-mentioned research methods based on English cannot be directly used. In this paper, through the word pair method, the key-phrases are extracted through grammatical, semantic rules and information entropy knowledge. A phrase database is set by using the text data, which is transferable in the semantic categories and normative texts. With the key-phrase as the basis of the LDA model, the topic noise and semantic ambiguity are reduced, and the accuracy of keywords are improved.

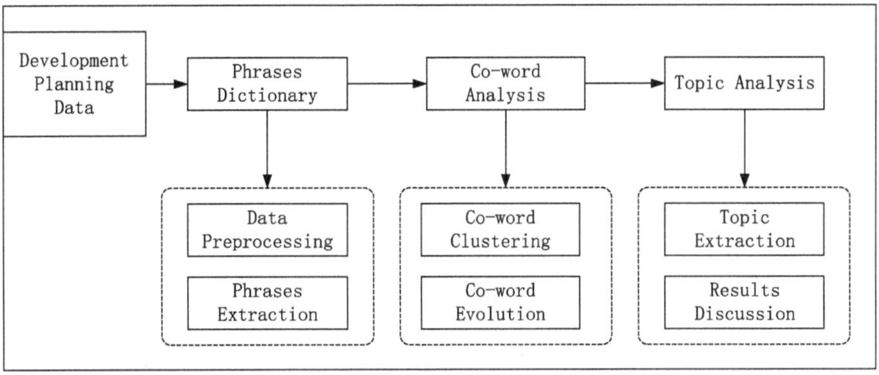

Fig. 23.1 Flowchart of analysis in this paper

23.3 Model Establishment

23.3.1 Research Ideas

Figure 23.1 shows the analysis of the "Thirteenth Five-Year Plan" and "Fourteenth Five-Year Plan". First, construct a phrase dictionary, preprocess the data and extract policy phrases. By calculating point-wise mutual information (PMI), grammatical rules, and similarity rules, the phrases are selected to generate a policy phrase dictionary. In this way, the unregistered new phrases can be obtained, and the newly generated phrases with contextual semantic features can satisfy the grammatical collocation rules of Chinese. Second, co-word analysis. The phrase dictionary is taken as a word segmentation library to obtain co-words in text data, and perform co-word clustering and evolution analysis. Perform co-word clustering through the community detection algorithm based on modularity, and the relationship among phrases is explicitly shown in a structured, clear and intuitive way. Then the co-word evolution in the two periods is analyzed in depth by using the clustering results. Third, topic analysis. With the phrases as the basic vocabulary unit of the model, the topic is analyzed through the LDA topic model that is widely used at present, improving the accuracy of the semantic expression of phrases. The high frequency words are filtered out through the word frequency ratio to improve the topic model quality.

23.3.2 Phrase Dictionary

According to the features of the Chinese language, the text can be divided into characters, words, phrases, sentences, paragraphs and chapters based on the granularity. Due to the vagueness of the meanings expressed by single character, words are often used as the basic analysis unit in text mining. Using words may lose the semantic relevance among words, and the same words can express different meanings in different contexts. Phrases not only retain the contextual semantic information to a certain extent, but also express the text content more clearly, with finer granularity and higher flexibility than sentences and paragraphs. Hence, it is necessary to establish a applicable phrase dictionary. A good phrase dictionary can reduce semantic ambiguity, have better text description capabilities, and provide an accurate word segmentation basis for in-depth mining. Based on the bigram proposed by Cheng et al. (2014), the phrases are extracted by using the statistics and part-of-speech rules. The phrase extraction process is shown in Fig. 23.2.

The construction of phrase dictionary includes data preprocessing and phrase extraction. Data preprocessing includes PDF document conversion, content extraction from image text, table content reading, and text preprocessing (word segmentation, deleting stop words, part-of-speech tagging, etc.). Phrase extraction includes calculating word relevance by using PMI value, filtering by grammatical rules, merging cross elements and deleting overlapping elements by similarity rules. Finally, a policy phrases dictionary is generated.

① Word relevance. PMI is used to measure the degree of interdependence between words in binary phrases and multiple groups, and its calculation formula is as follows:

$$\mathrm{PMI}(w_i, w_j) = \log\frac{p(w_i, w_j)}{p(w_i) * p(w_j)} \quad \text{(Binary phrases)} \tag{23.1}$$

$$\mathrm{PMI}\left(\sum_{i=1}^{n} w_i\right) = \log\frac{p\left(\sum_{i=1}^{n} w_i\right)}{\sum_{i=1}^{n} p(w_i)} \quad \text{(Multiple phrases)} \tag{23.2}$$

where $\mathrm{PMI}(w_i, w_j)$ in formula (23.1) is the PMI value of phrases w_i and w_j. $p(w_i, w_j)$ is the probability that the phrases w_i and w_j appear at the same time. $p(w_i)$ and $p(w_j)$ represent the probability that the phrases w_i and w_j appearing in the same article, respectively. From binary phrases to multiple groups, as shown in formula (23.2).

② Rule-based filtering. The word rules follow the Modern Chinese dictionary, combined with scholars' research on Chinese participles (Jiwei & Dongfan, 2010), key phrase extraction, and phrase structure rules (Haihua et al., 2022; Weidong, 2017). Which include common nouns, verbs, adjectives, pronouns and quantifier phrases, and specific phrase collocation is shown in Table 23.1. "a" represents adjectives, including adjective morphemes, adjective idioms, distinctive word idioms; "n" represents nouns, including adnouns, names, place names, transliterated place names, organization names, other special nouns, and ancient names; "v" represents verb, including noun-verbs, adverbs, formal verbs, and directional verbs; "r" represents pronouns, including personal pronouns and demonstrative pronouns.

③ Delete overlapping elements. The lower-order tuple can be either a single high-frequency word or a subset of a group of higher-order elements. In the former case, the low-order tuple needs to retain, and in the latter case, the low-order tuple is deleted as overlapping elements. Discriminant rule: If the frequency of low-order

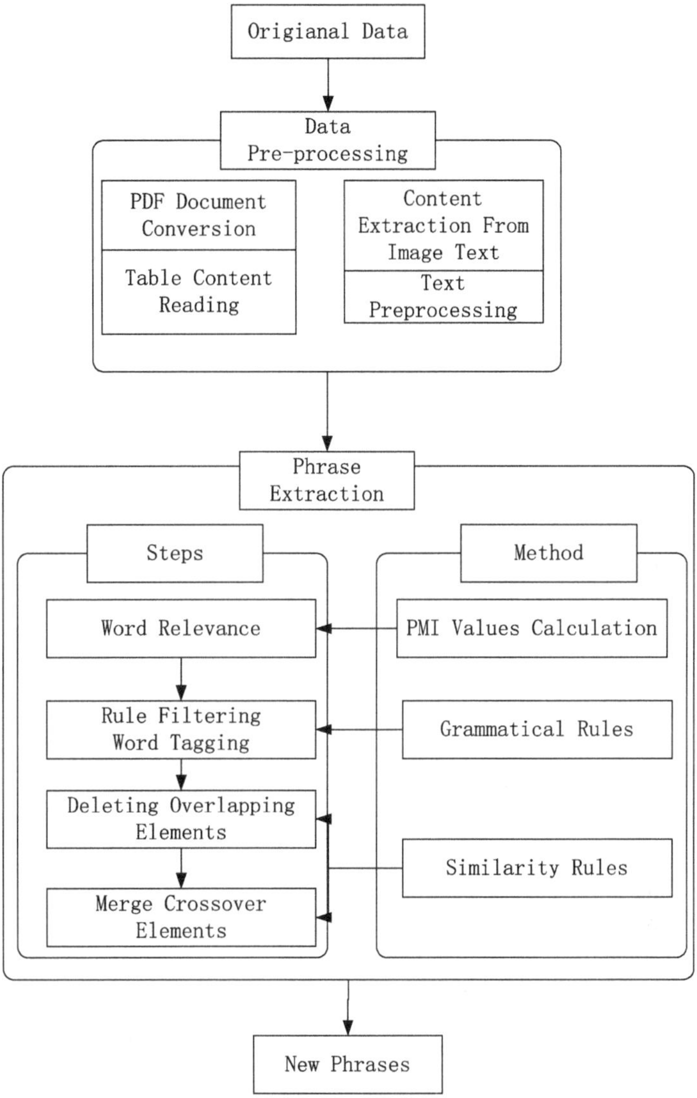

Fig. 23.2 Phrase extraction process

Table 23.1 Rules of part of speech

Phrase	Rules
Noun phrase	n+n; n+v; n+a n+n+n; n+v+v; n+v+n;
Verb phrase	v+r; v+n; v+a; v+q; v+v+n;
Adjective phrase	A+a; a+v; a+a; a+n;
Others	r+v; r+a; r+n; q+n;

tuples was close to high-order tuples, it means that low-order tuples in the text almost always appear as part of high-order tuples, and high-order tuples were the common or specialized words. Therefore, delete the low-order tuples from the dictionary. For example, the triple (w_i, w_j, w_k) contains two binary groups (w_i, w_j), (w_j, w_k). If Fre $(w'_i, w'_j) \approx \mathrm{Fre}(w_i, w_j)$ or $\mathrm{Fre}(w'_i, w'_j) \approx \mathrm{Fre}(w_j, w_k)$, delete (w_i, w_j), other wise retain it. Similarly, a subset of a unitary in a binary group can follow the word method and remove the overlapping elements.

④ Merge cross-elements. Unrelated low-order tuples have a low probability to generate related high-order words with a merger. Therefore, combining overlapped words can efficiently generate new strongly associated high-order tuples. Therefore, combining overlapped words can efficiently generate new strongly associated high-order tuples. At the same time, high-frequency and highly relevant new phrases are composed by borrowing the idea of linear programming for an optimal solution. The word overlap degree, lexical rules, and frequency statistical errors were used as constraints.

Similarly is calculated as follows:

If $A = \left\{ \sum_{m=1}^{n} w_{i,m} \right\}$; $B = \left\{ \sum_{j=1}^{n} w_{j,p} \right\}$; $c = \left\{ \sum_{q=1}^{n} w_{k,q} \right\}$.

When len(A) = len(B) = len(C)= 3

$$\text{If} \begin{cases} A \cap B = 2 \\ B \cap C = 2 \\ A \cap B \cap C = 1 \\ w_{i,m+1} = w_{j,p} \\ w_{j,p+1} = w_{k,q} \end{cases} \text{and} \begin{cases} A \cap B = 1 \\ B \cap C = 1 \\ A \cap B \cap C = \emptyset \\ w_{i,m+2} = w_{j,p} \\ w_{j,p+2} = w_{k,q} \end{cases}, \text{ similarly} = 1, \text{ or similarly} = 0. \tag{23.3}$$

When len(A), len(B), len(C) is not exactly 3,

$$\text{If} \begin{cases} A \cap B = 2 \\ B \cap C = 1 \\ A \cap B \cap C = \emptyset \\ w_{i,m+2} = w_{j,p} \\ w_{j,p+1} = w_{k,q} \end{cases}, \begin{cases} A \cap B = 1 \\ B \cap C = 1 \\ A \cap B \cap C = \emptyset \\ w_{i,m+1} = w_{j,p} \\ w_{j,p+2} = w_{k,q} \end{cases}, \begin{cases} A \cap B = 1 \\ B \cap C = 1 \\ A \cap B \cap C = \emptyset \\ w_{i,m+1} = w_{j,p} \\ w_{j,p+1} = w_{k,q} \end{cases}, \text{ similarly} = 1, \text{ or similarly} = 0. \tag{23.4}$$

When len(A), len(B)is not exactly 3

$$\text{If} \begin{cases} A \cap B = 2 \\ w_{i,m+1} = w_{j,p} \\ w_{i,m+2} = w_{j,p+1} \end{cases}, \begin{cases} A \cap B = 1 \\ w_{i,m+2} = w_{j,p} \end{cases}, \text{ similarly} = 1, \text{ or similarly} = 0. \tag{23.5}$$

$$\max(w_1, w_2, \ldots w_n)$$

$$\text{s.t.}\begin{cases} \text{similarly} = 1; \\ \frac{|Fre_A - Fre_B|}{\max\{Fre_A, Fre_B\}} + \frac{|Fre_A - Fre_C|}{\max\{Fre_A, Fre_C\}} + \frac{|Fre_B - Fre_C|}{\max\{Fre_B, Fre_C\}} < 3 * 0.1; \end{cases} \quad (23.6)$$

where the optimal qualified phrase is solved in formula (23.6) based on the idea of obtaining the optimal solution through linear programming. The optimal semantic phrase is formed through the overlapping between multiple phrases and the frequency constraint between phrase groups. The similarly represents the similarity between phrases, and it is calculated by similarity formulas (23.3, 23.4, 23.5). When similarly is 1, the similarity condition is satisfied. Fre_A, Fre_B, and Fre_C are the frequencies of sets A, B, and C.

23.3.3 Co-Word Analysis

According to the amount of text data, co-word analysis can be divided into co-word network analysis and cluster analysis. Co-word network analysis can visually present the relationship network and density of keywords when dealing with a small amount of data (Su et al., 2013). For a large amount of data, the cluster analysis can quickly obtain the classification results. Co-word analysis includes co-word clustering and co-word evolution. The community detection algorithm based on modularity optimization is adopted in co-word clustering (J, 2006; Javed et al., 2018). As one of the network clustering algorithms, it has good accuracy and stability. A collection of nodes with a class of the same characteristics is realized with the community as a module. Suppose that the complex network is undirected, that is, G = (V,E), V is the set of nodes in the entire network, and E is the set of sides in the entire network. Divide network G into N communities, $P = \{C_1, C_2, \cdots C_n\}$, and

$$Q(C) = \frac{1}{2m}\sum_{vw}\left(A_{vw} - \frac{k_v k_w}{2m}\right) * \delta * (c_v, c_w) = \sum\nolimits_{i=1}^{c}\left(e_{ii} - a_i^2\right) \quad (23.7)$$

$$e_{ij} = \sum\nolimits_{vw} \frac{A_{vw}}{m} \quad (23.8)$$

$$a_i = \frac{k_i}{2m} = \sum\nolimits_j e_{ij} \quad (23.9)$$

where A_{vw} is the sides between nodes v and w. When $A_{vw} = 1$, there are sides connected. k_v is the degree of node v, and m is the number of sides in the entire complex network. e_{ij} represents sides of the node in the community i and j. a_i is the degree of node in the community i.

The high-frequency phrases are extracted from the community classification results of co-word clustering by co-word evolution for frequency comparison in different periods.

23.3.4 Topic Model

Based on the LDA model, it is improved from the dimension of semantic association, and the Phrase-LDA model (Phrase-Latent Dirichlet Allocation Topic Model) is established. Suppose that there are multiple topics in each article, each word in the article selects a topic with a certain probability (Fig. 23.3). A word is selected from this topic with a certain probability to mine the underlying semantics of the vocabulary in the corpus, so the probability of each phrase can be expressed as:

$$p(Ph|Document) = \sum\nolimits_{Topic} p(Ph|Topic) * p(Topic|Document) \qquad (23.10)$$

The words in the LDA model are replaced with phrases by Phrase-LDA. The model structure is composed of two processes: "document-topic" and "topic-phrase". In the process of "document-topic", $\alpha \rightarrow \theta_m \rightarrow z_{m,n}$ is Dirichlet-Multionmial structure, that is, when the m-th document is generated, the document topic is obtained from the document-topic multinomial distribution θ_m. The topic number of the n-th phrase in the extracted document topic is $z_{m,n}$. In the "topic-phrase" process, $\beta \rightarrow \emptyset_k \rightarrow ph_{m,n}|k = z_{m,n}$ is also a Dirichlet-Multionmial structure. In the multinomial distribution $\emptyset_k$ of topic-phrase in k, the phrase $ph_{m,n}$ is obtained from $k = z_{m,n}$, that is, the n-th phrase in the m-th document in the generated corpus.

The joint probability of the SS-LDA model is shown in formula (23.11):

$$P(ph, z, |\alpha, \beta) = P(ph|z, \beta)P(z|\alpha) = \prod\nolimits_{k=1}^{K} \frac{\Delta(n_k + \beta)}{\Delta(\beta)} * \prod\nolimits_{m=1}^{M} \frac{\Delta(n_m + \alpha)}{\Delta(\alpha)} \qquad (23.11)$$

The documents and the phrases that make up each document are known, and the unknown variables θ_m and $\emptyset_k$ are obtained by utilizing the Collapsed Gibbs Sampling (Griffiths & Steyvers, 2004). For the specific phrase t, the unknown variable obtained is:

$$P(z_i = k|z_{-i}, ph) \propto P(z_i = k, ph_i = t|z_{-i}, ph_{-i}) = \hat{\theta}_{mk} * \hat{\emptyset}_{kt} \qquad (23.12)$$

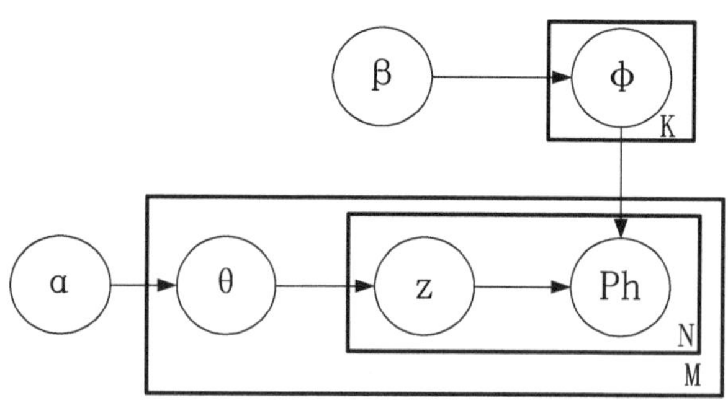

Fig. 23.3 Phrase-LDA model

$$\hat{\varnothing}_{kt}=\frac{n_{k,-i}^{(t)}+\beta_t}{\sum_{t=1}^{V}\left[n_{k,-i}^{(t)}+\beta_t\right]} \quad (23.13)$$

$$\hat{\theta}_{mk}=\frac{n_{m,-i}^{(k)}+\alpha_k}{\sum_{k=1}^{K}\left[n_{m,-i}^{(t)}+\alpha_k\right]} \quad (23.14)$$

Formulas (23.13) and (23.14) are substituted into formula (23.12) to obtain final Collapsed Gibbs Sampling formula:

$$P(z_i=k|z_{-i},ph)\propto\frac{n_{m,-i}^{(k)}+\alpha_k}{\sum_{k=1}^{K}\left[n_{m,-i}^{(t)}+\alpha_k\right]}*\frac{n_{k,-i}^{(t)}+\beta_t}{\sum_{t=1}^{V}\left[n_{k,-i}^{(t)}+\beta_t\right]} \quad (23.15)$$

23.4 Empirical Analysis

23.4.1 Text Data Source

In this paper, "Outline of the Thirteenth Five-Year Plan for National Economic and Social Development" and "Outline of the Fourteenth Five-Year Plan for National Economic and Social Development" (hereinafter referred to as "Development Plan") of the central, provincial, municipal and prefecture-level cities are taken as the research object. The selected policy text comes from the central, provincial, municipal and local public text materials, which are mainly searched and downloaded from the websites of local governments and development and planning committees. As of May 17, 2021, there are 395 "Thirteenth Five-Year Plan" and 347 "Fourteenth Five-Year Plan" texts, covering the national, provincial, and prefecture levels.

23.4.2 Phrase Dictionary

The key phrase extraction process is as follows: (1) Build a dictionary of source text phrases. The text is preprocessed with the help of Chinese word segmentation library jieba,[1] and then the phrase dictionary of the source text is generated according to the phrase dictionary algorithm. (2) Extract source text keywords. The source text phrase

[1] Jieba is a Chinese word segmentation, which combines two types of methods: rules and statistics. It supports simplified and traditional Chinese word segmentation. This word segmentation tool comes with a dictionary containing more than 20,000 words. In addition, it permits to customize dictionaries according to need. Jieba supports three-word segmentation methods: exact mode, full mode, and search engine mode.

dictionary is used for word segmentation, and a co-word matrix file is generated. The high-frequency phrases related to the topic are considered as keywords for co-word analysis and topic classification and are further analyzed in depth. According to the phrase dictionary extraction process, the generation of the phrase dictionary is divided into two key steps.

First, data preprocessing includes text classification and conversion, text preprocessing and content reading. ① Text classification and conversion. According to the title of the document, it is divided into "Thirteenth Five-Year Plan" and "Fourteenth Five-Year Plan". Part of the PDF document in the original document is converted to word format through Python. Read the contents of the table separately. Some of the original PDF documents are pictures, which need to be recognized and extracted. ② Text preprocessing and content reading. Unify the format of the generated text, such as Enter, space, page number etc. The word segmentation is conducted through jieba to generate unary, binary, and ternary phrases, and remove stop words and punctuation.

Second, the phrase extraction includes word relevance calculation, rule-based filtering, deleting overlapping elements, and merging cross elements. ① the word relevance is calculated by utilizing PMI (Formulas 23.1, 23.2). The first 90% of binary and ternary phrases are retained and saved to carry out the rule-based filtering step of the rule. ② filter the retained binary and ternary phrases, and delete the binary and ternary phrases that do not satisfy the part-of-speech rules (Table 23.1). ③ the cross relationship between binary and ternary phrases, and ternary phrases is obtained through formula (23.6), and merge the qualified phrases to generate new phrases. Read from the table with more phrases, and overlapping test is conducted for the unary and binary phrases, respectively. If the low-element phrase is completely contained in the higher-element phrase, the lower-element phrase is deleted. Multiple phrases are compared, and they are tested in descending order to iteratively generate a new phrase dictionary.

Table 23.2 is the extraction results from phrase dictionary in the periods of "Thirteenth Five-Year" and "Fourteenth Five-Year". With "ecology" as the key word, only 10 groups of phrases are extracted from the end of the dictionary for display. Compared with phrase dictionaries, common dictionaries (Jieba) mainly consist of single words, with concise expression, but they lack the semantic connection between contexts. In the process of text segmentation, it will cause polarization in text word frequency. Compared with commonly used dictionaries, phrase dictionaries have higher topic description capabilities, clearer semantic expressions, and short-distance contextual relevance, improving the expression accuracy. With the phrase dictionary as the basis of word segmentation, word frequency distribution interval of phrase segmentation is gentle. Comparing the "Thirteenth Five-Year" and "Fourteenth Five-Year" dictionaries, the generated phrase dictionaries are different due to the differences in text data. It shows the portability of the phrase dictionary construction. A phrase dictionary with strong relevance can be obtained according to the specific text content, which provides a more accurate word segmentation basis for subsequent analysis.

Table 23.2 Phrase extraction (partial)

Phrase dictionary		Jieba
Thirteenth Five-Year	Fourteenth Five-Year	
Ecological and Environmental Protection Planning	Ecosystem Protection And Restoration	Ecology, Environment, Protection, Planning, System, Restoration
Eco-environmental Protection Construction	Ecosystem Protection Effectiveness	Ecology, Environment, Protection, Construction, System, Effectiveness
Ecosystem Restoration And Management	Living Eco-Space Layout	Ecology, Environment, Recovery, Governance, Life, Space, Layout
Ecological Civilization System Reform	Continuous Improvement Of Ecological Environment	Ecology, Civilization, Institution, Reform, Continuous, Improve, Environment
Publicity And Education On Ecological Civilization	Protecting The Ecology Of Basic Farmland	Ecology, Civilization, Advocacy, Education, Protection, Basic, Farmland
Environmental Governance And Ecological Protection	Implementation Of Ecological Environmental Protection	Environment, Governance, Ecology, Protection, Implement
Natural Ecosystem Stability	Sound Ecological Protection	Nature, Ecology, System, Stability, Sound, Protection
Eco-environmental Damage Compensation System	Green Ecological Unit GDP	Ecology, Environment, Damage, Compensation, regime, Green, Unit, GDP
Restrictive Ecological Construction Environmental Protection	Green Ecological Forest Cover	Restrictive, Ecology, Construction, Forest, Environment, Protection, Green, Coverage
Anticipatory Ecological Construction Environmental Protection	Rational Layout Of Natural Ecological Construction	Expectancy, Ecology, Construction, Environment, Protection, Nature, Layout, Reasonable

23.4.3 Co-Word Analysis

In the analysis of co-word phrases, the modular optimization community discovery algorithm is used for clustering. In the co-word network diagram, it is necessary to distinguish keyword topics and mark the keyword frequency. Therefore, the dot size is used to indicate the keyword frequency, and different colors are used to indicate the theme of the keyword. The webpage files of co-word network diagram in the two periods are 3.31 MB and 4.25 MB, respectively. The color block diagram represents the whole keyword network, and the topic tables are differentiated and compared.

The co-word networks during the 13th Five-Year Plan and 14th Five-Year Plan are shown in Figs. 23.4 and 23.5, including four categories: green environment, resources and energy, economic development concept, and pollution prevention and control. These four categories are discussed based on the analysis results. Tables 23.3, 23.4, 23.5, and 23.6 show the changes of key phrases of various topics in the two periods. Co-Word analysis is the interpretation of related high-frequency words in the two periods. The 13th Five-Year Plan and 14th Five-Year Plan are two closely related stages, and the plan of one period is continuity as well as development of the other. Therefore, the analysis of key phrases in the two periods is elaborated from two aspects: thematic importance (word frequency of the same phrase) and development (differentiated phrase). Differentiated phrases are marked in italics in Tables 23.3, 23.4, 23.5 and 23.6, and detailed in the paper. However, the low-frequency words with great differences are not shown in this paper because they are less important.

(1) Green environment

Green environment is a central cluster of co-word network clustering, including ecology, environmental protection and green. The difference in the ecological green development indicators between the two periods is small. Phrases such as restrictive ecological civilization and restrictive ecological environment during the "Thirteenth Five-Year Plan" period indicate that at this stage, ecological indicators are more restrictive than that during the "Fourteenth Five-Year Plan" period. The binding index

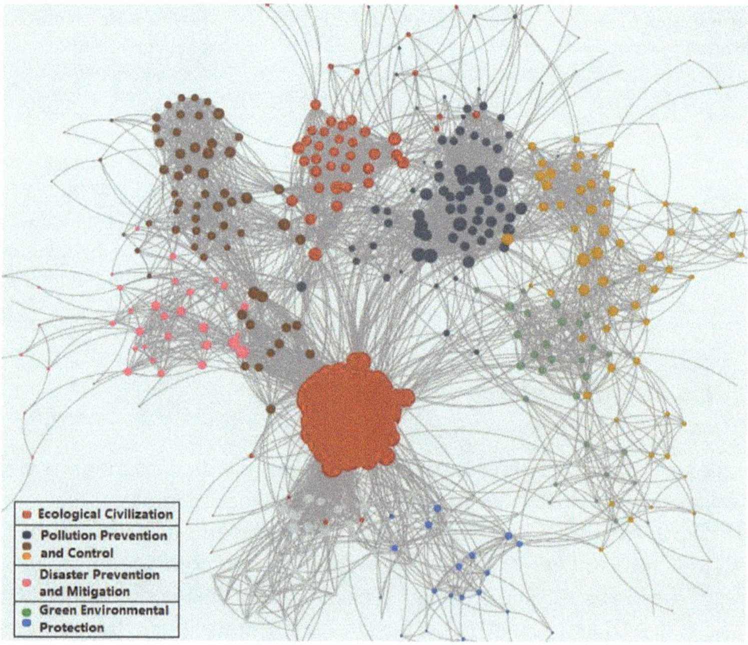

Fig. 23.4 "Thirteenth Five-Year Plan" co-word network

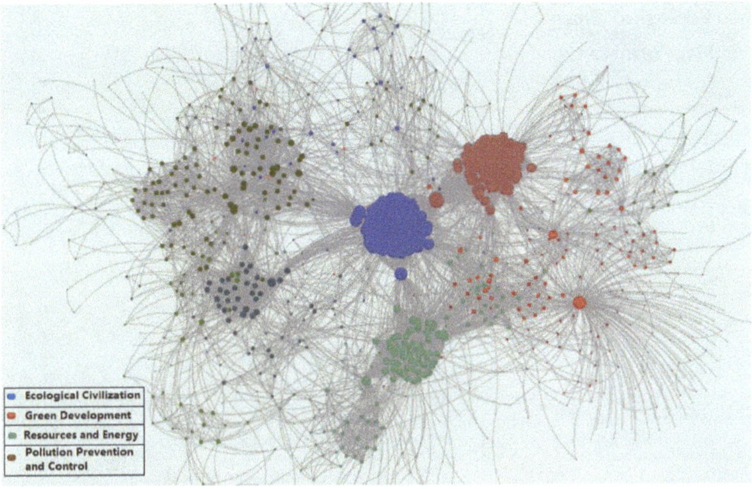

Fig. 23.5 "Fourteenth Five-Year Plan" co-word network

is the targets that cannot be breakthrough or must be achieved to achieve the planning goals within the planning period. The watershed ecological protection (1.12) of the "Fourteenth Five-Year Plan" is an important water ecological barrier. The Yangtze River basin and the Yellow River Basin are the two horizontal economic zones in China, which are the key areas for population gathering, agricultural cultivation and industry. Paying attention to the ecological protection of river basins is conducive to promoting coordinated regional development, promoting the harmonious coexistence between people and water, and achieving high-quality development. "Eco-industrialization (0.53)" and "improving the quality of ecosystem (0.39)" emphasize the integrity and systematization of the ecology. Green transformation (1.13), green travel (0.61), and green production and lifestyle (0.78) embody that green has become an important ecological concept of social production and life during the "Fourteenth Five-Year Plan" period. The 13th Five-Year Plan period focuses on improving the ecological environment, so this period sets more binding targets, including "binding ecological civilization", "binding ecological environment", "binding resources and environment", etc. The 14th Five-Year Plan, with the theme of promoting green development, promotes the height of ecological development from a holistic perspective. For example, "eco-industrialization", "upgrading ecosystem management", "ecological priority green development", etc. The construction of ecological civilization has achieved new progress, from quantitative accumulation to qualitative transformation.

(2) Resources and energy

The ecological environment is closely related to the development and application of energy resources. The increasing demand for energy intensifies ecological damage, constantly exceeding the carrying capacity of the environment, and the shortage of important energy sources will further restrict economic development concept.

Table 23.3 Ecological Green

Fourteenth Five-Year Plan			Thirteenth Five-Year Plan		
Phrases	Counts[1]	Frequency[2]	Phrases	Counts	Frequency
Ecology	7581	21.78	Ecology	9459	24.01
Ecological Environment	1553	4.46	Ecological Environment	1904	4.83
Ecological Protection	1068	3.07	Ecological Civilization	1782	4.52
Ecological Restoration	979	2.81	Ecological Protection	1638	4.16
Ecological Civilization Construction	745	2.14	Ecological Civilization Construction	1557	3.95
Eco-environmental Protection	659	1.89	Ecological Construction	1339	3.40
Eco-friendly	620	1.78	Eco-environmental Protection	791	2.01
Ecological Civilization	585	1.68	Ecotourism	789	2.00
Ecological Priority	404	1.16	Ecological Priority	126	0.32
Ecological Priority Green Development	99	0.28	Ecological Barrier	389	0.99
Ecological Barrier	424	1.22	Key Ecological Function Areas	588	1.49
Key Ecological Function Areas	176	0.51	Restrictive Ecological Civilization	469	1.19
Eco-industrialization	183	0.53	Restrictive Ecological Environment	158	0.40
Improving the quality Of Ecosystem	134	0.39	Restrictive Resources Environment	247	0.63
Watershed Ecological Protection	391	1.12			
Environmental Protection	1092	3.14	Environmental Protection	2535	6.43
Energy Conservation And Environmental Protection	576	1.66	Energy Conservation And Environmental Protection	1438	3.65
Energy Conservation And Environmental Protection Industry	156	0.45	Energy Conservation And Environmental Protection Industry	371	0.94
Energy Conservation And Emissions Reduction	226	0.65	Energy Conservation And Emissions Reduction	1176	2.98
Environmental Protection	499	1.43	Environmental Protection	2206	5.60

(continued)

Table 23.3 (continued)

Fourteenth Five-Year Plan			Thirteenth Five-Year Plan		
Phrases	Counts[1]	Frequency[2]	Phrases	Counts	Frequency
Environmental Governance	424	1.22	Environmental Quality	655	1.66
Green	4637	13.32	Green	5310	13.48
Green Development	1162	3.34	Green Development	1252	3.18
Green Transformation	392	1.13	Green Transformation	67	0.17
Green Travel	212	0.61	Green Travel	222	0.56
Green Production And Lifestyle	271	0.78	Restritive Green Development	116	0.29
Smart Green	174	0.50			
Significant Progress Has Been Made In The Green Transformation	114	0.33			

Notes 1. Counts: The number of occurrences of the phrase in the corpus. Same as below. 2. Frequency: The ratio of frequency to the number of texts in the corpus. Same as below

Therefore, ecological environment and energy constraints should be resolved to realize sustainable economic and social development. These problems can be alleviated by using clean, renewable energy and optimizing the energy structure. Through analyzing the characteristic words of ecological civilization in the two periods, it is found that the Fourteenth Five-Year Plan" has continued the utilization and promotion of clean energy, renewable energy, and new energy during the "Thirteenth Five-Year Plan" period. With low-carbon and green energy as the main direction of energy conservation and emission reduction, promote the green and high-quality development of the economy. Compared with the "Thirteenth Five-Year Plan" period, energy internet (0.3) and smart energy (0.64) appeared during the "Fourteenth Five-Year Plan" period. Meanwhile the proportion of primary energy consumption has decreased. The "13th Five-Year Plan" period focuses on new energy and related industries. At the same time, improve resource utilization efficiency to optimize resource allocation. The "proportion of disposable energy consumption" highlights the energy structure optimization to "reduce" and "control" as the main direction. The 14th Five-Year Plan, on the other hand, emphasizes systemic stability. For example, the "Energy Internet" can effectively integrate the industrial chain, and create interactive supply and demand transactions, thus realizing real-time dynamic optimization of energy. "Smart Energy" with continuous technological innovation, institutional change, triggering energy, energy industry structural upgrade, and promote ecological civilization development.

(3) Economic development concept

The high input, high energy consumption, and high pollution characteristics presented in the "factor accumulation mode" and "extensive mode of growth"

Table 23.4 Resources and energy

Fourteenth Five-Year Plan			Thirteenth Five-Year Plan		
Phrases	Counts	Frequency	Phrases	Counts	Frequency
Energy	2011	5.78	Energy	2705	6.87
New Energy	1662	4.78	New Energy	2139	5.43
New Energy Cars	913	2.62	New Energy Cars	1239	3.14
Clean Energy	856	2.46	Clean Energy	1161	2.95
Renewable Energy	373	1.07	Renewable Energy	574	1.46
Total Energy Consumption	150	0.43	Total Energy Consumption	275	0.70
New Energy Automotive Industry	110	0.32	New Energy Automotive Industry	160	0.41
Green Energy	178	0.51	Green Energy	138	0.35
New Energy Industries	214	0.61	New Energy Industries	361	0.92
Reduced Energy Consumption Per Unit Of Gross Regional Product	109	0.31	Energy Consumption	127	0.32
Lower Energy Consumption Per Unit Of GDP	65	0.19	Reduced Energy Consumption Per Unit Of Gross Regional Product	111	0.28
Energy Consumption	36	0.10	Lower Energy Consumption Per Unit Of GDP	109	0.28
Energy Internet	106	0.30	Lower Energy Consumption Of Gross Production	86	0.22
Primary Energy Consumption Share	85	0.24	Primary Energy Consumption Share	390	0.99
Smart Energy	224	0.64	Smart Energy	81	0.21
Wind Energy	146	0.42	Wind Energy	386	0.98
Photovoltaic Power Generation	200	0.57	Photovoltaic Power Generation	444	1.13
Wind Power Generation	112	0.32	Wind Power Generation	226	0.57
Biomass Power Generation	101	0.29	Biomass Power Generation	155	0.39
Domestic Waste Incineration Power Generation	55	0.16	Domestic Waste Incineration Power Generation	47	0.12
			Solar Power Generation	78	0.20
Mineral Resources	178	0.51	Mineral Resources	646	1.64
Saving Resources	97	0.28	Saving Resources	596	1.51
Comprehensive Utilization Of Resources	242	0.70	Comprehensive Utilization Of Resources	553	1.40

(continued)

Table 23.4 (continued)

Fourteenth Five-Year Plan			Thirteenth Five-Year Plan		
Phrases	Counts	Frequency	Phrases	Counts	Frequency
Resource Exploitation	184	0.53	Resource Exploitation	675	1.71
Resource Consumption	62	0.18	Resource Consumption	380	0.96
Resources Recycling	175	0.50	Resources Recycling	275	0.70
Resource Utilization Of Waste	133	0.38	Resource Utilization Of Waste	288	0.73
Recycling Of Renewable Resources	92	0.26	Renewable Resources	258	0.65

greatly threaten the ecological environment (Lin et al., 2016). Therefore, the high-quality development of the economy is a win–win path to realize the development of economic and ecological civilization construction. During the "Thirteenth Five-Year Plan" period, the concepts of green development (3.18), coordinated development (1.93), innovative development (3.64), and innovation-driven development (1.12) were put forward, promoting the ecological civilization construction that focuses on improving environmental quality. The report of at 19th CPC National Congress proposed that "accelerate the development of a green and low-carbon circular economic development concept system", and it guides the high-quality development (9.33) of the "Fourteenth Five-Year Plan". The high-quality development path of economy, industry, manufacturing, and service industry take ecological priority as a booster for the sustainable development of social economy. The development of "high-level opening up (0.74)" and "dual circulation of domestic economic cycle and international economic cycle (1.03)" deepen and expand win–win cooperation with the outside world. The "14th FYP" and the "13th FYP" follow the same development line. However, the development model has changed from a single industry development (real economy) to a comprehensive "green and high-quality development", and more esteem for the "sharing economy" and "balanced development" model. "Domestic and international dual cycle" is a positive exploration of the current international situation. So as to realize the economic circulation among various regions in China, China and other countries.

(4) Pollution prevention and control

Pollutants are caused by the social and economic development concept. Even if the resources and energy are used efficiently, the pollutants cannot be avoided. Therefore, the control and prevention of pollutants is still an important part of the ecological civilization construction. The pollution problems faced during the "Fourteenth Five-Year" and "Thirteenth Five-Year" periods are mainly water pollution and air pollution. For disaster problems, there are natural disasters, geological disasters, and meteorological disasters. Thus, the pollution and disaster control is a long

Table 23.5 Economic development concept

Fourteenth Five-Year Plan			Thirteenth Five-Year Plan		
Phrases	Counts	Frequency	Phrases	Counts	Frequency
Green Development	1162	3.34	Green Development	1252	3.18
Synergistic Development	1133	3.26	Synergistic Development	761	1.93
Innovation Development	1078	3.10	Innovation Development	1436	3.64
Coordinate Development	1044	3.00	Coordinate development	2306	5.85
Innovation-Driven Development	334	0.96	Innovation-driven Development	443	1.12
High-quality Development	3247	9.33	Real Economy	412	1.05
Promoting High Quality Development	728	2.09	Financial Services For The Real Economy	78	0.20
High-quality Development Of The Economy	346	0.99	Real Economy Development	61	0.15
High-quality Development Of Industry	160	0.46	Serving The Real Economy	48	0.12
High-quality Development Of manuFacturing Industry	158	0.45			
High-quality Development Of Service Industry	49	0.14			
Green High- Quality Development	63	0.18			
Circular Economy	296	0.85	Circular Economy	1188	3.02
Dual Circulation Of domestic Economic Cycle And International Economic Cycle	360	1.03	Recycling Development	319	0.81
Recycling Development	82	0.24	Vigorous Development Of Circular Economy	236	0.60
Vigorous Development Of Circular Economy	85	0.24			
Opening Up Development	260	0.75	Opening Up Development	612	1.55
High-level Of Opening up	256	0.74	Open Economy	544	1.38
High-level Of Open economy	163	0.47			
High-level Of Opening To The Outside World	150	0.43			
Digital Economy	1221	3.51			
Digital Transformation	539	1.55			

(continued)

Table 23.5 (continued)

Fourteenth Five-Year Plan			Thirteenth Five-Year Plan		
Phrases	Counts	Frequency	Phrases	Counts	Frequency
Digitalization Of Industry	332	0.95			
Real Economy	683	1.96			
Supporting The Real Economy	104	0.30			
The Focus Of Economic Development Is On The Real Economy	100	0.29			
Serving the Real Economy	88	0.25			
Financial Services For The Real Economy	59	0.17			
Real Economy Integration Development	52	0.15			
Balanced Development	363	1.04			
Sharing Economy	246	0.71			
Business Environment	1101	3.16	Business Environment	251	0.64
Optimizing The Business Environment	312	0.90			
Green Finance	209	0.60	Green Finance	65	0.16

game. The problems faced in the two periods remain unchanged, and the so does the prevention and control policies adopted. During the "Fourteenth Five-Year Plan" period, "joint prevention and control" was emphasized, indicating that the coordinated strategy adopted during the "Thirteenth Five-Year Plan" period has achieved good results. The joint prevention and control mechanism has strengthened the new path of ecological environment prevention and control of regional linkage, coordination and cooperation, and mutual benefit. The "carbon neutrality (0.75) and peak carbon dioxide emissions (0.51)" targets of the "Fourteenth Five-Year Plan" will become one of the important tasks in this period and the next stage, which puts forward new goals and requirements for atmospheric environmental governance in China. The "carbon" task of the development plan will play a positive gravitational role for ecological and environmental protection departments and local governments in formulating policies, decomposing targets, and supervising process. The carbon neutrality and peak carbon dioxide emission goals are closely related to economic growth and are important ways for high-quality and green development. They actively respond to ecological and environmental protection requirements, and promote the continuous optimization of energy and resource consumption structure and industrial structure.

Table 23.6 Pollution—Disaster—Prevention

Fourteenth Five-Year Plan			Thirteenth Five-Year Plan		
Phrases	Counts	Frequency	Phrases	Counts	Frequency
Pollution	839	2.41	Pollution	1581	4.01
Sewage Treatment Plant	597	1.72	Sewage Treatment Plant	1032	2.62
Sewage	574	1.65	Sewage	789	2.00
Pollution Prevention And Control	555	1.59	Pollution Prevention And Control	759	1.93
Pollution Control	463	1.33	Pollution Control	591	1.50
Sewage Treatment	445	1.28	Sewage Treatment	827	2.10
Air Pollution Prevention And Control	216	0.62	Air Pollution Prevention And Control	474	1.20
Water Pollution Prevention And Control	192	0.55	Water Pollution Prevention And Control	415	1.05
Soil Pollution Prevention And Control	181	0.52	Soil Pollution Prevention And Control	156	0.40
Joint Defense And Control	249	0.72	Joint Defense And Control	137	0.35
Joint Prevention And Control Mechanism	96	0.28	Joint Prevention And Control Mechanism	57	0.14
Joint Prevention And Control Of Atmospheric Pollution	57	0.16	Joint Prevention And Control Of Atmospheric Pollution	60	0.15
Regional Joint Prevention And Control	38	0.11	Regional Joint Prevention And Control	42	0.11
Carbon Sink	206	0.59	Carbon Sink	354	0.90
Reduced CO2 Emissions Per Unit Of GDP	94	0.27	Reduced CO2 Emissions Per Unit Of GDP	177	0.45
Gross Regional Product Per Unit Reduced Carbon Dioxide Emissions	122	0.35	Gross Regional Product Per Unit Reduced Carbon Dioxide Emissions	124	0.31
Carbon Dioxide Emissions	78	0.22	Carbon Dioxide Emissions	118	0.30
Carbon Cycle Development	120	0.34	Carbon Cycle Development	218	0.55
Carbon Neutrality	262	0.75			
Peak Carbon Dioxide Emissions	179	0.51			
Reach The CO2 Emission Peak	58	0.17			
Disaster	474	1.36	Disaster	537	1.36
Natural Disaster	282	0.81	Natural Disaster	399	1.01

(continued)

Table 23.6 (continued)

Fourteenth Five-Year Plan			Thirteenth Five-Year Plan		
Phrases	Counts	Frequency	Phrases	Counts	Frequency
Geologic Disaster	228	0.66	Geologic Disaster	348	0.88
Meteorological Disasters	96	0.28	Meteorological Disasters	203	0.52
Fire Disaster	129	0.37	Fire Disaster	133	0.34
Mountain Flood Prevention And Control	82	0.24	Mountain Flood Prevention And Control	122	0.31
Disaster Prevention And Mitigation	389	1.12	Disaster Prevention And Mitigation	588	1.49
Reserve Of Disaster Relief Materials	82	0.24	Reserve Of Disaster Relief Materials	150	0.38
Level Of Natural Disaster Preparedness	98	0.28	Disaster Prevention And Mitigation System	150	0.38
Disaster Monitoring And Early Warning	68	0.20	Disaster Monitoring And Early Warning	81	0.21

23.4.4 *Topic Classification*

According to the constructed phrase topic model, the open source gensim package in python is used to default hyper-parameters $\alpha = 0.0001$, $\beta = 0.00002$. Through perplexity calculation value, topic interpretability and topic distribution (Fig. 23.6), it is determined that the final topic number k is 7, the maximum number of iterations is 1000. The topic sets are shown in Table 23.7.

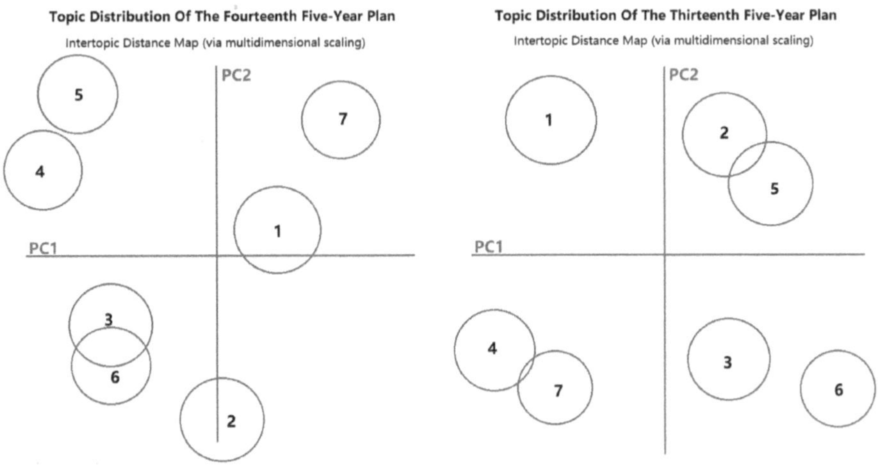

Fig. 23.6 Topic distribution map

Table 23.7 Topic set

Thirteenth Five-Year Plan	
T1 Industrial Development	Topic Feature Phrase: Industry Clusters(1802),[2] Equipment Manufacture(1235),Demonstration Base(1134), Strategic Emerging Industries(1176), Energy Conservation And Environmental Protection(1211), New Energy Cars(1040), Level Of Informatization(506), Industry Upgrade(465), Productive Service Industry(736), Industrial Layout(587)
	Ecological Civilization Feature Phrase: Strategic Emerging Industries(1176), Energy Conservation And Environmental Protection(1211), New Energy Cars(1040), Information Services(912), Level Of Informatization(506)
T2 Employment and Entrepreneurship	Topic Feature Phrase: Senior High School Education Universal(629), Vocational Skills Training(396), Practical Training Base(313), Employment And Entrepreneurship(457), Career Development(376), Ability To work(273), National Dducation(139), Entrepreneurship Promotes Employment(269), Professional Technical Personnel(291), Employment Services(256)
	Ecological Civilization Feature Phrase: Circular Economy(1188), Harmonious Development(186)
T3 Ecological Protection	Topic Feature Phrase: Eco-environmental Protection(791), Ecological Construction(994), Energy Conservation And Emissions Reduction(1176), Key Ecological Function Areas(496), Forest Coverage(611), Tangible Cultural Heritage(641), Ecological Restoration(662), Wetland Protection(347), Ecological Space(510), Pollution Prevention And Control(526)
	Ecological Civilization Feature Phrase: Ecological Protection(1638), Cultural Tourism(1430), Eco-environmental Protection(791), Energy Conservation And Emissions Reduction(1176), Key Ecological Function Areas(496), Forest Coverage(611), Tangible Cultural Heritage(641), Ecological Restoration(662), Wetland Protection(347), Ecological Space(510), Water Containment(335), Pollution Prevention And Control(759), Wetland Park(693)
T4 Social Development Strategy	Topic Feature Phrase: Economic And Social Development(3159), Building A Moderately Prosperous Society(2276), Ecological Civilization Construction(1299), Open Development(484), Scientific Development(675), Innovative Development(1124), Green Development(1065), Coordinated Development(1597), Scientific Development Of View(478), Structural Reform(774)
	Ecological Civilization Feature Phrase: Ecological Civilization Construction(1299), Ecological And Environmental Protection(651), Green Development(1065), Ecological Civilization(1782)

(continued)

[2] The numbers in parentheses are phrase frequencies.

Table 23.7 (continued)

T5 Internet Integration Development	Topic Feature Phrase: Internet Plus(1865), Logistics Center(622), Entrepreneurship And Innovation(641), Adjustment of Industrial Restructuring(402), Service Platform(1240), School-Enterprise Cooperation(388), Enabling Strength(401), Service System Construction(371), Human Resource Market(336), Informatization Construction(679)
	Ecological Civilization Feature Phrase: Tourist Attractions(917), Travel Products(430), Clean Energy(1161)
T6 Social Security	Topic Feature Phrase: Support For Development(214), Greening Projects(186), New Employment(224), Balanced Population Development(255), Urban And Rural Low Income Insurance(152), Social Security Pocket(111), Medical Insurance System(102), Health Industry(346), Drug System(253), Cultural And Creative Industrial Park(262)
	Ecological Civilization Feature Phrase: Sewage Treatment Plant(1032), Disaster Prevention and Mitigation(588), Binding Indicators(653), Environmental Renovation(555), Scenic Spots(378)
T7 Well-being Of The People's	Topic Feature Phrase: Improving People's Livelihood(823), Shanty Town Rehabilitation(668), Sheltered Housing(587), Ecological Function Area(270), Equalization Of Public Services(950), High School Education(272), Rural Drinking Water(304), Public Rental Housing(238), Leisure Vacation(424)
	Ecological Civilization Feature Phrase: Leisure Travel(692), Resource Advantages(668), Eco-tourism(789), Saving Resources(596), Photovoltaic Power Generation(44)
Fourteenth Five-Year Plan	
T1 Development And Construction	Topic Feature Phrase: Development Pattern(2789), Industrial Cluster(2207), Digital Economy(1193), Integration Development(2041), Advanced Manufacturing Industry(859), Strategic Emerging Industries(785), Real Economy(652), Smart Manufacturing(967), High-quality Development(3245), Structural Reform(967)
	Ecological Civilization Feature Phrase: High-quality Development(3245), Strategic Emerging Industries(785), Tangile Cultural Heritage(437), Development Quality(440)
T2 Innovation And Entrepreneurship	Topic Feature Phrase: Eco-friendly(611), Energy Saving And Environmental Protection(527), Innovation And Entrepreneurship(903), Science And Technological Innovation(1655), College graduates(335), Fundamental Research(516), Employment and Entrepreneurship(515), High-level Talent(484), New Research And Development Organization(447), Entrepreneurship-led Employment(229)
	Ecological Civilization Feature Phrase: Eco-friendly(611), Energy Saving And Environmental Protection(527), Sewerage Pipe Network(310), Disaster prevention And Mitigation(389), Value Of Ecological Products(277)

(continued)

Table 23.7 (continued)

T3 Ecological Environment	Topic Feature Phrase: Clean Energy(728), Pollution Prevention And Control(487), Territorial Space Planning(446), Forest Coverage(315), Household Waste Classification(364), Eco-environmental Quality(278), Biodiversity Conservation(261), Carbon Sink(206), Pollution Treatment(384), Carbon Neutrality(261)
	Ecological Civilization Feature Phrase: Clean Energy(728), Pollution Prevention And Control(487), Household Waste Classification(364), Forest Coverage(315), Pollution Control(384), Eco-environmental Governance(208), Biodiversity Conservation(261), Smart Tourism(315), Tourist Attractions(731), Rural Tourism(730), Territorial Spatial Planning(444), Carbon Sink(206), Carbon Neutrality(261)
T4 High-quality Development	Topic Feature Phrase: Ecological Protection(1034), Green Development(1110), Ecological Priority(405), High-quality Development(3245), Coordinated Development(890), Ecological Civilization Construction(695), Modernization Construction(696), Gold And Silver Mountains(472), New Industrialization(375), Pollution Control(384)
	Ecological Civilization Feature Phrase: Ecological Protection(1034), Green Development(1110), Ecological Priority(405), Ecological Environment Protection(659), Ecological Civilization Construction(695), Promoting High-quality Development(728), High-quality Development(3247), Water Containment(255), Ecological Civilization(585), Gold And Silver Mountains(472), Ecological Control(252)
T5 Innovative Development	Topic Feature Phrase: Strengthen The Main Position Of Enterprise Innovation(204), High And New Technology Industrial Development Zone(219), Technology Innovation Center(332), Industry-education Integration(525), Economic And Technological Development Zone(425), Key Laboratories(505), Demonstration Base(849), Equipment Manufacturing(910), Science And Technology Transformation(687), Industrial Transfer(379)
	Ecological Civilization Feature Phrase: Watershed Ecological Protection(391), Integrated Watershed Management(158), Rural Culture(200)
T6 Open To The Public	Topic Feature Phrase: Business Environment(1101), One Belt One Road(1966), Cross-border E-commerce(704), High-tech Enterprise(666), Technology-based SMEs(423), Comprehensive Free Trade Zone(534), Open Cooperation(338), Synergistic Development(1136), Linkage Development(675), Multimodal Transport(483)
	Ecological Civilization Feature Phrase: Ecological Protection And Restoration(278), High-quality Development(3247)
T7 Rural Revitalization	Topic Feature Phrase: Expanding The Results Of Poverty Eradication(649), Vocational Skills Training(422), Industrial Transformation And Upgrading(459), Ecological Industry(160), Casting A Firm Sense Of Chinese National Community(210), Helping Mechanism(378), Rural Revitalization(1937), Rescue System(374), Counterpart Support(252), Ecological Tourism(374)
	Ecological Civilization Feature Phrase: Ecological function(224), Ecological Tourism(471), Ecological Industry(201)

The top 30 key-phrases of topic are highly relevant to the topic, so the topic set in Table 23.7 only shows the top 10 phrases and their frequencies that are highly relevant to the topic. From the categories of the topic set, it can be found that the topic of the development plans in the two periods are relatively similar, covering overall development, industrial layout, green ecology, living standard, innovation and entrepreneurship, etc. However, there are great differences in topic feature words.

T1: the "Thirteenth Five-Year Plan" focuses on industrial development, including the layout of modern industries (1.49), clusters (4.57), and informatization (1.02). From the perspective of the overall high-quality development of the economy and society, the "Fourteenth Five-Year Plan" emphasizes the digital economy (3.43), the real economy (1.87) and structural reforms (2.78).

T2: the topic of feature words the "Thirteenth 13th Five-Year" is employment and entrepreneurship (1.16), which emphasizes basic employment stability, skills, occupational employment and services. The topic of the "Fourteenth Five-Year" feature words is innovation and entrepreneurship and employment, and the "Fourteenth Five-Year" plan focuses on higher education (0.98), high-level talents (1.39), innovation and entrepreneurship (2.59), which is in line with the high-quality development of human capital and employment and entrepreneurship.

T3: the "Thirteenth Five-Year Plan" focuses on ecological protection and restoration to achieve ecological civilization construction. In the "Fourteenth Five-Year Plan", clean energy (2.09), carbon sinks (0.59) and carbon neutrality (0.75) are incorporated in the ecological civilization construction.

T4: The feature words of the development strategy in the "Thirteenth Five-Year" include economic and social development (8.02), building a moderately prosperous society (5.78), innovation, green, and coordinated development (2.85, 2.70, 4.05). The topic feature words of the "Fourteenth Five-Year Plan" cluster focus on high-quality, ecological priority development (9.32, 1.16) strategy. The relationship between man and nature and the sustainability of development are highly valued at the new stage.

T5: Internet develops rapidly during the "Thirteenth Five-Year Plan" period, so the development and integration of Internet (4.73) receives greater attention. The topic feature words of the "Fourteenth Five-Year Plan" are innovation and development. The strategic emerging industries based on informatization and Internet have become the mainstay of innovation and development.

T6 and T7 of the "Thirteenth Five-Year Plan" are social security and improving people's livelihood. The overall development strategy in China always focuses on the "people-oriented", which clarifies the purpose of development, and emphasizes the essence of development. The T6 of the "Fourteenth Five-Year Plan" is opening up. In the face of the complex and changing development environment and the achievements of the "Belt and Road" (5.65), it emphasizes the diversity of cooperative development modes. T7: agriculture and rural development. The poverty alleviation during the Thirteenth Five-Year Plan period has been transformed into the comprehensive rural vitalization (5.57), and the depth and direction of development have changed greatly.

23.5 Conclusions and Prospect

Co-words and topic analysis play a very important role in interpreting text content and text classification. It is one of the most widely used methods in text mining. In this paper, combined the co-words and topic, 742 text data from the "Outline for National Economic and Social Development Plan" are utilized to understand the changing trends of development strategy goals of China in two different periods. The following conclusions are obtained:

(1) In the "Thirteenth Five-Year Plan" text, the institutional ecological security barriers are important task in the ecological environment planning. In the Fourteenth Five-Year Plan, safety barrier, control system, and supervising related ecological environmental protection system are implemented. Among them, the feature words of the "Thirteenth Five-Year Plan" contain 13 binding phrases, such as restrictive indices (1.66), restrictive ecological civilization (1.19), restrictive resource environment (0.63), restrictive ecological environment (0.4) etc. In the "Fourteenth Five-Year Plan", there are only 2 binding phrases, and the frequency of "binding index" is only 0.68. The red line (1.67) in the "Thirteenth Five-Year" means to delineate ecological red line, and formulate the ecological indicators. The red line (0.66) in the "Fourteenth Five-Year Plan" focus on ensuring that the red line in ecological protection in not crossed and implementing the ecological redline to implement and supervise the ecological protection indicators of the previous stage.

(2) Compared with the "Thirteenth Five-Year Plan", the text of the "Fourteenth Five-Year Plan" emphasizes the integrity and systematization of ecological civilization construction. The "Fourteenth Five-Year Plan" emphasizes high-quality development in the overall social and economic development concept, adheres to the ecological priority, and holds that lucid waters and lush mountains are invaluable assets. In the topic model, from the overall social development, T4 of the "Fourteenth Five-Year" focuses on high-quality development with ecological protection and green development as the center, and integrates development and ecological environmental protection to promote the sustainable development. The T4 of the "Thirteenth Five-Year" development strategy still focuses on economic and social development and building a well-off society, and it only emphasizes the green concept in the process of partial development and construction. The construction of ecological civilization in the "Fourteenth Five-Year Plan" takes "governance system (4.43)", "modernization of governance ability (2.34)" and "system governance (1.34)" as new missions. Compared with the higher-weighted "Environmental renovation (1.41)", "Environmental quality (1.66)" and "comprehensive environmental renovation (1.44)" in the "Thirteenth Five-Year Plan" for improving the ecological environment, it is more systematic.

(3) The ecological civilization construction plan in the "Fourteenth Five-Year" has higher initiative and scientific rationality. "Carbon neutrality (0.75) and peak carbon dioxide emissions (0.51)" are climate commitments from China for a

period of time in the future, and they have become important and clear ecological and environmental protection tasks in the "Fourteenth Five-Year" development plan, reflecting China's initiative and responsibility in the ecological civilization construction. The targets of carbon reduction (0.27), carbon cycle (0.55), and reduction of carbon dioxide emissions per unit of GDP (0.45) in the "Thirteenth Five-Year Plan" promote the carbon target development. The utilization of energy resources in the "Thirteenth Five-Year Plan" aimed at energy conservation (5.06), energy conservation and emission reduction (2.98), and energy conservation and environmental protection (3.65) has been transformed into the green transformation (1.13), smart energy (0.64), and energy Internet (0.30) during the "Fourteenth Five-Year" period, improving the utilization of resources and energy more accurately and scientifically.

There are similar regional plans or development strategies in other countries and regions. For example, the development strategy of intelligent manufacturing, Germany's "Industry 4.0", Japan's "Society 5.0" and the United States' "Industrial Internet". Mid- and long-term plans, such as the EU's "Digital Compass 2030: The Road to a European Digital Decade", the U.S. "Spatial Strategy 2050", and the UK's "Regional Economic Development Planning System". With text mining, it is possible to explore the development priorities and changing trends of these plans, which not only provides a new way of thinking for government public management, but also expands the application of these methods. Through mining the themes with key words and clusters from the five-year plan texts or other strategic documents, combined with the theories of ecological civilization construction, spatial data analysis and industrial development, our research will be important in the following three aspects: (1) Studying the historical stage of ecological civilization construction according to the evolution of high-frequency words, and forecasting the development of the situation can provide suggestions for policy optimization and implementation. (2) Analyzing regional differences according to regional high-frequency word forms and clustering differences can provide theoretical basis for regional development optimization, regional characteristic policy designation and implementation. (3) According to the co-word association between the country and a region and between one region and another, we can investigate the role of a region in the national development and the coordinated development strategy among regions.

Acknowledgements Shaoli He also made great contributions to this manuscript. We express our heartfelt thanks to them. This research also was supported by: National Natural Science Foundation of China (72474127) and National Social Science Fund Major Bidding Project of China (18ZDA052). Part of this chapter was published: He S, Wu X, Guo J. Features and Comparative Research on Ecological Civilization Vocabularies in the Five-Year Plan of China: Journal of Global Information Management 2023;31. https://doi.org/https://doi.org/10.4018/JGIM.330149. The publication has been conducted with copyright permission.

References

Allan, J. (2002). *Introduction to topic detection and tracking [Introduction to topic detection and tracking]*. Kluwer Academic Publishers (pp. 1–16). https://doi.org/10.1007/978-1-4615-0933-2_1

Baldwin, E., Cai, Y., & Kuralbayeva, K. (2018). To build or not to build? Capital stocks and climate policy. *Journal of Environmental Economic and Management, 100*, 1–25. https://doi.org/10.1016/j.jeem.2019.05.001

Blei, D. M., Ng, A. Y., & Jordan, M. I. (2003). Latent Dirichlet allocation. *Journal of Machine Learning Research*, 993–1022. https://doi.org/10.1162/jmlr.2003.3.4-5.993

Cai, W., Xu, W., Yang, N., Guo, X., & Ouyang, Z. (2021). Standards system for high-quality development of ecological civilization: Problems and implementation path. *Strategic Study of CAE, 23*(03), 40–45.

Calvo-Bartolomé, L., & Arenas-García, J. (2025). Topic models for decision-support systems, Part II: Expert-in-the-loop hierarchical topic models. *Engineering Applications of Artificial Intelligence, 159*, 1–22. https://doi.org/10.1016/j.engappai.2025.111508

Chang, D.-S., & Choi, K.-S. (2006). Incremental cue phrase learning and bootstrapping method for causality extraction using cue phrase and word pair probabilities. *Information Processing and Management, 42*(3), 662–678. https://doi.org/10.1016/j.ipm.2005.04.004

Cheng, X., Yan, X., Lan, Y., & Guo, J. (2014). Btm: Topic modeling over short texts. *IEEE Transactions on Knowledge and Data Engineering, 26*(12), 2928–2941. https://doi.org/10.1109/TKDE.2014.2313872

Du, H., Guo, Y., Lin, Z., Qiu, Y., & Xiao, X. (2021a). Effects of the joint prevention and control of atmospheric pollution policy on air pollutants—A quantitative analysis of Chinese policy texts. *Journal of Environmental Management.*

Du, W., Yan, H., Feng, Z., Yang, Y., & Liu, F. (2021b). The supply-consumption relationship of ecological resources under ecological civilization construction in China. *Resources Conservation and Recycling, 172*(1–2), Article 105679.

Eirini, P., & Grigorios, T. (2018). Local word vectors guide keyphrase extraction. *Information Processing and Management, 54*(6), 888–902. https://doi.org/10.1016/j.ipm.2018.06.004

Feng, F., & Croft, W. B. (2001). Probabilistic techniques for phrase extraction. *Information Processing and Management, 37*(2), 199–220. https://doi.org/10.1016/S0306-4573(00)00029-7

Griffiths, T. L., & Steyvers, M. (2004). Finding scientific topics. *Proceedings of the National Academy of Sciences of the United States of America, 101*, 5228–5235.

Group, R. (2020). Research on the experience of the past "Five-Year Plans" and the guideline for the "14th Five-Years Plan". *Economic Perspectives,* (4), 3–14.

Haihua, X., Xuefei, C., Yimin, D., Xiaoqing, L., & Zhi, T. (2022). A semi-supervised method for Chinese key phrase extraction based on statistical features and graph model. *Journal of Chinese Information Processing, 36*(4), 57–65.

He, X., Xu, H., Li, J., He, L., & Yu, L. (2017). FastBTM: Reducing the sampling time for biterm topic model. *Knowledge Based Systems, 132*, 11–20. https://doi.org/10.1016/j.knosys.2017.06.005

Hong, L., Zhou, N., Fridley, D., & Raczkowski, C. (2013). Assessment of China's renewable energy contribution during the 12th Five Year Plan. *Energy Policy, 62*, 1533–1543.

Huang, C., & Wu, C. (2021). Research on the synergetic effect of industrial green transformation and ecological civilization construction in the Yangtze River economic belt. *Resources and Environment in the Yangtze Basin, 30*(6), 1287–1297.

J, N. M. E. (2006). Modularity and community structure in networks. *Proceedings of the National Academy of Sciences, 103*(23), 8577–8582.

Javed, M. A., Younis, M. S., Latif, S., Qadir, J., & Baig, A. (2018). Community detection in networks: A multidisciplinary review. *Journal of Network and Computer Applications, 108*, 87–111.

Jean-Philippe, N. (2019). Emission reduction and profit-neutral permit allocations. *Journal of Environmental Economic and Management, 93*, 239–253. https://doi.org/10.1016/j.jeem.2018.12.001

Jiwei, Y., & Dongfan, Z. (2010). Disam biguation method in Chinese word segmentation based on phrase match. *Journal of Jilin University (Science Edition), 48*(3), 427–432. https://doi.org/CNKI:SUN:JLDX.0.2010-03-019

Lam, J. C. K., Cheung, L. Y. L., Wang, S., & Li, V. O. K. (2019). Stakeholder concerns of air pollution in Hong Kong and policy implications: A big-data computational text analysis approach. *Environmental Science and Policy, 101*, 374–382. https://doi.org/10.1016/j.envsci.2019.07.007

Lei, M., Huang, H., & Feng, C. (2021). Multi-granularity semantic representation model for relation extraction. *Neural Computing and Applications, 33*, 1–11.

Lin, M. (2021). Medium-and long-term Planning is an important way for the communist party of China to Govern the country. *Economic Perspectives*, (05), 31–40.

Lin, W., Su, J., & Zhou, Y. (2016). Economic development: Accelerating or slowing down measurement and analysis based on energy and environment perspective. *Economist, 206*(02), 35–43. https://doi.org/10.16158/j.cnki.51-1312/f.2016.02.005

Lo, K., & Wang, M. Y. (2013). Energy conservation in China's twelfth five-year plan period: Continuation or paradigm shift? *Renewable and Sustainable Energy Reviews, 18*, 499–507.

Lu, X., Zhang, S., Xing, J., Wang, Y., Chen, W., & Ma, M. (2020). Progress of air pollution control in china and its challenges and opportunities in the ecological civilization era. *Engineering*, (6), 1423–1431. https://doi.org/10.1016/j.eng.2020.03.014

Lu, Y., Bing, X., Stefan, S., Heiko, T., & Guotian, C. (2019). Indicators for energy transition targets in China And Germany: A text analysis. *Ecological Indicators, 111*, Article 106012. https://doi.org/10.1016/j.ecolind.2019.106012

Meng, F., Guo, J., Guo, Z., Lee, J. C. K., Liu, G., & Wang, N. (2021). Urban ecological transition: The practice of ecological civilization construction in China. *Science of the Total Environment, 755*, Article 142633. https://doi.org/10.1016/j.scitotenv.2020.142633

Michal, R.-Z., Chaitanya, C., Thomas, G., Padhraic, S., & Mark, S. (2010). Learning author-topic models from text corpora. *ACM Transactions on Information Systems, 28*(1). https://doi.org/10.1145/1658377.1658381

Onan, A., Korukoğlu, S., & Bulut, H. (2016). Ensemble of keyword extraction methods and classifiers in text classification. *Expert Systems with Applications*. https://doi.org/10.1016/j.eswa.2016.03.045

Papagiannopoulou, E., & Tsoumakas, G. (2018). Local word vectors guiding keyphrase extraction. *Information Processing and Management, 54*(6), 888–902. https://doi.org/10.1016/j.ipm.2018.06.004

Park, J. Y., Mistur, E., Kim, D., Mo, Y., & Hoefer, R. (2022). Toward human-centric urban infrastructure: Text mining for social media data to identify the public perception of COVID-19 policy in transportation hubs. *Sustainable Cities and Society*, 76.

Ran, L. (2020). Text analysis of Chinese government's Green Governance Policy (1949–2020): Its logic change and basic experience. *Journal of Shenzhen University: Humanities and Social Sciences, 37*(04), 46–55.

Su, J., Xu, X., & Li, X. (2013). Research on the structure and relationship of China's technological innovation policies based on co-word analysis. *Science and Technology Progress and Policy, 30*(9), 110–115.

Sung, Y. Y., & Kim, S. B. (2019). Topical keyphrase extraction with hierarchical semantic networks. *Decision Support Systems, 128*, Article 113163. https://doi.org/10.1016/j.dss.2019.113163

Tahir, H., Gelani, H. E., & El-Ferik, S. (2025). Electric vehicles and renewable energy for sustainable mobility: Insights from bibliometric, topic modeling, and societal impact analyses. *Energy Strategy Reviews, 61*, 1–21. https://doi.org/10.1016/j.esr.2025.101827

Thuestad, I. E. (2020). Have international pollution protocols made a difference? *Journal of Environmental Economic and Management, 103*, 1–24. https://doi.org/10.1016/j.jeem.2020.102358

Tong, Y., Chen, X., & Xinyu, L. (2021). Evaluation and analysis of new-energy vehicle industry policies in the context of technical innovation in China. *Journal of Clearner Production, 281*, Article 125126. https://doi.org/10.1016/j.jclepro.2020.125126

Vega-Oliveros, D. A., Gomes, P. S., Milios, E. E., & Berton, L. (2019). A multi-centrality index for graph-based keyword extraction. *Information Processing & Management, 56*(6), Article 102063.

Vikstrom, H., Mårald, E., & Junkka, J. (2025). Policy silences in Nordic-Arctic energy transitions: A topic-modelling study. *Energy Policy, 206*, 1–11. https://doi.org/10.1016/j.enpol.2025.114746

Wang, R., Zhou, D., & He, Y. (2019). ATM: Adversarial-neural topic model. *Information Processing and Management, 56*(6), Article 102098. https://doi.org/10.1016/j.ipm.2019.102098

Weidong, Z. (2017). On theoretical issues in building a knowledge database of Chinese constructions. *Journal of Chinese Information Processing, 31*(1), 230–238.

Wu, M., Liu, Y., Xu, Z., Yan, G., Ma, M., Zhou, S., & Qian, Y. (2020). Spatio-temporal dynamics of China's ecological civilization progress after implementing national conservation strategy. *Journal of Cleaner Production, 258*, Article 124886. https://doi.org/10.1016/j.jclepro.2020.124886

Xu, Y., Fan, X., & Long, Z. (2025). Unveiling synergy in China's new energy vehicle policies: A comprehensive text mining-based quantitative analysis. *Transport Policy, 4*, Article 103891. https://doi.org/10.1016/j.tranpol.2025.103891

Xu, Y., Yin, J., Huang, J., & Yin, Y. (2018). Hierarchical topic modeling with automatic knowledge mining. *Expert Systems with Applications, 103*, 106–117. https://doi.org/10.1016/j.eswa.2018.03.008

Yan, J., & Su, B. (2020). What drive the changes in China's energy consumption and intensity during 12th five-year plan period? *Energy Policy, 140*, Article 111383. https://doi.org/10.1016/j.enpol.2020.111383

Zhang, H., Xu, Z., Sun, C., & Ehsan, E. (2018). Targeted poverty alleviation using photovoltaic power: Review of Chinese policies. *Energy Policy, 120*, 550–558. https://doi.org/10.1016/j.enpol.2018.06.004

Zhang, X., Wang, Y., Qi, Y., Wu, J., Liao, W., & Shui, W. (2016). Evaluating the trends of China's ecological civilization construction using a novel indicator system. *Journal of Cleaner Production*, 910–923. https://doi.org/10.1016/j.jclepro.2016.06.034

Zhao, X., Wang, D., Zhao, Z., & Liu, W. (2021). A neural topic model with word vectors and entity vectors for short texts. *Information Processing and Management, 58*(2), Article 102455. https://doi.org/10.1016/j.ipm.2020.102455

FSC
www.fsc.org
MIX
Papier | Fördert
gute Waldnutzung
FSC® C083411

Kolibri 360 GmbH
Ferdinand-Jühlke-Straße 7
99095 Erfurt, Deutschland
produktsicherheit@kolibri360.de